B. S. Annigeri, K. Tseng (Eds.)

Boundary Element Methods in Engineering

Proceedings of the International Symposium
on Boundary Element Methods:
Advances in Solid and Fluid Mechanics
East Hartford, Connecticut, USA, October 2-4, 1989

Springer-Verlag
Berlin Heidelberg NewYork London
Paris Tokyo Hong Kong Barcelona

Dr. Balkrishna S. Annigeri
Dr. Kadin Tseng
United Technologies Research Center
East Hartford, Connecticut 06040
USA

ISBN-13: 978-3-642-84240-5 e-ISBN-13: 978-3-642-84238-2
DOI: 10.1007/978-3-642-84238-2

Softcover reprint of the hardcover 1st edition 1990

2161/3020-543210

INTERNATIONAL SYMPOSIUM ON BOUNDARY ELEMENT METHODS

sponsored by

United Technologies Research Center
East Hartford, CT 06108, USA

co-sponsored by

NASA Langley Research Center
and
International Association for
Boundary Element Methods (IABEM)

ORGANIZING COMMITTEE

LOCAL ORGANIZING COMMITTEE

Mr. R.E. Olson Dr. A.V. Srinivasan
Dr. B.S. Annigeri Dr. K. Tseng
United Technologies Research Center,
East Hartford, Connecticut, USA

Dr. R.B. Wilson
Pratt and Whitney Aircraft,
East Hartford, Connecticut, USA

Mr. A. Wanderlingh
Hamilton Standard,
Windsor Locks, Connecticut, USA

SCIENTIFIC ADVISORY COMMITTEE

Prof.	S.N.	Atluri	Prof.	P.K.	Banerjee
Dr.	C.	Brebbia	Prof.	M.P.	Cleary
Prof.	J.J.	Connor	Dr.	L.L.	Erickson
Dr.	F.	Farassat	Prof.	T.L.	Geers
Dr.	D.B.	Hanson	Dr.	G.K.	Haritos
Dr.	B.	Hunt	Prof.	O.A.	Kandil
Prof.	V.C.	Li	Prof.	G.	Maier
Prof.	P.	Mantegazza	Prof.	S.K.	Mukherjee
Prof.	R.	Piva	Prof.	F.J.	Rizzo
Prof.	S.R.	Sipcic	Dr.	J.	Sloof
Dr.	A.V.	Srinivasan	Prof.	M.	Tanaka
Prof.	N.	Tosaka	Dr.	R.B.	Wilson
Prof.	J.C.	Wu			

To Our Parents

Preface

The Boundary Element Method (BEM) has become established as an effective tool for the solutions of problems in engineering science. The salient features of the BEM have been well documented in the open literature and therefore will not be elaborated here. The BEM research has progressed rapidly, especially in the past decade and continues to evolve worldwide.

This Symposium was organized to provide an international forum for presentation of current research in BEM for linear and nonlinear problems in solid and fluid mechanics and related areas. To this end, papers on the following topics were included: rotary-wing aerodynamics, unsteady aerodynamics, design and optimization, elasticity, elastodynamics and elastoplasticity, fracture mechanics, acoustics, diffusion and wave motion, thermal analysis, mathematical aspects and boundary/finite element coupled methods. A special session was devoted to parallel/vector supercomputing with emphasis on massive parallelism.

This Symposium was sponsored by United Technologies Research Center (UTRC) , NASA Langley Research Center, and the International Association of Boundary Element Methods (IABEM). We thank the UTRC management for their permission to host this Symposium. In particular, we thank Dr. Arthur S. Kesten and Mr. Robert E. Olson for their encouragement and support. We gratefully acknowledge the support of Dr. E. Carson Yates, Jr. of NASA Langley, Prof. Luigi Morino, Dr. Thomas A. Cruse and Prof. Shoichi Kobayashi of IABEM for their enthusiastic support and assistance in the organization of this Symposium. We thank the members of the Local Organizing Committee and the Scientific Advisory Committee for their excellent support. We thank Mrs. Nancy E. Jestreby, the Symposium Secretary, for her dedication and enthusiasm in helping to bring the Symposium to its fruition. Many individuals at UTRC have helped in making this Symposium a success; our sincere thanks to them. Thanks also go to Springer Verlag for their effort in publishing this volume. Last but certainly not the least, we thank the authors for their excellent efforts in preparation and presentation of the papers.

We thank our families for their patience and support during the organization of this Symposium.

January, 1990

Balkrishna S. Annigeri, Kadin Tseng
United Technologies Research Center
East Hartford, Connecticut
U. S. A.

Contents

ELASTOPLASTICITY

ELASTODYNAMICS

FRACTURE

Boundary Integral Formulations of the Scalar and Vector Wave Equations with Arbitrary Moving Boundaries

Y.J. Lee and J.Y. Yang
Institute of Applied Mechanics,
National Taiwan University
Taiwan, Republic of China

Abstract

This paper presents an operational approach for deriving the boundary integral formulas of the scalar and vector wave equations with arbitrary moving boundary. Generalized derivatives are first introduced to the classical wave equation, the Ffowcs William–Hawkings (FW–H) equation and the Maxwell's equations. After analyzing the nonhomogeneous terms of the resultant distributional equations, it is found that the integral representations of the scalar and vector wave equations can be expressed in terms of three general operators which represent the surface source distribution, time derivative and spatial derivatives of the surface distribution, respectively. Explicit formulas for these general operators are then derived for some scalar and vector wave equations from aerodynamics, aeroacoustics and electro–dynamics.

Nomenclature

A	= acceleration of the surface point
B	= magnetic flux density
c	= wave speed ,speed of sound or speed of light
D	= electric flux density
E	= electric field intensity
f=0	= description of body surface; f>0 outside and f<0 inside the surface
H	= magnetic intensity
v	= velocity of boundary point
r	= vector **x**–**y**
r	= |**x**–**y**|; distance of source and observer
$\bar{r}$	= unit vector in r direction
R	= $\|1-M_r\| r$
ρ_0	= density of undisturbed fluid
x,t	= observer position, observer time
y,τ	= source position, source time
M	= |**v**|/c; Mach number
M_n	= **v**·**n**/c; Mach number in the normal direction

M_r = $v\cdot\bar{r}/c$; Mach number in the $\bar{r}$ direction
n = $\nabla f/|\nabla f|$, local unit outward normal on body $f = 0$
p = pressure field
ϕ = velocity potential
$\dot{\phi}$ = time derivative of ϕ with respect to surface moving point
Φ = surface source operator
μ = magnetic permeability
ϵ = electric permittivity
$\delta(f),\delta(g)$
= Dirac delta function
$\frac{\bar{\partial}}{\partial t}, \bar{\nabla}$,. = the bar over the differential operator denotes generalized

Introduction

It is known that most of the boundary integral methods for the wave equations are based on integral equation represented by the Green's third identity. Integral equations derived in this way,i.e., Kirchhoff's formula, however, can not be applied to problems with moving boundary.

A novel approach to the derivation of the boundary integral formulas without this limitation, originated from solving the Ffowcs Williams–Hawkings (FW–H) equation for the pressure field to predict the aerodynamic noise of advanced high speed propeller blades, has been developed by Farassat[1–3]. Considering a flow field outside a moving boundary which is described by f(x,t)=0, Farassat introduced the generalized derivatives

$$\bar{\partial}_t\phi = (\partial_t\phi) + \Delta\phi\, \partial_t f\delta(f) \tag{1}$$

$$\bar{\nabla}_x\phi = (\nabla_x\phi) + \Delta\phi\, \nabla_x f\delta(f) \tag{2}$$

into the Navier–Stokes equations, where $\delta(f)$ is the delta function and $\Delta\phi$ is the jump of ϕ across the moving boundary (discontinuity surface).The distributional form of the FW–H equation is then derived under the inviscid assumption as (Ref.[3]).

$$\bar{\Box}^2 p = \bar{\nabla}^2 p - \frac{\bar{\partial}^2}{c^2\bar{\partial}^2 t} p = -\frac{\bar{\partial}}{c\bar{\partial} t}\left[n\cdot M\rho_0 c^2|\nabla f|\delta(f)\right] + \bar{\nabla}\cdot\left[np|\nabla f|\delta(f)\right] \tag{3}$$

where $f(x,t) = 0$ describes the moving boundary, n denotes the unit normal vector at the boundary, ρ_0 is the density of the unperturbed air, c is the speed of sound and M is the dimensionless velocity (w.r.t. to speed of sound) of the boundary point. Several integral representations for the pressure field p were then developed by him from Eq.(3).

Recently Lee and Yang [4],[5] found that the derivation of the integral equation for the scalar wave equation (including the FW–H equation) can be generalized in a unified way by using the operational notions. A boundary integral method (panel method) for solving the linearized potential flow field (governed by the classical equation) was also implemented to demonstrate its capability to solve aerodynamic problems with rotary motions. In the present paper we shall extend the operational notions to the vector wave equation and derive the integral representation in a generalized sense.

Scalar and vector wave equations

Consider the following distributional equation

$$\Box^2 W = \left[h|\nabla f|\delta(f)\right] + \frac{\partial}{c\partial t}\left[n \otimes h_t |\nabla f|\delta(f)\right] + \nabla\circ\left[n\otimes h_x |\nabla f|\delta(f)\right] \tag{4}$$

where W, h, h_t and h_x can be either scalar or vector functions, $\nabla\circ$ denotes the gradient ∇, curl $\nabla\times$ or divergent operator and $\otimes$ represents dot product $\cdot$, cross product $\times$ or scalar product. It is noted that Eq.(4) is very general and encompasses many wave equations frequently encountered in engineering applications. For example, the FW–H equation (Eq.(3)), as well as the distributional form of the homogeneous wave equation

$$\bar{\Box}^2\phi = \left[(\nabla\phi\cdot n + M_n\frac{\partial\phi}{c\partial t})|\nabla f|\delta(f)\right] + \frac{\partial}{c\partial t}\left[n\cdot M\phi|\nabla f|\delta(f)\right] + \nabla\cdot\left[n\phi|\nabla f|\delta(f)\right] \tag{5}$$

derived in [3] and [4] are special cases of Eq.(4). The vector wave equations for the magnetic flux density and electric intensity field are also embedded in Eq.(4) and are given respectively by

$$\bar{\Box}^2 B = -\frac{\partial}{c\partial t}\left[(\Delta E\times n - M_n\Delta B)|\nabla f|\delta(f)\right] - \nabla\times\left[(\Delta B\times n + M_n\Delta E)|\nabla f|\delta(f)\right] + \nabla\left[\Delta B\cdot n|\nabla f|\delta(f)\right] \tag{6}$$

and

$$\bar{\Box}E = \frac{1}{c\partial t}\left[(\Delta B\times n + M_n\Delta E)|\nabla f|\delta(f)\right] - \nabla\times\left[(\Delta E\times n - M_n\Delta B)|\nabla f|\delta(f)\right] + \nabla\left[\Delta E\cdot n|\nabla f|\delta(f)\right] \tag{7}$$

The above vector equations are obtained by introducing generalized derivatives to the classical Maxwell's equations

$$\nabla\cdot D = 0,\ \nabla\times H - \frac{1\partial D}{a\partial t} = 0\ ,\ \nabla\times E + \frac{1\partial B}{a\partial t} = 0\ , \nabla\cdot B = 0$$
$$D = \epsilon E\ ,\ B = \mu H,\ c^2 = \frac{a^2}{\mu\epsilon} \tag{8}$$

and using the vector identity $\nabla\times\nabla\times A = \nabla(\nabla\cdot A) - \nabla^2 A$ after some manipulations. In Eq.(8), B,H,D,E are magnetic flux density, magnetic intensity,electric flux density and electric field intensity respectively; μ ,ϵ and a are magnetic permeability, electric permittivity and speed of light in vacuum respectively.

Integral representation of Eq.(4) can be obtained by taking convolution of the nonhomogeneous terms with the fundamental solution,$G(x,t) = -\frac{\delta(t-|x|/c)}{4\pi|x|}$. Using an operational notation, we have

$$W = \Phi[h] + \Phi_t[\otimes h_t] + \Phi_{\nabla\circ}[\otimes h_x] \tag{9}$$

where operators Φ_t and $\Phi_{\nabla\circ}$ are obtained from the operator Φ by

$$\Phi_t[\otimes h] = \frac{\partial}{c\partial t}\Phi[n\otimes h] \tag{10}$$

and

$$\Phi_{\nabla o}[\otimes h] = \nabla o \Phi[n\otimes h] \tag{11}$$

The operator Φ itself, by definition,is the convolution of a surface source distribution with the fundamental solution, that is

$$\Phi[h](x,y) = G^*\left\{h|\nabla f|\delta(f)\right\} = -\frac{1}{4\pi}\int h|\nabla f|\delta(f)\frac{\delta(\tau-t+r/c)}{r}dyd\tau \tag{12}$$

with $r = |x-y|$ denotes the distance between the observer **x** and the source point **y** at the boundary.

Formula (12) involves a 4–dimensional integral which can be reduced to a two dimensional integral after eliminating the delta functions, $\delta(f)$ and $\delta(\tau-t+r/c)$, by using the following relationships (Ref.[3]) :

$$dyd\tau=\frac{dfdSd\tau}{|\nabla f|}, \tag{13}$$

and

$$\delta(g(\tau))=\sum_i \frac{\delta(\tau-\tau'_i)}{|dg/d\tau|}\Big|_{\tau=\tau_i'} \tag{14}$$

where $g(\tau) = \tau-t+r/c = \tau-t+|x-y(\tau)|/c$ and τ_i' is the i–th root of equation $g(\tau) = 0$.

Since $dg/d\tau=1-dr/cd\tau = 1-\mathbf{M}\cdot(x-y)/r = 1-M_r$, formula (13) becomes

$$\Phi[h](x,y)=-\frac{1}{4\pi}\int\left[h\frac{1}{|1-M_r|r}\right]_{\tau=\tau'}dS \tag{15}$$

It is noted that we have neglected summation symbol for the sake of simplicity.

From Eqs.(10) and (15), it follows immediately

$$4\pi\Phi_t[\otimes h](x,t) = -\frac{\partial}{c\partial t}\int\frac{n\otimes h}{r|1-M_r|}_{\tau=\tau'}dS \tag{16}$$

One also has the following relations

$$\frac{\partial}{\partial t} = \frac{1}{(1-M_r)}\frac{\partial}{\partial \tau}, \quad \frac{\partial n}{\partial \tau}=\omega\times n \tag{17}$$

$$\frac{\partial}{c\partial\tau}\frac{1}{|r-\mathbf{r}\cdot\mathbf{M}|} = \frac{\mathbf{A}\cdot\mathbf{r}/c^2+M_r-M^2}{r^2(1-M_r)\,|1-M_r|} \tag{18}$$

and

$$\frac{\partial\bar{r}}{c\partial\tau}= (M_r\bar{r}-M)/r, \tag{19}$$

Using Eqs.(17) and (18), formula (16) can be explicitly written as

$$4\pi\Phi_t[\otimes h](x,t)=\int\Big[-\frac{(\omega\times n)\otimes h+n\otimes\frac{\partial h}{\partial\tau}}{cr(1-M_r)|1-M_r|}-\frac{A\cdot r/c^2+M_r-M^2}{r^2|1-M_r|^3}n\otimes h\Big]_{\tau=\tau'}dS \tag{20}$$

where A and ω are acceleration and angular velocity of the point at the moving boundary.

Before taking $\Phi_{\nabla o}$ into account, we find that

$$\nabla\left[\frac{\delta(g)}{r}\right]=-\frac{\partial}{c\partial t}\left[\frac{\bar{r}\ \delta(g)}{r}\right]-\frac{\bar{r}\ \delta(g)}{r^2} \tag{21}$$

where $\bar{r}$ is the unit vector in $r = x-y$ direction

Substituting Eqs.(21)and (15) into Eq.(11), one obtains

$$\begin{aligned}4\pi\Phi_{\nabla o}[\otimes h](x,t) &= \frac{\partial}{c\partial t}\int\frac{\bar{r}o(n\otimes h)}{r}\delta(g)\delta(f)|\nabla f|d\underline{y}d\tau\\ &+\int\frac{\bar{r}o(n\otimes h)}{r^2}\delta(g)\delta(f)|\nabla f|d\underline{y}d\tau\\ &=\frac{\partial}{c\partial t}\int\frac{\bar{r}o(n\otimes h)}{|r-r\cdot M|}dS+\int\frac{\bar{r}o(n\otimes h)}{r|r-r\cdot M|}dS\\ &=\int\frac{1}{(1-M_r)c\partial\tau}\frac{\partial}{}\frac{\bar{r}o(n\otimes h)}{|r-r\cdot M|}dS+\int\frac{\bar{r}o(n\otimes h)}{r|r-r\cdot M|}dS\end{aligned} \tag{22}$$

and with Eq.(19) one has

$$\begin{aligned}4\pi\Phi_{\nabla o}[\otimes h](x,t)&=\int\frac{A\cdot r/c^2+1\ \ -M^2}{r^2|1-M_r|^3}\bar{r}o(n\otimes h)dS\\ &+\int\frac{-Mo(n\otimes h)/r+\bar{r}o[(\omega\times n)\otimes h/c+n\otimes\frac{\partial h}{c\partial\tau}]}{r(1-M_r)|1-M_r|}dS\end{aligned} \tag{23}$$

Applications

As an example to test the validity of Eqs.(4),(9),(15),(20) and (23), we consider the homogeneous wave equation in distributional form (Eq.(5)). According to Eq.(9), its symbolical representation is

$$\phi=\Phi[\nabla\phi\cdot n+M_n\frac{\partial\phi}{c\partial\tau}]+\Phi_t[\cdot M\phi]+\Phi_{\nabla\cdot}[\phi] \tag{24}$$

Substitutes Eqs(15),(20),and (23) into Eq.(24), it becomes

$$4\pi\phi(x,y)=-\int\left[(\nabla\phi\cdot n+M_n\frac{\partial\phi}{c\partial\tau})\frac{1}{|1-M_r|r}\right]_{\tau=\tau'}dS$$
$$-\int\left[\frac{\phi(\omega\times n)\cdot M+n\cdot(A/c\phi+M\frac{\partial\phi}{\partial\tau})}{cr(1-M_r)|1-M_r|}\right.$$
$$\left.+\frac{A\cdot r/c^2+M_r-M^2}{r^2|1-M_r|^3}M_n\phi\right]_{\tau=\tau'}dS$$
$$+\int\frac{A\cdot r/c^2+1-M^2}{r^2|1-M_r|^3}\bar{r}\cdot(n\phi)dS$$
$$+\int\frac{-M\cdot(n\phi)/r+\bar{r}\cdot[(\omega\times n)\phi/c+n\frac{\partial\phi}{c\partial\tau}]}{r(1-M_r)|1-M_r|}dS \qquad (25)$$

The above equation is a generalized form of the Kirchhoff's formula and is valid for arbitrary moving boundary. Equation (25) was first derived in [4] and [5] and was applied to predict aerodynamic loading of the high speed propeller and helicopter blades by panel method (boundary integral method).

The integral representation of the FW–H equation used in [1] and [6] can also be reproduced using the present notions, that is

$$p=-\Phi_t\left[\cdot M\rho_0c^2\right]+\Phi_{\nabla\cdot}\left[p\right] \qquad (26)$$

The explicit form of Eq.(26) is given by

$$4\pi p(x,t)=\int\rho_0c^2\left[\frac{(\omega\times n)\cdot M+n\cdot A/c}{cr(1-M_r)|1-M_r|}+\frac{A\cdot r/c^2+M_r-M^2}{r^2|1-M_r|^3}M_n\right]_{\tau=\tau'}dS$$
$$+\int\frac{A\cdot r/c^2+1-M^2}{r^2|1-M_r|^3}\bar{r}\cdot npdS$$
$$+\int\frac{-M\cdot np/r+\bar{r}\cdot[(\omega\times n)p/c+n\frac{\partial p}{c\partial\tau}]}{r(1-M_r)|1-M_r|}dS \qquad (27)$$

which can be reduced to that appeared in [1] and [6] under the same assumptions.

Similarly the magnetic flux density is represented by

$$B=\Phi_t\left[(\times\Delta E+\cdot M\Delta B)\right]+\nabla_\times\Phi\left[(\times\Delta B-\cdot M\Delta E)\right]+\nabla\Phi\left[\cdot\Delta B\right] \qquad (28)$$

and its explicit form is

$$4\pi B=-\int\left[-\frac{1}{cr(1-M_r)|1-M_r|}\left[(\omega\times n)(\times\Delta E+\cdot M\Delta B)\right.\right.$$
$$\left.+n(\times\frac{\partial}{\partial\tau}\Delta E+\cdot\frac{\partial}{\partial\tau}M\Delta B)\right]$$
$$\left.-\frac{A\cdot r/c^2+M_r-M^2}{r^2|1-M_r|^3}n(\times\Delta E+\cdot M\Delta B)\right]_{\tau=\tau'}dS$$

$$+\int \frac{A\cdot r/c^2+1 \quad -M^2}{r^2|1-M_r|^3}\left[\bar{r}\times(n(\times\Delta B-\cdot M\Delta E))+\bar{r}(n\cdot\Delta B)\right]dS$$

$$+\int \frac{1}{r(1-M_r)|1-M_r|}\Big[-M\times(n(\times\Delta B-\cdot M\Delta E))/r-M(n\cdot\Delta B)/r$$
$$+\bar{r}\times[(\omega\times n)(\times\Delta B-\cdot M\Delta E)+n(\times\frac{\partial}{\partial\tau}\Delta B-\cdot\frac{\partial}{\partial\tau}(M\Delta E)]/c$$
$$+\bar{r}\,[(\omega\times n)\cdot\Delta B+n\cdot\frac{\partial}{\partial\tau}\Delta B]/c\Big]dS \qquad (29)$$

Although equation (29) is very complex but its derivation is rather straightforward using the present method.

Conclusions

The operatoral approach presented in this paper unifies the derivations of the integral formulas of the scalar and the vector wave equations into a generalized formula. Using this notions, the techniques used in deriving the integral formulas for the FW–H equation can be extended to other field problems such as electromagnetic field problems or aerodynamic problems. The present results, originated from Farassat's work in aeroacoustics, provide a theoretical background for the boundary integral method (panel method) of the wave equation with arbitrary moving boundary. Potential application of boundary integral method to electromagnetic phenomena is a subject of future study and is currently underway.

References

1. Farassat, F. and Brown, T.J.,"A New Capability for Predicting Helicopter Rotor and Propeller Noise Including the Effect of Forward Motion,"NASA TM –74037, 1977.
2. Farassat, F., "Discontinuities in Aerodynamics and Acroacoustics : The Concept and Applications of Generalized Derivatives," *Journal of Sound and Vibration*, Vol.55(2), pp. 165–193, 1977.
3. Farassat, F., "Linear Acoustic Formulas for Calculation of Rotating Blade Noise," *AIAA Journal*, Vol. 19, No. 9, pp 1122–1130, 1981.
4. Lee, Y. J. "On the Integral Formulations of Wave Equation for Arbitrary Moving Boundary and Its Applications to Aerodynamics and Aeroacoustics." Ph.D. Thesis (in Chinese), Institute of Applied Mechanics, National Taiwan University, Dec.,1988.
5. Lee, Y. J.and Yang, J.Y."A Panel Method for Arbitrary Moving Boundaries Problems. " to appear in *AIAA Journal Dec. 1989.*
6. Long, L.N.,"An Aerodynamic Theory Based on Time–Domain Aeroacoustics.", *AIAA Journal*, Vol. 23, No. 6, pp. 875–882, 1985.

Time and Frequency Domain Aerodynamics for Flutter of Helicopter Rotors in Hover

F. Mastroddi and L. Morino

Università di Roma "La Sapienza", Rome, Italy

The objective of this work is to assess the accuracy of the aerodynamic modeling used in the flutter analysis of helicopter rotors in hover. A boundary-element formulation for unsteady potential incompressible flows is introduced. The results obtained for the aerodynamic forces, for pitch and flap, are compared with those typically used in the flutter analysis of rotors, (which are obtained using a strip-theory extension of Loewy's formulation [3], although a boundary-element formulation has been used by Kwon [2] for a flutter analysis). The results demonstrate that Loewy's formulation [3] yields results that are considerably different from those obtained with the more accurate model introduced here.

1. Formulation

In the equations of the analysis of the classical linear flap-pitch flutter in frequency domain [1], the flap and pitch aerodynamic moments $\tilde{M}_\beta$ and $\tilde{M}_\theta$ [1] are assumed to be expressed as linear combinations of in the flap and pitch angles, $\tilde{\beta}$ and $\tilde{\theta}$,

$$\tilde{\mathbf{m}} = \begin{Bmatrix} \tilde{M}_\beta \\ \tilde{M}_\theta \end{Bmatrix} = \begin{bmatrix} \tilde{E}_{\beta\beta} & \tilde{E}_{\beta\theta} \\ \tilde{E}_{\theta\beta} & \tilde{E}_{\theta\theta} \end{bmatrix} \begin{Bmatrix} \tilde{\beta} \\ \tilde{\theta} \end{Bmatrix} = \tilde{\mathbf{E}}\,\tilde{\mathbf{q}} \tag{1}$$

In Eq. 1, the coefficient $E_{\theta\beta}$, for instance, is given by

$$\tilde{E}_{\theta\beta} = -\oiint_{S_R} \tilde{C}_{p_\beta}\, \mathbf{n}\cdot\mathbf{M}_\theta\, dS \tag{2}$$

where $\mathbf{M}_\theta$ is the pitch mode, $\tilde{C}_{p_\beta}$ is the pressure due to the flap motion, S_R is the surface of the rotor, and $\mathbf{n}$ is the outward normal at this surface. Similar definitions hold for the other entries of the matrix $\tilde{\mathbf{E}}$.

As mentioned above, in the past, the aerodynamic matrix $\tilde{\mathbf{E}}$ has been evaluated using Loewy's theory [3] (see for instance Ref. [8]). Here, we consider a method for the evaluation of the aerodynamic matrix. This method is closely related to that of Ref. [6], where details of the formulation are available.

[1] The overwritten $\tilde{}$ denotes the Laplace transform

The flow is assumed to be incompressible and potential. The governing equation is the continuity equation, $\nabla \cdot \mathbf{v} = 0$. For potential flows we have

$$\nabla^2 \phi = 0 \tag{3}$$

The boundary conditions are: (*i*) $\mathbf{v} \cdot \mathbf{n} = \mathbf{v}_R \cdot \mathbf{n}$ (impermeability) or $\partial\phi/\partial n = \mathbf{v}_R \cdot \mathbf{n}$ on $\mathcal{S}_R$; (*ii*) $\phi = 0$ at infinity (air frame of reference); (*iii*) $\Delta(\partial\phi/\partial n) = 0$ on the wake surface $\mathcal{S}_W$; (*iv*) $\Delta\phi$ =constant following a wake point (where $\Delta\phi$ denotes the discontinuity of the velocity potential across the wake surface); and (*v*) $\Delta\phi = \Delta\phi^{TE}$ at the trailing edge (where $\Delta\phi^{TE}$ denotes the difference of the trailing-edge values of ϕ on the upper and lower side of the body). Using the direct integral-equation approach, one obtains (see, *e.g.*, Ref. [6])

$$E(\mathbf{x}_*)\phi(\mathbf{x}_*) = \oiint_{\mathcal{S}_R} \left(\frac{\partial \phi}{\partial n} G - \phi \frac{\partial G}{\partial n} \right) d\mathcal{S} - \iint_{\mathcal{S}_W} \Delta\phi \frac{\partial G}{\partial n} d\mathcal{S} \tag{4}$$

where $G = 1/4\pi\|\mathbf{x} - \mathbf{x}_*\|$ is the fundamental solution of the three-dimensional Laplace equation. In Eq. 4, $E(\mathbf{x}_*) = 1$ if $\mathbf{x}_*$ is in the fluid volume, whereas $E(\mathbf{x}_*) = 0$ if $\mathbf{x}_*$ is inside $\mathcal{S}_R$. If $\mathbf{x}_*$ is on a smooth point of $\mathcal{S}_R$, then $E(\mathbf{x}_*) = 1/2$ and Eq. 4 is an integral equation for the unknown ϕ.

Note that the boundary conditions (*iv*) and (*v*) given above may be combined to yield $\Delta\phi(\mathbf{x}_W, t) = \Delta\phi^{TE}(t - \tau)$, where τ is the convection time, *i.e.*, the time necessary for a fluid point to be convected from $\mathbf{x}_{TE}$ to $\mathbf{x}_W$. Neglecting higher-order terms, this convection time may be expressed for a rotor in hover as $\tau = \alpha/\Omega$, where Ω is the angular velocity of the rotor and α is the azimuth angle along the wake. Combining with Eq. 4, assuming $\mathbf{x}_*$ to be a smooth point of $\mathcal{S}_R$, and taking the Laplace transform, one obtains,

$$\frac{1}{2}\tilde{\phi}(\mathbf{x}_*) = \oiint_{\mathcal{S}_R} \left(\frac{\partial \tilde{\phi}}{\partial n} G - \tilde{\phi} \frac{\partial G}{\partial n} \right) d\mathcal{S} - \iint_{\mathcal{S}_W} \Delta\tilde{\phi}^{TE} \, e^{-s\alpha/\Omega} \, \frac{\partial G}{\partial n} d\mathcal{S} \tag{5}$$

Equation 5 is the transform of the time-domain integral equation used in Ref. [6] for rotor. This equation is discretized using a zeroth-order boundary-element method. Dividing $\mathcal{S}_R$ and $\mathcal{S}_W$ into N_R and N_W elements, respectively, and assuming $\tilde{\phi}$, $\tilde{\psi} = \partial\tilde{\phi}/\partial n$, and $\Delta\tilde{\phi}$ to be constant within each element one obtains

$$\frac{1}{2}\tilde{\phi}_k = \sum_{h=1}^{N_R} \left(B_{kh}\tilde{\psi}_h + C_{kh}\tilde{\phi}_h \right) + \sum_{n=1}^{N_W} F_{kn} \, e^{-s\alpha_n/\Omega} \, \Delta\tilde{\phi}_n^{TE} \tag{6}$$

with

$$B_{kh} = \iint_{\mathcal{S}_h} G_k d\mathcal{S} \qquad C_{kh} = \iint_{\mathcal{S}_h} \frac{\partial G_k}{\partial n} d\mathcal{S} \qquad F_{kn} = \iint_{\mathcal{S}_n} \frac{\partial G_k}{\partial n} d\mathcal{S} \tag{7}$$

where $G_k = 1/4\pi\|\mathbf{x} - \mathbf{x}_k\|$ whereas $\mathcal{S}_h$ and $\mathcal{S}_n$ denote *hth* element on $\mathcal{S}_R$ and the nth on $\mathcal{S}_W$, respectively. The expressions for the evaluation of the coefficients B_{kh}, C_{kh} and F_{kn} are given in Ref. [7]. Once $\tilde{\phi}$ is known, the pressure is evaluated using Bernoulli's theorem and the aerodynamic matrix, $\tilde{\mathbf{E}}$, is obtained using Eq. 2 and related expressions.

2. Results

The results presented here are based on the work of Mastroddi [4], where additional results are available. The case considered here is that of the rotor considered in the flutter analysis of Ref. [5], *i.e.*, an isolated single-bladed rotor having rectangular planform, with $c/R = 0.05$. Our results are obtained using a symmetric airfoil with thickness ratio $\tau_R = 0.01$ and with a classical helicoidal wake with a wake spacing $h/c = 0.7$ as suggested in Ref. [1] for typical hovering conditions. For all the results the number of elements in the chordwise direction is 6, in the radial direction is 12, the number of elements used to represent each wake spiral in azimuth direction is 12 (this choice is based on the results of a convergence analysis, see [6]).

First consider an analysis of the effect of the number of spirals on the solution. The real and imaginary parts of the pitch moment coefficient due to pitch mode are shown in Figs. 1 to 3 as a function of the number of spirals, for $\omega/\Omega = 0.25$, 0.5, and 1.0, respectively (where ω is the frequency of oscillation). The pitching mode is around a quarter-chord axis. The results indicate that the eight-spiral solution is close to convergence. These values have been used for all the following results.

Next, consider a comparison between the time-domain and the frequency-domain solution. The pitch moment coefficient due to a pitch rigid mode is shown in Fig. 4 and 5 for 4 and 8 spirals, respectively. In the time domain solution, $\Delta\phi$ is assumed to be zero at time $t = 0$ for all the wake elements. In both cases $\omega/\Omega = 1$ and after the transient, the moment coefficient stabilizes itself around a steady amplitude. A comparison of the two time domain solutions of Figs. 4 and 5 is shown in Figure 6. The two curves coincide for the first four revolutions (*i.e.*, the first 48 time steps): this is to be expected because, in this case, $\Delta\phi \neq 0$ only on the first four spirals. It is apparent that the magnitude of the four-spiral case is considerably higher that those for the eight-spiral one (in agreement with the results of Fig. 3).

Finally, the real and the imaginary parts of the flap and pitch moment coefficients due to flapping and pitching modes as a function of ω/Ω are presented in Figs. 7 to 10. The results are compared with those obtained using Loewy's formulation within a strip theory approach. Note the difference in the ordinates. The largest scales are for the flap moments (due to both flap and pitch modes). The pitch moments (due to both flap and pitch modes) are smaller (the Loewy's theory [3] predicts a zero value in this case). Our results show the same trend as the results based on Loewy's formulation. However, quantitatively there is a considerable difference between the results. We believe that this disagreement is due to the simplifying approximations that are used in Loewy's formulation, but not in the more sophisticated formulation presented here.

3. Concluding remarks

Present work has examined an integral formulation for unsteady incompressible aerodynamics for hovering isolated rotors in the frequency and time domain. We successfully tested the algorithm against the time-domain solution used in Ref. [6]. The most important result that we obtained is that there exists considerable disagreement between our results and those based on the Loewy's formulation [3], which is widely accepted as the standard tool for flutter analysis. As mentioned above we believe that the disagreement is due to Loewy's simplifying approximations that are not in the more sophisticated three-dimensional unsteady formulation presented here. This results are to be considered preliminary: a more thorough convergence analysis appear to be desirable, especially in terms of numbers of spirals. If the present results are confirmed, the conclusion would be that Loewy's theory may be used to obtain qualitative results, but is inadequate for a quantitative analysis.

Applications to aeroelasticity are now underway. Preliminary results confirm that there exists a considerable discrepancy on the flutter boundaries between Loewy's and the present formulation. Current work includes also the extension of the present formulation to compressible flows and forward flight.

References

[1] Johnson, W., (1980), "Helicopter Theory," Princeton University Press, Princeton NJ.

[2] Kwon, O. J., (1988), "A Technique for the Prediction of Aerodynamics and Aeroelasticity of Rotor Blades," Ph.D. Thesis, Georgia Institute of Technology, Atlanta, Georgia.

[3] Loewy, R. G., (1957), "A Two-Dimensional Approximation of Unsteady Aerodynamics of Rotary Wings," Journal of Aeronautical Sciences, Vol. 24, No. 2, pp. 81-92, 144.

[4] Mastroddi, F., (1989), "Metodo degli Elementi al Contorno per l'Analisi Aerodinamica ed Aeroelastica di Ali e Rotori in "Hover" nel Dominio della Frequenza," Tesi di Laurea, Università degli Studi "La Sapienza", Roma, Italy.

[5] Miller, R. H., and Ells, C. W., (1956), "Helicopter Blade Vibration and Flutter," Journal of American Helicopter Society, Vol. 1, No. 3, pp. 19-38.

[6] Morino, L., Bharadvaj, B. K., "A Unified Approach for the Potential and Viscous Free-Wake Analysis of Helicopter Rotors," Vertica, Vol. 12, No. 1/2, pp. 147-154, 1988.

[7] Morino, L., Chen, L. T., and Suciu, E. O., (1975) "Steady and Oscillatory Subsonic and Supersonic Aerodynamic around Complex Configuration," AIAA Journal, Vol. 13, pp. 368-374.

[8] Venkatesan, C., and Friedmann, P. P., (1985), "New Approach to Finite-State Modeling of Unsteady Aerodynamics," AIAA Journal, Vol. 24, No. 12, pp. 1889-1897.

Acknowledgments

This work was partially supported by the Army Research Office (Contract No. DAAL-03-87-K-0022 to Boston University).

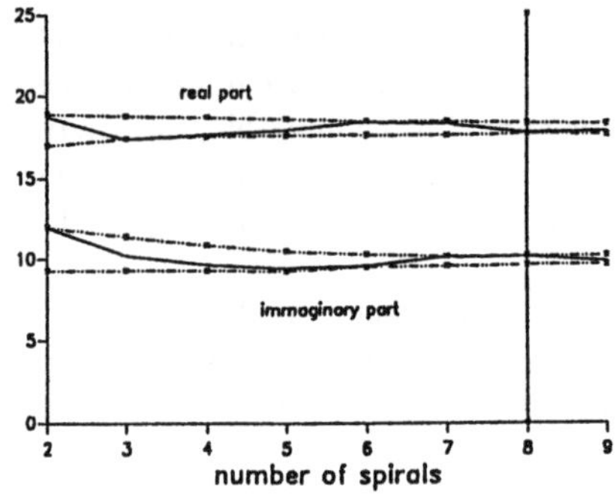

Figure 1. Effect of the number of spirals on $\tilde{E}_{\theta\theta}$ for $\omega/\Omega = 0.25$.

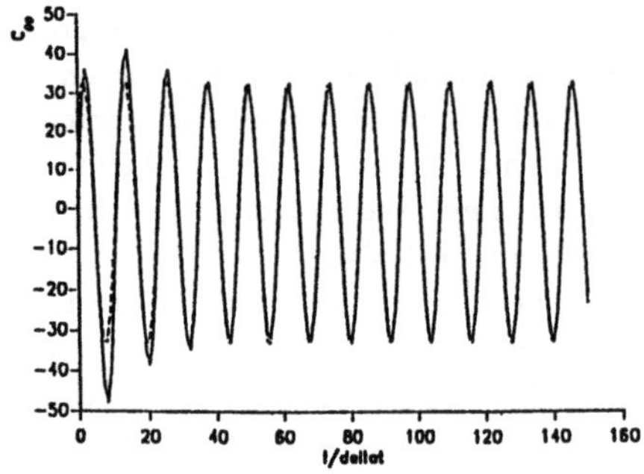

Figure 4. Comparison of frequency (---) and time (—) domain solution for 4 wake spirals: $\omega/\Omega = 1$.

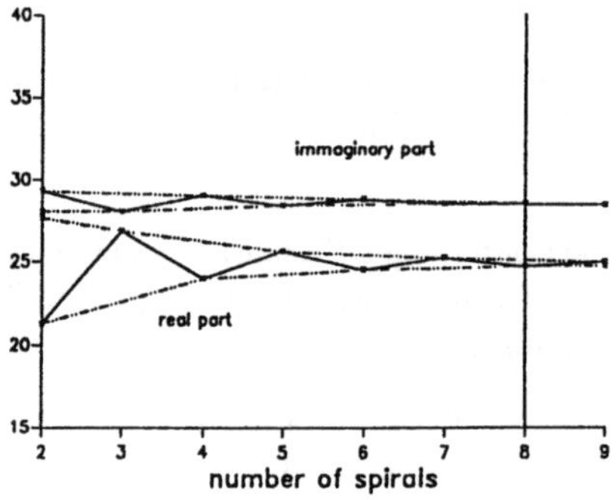

Figure 2. Effect of the number of spirals on $\tilde{E}_{\theta\theta}$ for $\omega/\Omega = 0.5$.

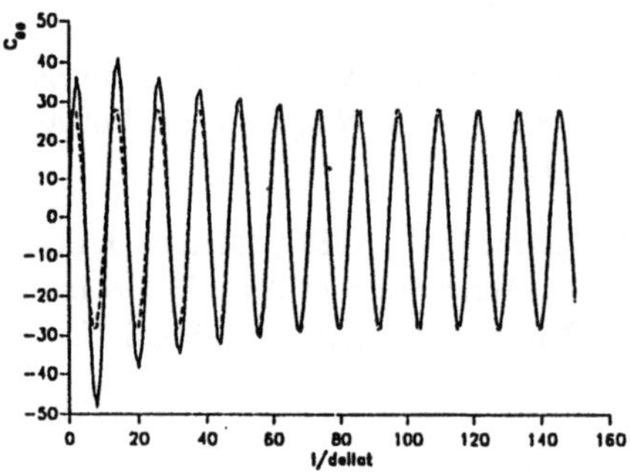

Figure 5. Comparison of frequency (---) and time (—) domain solution for 8 wake spirals: $\omega/\Omega = 1$.

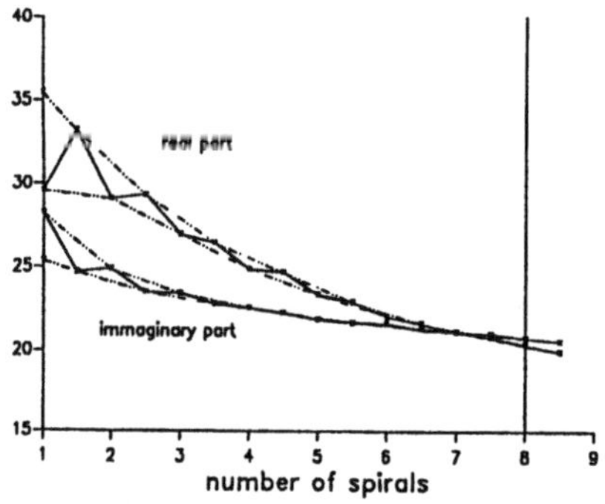

Figure 3. Effect of the number of spirals on $\tilde{E}_{\theta\theta}$ for $\omega/\Omega = 1$.

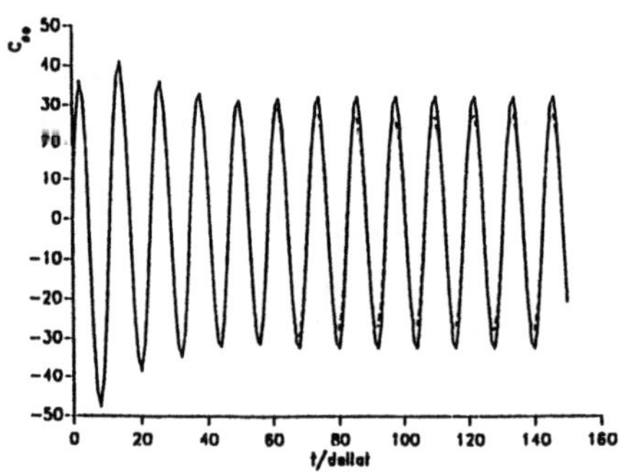

Figure 6. Time domain solution for 4 (—) spirals and 8 (---) spirals: $\omega/\Omega = 1$.

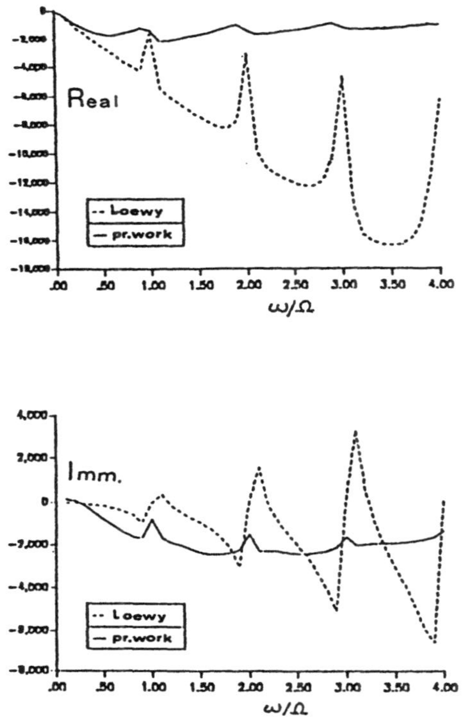

Figure 7. Real and immaginary parts of $\bar{E}_{\beta\beta}$, flap coefficient due to flapping mode.

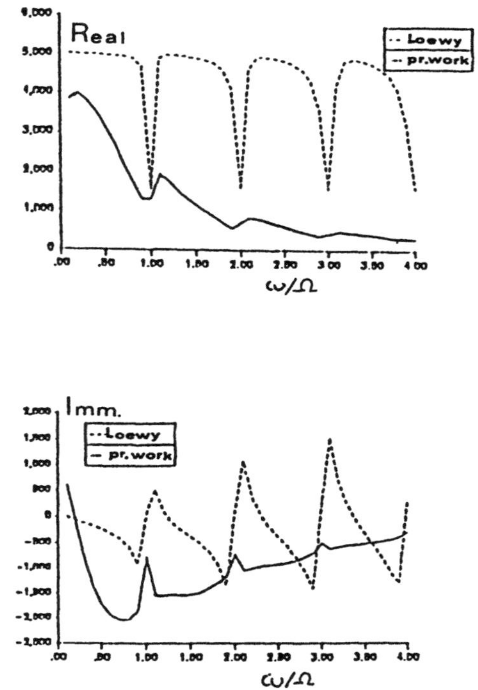

Figure 8. Real and immaginary parts of $\bar{E}_{\beta\theta}$, flap coefficient due to pitching mode.

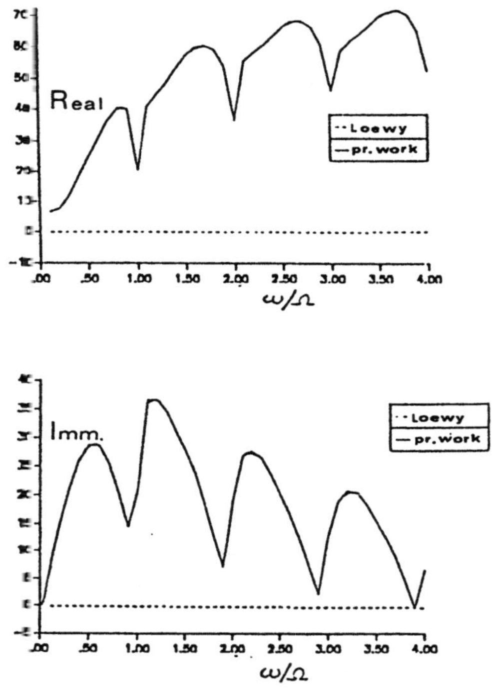

Figure 9. Real and immaginary parts of $\bar{E}_{\theta\theta}$, pitch coefficient due to pitching mode.

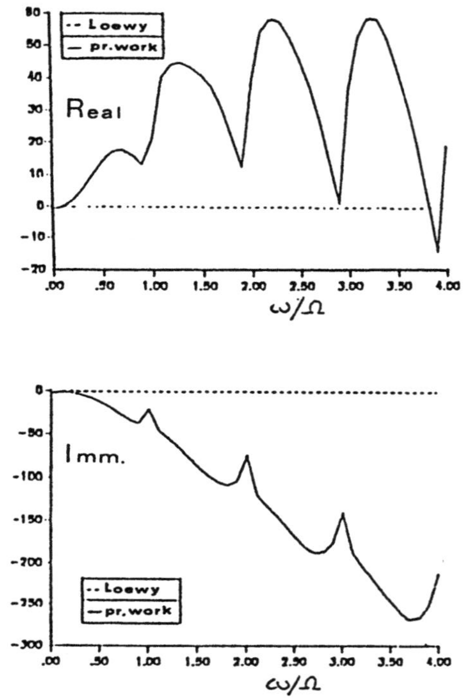

Figure 10. Real and immaginary parts of $\bar{E}_{\theta\beta}$, pitch coefficient due to flapping mode.

Application of a Panel Method to Aeroelastic Stability of Rotors in Hover

OH JOON KWON, DEWEY H. HODGES,
DAVID A. PETERS and LAKSHMI N. SANKAR

School of Aerospace Engineering
Georgia Institute of Technology
Atlanta, Georgia

Summary

A panel method based on the piecewise constant source and doublet singularities is applied to helicopter rotors in the hovering flight condition to investigate the effects of three-dimensional unsteady aerodynamics on coupled flap-lag-torsion aeroelastic stability using a realistic contracting wake model. The structural model of the elastic blade is a geometrically nonlinear beam model which allows for moderately large deflections. The results are compared with traditional two-dimensional airfoil theory and with experimental data where available.

Aerodynamic Model

For the case of an incompressible, inviscid, irrotational fluid, the Navier-Stokes equation can be reduced to the classical Laplace's equation. From Green's third identity, a solution to the Laplace's equation in the flow field and inside the body can be expressed in an integral form over the boundary surface. If the internal free stream condition is applied, the integral equation becomes

$$0=\iint_{S-P}\varphi\vec{n}\cdot\vec{\nabla}(\frac{1}{r})dS-2\pi\varphi_P+\iint_{W}(\varphi_u-\varphi_l)\vec{n}\cdot\vec{\nabla}(\frac{1}{r})dW-\iint_{S}\frac{1}{r}\vec{n}\cdot\vec{\nabla}\varphi dS \quad (1)$$

where φ is the perturbation potential and r is the distance from the blade surface element to a point on the blade, P. The first and second terms represent the unknown doublet distributions on the blade surface, S. The third term represents the doublet distribution on the wake, W, which can be defined in the limiting sense as the difference between the doublet strengths on the upper and lower surfaces at the blade trailing edge. The fourth term represents the contribution of the sources on the blade surface, the strengths of which are directly determined from the external Neumann boundary condition of zero penetration.

The far wake and the effect of preceding blades are assumed to influence the primary blade as an induced velocity which is added to the free stream velocity in determining the source strengths on the blade surface. In the present work, a prescribed-wake, lifting-line technique is used to provide the information concerning the near wake geometry, the effect of the far wake, and the influence of preceding blades. For a detailed description of the aerodynamic model, the reader is referred to [1].

Structural Model

The equations of motion of the blade are a coupled set of nonlinear integro-partial differential equations [2, 3] which are second-order in time and fourth-order in space. They govern the coupled flap bending, chordwise bending, and torsion of a slender, homogeneous, transversely isotropic beam with constant spanwise mass and stiffness properties. The bending and torsional deflections u, v, w, and ϕ, are illustrated in Fig. 1.

These equations of motion are reduced to ordinary differential equations in time by expansion of the flap bending, chordwise bending, and torsion deflection in terms of a series of generalized coordinates and mode shape functions and application of Galerkin's method. The nonlinear, nonhomogeneous, constant-coefficient ordinary differential equations are then linearized for small perturbation motions about the equilibrium operating condition by a separation of the time-dependent generalized coordinates into steady equilibrium quantities and small unsteady perturbation quantities [2].

First, if one substitutes only the steady equilibrium quantities into the nonlinear modal equations, then all of the time derivative terms disappear. The remaining terms constitute nonlinear algebraic equations which govern the equilibrium deflections. The second set of equations is obtained by a substitution of the steady equilibrium and dynamic perturbation quantities, followed by subtraction of the steady equilibrium equations. All nonlinear products of dynamic perturbation quantities are discarded. These linear, non–homogeneous, constant–coefficient ordinary differential perturbation equations define the unsteady blade motion near the equilibrium operating condition.

Numerical Procedure

Since the integrals appearing in equation (1) are over the blade surface and the near wake, these are replaced by sums of elemental integrals over panels which cover the blade surface and the wake. Equation (1) is satisfied simultaneously at a control point on each panel and thus can be represented by a system of algebraic equations for the doublet strength. If there are N_s panels representing the blade surface and N_w panels for the wake, equation (1) becomes

$$\sum_{k=1,k\neq j}^{N_s} (\mu_k C_{jk}) - 2\pi\mu_j + \sum_{k=1}^{N_w}(\mu_{w_k} D_{jk}) = -\sum_{k=1}^{N_s} \sigma_k B_{jk} \qquad (j = 1, ..., N_s) \qquad (2)$$

Here, μ_k is the unknown doublet value on the *k-th* panel and is equal to the perturbation velocity potential φ_k. Also, σ_k is the source strength on the *k-th* panel. The aerodynamic influence coefficients, B_{jk}, C_{jk}, and D_{jk}, are only functions of blade and wake geometries.

The calculations proceed as follows. Initially, a prescribed-wake, lifting-line code is run to obtain the induced downwash distribution due to the far wake and the preceding blades for a given flight condition. At first, the blade is assumed to be undeformed. From the known blade geometry and the flight condition, a system of equations is solved for the strengths of the blade surface doublets and the aerodynamic loads are then determined. A Newton-Raphson iteration of the non–linear structural equations provides the deflections of the blade which then define a new blade geometry. Next, new aerodynamic forces and moment are obtained for the current state of blade deformation. This iterative process is repeated until a satisfactory convergence is obtained for the equilibrium deflections.

After the equilibrium deflections are obtained, the unsteady equations of motion are integrated in time using a *5th* or *6th* order Runge-Kutta-Verner method. At each time step, a new set of aerodynamic loads is obtained from the known blade geometry and blade motion. The generalized force is calculated for these aerodynamic loads, which allows one to use the Runge-Kutta scheme to predict new deflections at the next time step. This process is continued for several rotor revolutions to provide the time history of blade motion. Once the time history of blade motion is known, the modal damping and frequency of any desired mode can be determined from a Fourier analysis.

For a detailed description of the numerical procedure, the reader is referred to [4].

Discussion of Results

Initially, the panel method is applied to a rigid rotor and the aerodynamic load distributions are compared with the experimental data [5] for a two-bladed hovering rotor. The model rotor had a solidity of 0.1061 with NACA 0012 airfoil section. The rotor blade had an untwisted, rectangular planform shape and was set at a collective pitch angle of 8.0 degrees. Fig. 2 shows the correlation of the calculated chordwise pressure distribution with the experimental data at several spanwise stations. The comparisons show good agreement over most of the chord for all the spanwise sections from inboard to outboard.

The influence of the present aerodynamic model on the steady equilibrium deflections has been discussed in detail in [4] and is briefly described here. The analysis, based on three-dimensional panel method aerodynamics, predicts slightly smaller flap deflection compared to the two-dimensional theory. This loss in flap deflection is concomitant with a decrease in thrust due to tip loss phenomena. Even though there exists a large difference between the two aerodynamic models for the induced drag distribution along the span, a small difference in lead-lag deflection is observed. The torsional deflection by the present panel method is about two thirds of what the two-dimensional theory predicts.

Once the steady equilibrium deflections are obtained, the blade is perturbed by an amount equal to 10% of the equilibrium deflections and released. The unsteady blade motion under the interaction of elastic blade and applied aerodynamic loads is then traced in time. Fig. 3 shows typical perturbation motions and the loads of a stiff inplane hingeless rotor blade [6]. The deflections shown in the figures are nondimensional tip deflections. The applied aerodynamic loads are shown for the first mode of each motion in the form of the generalized perturbation modal forces and moment. The results of the two-dimensional quasi–steady Greenberg theory are also shown for comparison.

Blade lead-lag, flap, and torsion are strongly coupled with each other so that the behavior of all three components shows up in each mode. In Fig. 3 (d), it is shown that the lift force predicted by the Greenberg theory is much larger than that of the panel method when the blade initially starts the oscillatory motion. Thus, the flapping motion obtained by this theory is damped out faster than the motion with the panel method aerodynamics as shown in Fig. 3 (c). The effect of lead-lag and torsion is also superimposed on the flapping motion at this early stage of motion, but is not seen separately. After about one and half revolutions of the blade, the flapping motion of the blade dies down except the motion excited by lead-lag through the structural coupling. This remaining motion due to coupling occurs at the same frequency as the lead-lag motion.

The lead-lag perturbation motion and the associated force in that direction are shown in Figs. 3 (a) and (b). The force here is the total induced drag in generalized form. The viscous profile drag is not shown because it is based on the constant drag coefficient and is the same as used in the two-dimensional theory. At the initial stage of blade motion, the magnitude of the lead-lag motion is quickly reduced for the Greenberg theory, which is due to the structural coupling under the effect of large flap damping. After one and half revolutions, the influences of flapping and torsional oscillations have died down, but the pure lead-lag motion persists. During the early portion, the induced drag is dominated by torsional and flapping motions and varies at the torsional frequency with respect to time.

Similar trends are also observed for the torsional perturbation motion as shown in Figs. 3 (e) and (f). It is seen that the generalized aerodynamic moment of the present panel method has a different magnitude and phase from the quasi–steady Greenberg theory for which the sectional moment is only a function of time change of the collective pitch angle. After about one and half revolutions of the blade, the torsional perturbation has mostly died down except the one due to the coupling between torsion and lead-lag motions which induces an oscillatory torsional motion at the lead-lag frequency.

The results of lead–lag damping and frequency values obtained from the unsteady blade motion are shown in Fig. 4 for a stiff-inplane hingeless model rotor with soft pitch flexure, zero precone, and droop for various collective pitch angles [7]. At low collective pitch angles, the aerodynamic force which contributes to lead-lag damping is mostly the viscous profile drag; and, thus, the two models predict approximately the same damping. However, as the absolute value of the pitch angle is increased, the two-dimensional theory predicts lead-lag damping that is approximately twice as high as the experimental value; while the results based on the present three-dimensional, unsteady aerodynamic model agree with the experimental results much better, falling closer to the scatter of the data. The improved predictions are attributed to the use of unsteady wake inflow and to the three–dimensional tip effect, which the two–dimensional theory lacks [8]. The lead–lag frequencies are not significantly affected by the present aerodynamic model. It is expected that blade stall effects will account for the discrepancy between theory and experiment in accordance with the discussion in [7].

Conclusion

A three–dimensional, unsteady, panel method aerodynamic model has been coupled to a structural dynamic model of a hingeless rotor with elastic blades. The resulting analysis has been applied to a two–bladed rotor in hover to investigate the effects of three-dimensional panel method aerodynamics on coupled flap–lag–torsion aeroelastic stability. The present results show the usefulness of a fully unsteady, three–dimensional, aerodynamic model with a realistic contracting wake which accounts both the unsteady inflow dynamics the unsteady tip relief effects for the accurate prediction of low–frequency aeroelastic stability in hover.

Acknowledgments

The authors gratefully acknowledge the support of this work by the U.S. Army Research Office as part of the Center of Excellence in Rotary Wing Aircraft Technology, contract numbers DAAG29-82-K-0094 and DAAL3-88-C-0003, monitored by Dr. Robert Singleton.

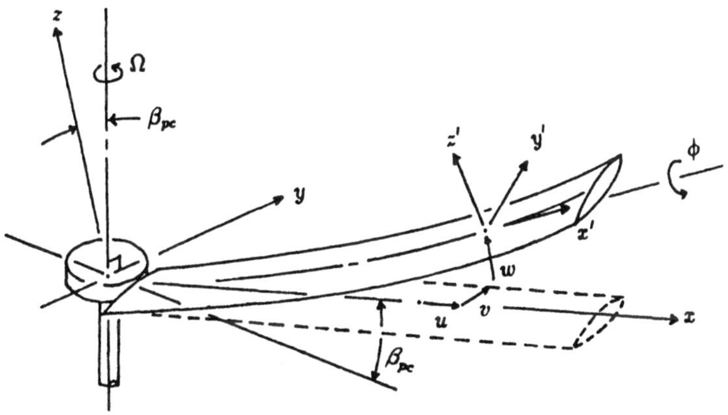

Figure 1: Rotor blade coordinate system and deflections (from Ref. 2).

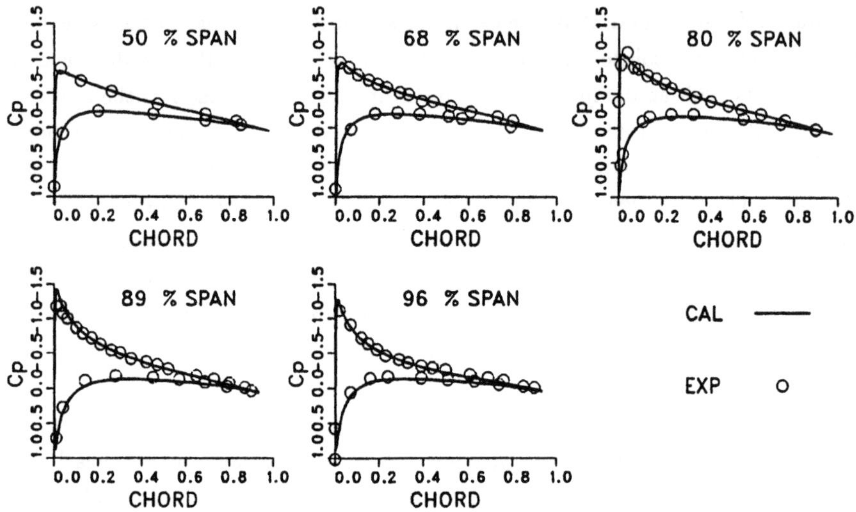

Figure 2: Comparison of chordwise pressure distributions of a rigid rotor.

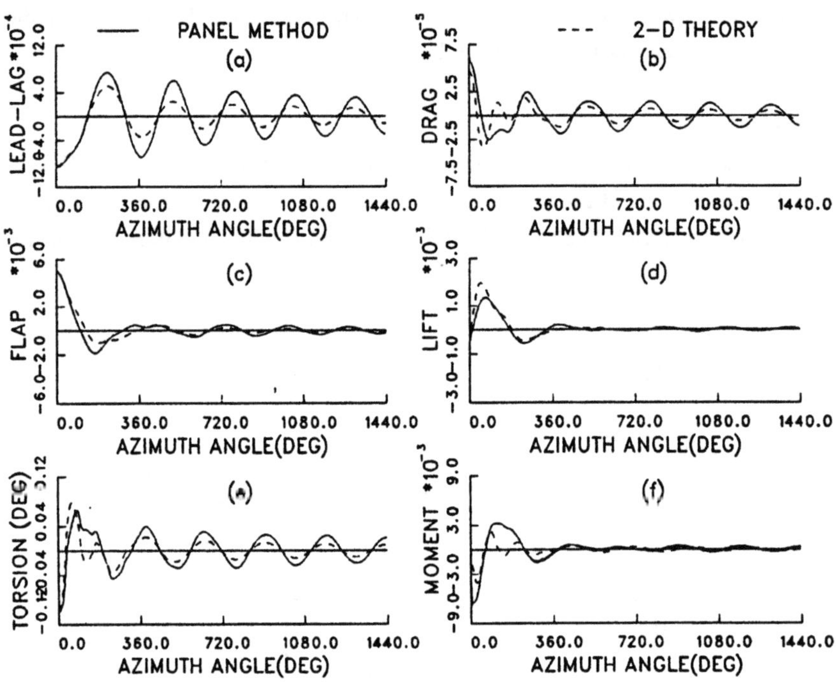

Figure 3: Typical unsteady perturbation motion of the blade and aerodynamic loads at 8 deg. collective pitch angle.

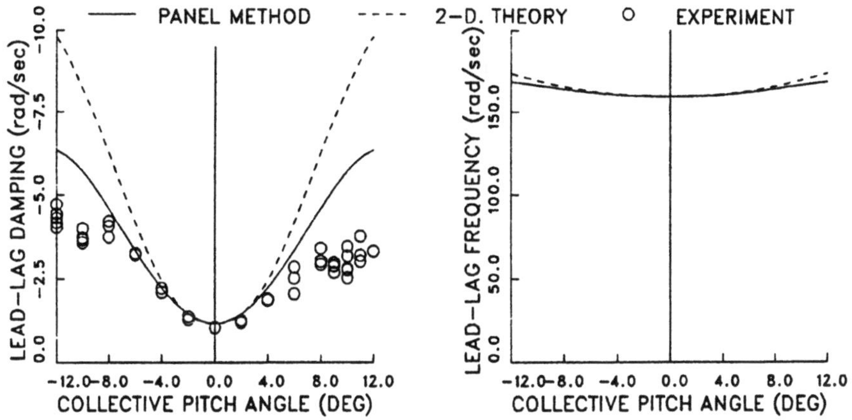

Figure 4: Comparison of lead-lag damping and frequency of a stiff inplane rotor with zero precone and droop angles.

References

1. Kwon, Oh Joon, *A Technique for the Prediction of Aerodynamics and Aeroelasticity of Rotor Blades*, Ph.D. Thesis, School of Aerospace Engineering, Georgia Institute of Technology, December 1988.

2. Hodges, D. H., and Ormiston, R. A., "Stability of Elastic Bending and Torsion of Uniform Cantilever Rotor Blades in Hover with Variable Structural Coupling," NASA TN D-8192, May 1976.

3. Hodges, D. H., "Nonlinear Equation of Motion for Cantilever Rotor Blades in Hover with Pitch Link Flexibility, Twist, Precone, Droop, Sweep, Torque Offset, and Blade Root Offset," NASA TM X-73,112, May 1976.

4. Kwon, O. J., and Hodges, D. H., "Application of Panel Method Aerodynamics to Rotor Aeroelasticity in Hover," AIAA Paper 89-1234.

5. Caradona, F. X. and Tung, C., "Experimental and Analytical Studies of a Model Helicopter Rotor in Hover," NASA CR-81232, 1981.

6. *Integrated Technology Rotor Accessment Workshop - Methodology Accessment Phase II*, NASA Ames Research Center, 1988, Task 86h.

7. Sharpe, D. L., "A Comparison of Theory and Experiment for Aeroelastic Stability of a Hingeless Rotor Model in Hover," Integrated Technology Rotor Assessment Workshop, NASA CP-10007, June 1983, pp. 7 - 26.

8. Hodges, D. H., Kwon, O. J., and Sankar, L. N., "Stability of Hingeless Rotors in Hover Using Three-Dimensional Unsteady Aerodynamics," *Proceedings of the 45th Annual Forum of the American Helicopter Society*, Boston, MA, May 22-24, 1989.

A General Purpose Fixed-Wing and Rotary-Wing Compressible Aerodynamics Analysis Method

KADIN TSENG

United Technologies Research Center
East Hartford, Connecticut 06108
U.S.A.

Summary

An existing boundary integral formulation for steady or unsteady subsonic flow for fixed-wing and rotary-wing applications is being recast into a form in which the wake and/or field influence is coupled into the linear Laplace solver through the boundary conditions. There are a number of advantages in using this approach. For instance, since the wake effect is "seen" as an induced velocity, any methodology that can provide this information may be used. From an engineering point of view, this is desirable for its flexibility and versatility. Furthermore, for transonic flow this formulation does not require a body-fitted field grid. This is of considerable advantage for flows around complex configurations. In addition, this new formulation is essentially based on the concept of velocity decomposition. Therefore, future extension to allow for viscous effects may be accomodated in a fairly straight-forward manner.

Introduction

The use of boundary element methods to solve inviscid irrotational flow may be considered standard practice. Various production codes have been developed over the years for routine applications within the airframe industry. Most of these boundary element methods may be categorized as Morino's Method (Ref. 1). For this type of flow, the unknown is the (scalar) velocity potential. For transonic flows, this technique may be extended to the field to include nonlinear effects such as shock waves (Refs. 2 – 7). For more complicated flow phenomena such as viscous flows, the concept of Helmholtz Decomposition has been adopted by various researchers – notably by Wu (*e.g.*, Ref. 8) and Morino (*e.g.*, Ref. 9).

An inviscid irrotational flow analysis computer code for fixed-wing as well as rotary-wing applications has been developed (Ref. 10). Although it is not neccessary to use the velocity decomposition concept for transonic flows, in order to later deal with viscous effects, the basic formulation has been recast to include the concept of velocity decomposition into an otherwise *linear* boundary element inviscid flow (Laplace) solver. For transonic flows, the nonlinear field effect is coupled-in through the boundary conditions. This concept has previously been demonstrated – within the context of integral methods – by Tseng (Ref. 11) to include the boundary-layer effect in transonic flow and more recently, by Sinclair (Refs. 12, 13) in using an indirect method to solve for fixed-wing transonic flows. In this paper, a generalized formulation which employs the techniques of transpiration boundary condition is presented. But first, the "classical" integral formulation (Ref. 3) is outlined in the following.

Transonic Integral Formulations

The method presented here is applicable to steady as well as unsteady flows. For simplicity, only the formulation for steady flow is discussed here. First, the equations governing the inviscid isentropic fluid flow will be presented. For inviscid fluid, the velocity field is irrotational. Hence, the perturbation velocity **v** may be expressed in terms of a scalar velocity potential ϕ as

$$\mathbf{v} = \nabla\phi \tag{1}$$

In a frame of reference fixed with the aircraft, the governing equation is

$$\nabla^2\phi = \chi \tag{2}$$

where χ represents a nonlinear function in ϕ and density ρ.

In this frame of reference, the boundary conditions associated with Eq. (2) are :

$$\begin{aligned}
\phi(\mathbf{x}) &= 0 && \textit{at } \infty \\
\frac{\partial\phi(\mathbf{x})}{\partial n} &= \mathbf{v}_b\cdot\mathbf{n} && \textit{on } \sigma_b \\
\frac{\partial\phi(\mathbf{x})}{\partial n} &= \mathbf{v}_w\cdot\mathbf{n} && \textit{on } \sigma_w \\
\Delta\phi(\mathbf{x}) &= \Delta\phi(\mathbf{x}_{te}) && \textit{on } \sigma_w
\end{aligned} \tag{3}$$

In the above, the subscripts b and w denote body and wake respectively. Hence, σ_b represents the body (*e.g.*, the propfan blade assembly) surface and σ_w the wake surface while $\mathbf{v}_b$ and $\mathbf{v}_w$ denote the velocity of the body and wake, respectively. Furthermore, **n** is the outwardly directed unit normal on the body (wake). For steady state flow, the potential jump $\Delta\phi$ across the wake sheet is constant along the wake strip and therefore the $\Delta\phi$ at any point **x** on the strip is equal to the $\Delta\phi$ at $\mathbf{x}_{te}$, the trailing edge where the wake strip emanates.

Note that the first boundary condition is obtained by requiring the perturbation velocity, defined by Eq. (1), to vanish at ∞. Impermeability of the body and wake surfaces lead to the second and third equations. Finally, the fourth boundary condition implies no pressure difference across the wake sheet.

Application of the Green's Function Method to invert Eq. (1) into an integral equation and upon introduction of space and time discretization yields (see Ref. 9)

$$\phi_i = \sum_{j=1}^{N_e} B_{ij}\psi_j + \sum_{j=1}^{N_e} C_{ij}\phi_j + \sum_{m=1}^{N_w} F_{im}\Delta\phi_m + \sum_{k=1}^{N_f} H_{ik}\chi_k \quad ; \quad i = 1,2,\ldots,N_e \tag{4}$$

where N_e is the number of body elements, N_w is the number of wake elements, N_f is the number of field elements and B_{ij}, C_{ij}, F_{im} and H_{ik} are defined as

$$B_{ij} = \iint_{\sigma_j} \frac{-1}{4\pi r}\, d\sigma \quad ; \quad C_{ij} = \iint_{\sigma_j} \frac{\partial}{\partial n}\left(\frac{1}{4\pi r}\right) d\sigma$$

$$F_{im} = \iint_{\sigma_m} \frac{\partial}{\partial n}\left(\frac{1}{4\pi r}\right) d\sigma \quad ; \quad H_{ik} = \iiint_{V_k} \left(\frac{-1}{4\pi r}\right) dV \tag{5}$$

In Eq. (5), "i" represents the control point, "j" denotes the element number on σ_b, "m" is the strip number on σ_w and "k" is the volume element number of the field volume V_f. Furthermore, $\Delta\phi$ is the potential discontinuity (jump) across the zero-thickness wake sheet while the normal-wash ψ is given by Eq. (3) and χ is a nonlinear function of ϕ and the density ρ. Note also that the present formulation is a low-order method, i.e., ϕ, ψ, $\Delta\phi$ and χ are assumed to be constant within each element. For instance, ϕ_j represents ϕ at the center of the element "j".

For the sake of convenience, Eq. (5) may be represented compactly as

$$\phi = B\psi + C\phi + F\Delta\phi + H\chi \tag{6}$$

Equation (6) given above was first presented in Ref. 2. It is referred to here as the Transonic Integral Equation; in contrast to the Transpiration Boundary Integral Equations, to be presented in the following.

Transpiration Boundary Integral Formulations

The underlying concept used in deriving the new formulation may be considered as a generalization of the formulations discussed in Refs. 11, 12 and 14. The objective of the approach is to represent Eq. (6) as a linear Laplace solver; with all the additional effects, such as the effect of the wake and the flow field, coupled through the boundary condition. Therefore, the new integral equations are referred to here as the Transpiration Integral Equations.

To this end, express ϕ as

$$\phi = \phi_b + \phi_w + \phi_f \tag{7}$$

Thus,

$$\mathbf{v} = \nabla\phi_b + \nabla\phi_w + \nabla\phi_f \tag{8}$$

where ϕ_b, ϕ_w and ϕ_f are the velocity potential induced by σ_b, σ_w and V_f, respectively.

Next, the impermeability boundary condition, $(\mathbf{v} - \mathbf{v}_b) \cdot \mathbf{n} = 0$, yields

$$\psi_b = \nabla\phi_b \cdot \mathbf{n} = (\mathbf{v}_b - \nabla\phi_w - \nabla\phi_f) \cdot \mathbf{n} \tag{9}$$

In addition,

$$\phi_b = B\psi_b + C\phi_b \tag{10}$$

while

$$\nabla\phi_w = (\nabla F)\Delta\psi_b - \mathbf{F}\Delta\phi_b \tag{11}$$

and

$$\nabla\phi_f = (\nabla H)\chi = \mathbf{H}\chi \tag{12}$$

In the above, the vector elements of the matrices **F** and **H** are defined as

$$\mathbf{F}_{im} = \iint_{\sigma_m} \nabla\left[\frac{\partial}{\partial n}\left(\frac{1}{4\pi r}\right)\right] d\sigma \quad ; \quad \mathbf{H}_{ik} = \iiint_{V_k} \nabla\left(\frac{-1}{4\pi r}\right) dV \tag{13}$$

Note that the gradient operator, ∇, in Eq. (11) and Eq. (12) are understood to be with respect to the control points "i".

Equations (9) through (13) above formed the Transpiraton Boundary Integral Formulations for the fluid flow problem governed by Eq. (2). This formulation in general will not offer any improvement in computational efficiency. Nonetheless, from a program development point of view, the new formulation lends itself to a modular programming environment. The induced velocity due to the wake and the field may be considered as separate entities – though a function of the solution of Eq. (9). As may be inferred from Eq. (8), additional effects such as velocity induced by the viscosity may be included. In fact, with the viscous induced velocity term, Eq. (8) essentially become a statement of the Helmholtz Decomposition Theorem. Application of the Helmholtz Decomposition, in conjunction with the boundary element method, to solve viscous flow problems have been pioneered by Wu (Ref. 11). More recently, Morino (Ref. 12) introduced the Poincaré Decomposition concept. It is akin to the Helmholtz Decomposition , though the rotational component of the velocity field is not required to be solenoidal and hence may be more suitable for compressible flows. Although no attempt is being made by the author to address the viscous flow effects at the present time, it is however the impetus for rewriting the velocity potential formulation for transonic flows in the present form. Another main advantage of the present formulation is the fact that for transonic flows, the field volume integral includes the entire flow field, i.e., it includes the volume outside as well as inside of the aircraft. This is in contrast to the classical method presented in Ref. 3 wherein the field volume excludes the portion enclosed by the aircraft. This is of significant importance because this simple field volume remove the need for any non-trivial grid-generation algorithm. For complex aircraft configurations, this is a major advantage.

Numerical Results

The objective here is to demonstrate the validity of the formulation and understand the effects of the various components of ϕ. Hence, the numerical analysis is restricted to subsonic (linear) flows only. In other words, while the wake influence will be coupled through the boundary condition, the flow field effect will not be included in the present study. Under this condition, the case of a United Technologies Hamilton Standard SR7-L propfan in axial flow is analyzed for which extensive experimental surface pressure data is available (Ref. 15). Furthermore, numerical results of the SR7-L using the classical method are available for comparison (Ref. 10). As in Ref. 10, the propfan is spinning at 1651 rpm with a steady axial flight Mach number of $M_n = 0.2$. At this condition, the flow is assumed incompressible so that the integral formulation presented in the last section applies. As expected, the present method yields exactly the same solution as the classical method and hence compares favorably with experiment (Ref. 15) as shown in Figure 1.

Shown in Figure 2 are ϕ_b, ϕ_w and ϕ as defined in the previous section. The symbols "l " and "u " denote lower and upper surfaces of the propfan, respectively. With a tip radius of 53.64 inches, the figure shows the chordwise potential distribution at the 18% spanwise station. A comparison of the velocity potential ϕ betweeen the classical method and the present method is made in Figure 3. Similar to Figures 2 and 3, Figures 4 and 5 show the velocity components and comparison for the 43% spanwise station.

Conclusions

A generalized transpiration boundary integral formulation for general aircraft configurations is presented. For simplicity, only the steady-state formulation is presented. The concept, however, is applicable to unsteady flows. For transonic flows, this new formulation resulted in a simple flow field and therefore alleviates greatly the complicated treatment of the field grid. Furthermore, additional effects such as boundary-layer and viscosity may be accomodated for in a straightforward manner.

Acknowledgement

Thanks are due Prof. Luigi Morino for providing the author with his work (Ref. 14) as well as the helpful discussions on the subject matter. The author is also indebted to Mr. T. Alan Egolf of the United Technologies Research Center for his valuable suggestions.

REFERENCES

1. Margason, R.J.; Kjelgaard, S.O.; Sellers, W.L. and Morris, C.E.K., "Subsonic Panel Methods — A Comparison of Several Production Codes," AIAA-85-0280.
2. Morino, L. and Tseng, K., "Time-domain Green's Function Method for Three-Dimensional Nonlinear Subsonic Flows," AIAA-78-1204.
3. Tseng, K., "Application of the Green's Function Method for 2- and 3-Dimensional Steady Transonic Flows," AIAA Paper 84-0425.
4. Piers, W.J. and Slooff, J.W., "Calculation of Transonic Flow by Means of a Shock-capturing Field Panel Method," AIAA-79-1459.
5. Chu, L., "Integral Equation Solution of the Full Potential Equation for Three-dimensional, Steady, Transonic Wing Flows," Ph.D. Dissertation, Dept. of Mechanical Engineering and Mechanics, Old Dominion University, 1988.
6. Hu, H., "Full Potential Solutions for Steady and Unsteady Transonic Airfoils With and Without Embedded Euler Domains," Ph.D. Dissertation, Dept. of Mechanical Engineering and Mechanics, Old Dominion University, 1988.
7. Samant, S.S. et. al., "TRANAIR: A Computer Code for Transonic Analyses of Arbitrary Configurations," AIAA-87-0034.

8. Wu, J.C., "Zonal Solution of Unsteady Viscous Flow Problems," AIAA-84-1637.
9. Morino, L. and Beauchamp, P., "A Potential-Vorticity Decomposition for the Analysis of Viscous Flows," *Boundary Element Methods in Applied Mechanics*, Editors: M. Tanaka and T.A. Cruse, Pergamon Press, 1988.
10. Tseng, K., "Rotary Wing Aerodynamics Research at UTRC Using the Boundary Element Method," *Boundary Element Methods in Applied Mechanics*, Editors: M. Tanaka and T.A. Cruse, Pergamon Press, 1988.
11. Tseng, K., "A Green's Function Method for Viscous Transonic Flows," ICARUS Inc., Rept. 83-02, 1983.
12. Sinclair, P.M., "An exact integral (field panel) method for the calculation of two-dimensional transonic potential flow around complex configurations," Paper No. 1394, Aeronautical Journal, June/July 1986, pp. 227 – 236.
13. Sinclair, P.M., "A three-dimensional field-integral method for the calculation of transonic flow on complex configurations – theory and preliminary results," Paper No. 1482, Aeronautical Journal, June/July 1986, pp. 235 – 241.
14. Morino, L. and Bharadvaj, B., "Two Methods for Viscous and Inviscid Free-wake Analysis of Helicopter Rotors," Boston University Report CCAD-TR-85-02, Boston, Massachusetts, August 1985.
15. Bushnell, P., "Measurements of Surface Pressure Distributions of a Single Rotation Large Scale Advance Propfan Blade of Mach Numbers from 0.03 to 0.78," NAS3-23051, 1987.

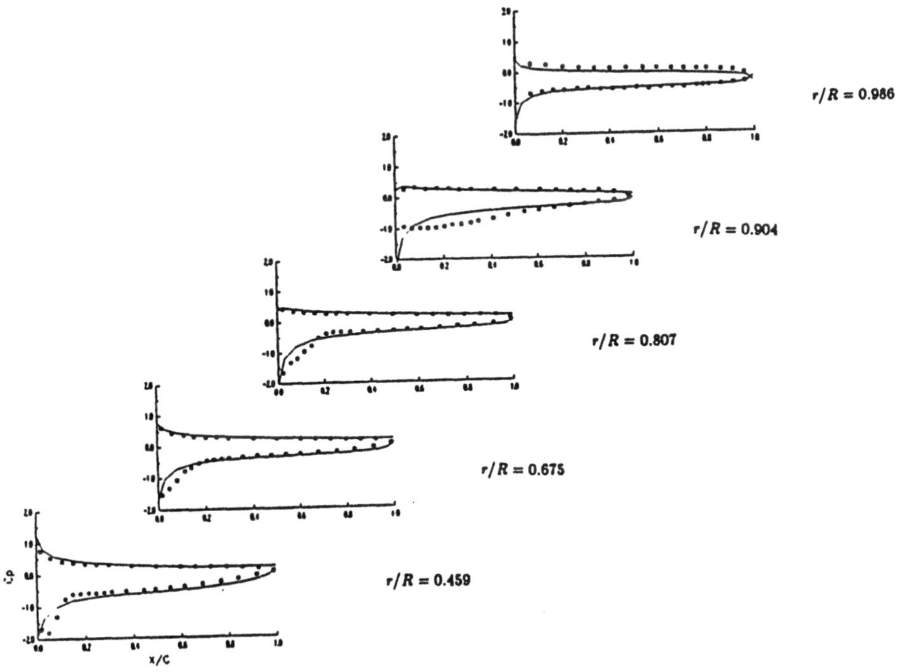

Fig. 1 SR7-L Propfan Surface Pressure Distributions
($\Omega = 1651 RPM, M_n = 0.2, J = 0.88$)

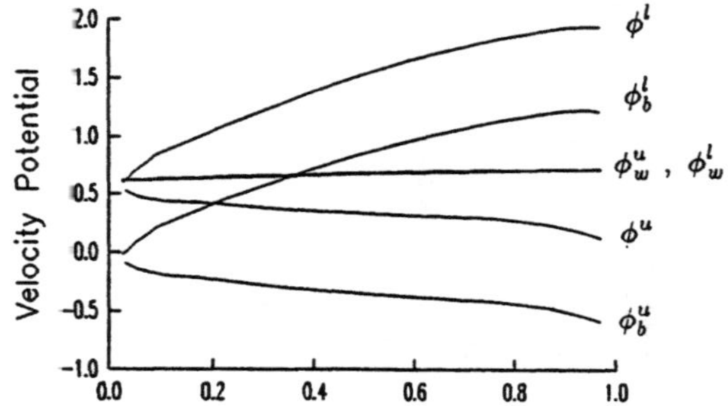

Fig. 2 Velocity potential at the 18% spanwise station of the SR7-L Propfan

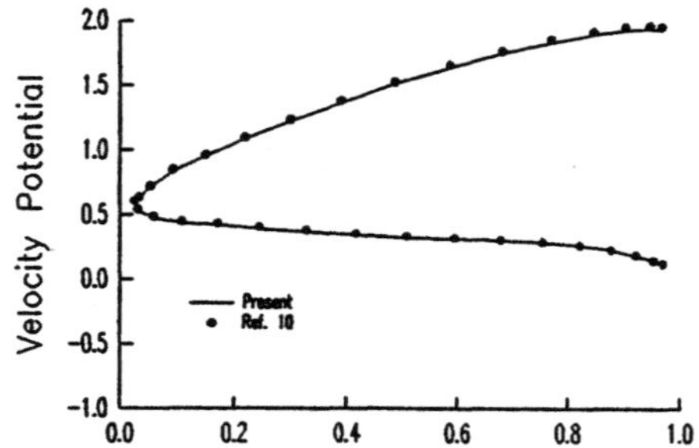

Fig. 3 Velocity potential at the 18% spanwise station of the SR7-L Propfan

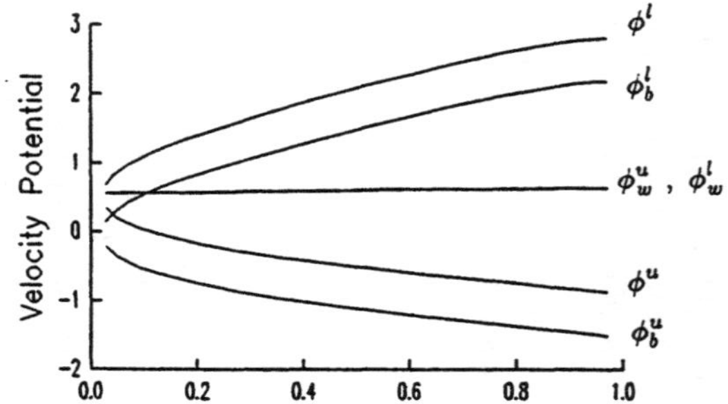

Fig. 4 Velocity potential at the 43% spanwise station of the SR7-L Propfan

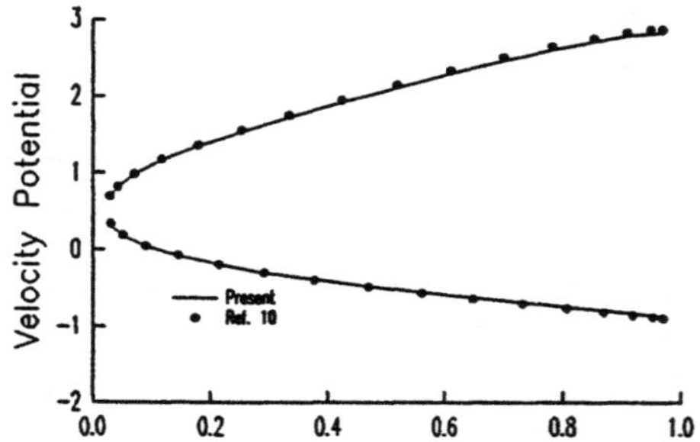

Fig. 5 Velocity potential at the 43% spanwise station of the SR7-L Propfan

Turbomachinery Blade Loading Prediction Using the Panel Method

SHIH H. CHEN

Rocketdyne Division
Rockwell International Corporation
Canoga Park, CA

ABSTRACT

A 2D frequency domain source-doublet based potential paneling formulation is used to predict the steady and unsteady loadings on turbomachinery blades. The method employed blade-to-blade periodic boundary condition and interblade phase angles to simulate the complex blade-viscous wake interaction in a relative motion. For centrifugal turbomachines, a quasi-3D flow is calculated along the stream surfaces, using appropriate coordinate transformation and stream tube thickness variations. Results obtained from this method agreed well with measured data and other predicting methods. This method is accurate and computational efficient compared to finite difference methods. In this paper, the method of frequency domain computation is discussed. Selected results from 2D and quasi 3D flow in steady, and unsteady states are presented.

INTRODUCTION

Despite the move toward finite difference methods in the fluid dynamics community, the panel method remains one of the most widely used and powerful methods for design. This is not only because of its reliability, accuracy and computational efficiency, but also because, in most cases, it can be easily modified to solve problems of complex geometries with complex motions.

The panel method has been extensively used to solve all kinds of fluid problems, however, few methods were developed for turbomachinery. Among them, the Douglas-Neumann method is based on the theory of Hess [1], McFarland [2] solved velocity directly instead of velocity potential on the blade boundary, and Fujinami [3] used a free-vortex method to simulate separated flow in a cascade.

Due to increasing concerns about turbomachinery blade cracks during development or operation, the ability to predict the unsteady aerodynamic forcing function for structural failure analysis is becoming essential. This not only gives an insight into the dynamic loading on turbomachinery blades due to blade-to-blade or blade-to-wake interactions in relative motions, but also gives better predictions of structural fatigue margins.

Rai [4] simulates the unsteady rotor-stator interaction using a finite difference method. However, in order to reduce the computer storage and computational time, the blade numbers are modified to be simple arithmetic ratios (e.g. 1:1 or 2:3, etc), and the blade geometries are enlarged (or reduced) accordingly to account for continuity. Nevertheless, the computational time remains tremendous, ranging from 2 hours to 20 hours of Cray time depending on the complexity of the problem. Liu [5] uses a simpler model to study rotor-stator interactions, but the tilted time method he uses to handle the unequal pitch of the rotor and stator is limited to a narrow range of values. A 2D frequency domain paneling method [6] has been successfully used in calculating blade-wake interactions in turbomachines. This method can simulate arbitrary blade geometry with thickness, camber, stagger, and any rotor-stator pitch ratio. The calculation has also proved to be very efficient, with CPU times 2 orders of magnitude faster than that of the finite difference methods. The method can therefore be effectively used to perform design optimization studies.

THEORY

The 2-dimensional frequency domain source-doublet based potential paneling method is similar to Morino's Green's function paneling method [7]. In the present analysis, the flow is assumed to be incompressible, inviscid, and irrotational. The governing equation is the Laplace's equation

$$\nabla \phi^2 = 0 \tag{1}$$

where ϕ is perturbation velocity potential. When specializing to the blade surface and using the Green's theorem, the velocity potential at any instant in time t, and any field point $\vec{x}$, can be expressed as a distribution of source and doublet over the boundary surface of blades and their wakes as

$$\phi(\vec{x},t) = \frac{1}{\pi}\int \frac{\partial \phi}{\partial n} lnR ds - \frac{1}{\pi}\int \phi \frac{\partial}{\partial n}(lnR) ds \tag{2}$$

where R is the distance between any two points and n is the unit surface normal.

In a 2D cascade calculation, The blade surfaces and their wakes are divided into smaller elements, Each element is distributed with a constant value of singularity. The singularity distribution is blade to blade periodic with a interblade phase angle

$$\sigma = 2\pi m \frac{(nb_1)}{(nb_2)} \tag{3}$$

where m is the harmonic index, nb1 is the upstream number of blades, and nb2 is the downstream number of blades. The wake doublet strength is sinusoidal with phase between wake elements

$$\Delta\varphi = m\omega_o \Delta t \tag{4}$$

where ω_o is the fundamental frequency, Δt is the specified time step.

The solid wall nonpenetrating boundary condition yields the surface normal velocity

$$v_n = -[U_x + \Delta u_x e^{i(\Delta\theta_x + m\omega_o\tau \pm (l-1)\sigma)}]n_x - [U_y + \Delta u_y e^{i(\Delta\theta_y + m\omega_o\tau \pm (l-1)\sigma)}]n_y \tag{5}$$

where U_x, U_y are the mean velocities, Δu_x, Δu_y and $\Delta\theta_x$, $\Delta\theta_y$ are the unsteady velocities and related phase angles, l is the blade index, τ is the retarded time ($=t - x_j/U_x$). The Kutta condition is determined by the potential across the wake and is expressed numerically as

$$\Delta\phi_w = (1 - \frac{im\omega_o\Delta x}{U_x})(\Delta\phi_{te}) \tag{6}$$

for steady and unsteady conditions, where Δx is the length of wake elements, $\Delta\phi_{te}$ is the potential difference across the blade surface at the trailing edge. Thus, all the velocity potentials can be related to those on one blade (the reference blade) and equation (2) can be written as

$$[\delta_{ij} - \bar{C}_{ij}]\{\bar{\phi}_j\} - [\bar{W}_{i1}]\{\Delta\bar{\phi}_w\} + [\delta_{ij} - \tilde{c}_{ij}]\{\tilde{\phi}_j\} - [\tilde{w}_{i1}]\{\Delta\tilde{\phi}_w\} = [\bar{B}_{ij}]\{\bar{v}_n\} + [\tilde{b}_{ij}]\{\tilde{v}_n\} \tag{7}$$

where δ_{ij} = Kronecker Delta function

$\bar{C}_{ij} = \frac{-1}{\pi}\sum_l \int \frac{\partial}{\partial n} lnRds$

$\tilde{c}_{ij} = \frac{-e^{im\omega_o t}}{\pi}\sum_l (e^{\pm i(l-1)\sigma}) \int \frac{\partial}{\partial n} lnRds$

$\bar{W}_{i1} = \frac{-1}{\pi}\sum_l \int \frac{\partial}{\partial n_w} lnRds_w$

$\tilde{w}_{i1} = \frac{-e^{im\omega_o t}}{\pi}\sum_l (e^{\pm i(l-1)\sigma} e^{-i(k_{te}-k)\Delta\varphi}) \int \frac{\partial}{\partial n_w} lnRds_w$

$\bar{B}_{ij} = \frac{1}{\pi}\sum_l \int lnRds$

$\tilde{b}_{ij} = \frac{e^{im\omega_o t}}{\pi}\sum_l e^{\pm i(l-1)\sigma} \int lnRds$

The unsteady solution $\tilde{\phi}_j$ can be mathematically decomposed from the steady solution $\bar{\phi}_j$ in equation (7). The overall velocity potential is simply the superposition of steady and unsteady solutions, i.e.

$$\phi_j = \bar{\phi}_j + \sum_m \tilde{\phi}_j e^{im\omega_o t} \tag{8}$$

The pressure is calculated using the unsteady Bernoulli's equation once the velocity potential is obtained,

$$\frac{p'}{\rho} = -\frac{\partial \phi}{\partial t} - \vec{U}_\infty \cdot \nabla\phi - \frac{1}{2}|\nabla\phi|^2 \tag{9}$$

RESULTS AND DISCUSSIONS

In this section, three cascade solutions will be discussed. Each one represents a particular application of the method in different flow fields. They are the 2D steady Gostelow cascade, a Quasi-3D impeller flow, and a Blade-Viscous Wake interaction.

* Gostelow Cascade

The Gostelow cascade is a typical turbomachinery blade test case. The airfoil of the cascade has a cusp trailing edge. The cascade has a stagger of 37.5 degrees, inflow angle 53.5 degrees and solidity 1.01. Figure 1 shows the predicted pressure distributions on the airfoil surfaces and compared with analytical results obtained by Gostelow [8]. The correlation shows good agreement in pressure distribution between the two methods. The obtained inviscid solutions can be easily coupled with simple loss models to reflect the boundary layer effect in true turbomachinery blade passages.

* Quasi-3D Flow

The 2D cascade calculation can be applied to a centrifugal flow blade-to-blade stream surface with an appropriate coordinate transformation and with streamtube thickness variations. Coriolis force effect and angular momentum correction are also implemented to account for the 3-dimensional nature of the centrifugal flow field.

The Eckardt impeller is one of the very few test cases available for anchoring predictions for centrifugal geometries. Even though it contains a region of separation and the flow in the passage has high subsonic Mach numbers, it is used here to test the quasi-3D formulation. Figure 2 shows the comparison of Mach number distribution along the blade shroud. The predicted mean pressure, Figure 3, is compared with measured [9] and a 3D prediction [10]. The present method predicted generally higher average pressure than the measured data. This could be due to the fact that the present method is based on incompressible, potential flow theory. Nevertheless, the prediction correlated well in the trend of pressure rise from the leading edge to the trailing edge.

* Blade-Viscous Wake Interaction

The fully unsteady scheme is tested on the diffuser of the space shuttle main engine high pressure Lox pump to simulate the blade and viscous wake interaction. The wake field velocities upstream of the blades are measured and Fourier decomposed. The four most significant harmonics with magnitudes and phase angles were picked. Each harmonic is considered as an sinusoidal oblique gust to the blades. The response to each gust is calculated and reconstructed at the end to obtain a complete unsteady response characterictics. Ideally, this method can simulate unsteady blade loading of arbitrary blade geometry with thickness, camber, stagger, rotor-stator pitch ratio, and in any excitation frequency (as long as the panel size is sufficiently small compared to the disturbing wave length).

Figures 4 and 5 show the predicted and measured unsteady pressure history for one blade revolution at around 10% from the leading edge on suction and pressure surfaces. The correlation indicates that the wake induced responses can be predicted well using the present method. For this case with 15 blades, 80 panels on each blade, and 100 wake elements/blade, the computational time on a DN4000 Apollo workstation is about 10 minutes for each individual frequency.

CONCLUSIONS AND FUTURE PLANS

The Green's function frequency domain potential paneling method has been developed for the purpose of efficiently predicting the unsteady interactions in turbomachinery blade passages. This method is also capable of predicting the steady loading on a 2D or quasi-3D blade cascade. The method is generally accurate and computationally efficient compared with other finite difference schemes.

The current analysis is based on a rigid blade assumption. Future plans will extend the current method to include blade vibration and fluid-structure interactions. Hopefully, this will provide a more complete turbomachinery blade unsteady load prediction in subsonic flow.

REFERENCES

[1] Hess, J. L., "Higher Order Numerical Solution of the Integral Equation for the Two-Dimensional Neumann Problem," Comp. Methods in Appl. Mech. and Eng., Vol. 2, p.1-15, 1973.

[2] McFarland, E. R., "A Rapid Blade-to-Blade Solution for use in Turbomachinery Design," J. of Engineering for Gas Turbines and Power, Vol. 106, p.376-382, 1984.

[3] Fujinami, T., "Computation of Unsteady Separated Compressible Flows Using Free-Vortex Method," M.S. thesis, Dept. of Aerospace Engineering, U. of Texas, Austin, 1985.

[4] Rai, M. M., "Navier-Stokes Simulations of Rotor-Stator Interaction Using Patched and Overlaid Grids," AIAA paper no. 85-1519.

[5] Liu, J. and Sockol, P., "Unsteady Euler Cascade Analysis," AIAA paper no. 89-0322.

[6] Chen, S. H., "Unsteady load prediction in a Turbomachinery Cascade Due to Nonuniform Upstream Flow Field," AIAA paper no. 89-0450.

[7] Morino, L., "A General Theory of Unsteady Compressible Potential Aerodynamics," NASA CR-2464, 1974.

[8] Gostelow, J. P., "Potential Flow through Cascades - A Comparison between exact and Approximate Solutions," ARC CP-807, 1965.

[9] Eckardt, D., "Detailed Flow Investigations with a High-Speed Centrifugal Compressor Impeller," J. of Fluids Engineering, p.390-399, 1976.

[10] Prince, T. C., and Bryans, A. C., "Three-Dimensional Inviscid Computation of an Impeller Flow," J. of Engineering for Gas Turbine and Power, Vol. 106, p.523-528, 1984.

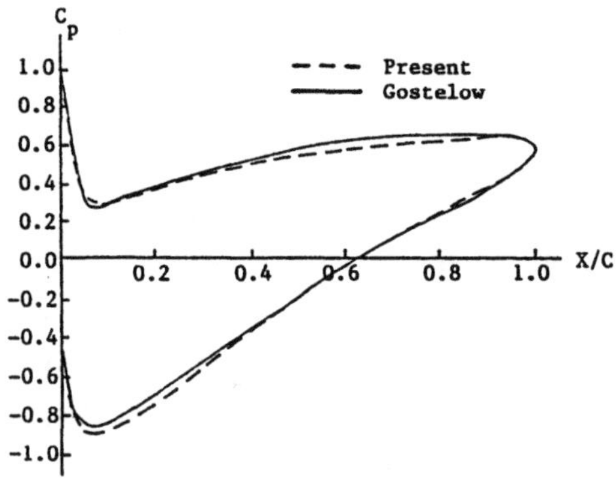

Figure 1 Pressure distribution on Gostelow cascade.

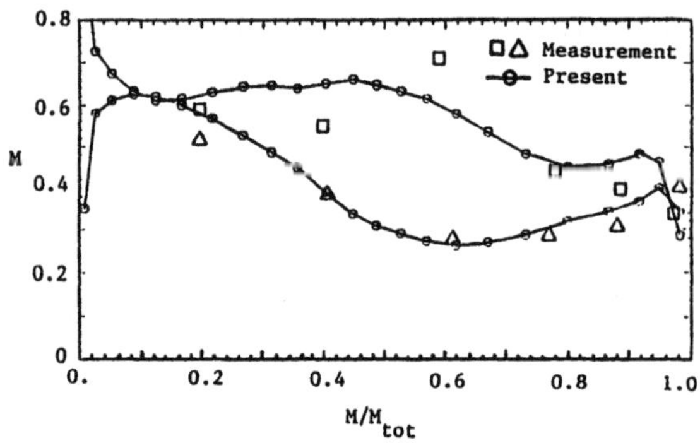

Figure 2 Velocity distribution on impeller tip.

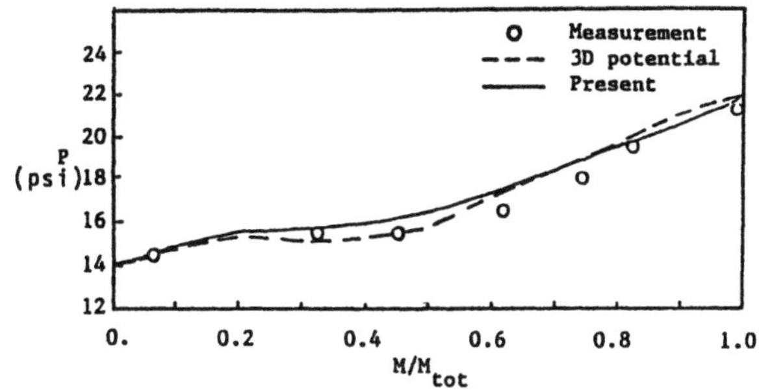

Figure 3 Tip mean pressure distribution.

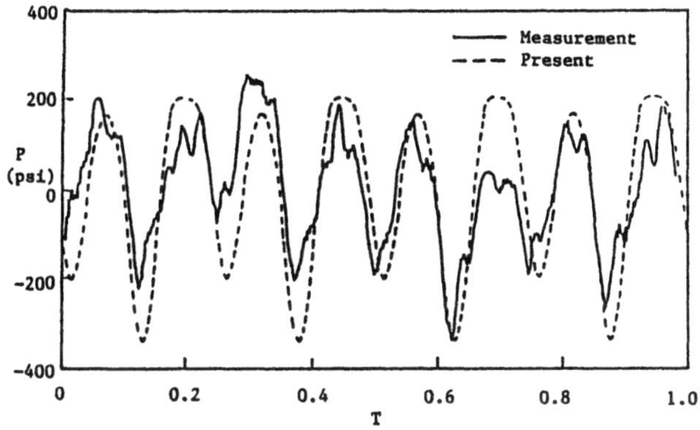

Figure 4 Unsteady pressure on suction surface.

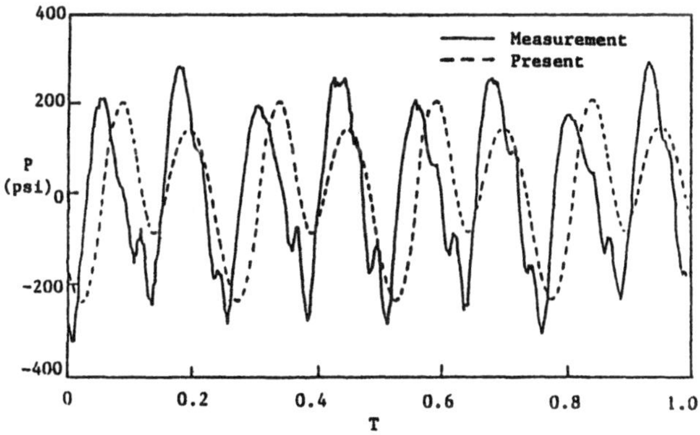

Figure 5 Unsteady pressure on pressure surface.

Panel Methods for Multiple-Rotor Aerodynamics

Marc H. Williams
School of Aeronautics and Astronautics
Purdue University, West Lafayette Ind.

Abstract

Two methods will be discussed for predicting the three dimensional unsteady aerodynamic loads which are exerted on blade assemblies in relative rotation. The earlier of the two is a full surface, time marching panel scheme based on Morino's formulation. It has been implemented only for low speed (incompressible) flow. The second scheme uses an iterative coupling of the rotor velocity fields, decomposed into harmonics. Each field harmonic is computed using linearized compressible aerodynamic theory. Both schemes are capable of handling axial or weakly nonaxial flow configurations, with either rigid or flexible blades. The principle applications are to the prediction of performance and aeroelastic response characteristics of counter-rotating systems.

Introduction

Many propulsion systems contain multiple lifting surfaces in relative rotation but with a common translational velocity, so that the system configuration is periodic. The flows about such systems are intrinsically unsteady, but periodic. Examples are rotor-stator, and rotor-wing-body interference. Computational methods are required for predicting the unsteady loads, as well as the vibration and noise that those loads produce. The main objective of the present work is to develop aerodynamic analysis tools which can be used to study the aeroelastic behavior of counter-rotating propfans.

One possible approach to the problem would be to develop a direct time marching numerical solution of the Euler/Navier-Stokes equations. Such schemes have been applied to rigid body motion [1], but are, at present, prohibitively expensive, even with that restriction, and would be more so if elastic deformations were to be included. Moreover, though such methods may be necessary for some applications (eg stall flutter prediction), forced response and stability analyses at or near cruise conditions, for a well designed rotor, should be possible with a lower order aerodynamic model. In the present work we have presumed that linear aerodynamics can be used. This is the lowest level possible, and, of course, is incapable of describing certain flow physics: transonic shocks , flow separation, and wake roll-up in particular. However, the advantages in terms of computational efficiency are great, and the results should be useful when those phenomena are absent or negligible.

For a single isolated rotor, any simple harmonic blade vibration will induce a simple harmonic load fluctuation on the blades with the same frequency. This is not true for interacting rotors. Even if the blades are rigid, the loads will vary periodically. If the blades undergo a forced simple harmonic vibration, the loads will have an infinite, but discrete, spectrum of frequencies. Therefore the problem cannot be simplified by assuming simple harmonic time dependence, as is commonly done for isolated rotors or fixed wings.

Solutions of the linearized equations can be found by a variety of methods. One approach, again, would be to solve the governing PDE's directly in the time domain using finite differences or like technique. This would probably be nearly as costly as solving the full nonlinear Euler equations, since the cost is set mostly by the number of grid points, not the nonlinearities in the equations. The linearized equations, though, possess known elementary solutions and a principle of superposition. Therefore the flow field can be constructed by distributing singularities over the blades, and adjusting their strengths to match the blade boundary conditions. Even here, though, there are numerous alternatives: The distributions can be placed on the undeformed blade outer surface, or on some reference surface within the blade. The singularities can be made up of sources, doublets, forces, or vorticity. Finally the problem can be posed in the time domain, and solved by time marching; or the known periodic nature of the solution can be used to project the problem into Fourier space, where it can be solved simultaneously for the interacting harmonics.

Time Domain Method

Chen [2,3], presented a full surface Green's Theorem based panel method for counter-rotating propellers in incompressible flow. The wakes were simply convected along the helical trajectories of the blade trailing edges. The potential is represented via Green's relation,

$$4\pi\Phi(\vec{x},t) = \iint_S \left[\Phi \frac{\vec{n}\cdot\vec{r}}{r^3} - \frac{u_n}{r}\right] dA_0 \tag{1}$$

where S is the surface of all blades and wakes, and $\vec{r}$ is the position vector $\vec{x}-\vec{x}_0$. The normal velocity, u_n is known on the blade surface and vanishes on the wake, so the task is simply to solve this equation for Φ on the blade surface. The wake doublets are fixed by imposing a Kutta condition at the trailing edges and convecting the results downstream. The integrals were discretized with a low order panel scheme. The potential on the blades (which give the pressures through Bernoulli's equation) was obtained by time marching until the solution became periodic. This procedure requires recalculation of the wake induction and of the rotor interaction influence coefficients as well as a new inversion of the full matrix at each time step. The entire calculation, therefore, becomes fairly expensive, consuming 2 to 3 minutes of Cyber 205 CPU time per step. The basic reason for this expense was that the full surface scheme puts unknowns on both the upper and lower surfaces (around 140 per blade) so the full matrix coupling the front and rear rotors is very large, which increases both the set-up and inversion costs.

A further problem with this scheme is that as the vortex wake shed by the forward rotor cuts across the rear blade it induces a large local disturbance. Since the wake is modeled with discrete doublets, the control points on the rear blade would see 'naked singularities' periodically as the wake sweeps past, which would cause large local discretization errors. The wake was, therefore, dynamically deformed around the control point to avoid the difficulty. This procedure, which adds modestly to the computational cost, eliminated most, but not all, of the numerical noise associated with blade-wake cutting. A typical result is shown in Fig(1), which gives the net power time history on the front row of a 4 by 4 system of straight HS3155 blades. (The 'quasi-steady' result is obtained by setting the wake doublets equal to their trailing edge values at each time step, rather than convecting them. This approximation was used by Lesieutre,[4] in a vortex lattice discretization, with similar results.)

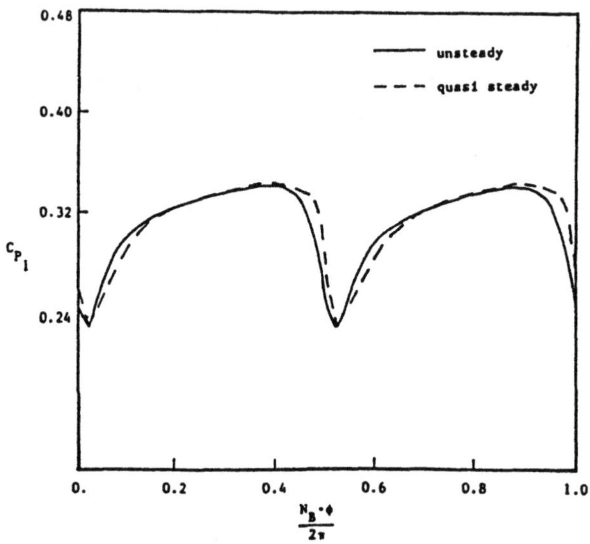

Fig.(1) Unsteady Power on Front Row of 4×4 HS3155 Rotor

Frequency Domain Method

We have, as well, pursued an alternative scheme which makes use of the fact that in most counter-rotation configurations the separation between blade rows is large enough that the aerodynamic coupling is weak. In particular, blade thickness effects, which have dipole decay, are essentially negligible. This is illustrated in Fig(2), which shows the sectional thrust distribution on an SR3 rotor for various assumed blade thicknesses. Note that Chen's method cannot do the zero thickness case, since the equation is degenerate in that limit. The zero thickness result shown was obtained from the single rotation lifting surface panel method of Williams [5,6]. (However, in counter-rotation, blade thickness does, in principle, contribute to the unsteady loads.) This suggests that a mean surface approximation, in which only blade loading is included and half the number of panels are used, would be as accurate and more efficient.

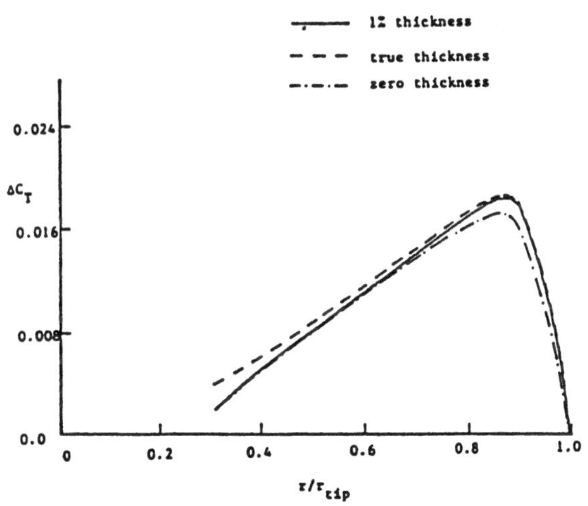

Fig (2) Blade Thickness Effect on SR3 Sectional Thrust

Weak coupling between rows also suggests that the spectrum of the induced loads is likely to be narrow, so that only a few frequencies are important, and that an iterative, rather than a direct solution is appropriate. These notions have been implemented by Cho [7] and applied to both counter-rotation propellers [8] and propeller-wing interaction [9]. This method appears to be about an order of magnitude faster than Chen's, and does not presume low Mach numbers.

The main principles of the model will be outlined. Let the two rotors be denoted by subscripts 1 and 2, and the pressure difference across the blades by Δp. The net velocity field generated by the rotors is linearly related to the past history of the loads on all blades, (summation convention on Roman subscripts is used unless otherwise indicated)

$$\vec{u}(\vec{x}, t) = \vec{A}_i \cdot \Delta p_i \tag{2}$$

where $\vec{A}_i$ is a space-time integral operator over the rotor i blade surfaces and past history. The normal component of velocity on each rotor is specified as V_i, giving a pair of integral equations,

$$A_{ij} \cdot \Delta p_j = V_i \tag{3}$$

where A_{ij} are normal projections of $\vec{A}_i$. The kernels of the diagonal operators $A_{\alpha\alpha}$ are translationally invariant in time, since they correspond to field and source points fixed in the same frame. The off-diagonal operators, though, are periodic in source time for fixed delay time, because of the relative rotation.

Let N_i and Ω_i be the number of blades and rotational speeds for the i'th rotor. We suppose that the normal velocity on one of the rotors, α, is simple harmonic in time with frequency ω_0 and constant interblade phase m_0. On the other rotor, β, we have $V_\beta = 0$. The periodicity of the aerodynamic interaction between rotors implies that the loads on the k'th blade of the j'th rotor must have a spectral decomposition,

$$\Delta p_{j_k} = \sum_n P_{jn}\, e^{i(\omega_{jn} t - k\, m_{jn} \Delta\theta_j)} \tag{4}$$

where $\Delta\theta_j = 2\pi/N_j$. The frequencies and phases are generally different on the driver and driven rotors,

$$\begin{aligned} \omega_{\alpha n} &= \omega_0 + nN_\beta\Omega_0 \quad , \quad \omega_{\beta n} = \omega_0 + [nN_\alpha - m_0]\Omega_0 \\ m_{\alpha n} &= nN_\beta + m_0 \quad , \quad m_{\beta n} = nN_\alpha - m_0 \end{aligned} \tag{5}$$

where $\Omega_0 = \Omega_\alpha - \Omega_\beta$ is the difference in rotor speeds. In general the harmonic index n should span $(-\infty,+\infty)$. The fundamental term on the α rotor is n=0. All other terms come from the interference with rotor β. All loads on the β rotor are induced by the interaction with α.

Our objective is to find the complex amplitudes P_{jn} of the loads, which represent simple harmonic excitations with fixed interblade phase. Each such term induces a velocity field which has the same frequency and phase (in the same frame.) The equations for P_{jn} can be written in analogy to the single rotation isolated rotor case,(with no summation on j)

$$\overline{A}_{jj}(\,\omega_{jn}\,,\,m_{jn}\,)\cdot P_{jn} = \overline{V}_{jn} \qquad (6)$$

Where $\overline{A}_{jj}$ is the isolated rotor operator for simple harmonic motion, or, equivalently, the harmonic amplitude of the temporal operator A_{jj}. The interaction with the other rotor has been absorbed into the right hand side $\overline{V}$, which therefore includes both the complex amplitude of the forcing terms V_j (in the n = 0 term), and of the interference velocities.

The basic solution of Eq(6) for a given right hand side is performed with the single rotation panel code described in reference [6],(though any frequency domain single rotation scheme could be used.) Interference effects are included iteratively. That is, on the first pass, interference effects are neglected and just the n=0 load on the α rotor is found. Given the loads on rotor α, the velocity field that it induces in the volume swept out by the β rotor is computed, using the same panel method. It is sufficient to do this in one blade passage.) This induced velocity field is then Fourier decomposed in circumferential harmonics, transferred to the β frame, and distributed to the appropriate terms $\overline{V}_{\beta n}$. The loads $P_{\beta n}$ are then found by solving Eq(6) for $j = \beta$, and all values of n. Each of these harmonic loads , then, generates an induced velocity field in the volume swept by the α rotor, which is computed, Fourier decomposed and distributed to the appropriate elements $\overline{V}_{\alpha n}$. This completes one pass of the iteration.

If three frequencies are retained in the load spectrum on each rotor, then a total of six load distributions will be found. This requires six isolated rotor influence coefficient matrices, which are computed, inverted, and stored at the start of the calculation to reduce cost. Most of the effort is expended in computing the induced velocity fields for each harmonic. However, although each $\overline{V}_{jn}$ receives contributions from each P_{ik}, these contributions are highly unequal. For example, the dominant contribution to $\overline{V}_{\beta n}$ comes from the fundamental $P_{\alpha 0}$. Therefore, the computational cost can be significantly reduced by only computing the velocity fields from a subset of the load harmonics (eg. $P_{\alpha 0}$ and $P_{\beta 0}$ only.)

Note that this procedure has a built in high frequency cut off in the spectrum truncation. This means that any sharp spikes in the rear blade load time histories at wake passing will be smoothed out. Further, the field points at which the induced velocities are computed are selected based on the source rotor discretization (not the receiving rotor) so the 'naked singularity' problem is gone.

Only two results will be shown, both taken from Ref[7]. The sectional thrust at 3/4 radius on a 4×4 HS3155 rotor front row is shown in Fig(3), comparing the time domain (Chen) and frequency domain methods. The two are in generally good agreement on the fluctuation amplitudes and waveform. This agreement gives some confidence in the correctness of both schemes. Finally, in Fig(4), the time history of crossflow velocity just behind the rear row of a 2× 2 CR system is shown. Half of the figure was computed from the frequency domain method, the other half contains LDV measurements made by Harrison [10]. The excellent match between these results indicates that the linearized theory is adequate for such systems.

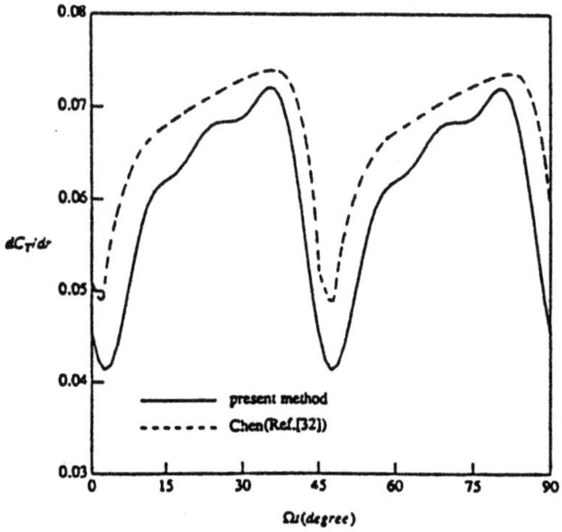

Fig (3) Sectional Thrust on front row of 4× 4 HS3155 Rotor at .75 Radius

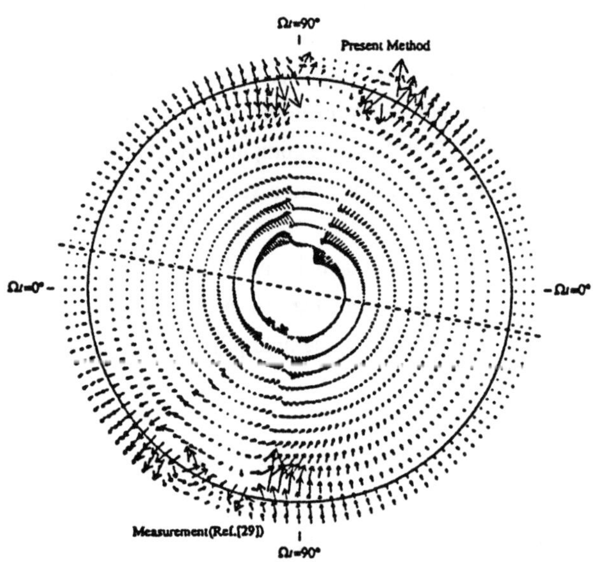

Fig (4) Cross-Flow Time History behind a 2× 2 Purdue Model Rotor

Conclusions

The frequency matching panel method appears to be more computationally efficient than the time marching scheme (for comparable accuracy in the results.) Much of this improvement comes from using half as many panels and doing the solution iteratively. However, the weak coupling that makes iteration desirable could also be used to speed up the time marching scheme.

References

1. Whitfield,D.L.;Swafford,T.W.;Janus,J.M.;Mulac,R.A. and Belk,D. "Three-Dimensional Unsteady Euler Solution for Propfans and Counter-Rotating Propfans in transonic Flow", AIAA Paper no. 87-1197.

2. Chen, S. H., "Panel Method for Counter Rotating Propellers," Ph.D. Thesis, Purdue University, December 1987.

3. Chen, S. H.; Williams M.H.,"Panel Method for Counter Rotating Propellers," AIAA Paper No. 87-1890

4. Lesieutre,D.J.;Sullivan J.P., "The Analysis of Counter-Rotating Propellers", SAE Paper no. 850869, 1985

5. Williams, M. H., "An Unsteady Lifting Surface Theory for Single Rotation Propellers," Purdue University Report, June 1985. To appear as a NASA CR.

6. Williams, M.H.; Hwang, C.C., "Three Dimensional Unsteady Aerodynamics and Aeroelastic Response of Advanced Turboprops," AIAA Paper No. 86-0846, May 1986.

7. Cho, J., "Frequency Domain Aerodynamic Analysis of Interacting Rotating Systems," Ph.D. Thesis, Purdue University, December 1988.

8. Cho, J.; Williams M.H., "Counter-Rotating Propeller Analysis using a Frequency Domain Panel Method" to appear in the AIAA Journal of Propulsion and Power

9. Cho, J.; Williams M.H., "Propeller-Wing Interaction using a Frequency Domain Panel Method" to appear in the AIAA Journal of Aircraft

10. Harrison,G.L., and Sullivan J.P. "Measurement of a Counter Rotation Propeller Flowfield Using a Laser Doppler Velocimeter", AIAA Paper No. 87-0008, January 1987.

BEM for the Analysis of Unsteady Aerodynamics of Windmill Rotors in the Presence of Yaw

G. Arsuffi
ENEA - C.R.E. Casaccia, Rome, Italy
G. Guj and L. Morino
Università di Roma "La Sapienza", Rome, Italy

The objective of this paper is to present a method for the evaluation of pressure and velocity fields for a horizontal-axis wind turbine (HAWT) in the presence of yaw. The flow is assumed to be potential and incompressible. The formulation used is general and is applicable to unsteady incompressible potential flows around wind turbines, as well as aircraft and ship propellers. The main problem in the aerodynamic analysis of HAWT is to define the wake geometry, because no systematic studies (such as those for helicopter rotors) exist in this field [1]. In the past, both numerical and experimental approaches were used to study this problem [2, 3]. Because of this lack of data on the wake geometry, the approaches used in the non-axial flow analysis are typically experimental [4, 5, 6]: a free-wake analysis is included in this paper. The formulation used is closely related to that of Morino and Bharadvaj [8].

1. Formulation

As mentioned above, the flow is assumed to be potential and incompressible. The equation for the velocity potential, obtained from the continuity equation, is $\nabla^2\phi = 0$. The boundary conditions, in the air frame of reference (*i.e.*, a reference frame fixed with the undisturbed air), are: (i) $\partial\phi/\partial n = \psi = \vec{n}\cdot\vec{v}_R$ on the rotor surface S_R (where $\vec{v}_R$ is the velocity of a point on the rotor surface and $\vec{n}$ the outward normal); (ii) $\phi = 0$ at infinity; (iii) potential discontinuity $\Delta\phi$ = constant, following a material wake point; (iv) $\Delta(\partial\phi/\partial n) = 0$ on the wake surface S_W; and (v) at the trailing edge, $\Delta\phi$ on the wake is equal to the potential discontinuity on the body. The corresponding boundary integral representation is

$$E_*\phi(\vec{x}_*, t_*) = \oiint_{S_R} \left(\frac{\partial\phi}{\partial n}G - \phi\frac{\partial G}{\partial n}\right) dS - \iint_{S_W} \Delta\phi\frac{\partial G}{\partial n}\, dS \tag{1}$$

where $G = -1/4\pi\|\vec{x} - \vec{x}_*\|$, with $\vec{x}$ denoting the point of integration, whereas $E_* = 1$ if $\vec{x}_*$ is outside S_R, and $E_* = 1/2$ if $\vec{x}_*$ is a smooth point of S_R.

The above equation is discretized by dividing S_R and S_W in quadrilateral elements (σ_m and σ_n, respectively), assuming ϕ, ψ, and $\Delta\phi$ to be constant within each element, and using the collocation method. Using M elements on S_R and N elements on S_W, one obtains

$$E_k\phi_k = \sum_{m=1}^{M} B_{km}\psi_m + \sum_{m=1}^{M} C_{km}\phi_m + \sum_{n=1}^{N} F_{kn}\Delta\phi_n \tag{2}$$

with

$$B_{km} = \iint_{\sigma_m} G_k\, d\sigma \qquad C_{km} = -\iint_{\sigma_m} \frac{\partial G_k}{\partial n}\, d\sigma \qquad F_{kn} = -\iint_{\sigma_n} \frac{\partial G_k}{\partial n}\, d\sigma \tag{3}$$

where $G_k = -1/4\pi\|\vec{x} - \vec{x}_k\|$. The expressions for the evaluation of the coefficients are given in [7]. Two types of analysis are considered: prescribed- and free-wake analysis. In the prescribed-wake analysis the wake is assumed to be a helicoidal surface (without skew), with the pitch calculated according to the velocity at the rotor disk, as given by the blade-element theory. The unskewed-wake assumption is acceptable for small yaw angles, but, as the yaw angle increases, it becomes questionable. In order to assess the importance of the wake geometry, we performed a free-wake analysis. In this case, using the air frame of reference, each wake node is moved according to $\vec{x}_k(t+\Delta t) = \vec{x}_k(t) + \vec{v}_k(t)\Delta t$ where $\vec{v}_k$ is the velocity of the node $\vec{x}_k$. For the first few spirals $\vec{v}_k$ is evaluated by taking the gradient of Eq. 1 and discretizing (near-wake model). For the remaining spirals, $\vec{v}_k$ is prescribed (intermediate-wake model). A source disk is used for the far-wake model. Details are similar to those given in [8] for helicopter rotors.

Next, consider the boundary conditions for non-axial flows. The normalwash may be expressed as $\psi = \vec{n} \cdot \vec{v}_R = \vec{n} \cdot (-\vec{v}_\infty + \vec{\omega} \times \vec{x} + \vec{\Omega} \times \vec{x})$, where $\vec{v}_\infty$ is the free-stream velocity, $\vec{\omega}$ is the angular velocity of the rotor, and $\vec{\Omega}$ is the angular velocity due to the yaw angle δ ($\Omega = d\delta/dt$). The vectors in the above equation are assumed to be expressed in the same frame of reference (*i.e.*, either air or body). However, for the actual computations, it is convenient to express $\vec{n}$ and $\vec{x}$ in the body frame, and $\vec{v}_\infty$, $\vec{\omega}$, and $\vec{\Omega}$ in the air frame. In order to accomplish this, we introduce the following notations: a lower-case boldface letter will denote column matrices (*e.g.*, the components of a vector expressed in a specified frame of reference, either body or air). Thus, the same vector will be identified by different symbols depending upon the frame of reference in which it is considered. In particular, the components of $\vec{x}$ in the body frame will be identified with $\mathbf{x}$, those in the air frame with $\mathbf{z}$. Capital boldface letters will denote square matrices (*e.g.*, transformation matrices relating the column matrices).

In order to obtain the matrix relating $\mathbf{x}$ to $\mathbf{z}$, consider the coordinate systems $(x_1,\ x_2,\ x_3)$ and $(z_1,\ z_2,\ z_3)$, as well as an intermediate system $(y_1,\ y_2,\ y_3)$, which is rotating with angular velocity $\vec{\Omega}$ with respect to $(z_1,\ z_2,\ z_3)$, and with respect to which the system $(x_1,\ x_2,\ x_3)$ rotates with the angular velocity $\vec{\omega}$. Choosing the z_3-axis to be parallel to $\vec{v}_\infty$, the y_2-axis and the z_2-axis to be parallel to $\vec{\Omega}$, and the x_3-axis and the y_3-axis to be parallel to $\vec{\omega}$, one obtains $\mathbf{z} = \mathbf{U}_1\mathbf{y}$ and $\mathbf{y} = \mathbf{U}_2\mathbf{x}$, or $\mathbf{z} = \mathbf{U}_1\mathbf{U}_2\mathbf{x}$, with

$$\mathbf{U}_1 = \begin{bmatrix} \cos\delta & 0 & \sin\delta \\ 0 & 1 & 0 \\ -\sin\delta & 0 & \cos\delta \end{bmatrix} \qquad \mathbf{U}_2 = \begin{bmatrix} \cos\omega t & -\sin\omega t & 0 \\ \sin\omega t & \cos\omega t & 0 \\ 0 & 0 & 1 \end{bmatrix} \tag{4}$$

Note that the derivatives of the matrices $\mathbf{U}_1$ and $\mathbf{U}_2$ have the form $d\mathbf{U}_1/dt = \mathbf{A}_1\mathbf{U}_1 = \mathbf{U}_1\mathbf{A}_1$ and $d\mathbf{U}_2/dt = \mathbf{A}_2\mathbf{U}_2 = \mathbf{U}_2\mathbf{A}_2$, where

$$\mathbf{A}_1 = \begin{bmatrix} 0 & 0 & \Omega \\ 0 & 0 & 0 \\ -\Omega & 0 & 0 \end{bmatrix} \qquad \mathbf{A}_2 = \begin{bmatrix} 0 & -\omega & 0 \\ \omega & 0 & 0 \\ 0 & 0 & 0 \end{bmatrix} \tag{5}$$

Then, setting $\mathbf{v}_\infty^T = [0,\ 0,\ V_\infty]$, the normalwash ψ is given by

$$\begin{aligned} \psi &= \mathbf{n}^T\left(\mathbf{U}_2^T\mathbf{U}_1^T\frac{d\mathbf{z}}{dt} + \mathbf{U}_2^T\mathbf{U}_1^T\mathbf{v}_\infty\right) = \mathbf{n}^T\left(\mathbf{U}_2^T\mathbf{A}_1\mathbf{U}_2\mathbf{x} + \mathbf{A}_2\mathbf{x} + \mathbf{U}_2^T\mathbf{U}_1^T\mathbf{v}_\infty\right) \\ &= \begin{bmatrix} n_1 \\ n_2 \\ n_3 \end{bmatrix}^T \left(\Omega\begin{bmatrix} x_3\cos\omega t \\ -x_3\sin\omega t \\ x_2\sin\omega t - x_1\cos\omega t \end{bmatrix} + \omega\begin{bmatrix} -x_2 \\ x_1 \\ 0 \end{bmatrix} + V_\infty\begin{bmatrix} \sin\delta\cos\omega t \\ -\sin\delta\sin\omega t \\ \cos\delta \end{bmatrix}\right) \end{aligned} \tag{6}$$

2. Numerical Results

Some numerical cases have been obtained in order to assess the validity and the efficiency of the above presented formulation. The following notations are used: R for the rotor radius, $\lambda = \omega R/V_\infty$ for the tip speed ratio, θ for the blade pitch angle, AR for the aspect ratio, N_1 and N_2 for the number of elements (on one side of S_R), in chordwise and spanwise direction, respectively, N_W for the number of wake elements in the azimuthal direction, N_S for the number of elements per spiral, and finally N_F for the number of the free spirals in the free-wake analysis. Twelve time steps per revolution were used in all the computations.

Two rotors were used in the calculations. The first, used in the experimental work of Guj, Terzitta, Arsuffi, and Morino [10], is a single-bladed rotor, having rectangular planform, with $R = 0.14\ m$, $AR = 4.67$, and $\theta = 12°$. The second, used in the experimental work of Wagner and Deppe [11], is a two-bladed rotor, having tapered and twisted blades, with $R = 2.0\ m$, $AR = 9.97$, λ=10.5, and $\theta_{.75} = 2°$.[1] They will be referred to as 'rotor A' and 'rotor B', respectively.

Consider first the case without yaw. A preliminary study was performed, using 'rotor A', in order to check for the convergence of the algorithm. Figure 1 shows the lift coefficient along the span for different S_R paneling ($N_1 \times N_2$). The results indicate that the case 7×14 is close to convergence. Next, consider the comparison with the experimental results of Wagner and Deppe [11]. The results are obtained using 'rotor B', described above. Figure 2 shows the dimensionless axial velocity *vs* blade azimuth. The results have the same trend, but there appear to be a vertical shift between experimental and numerical results. Next, consider a comparison with the experimental results of Guj, Terzitta, Arsuffi, and Morino [10] ('rotor A', defined above). This gives the velocity at a point that is fixed in the air reference of frame and is located upstream of the rotor plane, with $x_3/R = 0.0572$ (this quantity was used for the comparison because of the simplicity of the experimental data acquisition). The comparison is presented in Figs. 3 and 4, which show the dimensionless axial velocity *vs* blade azimuth for $\lambda = 4.42$. The agreement appears to be satisfactory.

Next, consider the results in the presence of yaw. First, consider a comparison with the experimental results of Wagner and Deppe [11]. The results are obtained using 'rotor B', described above. Figure 5 presents the power coefficient *vs* yaw angle. The trend is captured correctly, and the agreement between numerical and experimental results is satisfactory. Next, in order to assess the method on the basis of 'local' quantities, we considered a comparison with the experimental results of Guj, Terzitta, Arsuffi, and Morino [10]. These were obtained without a control on the angular speed (the rotor is allowed to assume the 'equilibrium' angular speed). Therefore, the experimental results exibit the simultaneous effects of two parameters: the yaw angle, δ, and the tip speed ratio, λ. These effects are presented in Fig. 6, which shows the dimensionless axial velocity as a function of the azimuth angle, for three different cases: (*i*) $\delta = 0°$ and $\lambda = 4.42$ (as in Fig. 3), (*ii*) $\delta = 0°$ and $\lambda = 3.90$, and (*iii*) $\delta = 15°$ and $\lambda = 3.90$ (as in the experimental results, see Fig. 7). We may observe that, for the case under consideration, the two parameters have contrasting effects on the peak values: as λ decreases, the peak values become more pronounced, whereas, as δ increases, the upper peak decreases (along with most of the curve), while the lower peak remains almost unchanged. Thus, the combined effect of δ and λ is not strong. This is confirmed by the experimental results, presented in Figs. 7 and 8, which show the dimensionless axial and tangential velocity as a function of the azimuth angle for $\delta = 15°$ and $\lambda = 3.90$. The comparison with the experimental results is satisfactory. Next, we compared the results with those obtained using an independently-developed frequency-domain code based on the same formulation [9]. The comparison is presented in Fig. 9, which shows the value of ϕ at the blade element No. 30 (*i.e.*, at the trailing-edge element where the section lift coefficient is higher, I_1=6, I_2=5), for $\delta = 15°$ and $\lambda = 4.5$: not surprisingly, the results are in excellent agreement with each other (the difference is due to the zero initial conditions for the time-domain solution).

Next, in order to verify that the discrepancy between theoretical and experimental results is not due to the assumption on the wake, we performed a free-wake analysis. In order to assess better the impact of the free-wake analysis, we considered a case that is particularly sensitive to the wake geometry, *i.e.*, we used a higher value for the tip speed ratio ($\lambda = 14$ with $\theta = 2°$). The results for the wake geometry are presented in Fig. 10. Those for the pressure coefficient as a function of x_1/c (for $r/R = 0.8$ and for different values of time) are presented in Figs. 10

[1]The twist, not available in the data provided by Wagner and Deppe [11], was assumed to be linear and determined by $\theta_{tip} = -2°$. This yields $\theta_{tip} - \theta_{root} = -9.86°$, typical for HAWT.

and 11, for prescribed and free wake, respectively. These results indicate that even for this case that is designed to emphasize the free-wake effect the results are relatively small. Therefore, we believe that the numerical results presented above are an accurate representation of the potential solution. The discrepancy between numerical and experimental results may be due to (1) the experimental error (estimated to be of the order of 4%, [10]), (2) to the the effects of viscosity (not included in the numerical formulation), and (3) to the truncation error (see, however, the convergence analysis of Fig. 1).

Finally, in order to show the generality of the proposed method, consider a case with nonconstant yaw (*i.e.*, with $\Omega \neq 0$). Rotor A (defined above) was used with $\lambda = 14$ and $\theta = 2°$. We assumed $\delta = 0$ for $t < t_0$ (where t_0 corresponds to 15 revolutions, in order to reach steady state), $\delta = \delta_1 = 15°$ for $t > t_1$ (where t_1 corresponds to 16 revolutions), and

$$\delta = \delta_1 sin^2\left(\frac{\pi}{2}\frac{t-t_0}{t_1-t_0}\right) \qquad \text{for } t_0 < t < t_1 \tag{7}$$

Three extra revolutions after t_1 were included in the calculations in order to better identify the transient region in the solution. Figure 13 shows ϕ at a given panel *vs* azimuthal angle. Figures 14 and 15 show the first two spirals wake of the geometry at $t = t_0$ and $t = t_1$. It may be noted that the wake geometry changes to reach the steady configuration of Fig. 10. Figure 16 shows the C_p distributions at $t = t_0$, as well as one and two revolutions thereafter.

3. Conclusions

A method for prescribed- and free-wake unsteady aerodynamic analysis of rotors in the presence of yaw has been presented. The results obtained for HAWT rotors are in satisfactory agreement with existing experimental results. The free-wake results indicate that wake distortion due to yaw should be taken into account only for high yaw angles: in this case, however, the potential-aerodynamic assumption is questionable. The proposed method is very general and it is applicable a wide range of unsteady aerodynamics applications.

References

1. Montgomerie, B., "Performance Methods (Again!)," Proc. 2nd Symposium on Aerodynamics of Wind Turbine, IEA R&D WECS Joint Action, Lyngby, Denmark, Nov. 21-22, 1988.

2. Gohard, J. C., "Free Wake Analysis of Wind Turbine Aerodynamics," ASRL TR-184-14, 1978.

3. Alfredsson, P. H., Dahlberg, J. A., Bark, F. A., "Some Properties of the Wake Behind Horizontal Axis Wind Turbines," Proc. 3rd International Symposium on Wind Energy Systems, Lyngby, Denmark, pp. 469-484, Aug. 26-29, 1980.

4. Anderson, M. B., Milborrow, D. J., Ross, J. N., "Performance and Wake Measurements on a 3m Horizontal Axis Wind Turbine - Comparison of Theory, Wind Tunnel and Field Test Data," Proc. 4th International Symposium on Wind Energy Systems, Stockholm, Sweden, pp. 113-135, Sep. 21-24, 1982.

5. Clayton, B. R., Filby, P., "Measured Effects of Oblique Flows and Change in Pitch Angle on Performance and Wake Development of Model Wind Turbine," Proc. 4th BWEA Wind Energy Workshop, Cranfield, Bedford, United Kingdom, Mar. 24-26, 1982.

6. Atkinson, G. T., Wilmshurst, S. M. B., Wilson, D. M., "The Aerodynamics of Rotors in Yaw," Proc. 9th BWEA Conference, Edinbourgh, United Kingdom, pp. 71-78, Apr. 1-3, 1987.

7. Morino, L., Chen, L. T., Suciu, E. O., "Steady and Oscillatory Subsonic and Supersonic Aerodynamics Around Complex Configurations," AIAA Journal, Vol. 13, pp. 368-374, 1975.

8. Morino, L., Bharadvaj, B. K., "A Unified Approach for the Potential and Viscous Free-Wake Analysis of Helicopter Rotors," Vertica, Vol. 12, No. 1/2, pp. 147-154, 1988.

9. Mastroddi, F., "Metodo degli Elementi al Contorno per l'Analisi Aerodinamica ed Aeroelastica di Ali e Rotori in "Hover" nel Dominio della Frequenza," Tesi di laurea, Università di Roma "La Sapienza," 1989.

10. Guj, G., Terzitta, M., Arsuffi, G., and Morino, L., "Unsteady Aerodynamics of Rotors: Experimental Measurements of Velocity Field," Atti del X Congresso Nazionale dell'Associazione Italiana di Aeronautica e Astronautica, Pisa, Italy, Oct. 16-20, 1989.

11. Wagner, W. J., Deppe, L., "Experimental Investigation of the Rotor Loads and the Rotor-Induced Flow of a Two-Bladed Wind Turbine Model in the Low Speed Wind Tunnel of DLR," Proc. 2nd Symposium on Aerodynamics of Wind Turbine, IEA R&D WECS Joint Action, Lyngby, Denmark, Nov. 21-22, 1988.

Acknowledgements: This work was partially supported by the Italian Commission for Nuclear and Alternative Energy Sources (ENEA) under contract No. 13984 to the University of Rome "La Sapienza".

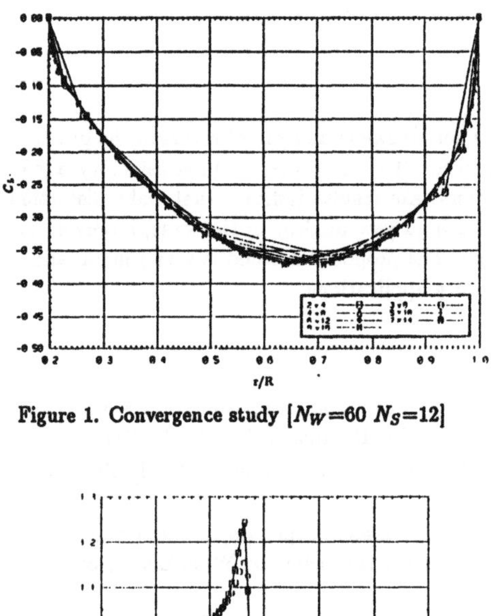

Figure 1. Convergence study [N_W=60 N_S=12]

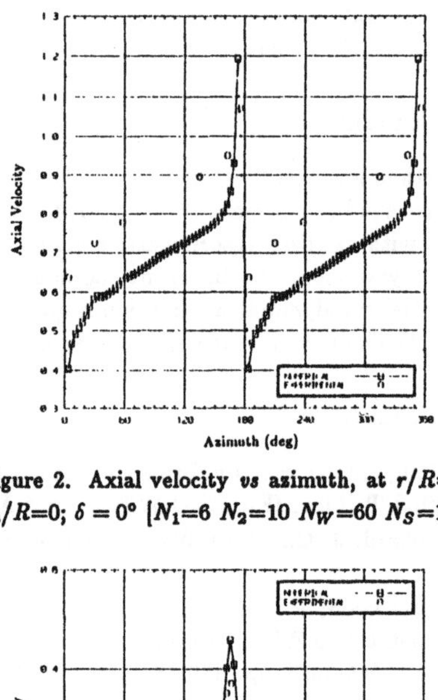

Figure 2. Axial velocity *vs* azimuth, at r/R=0.8, x_3/R=0; $\delta = 0°$ [N_1=6 N_2=10 N_W=60 N_S=12]

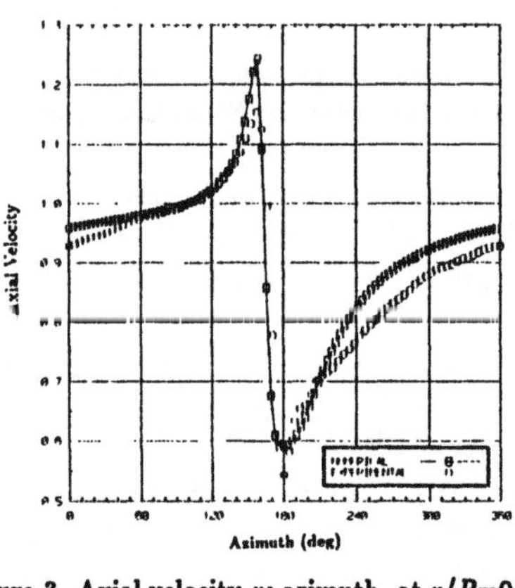

Figure 3. Axial velocity *vs* azimuth, at r/R=0.357, x_3/R=0.0572; $\delta = 0°$ [N_1=6 N_2=10 N_W=60 N_S=12]

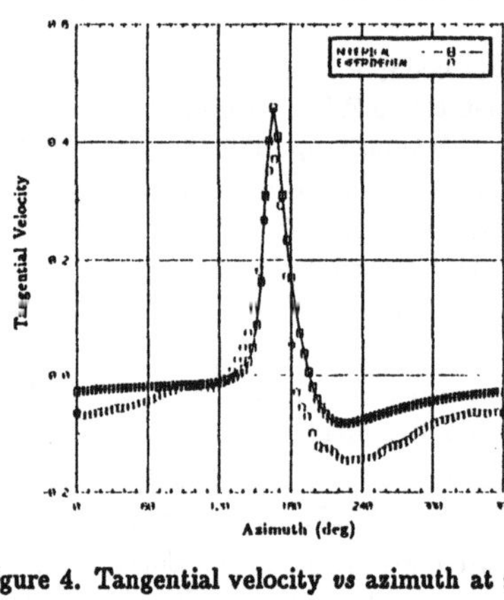

Figure 4. Tangential velocity *vs* azimuth at r/R = 0.357, x_3/R=0.0572; $\delta = 0°$ (prescribed wake) [N_1=6 N_2=10 N_W=60 N_S=12]

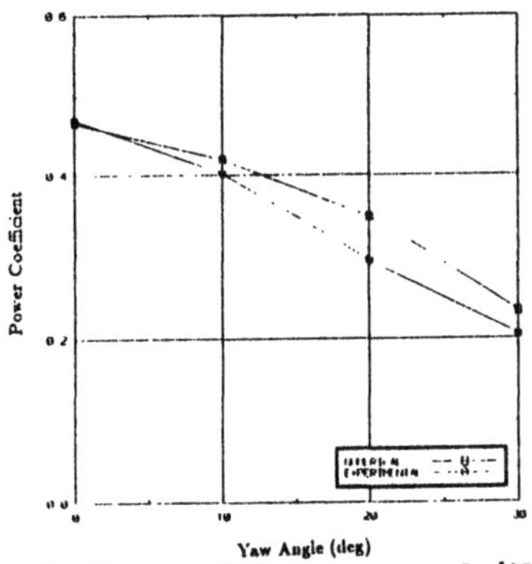

Figure 5. Power coefficient *vs* yaw angle [N_1=6 N_2=10 N_W=60 N_S=12]

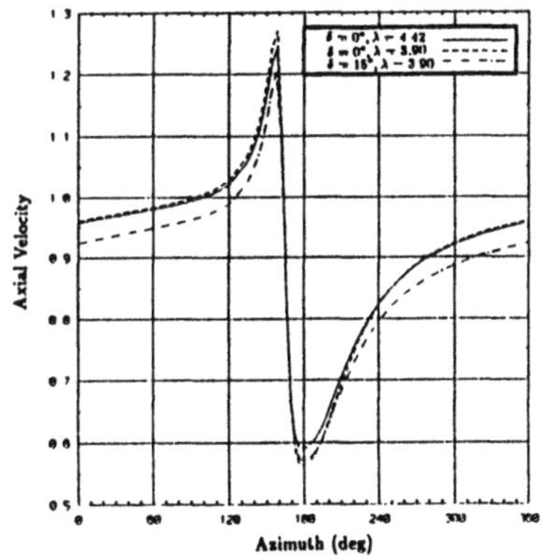

Figure 6. Axial velocity *vs* azimuth at r/R=0.357, x_3/R=0.0572; for different values of δ and λ (prescribed wake) [N_1=6 N_2=10 N_W=60 N_S=12]

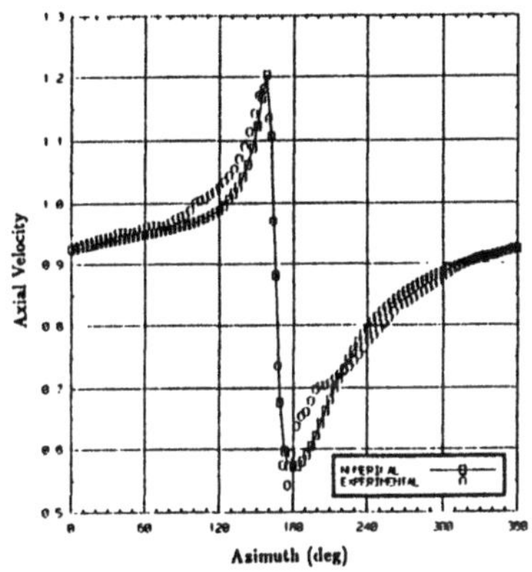

Figure 7. Axial velocity *vs* azimuth at r/R=0.357, x_3/R=0.0572; δ = 15° (prescribed wake) [N_1=6 N_2=10 N_W=60 N_S=12]

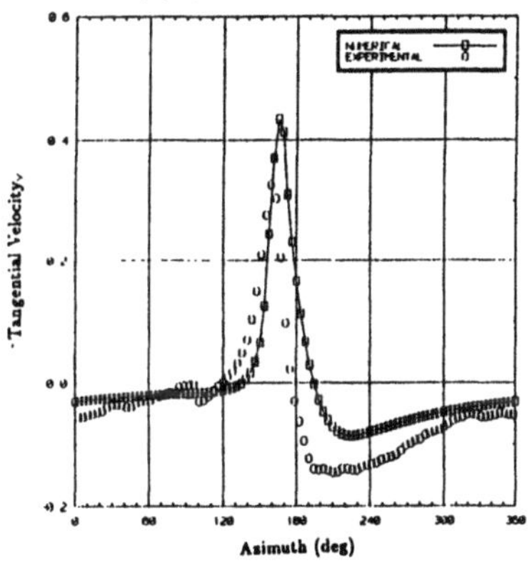

Figure 8. Tangential velocity *vs* azimuth at r/R = 0.357, x_3/R=0.0572; δ = 15° (prescribed wake) [N_1=6 N_2=10 N_W=60 N_S=12]

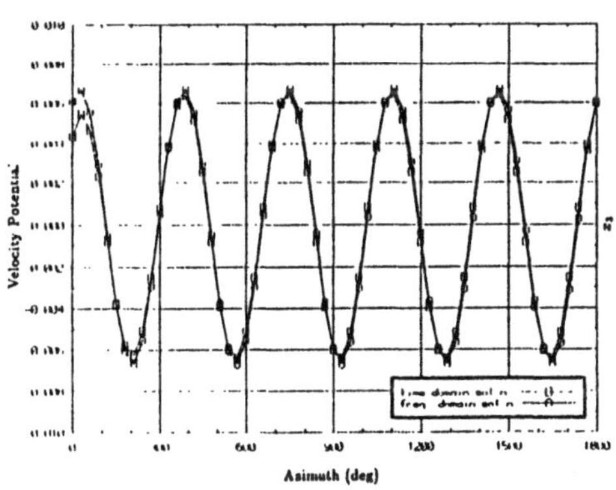

Figure 9. Time and frequency domain solutions: ϕ_{30} (r/R = .756) *vs* azimuth (prescribed wake) [N_1=6 N_2=10 N_W=60 N_S=12]

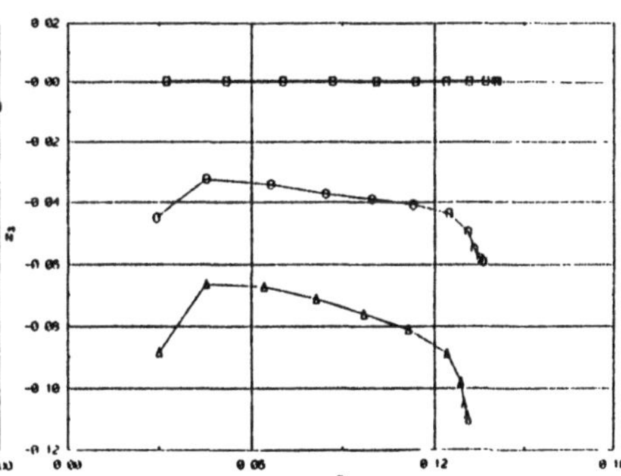

Figure 10. Wake geometry (first two spirals) δ = 15° [N_1=6 N_2=10 N_W=84 N_S=12 N_F=2]

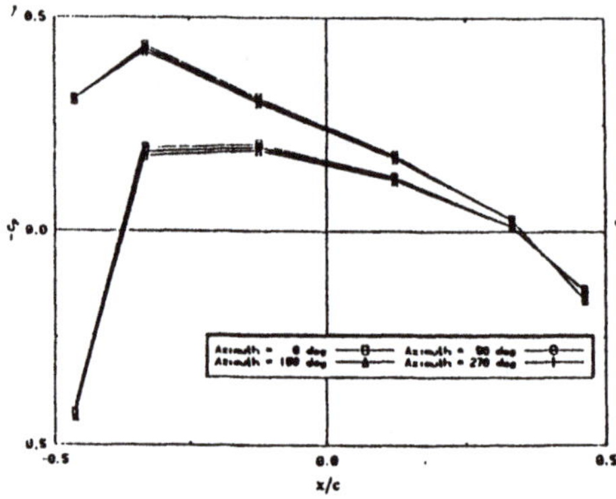

Figure 11. c_p *vs* x_1/c at $r/R = 0.8$; $\delta = 15°$ (prescribed wake) [N_1=6 N_2=10 N_W=84 N_S=12 N_F=2]

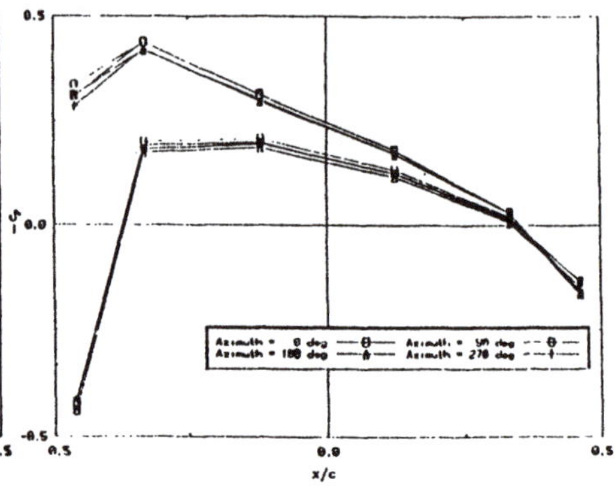

Figure 12. c_p *vs* x_1/c at $r/R = 0.8$ (free wake); $\delta = 15°$ [N_1=6, N_2=10 N_W=84 N_S=12 N_F=2]

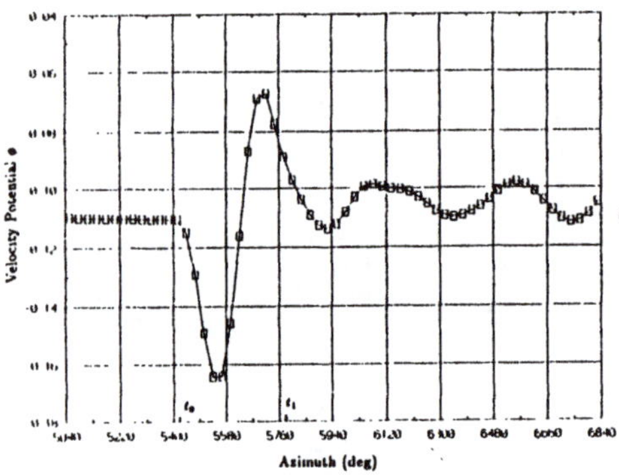

Figure 13. ϕ_{30} *vs* azimuth (r/R=0.756) [N_1=6 N_2=10 N_W=84 N_S=12 N_F=2]

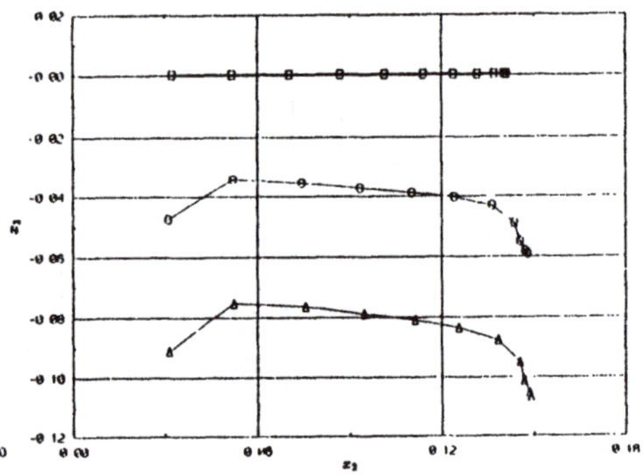

Figure 14. Wake geometry at $t = t_0$ [N_1=6 N_2=10 N_W=84 N_S=12 N_F=2]

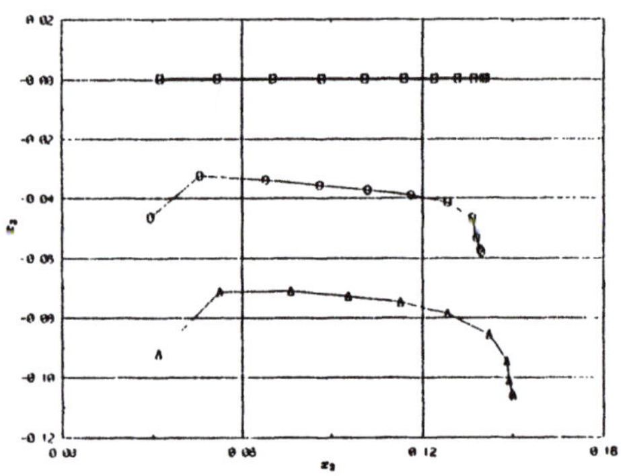

Figure 15. Wake geometry at $t = t_1$ [N_1=6 N_2=10 N_W=84 N_S=12 N_F=2]

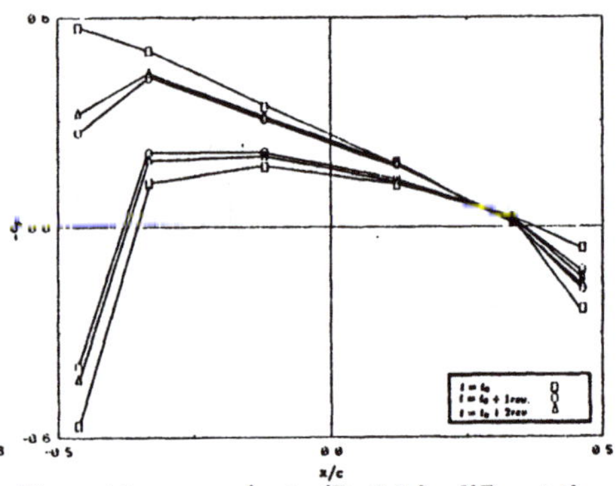

Figure 16. c_p *vs* x_1/c at r/R=0.8 for different time steps; [N_1=6 N_2=10 N_W=84 N_S=12 N_F=2]

A Vortex Method for Predicting Unsteady Flow and Boundary Layer Development Around an Airfoil

K.KAMEMOTO and T.SUZUKI

Department of Mechanical Engineering
Yokohama National University, Yokohama, Japan

Summary

This paper presents a two-dimensional vortex method for simulation of boundary layer development and unsteady flow around an airfoil, in which both laminar and turbulent boundary layers are treated. By using even a rather small computer, reasonable prediction of starting flows around an airfoil of NACA23012 are obtained for various angles of attack either with or without boundary layer separation. The emphasis in the following article will be on the explanation of the mathematical model and the presentation of the predicted flow characteristics.

Introduction

From a viewpoint of inexpensive design application, existing analyses of two-dimensional, unsteady and separated flow even using the appropriate Navie-Stokes equations still are not satisfiable, because they are not only costly but also usually provide results no better than simpler methods. Therefore, it is desirable to have a simple but reliable predictive technique which is convenient especially for engineering usage. Vezza and Galbraith [1] developed a new method for the prediction of unsteady, fully-attached potential flow about an airfoil, in which the airfoil is represented by vortex panels, and the vorticity shed from the trailing edge is represented by discrete vortices. Extending their inviscid model, Vezza and Galbraith [2] obtained encouraging results in the prediction of unsteady, incompressible airfoil flow with fixed upper surface separation. Recently, Coton and Galbraith [3] presented a direct viscid-inviscid interaction scheme for the prediction of two-dimensional airfoil lift and pitching moment in incompressible and steady flow. The scheme is based on a vortex panel method with boundary layer corrections.

In most cases, however, separated flow over an airfoil has unsteady features of vortical flow structures. Therefore, for much more appropriate prediction of such separated flow, not only the separating shear layers but also the boundary layers over the body surface should be calculated as unsteady flow problems. From this point of view, in this paper, a new vortex method of a direct viscid-inviscid interaction scheme is proposed for predicting unsteady flow around an airfoil either with or without boundary layer separation.

Modelling of boundary layer with vortex panels

The circulation γ due to the total vorticity existing in the boundary layer per the unit length of x along the body surface is expressed as γ = U, where U denotes the inviscid flow velocity outside the boundary layer. On the other hand, as explained by Kamemoto and Kawamata [4], the center of gravity yG of the vorticity distribution per the unit length of x is coincident with the displacement thickness δ^* over the surface. In the present method, therefore, the boundary layer displacement effect is accounted for by the replacement of instantanenous profile formed with displacement thickness by vortex panels as shown in Fig.1.

From the two-dimentional unsteady, incompressible boundary layer equation and the continuity equation, the displacement velocity, with which local displacement thickness varries with time, can be defined as

$$\frac{\partial\delta^*}{\partial t} = -\frac{\delta^*}{U}\frac{\partial U}{\partial t} - \left(1+\frac{2}{H}\right)\delta^*\frac{\partial U}{\partial x} + U\left(\frac{\delta^*}{H^2}\frac{\partial H}{\partial x} - \frac{1}{H}\frac{\partial\delta^*}{\partial x} + \frac{1}{2}C_f\right) \quad (1)$$

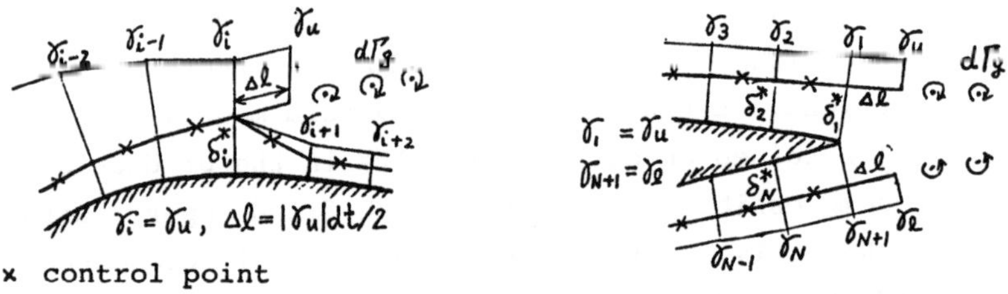

(a) Separation point (b) Trailing edge

Fig.1 Modelling with vortex panels and discrete vortices

where the x co-ordinate, time t, displacement thickness δ^* and momentum thickness θ are non-dimensional values based on the approaching flow velocity U_∞ and the cord length C of the airfoil. The shape factor H and friction coefficient Cf are defined as δ^*/θ and $\tau_w/(\rho U_\infty^2/2)$ respectively, where τ_w denotes the surface stress and ρ is density of fluid. The unsteady boundary layer correction to the inviscid calculation is introduced by displacing each panel corner point step by step in the direction normal to the airfoil surface by the movement during a small time interval dt, due to the displacement velocity defined by Eq.(1). In the present calculation for laminar boundary layers, Pholhausen's velocity approximation was used, and the local displacement velocity was estimated by solving Eq.(1) numerically. The occurrence of laminar boundary layer separation was estimated by the condition of Pholhausen's shape factor $\Lambda \leq -12$. In the case of turbulent boundary layers, the friction coefficient in Eq.(1) was approximately described by the empirical formula of Ludwieg-Tillman [5], and variation of the velocity shape factor H was represented by the entrainment equation [6]. In order to simplify the problem, it was assumed in the present study that the tubulent boundary layer begins from the stagnation point without transition, and the condition of turbulent boundary layer separation is assumed as $H \geq 2.0$

Discrete vortex model of separating shear layer

The present discrete vortex model representing the separation of free shear layer is almost similar to the algorithm of Vezza and Galbraith [1][2]. However, in the present method, the continuous leaving of either upper or lower boundary layer is individually represented by corresponding discrete vortices for either case of fully-attached or separated flow as illustrated in Fig.1. An extra panel of length $\Delta\ell$ and constant vorticity strength is attached to the downstream corner of the vortex panel which represents the displacement thickness at a trailing edge or a separation point, in the direction of extending the displacement thickness panel. In subsequent time, the extra panel is replaced by a discrete vortex which convects down stream according to the induced velocity. The strength of the discrete vortex is given as $d\Gamma = 0.5\gamma|\gamma|\,dt$.

In order to consider the effect of viscocity on vorticity diffusion, a viscous vortex core is introduced so that the vortex has the feature of a forced vortex within the core radius instead of that of a free vortex. The initial core radius for each vortex is given to be the displacement thickness at its separating point.

Calculation procedure

With the present method, the unsteady flow around an airfoil is solved at successive intervals of time starting with an initial flow at t=0. The initial values of displacement thickness are assumed as δ_0^*, and the profile formed with displacement thickness is represented by N vortex panels. In order to obtain a solution for the unknown bound vortex sheet strength at t, the boundary condition of normal flow due to the displacement velocity given by Eq.(1), is applied at the control points of the panels resulting in the following system of equations

$$\mathbf{U}\cdot\mathbf{n}_i + \sum_j^N A_{ij}\gamma_{ij} + A_{iu}\gamma_u + A_{il}\gamma_l + \sum_g^M B_{ig}\,d\Gamma_g - \mathbf{V}\cdot\mathbf{n}_i = k\frac{\partial\delta^*}{\partial t} \quad (2)$$

The second, third and fourth terms are the normal velocities induced at the i-th control point due to the bound vortex sheet and the two separating panels at time t respectively. These terms contain the unknown vortex strengths, whereas the first, fifth and sixth terms are the normal velocities due to the free stream, all discrete vortices and the moving boundary respectively and they can be completely evaluated. The coefficient k in the right hand term is a relaxation factor for numerical stability, and its value is in the range of $0 \leq k \lesssim 1$. The strength of the sheets of extra panels is determined by making use of Helmholtz's theorem of continuity of vorticity as shown by Vezza and Galbraith [2]. When boundary layer separation occures, the displacement thickness in the separated flow region is assumed to be the initial thickness, because it is thought to have little effect on the inviscid flow calculation.

The calculation of vortex strength on panels gives a new instantanenous velocity distribution for use with both the unsteady boundary layer calculation and the estimation of convective motion of discrete vortices. Then, the unsteady boundary layer calculation yields information on displacement corrections to the subsequent inviscid flow calculation.

Results and discussion

The present method was applied to the calculation of unsteady flow around an airfoil of NACA23012 impulsively starting with a constant speed U_∞ from rest. Both laminar and turbulent boundary layers were treated for Re=5000 and $5x10^6$, respectively, and the initial displacement thickness was assumed uniformelly around the airfoil to be 0.005 and 0.001 of its cord length, also respectively. Non-dimensional time interval dt was 0.05, and the number of panels was 40.

Fig.2-(a) shows the leaving of laminar boundary layers from trailing edge without flow separation for the attack angle 5° at t=10, and Fig.2-(b) shows a fully separated flow for the attack angle 20° at t=10. Fig.2-(c) shows velocity vectors of the same flow as shown in Fig.2-(b).

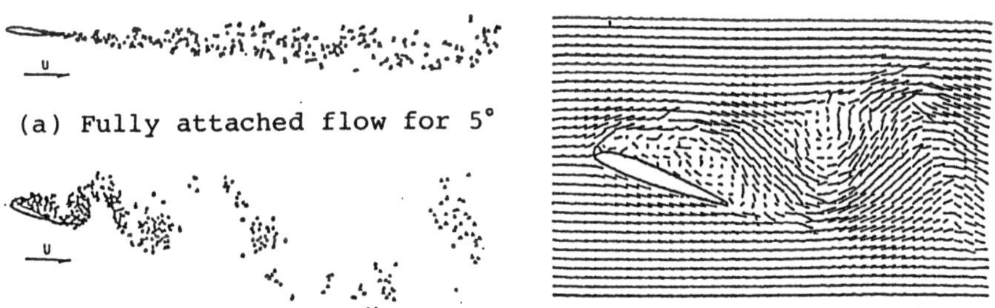

(a) Fully attached flow for 5°

(b) Separated flow for 20°

(c) Velocity vectors for 20°

Fig.2 Predicted flows around an airfoil of NACA23012
(Re = 5000, t = 10.0)

Fig.3-(a) and (b) respectively show the developments of displacement thickness with time for both laminar and turbulent boundary layers for the attack angle 0° where the point of x=0 denotes the leading edge and the region of positive and negative values of x correspond to the upper and lower surfaces respectively.

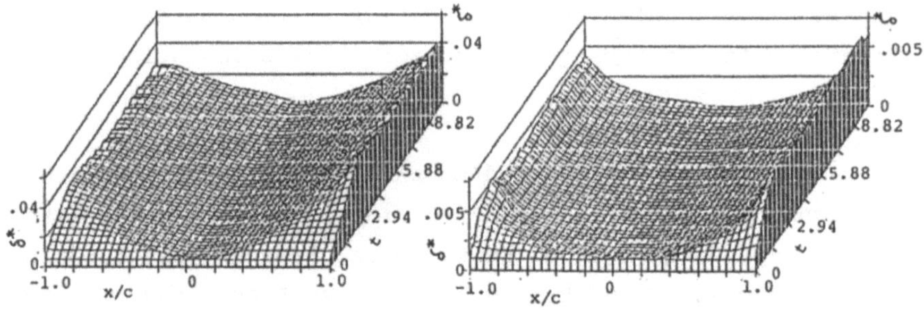

(a) Laminar boundary layer (Re=5000)　　(b) Turbulent boundary layer (Re=5×10^{6})

Fig.3 Development of displacement thickness around an airfoil of NACA23012 for attack angle 0 deg.

Fig.4 shows the time averaged values of lift and drag coefficients for various attack angles in comparison with experimental data [7]. It is clear that the good agreement between the predicted values of Cl and the experimental data has been obtained not only in the attack angle range of fully

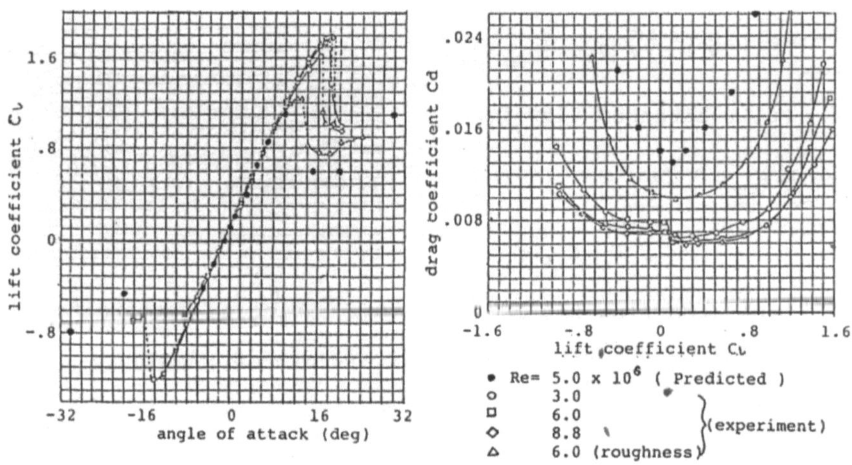

(a) Lift coefficient C_L　　(b) Drag coefficient Cd to C_L

Fig.4 Lift and drag coefficients of an airfoil of NACA23012

attached flow but also for the angles of fully separated flows, whereas the agreement on Cd is poor because of the ignorance of boundary layer transition. It should be noted here that the calculations were considerably unstable for the attack angles around ±16° due to violent change of separation points.

Conclusions

A new vortex method of a direct viscid-inviscid interaction scheme is proposed for predicting two-dimensional, incompressible, unsteady flow around an airfoil either with or without boundary layer separation. It has been confirmed that the flow patterns, displacement thickness distributions, lift and drag coefficients can be predicted reasonably by the present method. It is one of the most interesting points that the computation of all cases presented in this paper were carried out by using a 16 bits personal computer. From this fact, it may be concluded that the present algorithm is regarded as being very usueful especially in the field of engineering use for inexpensive design application.

The work was supported by RC-78 Research Committee of Japan Society of Mechanical Engineers.

References

1. Vezza,M. and Galbraith,R.A.Mcd.: A method for predicting unsteady potential flow about an aerofoil. Int. J. Numer. Meth. in Fluids, 5 (1985) 347-356.
2. Vezza,M. and Galbraith,R.A.Mcd.: An inviscid model of unsteady aerofoil flow with fixed upper surface separation. Int. J. Numer. Meth. in Fluids, 5 (1985) 577-592.
3. Coton,F.N. and Galbraith,R.A.Mcd.: A direct viscid-inviscid interaction scheme for the prediction of two-dimensional aerofoil lift and pitching moment in incompressible flow. Aeron. J., 93-924 (1989) 132-140.
4. Kamemoto,K. and Kawamata,Y.: Representation of two-dimentional boundary layers by discrete vortices. Computational Mechanics'86, Springer-Verlag, (1986) VII 261-266.
5. Ludwieg,H. and Tillmann,W.: Untersuchungen über die Wandschubspannung in turbulent Reibunsschichten. Ing-Arch. 17 (1949) 288-299.
6. Head,M.R.: Entrainment in the turbulent boundary layer. ARC RM 3152 (1958).
7. Abbott,I.H. and Von Doenhoff,A.E.; Theory of wing sections. New York: Dover 1959.

A General Approach to Unsteady Flows in Aerodynamics: Classical Results and Perspectives

Casciola C.M.,* Lancia M.R.,† Piva R. ‡

1 Introduction

Several methodologies and computational techniques have been developed for the analysis of flow problems in unsteady aerodynamics. Most of them are based on the integral representation for the velocity potential [1, 2], which leads to the numerical solution of boundary integral equations. These equations may be of different type depending either on the formulation (e.g. single or double layer) or on the choice of the unknown variables (e.g. direct or indirect methods). Up to recently, a preference was given to Fredholm integral equations of second kind due to the well established theory and the computational procedure which are available for them. In the framework of these potential methods the unsteady features of the flow are accounted for by the generation of the wake, which develops as a discontinuous surface within the flow field, to be included as an extra piece of the boundary in the discrete model. Any extension to a more general viscous flow field may be obtained by performing an iterative procedure with a boundary layer type solution. This is not an easy goal to be reached in many flows of interest, especially when separation occurs.

Another interesting approach, leading to a second kind Fredholm equation, is instead aimed to determine a correct boundary value for the vorticity (or its flux) to be applied as a boundary condition to the vorticity transport equation. To this purpose, different mechanisms for the vorticity generation have been proposed by several authors [3,4]. The dynamic equation for the vorticity is then usually solved by a finite difference technique, or again by a boundary/finite element methodology. The crucial point of such methods stays in the decoupling between the linear kinematical part and the nonlinear dynamical part of the equations. This leaves most difficulties to the solution of the vorticity equation, which, on the other hand, is restricted to the boundary layer and the wake region.

A more general approach is based on the integral representation of the solution for the unsteady Navier Stokes equations in primitive variables [5] Starting from the boundary integral equations provided by this representation, a general solution procedure has been recently proposed for large Reynolds number flows [6]. The above formulation which

*I.N.S.E.A.N., Roma
†I.A.C., C.N.R., Roma
‡Universitá di Roma, "La Sapienza"

involves first kind Fredholm equations presents, however, some computational difficulties, mainly related to the tendential singularity of the kernel for $Re \to \infty$. In particular it has been shown [7] that in the limiting case of Reynolds infinity, we obtain, after some analytical manipulation, a nice integral representation for inviscid flows, still including, coupled together, the dynamics and kinematics of the flow. The overall procedure leading to this final equation is shortly reported in section 2. By performing a decoupling of the dynamical part, we recover, as shown in section 3, a model consistent with the well known Poincaré formula, which gives a kinematical description for the most general flow field. It is important to notice that this integral representation has essentially the same structure of the original one, but the simplified conditions allow for a better investigation of the mathematical properties and the computational behaviour of the integral equation. In particular we analyze in section 4 the two integral equations obtained, in the two dimensional case, by the tangential and the normal projections, although only one of them should be solved in this case. In fact, the first one results to be a Fredholm equation of second kind with a weakly singular kernel, while the second one reduces to a first kind integral equation with a singular Cauchy type kernel. Several procedures for the solution of these equations are discussed and compared, by analyzing the results for both non lifting and lifting bodies.

In the last case, a model for the wake generation based on the incompressible Euler equations with zero jump on both pressure and normal velocity components, leads to a Lagrangian expression for the jump of the tangential velocity. This should be considered as an unsteady Kutta condition; in fact it is equivalent to a conservation principle for the vorticity which, leaving the boundary layer, is going to be located in the wake region.

In the authors' opinion, an accurate study of the previous simplified condition is essential to better approach and control the numerical solution of the full Navier Stokes equations for attached flows at large Reynolds numbers. In this case both equations obtained by the normal and tangential components have to be solved simultaneously.

A satisfactory solution of the Navier Stokes equations would still require a further investigation for the treatment of the diffusive phenomena.

2 From viscous to inviscid flows

Let us consider,for the sake of semplicity, the undisturbed fluid in uniform translation with velocity U_∞ with respect to the body frame of reference. The absolute velocity u is expressed as $U = U_\infty + u$ where u is a perturbation velocity. We also introduce a perturbation stagnation pressure P. The governing equations are

$$\nabla \cdot u = 0 \tag{1}$$

$$\mu \nabla^2 u - \nabla P - \rho \frac{\partial u}{\partial t} = \chi = -\rho U \times \varsigma \tag{2}$$

The integral representation of the velocity is given by [5]

$$u_k(x_*, t_*) = \int_0^{t_*} \int_{\partial\Omega} \left(u_j t_j^{(k)} - t_j u_j^{(k)} \right) dS\,dt + \int_0^{t_*} \int_\Omega \chi_j u_j^{(k)} dV\,dt - \int_\Omega \rho u_j u_j^{(k)} dV |_{t=0} \tag{3}$$

For the problem under investigation the boundary conditions are given by the velocity components (both normal and tangential). It follows, for x_* going to the boundary, a Fredholm integral equation of the first kind. At increasing Reynolds numbers, the kernel of this equation tends to become singular as shown by the expression of the fundamental solutions $u_j^{(k)}$ and $t_j^{(k)}$ listed below for the two-dimensional case (on the left hand side) together with the corresponding expressions in the limiting case of $Re \to \infty$ (on the right hand side).

$$u_j^k = \delta_{ij} F - \frac{\partial^2 E}{\partial x_j \partial x_k} \qquad u_j^{(k)} = -\frac{1}{\rho} H(t_* - t)\left(\delta(x_* - x)\delta_{jk} - \frac{\partial^2 G}{\partial x_j \partial x_k}\right)$$

$$p^{(k)} = -\frac{\partial G}{\partial x_k}\delta(t_* - t) \qquad p^{(k)} = -\frac{\partial G}{\partial x_k}\delta(t_* - t)$$

$$t_j^{(k)} = -p^{(k)} n_j + \mu\left(\frac{\partial u_j^{(k)}}{\partial x_i} + \frac{\partial u_i^{(k)}}{\partial x_j}\right) n_i \qquad t_j^{(k)} = -p^{(k)} n_j$$

with

$$G = \frac{1}{2\pi} lnr \qquad G = \frac{1}{2\pi} lnr$$

$$F = \frac{1}{\rho} H(t_* - t)\frac{e^{-r^2/4\nu(t_*-t)}}{4\pi\nu(t_* - t)} \qquad F = -H(t_* - t)\delta(x_* - x)$$

$$E = \frac{1}{4\pi\rho} E_1\left(\frac{r^2}{4\nu(t_* - t)}\right) - G \qquad E = -H(t_* - t)G$$

$$\nabla^2 E = F \qquad \nabla^2 G = \delta(x_* - x)$$

Combining the fundamental solutions for $Re \to \infty$ with the general representation formula we obtain for an inviscid rotational flow [7]

$$\begin{aligned} u(x_*, t_*) \;=\; & \int_{\partial\Omega} (u \cdot n)\nabla G \, dS + \int_t^{t^*}\int_{\partial\Omega} \frac{1}{\rho}\left(P\, n \cdot \nabla\nabla G - P n \nabla^2 G\right) dS \\ & + \int_0^{t^*}\int_\Omega \nabla_* \times \nabla_* G \times \chi \, dV dt - \int_\Omega \nabla_* \times \nabla_* G \times u_0 \, dV \end{aligned} \qquad (4)$$

where the folllowing vector identity has been considered

$$(b \cdot \nabla)\nabla G - b\nabla^2 G = \nabla_* \times \nabla_* G \times b$$

Assigning as boundary condition the normal component of the perturbation velocity, when x_* goes to the boundary it follows an integral equation for the stagnation pressure P. The kernel of this equation is composed of two terms which are both singular. The first one is hypersingular, the second one is a Dirac delta function on the boundary itself. Combining the two terms together with a few vectorial identities and the Stokes theorem, as suggested by Hsiao [8] and Nedelec [9] for potential flow, the second surface integral may be expressed in a form which corresponds to a vortex layer of density $\gamma = (\nabla P \times n)$. Afterwards the kernel doesn't show the same singularity as in the original form. By analogous manipulations on the volume integral terms and a combination with the vorticity transport equation we finally obtain

$$u(x_*,t_*) = \int_{\partial\Omega}(u{\cdot}n)\nabla G dS + \int_{\partial\Omega}(\nabla\tilde{P}\times n)\times\nabla G dS - \nabla_*\times\int_{\Omega} G\varsigma dV + \nabla_*\times\int_{\partial\Omega} Gn\times(u_0-\tilde{\chi})\, dS \tag{5}$$

where $\tilde{P}$ and $\tilde{\chi}$ stay respectively for the functions P and χ integrated in time. It is worth noticing that in the representation formula (5) the dynamical and the kinematical part of the equations are still coupled together.

3 A purely kinematical representation and its projections

By a back substitution of the Euler equation integrated in time

$$u - u_0 + \tilde{\chi} = -\nabla\tilde{P}$$

we may recover from eq.(5) a purely kinematical representation for the velocity vector

$$u(x_*) = \int_{\partial\Omega}(u\cdot n)\nabla G dS - \int_{\partial\Omega}(u\times n)\times\nabla G dS - \nabla_*\times\int_{\Omega}\varsigma G dV \tag{6}$$

which reduces to the well known Poincaré formula, for $\nabla\times u = \varsigma$ and $\nabla\cdot u = 0$. It is worth noticing, that the representation (6) has essentially the same structure of the original one (5) valid for coupled models. In this section we will examine more deeply the simplified kinematical formula to investigate the mathematical and the computational features of the related integral equations. In particular let us assume, for the sake of semplicity, zero vorticity in the field, hence the velocity representation (6), now given in terms of surface integral, is rewritten as

$$u(x_*) = -\nabla_*\int_{\partial\Omega}(u\cdot n)G dS - \nabla_*\times\int_{\partial\Omega}(u\times n)G dS \tag{7}$$

To study a two dimensional case we consider a cylindrical wing in a three dimensional space. We assume a set of Cartesian coordinates defined at each point of the boundary by the unit tangential vector τ, the unit normal vector n, both in the cross section, and the unit vector $k = \tau\times n$ normal to it. The normal velocity component is assigned on the boundary by $u\cdot n = -U_\infty\cdot n$. For x_* going to the boundary, we obtain two integral equations by performing the tangential and the normal component of eq.(7).

Accounting for the limit value of the double layer integral which gives a jump across the boundary, the tangential component may be expressed as

$$\frac{1}{2}u_\tau^* + \int_{\partial\Omega} u_\tau\frac{\partial G}{\partial n^*}dS = -\frac{\partial}{\partial\tau^*}\int_{\partial\Omega} u_n G dS \tag{8}$$

that is a Fredholm equation of second kind in the variable u_τ. By an analogous procedure the normal component gives

$$\frac{1}{2}u_n^* + \int_{\partial\Omega} u_n \frac{\partial G}{\partial n^*} dS = \frac{\partial}{\partial \tau^*} \int_{\partial\Omega} u_\tau G dS \tag{9}$$

which has the same form as (8), but now for the variable u_τ it is a integral equation of first kind. In fact we immediately recognize that the singular kernel $\frac{\partial G}{\partial \tau^*}$ implies a Cauchy type integral, which gives a continuous function across the boundary. It is clear at this point that only one of the two equations (8) and (9) has to be solved in the present case for the only unknown u_τ: we may therefore choose the easiest one from a computational point of view. However, for a more general application of this methodology, as mentioned in the previous sections, we are going to need the simultaneous solution of both of them. Let us analize briefly the main features of the two equations for the external flow problem.

The tangential component exactly corresponds to a Neumann internal problem for the potential flow with a simple layer representation. We have therefore a complete equivalence with regard to existence and uniqueness of the solution and compatibility conditions for the known term. In particular the solution is not unique, the compatibility condition is given by

$$\int_{\partial\Omega^*} \frac{\partial}{\partial \tau^*} \int_{\partial\Omega} u_n G \, dS \, dS^* = 0 \tag{10}$$

which is identically satisfied, giving the existence of the solution for any assigned distribution of normal velocity. The solution is expressed in the form

$$u_\tau = u_\tau^p + \alpha u_\tau^0 \tag{11}$$

where u_τ^p is a particular solution and u_τ^0 is the eigensolution, that satisfies the homogeneous equation for the Neumann problem

$$\frac{1}{2}u^0 + \int_{\partial\Omega} u^0 \frac{\partial G}{\partial n^*} \, dS = 0 \tag{12}$$

or the equivalent Dirichlet problem

$$\int_{\partial\Omega} u^0 G \, dS = c \tag{13}$$

which gives u_0 as the simple layer density for the related Robin potential. It is worth recalling that in this case

$$\int_{\partial\Omega} u^0 \, dS \neq 0 \tag{14}$$

We may then apply the condition on the value of the circulation, which is evaluated by imposing the conservation of the total vorticity,

$$\Gamma = \int_{\partial\Omega} u_p dS \tag{15}$$

in order to obtain the value of α and determine uniquely the solution (11).

Let us consider now the normal component (9). The integration of equation (9) along the tangential direction leads to

$$\int_{\partial\Omega} u_\tau G dS = F^* + c \tag{16}$$

with

$$F^* = \int_0^{s^*} (\frac{1}{2}u_n^* + \int_{\partial\Omega} u_n \frac{\partial G}{\partial n^*} dS) d\tau \tag{17}$$

having a unique value for the integral condition

$$\int_{\partial\Omega^*} (\frac{1}{2}u_n^* + \int_{\partial\Omega} u_n \frac{\partial G}{\partial n^*} dS) dS^* = 0 \tag{18}$$

which is automatically satisfied for an external domain. Eq. (16) is a first kind Fredholm equation and a variational formulation as indicated in [8] and [9] is advisable for discussing the existence and the uniqueness of the solution. We can always put the solution in the general form

$$u_\tau = u_\tau^p + \alpha u_\tau^0 \tag{19}$$

where the eigensolution u_τ^0 satisfies the equation

$$\int_{\partial\Omega} u_\tau G dS = c \tag{20}$$

that exactly corresponds to the one (13) previously considered for the second kind equation.

4 Present results and developments

The numerical solution of the second kind integral equation (8) has been performed for a lifting airfoil (NACA 0015) in unsteady flow. A simple model for the wake leads to a Lagrangian expression for the jump of the tangential velocity across the wake. Fig.1 shows the wake configuration at successive times after a sudden start of the airfoil at an angle of attack of 12^0 degrees. The time evolution of the vorticity released at the trailing edge is reported in fig.2. The numerical solutions obtained by the integral equation (8) and (9) are compared in fig. 3 and 4. In particular fig.3 shows the results for the steady flow about a cylindrical body with an elliptical cross section and an assigned circulation $\Gamma = \pi$. A perfect coincidence of the two eigensolutions, and a good agreement of the general solutions also with the analytical one, are shown respectively in *a*) and *b*). Fig.4 reports analogous results for the airfoil profile at angle of attack of 5^0 degrees. A better agreement of the results is exspected by adopting a variational approach for the solution of the first kind integral equation (9). This approach, in fact leads to a bilinear form which is very convenient for the numerical solution of the equation. As a concluding remark the above discussion and comparison concerning the tangential and normal components of the integral equation should be very enlightening to approach the numerical solution for the viscous flow, where both components have to be solved simultaneously.

Acknowledgements

This work was partially supported by NASA Grant no. NAGW 1313 to the University of Rome.

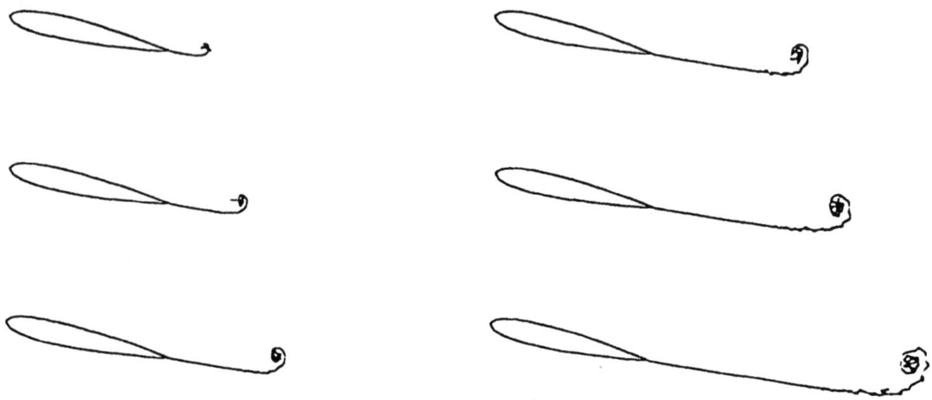

Fig.1 Wake configuration at successive times after a sudden start for a Naca 0015 profile, angle of attack $\alpha = 12$ degrees.

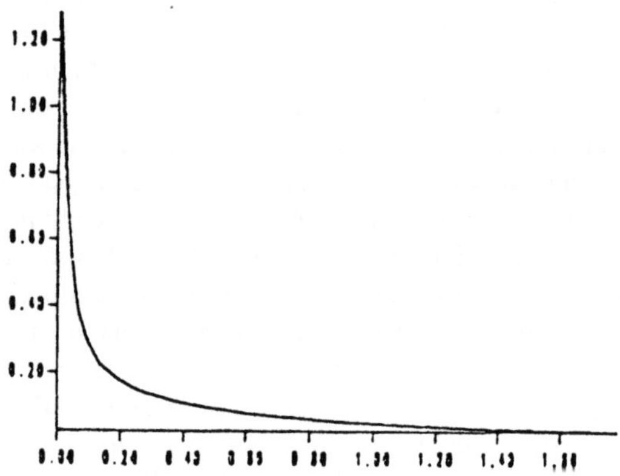

Fig.2 Time evolution of the vorticity released at the trailing edge.

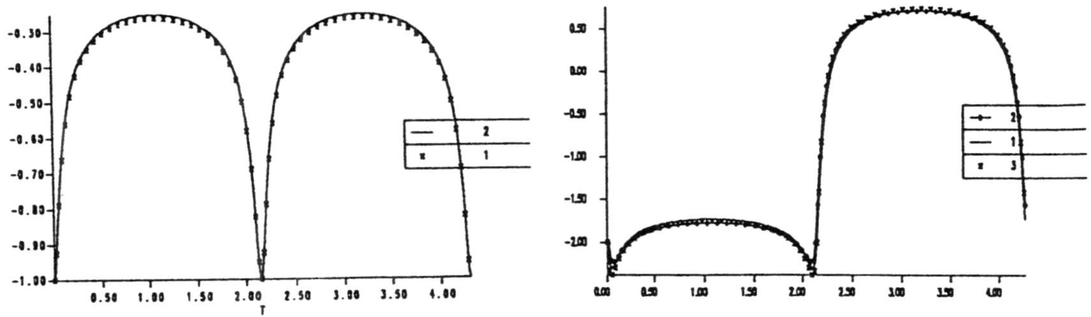

a) b)

Fig.3 Elliptic body with circulation $\Gamma = \pi$: eigensolution (a) and velocity distribution (b) given by analytical solution (3), first kind (2), second kind (1) integral equation.

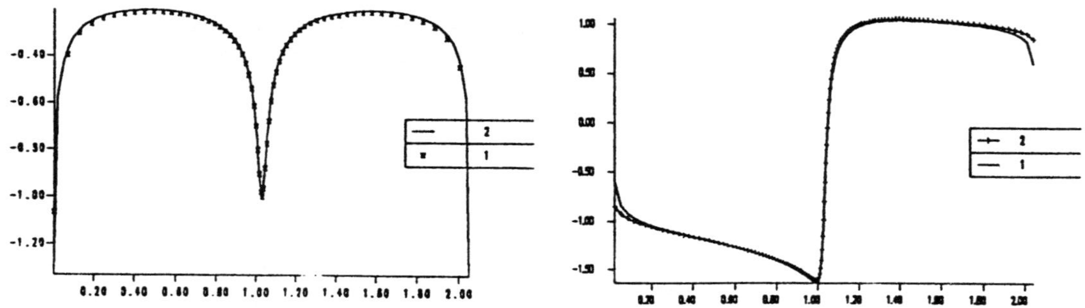

a) b)

Fig.4 Airfoil profile NACA 0015: eigensolution (a) and velocity distribution (b) given by first kind (2), second kind (1) integral equation.

References

[1] B. Hunt,"The Panel Method for Subsonic Aerodynamic Flows: a Survey of Mathematical Formulations and Numerical Models and an Outline of the New British Aerospace Scheme", VKI Lecture Series 1978-4 on Computational Fluid Dynamics, (March 13-17, 1978), ed. W Kolmann, Hemisphere Publ. 1980.

[2] L. Morino, "Helmholtz Decomposition Revisited: Vorticity Generation and Trailing Edge Condition", Computational Mechanics (1986) 1, 65-90.

[3] R. Kinney, "Analysis of Unsteady Viscous Flow Past an Airfoil: Part I – Theoretical Developments", AIAA Journal.

[4] J. C. Wu, " Numerical boundary conditions for viscous flow problems", AIAA Journal.

[5] R. Piva, L. Morino, "Vector Green's Function Method for Unsteady Navier-Stokes Equations", Meccanica, Vol. 22, No. 2, June 1987.

[6] R. Piva, G. Graziani, L. Morino, "Boundary Element Method for Unsteady Viscous and Inviscid Flows", ed. Th. A. Cruse, Advanced Boundary Element Methods, IUTAM Symposium San Antonio, Texas, 1987. Springer Verlag 1988.

[7] R. Piva, "The Boundary Integral Equation Method for Viscous and Inviscid Flows", Proc. Int. Symposium on Comp. Fluid Dynamics, Nagoya (Japan) 1989.

[8] Hsiao, G.C. (1986): "On the Stability of Integral Equations of the First Kind with Logarithmic Kernels", Archive for Rational Mechanics and Analysis, vol. 94, 2, pp. 179-192, Springer Verlag, Berlin.

[9] Nedelec, J.C., (1977): "Integral Equations with Non Integrable Kernels", Integral Equations and Operator Theory, Vol. 5, pp. 562–572.

[10] C.M. Casciola, R. Piva, " A Boundary Integral Formulation for Free Surface Viscous and Inviscid Flows about Submerged Bodies", Proc. 5th Int. Conf. on Numerical Ship Hydrodynamics, Hiroshima Sept. 25-29, 1989.

A Unified Approach for Potential and Viscous Flows in Fixed-Wing and Rotary-Wing Aerodynamics

Luigi Morino
Università di Roma "La Sapienza", Rome, Italy

The objective of this paper is to present an elementary introduction of the work of this author and his collaborators in the field of compressible potential aerodynamics and to introduce an extension of the formulation to compressible viscous flows. For the sake of simplicity, details of the formulation are given only for incompressible flows. For compressible flows, the formulations are only outlined. Also, because of space limitations no numerical results are presented and no attempt is made to give a review of related work in the field. For details of the compressible formulations and the discussion of numerical results, the reader is referred to the original papers, cited in the references. Reviews of boundary-integral formulations are given in Morino and Tseng (1990) for potential flows, in Morino (1990) for viscous flows (potential-vorticity decomposition), and in Piva and Morino (1990) for viscous flows (primitive-variable formulation).

1. Incompressible Potential Flows

The formulation presented here for potential flows was introduced by Morino (1974) and successively refined and implemented by several of his collaborators. Recent developments and details of the formulation are presented in Morino, Bharadvaj, Freedman, and Tseng (1989) and Morino and Tseng (1990), which also include recent numerical results for airplanes, helicopter rotors and propellers. Very recent results are included in other papers in this volume.

Consider an incompressible inviscid attached flow around a body (*e.g.*, a wing or a rotor). The flow is governed by the continuity equation, $\nabla \cdot \mathbf{v} = 0$, and the Euler equations, $\rho D\mathbf{v}/Dt = -\nabla p$ (where D/Dt denotes the substantial derivative). Combining the above equations one obtains Kelvin's theorem: the circulation Γ over a material contour remains constant in time. If, at $t = 0$, the flow is irrotational (*i.e.*, at rest), then $\Gamma = 0$ at all times, for all the material contours that remained in the flow at all times. Thus, recalling Stokes theorem, the flow remains irrotational (and hence potential, *i.e.*, the velocity is given by $\mathbf{v} = \nabla\phi$) at all times and at all points, except for those that came in contact with the body (which form a surface called the wake). Then, the Euler equations may be integrated to yield Bernoulli's theorem for unsteady incompressible potential flows[1]

$$\frac{\partial \phi}{\partial t} + \frac{1}{2}v^2 + \frac{1}{\rho}\,p = \frac{1}{\rho}\,p_\infty \tag{1}$$

Also, combining $\mathbf{v} = \nabla\phi$ with the continuity equation, one obtains

$$\nabla^2\phi = 0 \qquad \mathbf{x} \text{ outside } S \tag{2}$$

where S is a surface that surrounds both the surface of the body, S_B, and the surface of the wake, S_W. The following boundary conditions complete the problem. On the body we assume the surface to be impermeable, so that $(\mathbf{v} - \mathbf{v}_B) \cdot \mathbf{n} = 0$, *i.e.*, $\partial\phi/\partial n = \mathbf{v}_B \cdot \mathbf{n}$ where $\mathbf{v}_B$ is the velocity of a point of S_B. At infinity, we have $\phi = 0$. Next, consider the conditions on the wake. These are obtained from the principles of conservation of mass and momentum and are given by $\mathbf{v} \cdot \mathbf{n} = v_W$ (where v_W is the velocity of the wake surface in the direction of its normal) and

[1] Throughout this paper, the frame of reference is assumed to be connected with the undisturbed air (air frame of reference), unless otherwise stated.

$\Delta p = 0$ (where Δ denotes the discontinuity across S_W). For potential flows, the first equation gives $\Delta(\partial\phi/\partial n) = 0$. Also, combining $\Delta p = 0$ with Bernoulli's theorem, Eq. 1, one obtains

$$\frac{\partial \phi_1}{\partial t} + \frac{1}{2}v_1^2 - \frac{\partial \phi_2}{\partial t} - \frac{1}{2}v_2^2 = \frac{\partial \Delta\phi}{\partial t} + \frac{1}{2}(\mathbf{v}_1 + \mathbf{v}_2)\cdot(\mathbf{v}_1 - \mathbf{v}_2) = 0 \tag{3}$$

or $D_W(\Delta\phi)/Dt = 0$, with $D_W/Dt = \partial/\partial t + \mathbf{v}_W \cdot \nabla$, where $\mathbf{v}_W = \frac{1}{2}(\mathbf{v}_1 + \mathbf{v}_2)$. The above equation may be integrated to yield $\Delta\phi$ =constant following a wake point $\mathbf{x}_W$ (*i.e.*, a point having velocity $\mathbf{v}_W$). Finally, we assume that no concentrated vortices exist at the trailing edge (see Morino, 1986, for an analysis of this issue). This implies that, at the trailing edge, $\Delta\phi$ on the wake is equal to $\Delta\phi$ on the body.

Next, consider the integral formulation. Recall that any function that satisfies the Laplace equation outside S satisfies also the relationship

$$E(\mathbf{x}_*)\ \phi(\mathbf{x}_*) = \oiint_S \left(\frac{\partial\phi}{\partial n}\, G - \phi\, \frac{\partial G}{\partial n}\right) dS \tag{4}$$

where $G = 1/4\pi\|\mathbf{x}_* - \mathbf{x}\|$ ($\mathbf{x}$ being the point of integration on S), and $\mathbf{n}$ is the outer normal, whereas $E_* = 1$ if $\mathbf{x}_*$ is outside S, $E_* = 0$ if $\mathbf{x}_*$ is inside S, and $E_* = 1/2$ if $\mathbf{x}_*$ is a smooth point of S. In this last case, Eq. 4 is a compatibility condition between ϕ and $\partial\phi/\partial n$, which may be used to evaluate ϕ on S, if $\partial\phi/\partial n$ is given (and *vice versa*). Once ϕ on S has been evaluated, Eq. 4 may be used to evaluate ϕ in the field.

In the specific application of interest here, S surrounds S_B and S_W. Let two sides of the portion of S that surrounds S_W become infinitesimally close to S_W. In the limit, recalling that $\Delta(\partial\phi/\partial n) = 0$, one obtains

$$E(\mathbf{x}_*)\ \phi(\mathbf{x}_*) = \oiint_{S_B} \left(\frac{\partial\phi}{\partial n}\, G - \phi\, \frac{\partial G}{\partial n}\right) dS - \iint_{S_W} \Delta\phi\, \frac{\partial G}{\partial n} dS \tag{5}$$

If $\mathbf{x}_*$ is on S_B, Eq. 5 is an integral equation for the unknown ϕ ($\partial\phi/\partial n$ is known from the boundary condition given above, whereas the contribution of the surface is known from preceding time steps, as indicated below). After discretization, Eq. 5 may be used to evaluate ϕ on the S_B. Once ϕ on S_B is known, the same equation may be used to evaluate ϕ, the velocity $\mathbf{v} = \nabla\phi$, and hence the pressure (using Bernoulli's theorem). From the velocity of the wake points, one may evaluate the geometry of the wake at $t + \Delta t$, as $\mathbf{x}_p(t + \Delta t) = \mathbf{x}_p(t) + \mathbf{v}_p(t)\Delta t$. Recall that $\Delta\phi$ remains constant following a wake point $\mathbf{x}_W$, and is equal to value that it had when $\mathbf{x}_W$ left the trailing edge. Thus, the wake contribution at $t + \Delta t$ is known, and the process may be repeated.

The above formulation has been used, for instance, for the free-wake analysis of helicopter rotors in hover (see Morino, Bharadvaj, Freedman, and Tseng, 1989). Very recent unsteady-flow results for helicopter rotors in hover (unsteady aerodynamic forces due to flap and pitch) and windmills (effect of yaw) are included in other papers in this volume.

2. Compressible Potential Flows

Next, consider compressible potential flows (for details, see Morino, Bharadvaj, Freedman, and Tseng (1989). In analogy to incompressible flows, we assume that the fluid is inviscid and non-conducting. The flow is governed by the continuity equation $\partial\rho/\partial t + \nabla \cdot (\rho\mathbf{v}) = 0$, the Euler equations, given above, and the entropy evolution equation, $DS/Dt = 0$. Then, if the flow is initially isentropic and irrotational (*e.g.*, at rest and in thermodynamic equilibrium), it remains isentropic and irrotational (*i.e.*, potential) at all times and at all points (except for the wake points). Recalling that $dh = \frac{1}{\rho}dp + TdS$ (where h denotes the enthalpy), the Euler equations may be integrated to yield Bernoulli's theorem for unsteady compressible potential inviscid isentropic flows

$$\frac{\partial \phi}{\partial t} + \frac{1}{2} v^2 + h = h_\infty \tag{6}$$

Indicating with $a = (\partial p / \partial \rho |_S)^{1/2}$ the isentropic speed of sound, one obtains the equation for ϕ, $a^2 \nabla^2 \phi = D/Dt(\partial \phi / \partial t + \frac{1}{2} v^2)$. For ideal gases, the speed of sound is related to the enthalpy h by the relationship $a^2 = \gamma p / \rho = (\gamma - 1) h$, with h obtained from Eq. 6. Combining and isolating the linear terms, one obtains

$$\nabla^2 \phi - \frac{1}{a_\infty^2} \frac{\partial^2 \phi}{\partial t^2} = \chi \tag{7}$$

where χ denotes the non-linear terms, which are important in the transonic regime. Following a procedure equivalent to that used for incompressible flows, one obtains, for fixed-wing aerodynamics (assuming S to be rigid)

$$\begin{aligned} E(\mathbf{x}_*)\, \phi(\mathbf{x}_*, t_*) \; = \; & \oiint_S \left[\frac{-1}{4\pi \hat{\varrho}} \frac{\partial \phi}{\partial \check{n}} - \frac{\partial}{\partial \hat{n}} \left(\frac{-1}{4\pi \hat{\varrho}} \right) \phi + \frac{-1}{4\pi \hat{\varrho}} \frac{\partial \phi}{\partial t} \left(\frac{\partial \theta}{\partial \hat{n}} + 2 \frac{\mathbf{M} \cdot \mathbf{n}}{a_\infty} \right) \right]_{t_* - \theta} dS \\ & + \iiint_{\mathcal{V}} \left[\frac{-1}{4\pi \hat{\varrho}} \chi \right]_{t_* - \theta} d\mathcal{V} \end{aligned} \tag{8}$$

where $r_\beta = \sqrt{(x - x_*)^2 + \beta^2 (y - y_*)^2 + \beta^2 (z - z_*)^2}$, $\theta = [r_\beta + M_\infty (x - x_*)] / a_\infty \beta^2$ and $\partial / \partial \hat{n} = \partial / \partial \check{n} = \partial / \partial n - M_\infty^2 n_x \partial / \partial x$ (here, $\mathbf{x} = x,\ y,\ z$, is in the body frame of reference). The same integral representation is also valid for fixed-wing aerodynamics, with similar but more complicated expressions for r_β, θ, $\partial / \partial \hat{n}$, and $\partial / \partial \check{n}$.

Numerical results for wings (transonic flows), helicopter rotors in hover (compressible fixed-wake analysis), and propfans are presented in Morino, Bharadvaj, Freedman, and Tseng (1989) and Morino and Tseng (1990). Very recent results on hovering rotors are presented in a separate paper in this volume (which includes also applications of the formulation to aeroacoustics of propellers).

3. Poincaré Decomposition

In this section we present some general considerations on the extension to viscous flows of the potential-flow formulation presented above (for details see Morino, 1990). The key to this extension is the potential-vorticity decomposition: any vector field, in particular the velocity field, may be decomposed as

$$\mathbf{v} = \nabla \phi + \mathbf{w} \tag{9}$$

where ϕ is called the potential, $\nabla \phi$ the potential velocity, and $\mathbf{w}$ is called the vortical velocity and is a particular solution of

$$\nabla \times \mathbf{w} = \boldsymbol{\zeta} \tag{10}$$

where $\boldsymbol{\zeta} = \nabla \times \mathbf{v}$ is the vorticity.[2]

It may be noted that the classic Helmholtz decomposition is a particular case of Eq. 9, with $\mathbf{w} = \nabla \times \boldsymbol{\pi}$ and $\mathbf{v} = \nabla \phi + \nabla \times \boldsymbol{\pi}$. This decomposition is discussed in details in Morino (1986). Only some relevant aspects are discussed here. Imposing the arbitrary but legitimate condition

[2]Proof: Eq. 10 is necessarily satisfied (curl of Eq. 9). On the other hand, Eq. 10 implies $\nabla \times (\mathbf{v} - \mathbf{w}) = 0$. Hence, there exists ϕ such that $\mathbf{v} - \mathbf{w} = \nabla \phi$, in agreement with Eq. 9.

$\nabla \cdot \boldsymbol{\pi} = 0$, Eq. 10 yields $\nabla^2 \boldsymbol{\pi} = -\boldsymbol{\zeta}$. Also, taking the divergence of Eq. 9, with $\mathbf{w} = \nabla \times \boldsymbol{\pi}$, one obtains $\nabla^2 \phi = \theta$, where $\theta = \nabla \cdot \mathbf{v}$. The equations for $\boldsymbol{\pi}$ and ϕ may be integrated to yield

$$\begin{aligned} E(\mathbf{x}_*)\, \mathbf{v}(\mathbf{x}_*) &= \nabla_* \iiint_{\mathcal{V}} G\, \nabla \cdot \mathbf{v}\, d\mathcal{V} - \nabla_* \times \iiint_{\mathcal{V}} G\, \nabla \times \mathbf{v}\, d\mathcal{V} \\ &+ \nabla_* \oiint_{\mathcal{S}} G\, \mathbf{n} \cdot \mathbf{v}\, d\mathcal{S} - \nabla_* \times \oiint_{\mathcal{S}} G\, \mathbf{n} \times \mathbf{v}\, d\mathcal{S} \end{aligned} \tag{11}$$

which is the basis of Wu's method (see, *e.g.*, Wu, 1984).

The decomposition presented here differs from that of Helmholtz, in that $\mathbf{w}$ is obtained by direct integration of Eq. 10. Such integration is possible because of the Poincaré lemma on the integrability of differential form (see below). Thus, the decomposition will be referred to as the Poincaré decomposition. The reason for introducing this decomposition is briefly outlined here. In the Helmholtz decomposition, the vortical velocity $\mathbf{w} = \nabla \times \boldsymbol{\pi}$ is solenoidal. This implies that any signal propagates at infinite speed. This is consistent with the physics of the phenomenon for incompressible flows, but not for incompressible flows. The implication is that the equation for the potential ϕ for compressible flows contains field terms that yield a contribution equal and opposite to that of the vortical velocity (see the section on Compressible Viscous Flows). The presence of field terms makes the integral approach inefficient. As we will show, this problem is minimized by the use of the Poincaré equation.

4. Incompressible Viscous Flows

In this section we present an extension of the potential formulation to incompressible viscous flows (see Morino, 1990, for details). The governing equations are the continuity equation and the Navier-Stokes equations, $D\mathbf{v}/Dt = -\nabla p/\rho + \nu \nabla^2 \mathbf{v}$.

As mentioned above, the Poincaré decomposition is based on the direct solution of Eq. 10. For the sake of clarity, consider first the solution in Cartesian components. In order to obtain the solution, we make the arbitrary but legitimate choice $w_1 = 0$. Then, Eq. 10 becomes

$$\frac{\partial w_3}{\partial x_2} - \frac{\partial w_2}{\partial x_3} = \zeta_1 \qquad -\frac{\partial w_3}{\partial x_1} = \zeta_2 \qquad \frac{\partial w_2}{\partial x_1} = \zeta_3 \tag{12}$$

Last two equations may be integrated to yield

$$w_2(x_1, x_2, x_3) = \int_{-\infty}^{x_1} \zeta_3(x_1', x_2, x_3) dx_1' \qquad w_3(x_1, x_2, x_3) = -\int_{-\infty}^{x_1} \zeta_2(x_1', x_2, x_3) dx_1' \tag{13}$$

Recalling that the vorticity is solenoidal (integrability condition), it is easy to verify that the first equation in Eq. 12 is automatically satisfied. Equation 13, combined with $w_1 = 0$, may be expressed as $\mathbf{w} = \int_{\mathcal{L}} \boldsymbol{\zeta}(\mathbf{x}') \times d\mathbf{x}'$ where $\mathcal{L}$ is a x_1-line from $-\infty$ to $\mathbf{x}$. In order to illustrate the nature of this solution, consider the case in which the vorticity distribution is equal to zero outside a sphere, $\mathcal{V}_0$, and arbitrary within $\mathcal{V}_0$. It is apparent from the above equation that $\mathbf{w} = 0$ except for the points of $\mathcal{V}_0$ and of its "shadow" in the direction of the positive x_1-axis. However, if (and only if)

$$\int_{X_1'}^{X_1''} \zeta_k(x_1', x_2, x_3) dx_1' = 0 \qquad (k = 2, 3) \tag{14}$$

(where X_1' and X_1'' are limits of integration on the boundary of $\mathcal{V}_0$) for all values of x_2 and x_3, then the Poincaré solution $\mathbf{w}$ is zero everywhere outside $\mathcal{V}_0$. In any event, in the shadow $\mathbf{w}$ is constant along the x_1-lines.

In practical applications, it is convenient to use body-fitted curvilinear coordinates, ξ^α. In this case, the solution for $\mathbf{w}$ is

$$w_\alpha(\xi^1, \xi^2, \xi^3) = \int_{\xi_0^1}^{\xi^1} \epsilon_{\alpha\beta 1}\, \zeta^\beta(\lambda^1, \xi^2, \xi^3)\, d\lambda^1 \tag{15}$$

where w_α are the covariant components of $\mathbf{w}$, ζ^β are the contravariant components of $\boldsymbol{\zeta}$, and $\epsilon_{\alpha\beta\gamma}$ are the covariant components of the permutation tensor. The issue of the shadow is similar to that for Cartesian components, but the shadow is now curvilinear (in the "direction" of the ξ^1-lines); we will refer to this as the "ξ^1-shadow".

Having obtained a solution for $\mathbf{w}$, we now turn to the solution for ϕ. The equation for ϕ is obtained from the continuity equation, which combined with Eq. 9 gives $\nabla^2\phi = \sigma_\mathbf{w} = -\nabla \cdot \mathbf{w}$. The boundary conditions on S_B are $\mathbf{v} = \mathbf{v}_B$. The normal component of this equation, combined with Eq. 9 gives $\partial\phi/\partial n = \mathbf{v}_B \cdot \mathbf{n} - \mathbf{w} \cdot \mathbf{n}$. The solution to the above equation is

$$E(\mathbf{x}_*)\ \phi(\mathbf{x}_*) = \oiint_{S_B} G\, \frac{\partial \phi}{\partial n}\, dS - \oiint_{S_B} \phi\, \frac{\partial G}{\partial n}\, dS + \iiint_{\mathcal{V}} G\, \sigma_\mathbf{w}\, d\mathcal{V} \tag{16}$$

As $\mathbf{x}_*$ approaches a point on the surface S_B, Eq. 16 yields an integral equation that may be used to solve for the potential, provided that the vortical velocity, $\mathbf{w}$, is known.

Having discussed the method of solution for $\mathbf{w}$ and ϕ, we now turn to the evaluation of vorticity. Note that the vorticity in the field may be obtained from the vorticity evolution equation $D\boldsymbol{\zeta}/Dt = (\boldsymbol{\zeta} \cdot \nabla)\mathbf{v} + \nu\nabla^2\boldsymbol{\zeta}$. Therefore, the only remaining problem is the evaluation of the vorticity at the boundary, and the related issues of the tangential boundary conditions and the vorticity generation. The treatment of this problem is similar to that introduced by Morino (1986) for the Helmholtz decomposition. Consider the case of a sudden change of velocity at $t = t_0$ (in particular, an impulsive start). Given the values of the vorticity at $t = t_0$, we may evaluate the vortical velocity $\mathbf{w}$ at $t = t_0$ (Eq. 15). Recall that, for $\mathbf{x}_*$ on S_B, Eq. 16 represents an integral equation, which is used to evaluate the unknown distribution of the potential on the surface of the body, still at $t = t_0$. The resulting solution, it should be emphasized, will present a velocity discontinuity between the fluid and the body, even for viscous flows (this is apparent in the case of an impulsive start). The implication of the above statement is that at $t = t_0$ there exits a zero-thickness layer of vorticity with intensity $\boldsymbol{\gamma} = \mathbf{n} \times (\nabla\phi + \mathbf{w} - \mathbf{v}_B)$. For viscous flows, $\boldsymbol{\gamma}$ immediately diffuses into the field, and the tangential boundary conditions are automatically satisfied (for, if they were not, there would be a vortex layer). At this point, still using Eq. 16 but with $\mathbf{x}_*$ in the field, we may evaluate the potential, ϕ. Then, using Eq. 9, one obtains the value of the velocity anywhere in the field, in particular at any point in the vortical region where it is needed for the vorticity evolution equation. Finally, using this equation we obtain the vorticity at time $t = t_0 + dt$ anywhere in the field and we may continue the process. Numerical results obtained using the above formulation are presented in Morino and Beauchamp (1988) and Beauchamp (1990).

It may be noted that the pressure, not necessary for the solution of the velocity field, may be evaluated from a Bernoulli-type theorem,

$$\frac{\partial \phi}{\partial t} + \frac{1}{2}v^2 + \frac{1}{\rho}\, p + \int_\infty^\mathbf{x} \left(\frac{\partial \mathbf{w}}{\partial t} + \boldsymbol{\zeta} \times \mathbf{v} - \nu\nabla^2\mathbf{w} \right) \cdot d\mathbf{x}' = \frac{1}{\rho}\, p_\infty \tag{17}$$

obtained by integrating the Navier-Stokes equations (with $D\mathbf{v}/Dt = \partial\mathbf{v}/\partial t + \frac{1}{2}\nabla v^2 + \boldsymbol{\zeta} \times \mathbf{v}$). Note that the integrand is irrotational (*i.e.*, the integral is path independent) and equal to zero outside the vortical region and its ξ^1-shadow. This facilitates the evaluation of the integral.

Finally, we want to show that, in the limit as the thickness of the vorticity layer goes to zero, the Poincaré representation reduces to the integral representation for potential flows, Eq. 5. In order to accomplish this, note that integrating Eq. 16 by parts yields

$$E(\mathbf{x}_*)\ \phi(\mathbf{x}_*) = \oiint_{S_B} G\, \psi\, dS - \oiint_{S_B} \phi\, \frac{\partial G}{\partial n}\, dS + \iiint_{\mathcal{V}_F} \mathbf{w} \cdot \nabla G\, d\mathcal{V} \tag{18}$$

where $\psi = \mathbf{v}_B \cdot \mathbf{n}$. In Eq. 18 ϕ_* is expressed in terms of the values of ϕ and ψ on S_B, and of a volume distribution of doublets having intensity $|\mathbf{w}|$ and direction equal to $\mathbf{w}$. As shown in

Morino (1990), as the thickness goes to zero, the volume integral over the wake region in Eq. 18 reduces to the wake doublet layer of Eq. 5. Similarly, the volume integral over the boundary-layer region, combined with the doublet integral over the surface S_B, yields, in the limit, the body doublet layer in Eq. 5. Therefore, the limiting case of the viscous-flow formulation based on the Poincaré decomposition (in the form given by Eq. 18) is Eq. 5. In this sense, such a formulation is an extension of the potential-flow formulation. It may be worth noting that even the wake-transport formulation may be recast in a form that is closely related to that of potential flows (Morino, 1990).

5. Compressible Viscous Flows

In this section we outline the potential-vorticity formulation for compressible viscous flows (for details the reader is referred to Morino, 1990). The governing equations are the continuity equation, given above, the compressible Navier-Stokes equations $\rho D\mathbf{v}/Dt = -\nabla p + \nabla \cdot \mathbf{V}$, and the entropy evolution equation $\rho\; T\; DS/Dt = \mathbf{V} : \mathbf{D} - \nabla \cdot \mathbf{q}$ ($\mathbf{V}$, $\mathbf{D}$, and $\mathbf{q}$ are the viscous-stress tensor, the strain-rate tensor, and the heat-flux vector, respectively). Recalling that $D\mathbf{v}/Dt = \partial \mathbf{v}/\partial t + \frac{1}{2}\nabla v^2 + \boldsymbol{\zeta} \times \mathbf{v}$ and $dh = \frac{1}{\rho}dp + TdS$ (where h denotes the enthalpy), the Navier-Stokes equations may be rewritten as

$$\frac{\partial \mathbf{v}}{\partial t} + \frac{1}{2}\nabla v^2 + \boldsymbol{\zeta} \times \mathbf{v} = -\nabla h + T\nabla S + \frac{1}{\rho}\nabla \cdot \mathbf{V} \tag{19}$$

In order to extend the potential-vorticity approach to compressible viscous flows, the formulation for the evaluation of $\mathbf{w}$, for ϕ, and for $\boldsymbol{\zeta}$ are required. The formulations for $\mathbf{w}$ and $\boldsymbol{\zeta}$ are similar to those for incompressible flows. Here, we concentrate on the formulation for ϕ. In order to extend the potential-flow formulation, we need an extension of Bernoulli's theorem, Eq. 6 (similar to Eq. 17 for incompressible viscous flows). Combining Eqs. 9 and 19, one obtains $\nabla\left(\partial\phi/\partial t + \frac{1}{2}v^2 + h\right) + \mathbf{f} = 0$, where $\mathbf{f} = \partial \mathbf{w}/\partial t + \boldsymbol{\zeta} \times \mathbf{v} - T\nabla S - \nabla \cdot \mathbf{V}/\rho$. By taking the curl of the above equation, we see that $\mathbf{f}$ is irrotational. Hence, the integral of $\mathbf{f}$ is path independent, and therefore, $\mathcal{F} = \int_\infty^{\mathbf{x}} \mathbf{f} \cdot d\mathbf{x}$ defines a function (multi-valued for multiply-connected regions) which is such that $\nabla\mathcal{F} = \mathbf{x}$. Combining and integrating yields

$$\frac{\partial \phi}{\partial t} + \frac{1}{2}v^2 + h + \mathcal{F} = h_\infty \tag{20}$$

Equation 20 is the desired generalization of Bernoulli's theorems, Eqs. 6 and 17.

Next, consider the differential equation for ϕ. Following the procedure used to obtain Eq. 7 for potential isentropic flows, we may combine the continuity equation with Eq. 9 to obtain $\nabla^2\phi = \sigma_\mathbf{w} - \frac{1}{\rho}D\rho/Dt$, where $\sigma_\mathbf{w} = -\nabla \cdot \mathbf{w}$. Next, assuming h and S as fundamental variables of state, one obtains $D\rho/Dt = \rho_h Dh/Dt + \rho_S DS/Dt$, where $\rho_h = \partial\rho/\partial h|_S$ and $\rho_S = \partial\rho/\partial S|_h$. Note that $\rho_h = \rho/a^2$ (where a is the isentropic speed of sound, defined above), whereas for ideal gases, $\rho_S = -\rho/\mathcal{R}$ (where $\mathcal{R}$ is the ideal-gas constant). Combining yields

$$\frac{1}{\rho}\frac{D\rho}{Dt} = \frac{1}{a^2}\frac{Dh}{Dt} - \frac{1}{\mathcal{R}}\frac{DS}{Dt} \tag{21}$$

or, combining with $\nabla^2\phi = \sigma_\mathbf{w} - \frac{1}{\rho}D\rho/Dt$ and using Eq. 20 to eliminate h, one obtains

$$\nabla^2\phi = \sigma_\mathbf{w} + \frac{1}{a^2}\frac{D}{Dt}\left(\frac{\partial\phi}{\partial t} + \frac{1}{2}v^2 + \mathcal{F}\right) + \frac{1}{\mathcal{R}}\frac{DS}{Dt} \tag{22}$$

Note that the linear portion of Eq. 22 (excluding the contribution of ϕ to $\mathbf{V}$ (which is linear, but is treated as a field term), coincides with that of Eq. 7. Thus, the boundary-integral formulation for the solution of Eq. 7 (*i.e.*, Eq. 8) may be used for Eq. 22 as well.

In the remainder of this section we examine the advantages of the Poincaré decomposition over the Helmholtz decomposition for unsteady compressible viscous flows. For simplicity, consider the flow around an isolated wing. The differences between the two approaches are best understood by examining Eq. 22 in the outer region (*i.e.*, outside the boundary layer and wake), where, at least from a numerical point of view, we may assume $\boldsymbol{\zeta} = \mathbf{V} = 0$, and S=constant. Hence, for both cases, in the outer region we have $\mathbf{f} = \partial \mathbf{w}/\partial t$ and hence $\mathcal{F} = \int_\infty^x (\partial \mathbf{w}/\partial t) \cdot d\mathbf{x}$. Consider first the Poincaré decomposition. Let $\mathcal{V}_E$ indicate the region outside the rotational region and its ξ^1-shadow. In this region we have $\mathbf{w} = 0$, $\mathbf{v} = \nabla\phi$, and $\mathcal{F} = 0$. Hence, the additional field terms are limited to the rotational region and its ξ^1-shadow. On the other hand, still in the outer region, for the Helmholtz decomposition we have $\mathbf{w} = \nabla \times \boldsymbol{\pi}$. Hence, in this case $\sigma_\mathbf{w} = -\nabla \cdot \mathbf{w} = -\nabla \cdot \nabla \times \boldsymbol{\pi} = 0$. However, for unsteady flows $\mathcal{F} \neq 0$ outside the vortical region, requiring a much larger volume for the field-term integral. Finally, note that in the Poincaré decomposition, akin to incompressible flows, the term $\sigma_\mathbf{w} = -\nabla \cdot \mathbf{w}$ in the wake yields a source layer that integrated by parts yields a volume integral that is numerically close to the compressible doublet distributions obtained for compressible potential flows (see Morino, Bharadvaj, Freedman, and Tseng, 1988). No numerical results based on the above formulation for compressible viscous flows have been obtained.

References

Beauchamp, P. P., (1990): "A Potential-Vorticity Decomposition for the Boundary Integral Equation Analysis of Viscous Flows," Ph D Thesis, Boston University.

Morino, L., (1974): "A General Theory of Unsteady Compressible Potential Aerodynamics," NASA CR-2464.

Morino, L., (1986): "Helmholtz Decomposition Revisited: Vorticity Generation and Trailing Edge Condition, Part 1, Incompressible Flows," Computational Mechanics, Vol. 1, pp. 65-90.

Morino, L., (1990): "Helmholtz and Poincaré Vorticity-Potential Decompositions for the Analysis of Unsteady Compressible Viscous Flows," Eds.: P. K. Banerjee and L. Morino, *Developments in BEM, Vol.* 6: Elsevier Applied Science Publishers, Barking, UK.

Morino, L., and Beauchamp, P. P., (1988): "A Potential-Vorticity Decomposition for the Boundary-Element Analysis of Viscous Flows," Eds.: M. Tanaka and T. A. Cruse, *Boundary Element Methods in Applied Mechanics*, Pergamon Press, New York, NY, USA.

Morino, L., Bharadvaj, B. K., Freedman, M. I., and Tseng, K., (1989): "Boundary Integral Equation for Wave Equation with Moving Boundary and Applications to Compressible Potential Aerodynamics of Airplanes and Helicopters," Computational Mechanics, Vol. 4, pp. 231-243.

Morino, L., and Tseng, K., (1990): "A General Theory of Unsteady Compressible Potential Flows with Applications to Airplane and Rotors," Eds.: P. K. Banerjee and L. Morino, *Developments in BEM, Vol.* 6: Elsevier Applied Science Publishers, Barking, UK.

Piva, R., and Morino, L., (1990): "A Boundary Integral Formulation in Primitive Variables for Unsteady Viscous Flows," Eds.: P. K. Banerjee and L. Morino, *Developments in BEM, Vol.* 6: Elsevier Applied Science Publishers, Barking, UK.

Wu, J. C., (1984): "Fundamental Solutions and Numerical Methods for Flow Problems," International Journal for Numerical Methods in Fluids, Vol. 4, pp. 185-201.

Acknowledgments

This work was partially supported by the U. S. Army Research Office (Contract No. DAAL-03-87-K-0022 to Boston University) and by NASA (Grant No. NAG-1-934 to Boston University).

Linkage Between Potential and Viscous Flows

J. C. WU

School of Aerospace Engineering
Georgia Institute of Technology

Summary

The well-known boundary element method for computing potential flows is shown to be linked to a zonal method previously developed for computing viscous flows. This linkage permits the removal of conceptual difficulties and uncertainties resulting from the inviscid fluid idealization of the panel method. Extremely efficient and accurate boundary element methods are established for flows in the absence of significant separated regions.

Introduction

Flow problems are divisible into two main categories in accordance with the presence or absence of a significant region of flow separation. In the presence of a large scale flow separation, accurate simulation of the flow is achievable only by solving the full Navier-Stokes equations. This category of flows is referred to as the viscous flow in the present paper. In a flow containing no significant separated region, viscous effects are confined to small and well-defined parts of the flow. This category of flow is widely referred to as the potential flow since a great deal of information about the flow can be obtained by solving the much simpler potential flow equations.

Within the field of computational fluid dynamics, solution methods for the viscous flow problem have advanced along a different pathway from those for the potential flow problem. For the viscous flow, the finite difference approach has evolved into a reasonably well-established method. For the potential flow problem, the boundary element method, more commonly known as the panel method in fluid dynamics, today represents the most popular choice.

With the panel method, the familiar differential equation describing the flow is reformulated as an integral equation and then discretized and computed. In reformulating the flow problem, singularity elements (sources, doublets and vortices) are utilized and the fluid is viewed as being inviscid. Because of these underlying conceptions, the panel method is often regarded as unconnected to the viscous flow. The remarkable advantages offered by the panel method in solving the potential flow problem are thought to be not extendable to the viscous flow problem.

In this paper, it is shown that, although it is not practical to generalize the panel method for viscous flow computation, the panel method is readily interpretable as an approximation of a viscous flow solution method established previously by the present author and his co-workers [1]. This method, referred to as the zonal method, is developed mathematically rigorously on the basis of the full Navier-Stokes equations. The approximation of the zonal method, which yields the panel method, is justifiable on the basis of the physics of the viscous flow. The understanding of the linkage between the potential flow and the viscous flow offers important advantages in the further development of computation methods for both categories of flow. With the "bondage" of the panel method to the inviscid fluid idealization removed, it becomes practical to broaden the scope of application of the panel method. Accurate and reliable procedures can be established rigorously through viscous flow analyses. For viscous flow computation, the linkage between the two flows allows many well-established panel procedures to be adapted and used.

Flow Anatomy

The linkage between the potential flow and the viscous flow is best examined through the concept of vorticity. Real fluids are viscous. Viscous stresses, however, are important only in the flow region where the vorticity is present.

Consider a solid body immersed in a fluid and moving relative to the fluid. Both the solid and the fluid are initially at rest. At a certain instant of time, the solid is set into motion. Thereafter, the solid is kept moving and the induced unsteady motion of fluid is of interest. Before the solid is set into motion, the fluid is stationary and therefore free of vorticity. It is known [2] that vorticity can neither be created nor be destroyed in the interior of a fluid. Thus, the vorticity present in the interior of the fluid domain must originate from the fluid/solid interface. This vorticity, once generated, spreads into the interior of the viscous fluid by diffusion and, once there, is transported away from the solid by both convection and diffusion. Since the effective rates of diffusive and convective transport are finite, the region of significant vorticity is generally surrounded by a region in which no significant vorticity is present. This region is a potential flow region.

The region containing vorticity is the viscous region of the flow. This viscous region in general is composed of an attached boundary layer zone and detached flow zones. In two-dimensional flows, the detached viscous flow zones in general include recirculating flows (separation bubbles) and wakes. For three-dimensional flows, "tip vorticity are also present. The wake zone is in reality a continuation of the boundary layer and the recirculating flow. That is, the vorticity in the boundary layer and the recirculating flow is continually transported away from the solid body and feeds into the wake.

In most applications, the flow Reynolds number is high. In consequence, the thickness of the boundary layer is small compared to the characteristic length of the solid body. The detached viscous zones are much larger in comparision. The potential flow region, where the vorticity is absent, has an infinite length scale. In addition to the differences in length scales, the mathematical equations describing the flows in the several zones and in the potential flow region belong to different classifications. The coexistence of these diverse flow zones renders the task of analyzing the viscous flow problem an extremely difficult one.

For a flow containing no recirculating zone, only the wake zones exist in the detached viscous region. This wake is the continuation of only the boundary layer and is therefore thin. Wake layers trailing the solid body may roll up and form small clusters of concentrated vorticity, or vortices. In addition, in the case of a solid body undergoing rapid changes of motion, a large amount of vorticity is shed into the wake within a short time interval. This vorticity appears in a concentrated dosage and often becomes a prominent and well-defined feature in the flow. A well-known example of this wake feature is the starting vortex formed as an airfoil is set into motion suddenly.

Zonal Method for Viscous Flow

Wu and his coworkers [3] have shown that the mathematical formulation of the viscous flow problem can be partitioned into its kinetic and kinematic parts using the concept of vorticity. Specifically, the incompressible Navier-Stokes equations can be reexpressed as:

$$\frac{\partial \vec{\omega}}{\partial t} = \nabla \times (\vec{v} \times \vec{\omega}) + \nu \nabla^2 \vec{\omega} \tag{1}$$

$$\nabla \cdot \vec{v} = 0 \tag{2}$$

$$\nabla \times \vec{v} = \vec{\omega} \tag{3}$$

where $\vec{v}$ and $\vec{\omega}$ are respectively the velocity vector and the vorticity vector and ν is the kinematic viscosity of the fluid.

Equation (1), known as the vorticity transport equation, describes the kinetic processes of convection and viscous diffusion of vorticity. Equations (2) and (3) are kinematic equations describing the relationship between the vorticity field at any instant of time and the velocity field at the same instant. Equations (1), (2) and (3) are valid in the interior of the open fluid domain R. If an initial distribution of $\vec{\omega}$ is known in R and the velocity vector $\vec{v}$ is prescribed on the boundary B of R for a subsequent time interval, then the distributions of $\vec{v}$ and $\vec{\omega}$ in R for the subsequent time interval can be uniquely determined [3].

In the computation of a time-dependent viscous flow, the development of the vorticity field is followed. A computation loop that advances the solution from an old time level, at which the vorticity and velocity values are known in R and on B, to a new time level is composed of the following major steps:

(1) The computation of the new distribution of vorticity in the open domain R.

(2) The computation of the new distribution of vorticity on the boundary B.

(3) The computation of the corresponding new distribution of the velocity in the interior of the domain R.

Step (1) is the kinetic part of the loop. In this step, Eq. (1) is solved numerically. Step (2) is needed since vorticity is continually generated on the solid surface S through the no-slip condition [4]. New vorticity values on S are not obtainable by solving Eq. (1). Instead, they are needed to advance the solution to subsequent time levels. Wu [4] has shown that, with the vorticity distribution in the domain R and the velocity boundary values on B known, the boundary values of vorticity on the solid surface S is uniquely determined by the kinematics of the problem. Step (2) is therefore a kinematic part of the loop. Step (3) is obviously also a kinematic part of the loop.

The procedure just described can yield steady-state solutions asymptotically at large time levels. Alternatively, a steady state vorticity transport equation can be solved using an iterative procedure. The computation loop just described in that case advances the solution from an old iteration to a new one.

Wu and Thompson [3] have shown that, in steps (2) and (3), it is advantageous to use the concept of fundamental solution and to re-express Equations (2) and (3) as an integral representation as follows:

$$\vec{v}(\vec{r},t) = -\int_R \vec{\omega}_o \times \vec{Q}_o dR_o + \int_B [(\vec{v}_o \cdot \vec{n}_o)\vec{Q}_o - (\vec{v}_o \times \vec{n}_o) \times \vec{Q}_o] dB_o \qquad (4)$$

where $\vec{n}$ is a normal unit vector on B directed outward from R; the subscript "o" indicates that a variable or an integration is in the $\vec{r}_0$ space e.g., $\vec{\omega}_0$ = $\vec{\omega}_0(\vec{r}_0,t)$; the vector Qo is the gradient of the fundamental solution of ellipic differential equations, and is given by

$$\vec{Q}_o(\vec{r}_o,\vec{r}) = \frac{-(\vec{r}_o - \vec{r})}{2(D-1)\pi|\vec{r}_o - \vec{r}|^D} \qquad (5)$$

where D is the dimensionality of the problem, i.e. D = 2 for two-dimensional problem and D = 3 for three-dimensional problem.

With known vorticity distribution in R and on B and known velocity boundary condition on B, Eq. (4) allows velocity values to be computed explicity, one point at a time. This ability permits the velocity values on the demarcation boundary between the viscous and the potential regions to be computed independently of other computations. Once this is accomplished, the problem of computing the viscous region becomes independent of the potential region. Since the vorticity distribution is of interest, one need not be concerned with the computation of the potential region. The computation loop described earlier can be confined to the viscous region where the vorticity is non-zero.

The ability to explicitly compute velocity values can also be used to compute velocity values on the boundary of each viscous zone. Once this is accomplished, the computation of each viscous zone becomes independent of all other zones.

The advantages offered by the zonal method are drastic. The confinement of the solution field to the viscous region obviously reduces the needed number of grid points involved and leads to substantial savings in the amount of computation required. The independent computation of the coexisting viscous zones remove the difficulties causes by the diverse physical and mathematical properties of these zones. Different grids as well as different numerical procedures can be used in different zones. For example, in a boundary layer grid system with extremely close grid spacing in the direction normal to the solid surface can be used in the boundary layer zone. The boundary layer equations, instead of the Navier-Stokes equations, can be solved in this zone numerically using a non-iterative procedure that marches the solution station by station downstream.

Potential Flow Solution Method

In the absence of a recirculating flow zone, the viscous region contains only the boundary layer zone and the wake zone. Both these zones are confined to narrow corridors and justifiable approximations of the vorticity contents of these zones are available.

In the kinematic part of the computation, the vorticity in the boundary layer is justifiably approximated by a vortex sheet. The strength $\vec{\gamma}$ of the vortex sheet is given by

$$\gamma(s) = \int_0^\delta \bar{\omega}(s,n)dn \tag{6}$$

where s and n are respectively the tangential and the normal boundary layer coordinates and δ is the thickness of the boundary layer.

It is important to emphasize that this approximation does not imply an inviscid fluid idealization. It merely recognizes the fact that, since the boundary layer is thin, the distribution of vorticity across the thickness of the boundary layer is unimportant in the kinematic part of the analysis.

The vorticity in a thin wake layer is also justifiably approximated by a vortex sheet. Wake vorticity that exists in clusters, including the "rolled up" wake vortices and the starting vortices, are justifiably approximated by vortex filaments which are represented by point singularities in two-dimensional flows. These approximations also do not imply the inviscid fluid idealization.

To demonstrate the simplifications resulting from these approximations, consider a two-dimensional flow with the boundary layer and the wake vortices represented respectively by vortex sheets and vortex filaments. Equation (4) can now be expressed as:

$$\begin{aligned}\vec{v}(\vec{r},t) \; = \; & -\frac{1}{2\pi}\int_{S^+}\frac{\vec{\gamma}_o\times(\vec{r}_o-\vec{r})}{|\vec{r}_o-\vec{r}|^2}dB_o-\frac{1}{2\pi}\sum_k\frac{\vec{\Gamma}_k\times(\vec{r}_k-\vec{r})}{|\vec{r}_k-\vec{r}|^2}\\ & -\frac{1}{2\pi}\int_S\frac{(\vec{v}_o\times\vec{n}_o)\times(\vec{r}_o-\vec{r})}{|\vec{r}_o-\vec{r}|^2}dB_o \qquad (7)\\ & +\frac{1}{2\pi}\int_S\frac{(\vec{v}_o\cdot\vec{n}_o)(\vec{r}_o-\vec{r})}{|\vec{r}_o-\vec{r}|^2}dB_o+\vec{v}_\infty\end{aligned}$$

where Γ_k and $\vec{r}_k$ are respectively the strength and the position at the time t of the kth the vortex filament, the summation is over all wake vortices present, the surface S^+ indicates the location of the vortex sheet $\vec{\gamma}$ and $\vec{v}_\infty$ is the freestream velocity.

In Equation (7), the integral around S^+ and the summation containing Γ_k are the Biot-Savart contributions of the boundary layer vorticity and the wake vorticity, respectively, to the velocity field. They represent an approximation to the domain integral in Eq. (4). The boundary B in Eq. (4) is composed of the solid surface S and the external boundary S_∞ infinitely far from S. The contribution of S to the surface integral in Eq. (4) are given by the two surface integrals in Eq. (7) and the contribution of the surface S_∞ is simply the freestream velocity $\vec{v}_\infty$ (See [4] for a proof).

Consider the computation loop discussed earlier. The strength and the location of each wake vortex Γ_k are determined by the kinetics of the problem. Since the wake is convection dominated in a high Reynolds number flow, viscous diffusion of vorticity is negligible. The strength of each wake vortex at the new time level is taken to be the same as the strength at the old time level. That is, in step (1) of the loop, instead of using Eq. (1), each wake vortex Γ_k is allowed to convect with the local flow velocity $\vec{v}_k$ without changing its strength. The displacement of the wake vortex during a time interval Δt is simply $\vec{v}_k \Delta t$.

In addition to the convection of existing wake vortices, a nascent wake vortex is formed during each time interval due to the shedding of the vorticity from the boundary layer into the wake. In the literature, various models for the nascent vortex are proposed. In a recent article [6], Hsu and Wu presented a method for computing the nascent wake vortex based on a viscous flow analysis. This method has been shown to produce accurate simulation of unsteady flows.

The surface S^+ represents the location of the boundary layer. The thinness of the boundary layer suggests that S^+ can be placed placed infinitesimally close to S. S^+, however, must envelop and must not be coincident with S since, physically, the boundary layer surrounds the solid but is in the fluid domain.

The strength $\vec{\gamma}$ of the boundary layer is determined by the kinematics of the problem in Step (2). Specifically, one applies Eq. (7) to a point $\vec{r}_s$ on the solid boundary S. With prescribed solid motion, Equation (7) yields for $\vec{r} = \vec{r}_s$:

$$-\frac{1}{2\pi}\int_{S+}\frac{\vec{\gamma}_o \times (\vec{r}_o - \vec{r}_s)}{|\vec{r}_o - \vec{r}_s|^2}dB_o = \vec{G}(\vec{r}_s, t) \tag{8}$$

where $\vec{G}$ is a known function representing the left side of Equation (7) combined with the last three terms on the right side.

Equation (8) is a vector integral equation possessing a tangential component and a normal component. Wu [4] has shown that these two components are equivalent to each other. For the two-dimensional external flow problems under consideration, the fluid domain is multiply-connected. The strength γ is uniquely determined by Eq. (8) together with the principle of conservation of total vorticity [2]. This principle, with the approximations discussed, is expressible as:

$$\int_{S_+} \gamma dB + \int_S v_s dB + \sum_k \Gamma_k = 0 \tag{9}$$

In step (3), the results of steps (1) and (2) are placed into Eq. (7) to compute velocity values at any point in the flow. In order to advance the solution further in time, velocity values at the new time level are computed for each wake vortex Γ_k.

The procedure just outlined is a vortex panel method for unsteady flows. In the simplified computation loop just described, steps (2) and (3) constitute a boundary-element procedure. Namely, the problem of computing an unknown function (γ in the present case) is formulated as a general integral relation (Eq. 7 in the present case). The specialization of this integral relation to boundary points $\vec{r}_s$ yield a boundary integral equation for the unknown function. This boundary integral equation is solved numerically. The computed values of the unknown function is then used in the general integral relation to obtain information at points away from the boundary. It should be noted in computing viscous flows containing separated regions, a similar boundary element is used to compute the boundary values of vorticity [4,8].

The linkage between the potential flow and the viscous flow is most clearly revealed through an examination of a steady external two-dimensional flow. For this problem, the solid body can only undergo a constant translational motion. The integrated vorticity value across the wake layer is zero. In consequence, excepting the possible existence of a starting vortex, the wake vortices are not important. The starting vortex, being infinitely far from the solid body, does not contribute to the velocity field near the solid body. The significance of the starting vortex is manifest through the principle of conservation of total vorticity. For the present problem, the summation in Eq. (9) is replaced by simply Γ_s, the strength of the starting vortex. In a reference frame attached to the solid body, $\vec{v}_s = 0$ and the integral involving v_s in Eq. (9) vanishes. One thus has

$$\int_{S+} \gamma \, dB = -\Gamma_s \tag{10}$$

Equation (10) provides a physical interpretation of the concept of bound vortex, which is known to be equal to $-\Gamma_s$ in the context of an inviscid fluid. The bound vortex is considered as the circulation that remains "bound" to the solid body [7]. In the context of a viscous fluid, this bound vortex represents the total, integrated, vorticity in the boundary layers surrounding the solid. The vorticity in the boundary layers is not bound to the solid, but is continually transported downstream and feeds into the wake layer. At the same time, new vorticity is being generated on the solid surface. In a steady flow, the replenishment of vorticity and the transport of vorticity balance one another so that the total vorticity in the boundary layer does not change with time.

In order for the strength of the wake vortex sheet to be zero, the net rate of shedding of boundary layer vorticity into the wake must be zero. Hsu and Wu [6] has shown through a viscous flow analysis that, for an airfoil, the net rate of shedding of boundary layer vorticity is $1/2\ (\gamma_l^2 - \gamma_u^2)$, where γ_l and γ_u are the strengths of the vortex sheets representing the vorticity in the boundary layers respectively on the lower surface and the upper surface of the airfoil at the trailing edge. One thus conclude that $\gamma_l = -\gamma_u$ for a steady flow. This condition is equivalent to the well-known Kutta condition [7] and is often used in vortex panel methods in place of Eq. (10). In other words, with a vortex panel method, it is not necessary to know Γ_s directly.

Since there is no need to know the location or the strength of the starting vortex and the wake layer in a steady flow is not important, the kinetics of the flow problem is dissociated from in the analysis of steady two-dimensional potential flows. The kinematics of the flow, described by the following simplified version of Eq. (7) becomes the focal point of the solution procedure:

$$\vec{v}(\vec{r}) = -\frac{1}{2\pi} \int_{S+} \frac{\vec{\gamma}_0 \times (\vec{r}_0 - \vec{r})}{|\vec{r}_0 - \vec{r}|^2}\, dB_0 + \vec{v}_\infty \tag{11}$$

To determine the boundary layer vortex sheet strength $\vec{\gamma}$, one applies Eq. (11) to the solid boundary S, obtaining a boundary integral equation containing $\vec{\gamma}$ as the unknown function. This boundary integral equation, subject to the Kutta condition, is solved numerically. The computed values of $\vec{\gamma}$ can be placed into (11) to compute $\vec{v}$ at any point in the flow field.

The procedure just described is a vortex panel method for steady potential flows. In many problems of practical importance, it is only necessary to compute the velocity u_e at the edge of the boundary layer. It is easy to show, using Eq. (6), that $\gamma = -u_e$. This means that solution procedure can be confined to the boundary. The present two-dimensional flow problem is then solved by treating an one-dimensional integral equation. This remarkable advantage is of course well-known to the panel method.

For three-dimensional steady potential flows, the solution procedure can justifiably be confined to the solid-boundary provided that no tip vortices are present. This circumstance occurs only for a solid body which experiences no lift along any section of its span. In general, with the presence of tip vortices, a procedure similar to the one described for the unsteady potential flow must be used to compute the strengths and the locations of the tip vortices. The kinetics of the flow problem needs to be simulated.

Concluding Remarks

The present study demonstrates that the panel method for potential flows is interpretable as an approximation to a zonal method for viscous flows. There exist two levels of approximation for two major types of potential flows. In one type, the presence of wake vortices, including unsteady wakes and tip vortices, is important. In the second type, the presence of the wake is not important.

The ability of the zonal method to confine the solution field to the viscous region of the flow and to also compute the several coexisting viscous zones independently are available to the potential flow. Since no separation zone is present in the potential flow, the kinetic part of the computation for the potential flow is drastically simpler than that for the general viscous flow. In particular, for the type of potential flow where the wake is unimportant, the kinetic part of the computation procedure is not essential. The entire solution procedure reduces to a boundary element method for the kinematics of the flow.

At the present, the zonal method is fully established for incompressible laminar and turbulent flow computations. In a companion paper [5], selected results for two-dimensional and three-dimensional incompressible viscous flow problems is discussed. It has been found that remarkably accurate numerical results are generally obtainable using the zonal method. The reductions of the required amount of computation from those of the familiar finite difference and finite element methods are striking. Computations of two-dimensional viscous flows containing large scale separated regions can now be carried out using, instead of mainframe computers, the present generation of personal computers. Relatively recently, extensive results have been obtained for various unsteady potential flow problems where the wake effects are highly important. The problems studied range from biofluiddynamic flows associated with the flight of insects and the swimming of fishes to flows around oscillating airfoils. In general, it has been found that highly accurate results can be obtained using minimal computational efforts on the basis of the approximations discussed in the present paper. Additional information are presented in [8].

Acknowledgement

This work is supported by the U.S. Army Research Office.

References

1. Wu, J.C., Tuncer, J.H. & Wang, C.M.: Theoretical and Numerical Studies of Oscillating Airfoils; AIAA Paper # 89-0258, to appear in AIAA J., 1990.

2. Wu, J.C.: Theory of Aerodynamic Force and Moment in Viscous Flows. AIAA J., 19, 432-441, 1981.

3. Wu, J.C. & Thompson, J.R.: Numerical Solutions of Time-Dependent Incompressible Navier-Stokes Equations Using an Integro-Differential Formulation. J. Computers and Fluids, 1, 197-215, 1973.

4. Wu, J.C.: Numerical Boundary Conditions for Viscous Flow Problems. AIAA J., 14, 1042-1047, 1976.

5. Patterson, M.T., Wu, J.C. & Wang, C.M.: Boundary Element Methods for the Navier-Stokes Equations, Proc. of Internaional Symp. of Boundary Element Methods, 1989.

6. Hsu, T.M. & Wu, J.C.: Vortex Flow Model for the Blade-Vortex Interaction Problem. AIAA J., 26, 621-623, 1988.

7. Kuethe, A.M. & Chow, C.Y.: Foundations of Aerodynamics, Wiley, 1986.

8. Wu, J.C. & Wang, C.M.: Recent Advances in Solution Methods for Unsteady Viscous Flows. Chap. 7, Nonlinear Problems of Fluid Dynamics, Editors: P.K. Banerjee and L. Morino, Elsevier Applied Science Publishers, in print, 1988.

Boundary Element Analysis of Viscous Fluid Flow Problems Using the Time Splitting Method

K.KAKUDA and N.TOSAKA

Department of Mathematical Engineering,
College of Industrial Technology, Nihon University,
Narashino, Chiba 275, Japan

Summary

The effective attempt to solve unsteady viscous fluid flow with high Reynolds number is presented within the use of the boundary element method. This approach is based upon application of the time splitting scheme to the governing differential equations which are split into the advection equation and the Stokes equations. The advection solutions can be first obtained by means of both a simple iterative procedure and an interpolation with the cubic spline. By using these solutions, the Stokes equations are solved by the standard boundary element method. This technique is illustrated by several numerical results.

Introduction

The numerical solution of the Navier-Stokes equations for viscous fluid flow has been performed by many researchers using a finite difference method or a finite element method [1-3]. We have been developing a boundary-domain type integral equation analysis [4,5] to viscous fluid flow problems in terms of the velocity and the pressure as the unknown variables. The derived integral equation was discretized by employing reasonable space-time elements to the unknown variables. The final system of equations with a full coefficient matrix was computed effectively by using the standard Newton-Raphson iterative procedure. However, this scheme needs too much computation time and also many memory sizes.

In this paper, we present a new approach to solve the viscous fluid flow problems by means of the boundary element. This approach [6] is based upon splitting the governing set of equations into the nonlinear advection equation and the Stokes equations. The equation obtained by applying the theory of characteristics to the advection equation can be computed implicitly by means of both a simple iterative procedure and an interpolation with the cubic spline. By using the obtained advection solutions, the Stokes equations are solved explicitly by the standard boundary element method.

Numerical results of the Burgers' equation by this approach are compared to the analytical solutions [7] and the finite element solutions [8,9]. The driven cavity flow problem is also demonstrated through a comparison with the other existing solutions [2,3].

Time Splitting Approach

Let Ω be a bounded domain in Euclidean space R^2 with a piecewise smooth boundary Γ. The unit outward normal vector component to Γ is denoted by n_i. By applying the time splitting technique [6] to an incompressible viscous fluid flow problem, we obtain the two types of nonlinear system to be solved as follows :

$$\dot{u}_i + u_j u_{i,j} = 0 \tag{1}$$

$$\left.\begin{aligned} Re\, \dot{u}_i &= -Re\, p_{,i} + u_{i,jj} + u_{j,ij} \\ u_{i,i} &= 0 \end{aligned}\right\} \tag{2}$$

where Re denotes the Reynolds number of the flow.

By the use of the theory of characteristics, the solution of advection equation (1) can be given as follows :

$$u_i(\boldsymbol{x}, t_f) = u_i(\boldsymbol{x} - \boldsymbol{v}\Delta t_f, t_{f-1}) \quad , \ \boldsymbol{v} = u_j(\boldsymbol{x}, t_f)\boldsymbol{e}^{(j)} \tag{3}$$

where $\Delta t_f = t_f - t_{f-1}$.

Next, let us apply the boundary element method to solve the Stokes problems (2). The resulting boundary integral equations are given as follows [4] :

$$C(y)U_L(\boldsymbol{y},t_f) = \int_{t_{f-1}}^{t_f}\int_{\Gamma_i} u_l(\boldsymbol{x},t)\Sigma^*_{lL}(\boldsymbol{x},t;\boldsymbol{y},t_f)\,d\Gamma(\boldsymbol{x})dt - \int_{t_{f-1}}^{t_f}\int_{\Gamma_i} \tau_l(\boldsymbol{x},t)V^*_{lL}(\boldsymbol{x},t;\boldsymbol{y},t_f)\,d\Gamma(\boldsymbol{x})dt - \int_{\Omega_i} Re\; u_l(\boldsymbol{x},t_{f-1})V^*_{lL}(\boldsymbol{x},t_{f-1};\boldsymbol{y},t_f)\,d\Omega(\boldsymbol{x}) \quad (L=1,2,3\,;\, l=1,2) \tag{4}$$

where $C(y)$ denotes the shape coefficient matrix. Here, the fundamental solution tensor V^*_{lL} is determined from the following solution [4] :

$$\phi^*(\boldsymbol{x},t;\boldsymbol{y},t_f) = -\frac{H(t_f-t)}{2\pi Re}\{ln\, r - \frac{1}{2}E_i(-\alpha r^2)\} \tag{5}$$

where $E_i()$ is the integral exponential function, $\alpha = Re/4(t_f-t)$ and $r^2 \equiv (x_j-y_j)^2$. The pseud-traction tensor, Σ^*_{lL} , corresponding to the traction vector τ_l is also defined in terms of the fundamental solution tensor by

$$\Sigma^*_{lL} \equiv (-V^*_{3L}\delta_{lj} + V^*_{lL,j} + V^*_{jL,l})n_j \tag{6}$$

By using the standard boundary element techniques, we obtain the following discretized expression at the f-th time step for the boundary integral equation (4) :

$$\boldsymbol{H}\,\boldsymbol{u}^{(f)} = \boldsymbol{G}\,\boldsymbol{\tau}^{(f)} + \boldsymbol{C}\,\boldsymbol{u}^{(f-1)} \tag{7}$$

As the computational scheme, equation (3) corresponding to the advection equation can be first computed by the use of both a simple iterative procedure and an interpolation with the cubic spline. By using the solution determined with this procedure we can compute the equation set (7) for the Stokes problem.

Numerical Examples

In the following, we demonstrate the typical sample problems of viscous fluid flow in order to show an effectiveness and an adaptability of the proposed approach.

1.One-dimensional flow problems

As the first example, let us consider the following two types of initial-boundary value problems of the Burgers' equation as the one-dimensional version of the Navier-Stokes equations :

The Burgers' equation

$$Re(\dot{u} + u\,u_{,x}) = u_{,xx} \qquad (0<x<l) \tag{8}$$

Boundary conditions and initial conditions

$$\left.\begin{aligned} &\text{case 1}: u(0,t) = u(1,t) = 0 \\ &\qquad\quad u(x,0) = \sin \pi x \end{aligned}\right\} \tag{9}$$

$$\left.\begin{aligned} &\text{case 2}: u(0,t) = 1\ ,\ u(12,t) = 0 \\ &\qquad\quad u(x,0) = \begin{cases} 1 & , \ 0 < x < 5 \\ 6 - x & , \ 5 < x < 6 \\ 0 & , \ 6 < x < 12 \end{cases} \end{aligned}\right\} \tag{10}$$

The numerical solutions of case1 for $Re=10^2$ and 10^5 are shown in Figs.1 and 2, respectively. Our numerical solutions for $Re=10^2$ are in good agreement with the analytical solutions given in [7] (see Fig.1). The agreement between the numerical solutions from the present method and the finite element solutions[8,9] appears very satisfactory for $Re=10^5$ (see Fig.2).

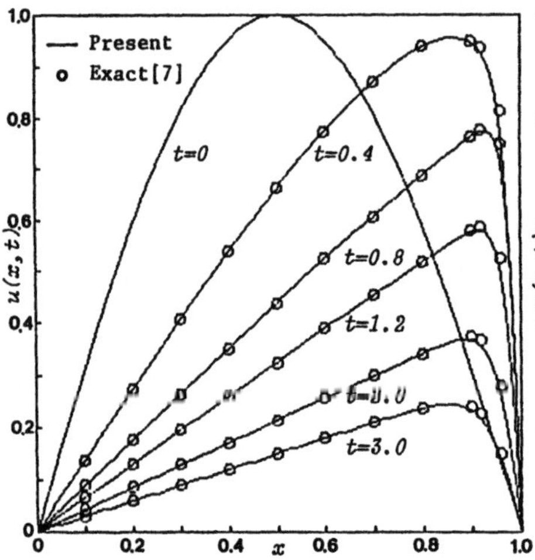

Fig.1 *case 1*; $Re=10^2, \Delta x=\Delta t_f=0.01$

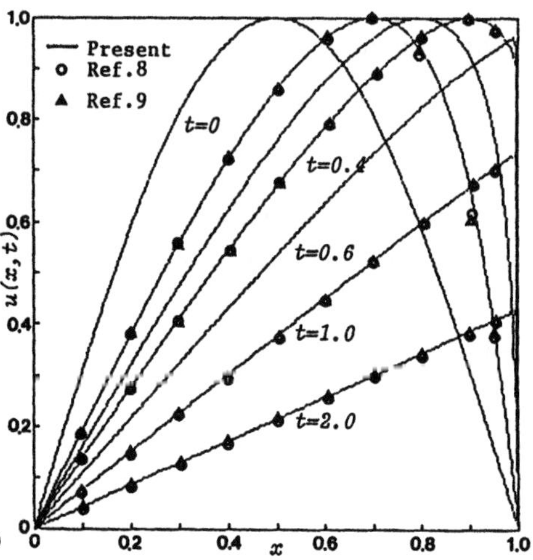

Fig.2 *case 1*; $Re=10^5, \Delta x=\Delta t_f=0.005$

Next, we demonstrate the numerical performances of the Burgers' equation under the conditions of case 2. Fig.3 shows that the initial front moves with the average initial velocity 1/2 for *Re=10*. The numerical solutions for $Re=10^2$ and 10^3 shown in Figs.4 and 5, respectively, demonstrate the formation of a very steep front at $t=1$. In Figs.3 and 4, our solutions are also compared with the finite element solutions obtained by Varoğlu and Finn [8].

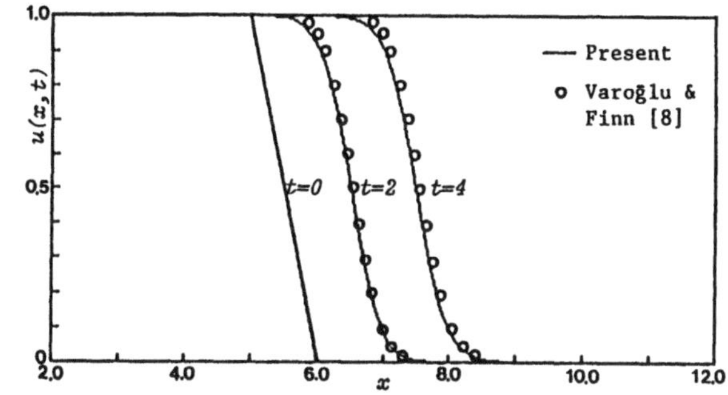

Fig.3 *case 2; Re=10* , $\Delta x=\Delta t_f=0.1$

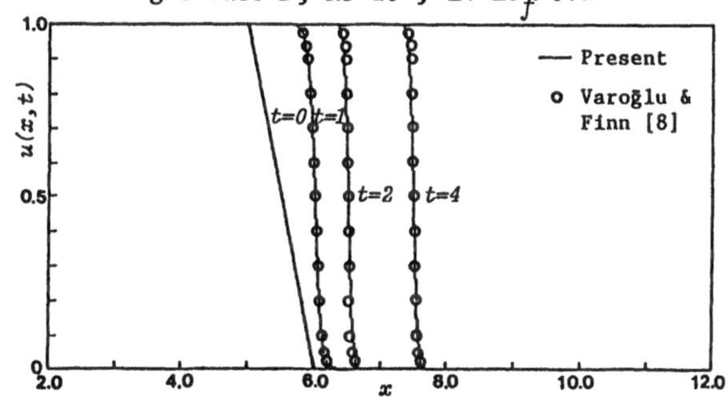

Fig.4 *case 2; Re=*10^2, $\Delta x=\Delta t_f=0.05$

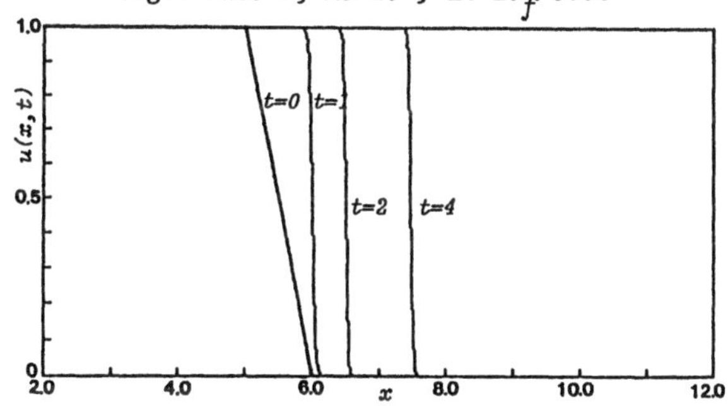

Fig.5 *case 2; Re=*10^3, $\Delta x=\Delta t_f=0.05$

2.Driven cavity flow problem

As a second example, we demonstrate an unsteady state flow in a square cavity driven by a lid sliding at a uniform velocity. We perform the numerical calculations for $Re=10^3$ with the time step Δt_f from 0.04 to 0.09. The velocity vector fields for each time step are shown in Fig.6. Figs.7 and 8 show the horizontal velocity profiles along vertical centreline in the cavity. In Fig.8, our results are also compared with the other existing results [2,3].

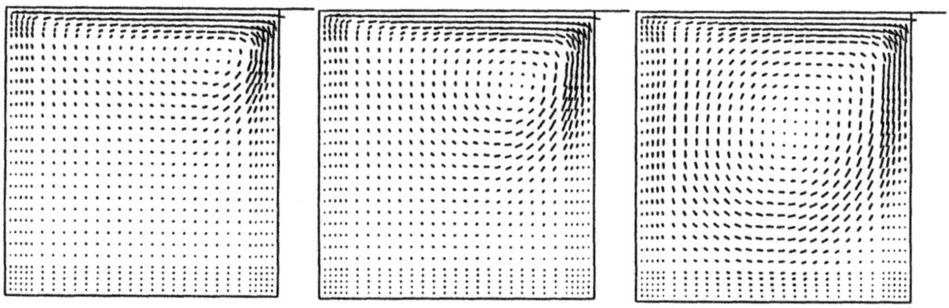

(a) $t=2.0$ (b) $t=4.0$ (c) $t=49.5$

Fig.6 Velocity vector fields for $Re=10^3$

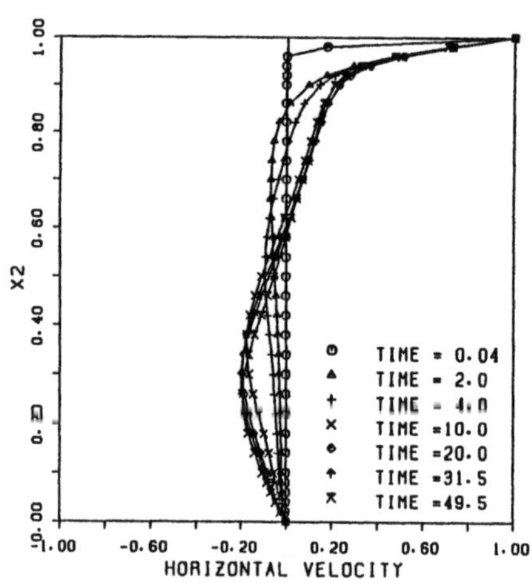

Fig.7 Horizontal velocity profiles along vertical centreline for each time step ($Re=10^3$)

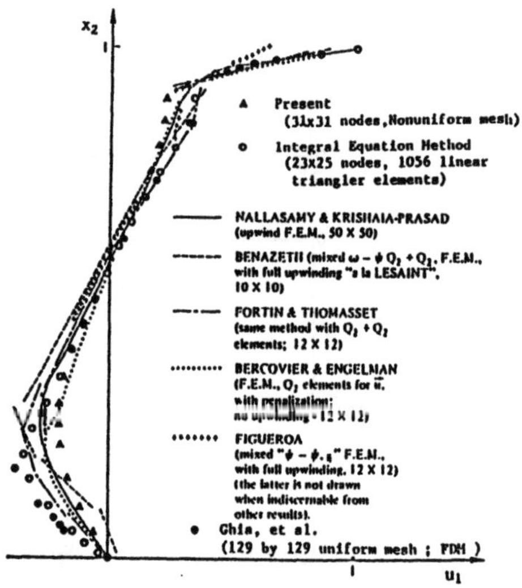

Fig.8 A comparison of the horizontal velocity profiles along vertical centreline for $Re=10^3$

Conclusions

We have presented the boundary element approach based upon the time splitting technique for an incompressible viscous fluid flow problem and its numerical implementations for typical examples. The numerical solutions of the Burgers' equation are computed by the approach with a simple iterative procedure. The accuracy and the effectiveness of the proposed method are discussed through a comparison with the analytical solutions and the finite element solutions. Moreover, the numerical performances for driven cavity flow problem are demonstrated for the Reynolds number $Re=10^3$.

References

1. Peyret,R.;Taylor,T.D.: Computational methods for fluid flow. New York, Heidelberg, Berlin: Springer-Verlag 1983.

2. Ghia,U.; Ghia,K.N.; Shin,C.T. : High-Re solutions for incompressible flow using the Navier-Stokes equations and a multigrid method. J.Comput.Phys. 48 (1982) 387-411.

3. Thomasset,F : Implementation of finite element methods for Navier-Stokes equations. New York, Heidelberg, Berlin : Springer-Verlag 1981.

4. Tosaka,N.: Integral equation formulations with the primitive variables for incompressible viscous fluid flow problems. Compt. Mech. 4 (1989) 89-103.

5. Tosaka,N.; Kakuda,K : Numerical simulations of the unsteady-state incompressible viscous flows using an integral equation. Boundary Elements(Ed.,Du Qinghua) (1986)163-172.

6. Benque,J.P.;Ibler,B.;Labadie,G.: A finite element method for Navier-Stokes equations. Numerical Method for Non-Linear Problems Vol.1 (1980) 709-720 : Pineridge Press.

7. Cole,J.D.: On a quasi-linear parabolic equation occurring in aerodynamics. Quart.Appl.Math. 9 (1951) 225-236.

8. Varoğlu,E.; Finn,W.D.L. : Space-time finite element incorporating characteristics for the Burgers' equation. Int.J.Num.Meth.Engng. 16 (1980) 171-184.

9. Nguyen,H.;Reynen,J.: A space-time finite element approach to Burgers' equation. Numerical Method for Non-Linear Problem Vol.2(Eds.,Taylor,C.,et al.)(1984)718-728, Pineridge Press.

Advanced Boundary Element Methods for Incompressible Thermoviscous Flow

P.K. Banerjee
G.F. Dargush
K. Honkala

Department of Civil Engineering
State University of New York at Buffalo
Buffalo, New York U.S.A.

ABSTRACT

A boundary element formulation is presented for moderate Reynolds number, incompressible, thermoviscous flows. The governing integral equations are written exclusively in terms of velocities and temperatures, thus eliminating the need for the computation of gradients. Furthermore, with the introduction of reference velocities and temperatures, volume modeling can often be confined to a small portion of the problem domain, typically near obstacles or walls. The numerical implementation makes use of all of the recent advances in boundary element technology, including higher order elements, adaptive integration, and multi-region capability.

INTRODUCTION

Integral equations governing flow of incompressible viscous fluids have been available since the early portion of this century. Oseen [1], in a most remarkable monograph, develops exact integral representations, along with the corresponding infinite space fundamental solutions for both two and three-dimensional flows.

In the present paper, current boundary element methodology is applied to Oseen's work on time-dependent incompressible viscous as well as thermoviscous flow. As a result, practical problems involving complicated geometry and boundary conditions can be solved. For Stokes flow, a boundary-only formulation is produced, thus reducing the dimensionality of the problem by one. In the moderate Reynolds number range, a nonlinear volume integral remains. However, in many cases, the domain discretization can be confined to a small portion of the flow field.

GOVERNING DIFFERENTIAL EQUATIONS

Initially, the governing equations for a general compressible, Newtonian fluid are presented. This set will provide the basis for the development of the boundary integral representation.

The conservation of mass in the absence of sources and sinks in the medium gives the equation of continuity:

$$\frac{\partial \rho}{\partial t} + \frac{\partial (\rho v_i)}{\partial x_i} = 0. \tag{1}$$

By introducing kinematics and the constitutive law for a Newtonian fluid with constant coefficients of viscosity, the familiar Navier-Stokes equations appear:

$$\rho\left(\frac{\partial v_i}{\partial t}+v_j\frac{\partial v_i}{\partial x_j}\right)=(\lambda+\mu)\frac{\partial^2 v_j}{\partial x_i\partial x_j}+\mu\frac{\partial^2 v_i}{\partial x_j\partial x_j}-\frac{\partial p}{\partial x_i} \tag{2}$$

In the above, v_i velocity vector, p pressure, t time, x_i Eulerian coordinate, ρ mass density, and λ,μ viscosity coefficients.

For non-Newtonian fluid, additional terms appear in (2). However, these terms can be conveniently considered as pseudo-body forces, exactly as done in an elastoplastic analysis of a solid (Banerjee and Butterfield [2]; Banerjee et al [3-6]).

Next, the balance expressed by the first law of thermodynamics in conjunction with Fourier's law of heat conduction gives the energy equation as

$$\rho C_v\left(\frac{\partial \Theta}{\partial t}+v_i\frac{\partial \Theta}{\partial x_i}\right)=k\frac{\partial^2 \Theta}{\partial x_i\partial x_i}-p\frac{\partial v_i}{\partial x_i}+Y \tag{3}$$

where Θ temperature, k thermal conductivity, c_v specific heat at constant volume, and Y viscous dissipation.

Note that in (3), the thermal conductivity has been assumed constant. Finally, the equation of state for an ideal fluid is introduced to relate temperature and pressure. That is,

$$p=\rho R\Theta \tag{4}$$

in which R is a gas constant.

In two-dimensions, equations (1-4) represent a coupled set of five equations with five unknowns, namely v_i,ρ,p and Θ. For the special case of incompressibility, ρ is constant and the continuity condition becomes simply

$$\frac{\partial v_i}{\partial x_i}=0, \tag{5}$$

while the equations of motion reduce to

$$\rho\left(\frac{\partial v_i}{\partial t}+v_j\frac{\partial v_i}{\partial x_j}\right)=\mu\frac{\partial^2 v_i}{\partial x_j\partial x_j}-\frac{\partial p}{\partial x_i}. \tag{6}$$

Then, (5) and (6) form a system of three equations in the unknowns v_i and p. The equations of energy and state are no longer required to determine fluid motion. However, under non-isothermal conditions, the fluid temperatures can be obtained from (3) after the velocities are established. The exception, to this two stage approach, is for buoyancy driven flow in which the body forces produced by temperature gradients are dominant. In this latter case, mass (5), momentum (6) and energy (3) conservation must be satisfied simultaneously.

INTEGRAL REPRESENTATIONS

One of the primary requirements of developing a boundary element formulation is that the fundamental solution of the governing differential equations must exist. Since the boundary element formulations could not be developed without these solutions, a substantial amount of effort has been devoted in the present work to successively derive more complete solutions of the differential equations. In essence, each advancement brings more of the physics of the problem into the fundamental solution. As a first approximation the compressibility terms in (2) were ignored and the complete fundamental solution for a transient body force and a transient heat source was utilized. Subsequently, these solutions were extended to include the effect of compressibility, although in the latter case, some approximation was necessary. Very recently, fundamental solutions have been derived which incorporate the effects of convection. Some of these were discussed by Oseen [1].

In any case, these fundamental solutions are used in conjunction with a reciprocal identity for fluid dynamics to produce the following integral representation for the velocity:

$$\begin{aligned} c_{\beta\alpha}(\xi)v_\beta(\xi,\tau) = &\int_S \left[\dot{G}_{\beta\alpha} * t_\beta(X,t) - \dot{F}_{\beta\alpha} * v_\beta(X,t)\right] dS(X) \\ &+ \int_V \left[\dot{G}_{\beta\alpha} * f_\beta(Z,t)\right] dV(Z) \end{aligned} \tag{7}$$

where in two-dimensions $v_\beta = [v_1 \quad v_2 \quad \Theta]^T, t_\beta = [t_1 \quad t_2 \quad q]^T$, and $f_\beta = [f_1 \quad f_2 \quad \psi]^T$.

The generalized body forces, f_β, appearing in (7), include the convective inertia forces, and in the compressible case, forces due to variable density. The time dependent functions $\dot{G}_{\beta\alpha}$ and $\dot{F}_{\beta\alpha}$ can be developed directly from the fundamental solution [14,15].

While the formulation presented above is perfectly valid, two additional modifications have proved quite beneficial. The first involves performing integration by parts on the convective body force. This releases a nonlinear surface integral, but also completely eliminates the need for calculating velocity gradients in incompressible flow. With compressibility, only the scalar dilatation is required. Then, for incompressible flow, (7) becomes [12]

$$\begin{aligned} c_{\beta\alpha}v_\beta = &\int_S \left[\dot{G}_{\beta\alpha} * t_\beta - \dot{F}_{\beta\alpha} * v_\beta\right] dS \\ &- \int_V \left[\dot{D}_{\beta\alpha i} * \rho v_i v_\beta\right] dV + \int_S \left[\dot{G}_{\beta\alpha} * \rho v_i v_\beta n_i\right] dS \end{aligned} \tag{8}$$

where $\dot{D}_{\beta\alpha i} = \dot{G}_{\beta\alpha,i}$, and n_i is the outward unit normal to the surface S.

The other modification involves the decomposition of the total velocity (v_i^T) into the free stream velocity (V_i) plus a velocity perturbation (v_i). Thus, let

$$v_i^T = V_i + v_i.$$

Substituting this decomposed form into the governing differential and integral equations leads to the following alternate form of (8):

$$c_{\beta\alpha}v_\beta = \int_S \left[\dot{G}_{\beta\alpha} * t_\beta - \dot{F}_{\beta\alpha} * v_\beta\right] dS - \int_V \left[\dot{D}_{\beta\alpha i} * \rho(V_i + v_i)v_\beta\right] dV$$
$$+ \int_S \left[\dot{G}_{\beta\alpha} * \rho(V_i + v_i)v_\beta n_i\right] dS. \tag{9}$$

As a result, one finds that volume integration is required only in portions of the flow field in which the total velocity differs from that of the free stream. In a practical sense, this means that, in many problems, volume discretization can be confined to a small region around an obstruction. Additionally, the free stream velocity can be selected on a region-by-region basis. This is helpful in limiting the magnitude of the nonlinear terms.

NUMERICAL IMPLEMENTATION

The numerical treatment of the equations in thermoviscous fluid dynamics follows very closely that described in references [7-13]. The geometry of each volume cell which is required for the nonlinear terms is defined by nodal points and quadratic shape functions. In two-dimensions, six and eight-noded cells are available. Meanwhile, either a linear or quadratic variation can be employed for the functional representation (Banerjee and Butterfield [2], Dargush and Banerjee [14,15]). For the complete nonlinear analysis boundary as well as interior collocation points must be included, and the resulting equations become highly nonlinear due to the convective terms. After the collocation process is complete, the final system of equations can be expressed in matrix form as

$$g^b(x, v) = A^b x + D^b \sigma^o - G^b t^o - B^b y = 0 \tag{10a}$$

for boundary points, and as

$$g^v(x, v) = v + A^v x + D^v \sigma^o - G^v t^o - B^v y = 0 \tag{10b}$$

for interior cell points, where the vectors σ^o and t^o have components defined by $\sigma^o_{i\beta} = \rho v_i v_\beta$ and $t^o_\beta = \sigma^o_{i\beta} n_i$ at each boundary and interior point. In (10), x and y are the known and unknown boundary quantities, while v is the interior velocity vector, and the matrices A, B, D and G are developed from the integrals of the kernel functions appearing in (9).

Initially, an iterative algorithm, along the lines of those traditionally used for BEM elastoplasticity (Banerjee and Butterfield [2]; Banerjee et al [3-6]), was employed to solve (10). However convergence is usually achieved only at low Reynolds number. More generally the interior equations must be brought into the system matrix along with the boundary equations, and a full or modified Newton-Raphson algorithm must be employed to obtain solutions at moderate or high Reynolds number. (Similar 'variable stiffness' algorithms have also been introduced by Banerjee and Raveendra [3] and Henry and Banerjee [16] for elastoplasticity.) Symbolically, at each iteration m,

$$\begin{bmatrix} \frac{\partial g^b}{\partial x} & \frac{\partial g^b}{\partial v} \\ \frac{\partial g^v}{\partial x} & \frac{\partial g^v}{\partial v} \end{bmatrix} \begin{Bmatrix} (\Delta x)^m \\ (\Delta v)^m \end{Bmatrix} = -\begin{Bmatrix} g^b(x^m, v^m) \\ g^v(x^m, v^m) \end{Bmatrix} \tag{11}$$

where

$$x^{m+1} = x^m + (\Delta x)^m$$

$$v^{m+1} = v^m + (\Delta v)^m$$

and the derivatives on the left-hand side of (11) are evaluated at (x^m, v^m). In the numerical implementation, the above equations are arranged to form a block banded system matrix for efficient multi-region solutions.

NUMERICAL EXAMPLES

Driven Cavity

This example involves the initiation of flow in a square cavity. An incompressible fluid of uniform density and viscosity is at rest within a unit square region. The velocities of the vertical sides and the bottom are fixed at zero throughout time. At time zero, the horizontal velocity of the top edge is suddenly raised to a value of 1000 and maintained at that level. A gradual transition of velocities is introduced near the top corners to provide continuity.

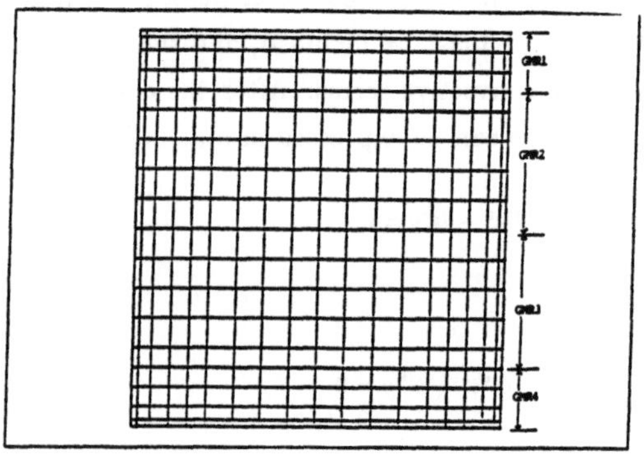

Fig. 1. Driven cavity: BEM model

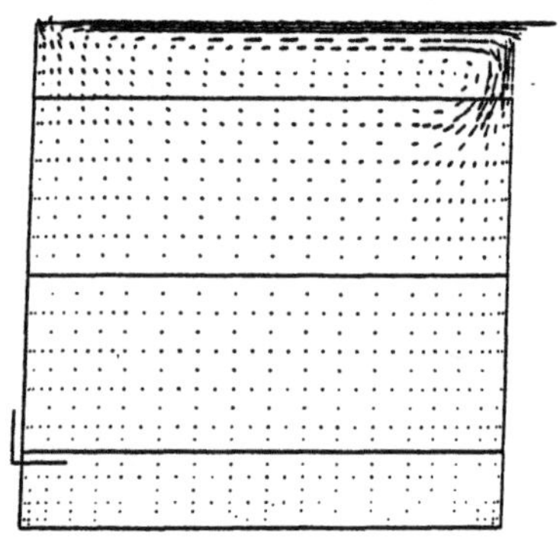

Fig. 2a. Driven cavity: Flow at time $t = 1.0$

The four region, 324 cell model shown in Figure 1 is employed for the boundary element analysis. The resulting velocity vector plots at several times are shown in Figure 2 for this case having a Reynolds number of 1000. The results lead directly to the steady solution after a sufficient number of time steps. This steady solution correlates closely with the results of Ghia et al [17], as presented in Figure 3. Notice that, in particular, the shift of the vortical center follows that described by Burggraf [18] in his classic paper.

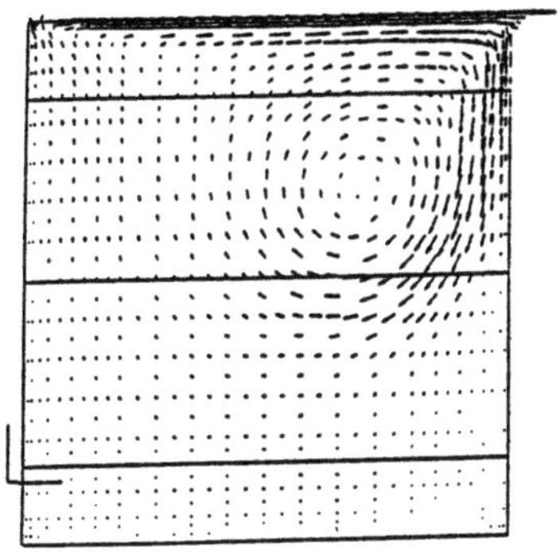

Fig. 2b. Driven cavity: Flow at time $t = 5$

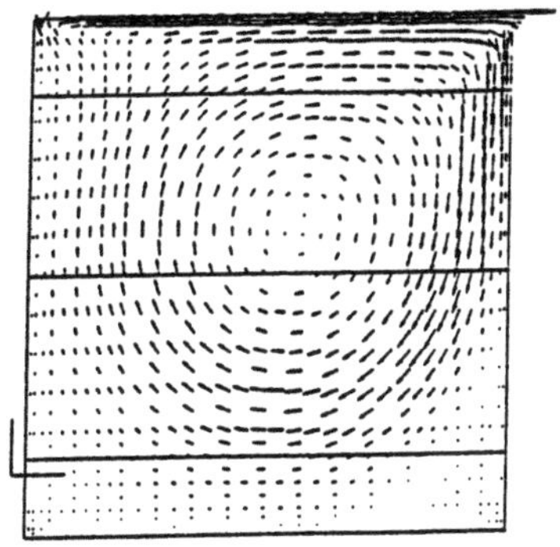

Fig. 2c. Driven cavity: Flow at time $t = 12.5$

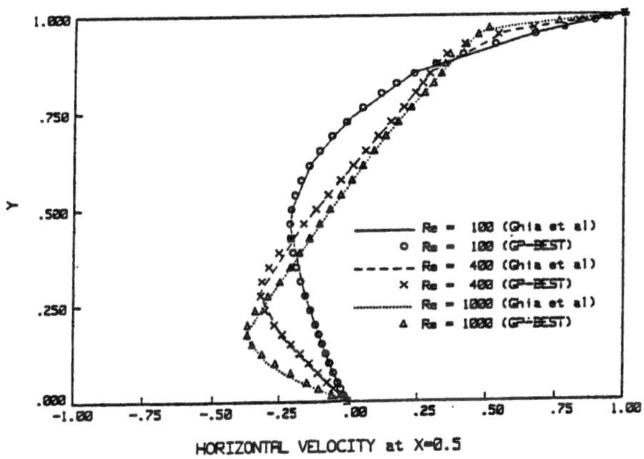

Fig. 3. Driven cavity: Comparison of velocity profile

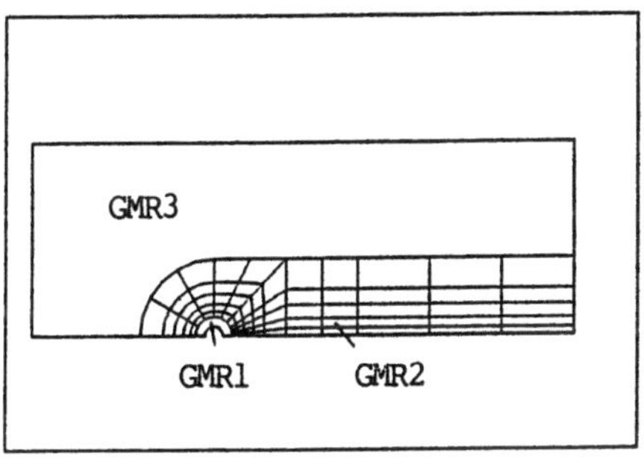

Fig. 4. Flow around a cylinder: Boundary element model

Flow Over a Cylinder

Next, an example of unconfined flow around an obstacle is considered. In particular, the oft-studied case of a unit diameter circular cylinder is examined. The boundary element mesh is illustrated in Figure 4. Notice that three distinct regions are evident. The smallest region, labeled GMR1, represents a thermoelastic thick-walled cylinder. Only the surface of the solid is discretized. The next region, GMR2, models a thermoviscous fluid in the vicinity of the cylinder. In GMR2, volumes cells are required due to convective body forces. However, sufficiently remote from the cylinder, these body forces become negligible and once again a boundary-only region, in this case GMR3, is valid.

Steady-state velocity vector plots are displayed in Figures 5 and 6 for $Re = 20$ and 40, respectively. The recirculating zone, behind the cylinder, is clearly visible.

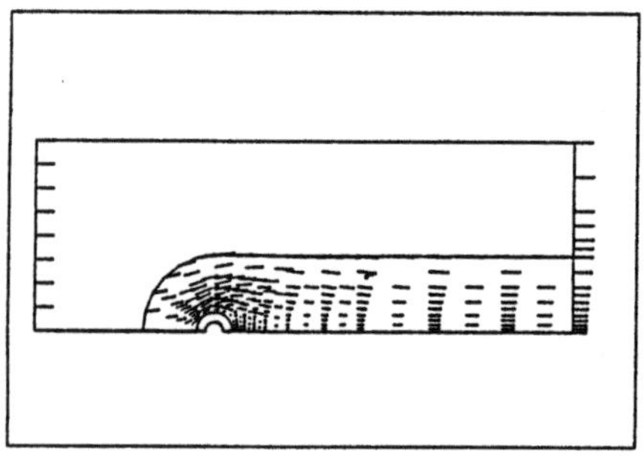

Fig. 5a. Flow around a cylinder: $Re = 20$

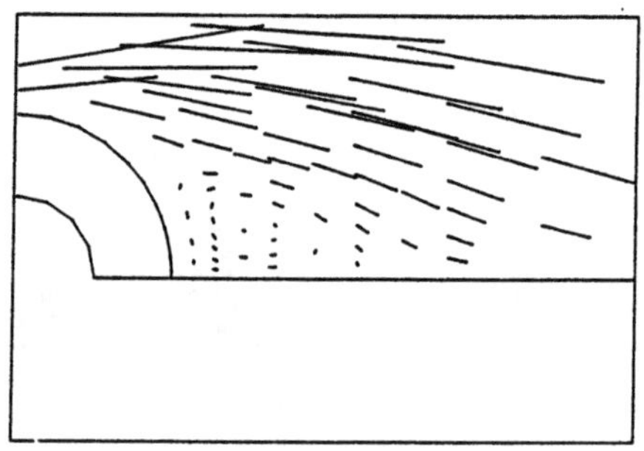

Fig. 5b. Flow around a cylinder: $Re = 20$

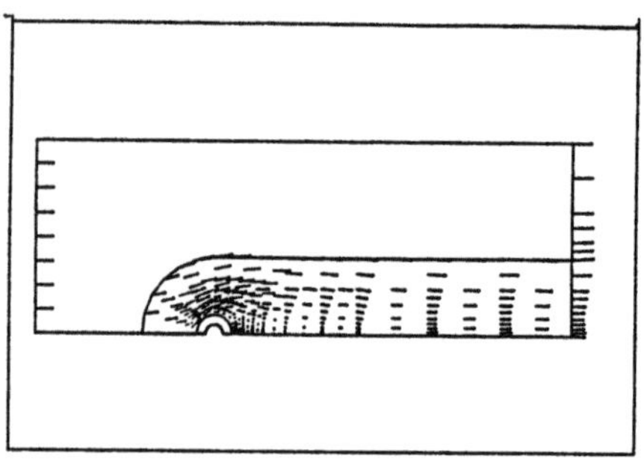

Fig. 6a. Flow around a cylinder: $Re = 40$

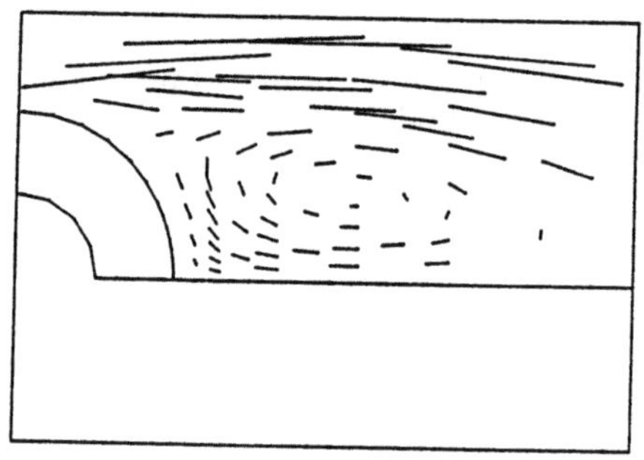

Fig. 6b. Flow around a cylinder: $Re = 40$

Additionally, the problem was extended to include thermal effects. The temperature of the fluid at inlet was specified as 1000°C, while that at the inner surface of the hollow cylinder was maintained at 0°C. The effective heat transfer coefficient between the fluid and solid can then be obtained from the resulting temperature and flux at the outer surface of the cylinder. The distribution of the nondimensional Nusselt number (Nu) around the circumference is plotted in Figure 7. These curves agree, at least, qualitatively with the experimental results of Eckert and Soehngen [19].

Flow Over an Airfoil

As a final example, the thermoviscous flow over a NACA 0018 airfoil is considered. The boundary element model shown in Figure 8 once again utilizes symmetry and employs the multi-region concept with cells confined to the vicinity of the airfoil. The airfoil is heated externally by a hot gas, flowing from left to right, at unit temperature, and cooled to zero on the surface of an internal cooling hole. The conductivity of the airfoil is set to one hundred times that of the hot gas, while unit values are assumed for the fluid density and viscosity.

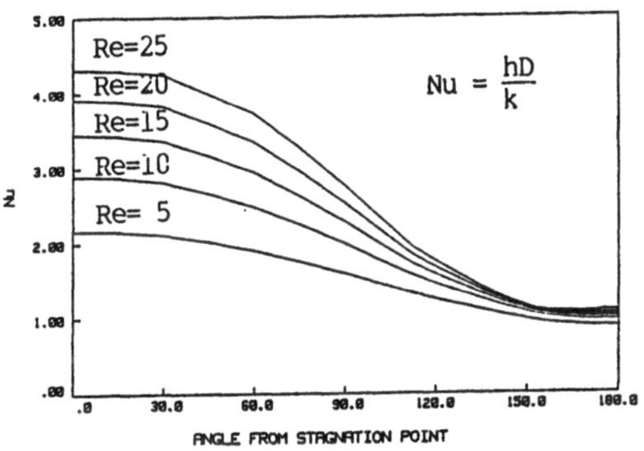

Fig. 7. Flow around a cylinder: Distribution of Nu on cylinder

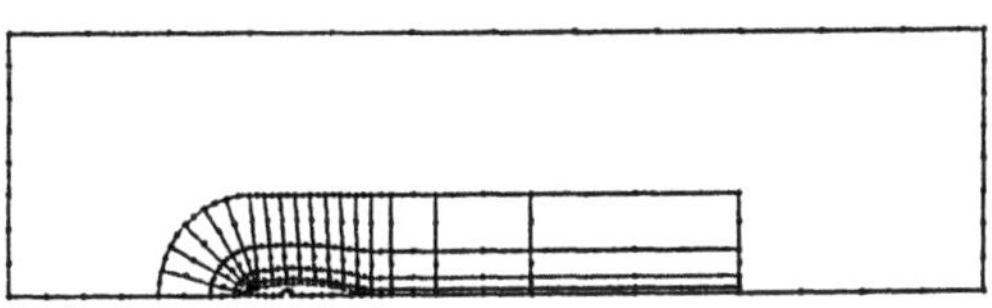

Fig. 8. Fig. 8. Flow over an airfoil: Boundary element model

The resulting steady-state velocity distribution at $Re = 100$ is displayed in Figure 9, and a shaded temperature contour plot of the region surrounding the airfoil is presented for $Re = 10$ in Figure 10. In the latter diagram, the hot regions are black, while lower temperature locations appear white. The effects of convection are visible downstream of the airfoil.

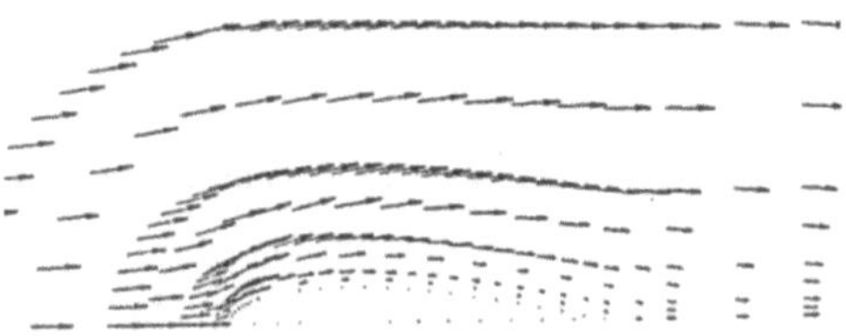

Fig. 9. Flow over an airfoil: Velocity vectors at $Re = 100$

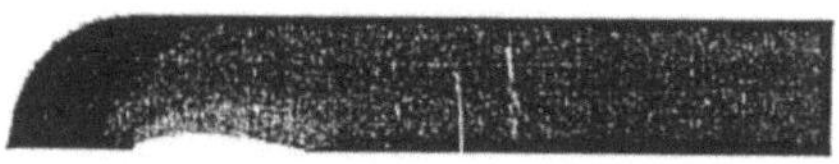

Fig. 10. Flow over an airfoil: Temperature field at $Re = 10, Pe = 10$

When the Reynolds number is elevated further, the convective terms begin to dominate, and the present formulation is not appropriate. The actual Re limit is highly problem dependent. In this higher Re flow regime, the physics of the problem demands that convected effects must be incorporated in the kernel functions. (Interestingly, this is analogous to the upwinding techniques that are required in finite difference and finite element analysis.) The thermal portion of this new work provides some interesting results. For example, Figure 11 depicts the temperature distribution around the airfoil in a uniform velocity field of 10. Notice the similarity between the images of Figures 10 and 11, which have a common Peclet number (Pe) of 10. Differences exist because the former includes the effects of viscosity, while the latter assumes a uniform velocity. As the magnitude of the velocity field increases further, convection becomes more evident. The thermal field at $Pe = 1000$ is depicted in Figure 12. Convection clearly dominates at this level.

Fig. 11. Flow over an airfoil: Temperature field at $Pe = 10$

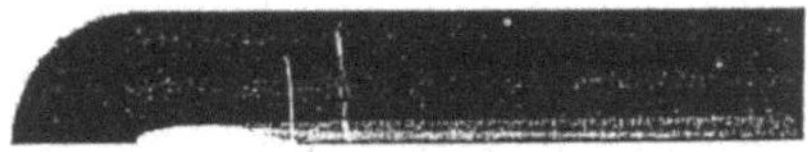

Fig. 12. Flow over an airfoil: Temperature field at $Pe = 1000$

ACKNOWLEDGMENT

This work was performed on grant NAG3-712 for NASA Lewis Center. The authors are indebted to Dr. C. Chamis for his support and encouragement.

REFERENCES

1. Oseen, C.W. Neuere Methoden und Ergebnisse in der Hydrodynamik, Akad. Verlagsgellschaft, Leipzig (1927).
2. Banerjee, P.K. and Butterfield, R., Boundary Element Methods in Engineering Science, McGraw-Hill, London (1981).
3. Banerjee, P.K. and Raveendra, S.T., 'A New Boundary Element Formulation for Two-dimensional Elastoplastic Analysis,' *Jour. of Engrg. Mech.*, ASCE, **V113**,(2), pp. 252-265 (1987).
4. Banerjee, P.K., Wilson, R.B. and Miller, N., 'Development of a Large BEM System for Three-dimensional Inelastic Analysis,' in *Advanced Topics in Boundary Element Analysis*, ed. T.A. Cruse, A.B. Pifko and H. Armen, **AMD-V72**, ASME, New York (1985).
5. Banerjee, P.K., Wilson, R.B. and Raveendra, S.T., 'Advanced Applications of BEM to Three-dimensional Problems of Monotonic and Cyclic Plasticity,' *Int. Jour. Mech. Sciences*, **V29**,(9), pp. 637-653 (1987).
6. Banerjee, P.K., Wilson, R.B. and Miller, N., 'Advanced Elastic and Inelastic Three-dimensional Analysis of Gas Turbine Engine Structures by BEM,' *Int. J. Num. Meth. Engrg.*, **V26**, pp. 393-411 (1988).
7. Ahmad, S. and Banerjee, P.K., 'Transient Elastodynamic Analysis of 3-D Problems by BEM,' *Int. Jour. Numerical Methods in Engineering*, **V26**(8), pp. 1560-1580 (1988).
8. Banerjee, P.K. and Ahmad, S., 'Advanced Three-dimensional Dynamic Analysis by Boundary Element Methods,' in *Advanced Topics in Boundary Element Analysis*, ed. T.A. Cruse, A.B. Pifko and H. Armen, **AMD-V72**, ASME, New York (1985).
9. Banerjee, P.K., Ahmad, S. and Manolis, G.D., 'Transient Elastodynamic Analysis of Three-dimensional Problems by Boundary Element Method,' *Earthquake Engineering and Structural Dynamics*, **V14**, pp. 933-949 (1986).
10. Dargush, G.F. and Banerjee, P.K., 'Development of a Boundary Element Method for Time-dependent Planar Thermoelasticity,' *Int. J. Solids Struct.* (in press) (1989).
11. Dargush, G.F. and Banerjee, P.K., 'Boundary Element Method in Three-dimensional Thermoelasticity,' *Int. J. Solids Struct.* (in press) (1989).

12. Dargush, G.F. and Banerjee, P.K., 'Development of an Integrated BEM for Hot Fluid-Structure Interaction,' *Transactions of the ASME*, Paper 89-GT-128 (1989).

13. Dargush, G.F. and Banerjee, P.K., 'Advanced Development of the Boundary Element Method for Steady-state Heat Conduction,' *Int. J. Num. Meth. Engrg.*, to appear (1989).

14. Dargush, G. and Banerjee, P.K. 'Advanced BEM for Steady Incompressible Thermoviscous Flow,' Chapter 2, *Developments in BEM*, **V5**, Ed. P.K. Banerjee and L. Morino, Elsevier Applied Science, London (1989).

15. Dargush, G. and Banerjee, P.K. 'A Time Dependent Incompressible Viscous BEM for Moderate Reynolds Numbers,' Chapter 3, *Developments in BEM*, **V5**, Ed. P.K. Banerjee and L. Morino, Elsevier Applied Science, London (1989).

16. Henry, D.P. and Banerjee, P.K., 'A Variable Stiffness Type Boundary Element Formulation for Axisymmetric Elastoplastic Media,' *Int. Jour. for Num. Methods in Engrg.*, **V25**, pp. 1005-1027 (1988).

17. Ghia, U., Ghia, K.N. and Shin, C.T., 'High-Re Solutions for Incompressible Flow Using the Navier-Stokes Equations and a Multigrid Method,' *J. Comp. Physics*, **V48**, pp. 387-411 (1982).

18. Burggraf, O., 'Analytical and Numerical Studies of the Structure of Steady Separated Flows,' *J. Fluid Mech.*, **V24**, Part 1, pp. 113-151 (1966).

19. Eckert, E.R.G. and Soehngen, E., 'Distribution of Heat-Transfer Coefficients Around Circular Cylinders in Crossflow at Reynolds Numbers From 20 to 500,' *Transactions of the ASME*, **V74**, pp. 343-347 (1952).

A Coupling of BEM and FEM for the Viscous Flow Problem

George C. Hsiao †
Dept. of Mathematical Sciences
University of Delaware
Newark, DE 19716, USA

and

John F. Porter ‡
Dept. of Mathematical Sciences
Ball State University
Muncie, IN 47306, USA

Summary

In this paper we present a combined boundary element method (BEM) and finite element method (FEM) scheme for treating the two-dimensional stationary viscous, incompressible fluid flow past an obstacle. The problem here can be formulated in terms of the dimensionless Navier-Stokes equations with a small parameter, the Reynolds number. This is a typical exterior singular perturbation problem in the sense that there is no solution to the reduced equations (that is, the equations when the parameter is zero) satisfying both the boundary condition and the condition at infinity. As is well known, because of the presence of small parameters, standard numerical schemes, without suitable modifications, are not, in general, applicable for single perturbation problems.

The essential idea of our approach is to divide the region under consideration into two regions, an inner and an outer region. By making use of the asymptotic behavior of the solution to the original problem, we solve actually a transmission problem for different linearized partial differential equations in the inner and outer regions,. In the inner region, Stokes equations (see (3)) will be employed, whereas in the outer region, the Oseen equations (see (4)) will be used. The transmission boundary conditions on an auxiliary boundary of the interface resemble the "matching conditions" in singular perturbation theory [3]. By employing Green's representation for the solution of the Oseen equations in the outer region, a *nonlocal* boundary condition [6] will be derived on the auxiliary boundary in terms of appropriate boundary integral operators. With this nonlocal boundary condition on the interface, we then apply the Galerkin method to the Stokes equations in the inner region. In particular, from the asymptotic expansions of the fundamental solution to the Oseen equations, an effective constructive procedure may be developed independent of the small parameter. In principle, our approach follows the works of [1] and [7], but we believe it is more appropriate for singular perturbation problems.

The Navier-Stokes Problem

We begin with the formulation of the slow viscous flow problem as the exterior boundary value problem for the Navier-Stokes equations in the dimensionless form:

$$-\Delta \mathbf{q} + \varepsilon(\mathbf{q}\cdot\nabla)\mathbf{q} + \nabla p = \mathbf{f} \text{ and } \nabla\cdot\mathbf{q} = 0 \text{ in } \Omega^c,$$

$$\mathbf{q} = 0 \text{ on } \Gamma, \tag{1}$$

$$\mathbf{q} \to \boldsymbol{\alpha} \text{ and } p \to 0 \text{ as } |x| \to \infty,$$

which we term the *Navier-Stokes Problem* (NS). Here Ω^c is the exterior domain to the smooth,

† The work of this author was supported by the Center for Advanced Study of the University of Delaware.
‡The work of this author was supported by a Ball State University new faculty grant.

simple closed curve Γ in the plane R^2. The unknowns are the velocity field $\mathbf{q}$ and the pressure field p, whereas $\mathbf{f}$ and $\boldsymbol{\alpha}$ are the given force field and constant unit velocity vector, respectively. We assume that $\mathbf{f}$ has a compact support in Ω^c. The parameter ε is the Reynolds number and is assumed to be small, $0 < \varepsilon \ll 1$. For later use, we introduce here the stress operator $\mathbf{T}[\mathbf{q}; p]$ defined by

$$\mathbf{T}[\mathbf{q}; p] := -p\mathbf{I} + (\nabla\mathbf{q} + \nabla\mathbf{q}') \tag{2}$$

where $\nabla\mathbf{q}'$ denotes the transpose of $\nabla\mathbf{q}$.

To reformulate the NS as a transmission problem, we let Γ_0 be a simple closed curve containing the supp $\mathbf{f}$. We denote the annular region between Γ and Γ_0 by Ω^-,the inner region and by $\Omega^+ := \Omega^c \setminus \bar{\Omega}^-$, the outer region. We approximate the Navier-Stokes equations by the *Stokes equations*,

$$-\Delta\mathbf{q}_s + \nabla p_s = \mathbf{f} \text{ and } \nabla\cdot\mathbf{q}_s = 0 \text{ in } \Omega^-, \tag{3}$$

and by the *Oseen equations*,

$$-\Delta\mathbf{q}_0 + \varepsilon(\boldsymbol{\alpha}\cdot\nabla)\mathbf{q}_0 + \nabla p_0 = 0 \text{ and } \nabla\cdot\mathbf{q}_0 = 0 \text{ in } \Omega^+. \tag{4}$$

The transmission problem (TP) then consists of (3), (4) together with the boundary conditions:

$$\mathbf{q}_s = 0 \text{ on } \Gamma, \tag{5}$$

$$\mathbf{q}_s^- = \boldsymbol{\alpha} + \mathbf{q}_0^+ \text{ and } \mathbf{T}[\mathbf{q}_s; p_s]\mathbf{n}^- = \mathbf{T}[\mathbf{q}_0; p_0]\mathbf{n}^+ \text{ on } \Gamma_0, \tag{6}$$

and the conditions at infinity,

$$\mathbf{q}_0 \to 0 \text{ and } p_0 \to 0 \text{ as } |x| \to \infty. \tag{7}$$

Here in the transmission conditions (6), $\mathbf{v}^{\pm}$ stands for the limits of the function on Γ_0 from $\Omega^{\pm}$, and $\mathbf{n}$ denotes the outer unit normal to Γ_0. Note that (3) and (4) represent different kinds of linearizations of the Navier-Stokes equations; the former is a linearization with respect to the zero velocity at the boundary Γ, and the latter with respect to the velocity $\boldsymbol{\alpha}$ at infinity. Thus, we define,

$$(\mathbf{q}, p) = \begin{cases} (\mathbf{q}_s, p_s) & \text{in } \Omega^- \\ (\boldsymbol{\alpha} + \mathbf{q}_0, p_0) & \text{in } \Omega^+, \end{cases} \tag{8}$$

as an approximate solution to the NS. For theoretical justification of this approximation, we refer to [3].

The Nonlocal Boundary Problem

We now reduce the TP to an equivalent nonlocal boundary problem. First we need to derive the so-called nonlocal boundary condition on the auxiliary boundary Γ_0. This can be achieved from the Green representation of the solution of (4):

$$\begin{aligned} \mathbf{q}_0(x) &:= \int_{\Gamma_0} \mathbf{q}_0^+(y)\cdot \mathbf{G}(x,y)\,ds_y - \int_{\Gamma_0} \mathbf{E}(x,y)\cdot \mathbf{T}[\mathbf{q}_0;p_0]\mathbf{n}(y)^+ ds_y \\ &\quad + \varepsilon \int_{\Gamma_0} \mathbf{q}_0^+(y)\cdot \mathbf{E}(x,y)\,(\boldsymbol{\alpha}\cdot \mathbf{n})\,ds_y, \quad x \in \Omega^+ \end{aligned} \tag{9}$$

$$\begin{aligned} p_0(x) &:= \int_{\Gamma_0} \mathbf{q}_0^+(y)\cdot \mathbf{g}(x,y)\,ds_y - \int_{\Gamma_0} \mathbf{e}(x,y)\cdot \mathbf{T}[\mathbf{q}_0;p_0]\mathbf{n}(y)^+ ds_y \\ &\quad + \varepsilon \int_{\Gamma_0} \mathbf{q}_0^+(y)\cdot \mathbf{e}(x,y)\,(\boldsymbol{\alpha}\cdot \mathbf{n})\,ds_y, \quad x \in \Omega^+. \end{aligned} \tag{10}$$

In the formulations (9) and (10), $\mathbf{E}(x, y)$ and $\mathbf{e}(x, y)$ are the fundamental velocity tensor and the fundamental pressure vector of the Oseen equations (4), whose explicit forms may be found in [4]; furthermore, the corresponding tensor $\mathbf{G}(x, y)$ and the vector $\mathbf{g}(x, y)$ are defined respectively by

$$\mathbf{G}(x,y) := \mathbf{T}_y[\mathbf{E}(x,y);\mathbf{e}(x,y)]\,\mathbf{n}(y) \text{ and } \mathbf{g}(x,y) := \mathbf{T}_y[\mathbf{e}(x,y);\mathbf{e}^*(x,y)]\,\mathbf{n}(y), \tag{11}$$

where we have introduced a "pressure"

$$\mathbf{e}^*(x,y) = -\frac{\varepsilon}{2\pi}\boldsymbol{\alpha}\cdot\nabla_y \log\frac{1}{|x-y|} \tag{12}$$

corresponding to the fundamental pressure vector field $\mathbf{e}(x, y)$.

From the representation formula (9), we let x approach the boundary and, by using the well-known jump relation of the double-layer potential, we arrive at the *boundary integral equation*:

$$\mathbf{q}_0^+ = \left(\frac{1}{2}\mathbf{I} + \mathbf{K}\right)\mathbf{q}_0^+ - \mathbf{V}(\mathbf{T}[\mathbf{q}_0;p_0]\mathbf{n}^+) + \varepsilon\mathbf{V}((\boldsymbol{\alpha}\cdot\mathbf{n})\mathbf{q}_0^+) \text{ on } \Gamma_0, \tag{13}$$

where the boundary integrals $\mathbf{V}$ and $\mathbf{K}$ are defined by

$$\mathbf{V}\varphi(x) := \int_{\Gamma_0} \mathbf{E}(x,y)\cdot\varphi(y)\,ds_y,\ x \in \Gamma_0 \text{ and } \mathbf{K}\phi(x) := \int_{\Gamma_0} \phi(y)\cdot\mathbf{G}(x,y)\,ds_y,\ x\in\Gamma_0. \tag{14}$$

Now we use the interface conditions from (6). We have $\mathbf{q}_0^+ = \mathbf{q}_s^- - \boldsymbol{\alpha}$, and if we put $\mathbf{T}[\mathbf{q}_s;p_s]\mathbf{n}^- = \boldsymbol{\sigma}$, then $\mathbf{T}[\mathbf{q}_0;p_0]\mathbf{n}^+ = \boldsymbol{\sigma}$. We substitute into (13) to obtain

$$\frac{1}{2}\mathbf{q}_s^- + \mathbf{V}\boldsymbol{\sigma} - \mathbf{K}\mathbf{q}_s^- - \varepsilon\mathbf{V}((\boldsymbol{\alpha}\cdot\mathbf{n})\mathbf{q}_s^-) = \boldsymbol{\alpha} - \varepsilon\mathbf{V}((\boldsymbol{\alpha}\cdot\mathbf{n})\boldsymbol{\alpha}) \text{ on } \Gamma_0. \tag{15}$$

This leads to the *nonlocal boundary problem: Find* $(\mathbf{q}_s, p_s, \boldsymbol{\sigma})$ *such that*

$$\begin{gathered} -\Delta\mathbf{q}_s + \nabla p_s = \mathbf{f} \textit{ and } \nabla\cdot\mathbf{q}_s = 0 \textit{ in } \Omega^-;\mathbf{q}_s = 0 \textit{ on } \Gamma, \\ \mathbf{T}[\mathbf{q}_s;p_s]\mathbf{n}^- = \boldsymbol{\sigma} \textit{ on } \Gamma_0, \\ \frac{1}{2}\mathbf{q}_s^- + \mathbf{V}\boldsymbol{\sigma} - \mathbf{K}\mathbf{q}_s^- - \varepsilon\mathbf{V}((\boldsymbol{\alpha}\cdot\mathbf{n})\mathbf{q}_s^-) = \boldsymbol{\alpha} - \varepsilon\mathbf{V}((\boldsymbol{\alpha}\cdot\mathbf{n})\boldsymbol{\alpha}) \textit{ on } \Gamma_0. \end{gathered} \tag{16}$$

We remark that the nonlocal boundary problem (16) is equivalent to the TP for the approximate NS in the sense that if $(\mathbf{q}_s, p_s, \sigma)$ is a solution of (16), and if we set

$$(\mathbf{q}, p) = \begin{cases} (\mathbf{q}_s, p_s) & \text{in } \Omega^- \\ (\alpha + \mathbf{q}_0, p_0) & \text{in } \Omega^+, \end{cases} \tag{17}$$

where $\mathbf{q}_0(x)$ and $p_0(x)$ are given by the representations (9) and (10) respectively with

$$\mathbf{q}_0^+ = \mathbf{q}_s^- - \alpha \text{ and } \mathbf{T}[\mathbf{q}_0; p_0]\mathbf{n}^+ = \sigma \text{ on } \Gamma_0,$$

then $(\mathbf{q}, p)$ is a solution of the approximate NS defined by (8) with $\mathbf{T}[\mathbf{q}_s; p_s]\mathbf{n}^- = \sigma$ on Γ_0.

The Variational Formulation

Our coupling procedure remains now to solve (16) by the FEM. For this purpose, we need the variational formulation of (16). In the following, let us introduce the function spaces:

$$\begin{aligned} \mathbf{H}(\Omega^-) &= \left\{\mathbf{q} \in (H^1(\Omega^-))^2 : \textstyle\int_{\Gamma_0} \mathbf{q}\cdot\mathbf{n}\, ds = 0\right\}, \quad \mathbf{Q}(\Omega^-) = \{\mathbf{q} \in \mathbf{H}(\Omega^-) : \mathbf{q}|\Gamma = 0\}, \\ \Pi(\Omega^-) &= \{p \in H^0(\Omega^-) : \textstyle\int_{\Omega^-} p\, dx = 0\}, \quad \Sigma(\Gamma_0) = \left\{\sigma \in (H^{-\frac{1}{2}}(\Gamma_0))^2 : \textstyle\int_{\Gamma_0} \sigma\cdot\mathbf{n}\, ds = 0\right\}, \end{aligned} \tag{18}$$

where $H^s(\Omega^-)$ and $H^r(\Gamma_0)$ denote the standard Sobolov spaces. Then, we see that for any $\mathbf{q}_s \in (H^2(\Omega^-)^2 \cap \mathbf{Q}(\Omega^-)$ with $\nabla\cdot\mathbf{q}_s = 0$ in Ω^- and for any $p_s \in H^1(\Omega^-) \cap \Pi(\Omega^-)$, the following Green's formula

$$\frac{1}{2}\int_{\Omega^-} \nabla\mathbf{q}_s : \nabla\phi\, dx - \int_{\Omega^-} p_s \nabla\cdot\phi\, dx = -\int_{\Omega^-} (\Delta\mathbf{q}_s - \nabla p_s)\cdot\phi\, dx + \int_{\Gamma_0} \mathbf{T}[\mathbf{q}_s; p_s]\mathbf{n}^-\cdot\phi\, ds$$

holds for all $\phi \in \mathbf{Q}(\Omega^-)$. Now, for given $(\mathbf{f}, \alpha) \in (H^0(\Omega^-))^2 \times R^2$, by a variational solution of the nonlocal boundary problem, we mean *the solution triples* $(\mathbf{q}_s, p_s, \sigma) \in \mathbf{Q}(\Omega^-) \times \Pi(\Omega^-) \times \Sigma(\Gamma_0)$ *satisfying the variational equations:*

$$\begin{aligned} a(\mathbf{q}_s, \phi) + b(\phi, p_s) - <\sigma, \phi^-> &= \ell_1(\phi) \quad \forall\phi \in \mathbf{Q}(\Omega^-), \\ b(\mathbf{q}_s, \psi) &= 0 \quad \forall\psi \in \Pi(\Omega^-), \\ <\mathbf{q}_s^-, \tau> + 2<\mathbf{V}\sigma, \tau> - 2<\mathbf{K}\mathbf{q}_s^-, \tau> &- 2\varepsilon <\mathbf{V}(\alpha\cdot\mathbf{n})\mathbf{q}_s^-, \tau> \\ &= \ell_2(\tau) \quad \forall\tau \in \Sigma(\Gamma_0). \end{aligned} \tag{19}$$

In this formulation, we have adopted the notation:

$$\begin{aligned} a(\mathbf{q}_s, \phi) &:= \frac{1}{2}\int_{\Omega^-} \nabla\mathbf{q}_s : \nabla\phi\, dx; \; b(\phi, p_s) := \int_{\Omega^-} p_s\nabla\cdot\phi\, dx, \\ &<\mathbf{q}_s^-, \tau> := \int_{\Gamma_0} \mathbf{q}_s^-\cdot\tau\, ds, \\ \ell_1(\phi) &:= \int_{\Omega^-} \mathbf{f}\cdot\phi\, dx; \; \ell_2(\tau) := 2<\alpha, \tau> + 2\varepsilon <\mathbf{V}(\alpha\cdot\mathbf{n})\alpha, \tau>. \end{aligned} \tag{20}$$

Alternatively, we may reformulate (19) in the form: *Given* $(\mathbf{f}, \boldsymbol{\alpha}) \in (H^0(\Omega^-))^2 \times R^2$, *find* $(\mathbf{q}_s, p_s, \sigma) \in \mathbf{Q}(\Omega^-) \times \Pi(\Omega^-) \times \Sigma(\Gamma_0)$ *such that*

$$\mathbf{A}((\mathbf{q}_s, p_s, \sigma), (\phi, \psi, \tau)) = \ell(\phi, \psi, \tau) \quad \forall((\phi, \psi, \tau) \in \mathbf{Q}(\Omega^-) \times \Pi(\Omega^-) \times \Sigma(\Gamma_0), \tag{21}$$

where $\mathbf{A}$ *is the bilinear form defined by*

$$\mathbf{A}((\mathbf{q}_s, p_s, \sigma), (\phi, \psi, \tau)) := a(\mathbf{q}_s, \phi) + b(\phi, p_s) - < \sigma, \phi^- > + b(\mathbf{q}_s, \psi)$$

$$+ < \mathbf{q}_s^-, \tau > +2 < \mathbf{V}\sigma, \tau > -2 < \mathbf{K}\mathbf{q}_s^-, \tau > -2\varepsilon < \mathbf{V}(\boldsymbol{\alpha} \cdot \mathbf{n})\mathbf{q}_s^-, \tau >$$

and ℓ *is a linear functional defined by*

$$\ell(\phi, \psi, \tau) := \ell_1(\phi) + \ell_2(\tau).$$

To establish the existence and uniqueness of the solution of (21), we now introduce the *associated variational problem: Find* $(\mathbf{q}_s, \sigma) \in \overset{\circ}{\mathbf{Q}}(\Omega^-) \times \Sigma(\Gamma_0)$, *such that*

$$\mathbf{B}((\mathbf{q}_s, \sigma), (\phi, \tau)) = \ell_0(\phi, \tau) \quad \forall(\phi, \tau) \in \overset{\circ}{\mathbf{Q}}(\Omega^-) \times \Sigma(\Gamma_0), \tag{22}$$

where $\mathbf{B}$ *is the bilinear form defined by*

$$\mathbf{B}((\mathbf{q}_s, \sigma), (\phi, \tau)) := a(\mathbf{q}_s, \phi) - < \sigma, \phi^- > + < \mathbf{q}_s^-, \tau > +2 < \mathbf{V}\sigma, \tau >$$

$$- 2 < \mathbf{K}\mathbf{q}_s^-, \tau > -2\varepsilon < \mathbf{V}(\boldsymbol{\alpha} \cdot \mathbf{n})\mathbf{q}_s^-, \tau >,$$

and ℓ_0 *is the linear functional given by* $\ell_0(\phi, \tau) := \ell_1(\phi) + \ell_2(\tau)$. Here $\overset{\circ}{\mathbf{Q}}(\Omega^-) = \{\mathbf{q}_s \in \mathbf{Q}(\Omega^-) : \operatorname{div} \mathbf{q}_s = 0\}$ is a closed subspace of $\mathbf{Q}(\Omega^-)$.

The existence and uniqueness of the solution of (21) can be easily established from that of (22). We note that the bilinear form $b(\cdot, \cdot)$ in (20) satisfies the Băbuska-Brezzi condition (see e.g., [4]. Hence we will have a similar existence and uniqueness result as in ([7]) for the Stokes equations. In principle, we may then solve (21) numerically and obtain an approximate solution for $(\mathbf{q}_s, p_s, \sigma)$. However, there are two disadvantages in doing so. First, due to the presence of the small parameter ε in the kernel of the integral operators in (21), it is difficult (if not practical) to numerically solve (21). Second, in view of the approximation (8) ([3]), there is no need to solve (21) more accurately than up to a certain order of ε. Indeed, this leads to the development of the following modified coupling procedure.

An Asymptotic Procedure

Following [4], we use the asymptotic series development of the fundamental tensor $\mathbf{E}(x, y)$ and rewrite $\mathbf{E}(x, y)$ in the form

$$\mathbf{E}(x, y) = \mathbf{E}_0(x, y) + \mathbf{M}(\varepsilon) + O(\varepsilon \log \varepsilon) \tag{23}$$

for x, y in a compact set. Here $\mathbf{E}_0(x, y)$ is the fundamental velocity tensor for the Stokes equations (3) and $\mathbf{M}(\varepsilon)$ is the diagonal matrix of the form

$$\mathbf{M}(\varepsilon) := \frac{1}{4\pi}\begin{pmatrix} \gamma_1 - \log\varepsilon & 0 \\ 0 & \gamma_2 - \log\varepsilon \end{pmatrix}, \tag{24}$$

where γ_1 and γ_2 are constants depending on the unit vector α in (1). By neglecting terms of $O(\varepsilon \log\varepsilon)$, we now approximate the boundary integral equation (15) and obtain

$$\frac{1}{2}\mathbf{q}_s^- + \mathbf{V}_0\sigma - \mathbf{K}_0\mathbf{q}_s^- + \mathbf{M}(\varepsilon)\int_{\Gamma_0}\sigma\,ds = \alpha \text{ on } \Gamma_0, \tag{25}$$

where $\mathbf{V}_0$ and $\mathbf{K}_0$ are the corresponding simple and double-layer potential boundary operators for the Stokes equations (3) (cf. (14)). Based on the nonlocal boundary condition (25), we now formulate the *asymptotic nonlocal boundary problem: Find* $(\mathbf{q}_s, p_s, \sigma, \omega)$ *such that*

$$-\Delta\mathbf{q}_s + \nabla p_s = \mathbf{f} \text{ and } \nabla\cdot\mathbf{q}_s = 0 \text{ in } \Omega^-;\ \mathbf{q}_s = 0 \text{ on } \Gamma,$$

$$\mathbf{T}[\mathbf{q}_s; p_s]\mathbf{n}^- = \sigma \text{ on } \Gamma_0;\quad \frac{1}{2}\mathbf{q}_s^- + (\mathbf{V}_0\sigma + \omega) - \mathbf{K}_0\mathbf{q}_s^- = \alpha \text{ on } \Gamma_0, \tag{26}$$

$$\int_{\Gamma_0}\sigma\,ds - \mathbf{M}^{-1}(\varepsilon)\omega = 0.$$

Note that in this formulation, we have introduced the auxiliary constant vector (unknown),

$$\omega := \mathbf{M}(\varepsilon)\int_{\Gamma}\sigma\,ds. \tag{27}$$

The variational formulation of (26) now reads: *Find* $(\mathbf{q}_s, p_s, \sigma, \omega) \in \mathbf{Q}(\Omega^-) \times \Pi(\Omega^-) \times \Sigma(\Gamma_0) \times R^2$ *such that*

$$\mathbf{A}_0((\mathbf{q}_s, p_s, \sigma, \omega), (\phi, \psi, \tau, \kappa)) + \Lambda_\varepsilon(\omega, \kappa) = \overset{\circ}{\ell}(\phi, \psi, \tau, \kappa) \tag{28}$$

$\forall(\phi, \psi, \tau, \kappa) \in \mathbf{Q}(\Omega^-) \times \Pi(\Omega^-) \times \Sigma(\Gamma_0) \times R^2$, *where* $\mathbf{A}_0$ Λ_ε, $\overset{\circ}{\ell}$ *are defined by*

$$\mathbf{A}_0((\mathbf{q}_s, p_s, \sigma, \omega), (\phi, \psi, \tau, \kappa)) := a(\mathbf{q}_s, \phi) + b(\phi, p_s) - \langle \sigma, \phi^- \rangle + b(\mathbf{q}_s, \psi)$$

$$+ \langle \mathbf{q}_s^-, \tau \rangle + 2\langle \mathbf{V}_0\sigma, \tau \rangle + 2\langle \omega, \tau \rangle - 2\langle \mathbf{K}_0\mathbf{q}_s^-, \tau \rangle + 2\langle \sigma, \kappa \rangle;$$

$$\Lambda_\varepsilon(\omega, \kappa) := -2\mathbf{M}^{-1}(\varepsilon)\omega\cdot\kappa$$

$$\overset{\circ}{\ell}(\phi, \psi, \tau, \kappa) := \int_{\Omega^-}\mathbf{f}\cdot\phi\,dx + 2\alpha\cdot\int_{\Gamma_0}\tau\,ds.$$

It can be shown that the solution of (28) admits the asymptotic representations:

$$\begin{aligned} \mathbf{q}_s &= \mathbf{q}_s^0 + c_1(\varepsilon)\mathbf{q}_s^1 + c_2(\varepsilon)\mathbf{q}_s^2; & p_s &= p_s^0 + c_1(\varepsilon)p_s^1 + c_2(\varepsilon)p_s^2; \\ \sigma &= \sigma^0 + c_1(\varepsilon)\sigma^1 + c_2(\varepsilon)\sigma^2; & \omega &= \omega^0 + c_1(\varepsilon)\omega^1 + c_2(\varepsilon)\omega^2; \end{aligned} \tag{29}$$

where $(\mathbf{q}_s^i, p_s^i, \sigma^i, \omega^i)$, $i = 0, 1, 2$ are solutions of

$$\mathrm{A}_0((\mathbf{q}_s^i, p_s^i, \sigma^i, \omega^i), (\phi, \psi, \tau, \kappa)) = \ell^i(\phi, \psi, \tau, \kappa) \tag{30}$$

with

$$\ell^0(\phi, \psi, \tau, \kappa) = \int_{\Omega^-} \mathbf{f}\cdot\phi\, dx + 2\alpha\cdot \int_{\Gamma_0} \tau\, ds,\ \ell^1(\phi, \psi, \tau, \kappa) = 2\mathbf{e}_1\cdot\kappa;\ \ell^2(\phi, \psi, \tau, \kappa) = 2\mathbf{e}_2\cdot\kappa,$$

and $\mathbf{e}_i$, $i = 1, 2$ are unit vectors along the x_i-axis. The coefficients $c_1(\varepsilon)$ and $c_2(\varepsilon)$ are the unique solutions of the algebraic system of equations:

$$(\mathbf{M}(\varepsilon)\mathbf{e}_1 - \omega^1)c_1(\varepsilon) + (\mathbf{M}(\varepsilon)\mathbf{e}_2 - \omega^2)c_2(\varepsilon) = \omega^0 . \tag{31}$$

We remark that the only dependence on ε in the solution $(\mathbf{q}_s, p_s, \sigma, \omega)$ is through the easily computable coefficients $c_1(\varepsilon)$ and $c_2(\varepsilon)$. Similar to (21), equation (30) is uniquely solvable; in particular, we see that in the present case, the map $\mathcal{A}$ defined by

$$\mathcal{A}(\sigma, \omega) := (\mathbf{V}_0\sigma + \omega, \int_{\Gamma_0} \sigma\, ds)$$

is *continuous* and *bijective* from $\overset{\circ}{\mathbf{H}}{}^s(\Gamma_0) \times R^2$ onto $\overset{\circ}{\mathbf{H}}{}^{s+1}(\Gamma_0) \times R^2$ for any $s \in R^2$, where

$$\overset{\circ}{\mathbf{H}}{}^s(\Gamma_0) = \{\mathbf{g} \in (H^s(\Gamma_0))^2 : \int_{\Gamma_0} \mathbf{g}\cdot\mathbf{n}\, ds = 0\} .$$

(See e.g. [4]; [5].) One may then consider the numerical implementation of (30) and error analysis in the usual manner.

References

1. Johnson, C.; Nedelec J.C.: On the coupling of boundary integral and finite element methods. Math. Comp., 35 (1980) 1063-1079.

2. Hsiao, G.C.: The Coupling of BEM and FEM - a brief review. Boundary Element X, Vol. 1, Brebbia,C.A.(ed.) Heidelberg and New York: Springer Verlag (1988) 431-445.

3. Hsiao, G.C.; MacCamy,R.C.: Singular perturbations for the two-dimensional slow viscous flow problem. Theory and Applications of Singular Perturbations (Eckharus,W.; De Jager,E.M.: Lecture Notes in Mathematics,942. Heidelberg and New York: Springer Verlag (1982) 229-244.

4. Hsiao, G.C.,;Porter, J.F.: The coupling of BEM and FEM for the two dimensional viscous flow problem. Appl. Anal. 27 (1988) 79-108.

5. Hsiao, G.C.; Wendland W.L.: A finite element method for some integral equations of the first kind. J. Math. Anal. Appl. 58 (1977) 449-481.

6. MacCamy,R.C.;Marin,S.P.: A finite element method for exterior interface problems. Int. J. Math. Sci. 3 (1980) 311-350.

7. Sequira, A.: The coupling of boundary integral and finite element methods for the bidimensional exterior steady Stokes problem. Math. Meth. Appl. Sci. 5 (1983) 356-375

Transient Flow Separation at High-Angle-of-Attack

George A. Watts and Jyh-Ching Juang
Lockheed Aeronautical Systems Company
Advanced Research Organization
P.O. Box 551; Burbank, CA, 91520

Summary

A 2-D boundary element, finite difference method is proposed for solving the laminar separated incompressible flow field about and forces on a slender forebody at high angle-of-attack (AOA) in steady or unsteady motion.

Introduction

The symmetric flowfield about the smooth, slender nose of a fighter can suddenly become asymmetric at a high critical AOA. The yawing moment accompanying this flow change can overpower the aircraft controls and force departure. Its magnitude and direction initially depend on small asymmetric errors built into the apex region and thus cannot be predicted. These are rapidly overcome by the much larger flow asymmetry induced by subsequent rolling and yawing. Resulting forebody forces and moments are repeatable aircraft to aircraft. A means of calculating the dependence of these on vehicle motions is the subject of this paper.

Method of Analysis

A slender body at high AOA is analyzed by considering that streaming air exists in thin slabs oriented normal to the body axis in its initial position. The body slices through a typical slab, as it translates past the body, as illustrated below, for coning motion about the wind axis.

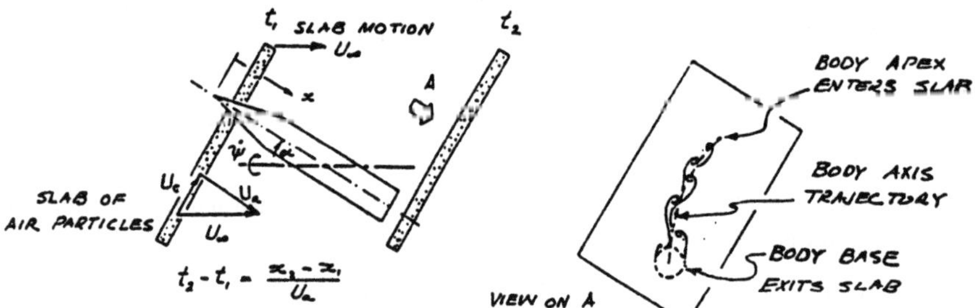

The space-time trajectory through the slab yields the histories of section diameter, radial and lateral velocities and accelerations.

Circumferential pressure distributions applied to time-varying body sections by the fluid in the slab may be integrated azimuthally into section force coefficients (lift and drag). These may then be resolved, for example, into components normal to the plane formed by the body axis and the wind axis to produce a time history of coning torque applied by one slab. If the body motions were unsteady, the trajectories and pressure histories in slabs starting at a succession of initial times would differ and the net torque could be found by combining the section coning moment coefficients along the axis of the body at a given instant of time. If steady, the pressures from a single run may be combined by transforming from the time-domain to x-domain, by $x = U_a t$, where x is body longitudinal position measured from its apex and U_a is the axial component of the wind velocity, to yield the net coning torque. Forebody force and moment coefficients needed for aircraft stability and control and produced by general motions may be found similarly.

Assumptions

If flow velocity perturbations parallel to the body axis are small and those in body section planes do not change rapidly in the body axis direction, slender body aerodynamics concepts apply. (The zero tangential axial velocity component boundary condition would clearly violate this requirement and is not imposed. At an AOA greater than 30 degrees this omission is expected to be insignificant). Therefore, pressures calculated on a cylinder of infinite length and in 2-D flow, due to rotation rate and time rate of change of radius, transverse velocity and acceleration through the fluid, approximate those developed on the body in corresponding motion. With flow Mach number, $M \ll 1.0$, and crossflow Reynolds number, $Re \ll 10^5$, incompressible laminar flow may be assumed.

If the body angular displacement is not large during the slab passage, assumptions that body sections remain circular and the body axis remains perpendicular to the slab are reasonable. The body may rotate about its axis.

Solution Mechanization

The above assumptions make a time-varying, incompressible, viscous, 2-D flowfield analysis appropriate. It is based on the incompressible Navier-Stokes equation transformed into the vorticity diffusion and transport equation, $\partial\omega/\partial t = -\mathbf{u}\cdot\nabla\omega + \nu\nabla^2\omega$, together with a Poisson's equation relating the stream function to flow vorticity, $\nabla^2\psi = -\omega$. The latter is derived from the vorticity definition, $\boldsymbol{\omega} = \nabla \times \mathbf{u}$, where ψ is the stream function, t is time and ω the vorticity.

Differentiation of the stream function provides flow velocity components, $u = e_r(1/r)(\partial\psi/\partial\phi) - e_\phi(\partial\psi/\partial r)$, needed for vorticity transport. The coefficient of kinematic viscosity, ν, of air is employed, unmodified.

The slab is assumed to move with variable velocity normal to an axis system that slides along the body center line, for flow kinematics analysis. For pressure coefficient analysis, the flow is transformed to an axis system moving in the slab at the fixed initial radial speed, U_c, of the body. For defining values of vorticity and velocity to be differentiated and inserted into the discretized vorticity transport and Poisson's equations, a grid of concentric circles divided into azimuthal segments is erected about the section. The polar coordinate grid is transformed to rectangular coordinates for numerical analysis as shown below.

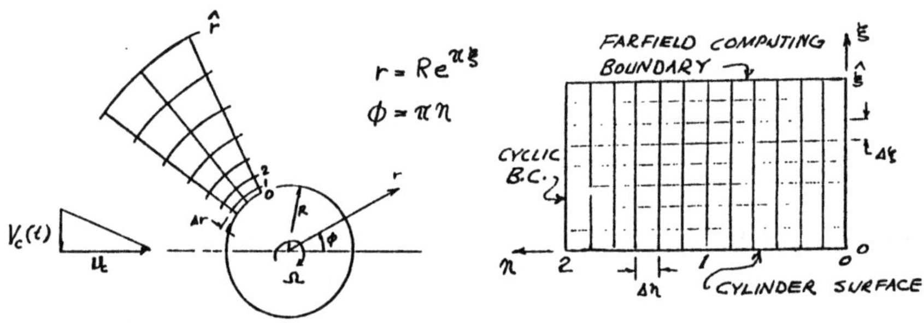

This results in five-element banded matrices that implicitly relate: a) the time derivative of vorticity to the terms of the transport and diffusion equation; and, b) the stream function to the vorticity at field points, off the body surface, at which the vorticity magnitude is significant. The ADI technique and upwind differencing are employed in numerical analyses.

Boundary Conditions

The boundary condition "no flow through the surface of the cylinder" is met by making the surface a streamline. The condition "no tangential flow relative to the surface" is imposed by expanding the stream function as a Taylor series in the radial direction near the surface. Evaluated at the first radial grid line it yields $\omega(R)=-(\partial^2\psi/\partial r^2)=-[\psi(R+\Delta r)-\psi(R)-\Omega R\Delta r]2/\Delta r^2$. The known stream function jump over the first radial increment, Δr, and the tangential velocity, ΩR, create the new surface vorticity needed to replace that drifted away, (factored by $\Delta r/2$, it becomes circulation per unit circumference).

Body dilation is provided by allowing the cylinder to remain at fixed radius for vorticity transport calculation; then it instantly increases to the next incremental radius. Field vorticity is squeezed into thinner radial elements by a line-source transformation, ie., inner and outer radii moved by line-source velocity to new positions dictated by the time increment. This requires a vorticity interpolation to values at points on the original grid at the next axial station.

The limited number of cells available, and the requirement for sufficient cells in the "surface boundary layer", combine to force the farfield boundary too close to the body to give a good approximation to the stream function at infinity. To compensate this, the stream function farfield boundary value is set to correspond to that of attached inviscid flow about the cylinder, ie., $\hat{\psi} = [U_c \sin\phi - V_c(t)\cos\phi](\hat{r} - 1/\hat{r})$. This indicates a time-varying farfield boundary value for the lateral translational velocity component, V_c.

Flow perturbation velocities in incompressible flow depend only on the instantaneous infinite field velocity, (and vorticity and body geometry distributions) and not on flow or body accelerations. Therefore a time-varying farfield boundary value, for the flow moving relative to the body, is valid for flow kinematics analysis.

Pressure Calculation

Flow acceleration (in addition to kinematics) strongly affects pressures and is more easily evaluated relative to axes moving at constant speed. Surface pressures are obtained by line integrating the Navier-Stokes momentum equation radially from a point in the undisturbed upstream flow to the body surface, and then circumferentially to the point of interest. The data employed comes from previous kinematics calculations.

$$\nabla(p + \rho/2q^2) = \rho u \times \omega \quad + \quad \mu\nabla^2 u \quad - \quad \rho\partial u/\partial t$$

total head circulatory viscous acceleration

The change from body-fixed coordinates to those moving through the slab requires flow acceleration terms to be amended as follows:

$$\partial u/\partial t = (\partial u/\partial t)_{\text{body axes}} + (\Delta U)_{\text{mov. axes}} \cdot (\nabla u)_{\text{body axes}}$$

Example Problem

A cone-cylinder model mounted at 45° AOA on a shaft aligned with the wind at infinity and free to rotate, was wind tunnel tested by Yoshinaga, Tate and Inoue, AIAA-81-1899. The test was conducted at $M<.03$ and a section normal Reynolds number of approximately 2×10^4; thus incompressibility and laminar flow could safely be assumed. Its terminal angular velocity as a function of AOA and wind speed is shown below:

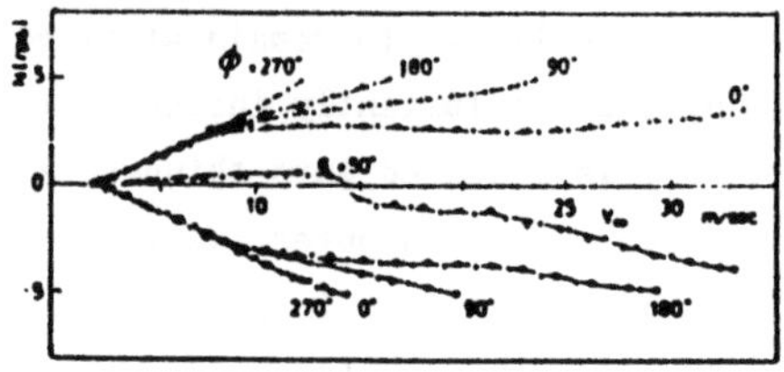

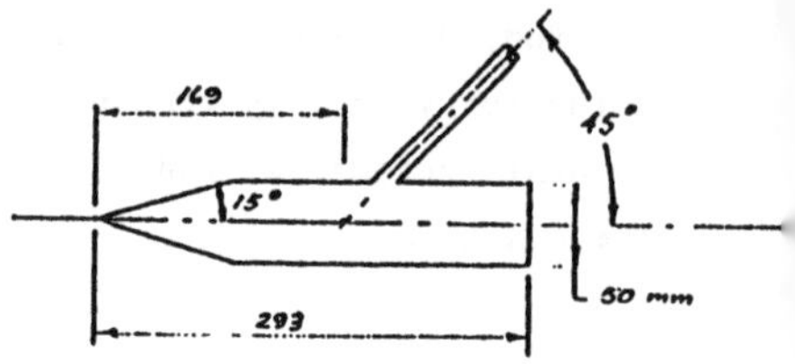

Body center line executes an almost circular path through the slab at U_∞=8.0 m/s, N=2.6 rev/s, as shown below.

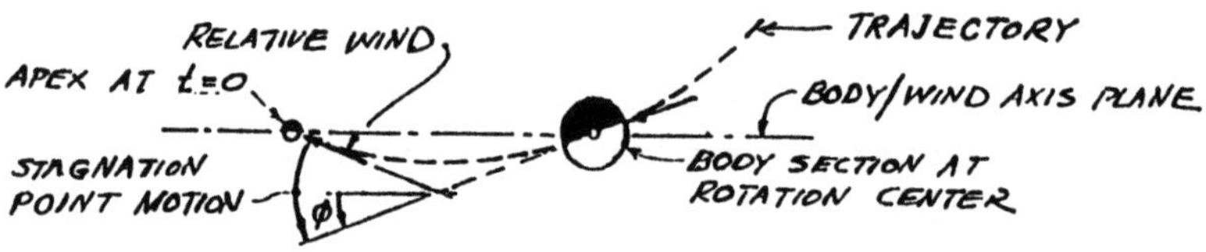

The stagnation point moves around the section at nearly twice the rotation rate to reverse the Magnus force into the propelling direction.

Computational Procedure Development

Initial flowfields have been calculated up to Re=200, equivalent to 1% of test velocity, to conserve grid points. Analysis to τ=10 ($\tau=U_c t/R$) is sufficient to capture the vortical flow shed at 45° AOA. Flowfields and surface pressures on a circular cylinder suddenly started into rectilinear motion, with and without rotation, are shown below for Re=40, 200.

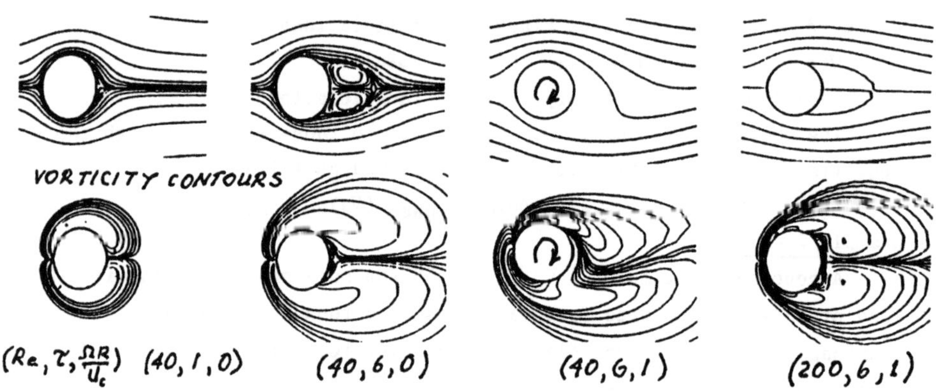

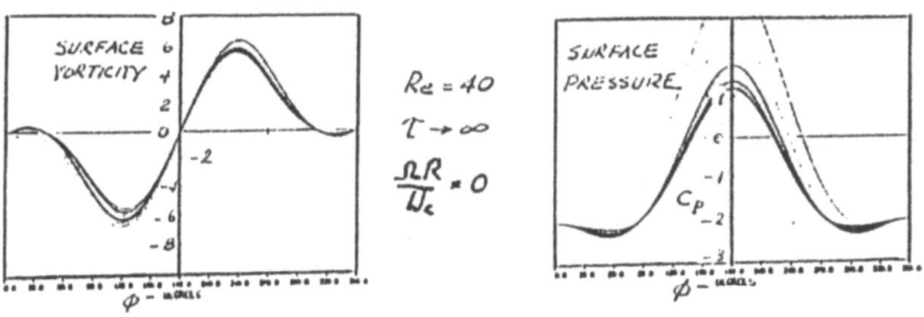

Boundary layer thickness reduction with increasing Re and severe flow asymmetry due to rotation are evident. A pointed body of revolution is simulated by a circular cylinder of radius increasing from zero at the rate, dR/dt. This produces the cone half-angle when combined with the axial flow. Circumferential surface flow velocity profiles produced by a 15^{0} cone at 45^{0} AOA, rotating about its axis at $\Omega R/U_c$=0.1, at Re=200, are shown below.

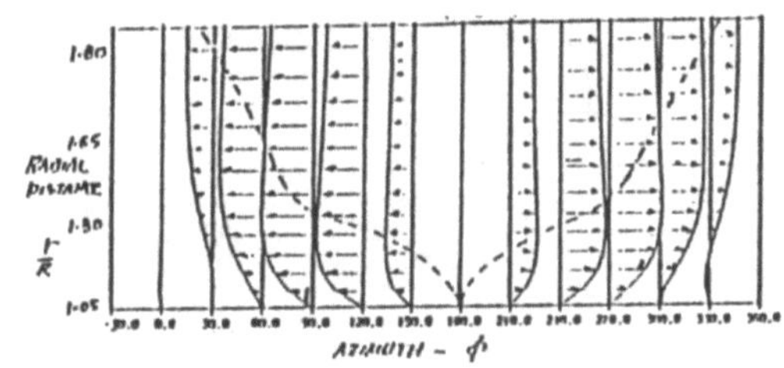

The approximate boundary layer and separated flow thickness is shown as a dashed line. The cone acts to flatten the boundary layer displacement thickness to 0.12R at ϕ=90° for Re=200, or provides an effective inviscid body radius about 10% greater than that of the test.

Example Problem Preliminary Results

Streamlines and constant vorticity contours are shown at four sections through the cone-cylinder body below.

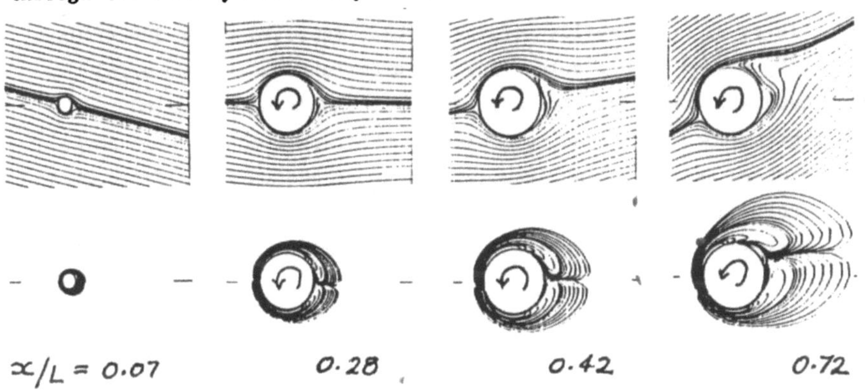

Azimuthal pressure coefficient distributions at the indicated stations and integrated lift and drag force distributions are shown below.

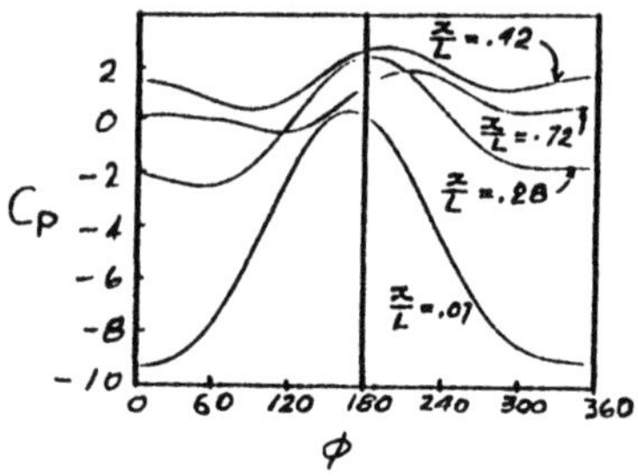

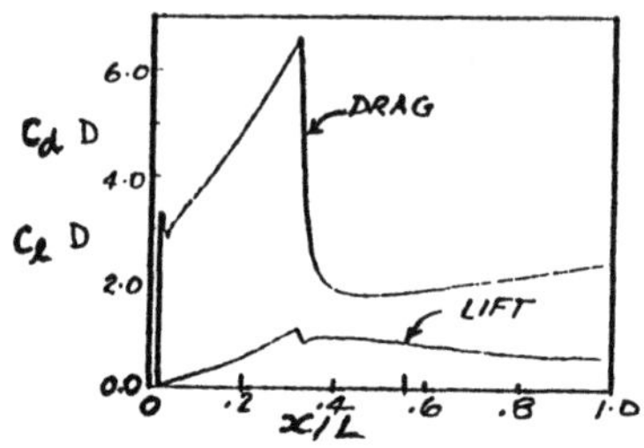

Lift and drag forces resolved normal to the plane through the body and wind axes and multiplied by distance to the wind axis give contributions to the coning moment.

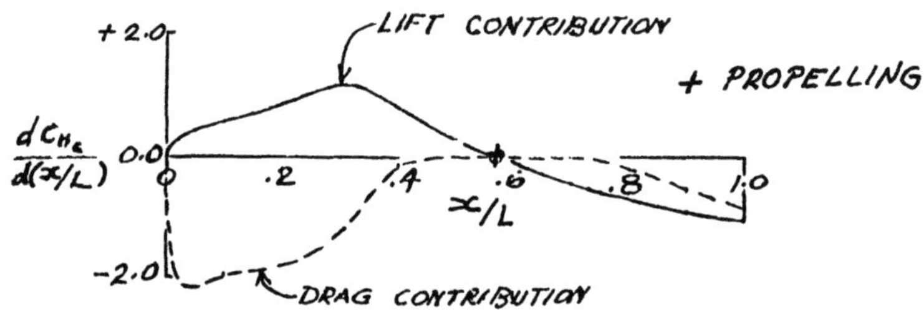

Even at 1% of the tested Reynolds number a large propelling moment is seen, though not enough to balance the drag contribution to the net coning moment at the chosen rotation speed.

Concluding Remarks

- o The azimuthal lag of the vortical structure and separation points behind the external pressure-producing flow appears to cause the propelling moment.
- o Limits of applicability of the method, expected to be 30° < AOA < 70°, caused by neglected flow perturbations due to the axial boundary condition and body bluntness should be evaluated.
- o This method, extended to arbitrary cross-sections, higher Reynolds numbers and turbulent flow, should be of use in aircraft forebody preliminary design.
- o This approach was suggested by the work of L.E. Ericsson.

A Boundary Element Model for the Taylor-Couette Instability

Graziani G. and Piva R.

Dipartimento di Meccanica e Aeronautica
Università di Roma "La Sapienza"

1. Introduction

When a viscous incompressible fluid is contained between two infinitely long concentric cylinders rotating at different angular velocities a large variety of flow conditions occurs. Such phenomenon is usually referred to as the Taylor–Couette instability [1]. For low values of the relative angular speeds of the cylinders the only non zero-velocity component is the azimuthal velocity u_θ and the streamlines are concentric circumferences in a plane to which the axis of rotation is normal. As long as the Reynolds number exceeds a critical value, Re_c, multiple solutions are possible with non-zero velocity components in the radial planes. Actually, at Re_c, the Couette flow solution becomes unstable giving rise to the so-called Taylor vortices which are counter-rotating toroidal vortices regularly spaced along the axial direction. An infinite multiplicity of flow solutions is possible, each of them having a different wavelength of the vortices. At larger values of the Reynolds number a second non-axisymmetric instability modifies the toroidal vortices by means of an azimuthal oscillation which leads to travelling waves. The instability of the basic Couette flow, that was theoretically analyzed by Taylor [1], has been later confirmed by the experimental results and, more recently, by several numerical computations.
The purpose of this paper is to present a boundary integral approach to investigate the onset of the Taylor instability and the development of different solutions with rotational symmetry. This approach yields some interesting features that can be advantageously accounted for in the solution of this problem. In fact, the integral representation of the solution clearly highlights the coupling mechanism between the three components of the momentum equation as shown in section 2.

Moreover, the system of linear algebraic equations provided by the discretization procedure contains a known term multiplied by the Reynolds number. By assuming different values of this number a parametric analysis can be easily carried out by inverting only once the coefficient matrix which results to be small but fully populated. For this reason the boundary integral approach seems to be competitive with respect to other numerical techniques which lead to large banded matrices. However a particular care has to be paid to the computation of the volume integral which is going to be the most critical part of the methodology in terms of the computing effort.

In the next section the integral representation for a rotationally symmetric incompressible viscous flow is reported and briefly discussed together with some comments on the solution procedure. The discussion of the results and some remarks about the future development by means of parallel algorithms are given in Sec. 3.

2. Mathematical formulation and solution procedure

The integral representation for the velocity components can be written upon considering a cylindrical coordinate system that is particularly suitable for the analysis of rotationally symmetric flow fields. The circumferential velocity V given by the Couette flow solution has been subtracted from the Navier–Stokes equations to achieve an accurate description of the secondary flow in terms of the perturbation velocity u. By moving the non linear acceleration terms to the right hand side, the governing equations become:

$$\nabla^2 \underline{u} - Re\, \underline{\nabla p} = Re\, \underline{\chi} \tag{1}$$

$$\underline{\nabla} \cdot \underline{u} = 0 \tag{2}$$

where the component of the forcing term $\underline{\chi}$, under the assumption of steady state conditions and negligible body force, are:

$$\chi_r = u_r \frac{\partial u_r}{\partial r} + u_z \frac{\partial u_r}{\partial z} - 2\frac{V u_\theta}{r} - \frac{u_\theta^2}{r} \tag{3}$$

$$\chi_\theta = u_r \frac{\partial u_\theta}{\partial r} + u_z \frac{\partial u_\theta}{\partial z} + \frac{u_r u_\theta}{r} + u_r \left(\frac{dV}{dr} + \frac{V}{r} \right) \tag{4}$$

$$\chi_z = u_r \frac{\partial u_z}{\partial r} + u_z \frac{\partial u_z}{\partial z} \tag{5}$$

As reported in [2] the integral representation of the solution in a point of the domain, for the velocity components in cylindrical coordinates is :

$$u_r = \int_\Gamma (t_r^r u_r + t_z^r u_z - u_r^r t_r - u_z^r t_z)\, r\, d\gamma + Re \iint_\Omega (u_r^r \chi_r + u_z^r \chi_z)\, r\, dS \tag{6}$$

$$u_z = \int_\Gamma (t_r^z u_r + t_z^z u_z - u_r^z t_r - u_z^z t_z)\, r\, d\gamma + Re \iint_\Omega (u_r^z \chi_r + u_z^z \chi_z)\, r\, dS \tag{7}$$

$$u_\theta = \int_\Gamma \left(t_\theta^\theta u_\theta - u_\theta^\theta t_\theta \right) r\, d\gamma + Re \iint_\Omega u_\theta^\theta \chi_\theta\, r\, dS \tag{8}$$

where Ω is the flow domain bounded by the contour Γ.

The fundamental solution u_j^k and the related tractions t_j^k, derived in [2], account for the rotational symmetry of the flow, that is all the variables are independent of θ. It is worthwhile to notice that u_j^k and t_j^k are independent of Re.

When the point is going to the domain boundary the above integral representation yields a system of Fredholm integral equations for the unknowns u_i or t_i which are approximated by using some boundary element technique, as is explained in [3,4].

As is shown by equations (6-8), the nonlinear term χ gives the coupling between the equations which implies the onset of the instability. Since the term χ depends on the unknowns, an iterative procedure is required to obtain the solution.

A vanishing value of the volume integral is obtained unless an initial perturbation is assumed. Hence, by defining an arbitrary initial condition, it follows a non-zero acceleration term that starts the iterative solution procedure. The initial perturbation either is damped, yielding the basic Couette flow, or it is amplified up to a steady Taylor vortex solution. The last one is not uniquely defined, depending on the initial disturbance as well as on the value of the Reynolds number.

This approach, based onto the explicit inversion of the differential operator for the Stokes' problem, allows for the direct solution of the problem in terms of

$$\mathbf{x} = Re\, \mathbf{A}^{-1}\, \mathbf{b}(\mathbf{x}) \tag{9}$$

where $\mathbf{A}$ is the coefficient matrix and $\mathbf{b}$ is the field integral term that is itself function of the dependent variables.

It should be noted that the evaluation of the geometric quantities as well as the inversion of the matrix A are performed only once, since this operations are independent of the Reynolds number and the initial conditions. On the contrary, for other methods of solution a new problem has to be solved for each modification of the parameters.

3. Discussion of the numerical results and concluding remarks

The explicit dependence of the flow field unknowns upon the Reynolds number allows for a parametric investigation of the flow problem. For a given geometrical configuration of the physical domain, a wide range of values of Re has been investigated to determine whether the basic flow is stable or not.

By assuming different wavelenghts λ for the Taylor vortices, we obtain a stability diagram that shows the critical values of the Reynolds number (Re_c) as function of the axial wavelenght, *i.e.*, the stability region of the Couette flow.

The computed stability diagram compared with the predictions of some analytical theories [5,6] in Fig. 1, provides, for a given value of Re higher than the lowest critical value, multiple solutions of the governing equations corresponding to the generation of Taylor vortices of different wavelenght.

In the numerical computations we have to assume a finite axial extension of the physical domain, *i.e.* the height H of the concentric cylinders, even maintaining at the upper and lower boundaries the periodicity condition. Only a few values of λ are compatible with the assumed finite axial dimension ($\frac{H}{\lambda} = integer$). Hence only the solutions for those values of λ may be obtained numerically.

An example of multiple solutions is reported in Fig. 2 where the computed streamlines of the secondary flow respectively featuring five, six and seven Taylor vortices within the same axial interval are shown. As predicted by the stability analysis different solutions have been obtained, for the same Reynolds number and geometrical configuration, only by varying the assumed initial condition.

The numerical results of the proposed methodology show a good agreement when compared with the theoretical and the experimental results available in the literature for the analysis of Taylor vortex flows. As a preliminary conclusion the proposed approach has a computational effectiveness that is competitive with other methodologies when a parametric analysis has to be performed.

As a further point of interest, we like to mention that this methodology, presently under investigation, seems to be particularly appropriate for the implementation on massively parallel computers. In fact, the boundary integral approach provides the evaluation of each unknown independently from the values of the other ones. Hence the computation of the flow solution and in particular the computation of the volume integral may largely benefit from the adoption of parallel algorithms. In particular, in the framework of massively parallel computers, a single processor could be associated with every field point. In this way all the field variables could be computed at the same time by means of an explicit evaluation.

Bibliography

1. G. I. Taylor, Proc. Roy. Soc. A, **23**, (1923).
2. G. Graziani, Int. Journ. Engng. Scie., **27**, 7, (1989).
3. S. Y. Ma, G. Graziani and R. Piva, Meccanica, **19**, 294-299, (1984).
4. R. Piva, G. Graziani and L. Morino, Computational Mechanics '86, Springer-Verlag, (1986).
5. S. Chandrasekhar, *Hydrodynamic and Hydromagnetic Stability*, London : Oxford University Press, (1961).
6. S. Kogelman and R. C. Di Prima , Phys. Fluids, **13**, 1-11, (1970).

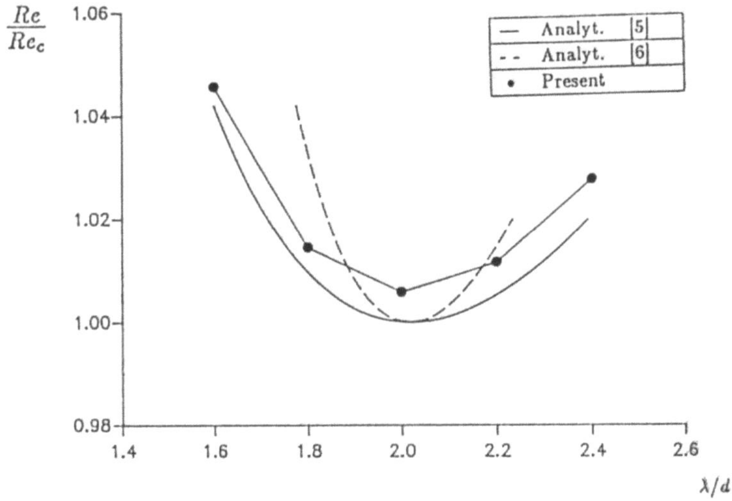

Fig. 1 - Re/Re_c as function of the nondimensional wavelenght λ/d.

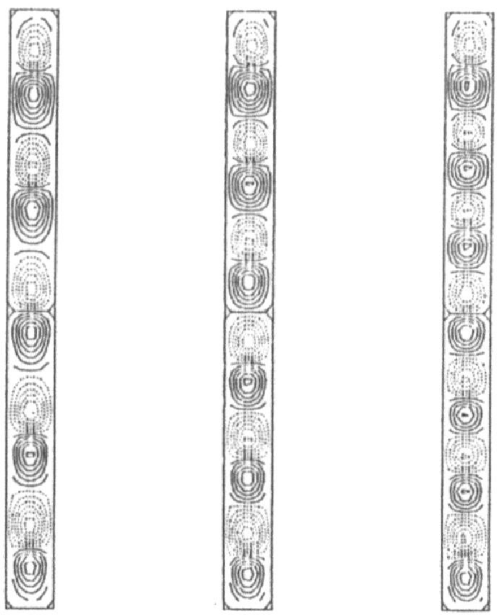

Fig. 2 - Contours of the secondary flow stream function for $Re = 150$ and different initial conditions.

Boundary Element Methods for the Navier Stokes Equations

Michael T. Patterson* James C. Wu† C. M. Wang‡

2 October, 1989

Summary

Numerical solutions to the full Navier-Stokes equations are appearing frequently within the fluid dynamics community. These solutions are usually determined with finite difference algorithms. These algorithms are providing excellent data, but at extrodinary computer costs.

Boundary element methods provide an alternative to finite difference methods. In this paper, a boundary element method for the Navier-Stokes equations is described. This method is very accurate and cost efficient when applied to two-dimensional, incompressible problems. Research is now being performed to extend the procedure to two-dimensional compressible flows and to three-dimensional incompressible flows. This research is reviewed in this paper.

Nomenclature

a	Speed of sound.
C_l, C_d, C_m	Lift, drag, and moment coefficients.
e	Total energy per unit volume.
n	Unit normal vector.
p	Pressure.
Pr	Prandtl number.
r	Position vector.
t	Time.
u, v, w	Velocity components in x, y, and z directions.
$\vec{v}$	Velocity vector.
x, y, z	Cartesian coordinates.
∂	Partial derivative.
γ	Ratio of specific heats.
∇, ∇^2	Del operator and Laplacian.
μ, λ, ν	Dynamic, bulk, and kinematic viscosities.
$\vec{\omega}$, $\vec{\Omega}$	Vorticity vectors for incompressible and compressible flow.
ρ	Density.

*Research Assistant, Georgia Institute of Technology
†Professor, Georgia Institute of Technology
‡Senior Research Engineer, Georgia Institute of Technology

1 Introduction

Boundary element methods are known to reduce computational requirements for the solution of aerodynamic problems over finite difference methods. This feature results mainly from the confinement of the solution procedure to some small region of the flow field.

For example, the widely used panel methods are boundary element procedures for inviscid flow. These methods confine the solution procedure to the surface of the body under investigation; the flow field itself is removed from the solution procedure.

Boundary element methods have also been applied to the viscous flow problem. In this paper, a method is presented which confines the solution procedure to the vortical (rotational) regions of the flow field. This feature is very attractive for both two and three-dimensional problems in that computer resource demands may be kept modest.

In this method, the (vector) conservation of momentum equation is differentiated to obtain a vorticity transport equation. Vorticity thus becomes the *primary* variable of the solution procedure and velocity becomes a *secondary* variable.

A finite difference procedure is applied to the vorticity transport equation and the velocity field is then determined by evaluating an integral relation which contains a domain integral in addition to the familiar boundary integral. Due to this mixture of finite differences and integral equations, this method has previously been called an *integro-differential* method. Because this method allows the partitioning of the flow field into zones (such as boundary-layers), the method has also been referred to as a *zonal* method.

2 Governing Equations

The boundary element method described in this paper is well suited for the solution of a solenoidal (incompressible) vector field. If incompressibility is assumed, the equations governing fluid motion may be formulated as

$$\begin{aligned} \nabla \cdot \vec{v} &= 0 \\ \nabla \times \vec{v} &= \vec{\omega} \\ \frac{\partial \vec{\omega}}{\partial t} = -(\vec{v} \cdot \nabla)\vec{\omega} &+ (\vec{\omega} \cdot \nabla)\vec{v} + \nabla^2(\nu\vec{\omega}). \end{aligned} \tag{1}$$

These equations are, respectively, the continuity equation, the definition of a vorticity vector, and the vorticity transport equation. The continuity equation and the vorticity vector definition represent the kinematic aspect of the method; this aspect describes the velocity field corresponding to a known vorticity field. The vorticity transport equation represents the kinetic aspect of the method; this aspect describes the redistribution, or transport, of vorticity.

The general solution procedure for these equations consists of three major steps which advance the vorticity-velocity field from an old time level to a new time level:

1. Using known vorticity and velocity values at the old time level, the vorticity transport equation is solved for new vorticity values in the fluid's interior;

2. New surface vorticity values are computed using the new interior vorticity values; and

3. New velocity values are computed using the vorticity values of steps 1) and 2).

Step 1) of this procedure may be performed using finite differences. For efficiency, the velocity field is lagged one time step during these computations. Note that the computations of this step may be confined to just the vortical region of the flow and that the attached and detached vortical regions may be treated independently, with the different length scales appropriately modeled (i.e., boundary-layer equations may be used when applicable).

Step 3) may be performed using either of two procedures. The first procedure requires rewriting the continuity equation and vorticity vector definition as

$$\nabla^2 \vec{v} \;=\; -\nabla \times \vec{\omega} \tag{2}$$

and then, with the vorticity field known, solving for the unknown velocity field with a Poisson solver. This procedure is inefficient, however, for several reasons. First, the Poisson's equation is elliptic and its solution at any point depends upon the solution at other points; the values of velocity at all grid points must be determined simultaneously. Second, a velocity boundary condition must be approximated at "infinity" by extending the grid points into the freestream to a distance far from the body of interest. Because of this, a large flow field must be included in the solution procedure even though only that portion of the field near the body may be of interest.

The second procedure re-expresses the continuity equation and vorticity definition through the use of fundamental solutions as an integral representation for the velocity vector at time t as

$$\begin{aligned} \vec{v}(\vec{r},t) \;&=\; \oint_{B_S} [(\vec{v}_0 \cdot \vec{n}_0) - (\vec{v}_0 \times \vec{n}_0)\times]\nabla_0 P \, dB_0 \\ &- \int_R \vec{\omega}_0 \times \nabla_0 P \, dR_0 \;+\; \vec{v}_\infty \end{aligned} \tag{3}$$

where the subscript 'o' denotes the variable of integration. In this equation, B_s denotes the body surface, and R denotes the interior region of the fluid domain, body surface included. The coordinate system is considered to be inertial and translating with the body. The fundamental solution of the Poisson's equation is defined by

$$P(\vec{r},\vec{r}_0) = \begin{cases} -\frac{1}{4\pi|\vec{r}-\vec{r}_0|} & \text{for three-dimensional problems} \\ -\frac{1}{2\pi}\ln\frac{1}{|\vec{r}-\vec{r}_0|} & \text{for two-dimensional problems.} \end{cases}$$

The velocity integral provides the following advantages over the Poisson's equation for the solution of velocity values:

- The velocity vector may be computed point by point. That is, the velocity vector at any point may be computed without knowledge of velocity values at other points.
- The far field boundary condition is satisfied analytically rather than approximated numerically.
- The region R is confined to only the regions where $\vec{\omega}$ is non-zero.

In general, any numerical method offering efficiency and accuracy may be used for this step of the solution procedure once the confinement of the solution field is achieved through use of the velocity integral. For example, a finite difference method based upon equation 2 may be used to compute the velocity field in the region R after equation 3 is employed to compute velocity values on the boundary of R.

The surface vorticity values of step 2) are determined by treating equation 3 as an integral constraint on the vorticity field. If this equation is applied at the body surface, boundary vorticity values may be computed explicitly because velocity values on the airfoil surface are known through the no-slip condition, and interior vorticity values are known from the solution of the vorticity transport equation, i.e., from step 1).

The three major steps of the solution procedure thus require, at each time step, one finite difference solution for interior vorticity values. one integral solution for surface vorticity values, and one integral solution for velocity values. The computations of these three steps may be confined to just the vortical region of the flow. This region includes the boundary-layers, any separated region, and the trailing wake. It is the confinement of the solution to the vortical region that distinguishes this computational method from any other.

It should lastly be mentioned that recent advances within the computer and aerodynamic industries may help to enhance the present boundary element method. Specifically,

1. This algorithm is well suited for parallel processing. The recent development of massively parallel machines is important in this respect.

2. Adaptive grid technology is becoming mature. This boundary element method will be very receptive to such gridding techniques.

3. Advances are being made in understanding the turbulence phenomenon. Part of this understanding is due to vorticity analyses. Use of vorticity as a primary variable may thus improve this modeling.

4. For three-dimensional flows. wake modeling is usually required and is often achieved using Biot-Savart laws. Because this boundary element method is an expression of these laws, incorporation of wake models is natural.

Further information about this method may be found in references [1] through [4].

3 Application To Two-Dimensional Flow

The assumption of two-dimensional flow simplifies the governing equations. Only one component of the vorticity vector exists, leaving only one scalar vorticity transport equation (as opposed to three for three-dimensional flow). The vorticity transport equation simplifies further in that the vortex stretching and rotation term $(\vec{\omega} \cdot \nabla)\vec{v}$ is identically zero.

The resulting vorticity transport equation is discretized implicitly in time on a conformal, body-fixed O-grid. Second order upwind differences are employed for the convection terms and central differences for the stress terms. The time derivative is approximated with a first order backward difference. For turbulent flows, the Baldwin-Lomax turbulence model is incorporated.

Utilization of a finite Fourier series in the azimuthal direction for each of the flow variables allows the analytic integration of the velocity integral in that direction. Evaluation of the velocity components thus requires only numerical quadrature in the radial direction. Again, this integration may be confined to the vortical region of the flow.

Figures 1 and 2 display the lift and drag coefficients for a NACA 0012 airfoil undergoing a rapidly pitching motion [4]. Reference [5] provides the experimental data. In this problem, the Reynolds number was one million and the reduced frequency was 0.10. The flow was considered fully turbulent.

4 Application To Three-Dimensional Flow

The main advantage of using the vorticity vector as the primary variable in three dimensions is that the solution procedure can be confined to the vortical region of the flow. Because the vorticity vector retains all three components in three-dimensional flow, the use of vorticity loses some attractiveness. However, confinement of the solution procedure to the vortical region offers greater computational savings in three dimensions than in two. This additional savings is attributed to the fact that a three-dimensional solution field is contracted along two directions while a two-dimensional solution field is contracted along only one direction.

Evaluation of the velocity integral (equation 3) is computationally demanding; this integral is thus evaluated only at the boundaries of the vortical region. Inside this region, the Poisson's equation (equation 2) is solved using a finite difference scheme.

An essential part of the present numerical procedure is the accurate determination of vorticity values on the solid surface. These vorticity values are again determined by an integral constraint and also by the vorticity divergence-free condition. The integral constraint is derived here from the satisfaction of the *normal* velocity boundary condition on the solid surface; in two dimensions, the tangential condition is used. Although conservation of vorticity is not required in three dimensions as it is in two, conservation of the spanwise component of vorticity is included during these computations to control numerical error. A least square technique is employed for these boundary vorticity computations.

Computation of flows over stationary (non-pitching) flat plate wings have been performed to demonstrate the abilities of this method. Figure 3 displays the distribution of the streamwise vorticity component at different downstream planes for a Reynolds number of 1000 and an angle of attack of 10 degrees. Very close to the trailing edge, this vorticity component has negative values inboard of and above the plate tip. Further downstream, only positive values are observed; the diffusion effect is apparent. Figure 4 displays streamlines near the wing tip. The roll-up of the streamlines coincides with the tip vortex.

Figure 5 presents the spanwise loading for this same problem at aspect ratios of 2 and 4. The hump appearing near the tip region on each curve results from the suction force associated with the tip vortex. A comparison of the lift coefficient at different aspect ratios with data from Abbott and Von Doenhoff [6] is shown in figure 6.

5 Application To Compressible Flow

As mentioned earlier, this boundary element method is well suited for incompressible flows. Compressible flows may be addressed by decomposing the momentum vector as

$$\rho\vec{v} = \rho\vec{v}_1 + \rho\vec{v}_2$$

where $\rho\vec{v}_1$ is a compressible, irrotational momentum field and $\rho\vec{v}_2$ is an incompressible, rotational field. By defining a new vorticity vector, both the kinematic and kinetic aspects may also be decomposed. The resulting equations are grouped into two sets and solved independently over a time step.

The first set of equations strongly resembles the equations addressed thusfar in this paper:

$$\begin{aligned} \nabla \cdot (\rho\vec{v}_2) &= 0 \\ \nabla \times (\rho\vec{v}_2) &= \vec{\Omega} \\ \frac{\partial\vec{\Omega}}{\partial t} + \nabla \cdot (\vec{v}\vec{\Omega}) &= \nabla^2 \left(\frac{\mu\vec{\Omega}}{\rho}\right) + \vec{S} \end{aligned}$$

where $\vec{S}$ is a source term describing both convection and diffusion processes. The remaining set of equations resembles the Euler equations and may be written in Cartesian coordinates as:

$$\partial_t\vec{q} + \partial_x\vec{E} + \partial_y\vec{F} = \partial_x\vec{R} + \partial_y\vec{S} \tag{4}$$

where the vectors are given in two dimensions by

$$\vec{q} = \begin{pmatrix} \rho \\ \rho u_1 \\ \rho v_1 \\ e \end{pmatrix} \quad \vec{E} = \begin{pmatrix} \rho u_1 \\ \rho u u_1 + p \\ \rho u v_1 \\ u(e+p) \end{pmatrix} \quad \vec{F} = \begin{pmatrix} \rho v_1 \\ \rho v u_1 \\ \rho v v_1 + p \\ v(e+p) \end{pmatrix} \quad \vec{R} = \begin{pmatrix} 0 \\ \tau_1 \\ 0 \\ R_4 \end{pmatrix} \quad \vec{S} = \begin{pmatrix} 0 \\ 0 \\ \tau_1 \\ S_4 \end{pmatrix}$$

The viscous stresses and constitutive equations are given by:

$$\begin{aligned} \tau_1 &= \lambda(u_x + v_y) & R_4 &= u\tau_{xx} + v\tau_{xy} + \mu \mathrm{Pr}^{-1}(\gamma-1)^{-1}\partial_x a^2 \\ \tau_{xx} &= \lambda(u_x + v_y) + 2\mu u_x & S_4 &= u\tau_{yx} + v\tau_{yy} + \mu \mathrm{Pr}^{-1}(\gamma-1)^{-1}\partial_y a^2 \\ \tau_{xy} &= \mu(u_y + v_x) & p &= (\gamma-1)[e - .5\rho(u^2+v^2)] \\ \tau_{yy} &= \lambda(u_x + v_y) + 2\mu v_y & a^2 &= \gamma(\gamma-1)[e/\rho - .5(u^2+v^2)]. \end{aligned} \quad (5)$$

By decomposing the momentum vector and equations as shown, the computational advantages of the boundary element method are retained; one has, however, the extra set of equations to resolve. This second set of equations is quite tractable, though. While some stress terms are yet present in the momentum equations, these are *normal* stesses and are easy to resolve. Furthermore, the stress terms present in the energy equation may be neglected for subsonic and transonic flows without much error. In fact, this entire set of equations may be approximated with unsteady potential flow equations.

Efforts are now directed at linking the boundary element method discussed in this paper with an Euler solver for the solution of these compressible flow equations.

6 Acknowledgments

This work is supported by the Army Research Office.

References

[1] Wu, J.C., and Gulcat, U., "Separate Treatment of Attached and Detached Flow Regions in General Viscous Flows," *AIAA Journal*, Vol. 19, No. 1. pp. 20 - 27, 1981.

[2] Wu, J.C., 1984, "Zonal Solution of Unsteady Viscous Flow Problems," *AIAA Paper* 84-1637.

[3] Wu, J.C., Wang, C.M., and Tuncer, I.H., 1986, "Unsteady Aerodynamics of Rapidly Pitched Airfoil," *AIAA Paper* 86-1105.

[4] Tuncer, I.H., 1988, "Unsteady Aerodynamics of Oscillating and Rapidly Pitched Airfoils," Ph.D. Thesis, Georgia Institute of Technology, Atlanta, Georgia.

[5] McAlister, K.W., Pucci, S.L., McCroskey, W.J., and Carr, L.W., "An Experimental Study of Dynamic Stall on Advanced Airfoil Sections, Pressure and Force Data," Vol. 2, NASA-TM-84245, 1982.

[6] Abbott, I.H., and Von Doenhoff, A.E., "Theory of Wing Sections," Dover Publications, New York, 1959.

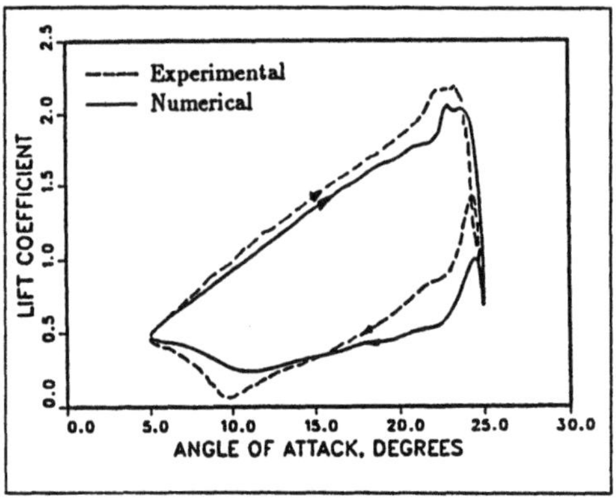

Figure 1: Two-Dimensional C_l

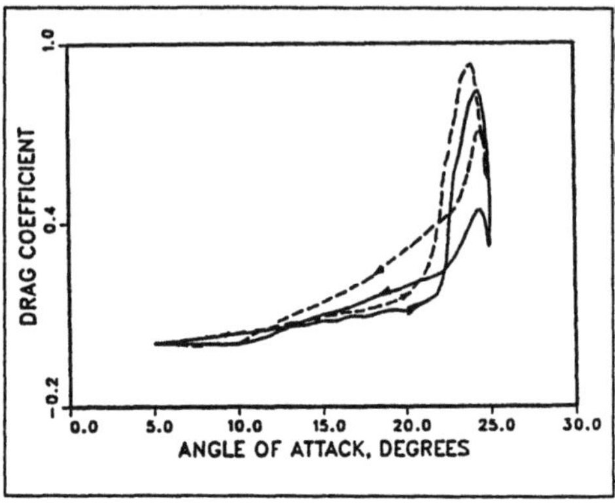

Figure 2: Two-Dimensional C_d

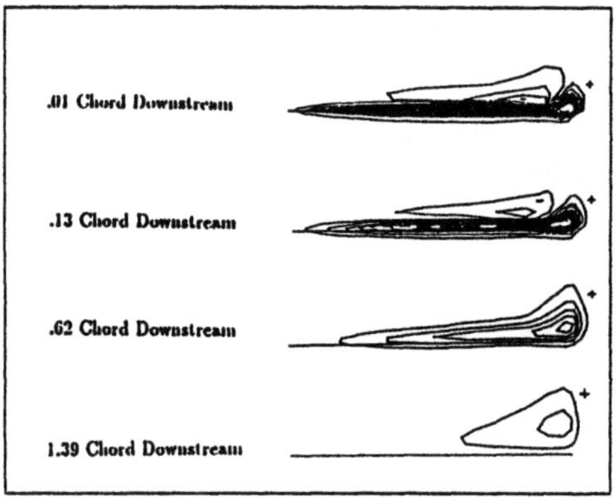

Figure 3: Streamwise Vorticity Component at Downstream Planes

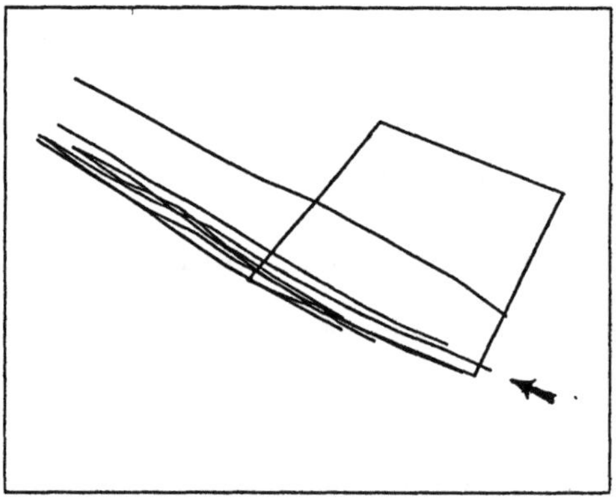

Figure 4: Streamlines at Plate Tip

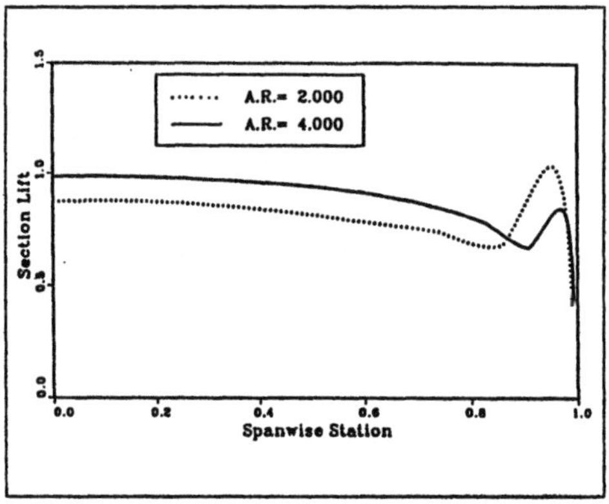

Figure 5: Flat Plate Spanwise Loading

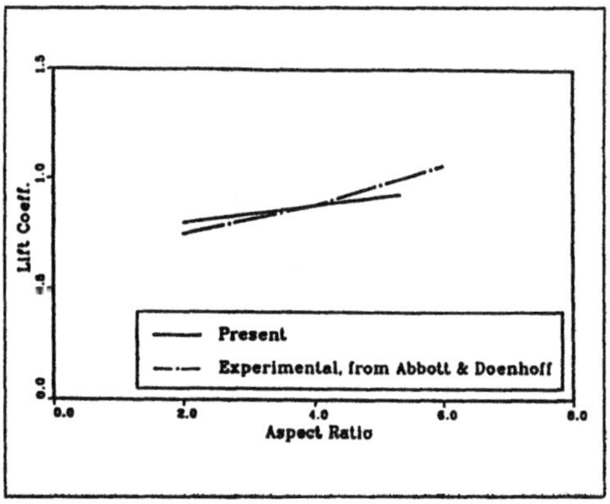

Figure 6: Three-Dimensional C_l

Boundary Element-Lagrangian Solution Method for Nonlinear Free Surface Problems

N. TOSAKA, R. SUGINO and H. KAWABATA

Department of Mathematical Engineering, College of Industrial Technology, Nihon University, Narashino, Chiba, 275, Japan.

Summary

A numerical approximation procedure for computing the unsteady irrotational motion of an inviscid and incompressible fluid with a free surface is developed within the boundary element approach. In order to track motion of fluid particles on the free surface which satisfy the well-known Laplace equation and nonlinear boundary conditions, the Lagrangian manner is introduced effectively. The solution procedure to be developed in this paper is the one which is constricted in aim at the use of the simplest boundary element, i.e., the constant element. The solution method is applied to several examples on the forced horizontal oscillation of a fluid in containers and generation, propagation, run-up and reflection on a vertical wall of a solitary water wave.

Introduction

The motion of fluid with a free surface is observed at sloshing of fluid in container, generation and propagation of solitary waves and so forth. The mathematical model of this motion is usually given as the initial-boundary value problem of the well-known Laplace equation. Although the governing equation is linear, the problem is highly nonlinear because the shape of free surface is unknown a priori and the boundary conditions on this surface is contained nonlinear terms. Therefore, some approximate solution procedure to solve this nonlinear problem is indispensable. There exist a number of researches on the problem by using the finite element method [1,2] and the boundary element method [3-7]. The boundary element approach is advantageous in solving the free surface problems of the potential flow which is governed by the Laplace equation because of its linearity.

In this paper, a procedure on the boundary element approach for solving nonlinear free surface fluid flow problems is developed with use of treatment in which the fluid particles on the free surface are moved in Lagrangian manner [6,7]. Especially, kinematical conditions on free surface are treated with Lagrangian manner. The approximate solution procedure is developed with the simplest element discretization with the constant element. The procedure developed in this paper is applied to solve numerically two dimensional problems of fluid flow with the free surface. Several test problems, which are nonlinear sloshing problems in containers and generation, propagation of a solitary wave in a rectangular tank of uniform depth, are examined.

Formulation

Let us consider two-dimensional motion with a free surface of the fluid which is assumed to be inviscid and incompressible. The flow is assmed to be irrotational. A fluid region Ω is bounded by boundary Γ with the outward unit normal n which is consisted by the free surface Γf and a wall Γw (Fig.1). The governing equation of the problems is given by the well-known Laplace equation,

$$\nabla^2\phi = \frac{\partial^2\phi}{\partial x^2} + \frac{\partial^2\phi}{\partial y^2} = 0 \qquad (\text{in } \Omega) \tag{1}$$

where the harmonic function $\phi(x,y,t)$ denotes the velocity potential. The boundary conditions to be prescribed are given as follows :

$$\frac{\partial\phi}{\partial t} + \frac{1}{2}\left\{\left(\frac{\partial\phi}{\partial x}\right)^2 + \left(\frac{\partial\phi}{\partial y}\right)^2\right\} + a\xi + g\eta = 0 \qquad (\text{on } \Gamma f) \tag{2}$$

$$\frac{D\xi}{Dt} = \frac{\partial\phi}{\partial x} = u \quad , \quad \frac{D\eta}{Dt} = \frac{\partial\phi}{\partial y} = v \qquad (\text{on } \Gamma f) \tag{3}$$

$$\frac{\partial\phi}{\partial n} = 0 \qquad (\text{on } \Gamma \omega) \tag{4}$$

where $a(t)$ denotes the forced horizontal acceleration applied to the container, g the acceleration of gravity and (ξ , η) the coordinates of horizontal and vertical positions of a fluid particle of a free surface, respectively. It is noteworthy that the kinematic boundary condition of the free surface is written by the Lagrangian expressions.

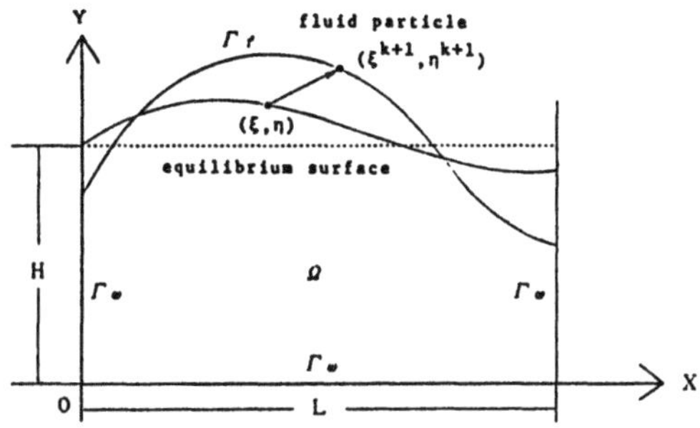

Fig.1 Geometry of the problem.

The boundary integral equation of the Laplace equation (1) under consideration of (4) can be derived as follows :

$$\frac{\alpha}{2\pi}\phi + \int_{\Gamma\omega}\phi q^* d\Gamma - \int_{\Gamma f}\frac{\partial\phi}{\partial n} v^* d\Gamma = -\int_{\Gamma f}\phi q^* d\Gamma + \int_{\Gamma\omega}\frac{\partial\phi}{\partial n} v^* d\Gamma \tag{5}$$

where the coefficient $\alpha/2\pi$ denotes the so-called shape coefficient. The well-known fundamental solution and its derivative are given, respectively, by

$$v^* = \frac{1}{2\pi}\ln\frac{1}{r}\ , \quad q^* = \frac{\partial v^*}{\partial n} = \frac{1}{2\pi}\frac{\partial}{\partial n}\left(\ln\frac{1}{r}\right) \tag{6}$$

in which r denotes the distance between the source point and the observation point.

Numerical Solution Procedure

Let us follow numerical solution procedure proposed in our previous papers [6,7] in solving the above-mensioned nonlinear problem. In order to solve numerically equation (5) under the nonlinear boundary conditions (2) and (3), we introduce two types of discretized scheme : (1) the discretization of unknowns ϕ and $\partial\phi/\partial n$ in (5) by means of the constant element (2) the approximation of the shape of free surface. Further we assume that the fluid particles on free surface are located at the nodal points on the approximated free surface. The unsteady shape of a free surface can be determined approximately by numerical solution of the boundary integral equation (5) under the time-dependent boundary conditions (2) and (3). Time integration of the problem is carried out the following schemes through a sequence of time increment Δt :

$$\begin{aligned}\phi(x+\Delta x, y+\Delta y, t+\Delta t) = \phi(x,y,t) &+ \Delta t\frac{\partial \phi}{\partial t}(x,y,t) \\ &+ \Delta x\frac{\partial \phi}{\partial x}(x,y,t) + \Delta y\frac{\partial \phi}{\partial y}(x,y,t)\end{aligned} \tag{7}$$

$$\begin{aligned}\xi_i^{k+1} &= \xi_i^k + \left\{\left(\frac{\partial \phi}{\partial x}\right)_i^{k+1} + \left(\frac{\partial \phi}{\partial x}\right)_i^k\right\}\frac{\Delta t}{2} \\ \eta_i^{k+1} &= \eta_i^k + \left\{\left(\frac{\partial \phi}{\partial y}\right)_i^{k+1} + \left(\frac{\partial \phi}{\partial y}\right)_i^k\right\}\frac{\Delta t}{2}\end{aligned} \tag{8}$$

where the superscript " k " indicates the time level $k\Delta t$ and the subscript " i " is the i th nordal point on the free surface element. Further, $(\partial\phi/\partial t)_i^k$ in (7) can be determined by using the dynamic boundary condition (2) and the Lagrangian coordinates (8). And, the derivatives $(\partial\phi/\partial x)_i^k$ and $(\partial\phi/\partial y)_i^k$ in (8) can be easily rewritten in terms of the normal and tangential derivatives $\partial\phi/\partial n$ and $\partial\phi/\partial s$ through the geometrical relation. Applying the above procedure to the problem, we can simulate numerically fluid motion of a free surface from the initial state to an arbitrary state.

Numerical Results

The solution procedure is applied to the test problems of nonlinear liquid sloshing in two-dimensional containers and of a solitary wave in a two-dimensional rectangular tank of uniform depth.

1. Sloshing in rectangular container [6,7]

The dimensions of container are H=0.6(m) and L=0.9(m). The boundary of the fluid region is divided into 50 constant elements. In particular, we divide the free surface into 20 segments. Two kinds of the forced horizontal acceleration with amplitude A and the frequency ω given by

$$a(t) = A\omega^2\sin(\omega t) \qquad \text{for } t \geq 0 \tag{9}$$

are selected : (1) Case I (A = 2.0 mm, ω = 5.5 rad/sec) (2) Case II (A =

10.0 mm, $\omega = 4.5$ rad/sec). The obtained results for cases I and II are shown in Fig.2 and Fig.3, respectively. It is clear that numerical results in Fig.3 show the so-called " beat " phenomenon.

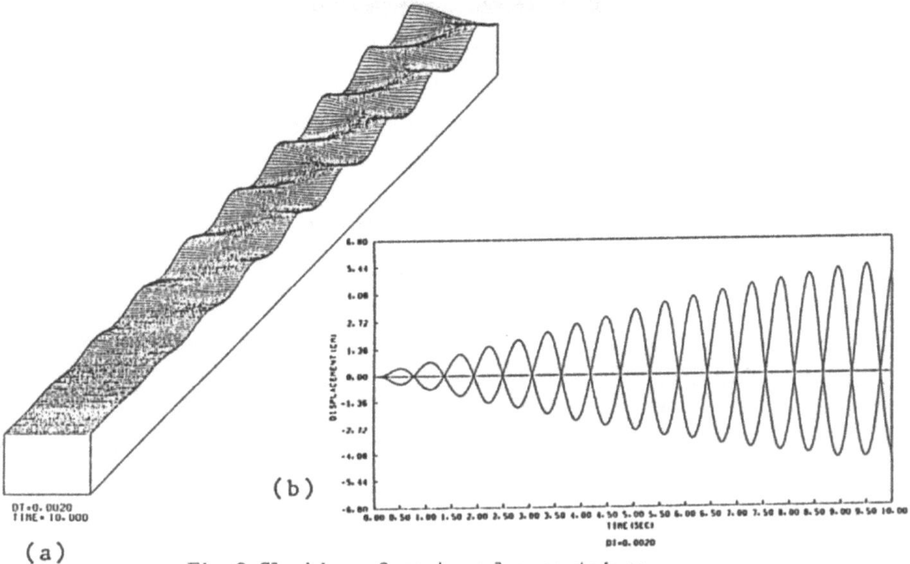

Fig.2 Sloshing of rectangular container.
(a) Growth of free surface.
(b) Time histories of $\eta(x,t)$ at x= 0, L.

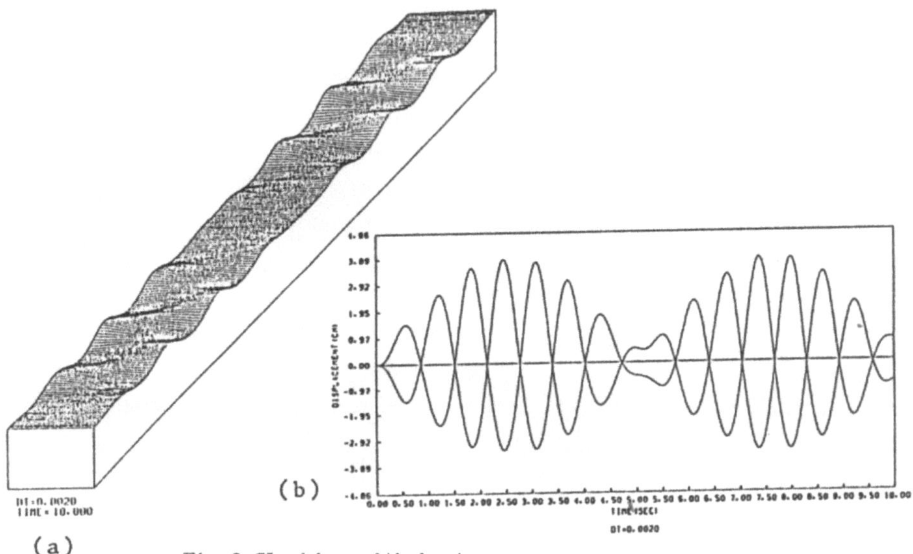

Fig.3 Sloshing with beat.
(a) Growth of free surface.
(b) Time histories of $\eta(x,t)$ at x= 0, L.

2. Sloshing in circular container

Next, let us consider liquid oscillation in a circular container subjected to a forced horizontal acceleration. In this calculation we adopt $R = 0.5$ m as the radius of container and $A = 10.0$ mm and $\omega = 5.0$ rad/sec as parameters on the foreced accleration.

The obtained free surface profiles at each time instant are shown in Fig.5.

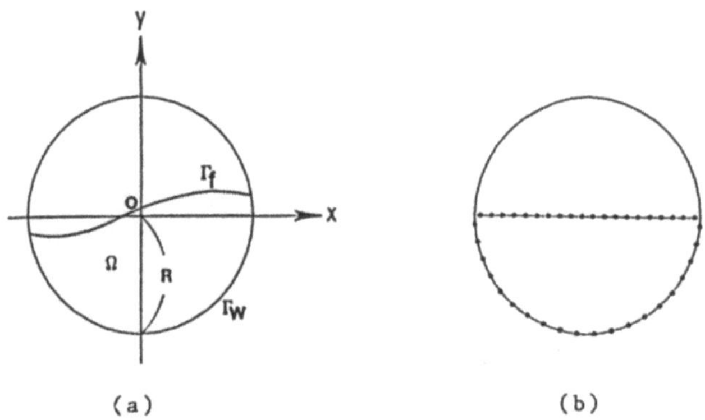

Fig.4 Sloshing in circular container.
(a) The geometry, (b) The boundary element mesh.

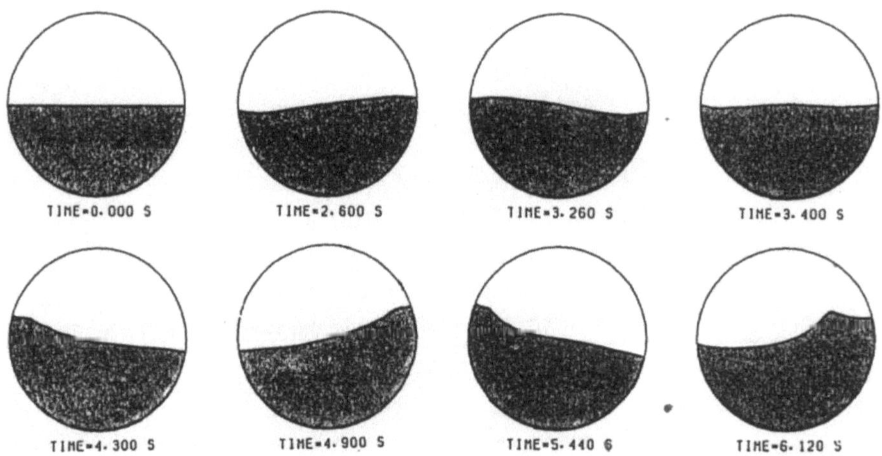

Fig.5 The motion of the fluid in circular container.

3. Motion of a solitary wave in a rectangular tank

As a third numerical example, we analyze the motion of a solitary wave in a rectangular tank of uniform depth shown in Fig.6. The tank has a width L and is filled with water to a height h in the stationary sate. A piston plate as a wave generator is located at the left-hand end of the tank. We assume that its forward movement follows the function $X_p(t)$ given by

$$X_p(t) = X_0 \tanh[\omega(t-t_c)] \qquad \text{for } 0 \leq t \leq T \tag{10}$$

in which X_0 equals the semistroke of the piston plate, ω and t_c are parameters characterizing the movement of the piston and the time T denotes the one at which the piston plate is stopped. Numerical calculations are carried out for the case given by numerical values of parameters of the wave generator, $L/h=18.0$, $X_0/h=0.6$, $\omega\sqrt{h/g}=0.5$, $t_c\sqrt{g/h}=5.0$ and $T\sqrt{g/h}=12.0$, and the element mesh shown in Fig.7. Fig.8 shows the time sequence of profiles of the solitary wave. In Fig.9 we show three phases of the motion of a solitary wave which are the generation, the propagation and run-up, and the reflection.

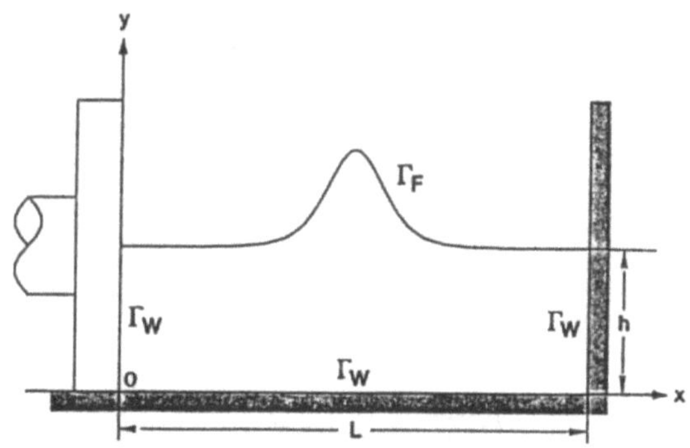

Fig.6 The water tank and wave generator.

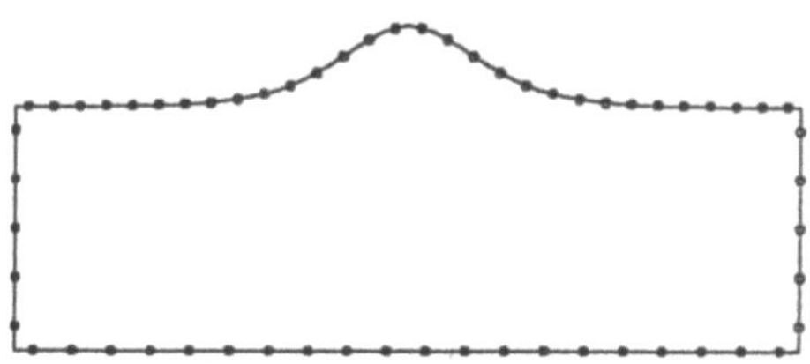

Fig.7 Boundary element mesh.

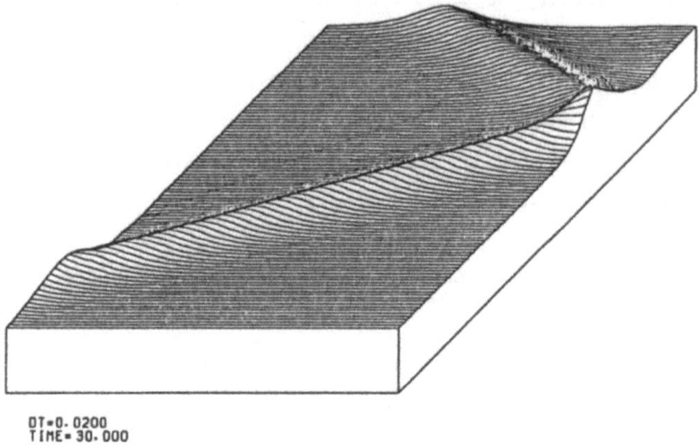

Fig.8 Profiles of a solitary wave.

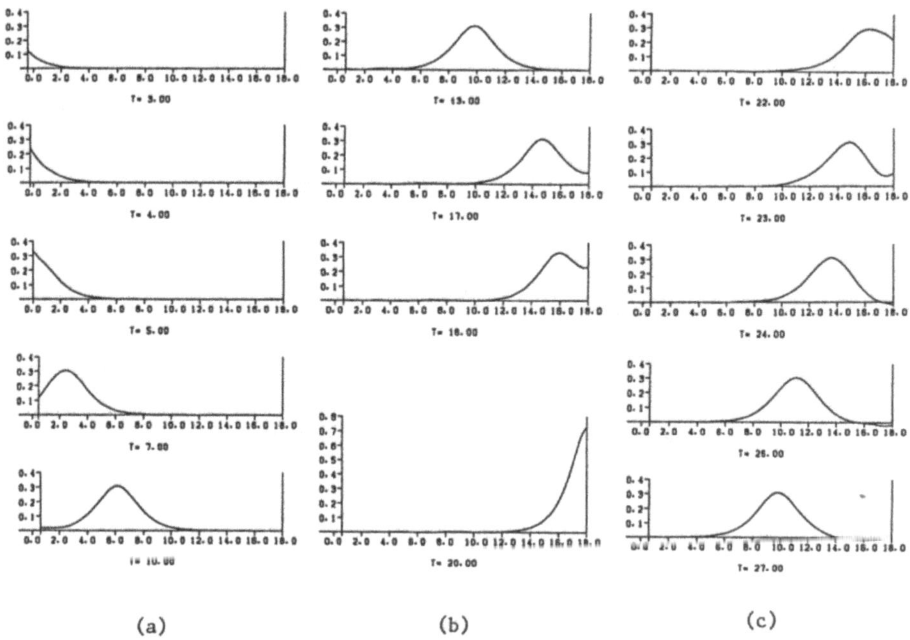

Fig.9 Three phases of the motion of a solitary wave.
(a) generation ; (b) propagation and run-up ; (c) reflection

Conclusions

The boundary element method with use of the manner of Lagrangian particle has been applied to numerical simulation of two-dimensional nonlinear water wave problems. The method is verified with nonlinear sloshing problems of container and the motion of a solitary wave in a rectangular tank. In solitary wave problem, three phases of the wave motion, which are the generation, the propagation and run-up on a vertical wall and the reflection, can be simulated by using our effective numerical solution procedure.

References

1. Ikegawa,M.: Finite element analysis of fluid motion in a container, Finite Element Methods in Flow Problems (eds.J.J.Oden et al.) UAH Press, pp.737-738,1974

2. Nakayama,T. and Washizu,K.: Nonlinear analysis of liquid motion in a container subjected to forced pitiching oscillation, Int.J.Num.Engrg,15 (1980) 1207-1200

3. Nakayama,T. and Washizu,K.: The boundary element method applied to the analysis of two-dimensional nonlinear sloshing problems , Int.J.Num. Mech.17 (1981) 1631-1646

4. Nakayama,T.: Boundary element analysis of nonlinear water wave problems, Int.J.Num.Mech.Engng.19 (1983) 953-970

5. Liu,P.L-F. and Liggett,J.A.: Boundary element formulations and solutions for some non-linear water wave problems, Developments in Boundary Element Methods-2 (Eds.P.K.Banerjee and R.P.Show), Applied Science, pp.37-67, 1982

6. Sugino,R. and Tosaka,N.: Boundary element analysis of non-linear free surface problems , Proceedings of Int.Symp.on Computational Fluid Dynamics Nagoya, pp.855-860, 1989

7. Sugino,R. and Tosaka,N.: Solution procedure for nonlinear free surface problems by boundary element approach, Theoretical and Applied Mechanics , Vol.38, University of Tokyo Press,(in press)

Unsteady Three-Dimensional Transonic Flow Computations Using Field Element Method

Hong Hu*
Hampton University
Hampton, Virginia 23668, U.S.A.

and

Li-Chuan Chu**
PRC Kentron Corp.
Hampton, Virginia 23665, U.S.A.

ABSTRACT

An unsteady field element (or called, integral equation) scheme for solving the full-potential equation has been developed for unsteady transonic wing flows. The unsteady full-potential equation is formulated in a moving frame of reference, and it is written in the Poisson's equation where unsteady term is included in nonhomogeneous term. Integral equation solution in terms of velocity field is obtained by the Green's theorem. The solution consists of a wing surface integral term of vorticity distribution, a wake surface integral term of free-vortex sheet and a volume integral term of compressibility and unsteadiness over a small limited domain around the wing. The solution is accomplished by a time-marching, iterative procedure. The Murman-Cole type-difference scheme is used to compute derivatives of density, and shocks are captured automatically. The scheme has been applied to the flow around a rectangular wing at transonic speeds undergoing acceleration motion. The time history of the wing surface pressure distributions has been presented.

INTRODUCTION

The problem of accurate and efficient prediction of the unsteady aerodynamic load over complex three-dimensional configurations at transonic speeds remains a challenging problem. Despite rapid advances in computing power available, the solution of the unsteady Navier-Stokes equations over complex three-dimensional configurations is still unrealistic. Experience has shown that rather accurate solutions can be obtained for many transonic flows using the inviscid modeling of the full-potential equation. For transonic flows without strong shocks and massive separations, the full-potential equation is an adequate approximation to the Navier-Stokes equations.

---

* Assistant Professor, Dept. of Mathematics
** Research Engineer, Aerospace Technology Division

Integral equation methods for transonic flows started few years ago. For steady, three-dimensional, inviscid transonic flows, integral equation solutions based on the full-potential equation formulation have been obtained for delta wing [1], aircraft [2], rectangular wing [3] and wing-body combination [4]. For unsteady, two-dimensional, inviscid transonic flows, an integral equation solution based on the full-potential equation formulation has been obtained for an airfoil undergoing pitching oscillation [5]. These solutions have been shown to be capable of treating transonic flows accurately and efficiently. Complex geometries were easily represented [2,4], since surface-conforming flow field grids were not used.

The objective of the present work is to combine the unsteady two-dimensional scheme [5] with the steady three-dimensional scheme [3], and extend them to unsteady three-dimensional transonic flows. In the present paper, the development of this unsteady three-dimensional scheme is presented.

FORMULATION OF THE PROBLEM

For a general unsteady motion of a body, the governing equations are simple to solve if a moving frame of reference formulation is used. This formulation does not require the grid-motion calculation since the grid is rigidly fixed in the moving frame of reference. In addition to the space-fixed frame of reference $OXYZ$, a moving frame of reference $oxyz$ is introduced as shown in Fig. 1. The moving frame of reference $oxyz$ is translating at a velocity of $\vec{V}_o(t)$ and rotating around a pivot point, $\vec{r}_p = (x_p, y_p, z_p)$, at an angular velocity of $\vec{\Omega}(t)$. The relation for the absolute velocity ($\vec{V}$), relative velocity ($\vec{V}_r$) and transformation velocity ($\vec{V}_o + \vec{\Omega} \times (\vec{r} - \vec{r}_p)$) is given by

$$\vec{V} = \vec{V}_r + \vec{V}_o + \vec{\Omega} \times (\vec{r} - \vec{r}_p) \tag{1}$$

where $\vec{r}$ is the position vector measured in the moving frame of reference.

The non-dimensional unsteady full-potential equation in the moving frame of reference has been derived as follows [6]:

$$\nabla^2 \Phi = G \tag{2}$$

with

$$G = G_1 + G_2 \tag{3}$$

$$G_1 = -\frac{\nabla \rho}{\rho} \cdot \vec{V}_r \tag{4}$$

$$G_2 = -\frac{1}{\rho}\frac{\partial' \rho}{\partial t} \tag{5}$$

and

$$\rho = \left\{1 + \frac{\kappa - 1}{2}\left[-V_r^2 + (\vec{V}_o + \vec{V}_e)^2 - 2\left(\frac{\partial' \Phi}{\partial t}\right)\right]\right\}^{\frac{1}{\kappa - 1}} \tag{6}$$

where the characteristic parameters, ρ_∞, a_∞ and l have been used; a is the speed of the sound, ρ the density, and l the length; Φ is the absolute velocity potential ($\vec{V} = \nabla\Phi = \nabla'\Phi$), G_1 the compressibility, G_2 the unsteadiness, κ the gas specific heat ratio, and t the time; the subscript, ∞, refers to infinity conditions; the prime ($'$) refers to the derivative with respect to the moving frame of reference. The boundary conditions are surface no-penetration condition, Kutta condition, wake kinematic and dynamic conditions and infinity condition.

Equation (2) is a Poisson's equation. By using the Green's theorem, the integral equation solution in terms of the absolute velocity field in the moving frame of reference is given by

$$\begin{aligned}\vec{V}(x,y,z,t) = {} & \frac{1}{4\pi}\int\!\!\int_g \frac{\vec{\gamma}_g(\xi,\eta,\zeta,t) \times \vec{d}}{d^3} ds \\ & + \frac{1}{4\pi}\int\!\!\int_w \frac{\vec{\gamma}_w(\xi,\eta,\zeta,t) \times \vec{d}}{d^3} ds \\ & - \frac{1}{4\pi}\int\!\!\int\!\!\int_V \frac{G(\xi,\eta,\zeta,t)}{d^2} \vec{e_d}\, d\xi\, d\eta\, d\zeta\end{aligned} \tag{7}$$

where $\vec{\gamma}$ is the surface vorticity distribution; the subscripts g and w refer to the wing and the wake surfaces, respectively; ds is the infinitesimal surface area; the vector $\vec{d}$ is given by $\vec{d} = (x - \xi)\vec{i} + (y - \eta)\vec{j} + (z - \zeta)\vec{k}$; and $\vec{e_d}$ is defined by $\vec{e_d} = \vec{d}/\mid\vec{d}\mid$.

COMPUTATIONAL SCHEMES

A sketch of the computational model with the computational volume size is shown in Fig. 2. The wing and its free-shear layer are represented by vortex sheets which are divided into triangular vortex panels. A rectangular parallelopiped type of cell is chosen to be used throughout the flow field. The solution procedure for solving unsteady three-dimensional flows follows the successful forms of the unsteady two-dimensional scheme [5] and the steady three-dimensional scheme [3], hence only a brief description is given. The solution procedure is a time-marching, iterative procedure. It can be described as follows:

Step 1: At Time Step ($n = 0$) - Steady Flow Computation:

In this step, $G_2^{(0)}$ and $(\partial'\Phi/\partial t)^{(0)}$ are set to be zero. Equations (3), (4), (6) and (7) with the boundary conditions are solved iteratively until the solution converges. Here, two loops are used. The inner loop is used to calculate and check the convergence of the nonlinear term, $G^{(0)}$. The outer loop is used to update and check the convergence of the wake shape and the wing surface pressure distribution. The details of this step are given in Ref. [3].

Step 2: At Time Step (n) - Unsteady Time March:

At this time step, $G_2^{(n)}$ is calculated numerically by a second-order accurate backward differencing, which is given by

$$G_2^{(n)} = -\left(\frac{1}{\rho}\frac{\partial'\rho}{\partial t}\right)^{(n)} = -\frac{1}{\rho^{(n)}}\frac{c_1\rho^{(n-2)} + c_2\rho^{(n-1)} + c_3\rho^{(n)}}{c_4} \tag{8}$$

where $c_1 = 1$; $c_2 = -[(\Delta t^{(n-1)} + \Delta t^{(n)})/\Delta t^{(n)}]^2$; $c_3 = c_2^2 - 1$; and $c_4 = -(\Delta t^{(n-1)} + \Delta t^{(n)}) - c_2\Delta t^{(n)}$. The time derivative term of the potential, $(\partial'\Phi/\partial t)^{(n)}$, can be numerically calculated by $\Phi^{(n)}$ and $\Phi^{(n-1)}$, and hence $\Phi^{(n)}$ and $\Phi^{(n-1)}$ must be calculated by integration of the velocity field numerically. In order to avoid numerical error when doing this numerical integration, Eq. (6) is used to compute $(\partial'\Phi/\partial t)^{(n)}$ distribution. Thus, Eq. (6) takes the form

$$\left(\frac{\partial'\Phi}{\partial t}\right)^{(n)} = \frac{1-\rho^{(n-1)^{\kappa-1}}}{\kappa-1} - \frac{1}{2}V_r^{(n)^2} + \frac{1}{2}(\vec{V}_o^{(n)} + \vec{V}_e^{(n)})^2 \tag{9}$$

With $G_2^{(n)}$ obtained from Eq. (8) and $(\partial'\Phi/\partial t)^{(n)}$ obtained from Eq. (9), Step 1 is repeated until the solution converges.

Step 3: At Time Step ($n + 1$):

Step 2 is repeated for time step ($n + 1$).

NUMERICAL EXAMPLE

Various cases for steady flows have been tested. The detailed results of those cases can be found in Ref. [3]. In order to improve the computational efficiency, a successive grid refinement technique is adopted. Figure 3 shows the comparison of the chordwise distributions of G and M_r for a steady flow case. As expected, when the grid size becomes finer the distributions of both values become sharper and stronger, and the shock region becomes narrower. For all the grid sizes the values of G and M_r drop asymptotically to zero and to M_o respectively. Moreover, the finer-grid values match the coarse-grid velues at the boundary of the reduced (finer-grid) computational domain. These results show clearly that this technique is feasible, efficient and capable of capturing sharp shock waves.

An unsteady transonic rectangular wing flow has been computed. In this numerical example, an zero-thickness rectangular wing with the aspect ratio of 2 is given an acceleration motion at an angle of attack of 5 degrees. The free stream Mach number, $M_o(t) = V_o(t)/a_\infty$, is given by $M_o(t) = 0.7+0.1t$, where t is given by $n\Delta t$; n is time step index (from 1 to 10) and Δt is the time-step size, which is chosen as 0.1. The half-span of the wing and wake is divided into 10×6 and 10×10 quadrilateral panels, respectively. Each quadrilateral panel consists of two triangular panels. One half of the computational volume is divided into $23 \times 9 \times 9$ volume cells in x, y and z directions, respectively. The time history of the surface pressure distributions over different wing across-sections, along with comparison of the steady flow solutions [3], is shown in Fig. 4. The results shows that this scheme can capture the shock and its motion correctly.

CONCLUDING REMARKS

An unsteady field element scheme for transonic wing flows has been developed. The scheme is capable of handling general unsteady motion, although it is applied to an acceleration motion in the present paper. The computational results show that the scheme is capable of capturing shock waves and their motion. This scheme is very stable, and the solution converges within 5 iterations for the first time step which starts from the uniform flow condition. The number of iterations required for convergence decreases with the increase of the time. The large time step ($\Delta t = 0.1$) allowed in the present scheme makes this scheme very efficient for unsteady computations. Further numerical tests and applications of this scheme are being carried out and compared with available experimental data and/or computational results to verify the accuracy of the scheme.

REFERENCES

1. Kandil, Osama A.; Yates, E. Carson, Jr.: Computation of transonic vortex flow past delta wings - integral equation approach. AIAA Journal, Vol.24, No.11, (1986), pp.1729-1736.

2. Erickson, L. L.; Madson, M. D.; Wou, A. C.: Application of the TranAir full-potential code to the F-16A. Journal of Aircraft, Vol.24, No.8, (1987), pp.540-545.

3. Chu, Li-Chuan: Integral equation solution of the full potential equation for three-dimensional, steady, transonic wing flows. Ph.D. Dissertation, Dept. of Mechanical Engineering and Mechanics, Old Dominion University, (1988).

4. Sinclair, P. M.: A three-dimensional field-integral method for the calculation of transonic flow on complex configurations - theory and preliminary results. Aeronautical Journal, June/July, (1988), pp.235-241.

5. Kandil, Osama A.; Hu, Hong: Unsteady transonic airfoil computations using the integral solution of full-potential equation. Boundary Elements X, Vol.2, edited by C. A. Brebbia. Springer-Verlag, (1988), pp.357-371.

6. Hu, Hong: Full-potential integral solutions for steady and unsteady transonic airfoils with and without embedded Euler domains. Ph.D. Dissertation, Department of Mechanical Engineering and Mechanics, Old Dominion University, Norfolk, Virginia, (1988).

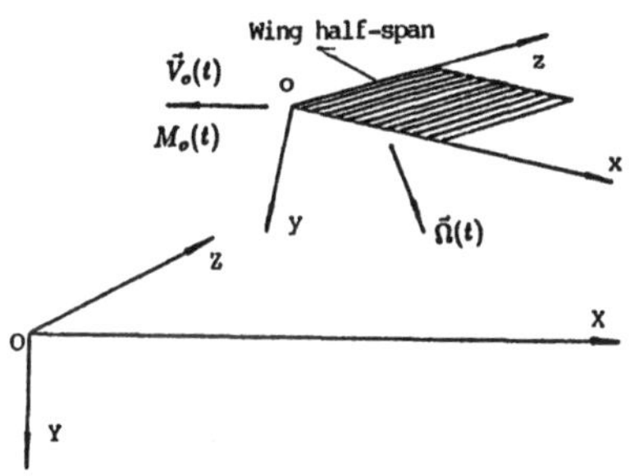

Fig. 1. Frames of reference

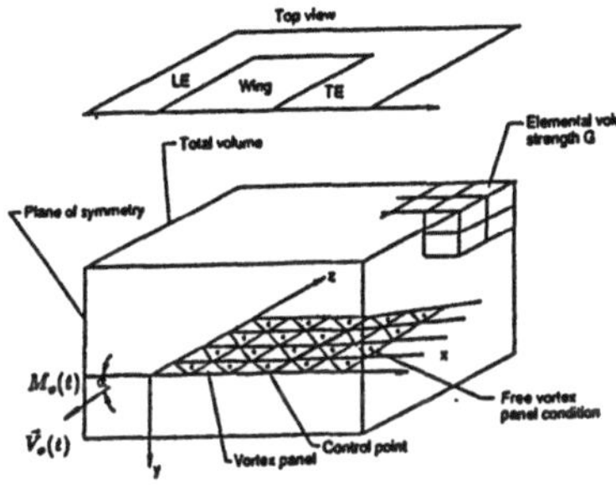

Fig. 2. Computational model

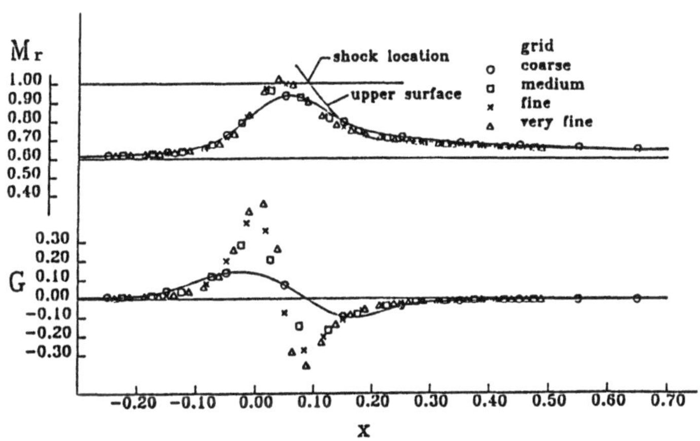

Fig. 3. Chordwise variation of G and M_r for $M_o = 0.6, \alpha = 10^o$, $z = 0.083, y = -0.042$ with various grid sizes

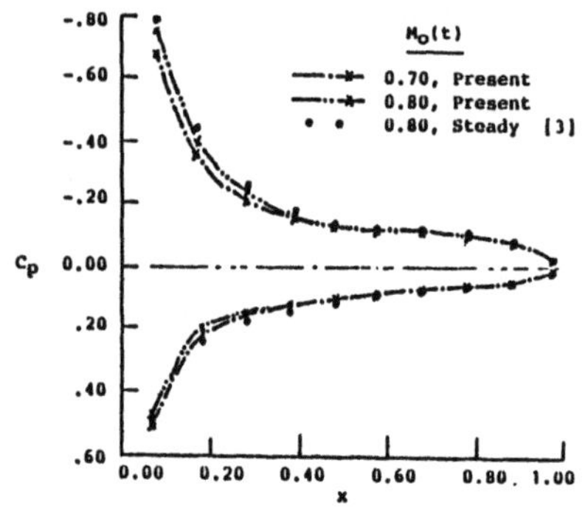

(a) Section I, z=0.111

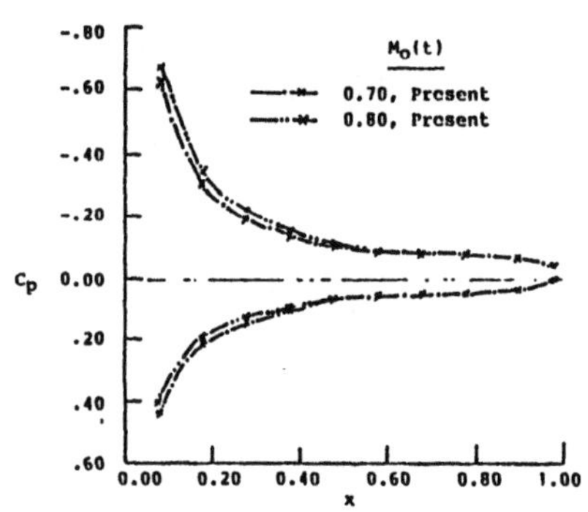

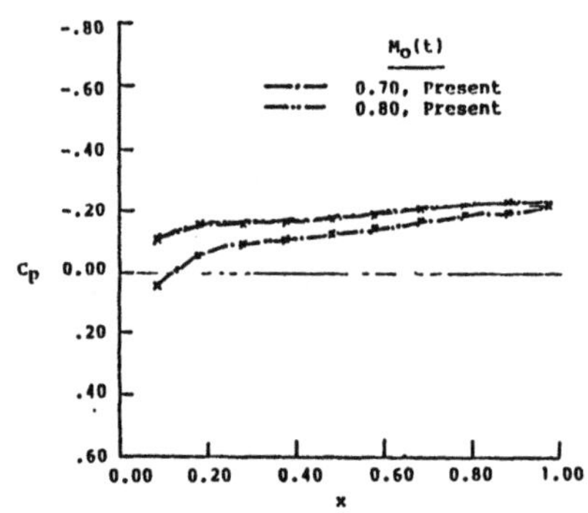

(b) Section II, z=0.444 (c) Section III, z=0.889

Fig. 4. Unsteady surface pressure distributions; rectangular wing, $AR = 2.0, \alpha = 5^o, M_o(t) = 0.70 + 0.1t$

Generalization of the Boundary Integral Method to Nonlinear Problems of Compressible Fluid Flow: The No-mesh Alternative – Part I

Barry Hunt and Ronald C. Plybon

GE Aircraft Engines, Cincinnati, Ohio, 45215

Summary

In this first part of the paper, we use a general identity for an arbitrary vector field to develop an integral equation relating the velocity in a compressible, rotational flow to a classical boundary integral plus field integrals involving the divergence and curl of the velocity. The field integrals giving the perturbation velocity due to this field source and vorticity are then reduced to standard boundary integral form, and then to a modified boundary condition for a pseudo-Laplacian problem. The numerical evaluation of these integrals is discussed.

1. The Mathematical Identity Underlying Field/Boundary Integral Methods

Consider an arbitrary 3-D domain V with boundary A. In the general case, this domain V may be considered to be the region of space external to a number of closed bodies and contained within some enclosing outer boundary. We confine our attention here to a single closed body in an infinite domain. The overall boundary A is then formed by the body boundary A_b plus an outer boundary A_∞ "at infinity." The unit normal vector $\hat{n}$, at any point on A, points into V.

Consider an arbitrary vector field $\underline{F}$ defined in V, and a second arbitrary vector field $\underline{F}'$ defined in the separate domain V′ in the interior of the body. We note that points in V′ are external to V, so that the vector $\hat{n}'$ pointing into V′ is equal to $-\hat{n}$. For an arbitrary point P in V, we have the following identities:

$$\underline{F}_P = \iint_A (\hat{n}\cdot\underline{F})\,\underline{K}\,dA + \iint_A (\hat{n}\times\underline{F})\times\underline{K}\,dA + \iiint_V (\mathrm{div}\,\underline{F})\,\underline{K}\,dV + \iiint_V (\mathrm{curl}\,\underline{F})\times\underline{K}\,dV \tag{1}$$

and

$$0 = \iint_A (-\hat{n}\cdot\underline{F}')\,\underline{K}\,dA + \iint_A (-\hat{n}\times\underline{F}')\times\underline{K}\,dA + \iiint_{V'} (\mathrm{div}\,\underline{F}')\,\underline{K}\,dV' + \iiint_{V'} (\mathrm{curl}\,\underline{F}')\times\underline{K}\,dV' \tag{2}$$

where $\underline{K}$ is the vector kernel function defined in 3-D as $\underline{r}_{QP}/4\pi r_{QP}^3$, and in 2-D as $\underline{r}_{QP}/2\pi r_{QP}^2$, with $\underline{r}_{QP}$ the position vector from a running point Q to the fixed point P. The symbol $\times$ denotes the vector product and $\cdot$ the scalar product of two vectors.

Let us express the vector $\underline{F}$ at the running point on A_b as the sum of two vectors, respectively normal and tangential to A_b:

$$\underline{F} = F_N\hat{n} + F_S\hat{s} \qquad \text{with} \quad F_N = \hat{n}\cdot\underline{F} \qquad \text{and} \quad F_S\hat{s} = \underline{F} - F_N\hat{n} \tag{3}$$

Confining our attention to the planar case, we define $\hat{s} \equiv \hat{j} \times \hat{n}$ where $\hat{j}$ is the unit vector oriented into the paper. We then have for $\underline{F}$:

$$\hat{n} \cdot \underline{F} = F_N; \qquad \hat{n} \times \underline{F} = F_S \hat{t}; \qquad \text{curl}\,\underline{F} = (\hat{t} \cdot \text{curl}\,\underline{F})\hat{t} \tag{4}$$

with similar relations for $\underline{F}'$. If the interior field $\underline{F}'$ is Laplacian, both div $\underline{F}'$ and curl $\underline{F}'$ are zero. Then, adding (1) to (2) gives:

$$\begin{aligned} \underline{F}_P = & \int_{A\infty} F_N \underline{K}\, dA + \int_{A\infty} F_S \hat{t} \times \underline{K}\, dA \\ & + \int_{Ab} (F_N - F_N') \underline{K}\, dA + \int_{Ab} (F_S - F_S') \hat{t} \times \underline{K}\, dA \\ & + \iint_V (\text{div}\,\underline{F}) \underline{K}\, dV + \iint_V (\hat{t} \cdot \text{curl}\,\underline{F}) \hat{t} \times \underline{K}\, dV \end{aligned} \tag{5}$$

where $(F_N - F_N')$, $(F_S - F_S')$ represent the normal and tangential components of the jump $(\underline{F} - \underline{F}')$ at any running point on A. We write $\sigma = (F_N - F_N')$ and $\omega = (F_S - F_S')$.

We now particularize to the case of compressible, rotational flow over a body in an infinite domain. $\underline{F}$ then becomes the velocity vector $\underline{Q}$ in the region external to the body. We assume that at every point on the outer boundary A_∞, $\underline{Q}$ in the presence of the body is identical to its unperturbed value, $\underline{Q}_\infty$. We assume also that the divergence and curl of the unperturbed field $\underline{Q}_\infty$ are zero. The integrals over A_∞ in (5) may then be interpreted as a representation of the case where the body is absent ($F_N' = Q_{N\infty}$ and $F_S' = Q_{S\infty}$). We can thus write:

$$\begin{aligned} \underline{Q}_P = \underline{Q}_\infty & + \int_{Ab} \sigma \underline{K}\, dA + \int_{Ab} \omega \hat{t} \times \underline{K}\, dA + \\ & \iint_V (\text{div}\,\underline{Q}) \underline{K}\, dV + \iint_V (\hat{t} \cdot \text{curl}\,\underline{Q}) \hat{t} \times \underline{K}\, dV \end{aligned} \tag{6}$$

When the "perturbed" velocity vector $\underline{Q}$ is also Laplacian, the field integrals in (6) vanish, leaving us with the general form of the familiar "indirect" boundary-integral equation for a Laplacian field. Various forms of "panel method" with different internal fields $\underline{F}'$, giving different jumps $(Q_N - F_N')$ and $(Q_S - F_S')$, are discussed in [1, 2, 3]. The terminology "indirect" introduced in [1] is used because $\underline{Q}$ is obtained indirectly, using the nonphysical functions σ and ω.

It is clear from (6) that both the nonphysical surface functions σ, ω and the "physical" spatial distributions div $\underline{Q}$, curl $\underline{Q}$, act like "singularity" distributions of source and vorticity, "inducing" a perturbation velocity field. This concept will be further developed in Part II.

We have thus progressed from the identity (1) to the integral equation (6) containing surface and volume singularity distributions. The field source and vorticity functions are considered "known"; they will be computed iteratively as functions of the local velocity, as discussed in Part II. The surface functions σ, ω are unknown; they will be computed from (6) by introducing boundary conditions.

The various types of boundary condition which can be applied at the boundary A_b are also discussed in [1, 2]. We shall confine our attention here to the case of Neumann boundary conditions, where the normal component of Q_P is prescribed at each boundary point P; say, $\hat{n}_P \cdot \underline{Q}_P = Q_{NP}$. The problem then reduces to finding the values of σ and/or ω at each point on the body boundary, which satisfy the following equation at each boundary point P:

$$\int_{Ab} \sigma \hat{n}_P \cdot \underline{K}\, dA + \int_{Ab} \omega \hat{n}_P \cdot \hat{t} \times \underline{K}\, dA = Q_{NP} - \hat{n}_P \cdot (\underline{Q}_\infty + \underline{Q}_F) \tag{7}$$

where

$$\underline{Q}_F = \iint_V (\text{div}\,\underline{Q}) \underline{K}\, dV + \iint_V (\hat{t} \cdot \text{curl}\,\underline{Q}) \hat{t} \times \underline{K}\, dV \tag{8}$$

It can be seen that this is equivalent to the standard form used for Laplace's equation, with the freestream velocity $\underline{Q}_\infty$ appearing in the r.h.s. of the latter now augmented by the velocity $\underline{Q}_F$ induced by the field distributions of source and vorticity. Thus if a panel method is available for solving the incompressible, irrotational flow problem (Laplacian field), that same method can be modified to solve the compressible, rotational flow problem.

To do this, we must establish a method of evaluating div $\underline{Q}$ and curl $\underline{Q}$; this will be deferred to Part II. We must also calculate the perturbation velocity $\underline{Q}_F$ induced by these distributions. We now address this second problem and show how the field integrals in (6) can be reduced to boundary integrals.

2. Reduction of the Field Integrals to Boundary Integral Form

2.1 Semi-Analytic Representation of the Field Source and Vorticity

Suppose we have evaluated div $\underline{Q}$ and curl $\underline{Q}$ and some number of their normal derivatives, at a set of points on the body boundary A_b. We shall see in Part II how these can be derived from (a current estimate of) the velocity vector $\underline{Q}$ and its tangential derivatives. The field equations used will be an idealized version of the compressible Navier-Stokes equation and a correspondingly idealized form of the physically dissipative energy equation.

We now wish to evaluate the two field integrals in (6). To reduce them to boundary integrals, we must introduce an approximation. We assume that we can replace the physical fields div $\underline{Q}$ and curl $\underline{Q}$ by semi-analytic source and vorticity distributions whose shape, as an analytical function of distance along some straight line roughly normal to A_b, is predefined. We denote by Σ the approximant of div $\underline{Q}$ and by Γ that of $\hat{\jmath} \cdot$ curl$\underline{Q}$.

These analytic shape functions are chosen to match the known value and normal derivatives of div $\underline{Q}$ and $\hat{\jmath} \cdot$ curl $\underline{Q}$ on the body boundary. In the present infinite-domain case, we choose functions which vanish at large distance. We consider here two types of function, the first suitable for low curvature regions (e.g. most of an airfoil surface, but not the nose), the second for regions which are approximately circular (e.g. an airfoil nose).

For low-curvature regions we select a reference plane A_{ref} near A_b; this could for example be an airfoil chord line. We construct a cartesian system (x, z), with x along A_{ref} and z normal to it, parameterizing P as (x_P, z_P). At each point on A_{ref} we define a set of "cartesian" shape functions $G_m(z)$:

$$G_m(z) = \exp(-c_m z) \tag{9}$$

with c_m a different positive constant for each m. The number of functions used depends on the number of normal derivatives to match on A_b.

For regions such as an airfoil nose, we select a circular reference surface A_{ref}, with center C and radius r_{ref} close to those of the local body surface A_b. We then construct a cylindrical polar system (r,θ), with r measured outwards from C and the angle θ positive clockwise about C. Any point P can then be parameterized as (r_P, θ_P). In this case we define a set of "polar" shape functions $G_m(r)$, again with c_m a different positive constant for each m:

$$G_m(r) = r^{-(c_m+2)} \tag{10}$$

For both the "cartesian" and "polar" case, we define the semi-analytic models $\Sigma(r, \theta)$ and $\Gamma(r, \theta)$ as a weighted sum of these shape functions. The weights $\Sigma_{ref,m}$, $\Gamma_{ref,m}$ (functions of θ defined on A_{ref} for each m) are chosen such that at each point P on the body boundary A_b (i.e. $r = r_P$), the value and normal derivatives of $\Sigma(r, \theta)$ and $\Gamma(r, \theta)$ match those supplied for the physically based values div $\underline{Q}$ and curl $\underline{Q}$. We confine our attention here to the canonical model, using only one value of m; it will then be possible to match only boundary values, and not normal derivatives. Dropping m, we thus write for this canonical case:

$$\Sigma(r, \theta) = \Sigma_{ref}\, G(r); \qquad \Gamma(r, \theta) = \Gamma_{ref}\, G(r) \tag{11}$$

with

$$\Sigma_{ref}(\theta) = \Sigma(r_P, \theta)/G(r_P) \quad \text{and} \quad \Gamma_{ref}(\theta) = \Gamma(r_P, \theta)/G(r_P) \tag{12}$$

We now wish to construct boundary source and vorticity functions on A_b, with an induced velocity field everywhere equal to that of Σ and Γ.

2.2 "Equivalent" Boundary Source and Vorticity Distributions

We have by simple rearrangement of (1), using the notation of (8):

$$\begin{aligned} \underline{Q}_F &= \iint_V (\text{div}\,\underline{F})\,\underline{K}\,dV + \iint_V (\text{curl}\,\underline{F}) \times \underline{K}\,dV \\ &= \underline{F} - \int_A (\hat{n}\cdot\underline{F})\,\underline{K}\,dA - \int_A (\hat{n}\times\underline{F}) \times \underline{K}\,dA \end{aligned} \tag{13}$$

This mathematical identity is valid for any vector field $\underline{F}$ defined in a domain V. It tells us that the velocity $\underline{Q}_F$ induced at any point in V, by field distributions of source of intensity div $\underline{F}$ and vorticity of intensity curl $\underline{F}$ in V, is equal to $\underline{F}$ plus the velocity induced by a surface source σ_F and a surface vorticity ω_F, defined at a running point Q on the boundary by:

$$\sigma_F = -\hat{n}_Q \cdot \underline{F} \quad \text{and} \quad \omega_F = -\hat{n}_Q \times \underline{F} \cdot \hat{\jmath} \tag{14}$$

It is thus clear that if we could construct a field $\underline{F}$ in V, with divergence and curl equal to those of the physical velocity field $\underline{Q}$, then the second line in (13) would be equal to the required velocity perturbation $\underline{Q}_F$ defined in (8). What we shall actually do is construct $\underline{F}$ with div $\underline{F}$ and curl $\underline{F}$ equal to the semi-analytic approximants (11). We shall see below that $\underline{F}$ will also be semi-analytic, with shape functions analogous to those of Σ, Γ.

The boundary on which the "equivalent" source and vorticity distributions σ_F and ω_F are located is the boundary A of the domain V, i.e. the closed body boundary A_b plus the boundary A_∞ at infinity. The shape functions chosen make σ_F and ω_F zero on A_∞. We thus need to place σ_F and ω_F only on A_b.

3. Construction of the Vector Field F

We shall discuss the construction of the field $\underline{F}$ for the "polar" case. The derivation is somewhat simpler for the "cartesian" case. In the cylindrical polar system (r, θ) based on a circle of radius r_{ref} centered at C, a vector $\underline{F}$ at some point P [i.e. (r, θ)] is defined in terms of its components F_r and F_θ as $\underline{F} = F_r\,\hat{r} + F_\theta\,\hat{\theta}$, where $\hat{r}$ is the unit vector oriented from C to P, and $\hat{\theta} = \hat{\jmath} \times \hat{r}$ is the unit circumferential vector at P. In this (r, θ) system, div $\underline{F}$ and curl $\underline{F}$ are defined by:

$$\text{div}\,\underline{F} = F_r/r + D_r\{F_r\} + D_\theta\{F_\theta\}/r \tag{15}$$

$$\hat{\jmath} \cdot \text{curl}\, \underline{F} = F_\theta / r + D_r\{F_\theta\} - D_\theta\{F_r\}/r \tag{16}$$

where the operator $D_r\{\}$ represents the partial derivative in the direction $\hat{r}$, and $D_\theta\{\}$ that in the direction $\hat{\theta}$, at the point (r, θ). Remembering we have used a shape function $r^{-(c+2)}$ for Σ and Γ, we postulate analogous forms for F_r and F_θ:

$$F_r(r,\theta) = F_{r,ref}\, r^{-(c+1)} \qquad F_\theta(r,\theta) = F_{\theta,ref}\, r^{-(c+1)} \tag{17}$$

Inserting (17) in (15), (16), replacing div $\underline{F}$ and curl $\underline{F}$ by their representations (11) and dividing the resulting equations by $r^{-(c+2)}$, we find that $F_{r,ref}$ and $F_{\theta,ref}$ must satisfy a coupled pair of first-order O.D.E's of the form:

$$D_\theta\{F_{\theta,ref}\} - c\, F_{r,ref} = \Sigma_{ref} \tag{18}$$

and

$$D_\theta\{F_{r,ref}\} + c\, F_{\theta,ref} = -\Gamma_{ref} \tag{19}$$

or differentiating (19) w.r.t. θ and substituting (18) into the result:

$$D_\theta^2\{F_{r,ref}\} + c^2 F_{r,ref} = -D_\theta\{\Gamma_{ref}\} - c\, \Sigma_{ref} = R, \text{ say.} \tag{20}$$

Equation (20) is a classical "forced simple-harmonic" equation. We shall discuss its solution in Section 4. We have thus shown that a vector field $\underline{F} = F_r \hat{r} + F_\theta \hat{\theta}$ in which F_r satisfies (20) and F_θ satisfies (19) at $r = r_{ref}$, will satisfy div $\underline{F} = \Sigma$ and $\hat{\jmath} \cdot$ curl $\underline{F} = \Gamma$. The values $F_r(r_P,\theta)$ and $F_\theta(r_P,\theta)$ on the body are then given by (17) with $r = r_P$. Finally, (14) yields σ_F and ω_F on A_b, which induce a velocity field identical to that induced by the semi-analytical models Σ and Γ of the physical field. We shall see in Section 4 how σ_F and ω_F may alternatively be used to define a modification to the boundary conditions for a pseudo-Laplacian panel method, analogous to the classical simulation of boundary layer displacement effects. The physical relevance of the various distributions appearing above will be discussed in Part II, where we shall also derive expressions for Σ and Γ, in terms of the local flow-speed, for a "physical" compressible flow.

4. Evaluation of the Overall Compressible Rotational Velocity Field

4.1 Construction of the Equivalent Boundary Source and Vortex Functions

We limit ourselves here to the "polar" case defined in Section 2; the "cartesian" case is somewhat simpler. We consider the physical domain V to be divided into a set of subdomains V_n. The initial and terminal radial lines bounding each V_n are chosen to coincide with a natural feature such as a trailing edge or a shock wave, or with the beginning or end of a "cartesian" or "polar" region as defined earlier. We shall now construct $\underline{F}$ in one such subdomain, in piecewise fashion. The part of the body boundary A_b in V_n is subdivided into a set of small panels, coinciding with those of the underlying panel method.

We consider the interval j of the reference circle, consisting of the radial projection of panel j onto the reference circle. We denote the value of θ at the front of interval j by θ_j, and employ a local measure θ' on interval j, defined as $\theta' = \theta - \theta_j$. Suppose the values of F_r and F_θ on the reference circle at the front of interval j (i.e. $\theta' = 0$) are known and defined by:

$$F_r(\theta_j) = F_{rj} \qquad \text{and} \qquad F_\theta(\theta_j) = F_{\theta j} \tag{21}$$

Suppose Σ_{ref} and Γ_{ref} are piecewise quadratic, and given on interval j by:

$$\Sigma_{ref} = \Sigma_0 + \Sigma_1 \theta' + \Sigma_2 \theta'^2; \qquad \Gamma_{ref} = \Gamma_0 + \Gamma_1 \theta' + \Gamma_2 \theta'^2 \tag{22}$$

Then $R(\theta')$ in (20) will also be piecewise quadratic. We introduce an analogous quadratic function $T(\theta')$, such that R and T can be written on interval j as:

$$R(\theta') = r_0 + r_1 \theta' + r_2 \theta'^2; \qquad T(\theta') = t_0 + t_1 \theta' + t_2 \theta'^2 \tag{23}$$

with

$$r_0 = -\Gamma_1 - c\,\Sigma_0; \qquad r_1 = -2\Gamma_2 - c\,\Sigma_1; \qquad r_2 = -c\,\Sigma_2 \tag{24}$$

$$t_0 = \Sigma_1 - c\,\Gamma_0; \qquad t_1 = 2\Sigma_2 - c\,\Gamma_1; \qquad t_2 = -c\,\Gamma_2 \tag{25}$$

The general solution of (20) is the sum of a complementary function and a particular integral. The former, $F_{\theta,CF}$, say, will satisfy the reduced form of (20), written using θ' as:

$$D^2_{\theta'}\{F_{r,CF}\} + c^2 F_{r,CF} = 0 \tag{26}$$

The solution of (26) at a point (r_{ref}, θ') on the reference circle is:

$$F_{r,CF}(r_{ref}, \theta') = K_1 \cos\phi + K_2 \sin\phi, \qquad \text{where } \phi = c\,\theta' \tag{27}$$

with K_1 and K_2 arbitrary constants. The field $\underline{F}_{CF}$ is Laplacian, with Σ and Γ both zero, inducing zero perturbation velocity. The perturbation-inducing part of $\underline{F}$ is the particular integral corresponding to the term R in (20). Inserting (23) in (20), a particular integral $F_{r,PI}$ of (20) can readily be seen to be:

$$F_{r,PI} = b_0 + b_1 \theta' + b_2 \theta'^2 \tag{28}$$

where

$$b_0 = r_0/c^2 - 2r_2/c^4; \qquad b_1 = r_1/c^2; \qquad b_2 = r_2/c^2 \tag{29}$$

Adding (27) to (28) to obtain F_r, then using (19) for F_θ, and selecting K_1 and K_2 to satisfy (21), we can write exact solutions for F_r and F_θ on interval j in the form:

$$F_{r,ref}(\phi) = F_{rj}\, f_c - F_{\theta j}\, f_s - \Gamma_0 f_1 + r_0 f_2 + r_1 f_3 + r_2 f_4 \tag{30}$$

$$F_{\theta,ref}(\phi) = F_{\theta j}\, f_c + F_{rj}\, f_s + \Sigma_0 f_1 + t_0 f_2 + t_1 f_3 + t_2 f_4 \tag{31}$$

where

$$\begin{aligned} f_c &= \cos\phi; \\ f_s &= \sin\phi; \\ f_1 &= [\sin\phi]/c; \\ f_2 &= [1 - \cos\phi]/c^2; \\ f_3 &= [\phi - \sin\phi]/c^3; \\ f_4 &= [\phi^2 - 2(1 - \cos\phi)]/c^4 \end{aligned} \tag{32}$$

Series expansions can be derived for each of these functions, in terms of ϕ, assumed small.

Equations (30) and (31) define F_r and F_θ on the reference circle. We can make $\underline{F}$ continuous in subdomain V_n by forcing the initial values F_{rj} and $F_{\theta j}$ on interval j to match the terminal values on interval j-1. The entire field $\underline{F}$ external to the reference circle can thus be constructed, starting from some assumed initial values F_{rI}, $F_{\theta I}$ at the front of the first interval j = 1, and ending with F_{rT}, $F_{\theta T}$ terminating the last interval. The value $\underline{F}$ at any point (r, θ) is then given by (17).

We have thus constructed $\underline{F}$ in a subdomain bounded by part of the reference circle, by the semi-infinite initial line I from the front of interval j = 1, by the semi-infinite line T terminating the last interval, and by part of the outer boundary A_∞. We are interested in $\underline{F}$ only in the "physical" region V_n external to A_b. We shall replace the velocity induced by Σ, Γ in V_n by that induced by distributions σ, ω on its boundary; i.e. on A_b, on the radial lines I and T, and on A_∞.

At a point Q on A_b, the functions σ_b, ω_b are given by (14) with $\underline{F}$ as in (17), the body normal $\hat{n}_Q$ pointing into V_n. On the initial line I, $\hat{n}_Q$ becomes the unit normal to I pointing into V_n, i.e. the vector $\hat{\theta}$ defined earlier. On line T, $\hat{n}_Q$ becomes $-\hat{\theta}$. On lines I and T, σ_I, ω_I and σ_T, ω_T decay in accordance with (17), and are zero on A_∞.

4.2 Evaluation of the Velocity Field Induced by the Boundary Functions

We have seen that the perturbation velocity due to the "physical" distributions div $\underline{Q}$ and curl $\underline{Q}$ may be approximated by that due to the model fields Σ and Γ, which is exactly equal to that induced by the boundary distributions σ_b, ω_b on A_b, σ_I, ω_I on the initial radial line I and σ_T, ω_T on the terminating radial line T.

The velocity due to σ_I, ω_I and σ_T, ω_T can be obtained analytically, with each line treated as a single, semi-infinite panel carrying analytic σ, ω. The velocity induced by σ_b, ω_b may be evaluated by using the "vector influence matrix" of the underlying panel method, though a word of warning is appropriate here. Most low-order panel methods are formulated such that σ, ω for a solid body are usually small, minimising errors due to curvature and source/vortex gradient. If the formulation is such that σ or ω is large, the resulting errors can also be large [1].

With the present method, σ_b and ω_b can take large values. Although these can be reduced by judicious choice of F_{rI}, $F_{\theta I}$ on the initial radial line I, the resulting numerical errors can still be large when a low-order panel method is employed. A considerable increase in accuracy can be obtained by using a higher order scheme with circular-arc panels carrying piecewise linear or quadratic σ, ω distributions. The influence functions can still be derived analytically. Extreme care is needed, moreover, with the end-point influences induced by σ, ω at the junction of the body boundary and the initial and terminal lines I and T.

4.3 Alternative Interpretation as Modified Boundary Conditions

We now propose an alternative formulation, almost eliminating the evaluation of extra boundary integrals. We first separate the contributions to the boundary integral in (13) into that from A_b and that from the radial lines I and T, which we designate as $\underline{Q}_\perp$. Inserting (13) in (6), using (14), replacing div $\underline{Q}$, curl $\underline{Q}$ by the approximants $\Sigma = \text{div}\,\underline{F}$, $\Gamma = \hat{\jmath}\cdot\text{curl}\,\underline{F}$, and writing $\sigma' = \sigma + \sigma_F$ and $\omega' = \omega + \omega_F$, we have for the overall velocity at a boundary point P in the subdomain V_n:

$$\underline{Q}_P = (\underline{Q}_\infty + \underline{Q}_\perp + \underline{F}) + \int_{A_b} \sigma' \underline{K}\, dA + \int_{A_b} \omega' \hat{\jmath} \times \underline{K}\, dA \qquad (33)$$

and (7) transforms into the following boundary integral equation:

$$\int_{Ab} \sigma' \hat{n}_P \cdot \underline{K}\, dA + \int_{Ab} \omega' \hat{n}_P \cdot \hat{T} \times \underline{K}\, dA = (Q_{NP} + \sigma_b) - \hat{n}_P \cdot (\underline{Q}_\infty + \underline{Q}_\perp) \tag{34}$$

It is thus apparent that the compressible, rotational flow equation (6) can now be interpreted as a pseudo-Laplacian boundary integral problem. The familiar: *"Use a Laplacian panel code to determine* σ, ω *which, in the presence of* $\underline{Q}_\infty$, *induce a resultant normal velocity* Q_{NP}," is replaced by: *"Use a Laplacian panel code to determine* σ', ω' *which in the presence of an onset flow* $(\underline{Q}_\infty + \underline{Q}_\perp)$ *induce a resultant normal velocity* $(Q_{NP} + \sigma_b)$."

The required velocity at any boundary point P will then simply be that from the Laplacian panel code with this modified onset flow and boundary condition, plus a correction equal to the value of $\underline{F}$ at P. Using (3) and (14), we can see that this restores the normal velocity to Q_{NP}, and increments by an amount F_S [see (3)] the tangential velocity component computed using the modified Laplacian code.

This generalizes Lighthill's proposal [4] for simulating boundary layer displacement by surface transpiration. The present technique eliminates the error in the latter due to omitting the tangential correction F_S, which can be significant near an airfoil trailing edge, affecting the lift prediction [5].

4.4 Synthesis of a Full-Domain Solution

The techniques in 4.2 and 4.3 apply to fields which are continuous throughout V_n. For such flows, we divide the entire domain V external to an airfoil into a set of adjacent subdomains V_n, with separate sets of subdomains for the upper and lower surfaces. For any subdomain V_n, the location of the initial radial line I and the value of $\underline{F}$ on that line, will be matched to those on the terminal line T of its preceding neighbor. A similar treatment will be applied between the first subdomains on the upper and lower surfaces. The values σ_I, σ_T and ω_I, ω_T on adjacent initial and terminal lines then cancel, leaving only those on the terminal lines of the last subdomains on the upper and lower surfaces. These will generally have coincident geometries but different σ_T, ω_T. Thus the overall contribution $\underline{Q}_\perp$ from the initial and terminal radial lines of all subdomains V_n collapses to the single contribution from this composite radial line at the rear of the body.

Shock discontinuities must be excluded from the subdomains V_n, and treated explicitly as a surface source distribution. The velocity perturbation induced by this explicit source distribution is incorporated into the definition of $\underline{Q}_\perp$ in (33) and (34). It is then still entirely proper to initialize the fictitious field $\underline{F}$, on the initial line I of the subdomain immediately downstream of this physical discontinuity, with the terminal value of that immediately upstream. This is analogous to treating a wake from a lifting wing as a vortex sheet embedded in an irrotational flow, its strength being determined iteratively. Further discussion of this shock treatment, and of that of the Σ, Γ fields, follows in Part II.

References

These appear at the end of Part II of the paper.

Generalization of the Boundary Integral Method to Nonlinear Problems of Compressible Fluid Flow: The No-Mesh Alternative – Part II

Barry Hunt and Ronald C. Plybon

GE Aircraft Engines, Cincinnati, Ohio, 45215

Summary

In this second part of the paper, we develop equations defining the field source and vorticity in terms of the local flow speed Q, for an idealized form of the compressible Navier-Stokes equation and dissipative energy equation. These equations naturally suppress the formation of expansion shocks, while indicating the incipient formation of compression shocks. We also propose an explicit treatment for compression shocks. We conclude with a brief discussion of the intuitive coupling between the boundary integral method and a knowledge-based system approach.

Note: The numbering sequence in this second part continues from Part I.

5. Source and Vorticity Distributions in a Physical Flow

5.1 Physical interpretation

We saw in Part I that div $\underline{Q}$ and curl $\underline{Q}$ act like source and vortex distributions, "inducing" a perturbation velocity field $\underline{Q}_F$ defined by (8). They are analogous to distributions and induced fields from other branches of physics very familiar to engineers. The source/velocity relation compares with Coulomb's law (electric charge and electric field); the Biot-Savart vorticity/velocity relation compares with Ampere's law (electric current and magnetic field).

In the general case, the surface functions σ, ω have no physical meaning. Note, however, that if we choose the internal field $\underline{F}'$ in (5) to be uniformly zero, then the surface source function ($\sigma = Q_N - F_N'$) vanishes if boundary A_b is solid, i.e. $Q_N = 0$. We then have only a surface vortex distribution, $\omega = Q_S$, which may be interpreted as the limit, as the viscosity μ goes to zero, of the local vorticity integral across a vanishingly thin boundary layer, across which the tangential velocity changes quasi-discontinuously from zero (i.e. $F_S' = 0$) on its inner face, to Q_S at its outer edge. This interpretation does not apply for other types of panel method. Further discussion appears in [3]. In a viscous flow, curl $\underline{Q}$ has direct physical interpretion as the vorticity in the boundary layer. Here, we consider only an idealized flow in which such shear effects are absent. Vorticity is limited to that generated by a shock.

The relevance of "divergence" is perhaps less intuitive. It arises from the fact that the volume of a given mass of fluid varies inversely with the density. In an integral formulation, we in effect model a compressible flow as an incompressible one in which fluid is locally injected or removed by local source or sink distributions. The source intensity is such that the mass of fluid in a given elemental volume in the incompressible flow is the same as that in the compressible flow.

5.2 Field Source in an Idealized Compressible, Rotational Flow

In a flow at high Reynolds number Re, the second viscosity coefficient λ plays a major role in the formation of compression shocks and suppression of expansion shocks, while the first coefficient μ plays only a secondary role, if we ignore shock/boundary-layer interaction. Here, we are not interested in shock structure, only in fixing its location. When this and the shock entry conditions are known, the Rankine-Hugoniot relations fix the shock strength and the entropy downstream.

We thus define an idealized steady flow, as follows. First, we hypothesize that the second coefficient of viscosity takes the value $\lambda = -\frac{2}{3}\mu$ consistent with a Rankine-Hugoniot shock [6]. We then express λ via $\lambda = -\frac{2}{3}\rho_\infty Q_\infty L/Re$, where ρ_∞ and Q_∞ are freestream density and speed, and L is a reference length. Second, having thus fixed λ, we now define our idealized fluid to have a first (shear) viscosity coefficient μ of zero. Thirdly, we define our idealized fluid to have zero thermal conductivity. Finally, we assume the fluid to be an ideal gas, with constant specific heats c_p and c_v of ratio γ and gas constant $R = c_p - c_v$. Thus idealized, the equations of momentum and energy [7] become, respectively:

$$\rho \underline{Q} \cdot \text{grad}\, \underline{Q} = -\text{grad}\, p + \text{grad}\, (\lambda\, \text{div}\, \underline{Q}) \tag{35}$$

and

$$c_v\, \rho\, \underline{Q} \cdot \text{grad}\, T = -p\, \text{div}\, \underline{Q} + \lambda\, (\text{div}\, \underline{Q})^2 \tag{36}$$

where p = pressure, ρ = mass density, and T = temperature. The last term in (36) is the "idealized" form of the viscous dissipation function. Taking the scalar product of (35) with $\underline{Q}$, noting that in 2-D we have $\underline{Q} \cdot \text{grad}\, \underline{Q} = \text{grad}\, \frac{1}{2} Q^2$, gives:

$$\underline{Q} \cdot \text{grad}\, p = -\rho\, \underline{Q} \cdot (\text{grad}\, \tfrac{1}{2} Q^2) + \lambda\, \underline{Q} \cdot \text{grad}\, \text{div}\, \underline{Q} \tag{37}$$

Mass conservation and the equation of state ($p = \rho RT$) respectively tell us:

$$\rho\, \text{div}\, \underline{Q} = -\underline{Q} \cdot \text{grad}\, \rho \tag{38}$$

and

$$c_v\, (\gamma - 1)\, \rho\, \underline{Q} \cdot \text{grad}\, T = \underline{Q} \cdot \text{grad}\, p - (\underline{Q} \cdot \text{grad}\, \rho)\, p/\rho \tag{39}$$

We introduce the Mach number $M = Q/a$, with the sonic speed a given by $\gamma p/\rho = a^2$, since on one side of a shock the flow is adiabatic and isentropic along a streamline. Inserting (36), (37), and (38) in (39) then gives, after some manipulation:

$$\text{div}\, \underline{Q} - \varepsilon\, D_s\{\text{div}\, \underline{Q}\} = M^2\, D_s\{Q\} - \varepsilon(\gamma - 1)(\text{div}\, \underline{Q})^2/Q \tag{40}$$

where $\varepsilon = -\lambda Q/\rho a^2 = \frac{2}{3} L\, M^2/Re\, \rho' Q'$, with $\rho' = \rho/\rho_\infty$ and $Q' = Q/Q_\infty$ and where $D_s\{\} = \underline{Q} \cdot \text{grad}\{\}/Q$ indicates partial differentiation along a streamline.

Note that if we take $\lambda = 0$, the second terms on each side of (40) vanish, leaving the relation for the (inviscid) Euler equation: $\text{div}\, \underline{Q} = M^2\, dQ/ds$. There are two important differences between (40) and the latter. First, since ε is a positive quantity, the l.h.s. of (40) gives the field source div $\underline{Q}$ at a point located a (variable) distance upstream of where the r.h.s. is evaluated. This offers a physical basis for "upwinding," as often applied in transonic Euler computations. Secondly, if an expansion shock were to attempt to form (steep positive velocity gradient, making div $\underline{Q}$ large and positive), the final term in (40) would become large and tend to make div $\underline{Q}$ less positive at the next iteration. This tends to suppress expansion shocks and, similarly, encourage compression shocks.

Equation (40) is the complete form for computing div $\underline{Q}$ and includes a physical mechanism for capturing shocks. In practice, the reduced Euler form will suffice for shock-free regions of the flow. The Mach number appearing in (40) is given by $M = Q/a$, where for a general point the local sonic speed a is given by:

$$a^2 = \tfrac{1}{2}(\gamma - 1)(Q^2_{max} - Q^2) \qquad \text{with} \qquad Q^2_{max} = Q^2_\infty \, [1 + 2/(\gamma - 1)M^2_\infty] \tag{41}$$

where M_∞ is freestream Mach number and Q_{max} is the maximum speed attainable in a flow with constant total temperature T_t, which we shall henceforth assume. Different relations are needed upstream and downstream of a shock for the density ρ appearing in (40). We shall first discuss an explicit shock treatment.

5.3 Rankine-Hugoniot Shock in Idealized Flow

Although (40) contains a physical mechanism for capturing compression shocks, the absence of thermal conductivity and shear viscosity prevents shocks "captured" in the solution of (40) from being fully realistic. On a more practical note, the streamline derivatives $D_s\{Q\}$ and $D_s\{\text{div}\,\underline{Q}\}$ could never be evaluated numerically inside a shock, because at Re values of interest the shock thickness is orders of magnitude smaller than any practicable "panel" length. Hence, in relation to shock modeling, we shall exploit (40) only to detect and locate compression shocks and to suppress expansion shocks.

Consequently, we are at liberty to fix ε in (40) by choosing a "local Re" which will cause the terms in (40) containing ε to take a significant value only when $D_s\{\text{div}\,\underline{Q}\}$ or $(\text{div}\,\underline{Q})^2$ become locally large, indicating an incipient compression or expansion shock. Elsewhere, we can set $\varepsilon = 0$. An incipient compression shock will be indicated by a large negative value of $D_s\{Q\}$. Once detected, we shall then model the shock explicitly as a sink surface producing a (negative) jump in the velocity component normal to that surface, consistent with the Rankine-Hugoniot value. This explicit shock will be located at the point in the incipient shock region where the Mach number M is decreasing, and passes through $M = 1.0$.

The shock strength is a function of M_1, the value of M on its upstream face. This value is not available directly in the solution with the incipient shock. We can estimate M_1 by extrapolating downstream onto the fitted shock, using values of M at points upstream of it in the incipient solution. In the next iteration (now also using the velocity perturbation from the explicit shock) the solution of (40) will produce a residual incipient shock to supplement the current explicit one, giving a new estimate of shock location. As successive iterates converge, we can use progressively shorter-range extrapolations to estimate M_1.

To convert Q to thermodynamic properties at boundary points downstream of a shock, we first recall the Rankine-Hugoniot jump conditions. Using subscripts 1 and 2 to denote the upstream and downstream shock faces, we first define the shock normal velocities $Q_{n1} = \hat{n}_s \cdot \underline{Q}_1$ and $Q_{n2} = \hat{n}_s \cdot \underline{Q}_2$, with $\hat{n}_s$ the unit normal to the shock surface, pointing downstream. The shock-normal Mach number is $M_1 = Q_{n1}/a_1$ on face 1, with the sonic speed defined generally by (41). We then introduce notation to indicate the ratio of properties across the shock: $Q_{21} = Q_{n2}/Q_{n1}$, $\rho_{21} = \rho_2/\rho_1$, $p_{21} = p_2/p_1$. For a Rankine-Hugoniot shock, these are given by:

$$Q_{21} = (1 + A\,M_1^2)/(B\,M_1^2); \qquad \rho_{21} = 1/Q_{21}; \qquad p_{21} = (\gamma\,M_1^2 - A)/B \tag{42}$$

with $A = \tfrac{1}{2}(\gamma - 1)$ and $B = \tfrac{1}{2}(\gamma + 1)$. The shock source intensity is then given by:

$$\sigma_s = Q_{n1}\,(Q_{21} - 1) \tag{43}$$

Expressions for the derivatives of σ_s along the shock will not be developed here. We recall that Q_{n1} has a theoretical maximum Q_{max} defined in (41), or roughly $3Q_\infty$ for transonic flows. Thus, σ_s has a maximum of $-2Q_{max}/(\gamma + 1)$, giving Q_{n2} a finite positive value of of $Q_{max}(\gamma - 1)/(\gamma + 1)$, or approximately $\frac{1}{2}Q_\infty$ for transonic flows.

Remembering that the total temperature T_t remains constant across the shock, the ratio p_{t21} of the total pressures across the shock must be equal to the ratio of the total densities ρ_{t21}. It is easy to show that:

$$p_{t21} = \rho_{t21} = [\rho_{21}^{\gamma} / p_{21}]^{1/(\gamma - 1)} \tag{44}$$

Since the total pressure, density, and temperature are conserved along a streamline downstream of a shock, it is easy to show that at any boundary point downstream of a shock the relative pressure and relative density are given by:

$$p' = p/p_\infty = p_{t21}\, T'^{\,\gamma/(\gamma - 1)} \qquad \text{and} \qquad \rho' = \rho/\rho_\infty = p'/T' \tag{45}$$

where the relative temperature T' can readily be shown to be given by:

$$T' = T/T_\infty = 1 + \tfrac{1}{2}(\gamma - 1)M_\infty^2(1 - Q'^2) \qquad \text{with} \qquad Q' = Q/Q_\infty \tag{46}$$

The same expressions apply for points upstream of the shock, but with $p_{t21} = 1.0$.

5.4 Field Vorticity Downstream of a Shock

In the absence of shear viscosity, vorticity will be created only in the process of traversing a shock. For Rankine-Hugoniot shocks, it can be shown that the vorticity curl $\underline{Q}|_2$ near the body surface immediately downstream of the shock is given by:

$$\hat{t} \cdot \text{curl}\, \underline{Q}|_2 = \kappa_1 \sigma_s (1 - M_1^2) \tag{47}$$

where κ = surface curvature. To evaluate this at boundary points downstream, we note that in an isentropic flow, the quantity (curl $\underline{Q}$)/p, where p = pressure, is conserved along a streamline (circulation is not conserved, as the flow is not barotropic). If the flow stays attached, the vorticity at boundary points downstream of the shock is thus:

$$\text{curl}\, \underline{Q} = (\text{curl}\, \underline{Q}|_2)p/p_2 = (\text{curl}\, \underline{Q}|_2)\, p'/(p'_1 p_{21}) \tag{48}$$

Certain important implications follow from (47) and (48). First, if Q_1 approaches its theoretical maximum Q_{max}, then curl $\underline{Q}|_2$ will become indefinitely large and positive, acting so as to reduce the speed of the approaching flow. Second, since p usually increases sharply near the trailing edge of an airfoil, (48) tells us that the vorticity streaming from the shock will do likewise, possibly enough to cause the flow velocity to become locally negative at some boundary point downstream of the shock, implying an inviscid shock-induced separation from the smooth surface. At a convex angular discontinuity, the flow speed in an attached flow, which would be infinite in the Laplacian case, will tend towards Q_{max}, giving the maximum shock strength σ_s and infinite vorticity immediately downstream, leading to a separation from that salient edge. This argument, which will not be pursued any further here, offers another interpretation of the Kutta condition at a trailing edge, usually attributed solely to the effects of shear viscosity.

canonical case in which only the values of div $\underline{Q}$ and curl $\underline{Q}$ are matched at the body boundary, but the arguments proposed extend readily to also matching a number of their normal derivatives.

We have also confined our attention to the case of an "inviscid" flow, retaining the expansion viscosity only to capture compression shocks and suppress expansion shocks. The resulting formulation offers a physical basis to the artificial viscosity and upwinding schemes used elsewhere. The general concepts applied here to compressible flows are also valid for viscous flows, both attached and stalled [3, 5].

The analysis developed here for the 2-D case is in fact fundamentally 3-D. Panel methods are now applied routinely in the airframe industry for the treatment of 3-D incompressible vortical flows over complex configurations, wherein the concept of a free vortex sheet takes on a physically intuitive meaning [8, 9]. The formulation proposed herein offers the prospect of extending the scope of the existing methods to treat the same problems in transonic flow, without significant increase in computational cost or difficulty of use. It is also shown in [9] how standard 3-D panel methods for steady flow can be adapted, with minimal effort, to treat configurations in uniform angular motion. Finally, [8, 9] show how a 3-D flow over a wing or blade can be decomposed exactly into a coupled set of pseudo-2D computations, with a significant increase in accuracy and decrease in computational cost.

We conclude by observing that a panel method with the above set of enhancements falls very naturally into the category of "knowledge-based systems" emerging from the field of Artificial Intelligence. Decisions such as whether a 3-D wing computation can be decomposed into a set of pseudo-2D problems, or whether a rapidly changing flow speed is to be interpreted as an incipient shock, require knowledge: some element of experience is required, beyond mere algorithms.

The knowledge-based approach also allows the user of the system to be involved in the modeling and the computation, to an unprecedented extent: the user can query the system, to gain information on why some particular conclusion was reached, or be offered an explanation of the underlying physics, or advised on the limitations of the model used. He can interactively be offered the opportunity to override defaults built into the system, or to make a selection from various possibilities recognized by the system during execution.

Perhaps most importantly, such a system can automate the bookkeeping involved in specifying the input to a complex computation, especially when a multiplicity of disciplines is involved – for example in fluid/structural interaction. A combination of all these elements offers the prospect of putting realistic computations, with acceptable accuracy, cost and time, on the designer's desk.

References

[1] Hunt, B.: The panel method for subsonic aerodynamic flows: a survey of mathematical formulations and numerical models and an outline of the new BAe scheme. VKI Lecture Series 1978-4.

[2] Hunt, B.: The mathematical basis and numerical principles of the boundary integral method for incompressible potential flow over 3D aerodynamic configurations. In: Numerical Methods in Applied Fluid Dynamics; Hunt, B. (ed.): Academic Press, London, 1980.

[3] Hunt, B., Hewitt, B.L.: The indirect boundary integral formulation for elliptic, hyperbolic and nonlinear fluid flows. In: Developments in Boundary Element Methods - 4; Banerjee, P.K., Watson, J.O. (eds.): Elsevier Applied Science Publishers, London, 1986.

[4] Lighthill, M.J.: On displacement thickness; J. Fluid Mech., 4, 383 (1958).

[5] Hunt, B.: Viscous/inviscid coupling; unpublished presentation to BAe and RAE, Pall Mall, London, January 1983.

[6] Dulikravich, G.S., Mortara, K., Marraffa, L.; Physically consistent models for artificial dissipation in transonic potential flow computations. AIAA paper 88-3653, 1988.

[7] Hughes, W.F., Brighton, J.A.: Schaum's outline of theory and problems of fluid dynamics. Schaum's Outline Series; McGraw-Hill, Inc., New York, 1967.

[8] Hunt, B.; Recent and anticipated advances in the panel method: the key to generalised field calculations? VKI Lecture Series 1980-5.

[9] Hunt, B. The role of computational fluid dynamics in high-angle-of-attack aerodynamics. AGARD Lecture Series 121, 1982.

Unsteady Field Element Method with Embedded Euler Domain for Transonic Pitching Airfoil Flows

Hong Hu*
Hampton University
Hampton, Virginia 23668, U.S.A.

and

Osama A. Kandil**
Old Dominion University
Norfolk, Virginia 23529, U.S.A.

ABSTRACT

An equation zonal scheme for unsteady transonic flow computations has been developed. In this scheme, the unsteady Euler equations in a moving frame of reference are solved in a small embedded domain (inner domain) around the airfoil using an implicit approximately factored finite-volume method. The unsteady full-potential equation in the same frame of reference is solved in the outer domain using the integral equation method to provide the boundary conditions for the inner Euler domain. The solution procedure is a time-marching procedure, where the outer boundary conditions are updated using the integral equation solution. The equation zonal scheme is applied to unsteady transonic flow around a pitching airfoil. The results have been compared with those of an implicit Euler equations solver throughout.

INTRODUCTION

For unsteady transonic flows, the unsteady Euler equations adequately model most of the flow features with the exception of viscous effects near solid walls, where they are substantial. Successful time-accurate solutions of the unsteady Euler equations have been obtained for a pitching airfoil [1]. However, the computation of the unsteady Euler equations in a large computational domain, whose outer boundary extends twenty to thirty airfoil chord length, is expensive. On the other hand, the integral equation (IE) solution of the potential equation represents an alternative to the finite-difference or finite-volume solution of the Euler equations for inviscid flows without shocks or with weak shocks. Recently, an integral equation solution of the full-potential equation has been

* Dept. of Mathematics, Assistant Professor
** Dept. of Mechanical Engineering and Mechanics, Professor

obtained for unsteady inviscid transonic flows [2]. Also, an integral equation with embedded Euler domain scheme has been developed for steady inviscid transonic flows with strong shocks [3]. In the latter method, the unsteady Euler equations were solved by using a central-differencing finite-volume scheme in a small embedded domain around the shock, while the integral equation solution of the full-potential equation was used in the region outside the Euler domain. This method has been shown to be efficient and accurate in treating transonic flows with strong shocks. The advantages of the small computational domain and coarse grid associated with the integral equation method have been utilized in that method. For unsteady transonic flows with strong shocks, this zonal scheme is highly desirable for further developments. In the present paper, the development of the scheme is presented, and the resulting scheme is applied to a NACA 0012 airfoil undergoing pitching oscillation.

FORMULATION OF THE PROBLEM

For a general unsteady motion of a body, the governing equations are simple to solve if a moving frame of reference formulation is used. This formulation does not require the grid-motion calculation since the grid is rigidly fixed in the moving frame of reference. In addition to the space-fixed frame of reference $OXYZ$, a moving frame of reference $oxyz$ is introduced as shown in Fig. 1. The moving frame of reference $oxyz$ is translating at a velocity of $\vec{V}_o(t)$ and rotating around a pivot point, $\vec{r}_p = (x_p, y_p, z_p)$, at an angular velocity of $\vec{\Omega}(t)$. The relation for the absolute velocity ($\vec{V}$), relative velocity ($\vec{V}_r$) and transformation velocity ($\vec{V}_o + \vec{\Omega} \times (\vec{r} - \vec{r}_p)$) is given by

$$\vec{V} = \vec{V}_r + \vec{V}_o + \vec{\Omega} \times (\vec{r} - \vec{r}_p) \tag{1}$$

where $\vec{r}$ is the position vector measured in the moving frame of reference.

Unsteady Euler Equations

The two-dimensional, conservative, unsteady Euler equations in the moving frame of reference are given by [1]

$$\frac{\partial' \vec{q}_r}{\partial t} + \frac{\partial' \vec{E}_r}{\partial x} + \frac{\partial' \vec{F}_r}{\partial y} = \vec{S} \tag{2}$$

where

$$\vec{q}_r = [\rho, \rho u_r, \rho v_r, \rho e_r]^t \tag{3}$$

$$\vec{E}_r = [\rho u_r, \rho u_r^2 + p, \rho u_r v_r, \rho u_r h_r]^t \tag{4}$$

$$\vec{F}_r = [\rho v_r, \rho u_r v_r, \rho v_r^2 + p, \rho v_r h_r]^t \tag{5}$$

and

$$\begin{aligned}\vec{S} =& [0, \rho(\ddot{\alpha}(y - y_p) + 2\dot{\alpha} v_r + \dot{\alpha}^2(x - x_p)), \\ & \rho(-\ddot{\alpha}(x - x_p) - 2\dot{\alpha} u_r + \dot{\alpha}^2(y - y_p)), \\ & \rho\ddot{\alpha}((y - y_p)u_r - (x - x_p)v_r - \dot{\alpha}(x - x_p)^2 + \dot{\alpha}(y - y_p)^2)]^t\end{aligned} \tag{6}$$

The energy and the rothalpy referred to the moving frame of reference are given by

$$e_r = \frac{p}{\rho(\kappa - 1)} + \frac{u_r^2 + v_r^2}{2} - \frac{\dot{\alpha}^2}{2}[(x - x_p)^2 + (y - y_p)^2] \tag{7}$$

and

$$h_r = \frac{\kappa p}{\rho(\kappa - 1)} + \frac{u_r^2 + v_r^2}{2} - \frac{\dot{\alpha}^2}{2}[(x - x_p)^2 + (y - y_p)^2], \tag{8}$$

respectively. The flow vector field $\vec{q}_r$, and the inviscid fluxes $\vec{E}_r$ and $\vec{F}_r$ are based on relative velocity, $\vec{V}_r = (u_r, v_r)$; and the source term, $\vec{S}$, is an algebraic term contributed from the motion of the moving frame of reference, which has been written for a pitching motion at an angular velocity and angular acceleration of $\vec{\Omega} = \dot{\alpha}\vec{k}$ and $\dot{\vec{\Omega}} = \ddot{\alpha}\vec{k}$. The p is the pressure, ρ the density, t the time, and κ the gas specific heat ratio. The subscript r refers to the relative values with respect to the moving frame of reference. The prime ($'$) represents the derivative with respect to the moving frame of reference.

Unsteady Full-Potential Equation

The unsteady full-potential equation in the moving frame of reference has been derived as follows [4]:

$$\nabla^2\Phi = G_1 + G_2 \tag{9}$$

with

$$G_1 = -\frac{\nabla\rho}{\rho} \cdot [\nabla\Phi - \vec{e}_o - \vec{\Omega} \times (\vec{r} - \vec{r_p})] \tag{10}$$

$$G_2 = -\frac{1}{\rho}\frac{\partial'\rho}{\partial t} \tag{11}$$

and

$$\begin{aligned}\rho = \Big\{1 + \frac{\kappa - 1}{2} M_\infty^2 \big[&- (\nabla\Phi - \vec{e}_o - \vec{\Omega} \times (\vec{r} - \vec{r_p}))^2 \\ &+ (\vec{e}_o + \vec{\Omega} \times (\vec{r} - \vec{r_p}))^2 - 2(\frac{\partial'\Phi}{\partial t})\big]\Big\}^{\frac{1}{\kappa - 1}}\end{aligned} \tag{12}$$

where Φ is the absolute velocity potential ($\vec{V} = \nabla\Phi = \nabla'\Phi$), $\vec{e}_o$ the unit vector of the translation velocity of the moving frame of reference ($\vec{e}_o = \vec{V}_o / \mid \vec{V}_o \mid$), and M_∞ the free stream Mach number.

Equation (9) is a Poisson's equation. By using the Green's third identity, the integral equation solution of Eq. (9) for two-dimensional flows, for the absolute velocity field in the moving frame of reference, is given by

$$\begin{aligned}\nabla\Phi(x,y,t) = & \frac{1}{2\pi}\oint_g \gamma_g(s,t)\frac{(y-\eta)\vec{i}-(x-\xi)\vec{j}}{(x-\xi)^2+(y-\eta)^2}ds \\ & + \frac{1}{2\pi}\int\int G_1(\xi,\eta,t)\frac{(x-\xi)\vec{i}+(y-\eta)\vec{j}}{(x-\xi)^2+(y-\eta)^2}d\xi d\eta \\ & + \frac{1}{2\pi}\int\int G_2(\xi,\eta,t)\frac{(x-\xi)\vec{i}+(y-\eta)\vec{j}}{(x-\xi)^2+(y-\eta)^2}d\xi d\eta \\ & + \frac{1}{2\pi}\int_w \gamma_w(s,t)\frac{(y-\eta)\vec{i}-(x-\xi)\vec{j}}{(x-\xi)^2+(y-\eta)^2}ds\end{aligned} \tag{13}$$

The first integral term of Eq. (13) is the contribution of the airfoil surface vorticity (γ_g). The second and the third integral terms are the contributions of the compressibility (G_1) and the unsteadiness (G_2), respectively. The last integral term is the contribution of the wake vorticity (γ_w) shed at the airfoil trailing edge. It should be noted that the integral equation solution, Eq. (13), automatically satisfies the far-field boundary condition. Therefore, only a small computational domain around airfoil is needed.

COMPUTATIONAL SCHEMES

A sketch of an Euler computational domain inside a full-potential integral equation computational domain is shown in Fig. 2. The domain sizes are also shown in that figure. In this zonal scheme, the unsteady Euler equations are solved by an implicit approximately factored finite-volume scheme. The outer boundary conditions are provided by the IE solution of the unsteady full-potential equation. The IE computation in the region outside the embedded Euler domain is carried out simultaneously with the unsteady Euler computation, but relatively larger time steps are used. The details of the implicit approximately factored finite-volume scheme were presented in Ref. [1], and will not be repeated here.

The solution procedures of this zonal scheme can be described as follows:

Step 1 - Euler Time Marching:

Starting with the initial conditions, which may correspond to a steady flow solution, the finite-volume Euler computation in the small embedded domain is carried out for several time steps.

Step 2 - IE Computation for B.C.:

The IE computation in the outer domain over coarse grid is then carried out to provide the boundary conditions for next several time steps of Euler computations. In this step, the airfoil surface vorticity, γ_g, is calculated by numerical integration around the airfoil along a closed path, while the wake vorticity, γ_w, is determined by the fact that the total circulation is conserved. Once the vorticities are determined and the Euler solutions are interpolated over the IE coarse grid, Eqs.(10) - (12) are used to calculate ρ, G_1 and G_2. Then, IE solution, Eq. (13), is used to compute the velocities in the outer IE domain. Finally, the boundary conditions for Euler computations are extropolated from IE solutions.

Step 3 - Euler Time Marching:

With initial conditions obtained from the previous Euler computation and boundary conditions extropolated from the IE computation, Step 1 is repeated for the next several time steps.

Step 4:

Steps 2 and 3 are repeated to march in time until a periodic solution is obtained.

NUMERICAL EXAMPLE

An unsteady transonic airfoil flow has been computed, and the results have been compared with the Euler results [1]. In this numerical example, an NACA0012 airfoil is given a pitching oscillation around an axis at the quarter-chord length at a free stream Mach number of 0.755. The angle of attack, $\alpha(t)$, is given by $\alpha(t) = 0.016^o + 1.255^o \sin(0.1632t)$, where the value of reduced frequency (0.1632) is based on the airfoil chord length. The comparison of the time history of the surface pressure distributions for one cycle is shown in Fig. 3. The comparison shows that this scheme can capture the shock and its motion correctly with the exception of under- or over- prediction of the peak pressure (less than 7%). However, it should be recalled that the Euler equations solution overpredicts the experimental peaks.

CONCLUDING REMARKS

An unsteady field element method with embedded Euler domain has been developed. Since the Euler equations are solved in a small computational domain, the present zonal scheme requires less computational time than other existing Euler schemes, which solve Euler equations in the whole computational domain. The zonal scheme has been applied to unsteady transonic flow around a pitching airfoil. The computational time required for the present application using the zonal scheme is 70% that of the computational time required for the same application using the Euler equations solver throuthout the whole computational domain. The reduction of the computational time is very promising and in particular for three-dimensional transonic flow problems.

REFERENCES

1. Kandil, Osama A.; Chuang, H. A.: Unsteady transonic airfoil computation using implicit Euler scheme on body-fixed grid. AIAA Journal, Vol. 27, No. 8, (1989), pp.1031-1037.
2. Kandil, Osama A.; Hu, Hong: Unsteady transonic airfoil computations using the integral solution of full-potential equations. Boundary Elements X, Vol.2, edited by C. A. Brebbia. Springer-Verlag, (1988), pp.357-371.
3. Kandil, Osama A.; Hu, Hong: Full-potential integral solution for transonic flows with and without embedded Euler domains. AIAA Journal, Vol.26, No.9, (1988), pp.1079-1086.
4. Hu, Hong: Full-potential integral solutions for steady and unsteady transonic airfoils with and without embedded Euler domains. Ph.D. Dissertation, Department of Mechanical Engineering and Mechanics, Old Dominion University, Norfolk, Virginia, (1988).

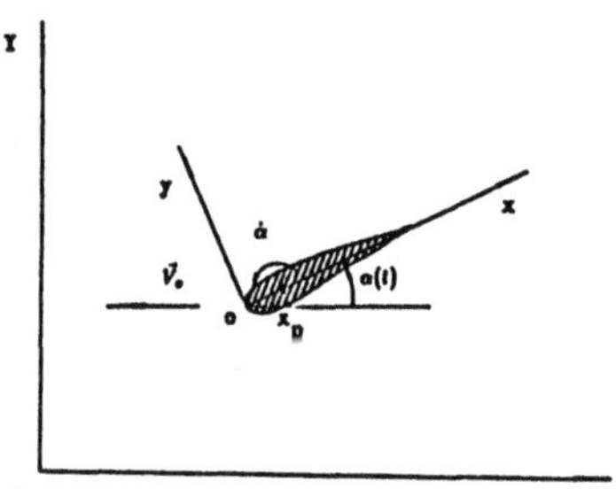

Fig. 1. Frames of reference

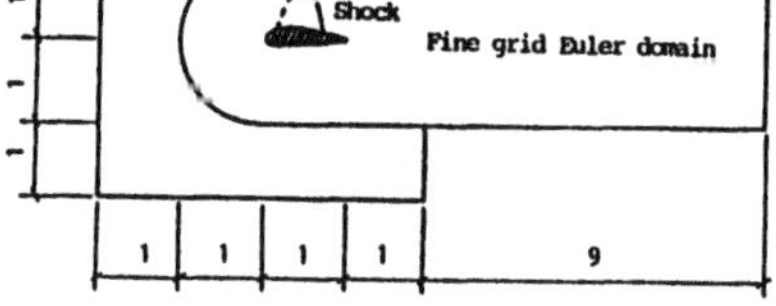

Fig. 2. Computational domains

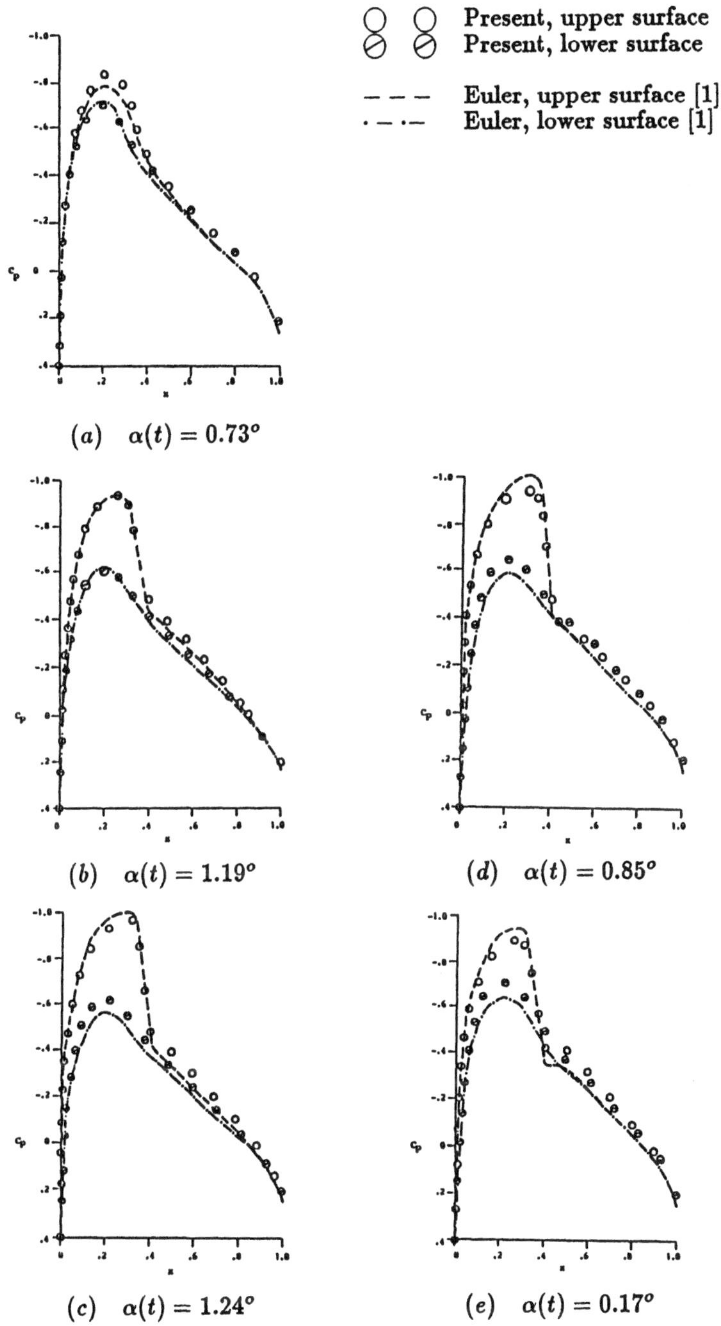

Fig. 3. Unsteady surface pressure distributions,NACA 0012, $M_\infty = 0.755, \alpha(t) = 0.016^o + 1.255^o \sin(0.1632t)$

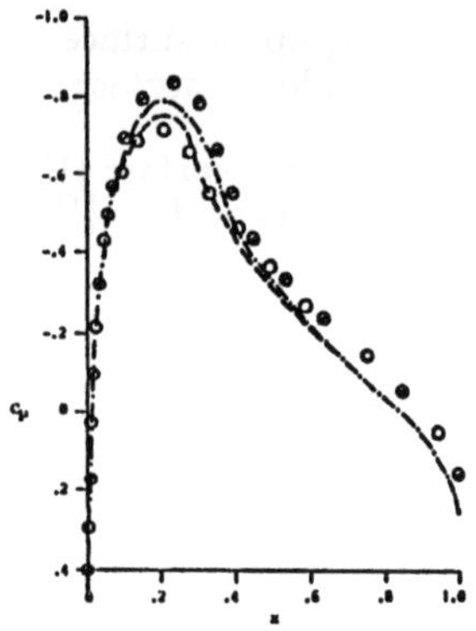

(*f*) $\alpha(t) = -0.70^o$

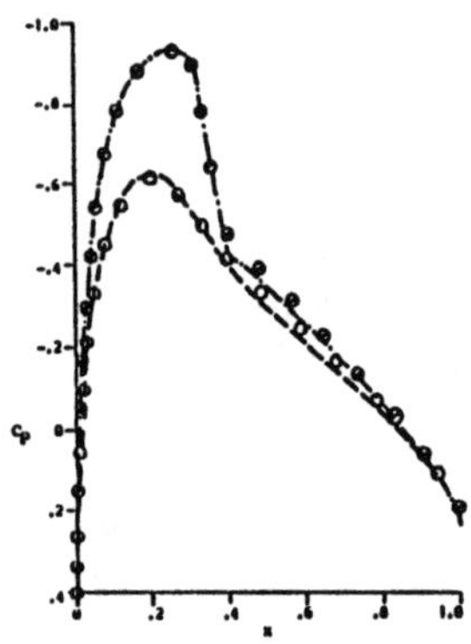

(*g*) $\alpha(t) = -1.16^o$

(*h*) $\alpha(t) = -1.20^o$

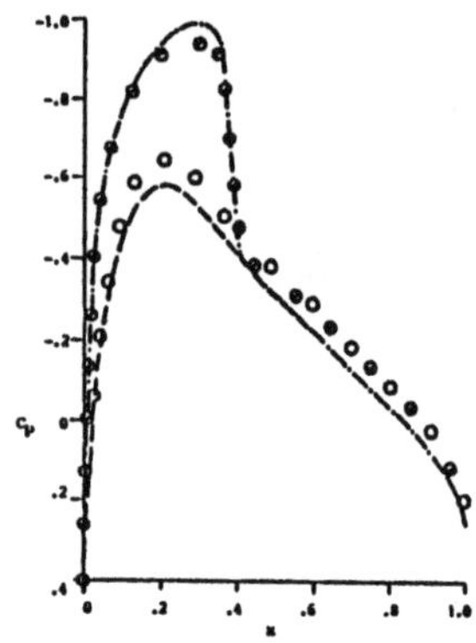

(*i*) $\alpha(t) = -0.82^o$

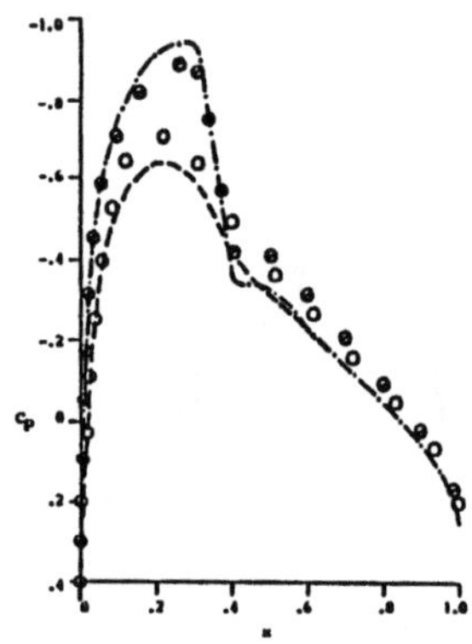

(*j*) $\alpha(t) = -0.14^o$

Fig. 3. (Continued)

Direct vs. Indirect Boundary Element Methods

JEFFREY L. O'BRIEN
THOMAS L. GEERS
Department of Mechanical Engineering
University of Colorado, Boulder, CO

Summary

Direct and indirect boundary-element formulations are compared in the context of calculating the fluid mass matrix for two-dimensional flow of an ideal fluid external to cylindrical surfaces. The accuracy of calculated eigenvalues for the associated fluid-boundary eigenproblem is used to assess performance. It is found that the indirect method is generally more accurate, which is attributed to the fact that boundary-element discretization directly constrains only the source field in the indirect formulation, whereas it directly constrains both the pressure and velocity fields in the direct formulation.

Introduction

Boundary element methods are widely used to solve engineering problems in fluid mechanics, solid mechanics, and electromagnetics. Two basic approaches are used to formulate the governing equations, which are frequently referred to as the direct and indirect formulations. The direct formulation makes use of generalized integration by parts to reduce the spatial dimensionality of the problem, while the indirect formulation utilizes source distributions to accomplish that purpose (see, *e.g.*, [1]).

Surprisingly, relatively little work has been done to compare the performance of the two formulations in terms of accuracy of solution as a function of mesh refinement. This paper reports on a modest effort to do just that for one of the simplest of boundary-element applications, the solution of Laplace's equation for two-dimensional ideal flow exterior to cylindrical surfaces.

The basis for comparison of the two formulations is the accuracy of computed eigenvalues for the fluid-boundary eigenvalue problem produced by finite-element surface discretization. This problem is defined by $\mathbf{Mu} = \lambda\mathbf{Au}$, where $\mathbf{M}$ is the full fluid mass matrix and $\mathbf{A}$ is the diagonal area matrix for the surface mesh. Both $\mathbf{M}$ and $\mathbf{A}$ are real, symmetric, positive-definite matrices, so all eigenvalues are positive real numbers. For the special case of the circular cylinder, the eigenvalues are compared with known exact values; for other cylindrical cross-sections, the exact eigenvalues are estimated by the extrapolation of computed values for meshes of varying refinement.

Problem Definition

Consider the two-dimensional, irrotational flow of a uniform, infinite, inviscid, incompressible fluid external to an infinite cylindrical body with cross-sectional contour C. The governing equations for such flow are (see, *e.g.*, [2])

$$\begin{aligned} &\nabla^2\phi = 0 && \text{in the domain D} \\ &p = \rho\phi_{,t} && \text{in the domain D} \\ &\boldsymbol{u} = -\nabla\phi && \text{in the domain D} \\ &-\phi_{,n} = u && \text{on the contour C} \\ &\nabla\phi \to 0 && \text{far from the contour C} \end{aligned} \tag{1}$$

in which ϕ is the velocity potential, p is the fluid pressure, ρ is the mass density of the fluid, the subscript t denotes differentiation in time, $\boldsymbol{u}$ is the fluid-particle velocity, and $\phi_{,n} = \boldsymbol{n}\cdot\nabla\phi$, where $\boldsymbol{n}$ is the unit normal to the surface (defined as positive going into the fluid), and u is the normal velocity of the body's surface (also defined as positive going into the fluid).

The geometry of the problem is depicted in Figure 1. There, P and Q are field and integration points, respectively, on the contour C. The surface normals $\boldsymbol{n}_P$ and $\boldsymbol{n}_Q$ are positive going into D. The angles $\theta_P(P,Q)$ and $\theta_Q(P,Q)$ are the angles between $\boldsymbol{R}(P,Q)$ and $\boldsymbol{n}_P$ and between $\boldsymbol{R}(P,Q)$ and $\boldsymbol{n}_Q$, respectively. For this geometry, the free-space (monopole) Green's function for Laplace's equation in two dimensions is (see, *e.g.*, [1])

$$\psi(P,Q) = \ln R^{-2}(P,Q) \tag{2}$$

where the scalar $R(P,Q) = |\boldsymbol{R}(P,Q)|$.

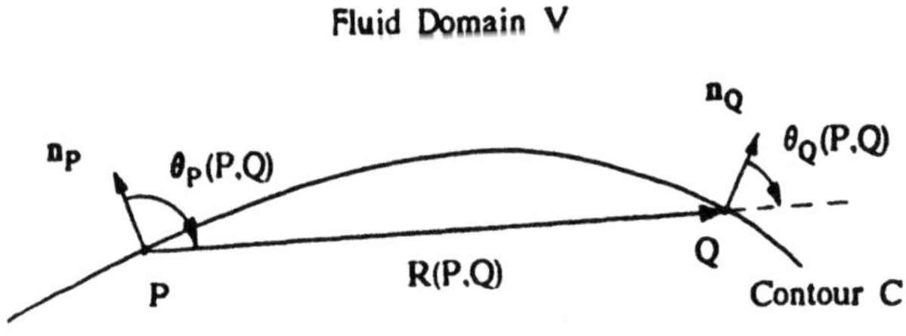

Figure 1. Problem geometry.

Indirect Approach

In the indirect approach, the fluid boundary is replaced by a distribution of monopole sources $\sigma(Q)$ such that

$$\phi(P) = \int_C \sigma(Q)\psi(P,Q)\, ds_Q \tag{3}$$

where s_Q is the contour coordinate for point Q. Differentiation with respect to the normal at P and analysis of the normal derivative of a simple source (see, *e.g.*, [1]) yields

$$\phi_{,P}(P) = -2\pi\sigma_P + \int_C \sigma(Q)\, \psi_{,P}(P,Q)\, ds_Q \tag{4}$$

where the singular contribution has been removed from the integral.

Discretization of the source distribution is readily effected by standard finite element procedures as

$$\sigma(Q) = \sum_j M_j(Q)\sigma_j \tag{5}$$

where the σ_j are source values at the nodes of the finite element mesh on C, and $M_j(Q)$ are global shape functions. In the interest of simplicity, the shape functions M_j are herein chosen as constant over element j and zero elsewhere, and the nodes are placed at the element centroids. Then the substitution of (5) into (3) and (4), followed by the enforcement of (3) and (4) at the nodes yields the matrix equations

$$\boldsymbol{\phi} = \mathbf{B}\boldsymbol{\sigma} \qquad \boldsymbol{\phi}_{,P} = -\mathbf{C}\boldsymbol{\sigma} \tag{6}$$

in which, through the utilization of (2), the matrix elements of **B** and **C** are given by

$$b_{ij} = -\int_{C_j} \ln R^{\,2}(P_i,Q)\, ds_Q \qquad c_{ij} = 2\pi\delta_{ij} - \int_{C_j} \frac{2\cos\theta_P(P_i,Q)}{R}\, ds_Q \tag{7}$$

where δ_{ij} is the Kronecker delta.

The boundary-element equations given in (6) permit a direct relation to be established between the velocity-potential vector $\boldsymbol{\phi}$ and the normal-fluid-particle-velocity vector $\mathbf{u} = -\boldsymbol{\phi}_{,P}$ [see the fourth of (1)]. By elimination of of the source vector, the desired relation for the indirect formulation is

$$\boldsymbol{\phi} = \mathbf{B}\mathbf{C}^{-1}\mathbf{u} \tag{8}$$

Now the kinetic energy of the external fluid, under the preceding discretization, is given by [3]

$$T = -\frac{1}{2}\,\rho\boldsymbol{\phi}_{,P}^{T}\mathbf{A}\boldsymbol{\phi} \tag{9}$$

where the superscript T denotes transposition. The introduction of $-\boldsymbol{\phi}_{,P} = \mathbf{u}$ and (8) into (9) then yields

$$T = \frac{1}{2}\,\rho\mathbf{u}^{T}\mathbf{A}\mathbf{B}\mathbf{C}^{-1}\mathbf{u} \tag{10}$$

Finally, because $\mathbf{u}^{T}\mathbf{S}\mathbf{u} = 0$ for $\mathbf{S}$ skew-symmetric, only the symmetric part of $\mathbf{ABC}^{-1}$ contributes to the kinetic energy. Therefore, the kinetic energy of the fluid is $T = \frac{1}{2}\mathbf{u}^{T}\mathbf{M}\mathbf{u}$, where the fluid mass matrix $\mathbf{M}$ is given by

$$\mathbf{M} = \frac{1}{2}\,\rho[\mathbf{ABC}^{-1} + (\mathbf{ABC}^{-1})^{T}] \tag{11}$$

Direct Approach

The foundation of the direct approach in two dimensions is the Green's identity

$$\int_{A}[\phi(Q)\,\nabla_Q^2\psi(P,Q) - \psi(P,Q)\,\nabla_Q^2\phi(Q)]\,dA_Q = -\int_{C}[\phi(Q)\,\psi_{,Q}(P,Q) - \psi(P,Q)\,\phi_{,Q}(Q)]\,ds_Q \tag{12}$$

Because both ψ and ϕ satisfy Laplace's equation, the left side of this equation vanishes. Also, from an analysis of a double source (see, *e.g.*, [1]), the singular contribution may be removed from the first term on the right side of (12) to give

$$-2\pi\sigma_P + \int_{C}\phi(Q)\psi_{,Q}(P,Q)\,ds_Q = \int_{C}\psi(P,Q)\phi_{,Q}(Q)\,ds_Q \tag{13}$$

Then, finite-element discretization of $\phi(Q)$ and $\phi_{,Q}(Q)$ in the manner of (5) with the shape functions M_j chosen as constant over element j and zero elsewhere, followed by the enforcement of (13) at the nodes, yields the matrix equations

$$-\mathbf{G}\boldsymbol{\phi} = \mathbf{F}\boldsymbol{\phi}_{,Q} \tag{14}$$

where, through utilization of (2), the matrix elements of $\mathbf{F}$ and $\mathbf{G}$ are given by

$$f_{ij} = -\int_{C_j}\ln R^2(P_i,Q)\,ds_Q \qquad g_{ij} = 2\pi\delta_{ij} + \int_{C_j}\frac{2\cos\theta_Q(P_i,Q)}{R}\,ds_Q \tag{15}$$

Premultiplication of (14) through by $\mathbf{G}^{-1}$ and the introduction of $\mathbf{u} = -\boldsymbol{\phi}_{,Q}$ yields the desired relation for the direct formulation

$$\phi = G^{-1}Fu. \tag{16}$$

Derivation of the fluid mass matrix for the direct formulation parallels that for the indirect formulation. The fluid mass matrix for the direct formulation is therefore given by

$$M = \frac{1}{2}\rho[AG^{-1}F + (AG^{-1}F)^T] \tag{17}$$

A comparison of (7) and (15) reveals that $b_{ij} = f_{ij}$, but that $c_{ij} \neq g_{ij}$ because, in general, $\cos\theta_P \neq -\cos\theta_Q$. However, for the circular cylinder, $c_{ij} = g_{ij}$ because $\theta_P = \pi - \theta_Q$. If, in addition, a uniform mesh is constructed on the circular cylinder, A is a scalar matrix and B = F and C = G are all symmetric. Hence, in this special case, the fluid mass matrices coincide and the two formulations yield identical fluid-boundary eigenvalues and eigenvectors.

Numerical Computations

The numerical work in this study was carried out with the FLUMAS Module in the USA Code [4]. As the original module is based on the indirect formulation, modifications were introduced to carry out direct-formulation calculations; additions were also introduced to handle contours not included in the original module. Computations were performed on the VAX/VMS system at the University of Colorado, Boulder.

As described in [3] and [4], FLUMAS surface modelling utilizes flat elements tangent to the surface at their centroids. For a convex surface, the model is external to the surface and contains inter-element gaps. Nonetheless, the technique is shown in [3] to yield highly accurate results. Here, all models were constructed with elements of equal length.

Three contour families were examined: ellipses, rectangles and triangles. Five levels of mesh refinement were constructed for each contour. Each level possessed twice the number of elements characterizing the previous level. A convergence trajectory for each eigenvalue was determined from the computed values for that eigenvalue as the mesh was refined. Some trajectories exhibited monotonic convergence and others exhibited oscillatory convergence. The circular cylinder is the only geometry for which the direct and indirect formulations yield identical results under the modelling used herein.

To the authors' knowledge, analytical solutions to the fluid-boundary eigenproblem exist only for the circular contour. Hence exact eigenvalues for the other contours were estimated by the extrapolation of convergence trajectories to ultimate mesh refinement. When a trajectory appeared to be monotonic, a decaying exponential function was fit to the last three calculated values. If monotonicity characterized both trajectories for corresponding direct-formulation and indirect-formulation eigenvalues, the exact value was taken as the average of the two extrapolated values. If only one trajectory was monotonic, the exact value was taken as the extrapolated value for that trajectory. If neither trajectory exhibited monotonic convergence, the exact value was taken as the average of the direct-formulation and indirect-formulation values produced by the most refined mesh.

Eight eigenvalues for twelve contours were calculated, producing 96 trajectories for each formulation. The direct method converged monotonically 67 times, and the indirect method converged monotonically 57 times. Only in three cases did the procedure of the preceding paragraph yield extrapolated values that differed from the "exact" value by more than 0.9%: those cases were omitted from the evaluation.

The accuracy of the extrapolation procedure may be assessed by comparing extrapolated eigenvalues for the circular contour with their exact counterparts produced by an analytical solution. This solution is readily obtained as described in [3] for the exact two-dimensional velocity potential

$$\phi(r,\theta) = -\sum_{n=1}^{\infty} \frac{a^{n+1}}{nr^n} (x_n \cos n\theta + y_n \sin n\theta) \tag{18}$$

where a is the radius of the circle, and x_n and y_n are modal normal-velocity components. The resulting exact eigenvalues are given by $\lambda_n/\rho a = n^{-1}$. The trajectory for the first eigenvalue exhibits monotonic convergence, producing a nondimensional extrapolated value that differs from unity by less than 0.003%. The next several trajectories exhibit oscillatory convergence, with extrapolated values differing from their exact counterparts by less than 0.3%.

Numerical Results

Plots of absolute error in calculated eigenvalues versus mesh refinement are shown in Figures 2-4. Errors in the eigenvalues calculated from the direct/indirect formulation are plotted above/below the abscissa. "Number of elements" pertains to discretization over the entire contour, even though all calculations were done for half-models that invoked positive or negative field symmetry. The error in a calculated eigenvalue is marked by the integer label for that eigenvalue.

Figure 2 shows four error plots for a family of ellipses. As it should, the plot for the circular contour exhibits perfect symmetry across the abscissa. Asymmetry increases, however, with increasing aspect ratio, with the indirect formulation generally producing eigenvalues that are slightly more accurate than those produced by the direct formulation. For some reason, the direct formulation appears to have difficulty producing an accurate fifth eigenvalue for ellipses with high aspect ratios.

Four error plots for a family of rectangles are shown in Figure 3. Here, the two formulations are quite comparable, as shown by the high degree of symmetry acrosss the abscissas. The situation is completely different, however, with the family of triangles, for which error plots are shown in Figure 4. Here, the indirect formulation exhibits clear superiority, with the direct formulation typically requiring twice as many elements as the indirect formulation requires to achieve the same accuracy.

Conclusion

The results from this limited study indicate that the indirect formulation for boundary-element analysis possesses a moderate accuracy advantage over the direct formulation. This might have been anticipated from the fact that finite-element discretization constrains only the source field in the former, whereas it constrains both the pressure and velocity fields in the latter. Why the difference in accuracy was more pronounced for triangular contours than for elliptical or rectangular contours is a question for future examination.

Acknowledgements

The authors express their appreciation to Dr. John DeRuntz for making the latest version of FLUMAS available for this study, and for providing helpful consultation on its use. This work was sponsored by the Defense Nuclear Agency with Drs. Kent Goering and Thomas Tsai as technical monitors.

References

1. Bannerjee, P.K. and Butterfield, R.; *Boundary Element Methods in Engineering Science*, McGraw-Hill Book Company, 1981.

2. Milne-Thomson, L.M.; *Theoretical Hydrodynamics*, MacMillan, New York, 1960.

3. DeRuntz, J. A. and Geers, T. L.; Added Mass Computation By The Boundary Integral Method. *International Journal For Numerical Methods In Engineering* 12 (1978) 531-549.

4. DeRuntz, J.A., *et al.*; The Underwater Shock Analysis Code (USA-Version 3). DNA5615F, Defense Nuclear Agency, Washington, DC (1980).

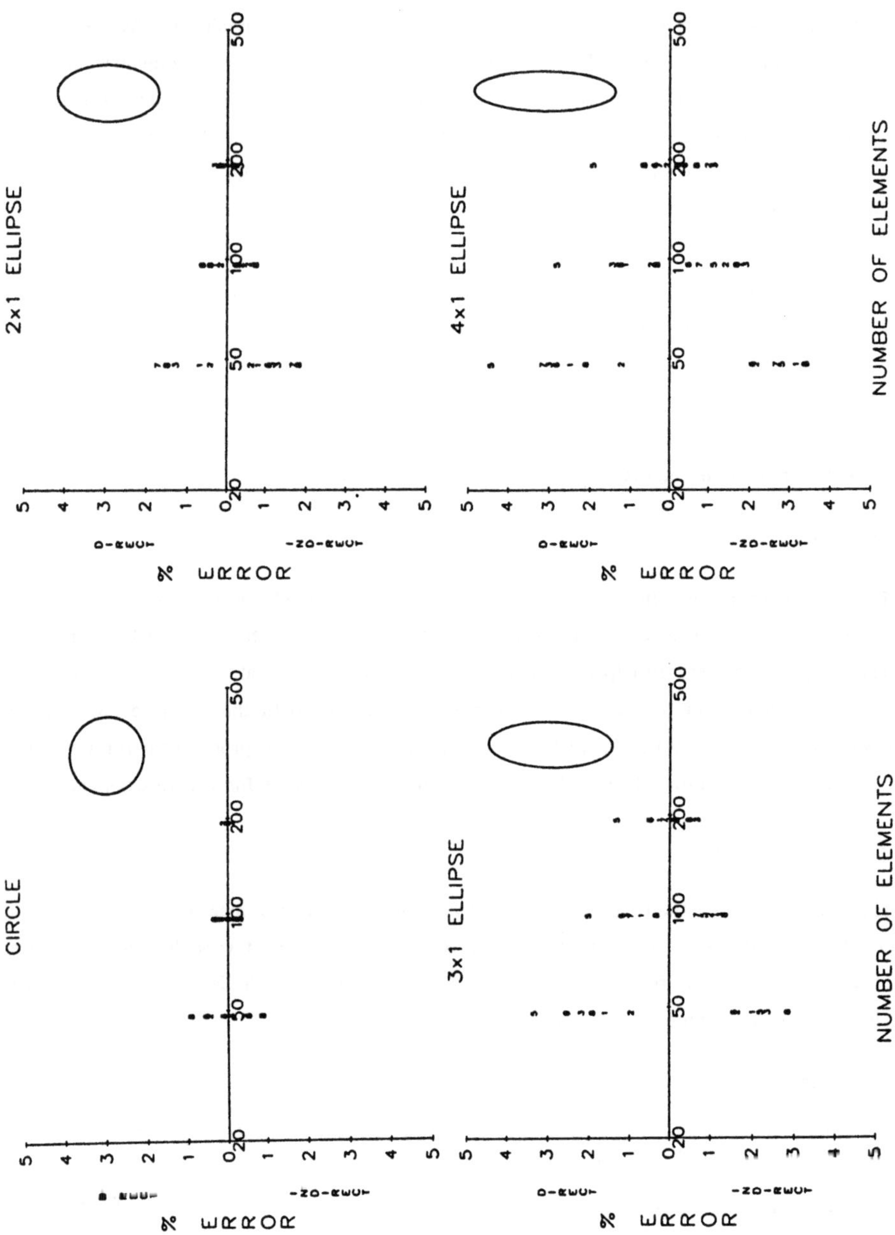

Figure 2. Errors in calculated eigenvalues for elliptical contours.

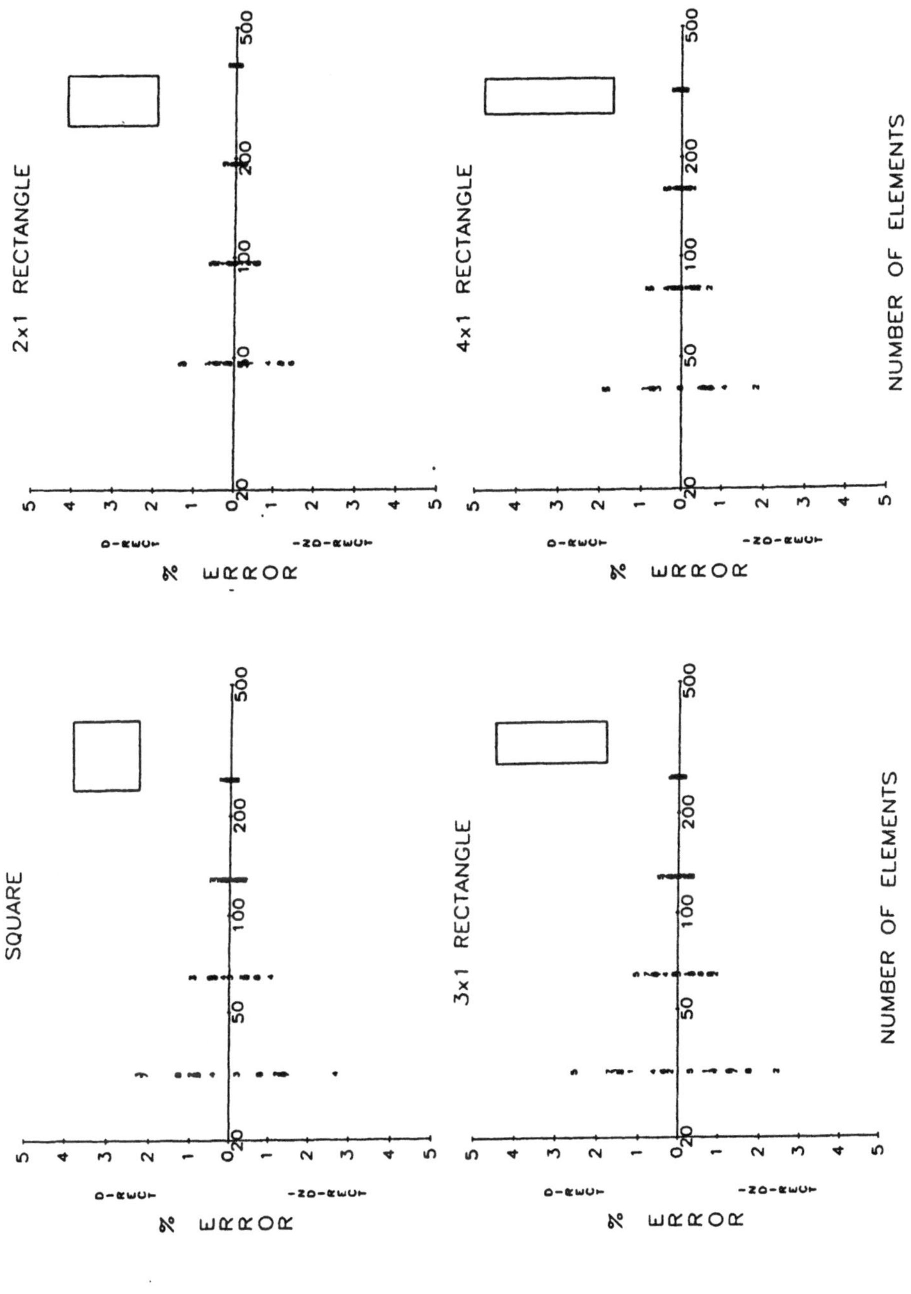

Figure 3. Errors in calculated eigenvalues for rectangular contours.

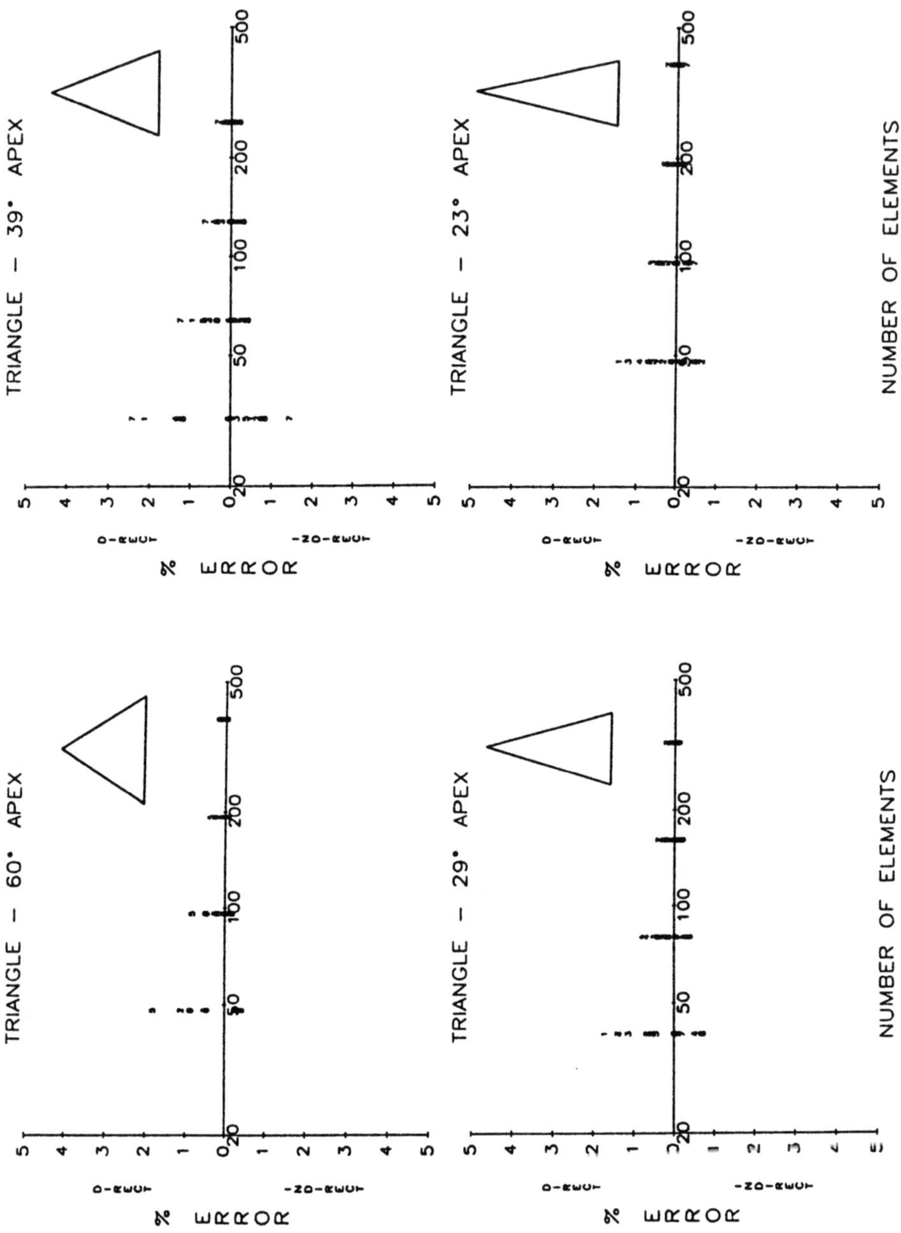

Figure 4. Errors in calculated eigenvalues for triangular contours.

Determination of the Maximum Thread Stress in Threaded Structures Using the Boundary Element Method

A. Wanderlingh
Applied Numerical Analysis Manager, Applied Mechanics Group, Hamilton Standard, Division of United Technologies Corporation, Windsor Locks, Ct. USA

INTRODUCTION

In the aerospace industry, use of threaded structures is widespread. In appearance, threaded structures can be as simple as a bolt or as complex as the jet engine fuel control housing shown in Figure 1. The primary function of threads is to join components together in a way that will ensure the safe operation of the assembled system for the duration of its structural fatigue life.

Structural fatigue life prediction of threaded structures relies on an accurate determination of the maximum thread stress. Determination of this parameter, however, poses significant challenges in both testing and analytical approaches. For instance, typical thread geometry and dimensions are so small that strain gage measurements, quite often, are virtually impossible to obtain. These same factors also make structural analysis using the finite element method very difficult. In this case, concerns regarding element distortion and the mesh density required to obtain accurate thread stresses, prevent this approach from being feasible.

As a result, designers over the years have had to rely on empirical methods, such as the one based on photo-elastic studies performed by Heywood [1] combined with the finite element method (FEM), to approximate the maximum thread stress. When compared to field data of existing structures, this empirical/FEM method has been shown, for some cases, to predict a maximum thread stress which is too high, thereby, resulting in a structural fatigue life prediction which is lower than the life the structures actually experience.

This conservatism can lead to new designs being manufactured unnecessarily with additional threads to reduce the load per thread and increased wall thickness to reduce the overall stress. Consequently, these modifications have the undesirable impact of increasing manufacturing costs and making the heavier structure potentially less competitive.

The purpose of this paper is to discuss how the boundary element method has been used, in a continuing structural analysis study, to identify improvements over the empirical/FEM method for determining the maximum thread stress.

APPROACH

An empirical approach combining Heywood's work and FEM was applied to the threaded jet engine fuel control housing section shown in Figure 2. The cylindrical section has eleven full threads and is made of aluminum. When assembled, the mating part loads ten of the eleven threads as illustrated.

This threaded section was also modeled and analyzed using the boundary element computer program BEASY [2]. The boundary element results were then used to evaluate the empirical/FEM method.

EMPIRICAL/FEM METHOD DESCRIPTION

The maximum thread stress is defined by the following empirical relationship developed by Heywood;

$$\sigma_{max} = (K_m)\, \sigma_{member} + A\ \sigma_{thread} \tag{1}$$

where:

- σ_{max} - The maximum thread stress resulting from thread and member loads.
- K_m - The member elastic stress concentration factor.
- σ_{member} - The elastic stress at the bottom center of the thread root due to loading of the member.
- A - The reduction factor which accounts for the location differences between σ_{member} and σ_{thread}.
- σ_{thread} - The elastic stress in the thread root (normally 30 deg. to the flank) due to bending and bearing action of the thread.

A description of how each term on the right hand side of Equation (1) is determined follows:

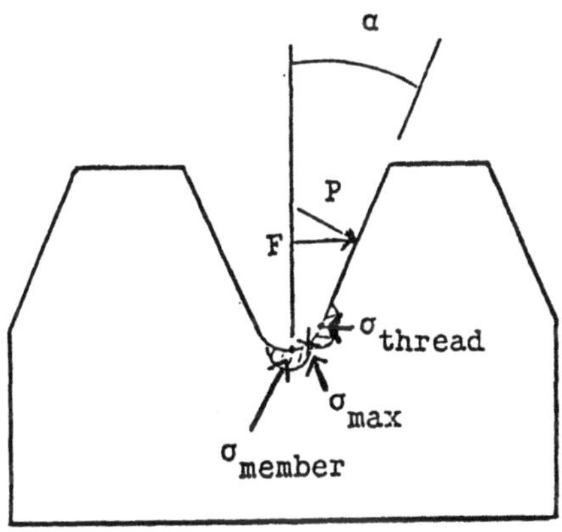

α- flank angle
P- applied thread load (lbs.)
F- axial component of thread load (lbs.)

The thread stress (σ_{thread}) represents that portion of the maximum stress due to the thread load only. Its magnitude is obtained from the following expression:

$$\sigma_{thread} = \frac{HKFNf}{Dn} \tag{2}$$

where:

H = thread load concentration factor due to fit and deformation of the mating threads.

F = axial load -lbs.

D = load diameter =~ pitch diameter-inches.

f = thread factor which depends on the specific thread geometry, i.e. thread form, size, tolerance class.

N = number of threads per inch.

n = number of engaged threads.

K = elastic stress concentration factor for the root fillet under thread bending.

Further,

$$K = 1 + .26\left(\frac{e}{R}\right)^{.7} \tag{3}$$

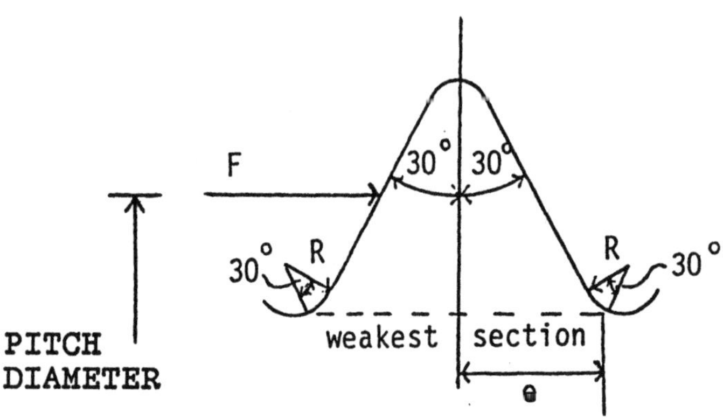

where:

e = half the thickness of the thread projection between points 30 degrees to the thread flank.

R = thread root radius.

A detailed discussion regarding Equation (2) can be found in Reference 1.

The member stress (σ_{member}) is obtained by performing a finite element analysis of the structure shown in Figure 3. The threads are replaced with equivalent elements and loaded with bending and shear tractions. The maximum surface stress is identified and used in Equation (1).

The factor A is used to combine the thread and member elastic stresses. For the threaded structure used in this study, the factor A was calculated for 60 degree threads using the following equation:

$$A = \frac{1}{1 + Co \left| \dfrac{\sigma_{member}}{\sigma_{thread}} \right|} \qquad (4)$$

where:

$$Co = \frac{\{60 \text{ deg.} - \alpha\}^2}{44 \text{ deg.}}$$

The member stress concentration factor (Km) for multiple notches is calculated using Neuber's Nomograph shown in Reference 1 on pages 200 and 201.

EMPIRICAL/FEM METHOD RESULTS

For the threaded structure shown in Figure 1, the following values were determined:

$$\sigma_{thread} = 10{,}850 \text{ psi}$$

$$\sigma_{member} = 12{,}994 \text{ psi}$$

$$A = .6424$$

$$K_m = 2.75$$

Inserting these values into Equation (1)

$$\sigma_{max} = \sigma_{member} (K_m) + A(\sigma_{thread})$$

$$\sigma_{max} = \underline{42{,}703 \text{ psi}}$$

This maximum thread stress occurs in fillet 12. The stress concentration factor (K_t) which combines the member and thread bending effects is determined as follows:

$$K_t = \frac{\sigma_{max}}{\sigma_{member}} = \frac{42{,}703}{12{,}994} = \underline{3.29}$$

BOUNDARY ELEMENT METHOD DESCRIPTION

The boundary element axisymmetric model of the aluminum threaded structure used in this study is shown in Figure 4. The same load distribution calculated in the empirical/FEM method approach was applied to the ten threads which contact the mating part. The loads were applied about the pitch diameter. Two hundred and thirty-one continuous and discontinuous quadratic elements were used to define the geometry. The structure was divided into four zones to reduce the solution CPU time.

BOUNDARY ELEMENT METHOD RESULTS

The boundary element method (BEM) solution confirmed the maximum thread stress location to be at fillet 12. However, the magnitude of the stress (σ_{max} = 25,872 psi) was found to be 39% lower than the empirical/FEM method prediction. The BEM solution provided a structural fatigue life prediction which more accurately reflects field experience.

Assuming Equation (1) is valid, the major deficiency in the empirical/FEM method appears to be in the calculation of the stress concentration factors used in determining σ_{max}. It should be noted that the use of Equation (3) and the Neuber Nomographs are approximate means of determining stress concentration factors. Judgements on the type of structure loading (i.e. bending, shear, membrane) as well as correction factors for the presence of multiple notches strongly influence the final determination of K_t. In contrast, no such approximations are necessary using the boundary element method because geometry and load distributions are much more precise.

A more accurate determination of K_t was made by using the BEM results for the threaded model and the finite element results for the no-thread model; see Figure 5. By dividing the un-notched surface stress into the notched surface stress a K_t of 1.80 was obtained. This value represents a 45% difference in the Kt value (3.29) obtained from the empirical/FEM method.

Another deficiency in the empirical/FEM method is the calculation of the thread bending and shear tractions which are used in the finite element analysis to determine the member stress. For example, a comparison of the normal thread traction applied to the finite element model with that obtained from the boundary element solution shows a significant difference in the distribution across the root of the thread; see Figure 6. A similar discrepancy occurs when a comparison of the shear traction is made. Application of the boundary element thread root normal and shear tractions to the finite element model produced a member stress that is 6% lower than that obtained in the empirical/FEM method approach.

CONCLUSION

It has been shown that use of BEM to determine the maximum thread stress is a clear improvement over the empirical/FEM method in that approximations of K_t and thread loading are not required. The more accurate calculation of the maximum thread stress results in an improved structural fatigue life prediction which correlates with field experience. The improved accuracy obtained with this method will ultimately translate into more competitive designs.

REFERENCES

1. Heywood, R.B., "Designing by Photoelasticity", First edition, Chapter 7, Stress Concentration in Screw Threads, Bolts and Nuts, 1952.

2. BEASY is a product of Computational Mechanics Inc., Southampton England

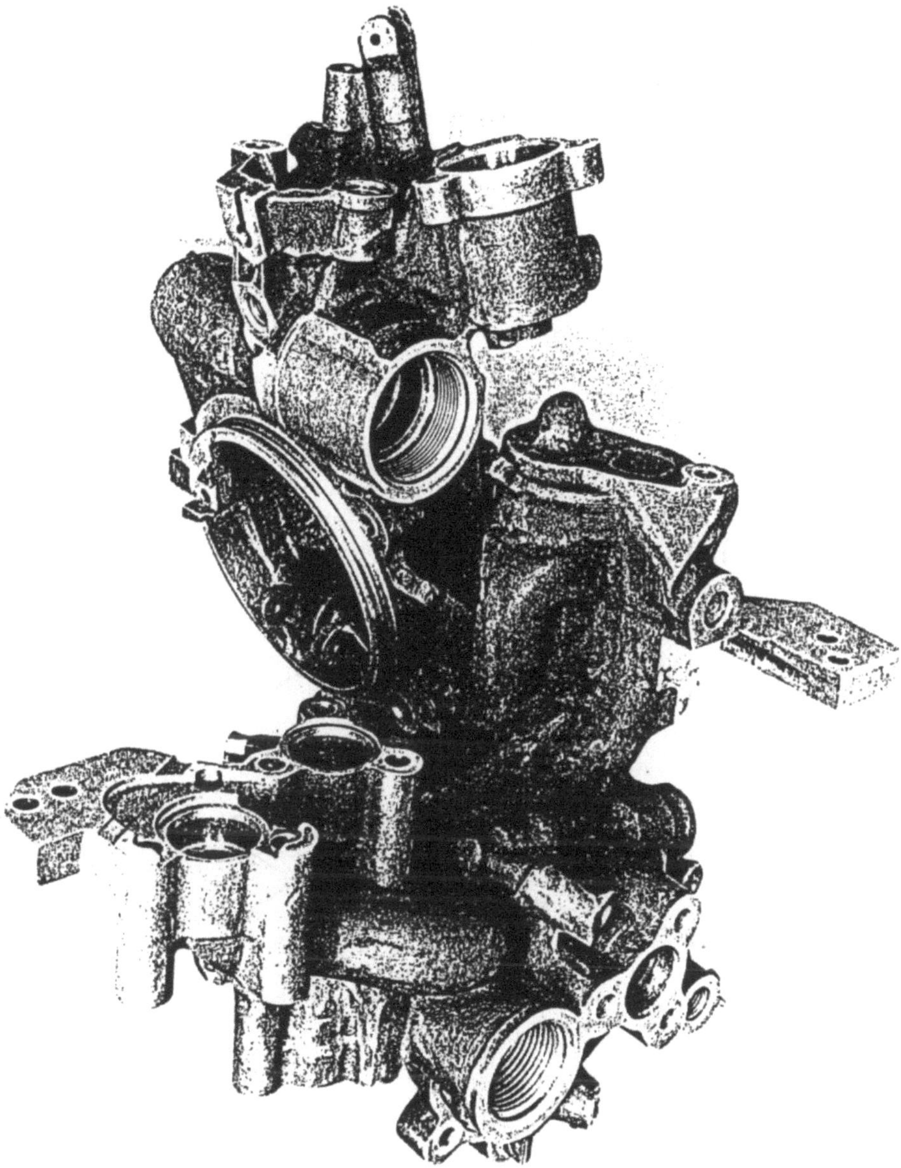

Figure 1 : Jet Engine Fuel Control Housing

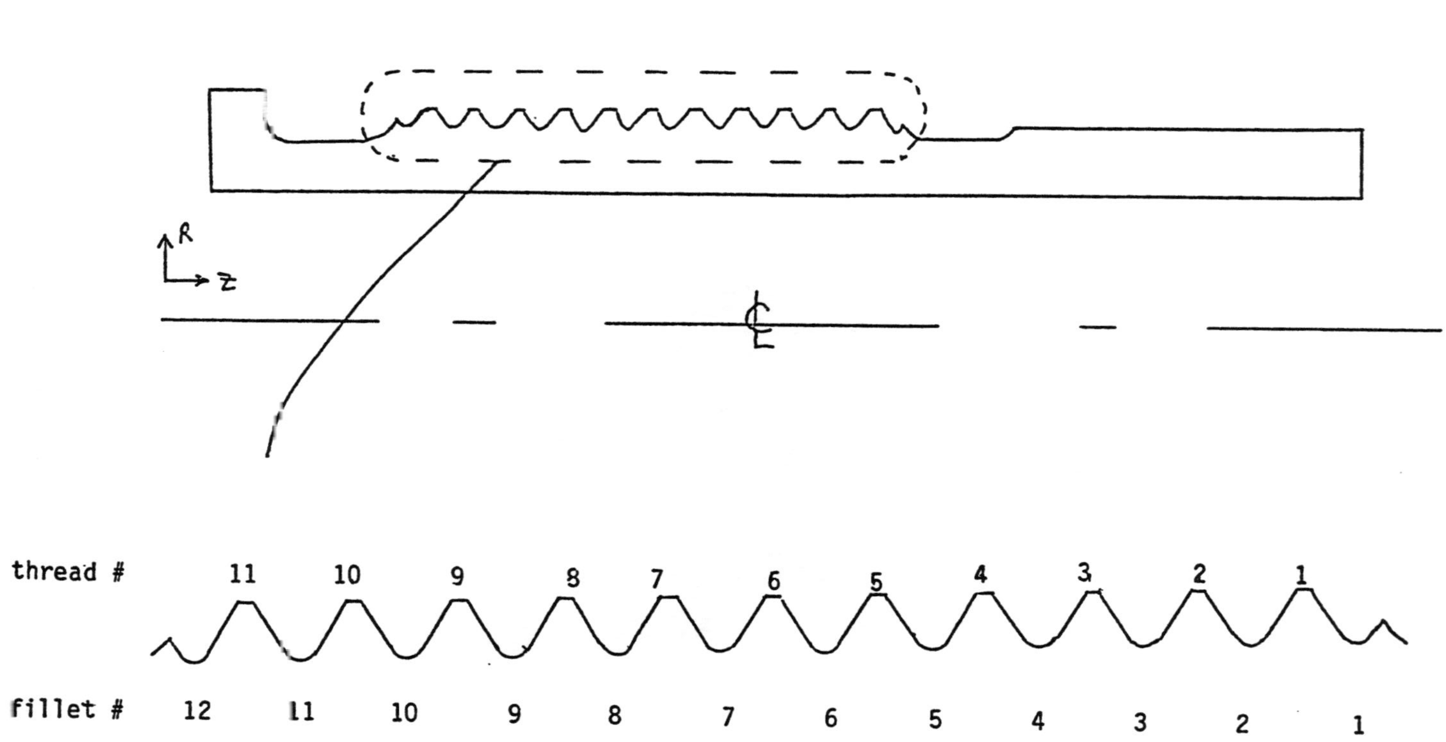

Figure 2 : Jet Engine Fuel Control Housing Section

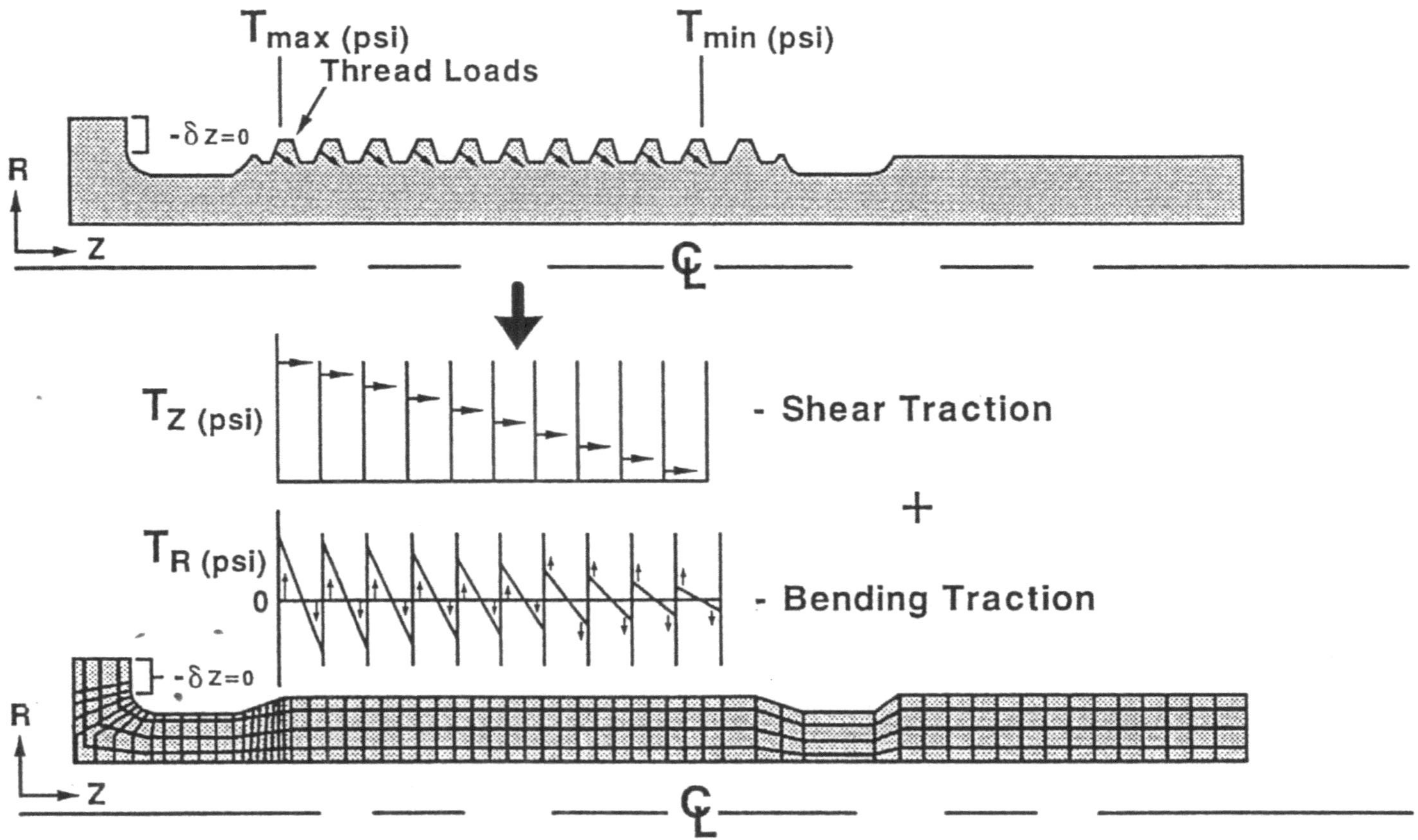

Figure 3: Finite Element Model used to Determine Member Stress

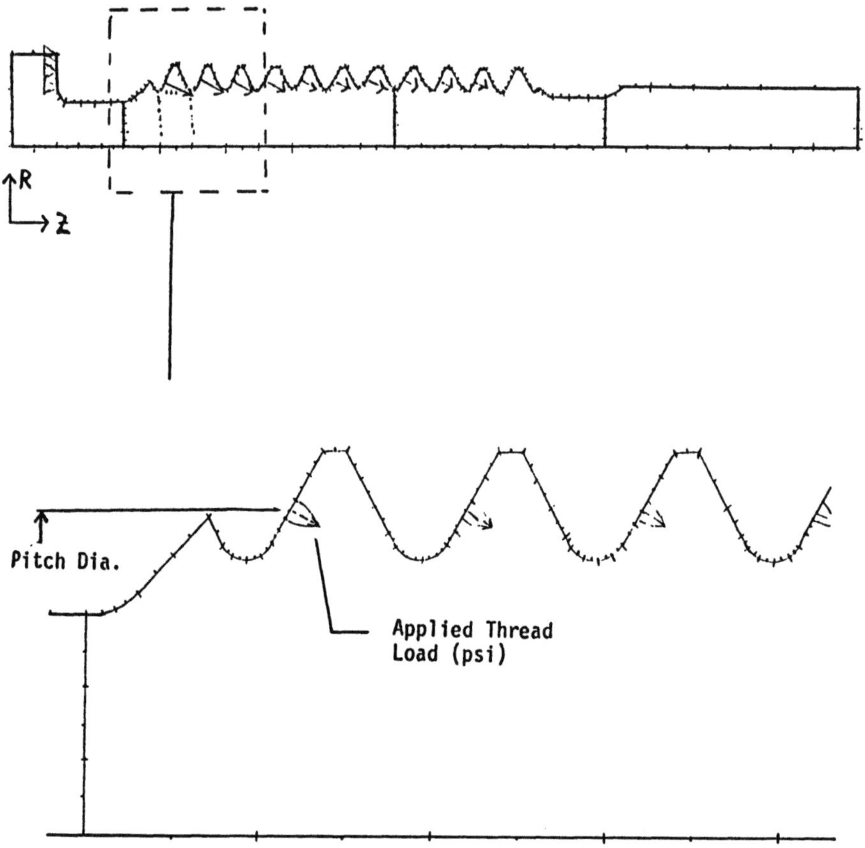

Figure 4 : Boundary Element Model used to Determine the Maximum Thread Stress

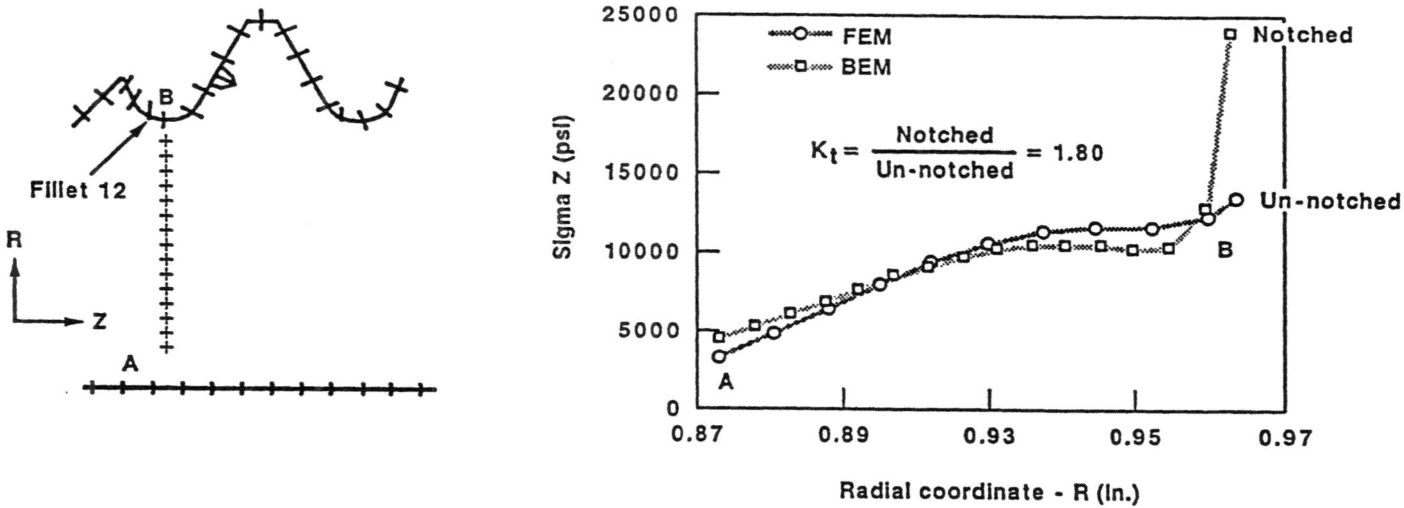

Figure 5: Determination of Stress Concentration Factor(K_t) using BEM and FEM Results

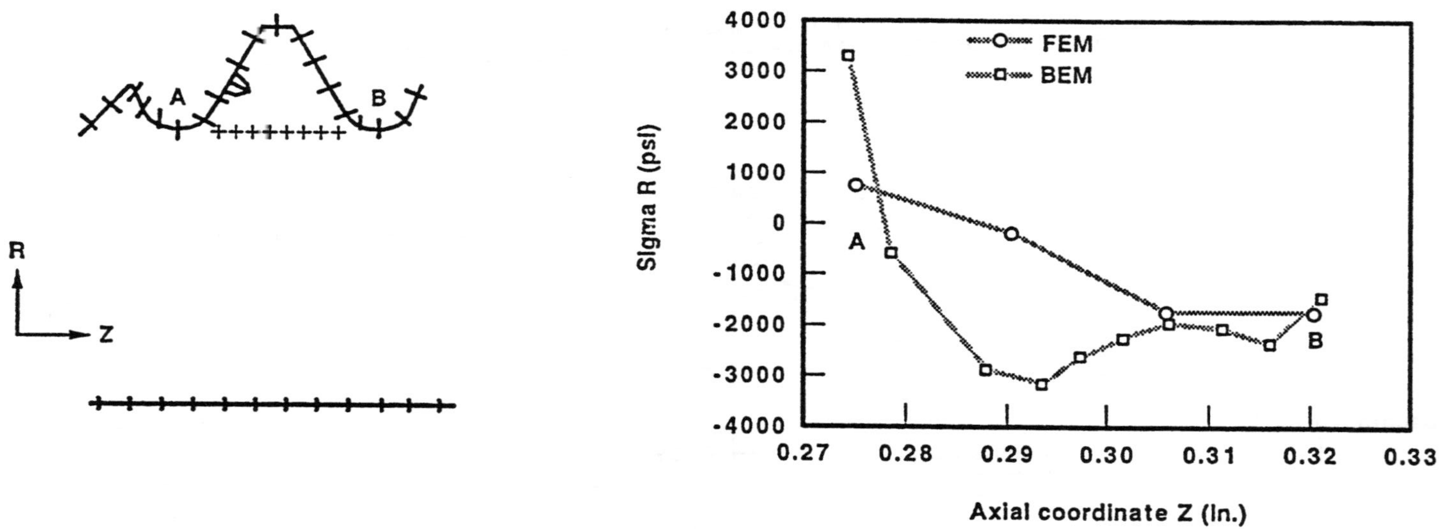

Figure 6 : Comparison of Bending Traction Between the Design Method and BEM

Absolute p-Refinement of Two-Dimensional Elasticity Problems in the Vicinity of Boundary Solution Singularities

Timothy J. Urekew and Joseph J. Rencis

Mechanical Engineering Department,
Worcester Polytechnic Institute,
Worcester, Massachusetts 01609 USA

ABSTRACT

A simple polynomial refinement formulation is developed to study the convergence of the collocation boundary element method for two-dimensional elasticity problems, in the vicinity of a boundary solution singularity. The boundary solution singularity considered here is created by discontinuous boundary conditions prescribed at a point on the boundary. A combination of analytic and numeric integrations are used in evaluating the boundary integrals for discontinuous, constant, and continuous, linear through cubic, solution distributions. The L_2 boundary error norm, relative percentage error, and solution norm ratio are used to study convergence of displacements and tractions at the crack tip of a plate fracture problem. The reference for convergence is the solution obtained from a fine mesh of quadratic boundary elements.

INTRODUCTION

Polynomial refinement (p-refinement) is a means of improving a numerical solution by increasing the order of the solution interpolation while holding the element size fixed. The refinement basis functions used in the absolute p-refinement strategy are of the same order, and from the same family of polynomials as the solution basis functions. This is in contrast to the relative or hierarchial p-refinement strategy [1,5] where the order of the solution basis functions remains fixed and refinement is carried out by adding bubble functions or relative degrees of freedom. Relative p-refinement can employ a variety of functions, such as Peano and Legendre, to effect the solution between the nodes of the initial mesh. Both forms of p-refinement have been thoroughly researched for the FEM [5, 11], and implemented into computer codes. Only recently has polynomial convergence been researched for the BEM.

In 1985, Alarcón et al. [1, 2] studied p-convergence of two-dimensional (2-D) potential and elasticity problems using the Peano family of refinement polynomials and the collocation BEM. Reverter et. al. [8] studied p-refinement of potential problems in conjunction with the development of indicators and estimators for adaptive refinement. This work also includes a study of the impact of relative p-refinement on a boundary containing a singularity created by discontinuous boundary conditions. A similar study was performed by Alarcón and Reverter [2], showing the impact of the Peano refinement functions on the convergence of a potential problem containing a boundary singularity. In 1987 Parreira [6] and Rank [7] reported that boundary solutions are improved during p-refinement if additional elements (h-refinement) are added in the vicinity of singular points. Rank [7] found that relative p-convergence of potential problems is exponential where the solution is smooth and is algebraic in the vicinity of singularities. An exponential rate of convergence was obtained however, even in the presence of singular points, through a combination of h- and relative p-refinement. Umetani [9] examined local convergence of the 2-D Galerkin boundary element formulation for elasticity using the Legendre family of polynomials for refinement. In this work Umetani concludes that a good initial mesh discretization is required for p-convergence and that convergence cannot always be attained through pure p-refinement.

The work presented in this paper uses the direct collocation BEM formulation for 2-D elasticity to examine absolute p-refinement. Constant through cubic interpolation functions from the Lagrange family of polynomials are used to improve the boundary element solution of a 2-D elasticity problem with a boundary solution singularity. Several error parameters are introduced to gage the local and global convergence of the absolute p-refinement strategy.

ABSOLUTE p-REFINEMENT STRATEGY

The governing equations for two-dimensional, isotropic, linear elasticity are the well-known Navier's equations. The solution for Navier's Equations in terms of the boundary integral equation (BIE), can be obtained from the direct formulation discussed in [3], and is given here, in the absence of body forces, as

$$c_{ij}\, u_j(\xi) + \int_{\Gamma} t^*_{ij}(\xi,x)\, u_j(x)\, d\Gamma(x) = \int_{\Gamma} u^*_{ij}(\xi,x)\, t_j(x)\, d\Gamma(x), \qquad i = 1,2;\ j = 1,2 \tag{1}$$

where c_{ij} is a coefficient matrix dependent upon local surface geometry, t^*_{ij}, u^*_{ij} are the fundamental solutions, and $u_j(x)$, $t_j(x)$ are the displacement and traction distributions, respectively, at the boundary. The integrals in Equation (1) are evaluated numerically when integrating over an element (straight) not containing the collocation point (source point), and analytically when integrating over an element containing the collocation point. After evaluating the integrals and applying boundary conditions Equation (1) can be rearranged into the linear system of algebraic equations $A_{ij}\, x_j = b_j$. The unknown nodal displacements and tractions are found by solving for x_j.

The absolute p-refinement strategy requires adding new degrees of freedom to affect the element distributions of $u_j(x)$ and $t_j(x)$ in Equation (1). This leads to complete higher-order, BEM solution distributions of the form

$$\hat{u}_i = \sum_{k=1}^{n} N_k\, u_k, \qquad i = 1,2, \tag{2}$$

$$\hat{t}_i = \sum_{k=1}^{n} N_k\, t_k, \qquad i = 1,2 \tag{3}$$

where n is the number of collocation points used to define the solution, and u_k, t_k are the displacements and tractions, respectively, at the collocating (nodal) points. Equations (2) and (3) are the solution functions used in Equation (1) to evaluate the unsymmetric coefficient matrix, A_{ij}. During mesh refinement, successively higher-order interpolation functions, N_k, from the Lagrange family of polynomials, are introduced until the numerical solution is within some acceptable range of error. In this work the geometry and boundary conditions are prescribed exactly, thereby introducing no error.

CONVERGENCE GAGING PARAMETERS

The L_2 boundary error norm, relative percent error, solution norm ratio and equilibrium residual are used to quantify errors in traction and displacement and to gage convergence. The first three parameters depend upon the error functions e_{u_i} and e_{t_i} associated with the displacement and traction components i, respectively. These functions are defined as

$$e_{u_i} = u_i - \hat{u}_i, \qquad i = 1,2, \tag{4}$$

$$e_{t_i} = t_i - \hat{t}_i, \qquad i = 1,2 \tag{5}$$

where u_i, t_i represent the exact solution of displacement and traction, respectively, and $\hat{u}_i$, $\hat{t}_i$ represent the p-refined, BEM solution of displacement and traction, respectively. The equilibrium residual, Er, is defined as the integral of the traction distribution over the entire mesh and is equal to 0.0 for the exact solution.

The exact L_2 boundary error norm for displacement is defined here, for each element, j, as

$$\| e_{u_i} \|_j = \left(\int_{\Gamma_j} (e_{u_i})^T (e_{u_i})\, d\Gamma \right)^{1/2}, \qquad i = 1,2 \tag{6}$$

for displacement components i. Also defined on the element level is the exact boundary solution norm given here as

$$\| u_i \|_j = \left(\int_{\Gamma_j} u_i^2\, d\Gamma \right)^{1/2}, \qquad i = 1,2, \tag{7}$$

for exact displacements, u_i. Similarly, the numerical solution norms are calculated for each element in the p-refined mesh from

$$\| \hat{u}_i \|_j = \left(\int_{\Gamma_j} \hat{u}_i^2\, d\Gamma \right)^{1/2}, \qquad i = 1,2. \tag{8}$$

The L_2 boundary error and solution norm ratios for traction are defined in a similar manner. The relative percent error quantifies the solution error as a percentage of the exact solution, and is calculated for each element, j, from

$$\eta^j_{u_i} = \frac{\| e_{u_i} \|_j}{\| u_i \|_j} x\, 100\%, \qquad i = 1,2 \tag{9}$$

for displacement components i. Lastly, the solution norm rations can now be defined for each element, j, as

$$\theta^j_{u_i} = \frac{\| \hat{u}_i \|_j}{\| u_i \|_j}, \qquad i = 1,2. \tag{10}$$

for displacement components i. The solution norm ratios approach 1.0 as the magnitude of the p-refined solution converges with the magnitude of the exact solution. A more detailed discussion of the error parameters presented here, as well as the global formulations, can be found in [10].

RESULTS

A plate with a centrally located crack, loaded in tension, was selected to study convergence in the vicinity of a boundary solution singularity. The geometry and boundary conditions for the quarter symmetry model are given in Figure 1. Two initial meshes, shown in Figures 2 and 3, are used to study local (element level) convergence in the vicinity of the crack tip and to study global convergence. This problem is examined by Umetani [9] in studying adaptive p-convergence of the Galerkin BEM formulation using Legendre polynomials for refinement. Cerrolaza et al. [4] also consider this problem in evaluating an adaptive p-refinement strategy using the collocation BEM and the Legendre family of refinement functions. The plate fracture problem is well-suited to adaptive refinement because of the non-smooth nature of the solution in the vicinity of the crack tip.

Convergence of the absolute p-refinement method is studied relative to an exact solution which is approximated by the solution from a boudary element mesh containing 240 quadratic elements. The exact displacement and traction distributions are presented in Figures 4 and 5, respectively, as a function of the problem geometry shown in Figure 1. The traction distribution shows a singularity at point C (crack tip), where the traction component in the y direction tends to infinity. Only traction distributions along sides BC and EA are shown in Figure 5 because traction is prescribed as a boundary condition on all other sides.

In Figures 6 through 9 local p-convergence, in the vicinity of the crack tip, is examined using the L_2 boundary error norm and the relative percent error. As can be seen in Figures 8 and 9, the greatest percentage of traction error resides in the element closest to the fracture tip in going from the 5 to the 9 element mesh. By grading the mesh towards the singularity, as was done with the 9 element mesh, the discretization error is concentrated in element I, thereby enabling adjacent elements V and VI to converge to a more accurate solution. This observation is in agreement with the findings in [4, 9].

Figures 10 and 11 show local, element level convergence of the L_2 boundary error norm for displacement. An element level improvement, similar to that observed for traction, is seen for displacement in going from a 5 to a 9 element mesh.

Figures 12 and 13 show global, Log-Log convergence for the 5 and 9 element meshes using the equilibrium residual and the global relative percent errors. In Figure 12 the convergence rates for the 5 element and 9 element meshes are the same. The net percentage of error is however lower for the 9 element mesh, due to the additional degrees of freedom.

CONCLUSIONS

A 2-D, BE elasticity program was developed to study absolute p-convergence of discontinuous constant, and continuous linear through cubic solution distributions. A convergence study of the plate fracture problem suggests that the BEM solution, in problems containing singularities, cannot be improved beyond a finite limit through absolute p-refinement up to cubic order, without decreasing element size h. In elastostatic problems involving singular points, a combination of h- and p-refinement may be a better alternative than pure p-refinement in improving the numerical solution.

ACKNOWLEDGEMENT

The authors would like to thank Professor William Durgin, Head of the Mechanical Engineering Department at Worcester Polytechnic Institute for providing the publication support for this work. In addition, the use of the Computational Mechanics Laboratory in the Mechanical Engineering Department at WPI is gratefully appreciated.

REFERENCES

1. Alarcón, E., Reverter, A., and Molina, J. Advances in Finite Element Technology - Hierarchial Boundary Elements, Computers and Structures, Vol. 20, pp. 151-156, 1985.

2. Alarcón, E., and Reverter, A., p-Adaptive Boundary Elements, International Journal for Numerical Methods in Engineering, Vol. 23, pp. 801-829, 1986.

3. Banerjee, P. K., and Butterfield, R., Two-Dimensional Problems of Elastostatics. Chapter 4, Boundary Element Methods in Engineering Science, pp. 177-215, McGraw-Hill, London, 1981.

4. Cerrolaza, M. and Alarcón, E., Further Applications of p-Adaptive Boundary Elements, Boundary Elements X, (Ed. Brebbia, C. A.), Computational Mechanics Publications, Southampton, 1988.

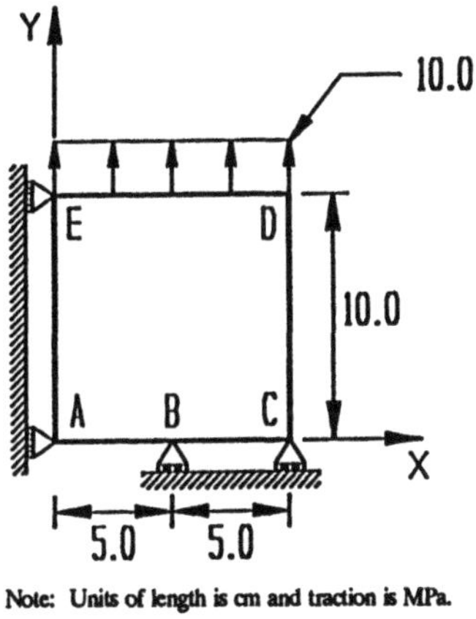

Figure 1 - Qaurter symmetry plate fracture model.

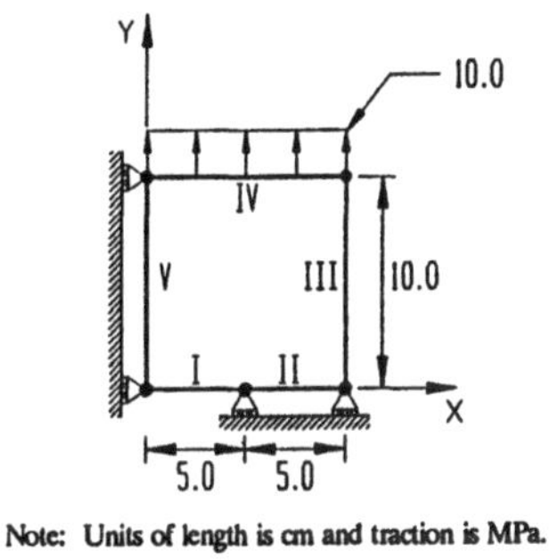

Figure 2 - Five element quarter symmetry mesh.

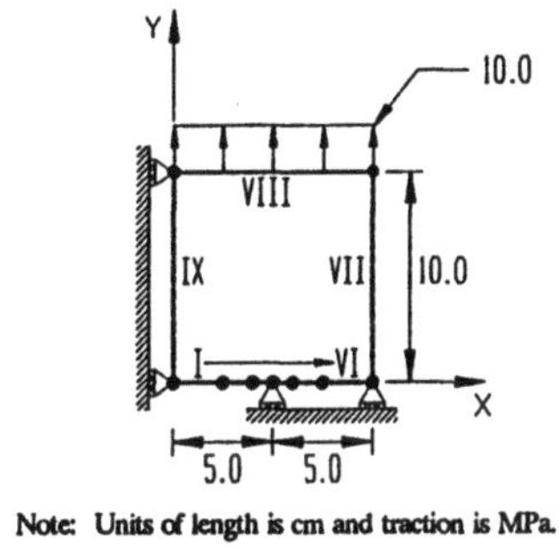

Figure 3 - Nine element quarter symmetry mesh.

5. Kardestuncer, H. and Norrie, D. H., Part 2 - Other Finite-Element Methods. Chapter 6, Finite Element Handbook, McGraw-Hill, New York, 1987.

6. Parreira, P., Self-Adaptive p-Hierarchical Boundary Elements in Elastostatics, Boundary Elements IX, (Eds. Brebbia, C. A., Wendland, W. L., Kuhn, G.), Computational Mechanics Publications, Southampton, 1987.

7. Rank, E., Adaptive h-, p-, and hp-Versions for Boundary Integral Element Methods, International Journal for Numerical Methods in Engineering, Vol. 28, pp. 1335-1349, 1989.

8. Reverter, A., Gónzalez, A.and Alarcón, E., Indicators and Estimators in p-Adaptive Boundary Elements, Boundary Elements VII, (Eds. Brebbia, C. A. and Maier, G.), Computational Mechanics Publications, Southampton, 1985.

9. Umetani, S-I., Adaptive Boundary Element Methods in Elastostatics, Computational Mechanics Publications, Southampton, 1988.

10. Urekew, T. J., and Rencis, J. J., Absolute p-Refinement of Boundary Elements for Two-Dimensional Elasticity Problems, Boundary Elements XI, (Eds. Brebbia, C. A. and Connor, J. J.), Computational Mechanics Publications, Southampton, 1989.

11. Zienkiewicz, O. C., Gago, J., and Kelly, D. W., The Hierarchical Concept in Finite Element Analysis, Computers and Structures, Vol. 16, pp. 53-65, 1983.

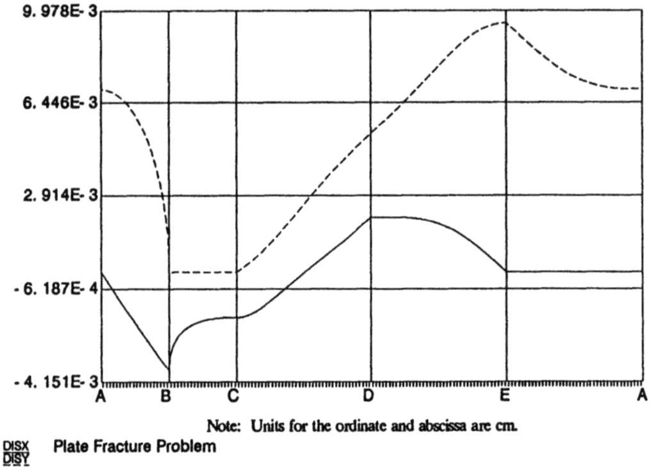

Figure 4 - Displacements from 240 element mesh.

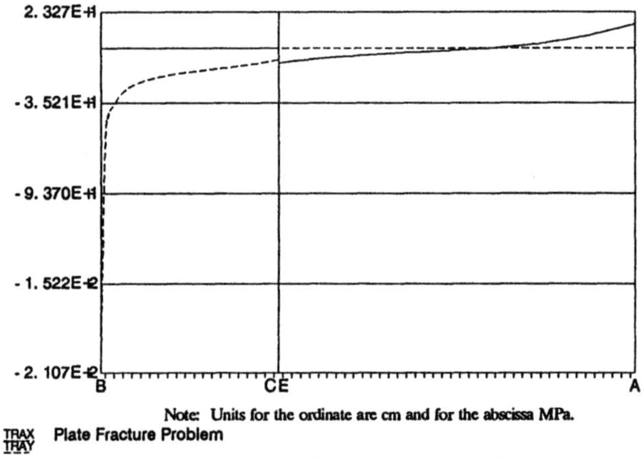

Figure 5 - Tractions from 240 element mesh.

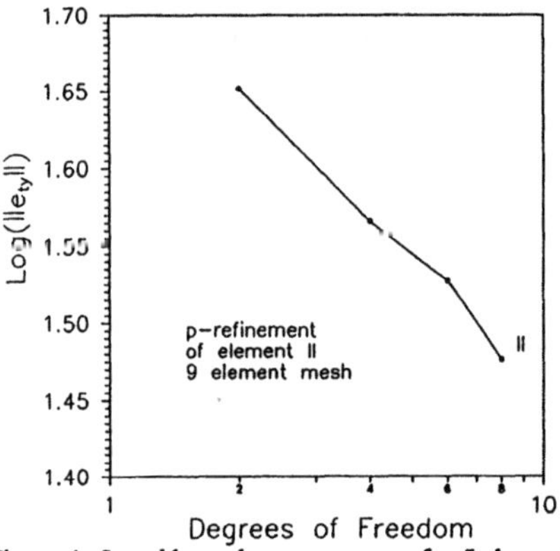

Figure 6 - Local boundary error norm for 5 element mesh.

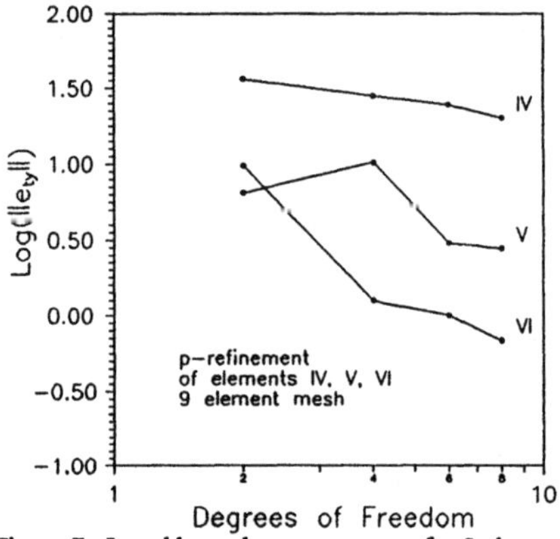

Figure 7 - Local boundary error norm for 9 element mesh.

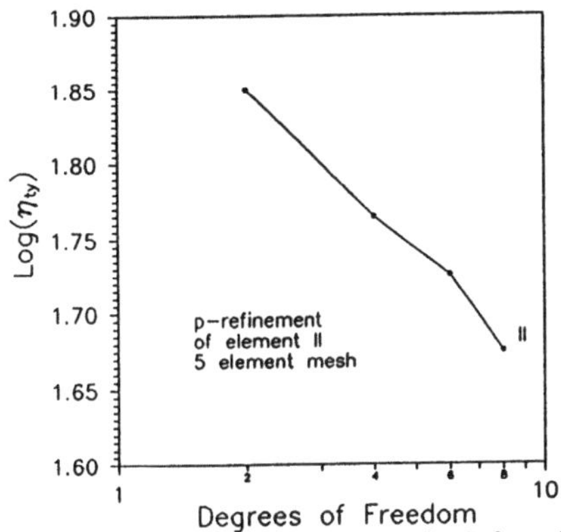

Figure 8 - Local relative percent error for 5 el. mesh.

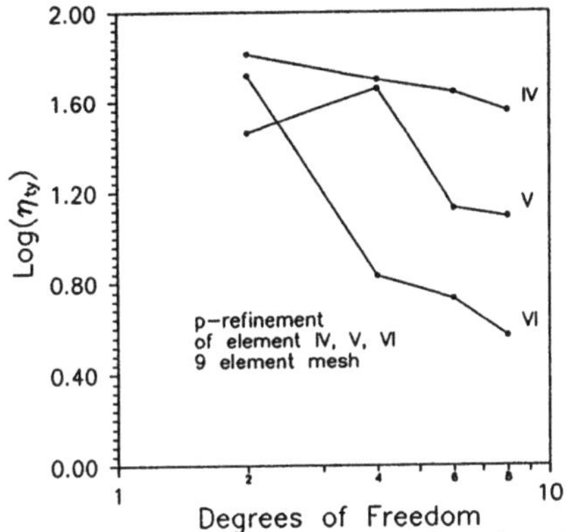

Figure 9 - Local relative percent error for 9 el. mesh.

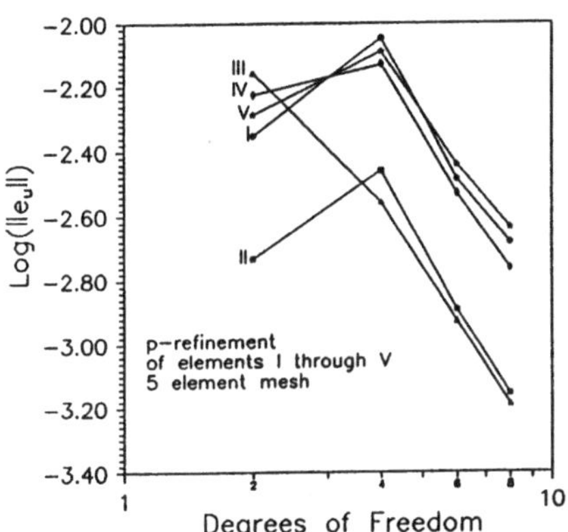

Figure 10 - Local boundary error norm for 5 element mesh.

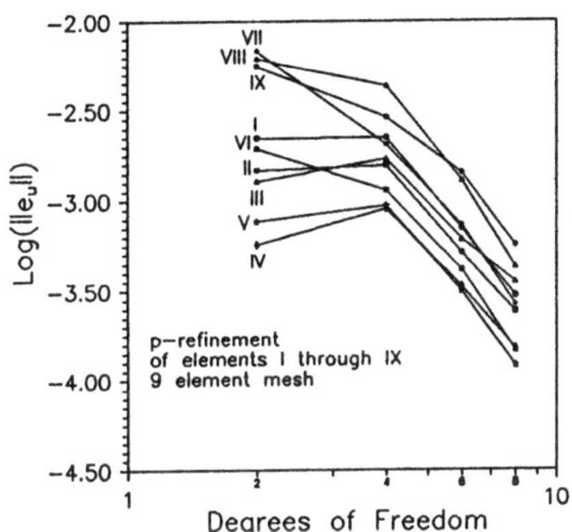

Figure 11 - Local boundary error norm for 9 element mesh.

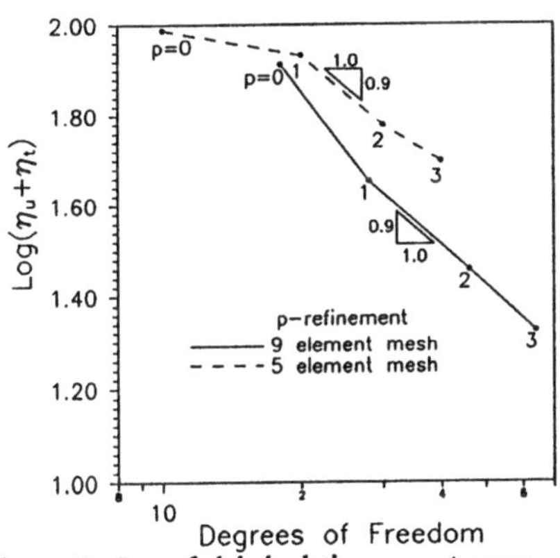

Figure 12 - Sum of global relative percent errors.

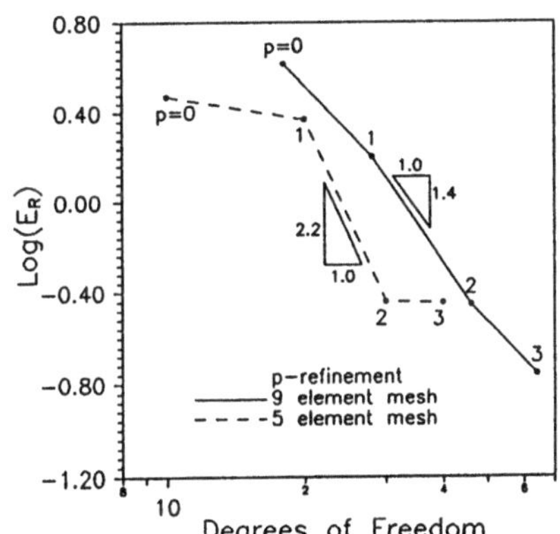

Figure 13 - Equilibrium residual estimator.

A Boundary Element Algorithm for Plate Bending Problems Based on Cauchy's Integral Formula

R. Piltner and R.L. Taylor
Department of Civil Engineering, University of California at Berkeley,
Berkeley, CA 94720, U.S.A.

SUMMARY

A representation for plate bending solutions (Kirchhoff-plate theory) in terms of two arbitrary complex valued functions is considered. The complex functions are chosen in the form of Cauchy integrals which are discretized along the boundary of the plate. Using the complex representation we a priori ensure the satisfaction of the homogeneous plate equation. An inhomogeneous (particular) solution is added. The constructed boundary functions for the deflection, the moments and the shear forces are compatible.

INTRODUCTION

The concept of using representations of elasticity solutions in terms of arbitrary functions has been used for finite element calculations [1-4] as well as for boundary element computations [5]. The functions in the numerical procedure are arbitrary in the sense that with any choice of functions of a characteristic complex variable the governing differential equations are satisfied. Representations of elasticity solutions in terms of arbitrary complex functions are given in references [6-10]. Using such representations we are able to construct compatible boundary displacements and stresses (or resultants) for the boundary element procedure. The use of compatible boundary functions is the main difference of the present algorithm compared to conventional boundary element procedures where independent displacements and tractions are used [11-13].

REPRESENTATION OF THE PLATE DEFLECTION AND THE RESULTANTS

The solution of the (Kirchhoff-) plate equation can be given as

$$w=\frac{1}{2}[\,z\overline{\Phi(z)} + \bar{z}\Phi(z) \;+\; \Psi(z) + \overline{\Psi(z)}] + w_p \tag{1}$$

where Φ and Ψ are functions of the complex variable $z = x + iy$ and w_p is a particular solution for a given load distribution. For a constant load q_0 we can choose the particular solution as

$$w_p = \frac{q_0}{48D}(x^4 + y^4) \tag{2}$$

where $D = Eh^3/[12(1-\nu^2)]$. For the moments and shear forces we have

$$\begin{aligned}
M_{xx} &= -D(w_{xx} + \nu w_{yy}) \\
M_{yy} &= -D(w_{yy} + \nu w_{xx}) \\
M_{xy} &= D(1 - \nu)w_{xy} \\
Q_{xx} &= -D\frac{\partial}{\partial x}\Delta w \\
Q_{yy} &= -D\frac{\partial}{\partial y}\Delta w
\end{aligned} \tag{3}$$

where

$$\begin{aligned}
w_{xx} &= \mathrm{Re}[\Phi'(z) + \overline{\Phi'(z)} + z\overline{\Phi''(z)} + \overline{\Psi''(z)}] + \frac{q_0}{4D}x^2 \\
w_{yy} &= \mathrm{Re}[\Phi'(z) + \overline{\Phi'(z)} - z\overline{\Phi''(z)} - \overline{\Psi''(z)}] + \frac{q_0}{4D}y^2 \\
w_{xy} &= \mathrm{Im}[z\overline{\Phi''(z)} + \overline{\Psi''(z)}] \\
\Delta w &= 4\mathrm{Re}[\Phi'(z)] + \frac{q_0}{4D}(x^2 + y^2) \\
\frac{\partial}{\partial x}\Delta w &= 4\mathrm{Re}[\Phi''(z)] + \frac{q_0}{2D}x \\
\frac{\partial}{\partial y}\Delta w &= -4\mathrm{Im}[\Phi''(z)] + \frac{q_0}{2D}y
\end{aligned} \tag{4}$$

For our boundary element procedure we choose the complex functions Φ and Ψ in the form of Cauchy integrals. As both functions are assumed in a similar way we write now f instead of Φ and Ψ in order to explain the form and discretization of the complex functions.

A complex function f(z) in a domain Ω with the boundary Γ can be expressed with the aid of its boundary values along Γ by the Cauchy integral formula [14-16]

$$f(z) = \frac{1}{2\pi i}\int_\Gamma \frac{f(\zeta)}{\zeta - z}d\zeta. \tag{5}$$

After the discretization of the boundary into boundary elements with the boundary portions Γ_j and the choice of a basis function $\hat{f}^j$ for every element we use Cauchy's integral formula in the discretized form

$$f(z) = \frac{1}{2\pi i}\sum_j \int_{\Gamma_j} \frac{\hat{f}^j(\zeta)}{\zeta - z}d\zeta. \tag{6}$$

THE CHOICE OF COMPLEX BASIS FUNCTIONS

The basis functions $\hat{f}^j$ have to be chosen such that all limit values on the boundary exist. If we would choose for example a linear basis function $\hat{f}^j$ we would get no limit values for $f'(z)$ and $f''(z)$ for the node points of the boundary elements. In order to get limit values for the deflection, the moments and the shear forces given by eqs. (1) and (3) we have to use at least fifth order basis functions.

Choosing straight line boundary elements we can perform the integration in (6) exactly. The use of straight line boundary elements is no particular restriction as we can choose all collocation points on the curved boundary if the straight line boundary element discretization encloses the real curved boundary. Using the discrete node values f_j, f_j' and f_j'' we can write the fifth order boundary element trial functions as

$$\hat{f}^j(\zeta(s)) = N_1(s)\, f_{j-1} + N_2(s)\, f_{j-1}'(z_j - z_{j-1}) + N_3(s)\, f_{j-1}''(z_j - z_{j-1})^2 + N_4(s)\, f_j + N_5(s)\, f_j'(z_j - z_{j-1}) + N_6(s)\, f_j''(z_j - z_{j-1})^2 \tag{7}$$

where

$$\begin{aligned} N_1(s) &= [\,2 - 20s^3 + 30s^4 - 12s^5\,]/2 \\ N_2(s) &= [\,2s - 12s^3 + 16s^4 - 6s^5\,]/2 \\ N_3(s) &= [\,s^2 - 3s^3 + 3s^4 - s^5\,]/2 \\ N_4(s) &= [\,20s^3 - 30s^4 + 12s^5\,]/2 \\ N_5(s) &= [\,-8s^3 + 14s^4 - 6s^5\,]/2 \\ N_6(s) &= [\,s^3 - 2s^4 + s^5\,]/2 \end{aligned} \tag{8}$$

and

$$s = \frac{\zeta - z_{j-1}}{z_j - z_{j-1}} \tag{9}$$

The local boundary coordinate s is chosen in the interval $0 \le s \le 1$ so that $s = 0$ is associated to $\zeta = z_{j-1}$ whereas $s = 1$ is associated to $\zeta = z_j$. After the substitution of (7) into the Cauchy integral formula we can integrate analytically and get the element contribution $f^j(z)$ for boundary element j in the form

$$f^j(z) = \frac{1}{4\pi i}\Bigg\{ f_{j-1}\left[-\frac{47}{30} - 3K - 9K^2 + 24K^3 - 12K^4 + 2N_1(K)\ln\frac{z_j - z}{z_{j-1} - z} \right]$$

$$+ (z_j - z_{j-1})f_{j-1}'\left[\frac{4}{5} - \frac{13}{6}K - 6K^2 + 13K^3 - 6K^4 + 2N_2(K)\ln\frac{z_j - z}{z_{j-1} - z} \right]$$

$$+ (z_j - z_{j-1})^2 f''_{j-1} \left[\frac{1}{20} + \frac{1}{4}K - \frac{11}{6}K^2 + \frac{5}{2}K^3 - K^4 + 2N_3(K) \ln\frac{z_j - z}{z_{j-1} - z} \right]$$

$$+ f_j \left[\frac{47}{30} + 3K + 9K^2 - 24K^3 + 12K^4 + 2N_4(K) \ln\frac{z_j - z}{z_{j-1} - z} \right]$$

$$+ (z_j - z_{j-1}) f'_j \left[-\frac{11}{30} - \frac{5}{6}K - 3K^2 + 11K^3 - 6K^4 + 2N_5(K) \ln\frac{z_j - z}{z_{j-1} - z} \right]$$

$$+ (z_j - z_{j-1})^2 f''_j \left[\frac{1}{30} + \frac{1}{12}K + \frac{1}{3}K^2 - \frac{3}{2}K^3 + K^4 + 2N_6(K) \ln\frac{z_j - z}{z_{j-1} - z} \right] \Bigg\} \qquad (10)$$

where

$$K = \frac{z - z_{j-1}}{z_j - z_{j-1}}. \qquad (11)$$

The approximation function f(z) is obtained by adding all element contributions $f^j(z)$. In our example of a fifth order trial basis function the approximation function f(z) has the form

$$f(z) = \frac{1}{2\pi i} \Big[\quad f_1 F_1(z) + f_1' G_1(z) + f_1'' H_1(z) + \cdots \qquad (12)$$

$$\cdots + f_j F_j(z) + f_j' G_j(z) + f_j'' H_j(z) + \cdots$$

$$\cdots + f_N F_N(z) + f_N' G_N(z) + f_N'' H_N(z) \quad \Big]$$

If z approaches a node point z_j on the boundary we get expressions of the form $\infty - \infty$ and $0 * \infty$. However, all limit values for f(z), f'(z), f''(z) exist for z approaching z_j. For example we get for f at the node point z_j the simple expression

$$\lim_{z \to z_j} f(z) = \frac{1}{4\pi i} \Bigg\{ \quad f_j \left[2\ln\frac{z_j - z_{j+1}}{z_j - z_{j-1}} \right] \qquad (13)$$

$$+ f_j' \left[\frac{4}{5} \left\{ (z_j - z_{j-1}) + (z_{j+1} - z_j) \right\} \right]$$

$$+ f_j'' \left[\frac{1}{20} \left\{ -(z_j - z_{j-1})^2 + (z_{j+1} - z_j)^2 \right\} \right]$$

$$+ \quad \text{remaining terms} \Bigg\}$$

Using complex functions as described above for the evaluation of the plate deflection and the resultants we have as unknowns the complex node values Φ_j, Φ_j', Φ_j'' and Ψ_j, Ψ_j', Ψ_j''. Requiring the plate boundary conditions to be satisfied for selected collocation points on the boundary, we obtain the system of equations for the unknowns. Collocation points include the boundary element nodes as well as points between the nodes. More details about the construction of complex approximation functions and their derivatives are given in reference [5].

CHOICE OF BRANCH CUTS OF THE COMPLEX LOGARITHMIC FUNCTIONS

A crucial point in the proposed boundary element algorithm is the right choice of branch cuts for the multivalued complex logarithmic terms. Since in our case the logarithmic terms contain a fraction we can rewrite these terms as

$$\ln\frac{z_{j+1}-z}{z_j-z} = \ln(z_{j+1}-z) - \ln(z_j-z) \tag{14}$$

$$= \ln r_{j+1} + i\,\phi_{j+1} - \ln r_j - i\,\phi_j$$

The angles ϕ_j and ϕ_{j+1} have to be chosen such that no branch cut crosses the solution domain under consideration because this would cause a jump of function values inside the domain.

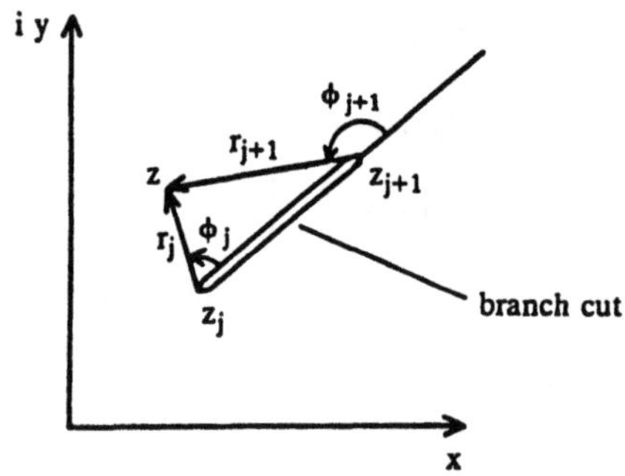

Figure 1. Choice of a branch cut for $\ln[(z_{j+1}-z)/(z_j-z)]$

The only generally applicable choice of branch cuts is to define cuts of finite length as shown in figure 1. In this case the branch cut lies on the boundary element. One surface of the branch cut is identical with the boundary portion Γ_j whereas the other surface lies outside the solution domain considered. For the complex logarithm we have

$$\ln\frac{z_{j+1}-z}{z_j-z} = \ln\left|\frac{z_{j+1}-z}{z_j-z}\right| + i\gamma \tag{15}$$

where

$$\gamma = \arg\left([z-z_{j+1}]e^{-i\alpha_{j+1,j}}\right) - \arg\left([z-z_j]e^{-i\alpha_{j+1,j}}\right) \tag{16}$$

$$\alpha_{j+1,j} = \arctan\frac{y_{j+1}-y_j}{x_{j+1}-x_j} \tag{17}$$

NUMERICAL EXAMPLES

For the numerical analysis of rectangular plates with simply supported and clamped boundary conditions the proposed boundary element procedure has been used. The results of two discretizations are given in Table I and Table II. Very good agreement with the exact solution which is also given in the tables was achieved using rather coarse discretizations.

Table I: Results at the center of a simply supported rectangular plate (a=4, b=6, E=1, ν=0.3, t=0.2, uniform load q_0=1.0)

	4 boundary elements	8 boundary elements	Timoshenko [17]
w	2699.54	2698.70	2697.60
M_{xx}	1.2985	1.2984	1.2992
M_{yy}	0.7976	0.7974	0.7984

Table II: Results for a clamped rectangular plate (a=4, b=6, E=1, ν=0.3, t=0.2, uniform load q_0=1.0 , $-a/2 \leq x \leq a/2$, $-b/2 \leq y \leq b/2$)

	4 boundary elements	8 boundary elements	Timoshenko [17]
w(x=0,y=0)	765.47	768.07	768.00
M_{xx}(x=a/2,y=0)	-1.2106	-1.2109	-1.2112

CLOSURE

A boundary element procedure with compatible displacements and resultants has been considered for plate bending applications. All integrations were performed analytically. For the results shown in this paper collocation points have been used to generate a (nonsymmetric) system of equations. However, since all boundary functions are compatible we can get also symmetric formulations. Hence, the complex functions presented may be used to obtain symmetric stiffness matrices for finite elements via boundary integral evaluations and have been applied to obtain plate bending finite elements [18]. A similar way to obtain symmetric finite element matrices via boundary integral evaluations has been successfully applied in [1-4].

ACKNOWLEDGEMENT

The first author gratefully acknowledges his support from the DFG (Deutsche Forschungsgemeinschaft).

REFERENCES

1. Piltner, R., Spezielle finite Elemente mit Löchern, Ecken und Rissen unter Verwendung von analytischen Teillösungen, Doctoral Thesis, Ruhr-Universität Bochum, Fortschr.-Ber. VDI-Z. Reihe 1 Nr. 96, VDI-Verlag, Düsseldorf, 1982.
2. Piltner, R. Special finite elements with holes and internal cracks, Int. J. Num. Methods Eng., 21, 1471 - 1485, 1985.
3. Piltner, R. Finite Elemente mit Ansätzen im Trefftz'schen Sinne, pp. 267 - 278, In: Finite Elemente - Anwendung in der Baupraxis, (Eds. H. Grundmann, E. Stein, W. Wunderlich), Verlag Ernst + Sohn, Berlin/ München/ Düsseldorf , 1985.
4. Piltner, R. Special finite elements for an appropriate treatment of local effects, pp. 299 - 314, in: Local Effects in the Analysis of Structures, (Ed. P. Ladeveze), Elsevier Science Publishers, 1985.
5. Piltner, R. and Taylor, R.L. A boundary element algorithm using compatible boundary displacements and tractions, Int. J. Num. Meth. Eng., (to appear).
6. Muskhelishvili, N.I., Some Basic Problems of the Mathematical Theory of Elasticity, Noordhoff, Groningen, Holland , 1953.
7. Piltner, R., The use of complex valued functions for the solution of three-dimensional elasticity problems, Journal of Elasticity, 18, 191 - 225, 1987.
8. Piltner, R., The representation of three-dimensional elastic displacement fields with the aid of complex valued functions for several curvilinear coordinates, Mechanics Research Communications, 15 (2), 79 - 85, 1988.
9. Piltner, R., The application of a complex 3-dimensional elasticity solution representation for the analysis of a thick rectangular plate, Acta Mechanica, 75, 77 - 91, 1988.
10. Piltner, R., On the representation of three-dimensional elasticity solutions with the aid of complex valued functions, Journal of Elasticity, (to appear).
11. Banerjee, P.K. and Watson, J.O. Developments in Boundary Element Methods - 4, Elsevier, London/ New York, 1986.
12. Brebbia, C.A., Telles, J.C.F. and Wrobel, L.C. Boundary Element Techniques, Springer, Berlin/ Heidelberg/ New York/ Tokyo, 1984.
13. Cruse, T.A. (Ed.), Advanced Boundary Element Methods, Proceedings of the IUTAM Symposium San Antonio, Texas, April 1987, Springer, Berlin/ New York/ Tokyo, 1988.
14. Henrici, P. Applied and Computational Complex Analysis, Wiley, New York/ London, 1986.
15. Spiegel, M.R. Complex Variables, Mc Graw-Hill, New York, 1964.
16. Marsden, J.E. and Hoffman, M.J. Basic Complex Analysis, Freeman and Company, New York, 1987.
17. Timoshenko, S., Theory of Plates and Shells, McGraw-Hill, New York/London, 1940.
18. Piltner, R. and Taylor, R.L., The evaluation of stiffness matrices for elasticity problems with the aid of boundary integrals, Proceedings of the NUMETA 90 Conference, (to appear).

Stepwise Linear Regression Particular Integrals for Uncoupled Thermoelasticity with Boundary Elements

Anil Gupta, Worcester Polytechnic Institute, Worcester, Massachusetts 01609
Jizu Cheng, Worcester Polytechnic Institute, Worcester, Massachusetts 01609
Sunil Saigal, Carnegie Mellon University, Pittsburgh, Pennsylvania 15213

Summary
This paper presents a simple technique for the efficient treatment of non-uniform body forces in boundary elements without the need for domain discretization. The distribution of the non-uniform forces in the domain is described using regression polynomials. Particular integrals corresponding to these polynomials are developed for the solution of the non-homogeneous differential equations of elasticity. The method allows the use of low order polynomials to model a general distribution of thermal effects in the domain. Numerical data is given to provide comparisons of boundary only data versus combined boundary and domain data based regression polynomials.

Introduction

The Boundary Element Method (BEM) primarily involves the discretization of only the boundary of the object for a linear elastic analysis. The initial attempts to include the effects of body forces [1] required the discretization of the domain into cells. The volume integrals involved may be reduced to equivalent surface integrals for conservative body forces [2,3]. A technique based on the particular integral method for the solution of non-homogeneous ordinary differential equations was presented in References [4,5] to account for conservative body forces through direct nodal calculations. Recently, Henry and Banerjee [6] developed a particular integral formulation for non-conservative body forces and presented the results for thermoelastic stress analysis. In this formulation the temperature distribution in each region of a multi-region boundary discretization is represented using global shape functions whose coefficients are determined based on the temperatures at boundary element nodes and some interior points. The particular integrals corresponding to the global shape functions are then developed independently for each sub-region for use in BEM analysis.

A different approach, also based on the particular integral method for the treatment of non-conservative body forces, is presented in this paper. This approach starts by representing the distribution of the body forces in a domain using polynomials that are based on the statistical procedure of linear regression. The process of linear regression forces the interpolation error to be normal to the domain in a least squared sense [7]. This results in a more accurate representation of the body force field compared to other available interpolations. Also, any general distribution can be approximated using a polynomial which retains the minimum number of terms required for an accurate interpolation. The number of terms retained does not have to be equal to the number of data points used to generate the regression polynomial. After obtaining the regression polynomials representing a general distribution of body forces, particular integrals corresponding to these polynomials are presented. The regression particular integral formulation is implemented in a multi-zone boundary element program. The results obtained indicate good accuracy.

Thermoelastic BEM with Particular Integrals

The linear, non-homogeneous differential equation of equilibrium for an elastic body is given as

$$\zeta(u_i) = f_i \quad ; \quad \zeta = (\lambda + \mu)u_{j,ji} + \mu u_{i,jj} \tag{1}$$

λ, μ are the Lame's constants, u_i is the displacement and f_i is the body force. For thermal loading, $f_i^{(T)} = \alpha(3\lambda + 2\mu) T_{,i}$, where α is the coefficient of thermal expansion and T is the temperature. The particular integral method proceeds by first decomposing the response variables into a complementary (c) and a particular integral (p) components as

$$u_i = u_i^c + u_i^p \quad ; \quad t_i = t_i^c + t_i^p \tag{2}$$

For thermoelastic displacements, u_i^c and u_i^p are given by the equations

$$\zeta(u_i^c) = 0 \quad \text{, and} \quad \zeta(u_i^p) = \alpha(3\lambda + 2\mu) T_{,i} \tag{3}$$

Following the theory of linear non-homogeneous differential equations, it is known that the particular integral solution u_i^p is not unique and any expression satisfying the non-homogeneous equation (3) is a valid particular integral. However the particular integral solution for computational purposes must be chosen carefully.

For the BEM, the integral equation due to Somigliana satisfies the homogeneous equation (3) and is used to describe the complementary solution as

$$C_{ij}(\xi)\, u_i^c(\xi) = \int_\Gamma \left[G_{ij}(x,\xi)\, t_i^c(x) - F_{ij}(x,\xi)\, u_i^c(x)\right] d\Gamma(x) \tag{4}$$

where G_{ij} and F_{ij} are the fundamental solutions for displacements and tractions, respectively; C_{ij} is the corner tensor; and Γ is the boundary of the object. Discretizing the integral equation (4) using boundary elements leads to a system of matrix equations given as $[F]\{u^c\} = [G]\{t^c\}$. Substituting for $\{u^c\}$ and $\{t^c\}$ from equation (2), and after some manipulation, we get,

$$[F]\{u\} = [G]\{t^{(e)}\} + [F]\{u^P\} - [G]\{\hat{t}^p\} \tag{5}$$

where $\{t^{(e)}\} = \{t\} - (3\lambda + 2\mu)\,\alpha T\,\{n\}$, $\quad \{\hat{t}^p\} = \{t^P\} - (3\lambda + 2\mu)\,\alpha T\,\{n\}$

$\{t^{(e)}\}$ is the externally applied traction and the specified traction boundary conditions are applied to this quantity, $\{n\}$ is the outward normal vector at the boundary collocation point. The matrix $\{\hat{t}^p\}$ is referred to here as the modified particular integral vector. It is noted that the vectors $\{u^p\}$ and $\{t^P\}$ may include the contributions due to the various types of body forces acting on the object, such as gravitational, centrifugal, and thermal.

Regression Model for Temperature Distributions

The following discussion holds for any non-conservative body force with a general distribution over the domain of the body. Only the case of thermal body force is discussed here. It is not generally possible to obtain the particular solutions $\{u^p\}$ and $\{t^P\}$ for a general distribution of body forces. To enable the determination of particular integral solutions for these body forces, their distribution is first represented in a form for which it is possible to obtain such particular integral solutions. In the present study, the distribution of body forces is represented using a linear regression model.

A simple regression model for a system defined by a single variable can be written as

$$y_i = \beta_0 + \beta_1 x_i + \varepsilon_i \tag{6}$$

where x_i are the values of the independent variables, for example, geometric coordinates; y_i are the values of the response variables, for example, temperature; β_0 and β_1 are the regression coefficients; and ε_i is an error term with mean $E(\varepsilon_i) = 0$. For the present case, it is required to determine the regression coefficients β_0 and β_1 based on the quantities (x_i, y_i) available at N data points. To find "good" estimators of the regression parameters, the method of least squares is employed for which β_0 and β_1 have the values b_0 and b_1, respectively, that minimize the quantity Q given by

$$Q = \sum_{i=1}^{N} (y_i - \beta_0 - \beta_1 x_i)^2 \tag{7}$$

To minimize Q, its partial derivatives with respect to the coefficients to be determined are set equal to zero. Thus, b_0 and b_1 can be obtained from the simultaneous equations

$$\sum_{i=1}^{N} y_i - N b_0 - b_1 \sum_{i=1}^{N} x_i = 0 \quad ; \quad \sum_{i=1}^{N} x_i y_i - b_0 \sum_{i=1}^{N} x_i - b_1 \sum_{i=1}^{N} x_i^2 = 0 \tag{8}$$

The above procedure describes a simple regression model for one independent variable. A similar procedure is applied for more than one independentvariables. For such cases not all terms of the regression model are retained. The decision for retaining or deleting a term corresponding to a variable or product of variables may be based on some specified criterion. In the present study, a stepwise regression scheme in conjunction with the F* statistics criteria [7] was employed in obtaining the regression model. The stepwise regression scheme proceeds by adding a new term to the regression model and thereafter applying F* criteria, term by term, to determine if the existing terms should be retained or deleted. This procedure thus results in a robust representation of the body force field being modeled. The regression models were obtained in this study using the subroutine library SPSSX [8].

Particular Integrals for Regression Models

The particular integral expressions for terms up to the third order in the regression model are presented below. Corresponding expressions for terms of order higher than three can be similarly developed. Consider the regression model given by the polynomial

$$T = b_1 + b_2 x + b_3 y + b_4 x^2 + b_5 y^2 + b_6 xy + b_7 x^3 + b_8 y^3 + b_9 x^2 y + b_{10} xy^2 \tag{9}$$

The particular integral expressions corresponding to the regression model in equation (9) are

$$u_x^p = -\frac{c_1 x^2 + c_2 y^2}{8.0\mu c_3} b_2 + \frac{c_2 xy}{4.0\mu c_3} b_3 + \frac{\mu x y^2}{c_1 c_2} b_4 - \frac{c_3 x y^2}{c_1 c_2} b_5 - \frac{y^3}{6.0\mu} b_6 - \frac{x^4}{4.0 c_1} b_7 - \left(\frac{x^3 y}{6.0\mu} + \frac{xy^3}{6.0 c_1}\right) b_9 - \left(\frac{y^4}{12.0\mu} - \frac{c_3 x^2 y^2}{4.0\mu c_1}\right) b_{10} \tag{10}$$

$$u_y^p = \frac{c_2 xy}{4.0\mu c_3} b_2 - \frac{c_2 x^2 + c_1 y^2}{8.0\mu c_3} b_3 - \frac{c_3 y x^2}{c_1 c_2} b_4 + \frac{\mu y x^2}{c_1 c_2} b_5 - \frac{x^3}{6.0\mu} b_6 - \frac{y^4}{4.0 c_1} b_8 - \left(\frac{x^4}{12.0\mu} - \frac{c_3 x^2 y^2}{4.0\mu c_1}\right) b_9 - \left(\frac{y^3 x}{6.0\mu} + \frac{x^3 y}{6.0 c_1}\right) b_{10} \tag{11}$$

where $c_1 = \lambda + 2\mu$, $c_2 = \lambda$, and $c_3 = 1 + \mu$. For the terms in equation (9) that are not selected by the stepwise regression procedure explained before, the corresponding coefficients are set equal to zero in equations (9)-(11). To obtain the particular integrals for tractions, the strains ε^P_{ij} corresponding to the displacement particular integrals in equations (10) and (11) are first determined from the strain-displacement relationship of elasticity. The stresses σ^P_{ij} are then obtained using Hooke's law. Finally, the traction solutions are obtained using the relation $t_i^P = \sigma_{ij}^P n_j$, where n_j is the outward normal. The particular integral expressions for the axisymmetric case can be similarly determined.

Data Points for Regression Model

To obtain a regression model representative of the distribution of temperatures in the domain, the model should be obtained based on a sufficient number of data points lying on the boundary or within the domain. The computation of temperatures at any location in the domain of the object, however, requires additional computation in the BEM approach. For the case of steady state thermoelasticity, since the extreme temperatures occur on the boundary, it is sufficient to use data points only on the boundary. The transient thermoelasticity problems may also be treated by considering only the boundary data points. This may not, however, provide a very accurate response especially for cases when high temperature gradients exist in the domain. To obtain an improved response, some internal points may be used in addition to the boundary points for the regression model. The inclusion of these internal points does not add to the number of unknowns but serves merely to improve the quality of the regression model. In case of multi-zone BEM analysis, a separate regression model may be developed for each zone to provide a better representation of the distribution of temperatures in the domain. For large domains it is recommended that these domains be broken up into zones as the regression polynomial representation of temperature field in a large domain may not lead to very accurate results.

Numerical Results

The above formulation was applied to obtain numerical results for both steady state and transient thermoelastic responses. A representative example is presented here to demonstrate the validity of the present formulation.

A rectangular strip with a circular hole subjected to a harmonic temperature distribution was examined. A half symmetry model of the strip along with the geometric data are shown in Figure 1. The material data used was: modulus of elasticity,E = 1000 psi; Poisson's ratio, υ = 0.3; Coefficient of thermal expansion, α = 0.02 in/in/°F. The harmonic temperature distribution on the strip was given by $T(x,y) = x^2 - y^2 - 4.0$. The half strip was modeled using 17 continuous quadratic boundary elements, all of equal length. The results were obtained for three cases. Case I : Using the surface integral formulation [2] for thermal body forces caused by harmonic temperature distribution ; Case II : Using the particular integral formulation based on exact temperature distribution given by T(x,y) ; and Case III : Using the particular integral formulation based on the regression models developed in this paper for representing the temperature distribution. It is noted that for Case III the regression model produced the exact distribution given by T(x,y) using only the boundary point information. However, these models were constrained so that the terms with x^2 and y^2 both have a coefficient of zero. Two different regression models were generated. The first model (Case III (a)) was based solely on the boundary points and the second model (Case III (b)) was based on the boundary points as well as 6 internal points as shown in Figure 1. The results for all these cases are shown in Table I and II . It is seen that for an exact representation as in Case II , the particular integral technique produces very accurate results. For approximate representations of the temperature field such as those using the regression models (Case III (a) and Case III (b)) an improved accuracy is obtained as the number of data points used to create the regression model are increased. Good agreement is seen for both the displacements and stresses due to thermoelastic behavior.

Concluding Remarks

The distribution of temperature within the domain of a body is represented using a regression model. The model is based on the temperature data at the boundary element nodes and a few interior points. A stepwise regression technique that ensures the selection of only the most effective terms is used to obtain the regression model. This model uses lower order polynomials and fewer terms to effectively represent the temperature distribution in a domain. Particular integrals are developed for the regression model obtained to account for the thermal body forces. The present technique avoids the need for subdividing the domain into cells for computing volume integrals for the treatment of non-uniform body forces. In this paper the non-uniform body force field is represented using a regression model which provides an approximate representation of this field. For large domain, this representation will provide only an approximate response but at the same time will highlight the high stress regions. A numerical example is presented to demonstrate the effectiveness of the present technique.

Acknowledgements
Anil Gupta was supported by the Wyman Gordon Company, Worcester, MA. Sunil Saigal was supported by the National Science Foundation grant DMC-8708254 with Dr. Tony Woo as the project monitor.

References

1. Banerjee, P.K.; Butterfield, R. Boundary Element Methods in Engineering Science, McGraw-Hill, London, 1983

2. Cruse, T.A.; Snow, D.W.; Wilson, R.B. : Numerical Solutions in Axisymmetric Elasticity. Computers and Structures. 7 (1977), 445-451.

3. Rizzo, F.J. ; Shippy, D.J. : An Advanced Boundary Integral Equation Method for Three-Dimensional Thermoelasticity. International Journal for Numerical Methods in Engineering.11 (1977), 1753-1768.

4. Henry, D.P., Jr. ; Pape, D.A. ; Banerjee, P.K. : A New Axisymmetric BEM Formulation for Body Forces Using Particular Integrals. Journal of Engineering Mechanics, ASCE. 113 (1987), 671-688.

5. Pape, D.A. ; Banerjee, P.K. : Treatment of Body Forces in Two-Dimensional Elastostatic BEM Using Particular Integrals. Journal of Applied Mechanics, ASME. 54 (1987), 866-871.

6. Henry, D.P., Jr. ; Banerjee, P.K. : A New Boundary element Formulation for Two- and Three- Dimensional Thermoelasticity Using Particular Integrals. International Journal for Numerical Methods in Engineering. 26 (1988), 2061-2078.

7. Neter, J. ; Wasserman, W. ; Kutner, M.H. : Applied Linear Statistical Models, 2nd Edition. R.D. Irwin, Inc., Homewood, Illinois, 1985.

8. SPSSX User's Guide. SPSS Inc., Suite 3300, 444 North Michigan Avenue, Chicago, IL, 1983.

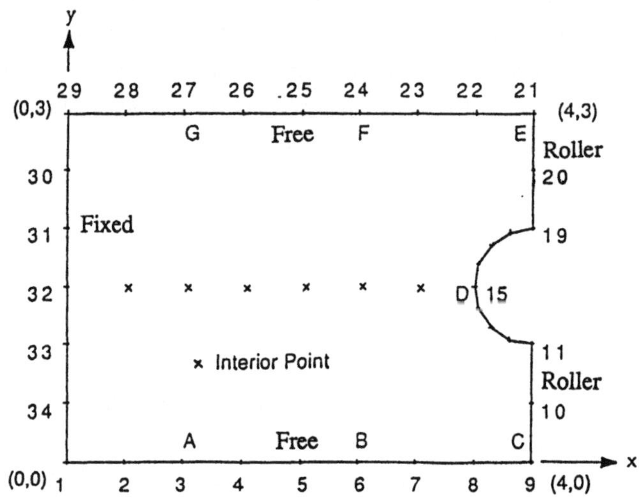

Figure 1. A Thick Strip With a Central Circular Hole

Table I. Comparison of Displacements in Two Dimensional Response Using Surface Integral Approach and Regression Particular Integral Approach for a Thick Strip With a Central Circular Hole.

Node	DISPLACEMENT (inches)							
	CASE I		CASE II		CASE III (a)		CASE III (b)	
	U_x	U_y	U_x	U_y	U_x	U_y	U_x	U_y
A	-1.492	1.486	-1.491	1.487	-1.442	1.479	-1.448	1.485
B	-2.923	-0.201	-2.922	-0.200	-2.918	-0.142	-2.927	-0.144
C	0.000	-2.552	0.000	-2.552	0.000	-2.536	0.000	-2.539
D	-1.287	0.554	-1.280	0.556	-1.282	0.549	-1.284	0.549
E	0.000	1.023	0.000	1.023	0.000	1.014	0.000	1.017
F	-3.062	-1.318	-3.061	-1.318	-3.056	-1.371	-3.065	-1.368
G	-1.686	-3.183	-1.686	-3.183	-1.636	-3.168	-1.641	-3.173

Table II. Comparison of Stresses in Two Dimensional Response Using Surface Integral Approach and Regression Particular Integral Approach for a Thick Strip With a Central Circular Hole.

Y - COORD. (inch)	STRESS AT FIXED END, σ_{xx} (psi)			
	CASE I	CASE II	CASE III (a)	CASE III (b)
0.00	164.11	164.21	163.07	163.80
0.50	-52.15	-52.13	-49.32	-49.38
1.00	-1.63	-1.60	-1.62	0.02
1.50	10.55	10.58	9.16	9.29
2.00	78.29	78.33	79.05	79.12
2.50	56.06	56.13	58.08	57.89
3.00	636.05	636.21	634.25	634.85

Boundary Element Method for Composite Bodies

Madhukar Vable and Yikang Zhang
Mechanical Engineering and Engineering Mechanics,
Michigan Technological University
Houghton, MI 49931, U.S.A.

1.0 Summary:

A general algorithm, based on Boundary Element Method for two dimensional composite bodies is presented. The composite body is viewed as a homogeneous orthotropic body. Unlike an isotopic body, an orthotropic body has 3 possible forms of fundamental solutions, depending upon the relationship of the four material constants of an orthotropic body. The structure of these fundamental solution is discussed briefly. Discretization of the problem results in integrals over straight line segments. The integrand of these integrals is a product of the fundamental solutions and a polynomial. Analytical expressions for any order of polynomial were found using an iterative scheme. The key to obtaining the analytical expressions is a set of transformation equations and the structure of the fundamental solution. These are discussed briefly in this paper. Results are presented for an elastostatic problem, a contact problem that arises in a pin fastened composite and a plate bending problem.

2.0 Introduction

Composite materials hold enormous promise and the impact of the composite material on gross national product, balance of trade and national employment could be dramatic, according to a report from the Office of Technology Assessment [1]. As in metals, design and manufacturing requirements and procedures introduce flaws, holes, material discontinuities etc., that result in large stress gradients. But unlike metals, composites are less forgiving because of the low strain to failure and hence require better accuracy in stress predictions. For isotopic bodies Boundary Element Method (BEM) has proven to have better resolution of stress gradients than other existing numerical methods. However, there is very little literature on boundary element method for composite bodies. A composite body by definition is made up of two or more constituents. Thus, a composite body is a non-homogeneous body at the micromechnics level. At a macro level the body is viewed as a homogeneous anisotropic body. The material constants of the homogeneous anistoropic body can be related to the individual constituents [2]. In this work the composite body will be viewed as a homogeneous orthotropic body. Thus the stress (σ_{ij}) and strain (ε_{ij}) are related by the following relationship

$$\begin{aligned}\varepsilon_{xx} &= C_{11}\,\sigma_{xx} + C_{12}\sigma_{yy}\\ \varepsilon_{yy} &= C_{12}\,\sigma_{xx} + C_{22}\sigma_{yy}\\ \gamma_{xy} &= C_{33}\sigma_{xy}.\end{aligned} \tag{1}$$

The use of BEM for an analysis of two-dimensional anisotropic elastostatic problem was first presented in 1972 by Sikarskie and Benjumea [3], who used an indirect formulation. Snyder and Cruse [4] published the direct formulation in 1975. In a latter work Cruse [5] improved the accuracy by using a linear variation in place of constant approximation for the unknown. All three papers [3-5] used complex variable representation for the fundamental solution. Zastrow [6-7] discussed the fundamental solution associated with different singularities for plane anisotropic material. In a later work [8] he presented the fundamental solution of plates and discussed the analogy between the plane elastostatic problem and the plate. The emphasis of the papers were on the tensorial character of the fundamental solution but there was no discussion on evaluation of integrals over boundary elements. Only two of the possible 3 material cases were discussed by Zastrow.

In the area of anisotropic plates Wu and Altiero [9] described a method in which the point sources were distributed outside the actual boundary at a finite number of points. Irschik [10] considered a transversely isotopic material which he solved by scaling the geometry. Shi and Bezine [11] described a direct boundary element formulation, in which they used constant approximation for the unknowns.

In this paper the work at Michigan Technology University in the area of BEM for composite bodies is described briefly. In all cases the fundamental solutions have been rederived using Fourier Transform Technique. This has resulted in a better understanding of the structure of the fundamental solution, which is elaborated in section 4.0. By exploiting the structure and developing a set of transformation equations [12,13] it has been possible to generate iterative formulas which resemble the formulas developed for isotropic problems presented in reference [14] by one of the authors. These iterative formulas are based on analytical integration and are applicable to any order of polynomial approximation of the unknown. In this paper results for linear and cubic approximation are discussed. The development of the algorithm and the presentation closely resembles to that presented for isotropic bodies in reference [14].

3.0 Formulation

Let $\sigma_I(Q)$ represent the quantity of interest at a field point Q. The subscript I is a generic subscript. For example if σ_I is a vector such as displacements in elastostatic or rotation in plates then I = i; if σ_I is a second order tensor such as stresses in elastostatic or curvatures in plates then I = ij. We assume σ_I can be represented by the following integral expression

$$\sigma_I(Q) = \S_B\ H_{Ij}(Q,P)\ f_j(P)ds \qquad j = x,y \tag{2}$$

where H_{Ij} is a known fundamental solution of the problem. f_j is an unknown strength of singularity and is determined from the specified boundary conditions. The integral expression in equation (2) is reduced to an algebraic expression through the following set of assumption.

Assumption 1: The boundary is assumed to be made up of M segments. In each of the mth segment f_j is approximated by a polynomial, $g_m(s)$ which satisfy certain continuity requirements.

Assumption 2: Each mth segment is further subdivided into N_m straight line segments. This assumption does not introduce any additional unknown but permits a better approximation of the boundary.

Assumption 3: In each of the straight line segments the function g_m is expanded about the mid point by Taylor series. Number of terms retained in the Taylor series is the same as the order of polynomial approximation. This assumption does not introduce any additional error. The purpose of the assumption is to permit a translation to a local coordinate with origin at the center of the line element as discussed in section 5.0.

The above assumptions reduce the integration problem on the boundary to an integration over straight line segments. Symbolically these integrals can be written as

$$T^{[q]}_{Ij}\ (Q,\ S_p) = \int_{S_p}^{S_{p+1}} (s - \bar{s}_p)^q\ H_{Ij}\ (Q,P)\ ds. \tag{3}$$

where $\bar{s}_p = (s_p + s_{p+1})/2$ and S_p is the value of s at the pth node of the pth line segment.

Before proceeding to the discussion of the fundamental solution, it should be pointed out that the bending rigidity constants D_{ij} in plates are related to the material constants C_{ij} as follows.

$$D_{ij} = \frac{h^3}{12}\ C_{ij} \tag{4}$$

where h is the thickness of the plate. Thus if C_{ij} satisfy a certain relationship then the same is true for D_{ij}

4.0 Fundamental Solution:

To develop the fundamental solution for orthotropic bodies the following characteristic equation has to be solved for plane elasticity as well as for plate problems:

$$C_{11}\mu^4 + 2\,(C_{12} + C_{33})\,\mu^2 + C_{22} = 0 \tag{5}$$

The roots of the equation are complex [15] and come in conjugate pairs. Two roots of the equation may be symbolically written as

$$\mu_k = i\lambda_k e^{i\delta_k} \qquad k = 1,2 \tag{6}$$

where $i = \sqrt{-1}$. The other two roots are the complex conjugate of the ones given in equation (6). The value of the constants λ_k and δ_k depends upon the relationship of the material constants C_{ij} and are discussed below: It clear from equation (5) that the roots will contain a radical. It is the term below the radical sign the dictates the nature of the roots.

Material Type 1

$$\left(\frac{C_{12} + C_{33}}{C_{11}}\right)^2 > \left(\frac{C_{22}}{C_{11}}\right)$$

For this case, $\delta_1 = \delta_2 = 0$, and (7)

$$\lambda_{1;2} = \sqrt{\frac{C_{12} + C_{33}}{C_{11}} \pm \sqrt{\left(\frac{C_{12} + C_{33}}{C_{11}}\right)^2 - \frac{C_{22}}{C_{11}}}}$$

The roots have no real part and are purely imaginary.

Material Type 2

(8)

$$\left(\frac{C_{12} + C_{33}}{C_{11}}\right)^2 > \left(\frac{C_{22}}{C_{11}}\right)$$

For this case, $\delta_1 = \delta_2 = 0$, and

$$\lambda_1 = \lambda_2 = \sqrt{\frac{C_{12} + C_{33}}{C_{11}}}$$

Materials belonging to this category are called transversely isotropic materials. Isotropic materials also belong to this category and correspond to $\lambda_1 = \lambda_2 = 1$. By scaling the y-coordinate (or x-coordinate) by the factor given in equation (6) the transversely isotropic material can be made to resemble an isotropic material.

Material Type 3:

$$\left(\frac{C_{12}+C_{33}}{C_{11}}\right)^2 > \left(\frac{C_{22}}{C_{11}}\right)$$

For this case, $\delta_1 = \delta_2 = \delta$, and $\lambda_1 = \lambda_2 = \lambda$ (9)

$$\text{where } \lambda\cos\delta = \sqrt{\frac{1}{2}\left[\sqrt{\frac{C_{22}}{C_{11}}} + \frac{C_{12}+C_{33}}{C_{11}}\right]}$$

$$\lambda\sin\delta = \sqrt{\frac{1}{2}\left[\sqrt{\frac{C_{22}}{C_{11}}} - \frac{C_{12}+C_{33}}{C_{11}}\right]}$$

The roots of the equation have a real and an imaginary part. The fundamental solution for all three types of material are a linear combination of the following 4 function.

$$\frac{r_{x_k}^m\, r_{y_k}^n}{r_k^{m+n+l}} \qquad r_{x_k}^m\, r_{y_k}^n \tag{10}$$

$$r_{x_k}^m\, r_{y_k}^n \ln r_k \qquad r_{x_k}^m\, r_{y_k}^n \tan^{-1}(r_{y_k}/\, r_{x_k})$$

where

$$r_{x_k} = r_x + r_y\,\lambda_k \sin\delta_k$$

$$r_{y_k} = r_y\,\lambda_k \cos\delta_k$$

$$r_k^2 = r_{x_k}^{\;2} + r_{y_k}^2 \tag{11}$$

$$r_x = x(Q) - x(P)$$

$$r_y = y(Q) - y(P)$$

and l, m and n are positive integers which satisfy the following condition `(l + m + n) is an even integer´. The integer l represents the order of singularity. In plate formulations the fundamental solution H_{lk} also contain direction cosines of the unit normal. However due to the straight line assumptions the direction cosines are constants and can be taken outside the integrals. Thus to evaluate the integrals in equation (3) one needs to evaluate the following 4 integrals over the pth straight line segment

$$IXY_k^{[l,m,n,q]} = \int_{s_p}^{s_{p+1}} (s - \bar{s}_p)^q \left(\frac{r_{x_k}^{\ m} \ r_{y_k}^{\ n}}{r^{l+m+n}} \right) ds$$

$$IL_k^{[m,n,q]} = \int_{s_p}^{s_{p+1}} (s - \bar{s}_p)^q \ r_{x_k}^{\ m} \ r_{y_k}^{\ n} \ \ln r_k \ ds$$

$$IT_k^{[m,n,q]} = \int_{s_p}^{s_{p+1}} (s - \bar{s}_p)^q \ r_{x_k}^{\ m} \ r_{y_k}^{\ n} \ \tan^{-1}\left(\frac{r_{y_k}}{r_{x_k}} \right) ds \tag{12}$$

$$IP_k^{[m,n,q]} = \int_{s_p}^{s_{p+1}} (s - \bar{s}_p)^q \ rx^m \ ry_k^n \ ds$$

To evaluate these integrals we consider the transformation equation given by one of the authors in references [12,13].

5.0 Transformation equation: Let the length of the pth segment be L as shown in figure 1 and the angle this segment makes with the x-axis be θ. Point A is the midpoint of the segment with coordinates $(\bar{x}, \bar{y})$.

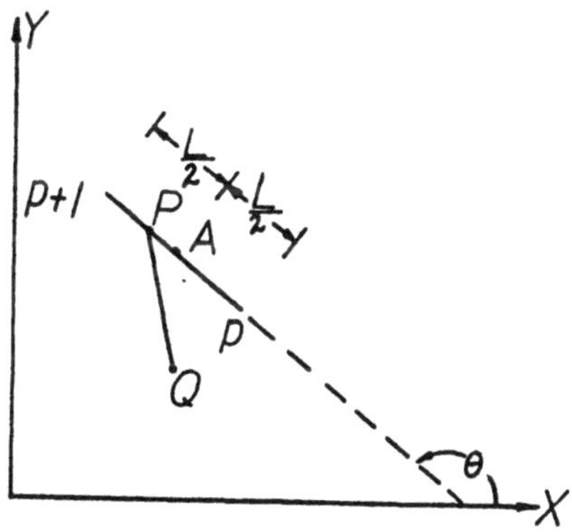

Figure 1. Geometery of the pth line segment.

We define the following quantities:

$$H_k^2 = \cos^2\theta + \lambda_k^2 \sin^2\theta + 2\lambda_k \sin\delta_k \sin\theta\cos\theta$$

$$\cos\gamma_k = (\cos\theta + \lambda_k \sin\delta_k \sin\theta)/H_k \tag{13}$$

$$\sin\gamma_k = (\lambda_k \cos\delta_k \sin\theta)/H_k$$

$$C_k = \{[(x - \bar{x}) + \lambda_k \sin \delta_k (y - \bar{y})] \cos \gamma_k$$

$$+ (y - \bar{y}) \lambda_k \cos \delta_k \sin \gamma_k$$

$$D_k = \{-[(x - \bar{x}) + \lambda_k \sin \delta_k (y - \bar{y})] \sin \gamma_k + (y - \bar{y})\lambda_k \cos \delta_k \cos \gamma_k\}/H_k$$

Using the above definition it can be shown that

$$r_x + \lambda_k \sin \delta_k r_y = -H_k [(t - C_k) \cos \gamma_k + D_k \sin \gamma_k]$$

$$r_k \cos \delta_k r_y = -H_k [(t - C_k) \sin \gamma_k + D_k \cos \gamma_k] \qquad (14)$$

It should be pointed out that for isotropic problems $\lambda_k = 1$ and $\delta_k = 0$, in which case $H_k = 1$, $\cos \gamma_k = \cos\theta$ and $\sin \gamma_k = \sin\theta$ and the transformation equations reduce to those given in reference [14]. Following the logic given in reference [14] and the above transformation the integral in equation (12) can be evaluated iteratively, the iterative formulas are very similar to those reported in reference [14].

6.0 Results:

Results are reported for a hole in an infinite plane, a pin in a coupon and for a square plate subjected to a constant moment. The first two problem are plane elastostatic problem. The fundamental solution for plane elastostatic problem has a first order singularity. This implies that the unknown functions must be continuous. Hence Lagrange polynomials of order 1 were choose for the approximation. The fundamental solution for plate problem has a second order singularity. This implies that corresponding unknown and its first denuative must be continuous. For this reason a cubic Hermite polynomial was chosen for approximation, in plate problems.

6.1 Example 1: A hole in an infinite orthotropic material is considered. A uniform tension along the material axis in the y-direction is applied at infinity. The stress concentration factor is computed and compared with the known analytical solution given by Lektinitski [13]. material properties were chosen to simulate the three types of material and are reported in Table 1. material type 2a represents transversely isotropic material, while type 2b represents isotropic material

Material	C_{11}	C_{12}	C_{22}	C_{33}
Type 1	1	-0.25	1	1.65
Type 2a	1	-0.25	1.96	1.65
Type 2b	1	-0.25	1	1.25
Type 3	1	-0.25	1	1.25

Table 1. Material properties for example 1.

In table 2 the analytical and computed values of stress concentration are reported. These values were first reported by one of the authors in reference [12]

Material	Type 1	Type 2a	Type 2b	Type 3
Analytical	3.191	2.690	3.000	2.565
Computed	3.053	2.604	2.894	2.604

Table 2: Result for stress concentration factor.

Results of Table 2 show a good correlation between analytical and computed values.

6.2 Example 2: A problem of mechanical fastening is considered next. The fastener is idealized as a pin. For comparison purposes isotropic material properties were simulated in the program and results compared with those reported by Hart Smith in reference [16]. The contact problem that has to be solved is described by one of the authors in reference [17]. The coupon geometry is shown in figure 2.

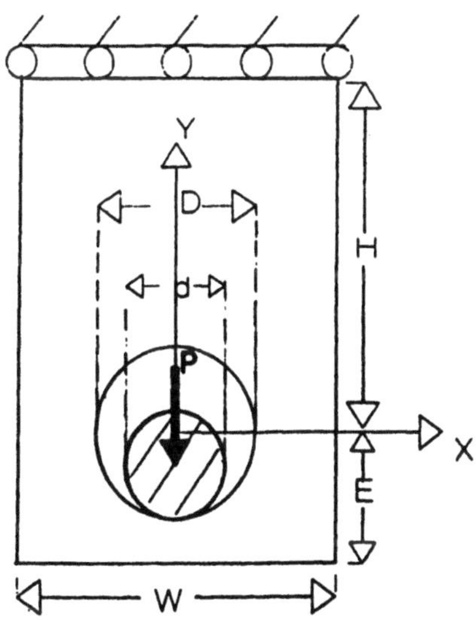

Figure 2: Geometry of a coupon.

In table 3 the stress concentration factor (K_E) determined experimentally [16] and the value computed (K_c) are reported for various value of W/D. The ratio E/W is kept constant at a value of 0.5 and D = d. The results of table 3 were reported by one of the authors in reference [17].

W/D	K_E Expt.	K_c Comp.
10	10.39	9.69
8	8.42	8.21
6	6.46	6.72
5	5.50	6.02
4	4.96	5.5

Table 3: Results for a pin in a coupon.

The decrease and then an increase in the difference between the two stress concentration factor is most likely caused from numerical modeling. To generate the data all dimensions were changed proportionately without changing the mesh. This is unlikely to yield the best mesh description. The matter is under further investigation.

<u>6.3. Example 3:</u> A 2 x 2 square is hinged at the corner is considered next. A constant moment of $M_x = -2\ (D_{11} - D_{12})$ and $M_y = -2\ (D_{12} - D_{22})$ are applied at $x = \pm 1$ and $y = \pm 1$ respectively. The analytical solution to the deflection of the plate is

$$w = (x^2 - y^2). \tag{15}$$

All four types of material were simulated. Figure (3a) (3b) (3c) and (3d) show the computed and the analytical value of the displacement correlate well. The analytical value of the stresses correspond to the applied moment. In all four material cases constant values for the stresses were found. For type 2a and 2b material the computed value and the analytical values were within 1%. However for type 1 and type 3 material the error in stresses was close to 20%. The cause of the problem is under investigation.

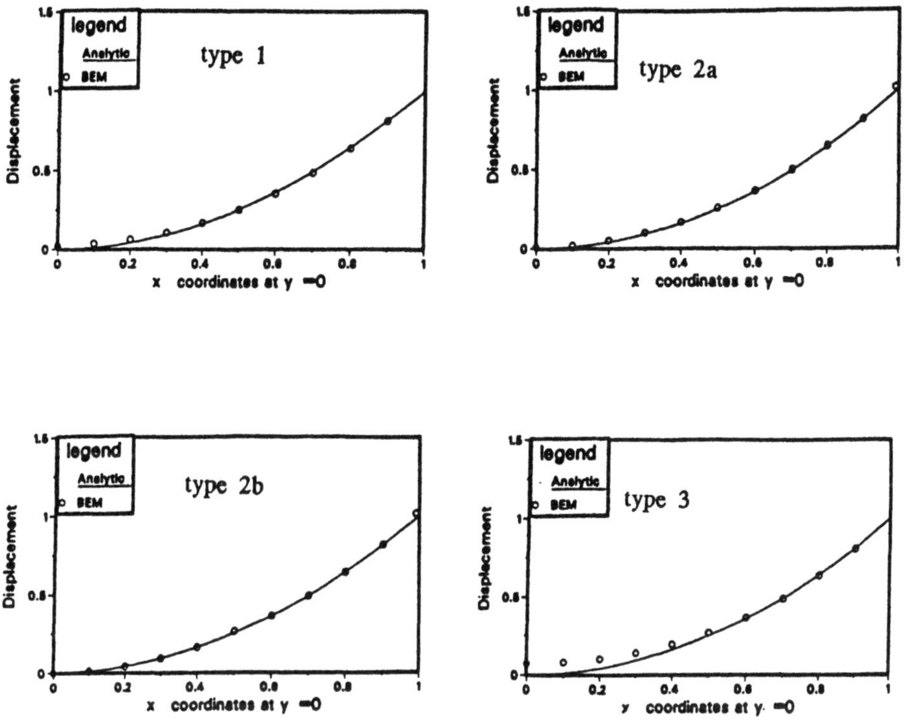

Figure 3: Displacement results for the orthotropic plate.

References:

1. U.S. Congress, Office of Technology Assessment (1986) "New structural materials technologies: Opportunities for the use of advanced ceramics and composites" - OTA-TM-E-32.

2. Tsai, S.W. and Hahn, H.T. (1980) "Introduction to composite materials." Technomic Publishing Co., Lancaster, PA.

3. Sikarskie, D.L. and Benjumea, R. (1972) "On the solution of plane orthotropic elastic problem by integral method". J. Appl. Mech. Trans. ASME, 94, 801-808.

4. Snyder, M.D. and Cruse (1975) "Boundary integral equation analysis of cracked Anisotropic plates. Int. J. Fracture Mech. II, 315-328.

5. Cruse, T.A. (1978) Two dimensional BIE fracture mechanics analysis. Appl. Math. Modelling 2, 287-293 (1978).

6. Zastrow, U. (1982) "Solution of plane anisotropic elastostatical boundary value problems by singular integral equations. Acta Mechanica 44, 59-71.

7. Zastrow, U. (1985) "On the formulation of the fundamental solutions for orthotropic plane elasticity". Acta Mechanica 57, 113-122.

8. Zastrow, U. (1985) "On the complete system of fundamental solutions for anisotropic slices and slabs: A comparison by use of the slab analogy" J. of elasticity 15, 293-318.

9. Wu, B.C. and Altiero, H.J. (1981) A new numerical method for the analysis of anisotropic thin plate bending problems". Computer Methods in Applied Mechanics and Engineering 25, 343-353.

10. Irsetik, H. (1984) "A boundary-integral equation method for bending of orthotropic plates" Int. J. Solids and structures V20, N3, 245-255.

11. Shi, G. and Bezine, G. (1988) "A general boundary integral formulation for the anisotropic plate bending problems". J. of Composite Materials V22, 694-716.

12. Vable, M. and Sikarskie, D.L. (1988) "Stress analysis in plane orthotropic material by the boundary element method". V24, N1, pp1-11.

13. Vable, M. and Sikarskie, D.L. (1987) "Boundary element method for mixed boundary value plane elastostatic orthotropic problems. BEM IX Ed. C.A. Brebbia, W.L. Wendland and G. Kuhn, CMP, Boston.

14. Vable, M. (1985) "An algorithm based on the boundary element method for problems in engineering mechanics". IJNME V21, 1625-1640.

15. Lekhnitski, S.G. (1963) "Theory of elasticity of an anisotropic elastic body". Hoden-Day, San Francisco, CA.

16. Hart-Smith, L.J. (1980) "Mechanically - Fastened Joints for advanced composites". - Phenomenological consideration and simple analysis" - Proceeding fourth conference on Fibrous composite structural design, 543-574.

17. Vable, M. (1989) "Boundary Element Analysis of Mechanically Fastened Composite". Boundary Element Techniques: Application in Engineering. Ed. C.A. Brebbia and M.G. Zamani, CMP, Boston.

Integration of 3D Boundary Elements for Static Elasticity

C. T. DYKA and A. M. REMONDI

General Dynamics, Electric Boat Division
Groton, Conn. 06340

H. R. MILLWATER

Southwest Research Institute
San Antonio, Texas 78280

Summary

Through the use of a local polar system and the shape functions, the strong singularities that are present in 3D static elasticity when the source point is within the element, are removed prior to numerical integration. A linear triangular and quadrilateral element are developed and used in an example problem. A procedure for a general subtriangle within any type of 3D boundary element is also presented.

Introduction

The boundary element method (BEM) has achieved a high level of success for two dimensional, axisymmetric and three dimensional (3D) problems in potential analysis and elasticity [1]. However, the general implementation of BEM requires the computation of surface integrals which contain the singularities - Ln R_i and/or $1/R_i$ where R_i is the distance from the source point i.

For 3D boundary elements a standard procedure [2] is to introduce subtriangles when the source point is within the element being integrated. The subtriangles allow the use of a polar coordinate system which will then remove the strong singularity $1/R_i$ in the integrand. However for the $\underline{H}$ matrix the remaining $1/R_i$ is not implicitly removed for off diagonal terms. In [1] a finite part numerical integration procedure is employed to perform this singular integration. For flat triangular elements, Cruse [3] employs Stokes's theorem to attain line integrals amenable to analytical integration.

In this paper we discuss a general approach that is similar to the one used in [4] in which the shape functions are employed to remove the remaining strong singularity. Linear triangular and quadrilateral elements are developed along with a procedure for a general subtriangle within any type of 3D boundary elements.

Basic Equations

The governing boundary element equations for 3D elasticity can be expressed as [1]:

$$\underline{\underline{C}}\,\underline{u} + \underline{\underline{H}}\,\underline{u} = \underline{\underline{G}}\,\underline{p} \tag{1}$$

where $\underline{u}$ and $\underline{p}$ are the global displacements and tractions and $\underline{C}$ is a diagonal matrix determined by rigid body considerations. For a specific element, the components of $\underline{\underline{H}}$ and $\underline{\underline{G}}$ are of the form:

$$H_{mk}^{iQ} = \int_{r=-1}^{r=1} \int_{s=-1}^{s=1} p_{mk}^{*}(r,s) \cdot h_Q(r,s) \cdot J(r,s)drds \tag{2}$$

and

$$G_{mk}^{iQ} = \int_{r=-1}^{r=1} \int_{s=-1}^{s=1} u_{mk}^{*}(r,s) \cdot h_Q(r,s) \cdot J(r,s)drds \tag{3}$$

in which i is the source node, r and s the parametric coordinates, $h_Q(r,s)$ the shape function associated with node Q, and J(r,s) the determinant of the Jacobian. The fundamental solutions p_{mk}^{*} and u_{mk}^{*} are:

$$u_{mk}^{*} = \frac{1}{16\pi G(1-\nu)\, R_i} \left[(3-4\nu)\, \delta_{mk} + \frac{\partial R_i}{\partial x_m} \frac{\partial R_i}{\partial x_k}\right] \tag{4}$$

$$p_{mk}^{*} = \frac{1}{8\pi(1-\nu)\, R_i^2} \left[\frac{\partial R_i}{\partial n} \left((1-2\nu)\, \delta_{mk} + 3 \frac{\partial R_i}{\partial x_m} \frac{\partial Ri}{\partial x_k}\right)\right.$$

$$\left. -(1-2\nu) \left(\frac{\partial R_i}{\partial x_m} n_k - \frac{\partial R_i}{\partial x_k} n_m\right)\right] \tag{5}$$

in which G and ν are the shear modulus and Poisson's ratio, δ_{mk} the Kronecker's delta, R_i the distance from the source point i to the field point, and n_k, n_m the components of the normal vector at the field point.

A Linear Triangle

Figure 1 describes the geometry of a continuous linear triangle (TRIL). The shape functions $h_k(r,s)$ associated with this element are:

$$h_1 = 1-r-s, \quad h_2 = r, \quad h_3 = s \tag{6}$$

When the source point i is located outside the element being integrated, the calculation of the $\underline{G}$ and $\underline{H}$ matrices is performed with Gaussian quadrature. However, when i is within the element in question, special techniques must be used to remove the strong singularities that are present. A polar system (R_i, $\emptyset$) is typically introduced [2] which will remove $1/R_i$ in the fundamental solutions. The removal of the second $1/R_i$ present in p^*_{mk} can be accomplished through the use of the shape functions. In Figure 1 the source point i is located at node 1. The vector $\underline{R}_i$ can be expressed in terms of the local ($\bar{x}$, $\bar{y}$) system as:

$$\underline{R}_i(r,s) = \bar{x}(r,s)\,\underline{\bar{i}} + \bar{y}(r,s)\,\underline{\bar{j}} \tag{7}$$

where $\underline{\bar{i}}$ and $\underline{\bar{j}}$ are the local base vectors. The coordinates $\bar{x}$ and $\bar{y}$ can be determined by interpolation of the local geometry to produce:

$$\bar{x}(r,s) = r \cdot L_{12} + s \cdot \bar{x}_3 \quad , \quad \bar{y}(r,s) = s \cdot \bar{y}_3 \tag{8}$$

The components of $\underline{R}_i$ can be expressed as:

$$\bar{x}(r,s) = R_i \cdot \cos\emptyset \quad , \quad \bar{y}(r,s) = R_i \cdot \sin\emptyset \tag{9}$$

Solving equations (8) and (9) for r and s and substituting in (6) generates the following expressions for the shape functions:

$$h_1(R_i,\emptyset) = 1 - R_i \cdot (F_1 + F_2)$$

$$h_2(R_i,\emptyset) = R_i \cdot F_1 \quad , \quad h_3(R_i,\emptyset) = R_i \cdot F_2 \tag{10}$$

where:

$$F_1 = (\cos\emptyset - \sin\emptyset \cdot \frac{\bar{x}_3}{\bar{y}_3})/L_{12} \quad , \quad F_2 = \sin\emptyset/\bar{y}_3 \tag{11}$$

In this form the shape functions h_2 and h_3 can now be used to remove the remaining $1/R_i$ which is present in $\underline{H}$. Equation (2) can then be directly integrated with Gaussian quadrature in the $\emptyset$ variable.

A Linear Quadrilateral Element

In Figure 2, a continuous 4 node quadrilateral element (QUAD) is described. For simplicity, we have assumed that the element is flat. However, the approach is directly applicable to warped geometries. The shape functions for the QUAD element are:

$$h_1 = \frac{1}{4}(1-r)(1-s) \quad , \quad h_2 = \frac{1}{4}(1+r)(1-s)$$
$$h_3 = \frac{1}{4}(1+r)(1+s) \quad , \quad h_4 = \frac{1}{4}(1-r)(1+s) \tag{12}$$

The source point i is assumed to be located at node 1. To remove the strong singularity in $\underline{G}$ and $\underline{H}$, two subtriangular regions A and B are introduced. A new triangular parametric coordinate system $(\overline{r},\overline{s})$ can be introduced. The corresponding shape functions for region A are in the same form as (6) with (r,s) replacing $(\overline{r},\overline{s})$. The original parametric coordinates (r,s) can then be interpolated in region A to yield:

$$r = -1 + 2\overline{r} + 2\overline{s} \qquad \text{and} \qquad s = -1 + 2\overline{s} \tag{13}$$

In a fashion similar to the TRIL element, (see [5]), $\overline{r}$ and $\overline{s}$ can be expressed as:

$$\overline{r} = R_i \cdot F_1 \qquad \text{and} \qquad \overline{s} = R_i \cdot F_2 \tag{14}$$

Substituting equations (13) and (14) into (12) then produces the original shape functions in the following form:

$$h_1(R_i,\emptyset) = [1 - R_i(F_1 + F_2)]\,(1 - R_i \cdot F_2)$$
$$h_2(R_i,\emptyset) = R_i(F_1 + F_2)\,(1 - R_i \cdot F_2)$$
$$h_3(R_i,\emptyset) = R_i^2(F_1 + F_2) \cdot F_2 \tag{15}$$
$$h_4(R_i,\emptyset) = R_i \cdot F_2 \cdot (1 - R_i(F_1 + F_2))$$

Equations (15) now allow the remaining strong singularity present in $\underline{H}$ to be removed within subtriangle A. An analogous procedure is followed in subtriangle B [5].

A General Subtriangle

Figure 3 describes a typical subtriangle which is contained within some 3D element. Point i is the source node, while points j and k may or may not represent node points on the element. The local parametric coordinates $(\bar{r},\bar{s})$ can be expressed as functions of R_i and Ø. This is followed by the interpolation of the elements parametric variables (r,s) with the local shape functions for the subtriangle. The parametric variables r and s can then be expressed as in the form:

$$r = r_i + R_i \cdot \bar{F}_3 \quad , \quad s = s_i + R_i \cdot \bar{F}_4 \tag{16}$$

where $\bar{F}_3$ and $\bar{F}_4$ are functions of Ø and the parametric coordinates of i, j, and k [5].

Using equations (16), the shape functions within the subtriangle of a 3D quadratic element (for example) are of the form:

$$h_n(R_i,\emptyset) = a_n + b_n R_i + c_n R_i^2 + d_n R_i^3 \tag{17}$$

In equation (17), n is the node number and a_n, b_n, c_n and d_n are constants. The constant a_n can be shown to be zero ($n \neq i$ where i is the source point) for quadratic elements. Thus, the remaining strong singularity $1/R_i$ can be removed from $\underline{H}$ prior to integration.

Example Problem

An example problem is presented to demonstrate the performance of the TRIL and QUAD elements. Standard quadrature rules are used when the source point is outside the element being integrated. When the source point is inside the element, the resulting subtriangles, which are introduced, are integrated with a 6 point quadrature rule for the Ø variable.

The problem investigated is a hollow sphere subjected to an inner and outer pressure. Figure 4 describes a 1/8 symmetric model composed of QUAD elements spaced every 15 degrees of arc, with TRIL elements employed at the pole (Case 1). A similar model using all TRIL elements is Case 2. Table 1 presents the boundary element results and the closed form calculations. The results are good for the coarse grids that are employed.

Conclusions

Through the use of the shape functions, the remaining strong singularity present in 3D elasticity has been removed for continuous linear triangles and quadrilateral elements. A procedure for the removal of the remaining strong singularity was presented which is applicable to a general subtriangle within any 3D element. The overall methodology of using the shape functions to remove the remaining strong singularity for 3D static elasticity would appear to be an important approach to the accurate and efficient integration of 3D boundary elements.

Acknowledgement. This research was sponsored by the Independent Research and Development (IR&D) program at the Electric Boat Division of General Dynamics.

References

1. C. A. Brebbia, J. C. F. Telles and L. C. Wrobel, Boundary Element Techniques, Theory and Applications in Engineering, Springer-Verlag, New York 1984.

2. J. C. Lachat and J. O. Watson, "Effective Numerical Treatment of Boundary Integral - Equations: A Formulation for Three Dimensional Elastostatics", Int. J. Num. Meth. Engng., Vol. 10, pp. 991-1005, 1976.

3. T. A. Cruse, "Numerical Solutions in Three Dimensional Elastostatics", Int. J. Solids Structures, Vol. 5, pp. 1259-1274, 1969.

4. C. T. Dyka and H. R. Millwater, "Formulation and Integration of Continuous and Discontinuous Quadratic Boundary Elements for 2D Potential and Elastostatics", Computers and Structures, Vol. 31, No. 4, pp. 495-504, 1989.

5. C. T. Dyka, A. M. Remondi and H. R. Millwater, "3D Boundary Elements and the Integration of Strong Singularities," Submitted to Int. J. Solids Structures.

Table 1 - Results for the Hollow Sphere

	Inner Radius R_I = 1 inch			Outer Radius R_O = 2 inches		
	U_r x10^{-6} in.	σ_{rr} psi	$\sigma_{\phi\phi}$ psi	U_r x10^{-6} in.	σ_{rr} psi	$\sigma_{\phi\phi}$ psi
Case 1						
• Pole (X=Z=o)	-.5246 (4.9)	-103.4 (3.4)	-262.5 (3.3)	-.6312 (.3)	-201.3 (.65)	-218.3 (1.4)
• Nodes along Y=0	-.5241 (4.8)	-105.7 (5.7)	-259.4 (4.4)	-.6181 (2.4)	-201.0 (.5)	-288.6 (2.9)
Case 2						
• Pole (X=Z=o)	-.5277 (5.5)	-103.4 (3.4)	-259.9 (4.2)	-.6299 (.5)	-201.5 (.75)	-225.7 (1.9)
• Nodes along Y=0	-.5191 (3.8)	-105.8 (5.8)	-265.0 (2.4)	-.6180 (2.4)	-200.5 (.25)	-214.3 (3.2)
Elasticity Solution	-.5000	-100.0	-271.4	-.6333	-200.0	-221.4

( ) - Percent error w.r.t. the elasticity solution.

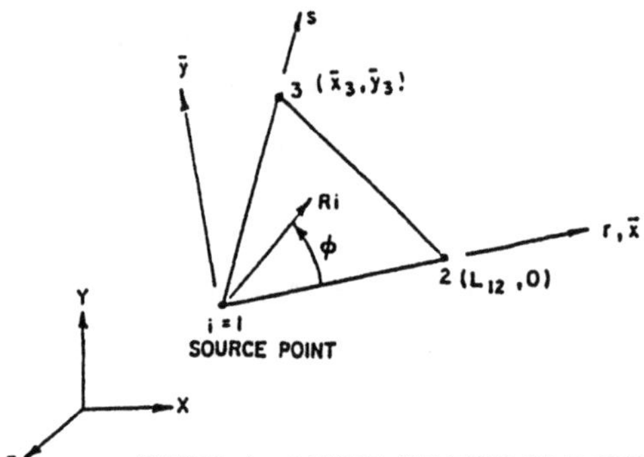

FIGURE 1 LINEAR TRIANGULAR ELEMENT

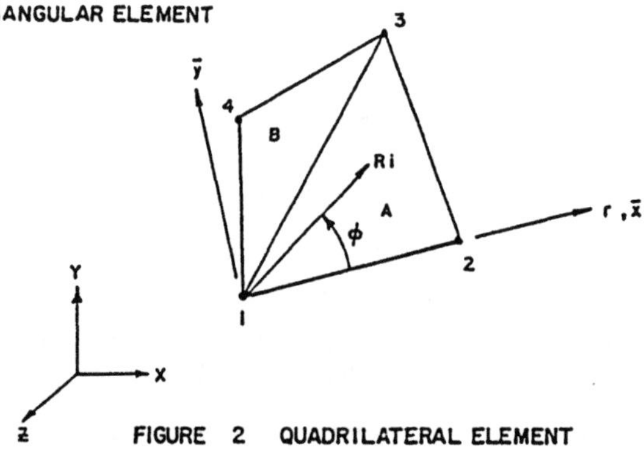

FIGURE 2 QUADRILATERAL ELEMENT

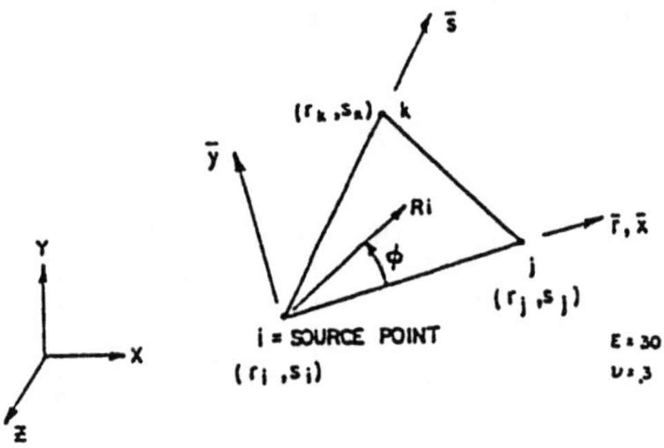

FIGURE 3 A GENERAL SUBTRIANGLE

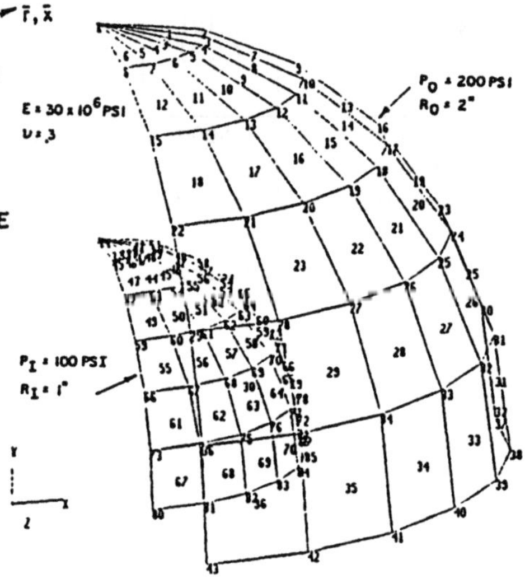

FIGURE 4 HOLLOW SPHERE

Alternative Methods for Substructures

F. Hartmann, P. Schoepp

Department of Civil Engineering
Gesamthochschule Kassel
35 Kassel, FRG

1. Introduction

Compound domains give rise to differential equations with discontinuous coefficients. Each subdomain requires a different fundamental solution and so the problem is usually solved by formulating the integral equations for the different subdomains separately and then combining the resulting systems of linear equations according to the coupling conditions at the interfaces.

In this paper we want to propose two alternative methods which require less complicated data structures, the number of unknowns on the interface is split into half.

2. The technique

To explain the technique we choose a compound domain which consists of two subdomains Ω_a and Ω_b which have thermal diffusity N_a and N_b respectively. The temperature jdistribution u induced by a continuous distribution of temperature sources p satisfies the equations

$$
\begin{aligned}
-N_a \Delta u &= p \quad \text{in } \Omega_a \\
-N_b \Delta u &= p \quad \text{in } \Omega_b \\
u &= 0 \quad \text{on } \Gamma \\
u_a - u_b &= 0 \quad t_a + t_b = 0 \qquad \text{on } \Gamma_i \text{ (the interface)}
\end{aligned}
$$

To set the fundamental solutions of the two subdomains apart we label them as follows

$$
g_0^a(\boldsymbol{y}, \boldsymbol{x}) = -\frac{1}{2\pi N_a} \ln r, \qquad g_0^b(\boldsymbol{y}, \boldsymbol{x}) = -\frac{1}{2\pi N_b} \ln r
$$

and similarly the different parts of Γ,

$$
\begin{aligned}
\Gamma_a &= \text{exterior boundary of } \Omega_a \\
\Gamma_b &= \text{exterior boundary of } \Omega_b \\
\Gamma_i &= \text{interface between } \Omega_a \text{ and } \Omega_b
\end{aligned}
$$

The influence function of the domain Ω_a is

$$c_a(\boldsymbol{x})u(\boldsymbol{x}) = \int_{\Gamma_a} [g_0^a t - N_a \frac{\partial}{\partial \nu} g_0^a u] ds_{\boldsymbol{y}} + \int_{\Gamma_i} [g_0^a t^a - N_a \frac{\partial}{\partial \nu_b} g_0^a u^a] ds_{\boldsymbol{y}} + \int_{\Omega_a} g_0^a p \, d\Omega_{\boldsymbol{y}} \tag{1}$$

and the influence function of the domain Ω_b is

$$c_b(\boldsymbol{x})u(\boldsymbol{x}) = \int_{\Gamma_b} [g_0^b t - N_a \frac{\partial}{\partial \nu} g_0^b u] ds_{\boldsymbol{y}} + \int_{\Gamma_i} [g_0^b t^b - N_b \frac{\partial}{\partial \nu_a} g_0^b u^b] ds_{\boldsymbol{y}} + \int_{\Omega_b} g_0^b p \, d\Omega_{\boldsymbol{y}} \tag{2}$$

where the functions $c_a(\boldsymbol{x})$ and $c_b(\boldsymbol{x})$ are the characteristic functions of Ω_a and Ω_b respectively. The indices on the normal derivative

$$\frac{\partial}{\partial \nu_a} \cdots \qquad \frac{\partial}{\partial \nu_b}$$

denote the direction of the normal vector which either points to Ω_a or to Ω_b. Because $N_a g_0^a = N_b g_0^b$ and because the normal vectors on the interface point into opposite directions we have

$$N_a \frac{\partial}{\partial \nu_b} g_0^a = -N_b \frac{\partial}{\partial \nu_a} g_0^b$$

so that if we multiply each influence function with the pertinent factor N and then add the two equations we obtain

$$(N_a c_a(\boldsymbol{x}) + N_b c_b(\boldsymbol{x}))u(\boldsymbol{x}) = \frac{1}{2\pi} \{ \int_{\Gamma_a} [-\ln r \, t + N_a \frac{r_\nu}{r} u] ds_{\boldsymbol{y}} + \int_{\Gamma_b} [-\ln r \, t + N_b \frac{r_\nu}{r} u] ds_{\boldsymbol{y}} \int_{\Gamma_i} (N_a - N_b) \frac{r_\nu}{r} u ds_{\boldsymbol{y}} + \int_{\Omega} \ln r \, p \, d\Omega_{\boldsymbol{y}} \} \tag{3}$$

where it is understood that the normal vector on the interface points into Ω_b. This sum ist the influence function for the compound domain. It contains as an additional unknown the trace u of the solution on the interface. This function can be determined simply by placing additional collocation points at the interface.

Our first idea was to simply add the two equations (1) and (2) which then leads to the expression

$$(c_a(\boldsymbol{x}) + c_b(\boldsymbol{x}))u(\boldsymbol{x}) = \int_{\Gamma_a} [g_0^a t - N_a \frac{\partial}{\partial \nu} g_0^a u] ds_{\boldsymbol{y}} + \int_{\Gamma_b} [g_0^b t - N_b \frac{\partial}{\partial \nu} g_0^b u] ds_{\boldsymbol{y}}$$

$$+ \int_{\Gamma_i} g_0^i t \, ds_{\boldsymbol{y}} + \int_{\Omega_a} g_0^a p \, d\Omega_{\boldsymbol{y}} + \int_{\Omega_b} g_0^b p \, d\Omega_{\boldsymbol{y}} \tag{4}$$

where

$$g_0^i = -\frac{1}{2\pi N_i} \ln r, \qquad N_i = \frac{N_a N_b}{N_b - N_a}$$

and t on Γ_i stands for

$$t = t^a = -t^b$$

The unknown interface function is now the slope t. To find this function t we must formulate an integral equation for the slope.

If we denote the boundary integrals over the outer boundary Γ_a by

$$B^a(\boldsymbol{x}) = \int_{\Gamma_a} [g_0^a t - N_a \frac{\partial}{\partial \nu} g_0^a u] ds_{\boldsymbol{y}}$$

and the domain integral by

$$P^a(\boldsymbol{x}) = \int_{\Omega_a} g_0^a p d\Omega_{\boldsymbol{y}}$$

(the functions $B^b(\boldsymbol{x})$ and $P^b(\boldsymbol{x})$ have the analogous meaning) then the influence function for the slope t in the domain Ω_a is

$$t = N_a \frac{\partial u}{\partial n}(\boldsymbol{x}) = N_a \{B_n^a + B_n^b + P_n^a + P_n^b\} + \int_{\Gamma_i} N_a \frac{\partial}{\partial n} g_0^i t \, ds_{\boldsymbol{y}} \tag{5}$$

where the index n denotes the normal derivative. The interface integral in Eq.(5)

$$I(\boldsymbol{x}) = \int_{\Gamma_i} N_a \frac{\partial}{\partial n} g_0^i t \, ds_{\boldsymbol{y}} = -\frac{1}{2} \frac{N_b - N_a}{N_b} \int_{\Gamma_i} \frac{1}{r} r_{,n} t \, ds_{\boldsymbol{y}}$$

is a potential of the second kind which is discontinuous at the interface. Its limit on approaching the interface from the left, from within Ω_a, is

$$\lim_{\boldsymbol{x} \to \Gamma} I(\boldsymbol{x}) = \frac{1}{2} \frac{N_b - N_a}{N_b} t(\boldsymbol{x}) - \frac{1}{2} \frac{N_b - N_a}{N_b} \int_{\Gamma_i} \frac{1}{r} r_{,n} \, ds_{\boldsymbol{y}}$$

so that, if we take the limit of both sides, we obtain

$$t(\boldsymbol{x}) = \frac{1}{2}\frac{N_b - N_a}{N_b}t(\boldsymbol{x}) + \ldots$$

or

$$\frac{1}{2}\frac{N_b + N_a}{N_b}t(\boldsymbol{x}) = \int\limits_{\Gamma_i} N_a \frac{\partial}{\partial n} g_0^i t \, ds_{\boldsymbol{y}} + N_a\{B_n^a + B_n^b + P_n^a + P_n^b\} \tag{5}$$

Equations (4) and (5) constitute a pair of coupled integral equations for the slope t on the outer boundary and the slope t at the interface.

The two techniques we introduced here differ by what is the unknown at the interface. In the first case (multiply with N) the unknown is the trace of the solution u on the interface and in the second case (naive approach) the unknown is the slope. The advantage of the N-approach is that the integral equation on the interface is the same integral equation as on the outer boundary. Its disadvantage is that the slope at the interface is not a system variable. So that for to calculate the slope on the interface you have to evaluate a strongly singular integral. This means that you run, only later, into the same difficulty as in the naive approach. The sequence of events is only reversed. In the naive approach the slope is a system variable, and to set up its integral equation we have to integrate the strongly singular kernels in the influence function (5) analytically, while the trace u at the interface is evaluated afterwards by employing the influence function which serves in the N-approach as integral equation. But the main and probably decisive disadvantage of the naive approach is that the interface integral in (5) does not exist at non-smooth points of the interface (if the layer has not a zero at the point), because it formulates a directional derivative. Though this can be overcome by introducing a double-node and by continuing the layers beyond this point by zero.

3. Examples

We use the N-approach mainly in our program BE-PLATE-BENDING for the solution of the fourth order biharmonic equation, $K\Delta\Delta w = p$. This program runs on an IBM-PC and the limited storage capacity of these computers under the operating system MS-DOS was the main reason for the development of these alternatives.

Considering the interface conditions

$$w_a - w_b = 0\,, \qquad \frac{\partial w_a}{\partial n} + \frac{\partial w_b}{\partial n} = 0\,, \qquad M_a + M_b = 0\,, \qquad V_a + V_b = f\,,$$

and the properties of the two fundamental solutions

$$K_a g_0^a = K_b g_0^b \qquad K_a \frac{\partial}{\partial \nu_b} g_0^a = -K_b \frac{\partial}{\partial \nu_a} g_0^b$$

$$V_\nu(g_0^a) = -V_\nu(g_0^b) \qquad M_\nu(g_0^a) = M_\nu(g_0^b)$$

the influence function for the solution of the biharmonic equation in the compound domain becomes

$$
\begin{aligned}
(K_a c_a(\boldsymbol{x}) + K_b c_b(\boldsymbol{x}))w(\boldsymbol{x}) = \int_{\Gamma} & K[\, g_0(\boldsymbol{y},\boldsymbol{x}) V_\nu(w)(\boldsymbol{y}) - \frac{\partial}{\partial \nu} g_0(\boldsymbol{y},\boldsymbol{x}) M_\nu(w)(\boldsymbol{y}) \\
& - V_\nu(g_0(\boldsymbol{y},\boldsymbol{x}))w(\boldsymbol{y}) + M_\nu(g_0(\boldsymbol{y},\boldsymbol{x}))\frac{\partial w}{\partial \nu}(\boldsymbol{y})]\, ds_{\boldsymbol{y}} + \int_{\Omega} K g_0(\boldsymbol{y},\boldsymbol{x}) p(\boldsymbol{y})\, d\Omega_{\boldsymbol{y}} \\
& + \int_{\Gamma_i} [M_\nu(g_0^a(\boldsymbol{y},\boldsymbol{x}))(K_a - K_b)\frac{\partial w}{\partial \nu}(\boldsymbol{y}) - V_\nu(g_0^a(\boldsymbol{y},\boldsymbol{x}))(K_a - K_b)w(\boldsymbol{y}) \\
& + K_a g_0^a(\boldsymbol{y},\boldsymbol{x}) f(\boldsymbol{y})]ds_{\boldsymbol{y}} + \sum_c K[\, g_0(\boldsymbol{y}^c,\boldsymbol{x})F(w)(\boldsymbol{y}^c) - w(\boldsymbol{y}^c)F(g_0)(\boldsymbol{y}^c,\boldsymbol{x})]
\end{aligned}
\tag{6}
$$

where f denotes the unknown support reactions at the interface (if there are any). For not to repeat each integral twice we simplified the notation somewhat: we did not mark integrals which refer to the domain Ω_a and integrals which refer to the domain Ω_b separately. In the same way the factor K on the right side takes on either the value K_a (if the integral extends over the edge of Ω_a) or K_b (if the integral extends over the edge of Ω_b) respectively.

The plate thickness of bridge structures or slabs can vary considerably, see Fig. 1 and Fig. 2, so the savings are considerable if we reduce by this technique the number of interface functions from four to two.

Our experience with this N-approach is quite satisfactory though we must make two remarks:

1) When we use this technique on two neighboring identical domains, same size, same coefficient $N_a = N_b = N$, as two unit-squares, then the resulting system matrix seems to become singular. But this is perhaps obvious.

2) In engineering analysis we are not only interested in the boundary values but the internal distribution of the bending moments and the shear forces as well. This requires the evaluation of the pertinent influence functions and as you are well aware the accuracy of the results depends on the distance from the boundary or, more generally speaking, from the source, from the boundary or interface layer. Therefore plate structures with many subdomains aggravate this problem. The interface layers — being of displacement type in the N-approach — are coupled with singular and hypersingular kernels which tend, if not integrated analytically, to spoil the higher derivatives of the solution near the interface and on the interface.

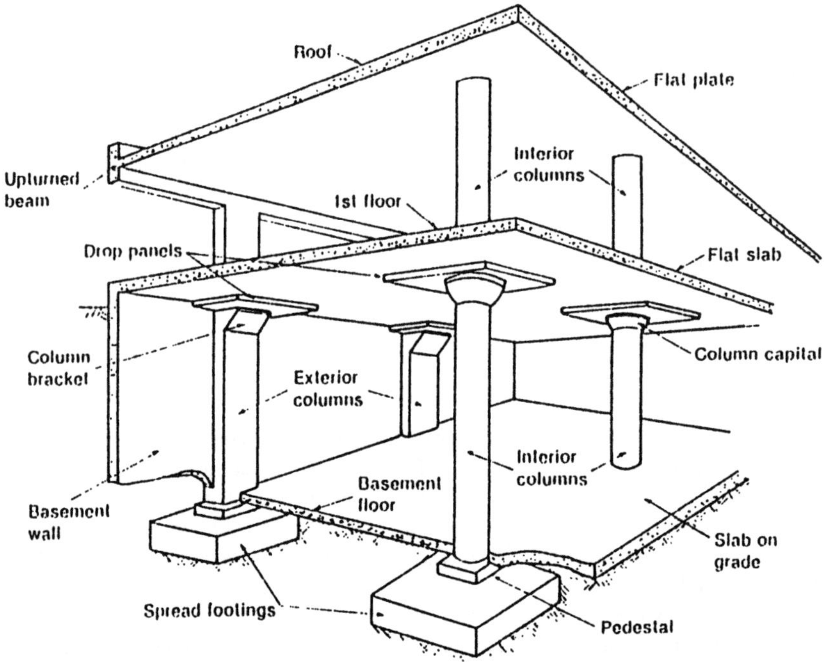
Roof
Flat plate
Interior columns
Upturned beam
1st floor
Flat slab
Drop panels
Column capital
Column bracket
Exterior columns
Interior columns
Basement wall
Basement floor
Slab on grade
Spread footings
Pedestal

Fig. 1

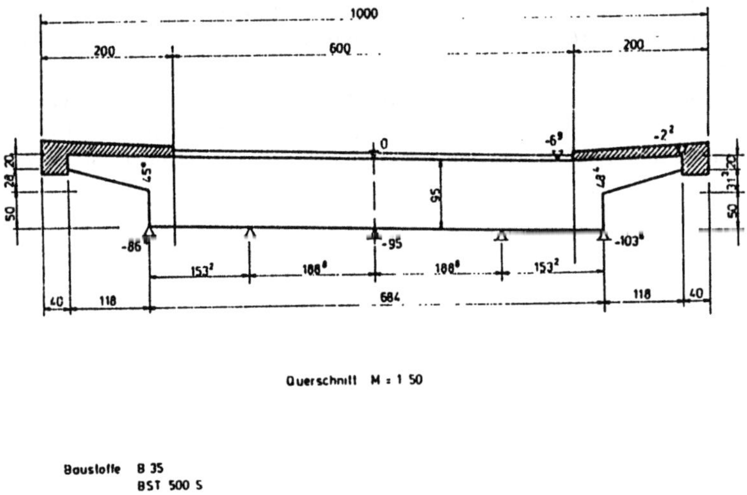
1000
200
600
200
0
-6
-2
95
-86
-95
-103
153
188
188
153
40
118
684
118
40
Querschnitt M = 1 50
Baustoffe B 35
BST 500 S

Fig. 2

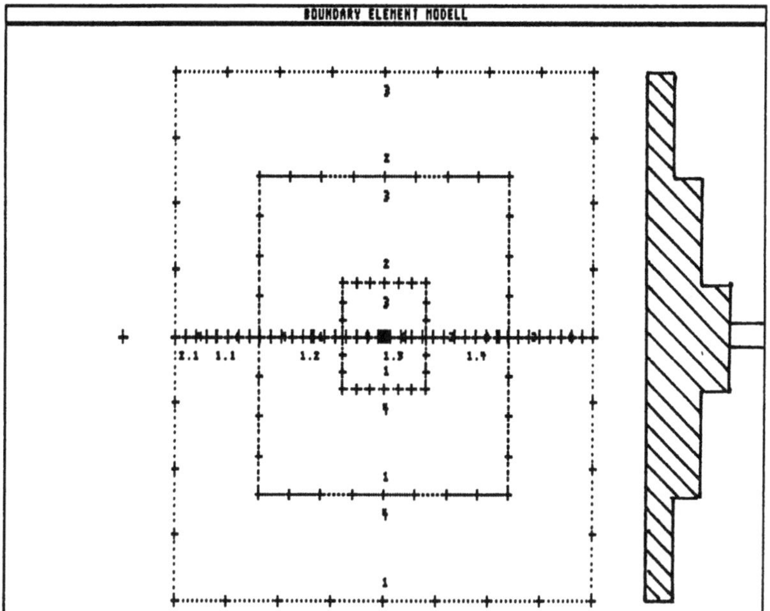

Fig. 3 Plate structure with varying thickness

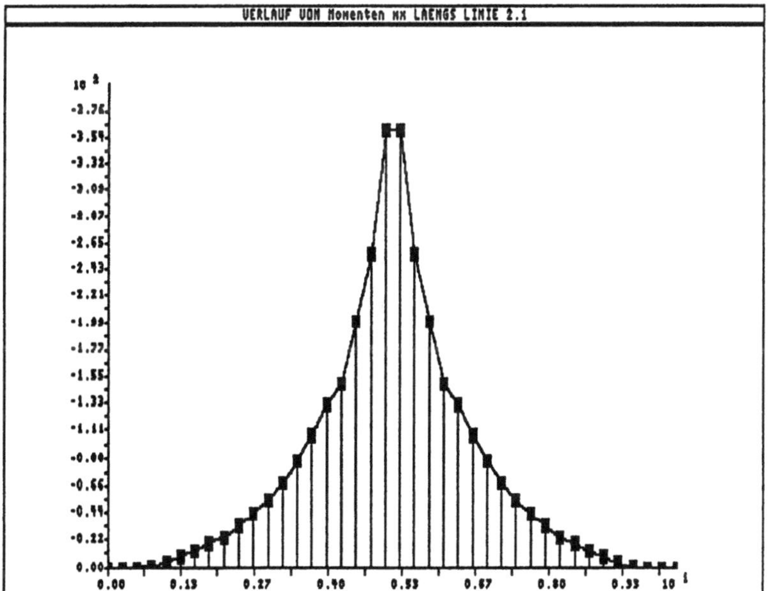

Fig. 4 Distribution of bending moments mx along the central axis

BEASY - An Analysis Tool for Design Engineers

Shan Nageswaran
Vice President
Computational Mechanics, Inc.
Billerica, MA 01821
(508) 667-5841

Abstract

Computer Aided design/analysis procedures have relied heavily on finite-element analysis over the last twenty years. In practice reliance only on FE codes for analysis usually required the design to be sent for manufacturing before detailed analysis results were available to guide the design. New designs were incrementally evolved from existing ones and detailed analysis was carried out only to validate designs which have been developed based on simple hand calculations. A more attractive approach based on the Boundary element method is proposed. This technique allows the design engineers to create analysis models quickly from their design geometry and perform analysis so that the results would be quickly available to help them to modify the designs and optimize them before detailed production drawings were made. **BEASY** is a widely used and validated commercial **software package** which provides the tools based on boundary element method to **design engineers** to explore new design concepts. BEASY is also a **cost effective** alternative to finite element systems.

Introduction:

Since the very first engineer proposed a solution to a problem the procedures we follow have not changed substantially. Design engineers initially create conceptual models to satisfy the basic objective function of the product and then they obtain basic geometry dimensions based on hand calculations or handbooks during the preliminary design stages. Most of the time the new designs are based on existing products with few minor modification. Next step is to send this design to analysts for detailed analysis. This process is often time consuming and during that time design engineers would have finished their detailed drawings for manufacturing. The impact of the results from the analysis is not felt immediately on the product design cycle. Even today the advanced computer aided techniques still follow the same procedure, but uses computer based programs to visualize the basic design geometry and to perform faster analysis. However the **distance between the design engineers and analysis engineers still remain the same.** The Analysis results are still used to validate the design. Designs are not modified quickly depending on the results. Therefore we cannot strictly call this procedure as computer aided design. Boundary Element Analysis System (**BEASY**) is specifically design to overcome these difficulties by integrating the design and analysis tasks together. The design engineers can perform their day to day analysis very easily and validate or modify their design before the manufacturing process starts. In this paper we discuss how this is done using BEASY design/analysis system which differs from other FE traditional based systems.

BOUNDARY ELEMENT METHOD Vs FINITE ELEMENT METHOD IN DESIGN

* Faster Model preparation with BEASY

Model building with BEASY is faster than in finite element method(FEM) because elements need only be placed on the boundary of the object. For a 2D or axisymetric problems elements will be lines drawing a " Silhouette of the components. For 3D problems elements will be surfaces covering the surface area of the object. Therefore the boundary information (lines, splines, arcs, or surface patches) are adequate for BEASY. At no stage are finite elements generated from boundary information. FE systems need additional mesh generators to create elements inside the domain. Even with todays advances in mesh generation and graphics it is not possible to automatically mesh a three-dimensional solid.

There are other advantages that spring from defining only boundary elements, particularly for 3D modeling. Like finite elements, boundary elements have to be small where conditions changes rapidly at a stress concentration, for instance. Also like FEA it would be uneconomical for use such small elements throughout the model. As a result, a mix of both small and large elements should be used. Infact, BEASY makes it easy to vary element size since the user is not forced to match the corners of adjacent elements fig 1. It is sufficient, within a few simple guide lines, to cover the surface of the object with no gaps or overlaps. This capability leads naturally to automatic mesh generation which is easier than FEA. For a 3D problem, automatic mesh generation of boundary elements is simpler than automatic mesh generation of finite elements for a 2D problem. BEASY has automatic mesh generation and connectivity correction facilities for 2D & 3D models.

* Quicker Design Modification using BEASY

Only the boundary needs to be defined for BEASY, so less work is required than a corresponding finite element model where volumetric elements must be defined (fig 2). On average this feature alone provides savings of 10:1

over finite-element modeling. Perhaps more important is ease of changing on existing model in BEASY. Modification of the model is serious and often overlooked draw back in FEA. Indeed, it is often easier to start all over than to change an existing finite-element model. In contrast, changes in a boundary element model are local and can this be made quickly (fig 3). This advantage provides even greater savings in 3D models.

* <u>Better Integration with CAD systems with BEASY</u>

Design engineers **are not** often trained to create FEA models and its associated properties such as element density and element types etc, which are often taken care by expert analysts. However they are more familiar with design geometries such as points, lines, arc's, splines, wireframes (fig 4) and surface patches. They like to associate the boundary condition and material properties to the geometry entities(Eg displacement constraints on a circular arc). BEASY system interacts with the design systems and design engineers at the geometry level. One can transfer the design geometry associated boundary conditions to BEASY directly or through CAD/IGES files. BEASY automatically creates elements on these geometry and perform analysis. The results are often displayed on the geometry entities. Use typically asks to plot the stresses on lines or arcs or surfaces. (Not on elements). BEASY provides **no element concept** to the designers. They are using the design geometry entities even inside the BEASY system. They can modify the geometry inside BEASY system depending on the analysis results and still take this modified geometry back to the CAD system via IGES files. In addition to all these benefits BEASY provides accurate, robust and faster analysis results. BEASY solution is almost as fast as simple hand calculations and approach accuracy of established FE codes. Large number of design iterations can be performed with BEASY in less time and cost to meet the weight and stress targets required by the design engineers.

Let us consider an analysis of a mechanical component already created by a designer using a CAD system. Figure 5 shows the design drawing from a CAD system where the part to be analyzed is shaded. Now the designer selects the component and writes the IGS description (points and lines only) and figure 6 shows how that component is picked up interactively by the BEASY system. The designer then uses BEASY to create mesh automatically only on the boundary lines which is much easier than creating FE element inside the domain. With the application of design loads BEASY analyzes the problem interactively as fast as simply hand calculations and gives the results in graphical form as shown in figure 7. The results are only on the boundary and designers can easily identify the maximum values and its locations quickly. Now the designer can modify the geometry using BEASY and re-analyze quickly to obtain the optimum shape. BEASY is capable of creating a new IGES description of the modified geometry which could be read by CAD systems. This is true Computer Aided Design (CAD).

The same principle is also applicable to three dimensional solid models. Figure 8 shows a cross sectional view captured from 2D CAD IGS file. This cross section is then rotated to create the surface model(figure 9). BEASY requires element only on the surface of the solid object and this is carried out automatically mapping the surface patches to elements (fig 10a). FE analysis requires brick element and designers will not be able to carry out this task automatically. Figure 10b shows the boundary elements created by BEASY and the displacement pattern from the analysis. In this example a turbine design engineer is estimating the stiffness of the casing system very quickly and accurately.

* Faster convergency and very accurate results.

Many engineers are surprised to find that even though boundary element method requires less information, the results turn out to be more accurate in general than FEA. There are a number of mathematical reasons for this, but intuitively it can be explained like this: every piece of information supplied to an analysis program represents an approximation to real life. With boundary elements, less information is supplied, there are fewer approximations, therefore more accurate.

The improved accuracy is particularly evident in problems involving stress concentrations (or hot spots in a heat transfer analysis). This is a large class of problems in today's engineering design, and one that requires particular care as accuracy is extremely important. The sharply peaking stress pattern which occurs in a crack tip, for instance, is simply not predicted by general finite elements. The boundary element formulation gives rise to highly accurate stress predictions in such cases. Boundary element results are not sensitive to mesh grading us FEM and the solution does not deteriorate at stress concentrations and discontinuity as is common with FEM.

The sensitivity of FE to mesh refinement is well known but not always appreciated by engineers. Specially in large 3D analysis engineers are often satisfied with one FE results and failed to verify that results by further mesh refinements or alterations. For example, in a recent study by P.R.Hearn (ref 1) showed the importance of mesh sensitivity analysis in FE. Fig 11a shows the first FE model compared to BEASY model (fig 12a) gave 15% lower peak stress value. When he refined the FE model further (fig 11b) the peak stress matched BEASY results. However when BEASY model was refined(fig 12b), he noticed very little change in the results showing the converged results.

Summarizing these solutions, the BEASY mesh (which basically represents the geometry of the problem) can provide locally accurate stresses with very simple modeling compared with the FEM results.

* <u>Results error indicator guides mesh refinement in BEASY</u>

It is essential to verify the results obtained by FEM or BEM before using them to alter the design. Use of FEM requires either to increase the mesh density (h convergency) or to increase the order of the elements (p convergency) and performing the full analysis to verify the results convergency. This procedure is very simple in BEASY. BEM also provides a guide line where to refine the mesh. Fig. 13 shows a typical BEM results when plotted along the boundary elements. If the full convergence is obtain, the results will be continuous and smooth between adjacent element. The discontinuity in results usually indicate the inadequacy of the mesh density at that location. Beasy system will plot you the error norm to indicate these functions. Depending on the error norm, new mesh refinement will be carried out. BEASY has the facility to perform h or p refinement or both. This procedure is valid for 3D as well. Lack of continuity on the element based contour plots means inadequate mesh density. BEM has very faster convergency property than FEM. The table below shows some comparison of results convergency using p version of elements in BEASY.

* <u>Excellent for fracture and stress concentration problems</u>

Because the boundary element method has been found to give highly accurate surface stresses, engineers are using BEASY more and more when they need to obtain accurate fatigue and life predictions. Crack tip stresses are found from the normal BEASY elements, and are very accurate when compared with typical finite element predictions(fig 14), in which the peak stress is often smoothed out.
BEASY is accurate for stress prediction around cracked surfaces for a number of reasons:

(a) The stresses are calculated at the same stage as the displacements (mixed formulation). They therefore have the same accuracy as the displacements. In most finite element programs, the displacements are calculated first, and these are then used to calculate the stresses- a step which involves differentiation, and therefore approximation and a loss in accuracy.

(b) BEASY calculates the results right on the boundary, where the peak stress is occurring. Finite element programs have to extrapolate from the integration points internal to each element.

(c) The discontinuous elements BEASY uses around features like cracks allow a discontinuous stress pattern which will occur. The peak stress is not smoothed out by averaging stresses with adjacent elements.

The boundary element method is more accurate generally than domain techniques, because the approximation is limited to the boundary. Fig 16 show the results of a 3D Piston model where BEASY predicts the presence of high stress concentration at the fillets accurately which were not correctly shown by FE programs. FE required finer mesh density at the fillet to predict good stress values which took large amount modeling time.

Conclusion:

BEM is a better design tool than FEM. It closes the gap between the design engineers and traditional Analysis engineers. Design engineers could perform the majority of the analysis tasks using BEM, which will help the analysis to perform difficult tasks where expert knowledge is required.

BENEFITS of using BEASY which is based on BEM is a design environment are:

===> **Better and faster design.**
Easy model creation, easy to change the geometry and quick and intelligent feed back of results to optimize the design.

===> **Lower Project Cost**
Lowering the project cost by reducing the model preparation and results reviewing time. Only boundary data and results are presented to the designers.

===> **Closer link to geometry and BEM models.**
BEM models are essentially the same as the design geometry.The geometry need not to be over simplified. Material properties and loads are not lined to elements. They can be created separately based on geometry entities.

===> **True mesh automation is possible.**
3D BEM meshing is the same as 2D FEM meshing.

Knowledge loading before mesh generation also helps intelligent auto mesh generation in BEASY. It is interesting to note that the final boundary element analysis model description contains all the initial inputs necessary for a FE auto mesh generation program.

===> **Accurate Results:**
BEM is very good for stress concentration problems and has faster convergency characteristics.

Finally
===> **Closer Designer-Analysis link. BEM is designer's analysis tool**

References:

1. BEASY used for optimization of rotor stress
BEM 11 .. 1989 Computational Mechanics
by Paul Hearn Hamilton Standard UTC.

2. **BEASY... Boundary Element Analysis System
by Computational Mechanics
25 Bridge Street
Billerica, MA 01821
TEL 508-667-5841**

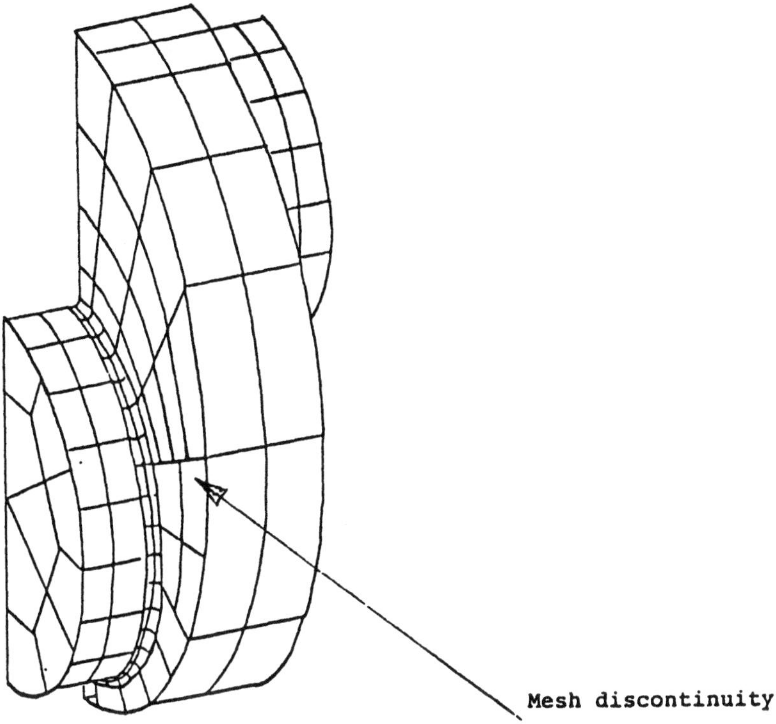

Figure 1 Surface element mapping is easy since BEM can handle mesh discontinuity.

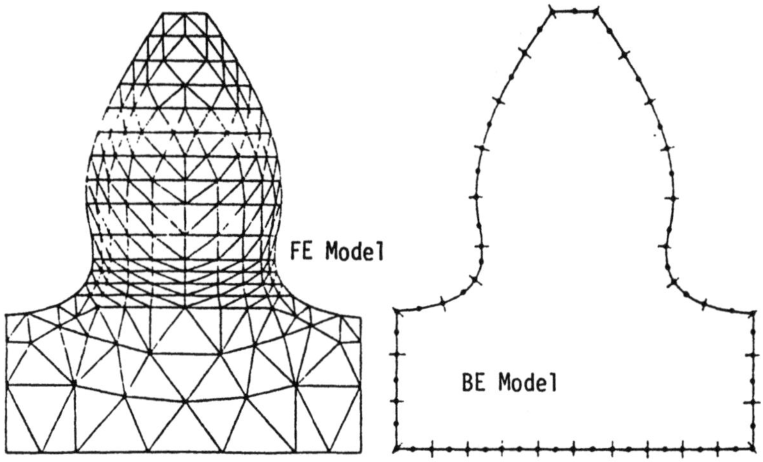

Fig 2. Model Creation is easy in BEM

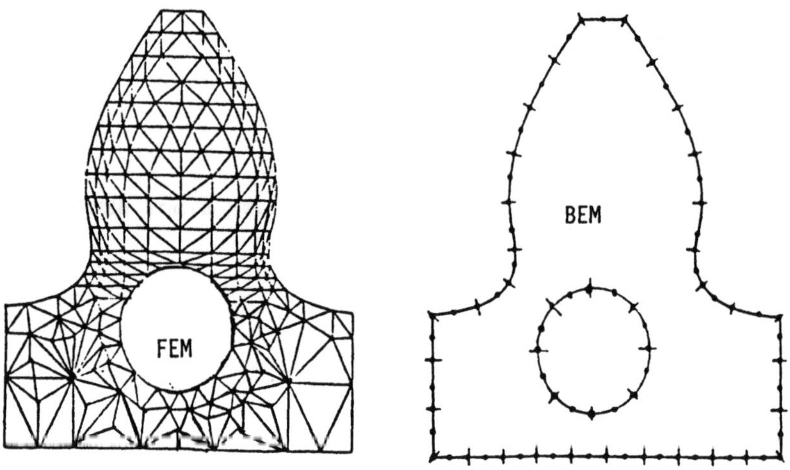

Fig 3. Model Modification is easy in BEM

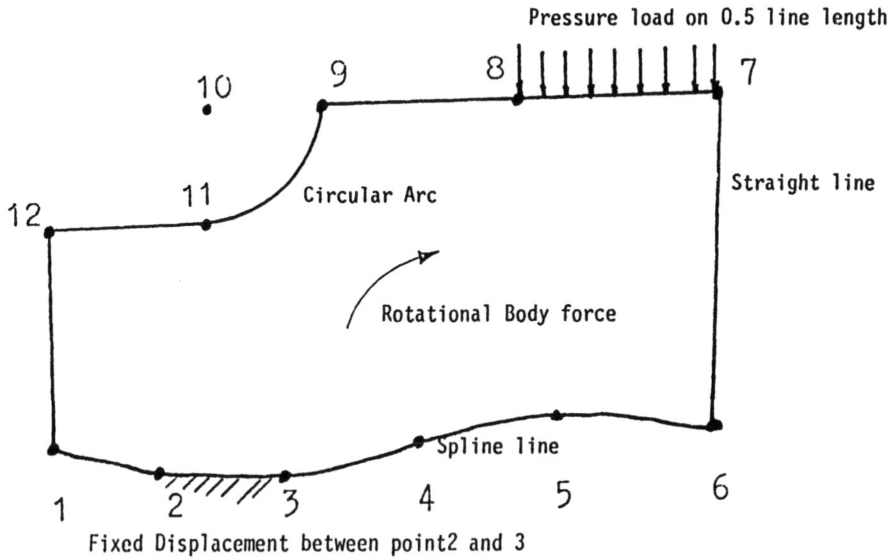

Design Engineer's Analysis Model. Fig 4.

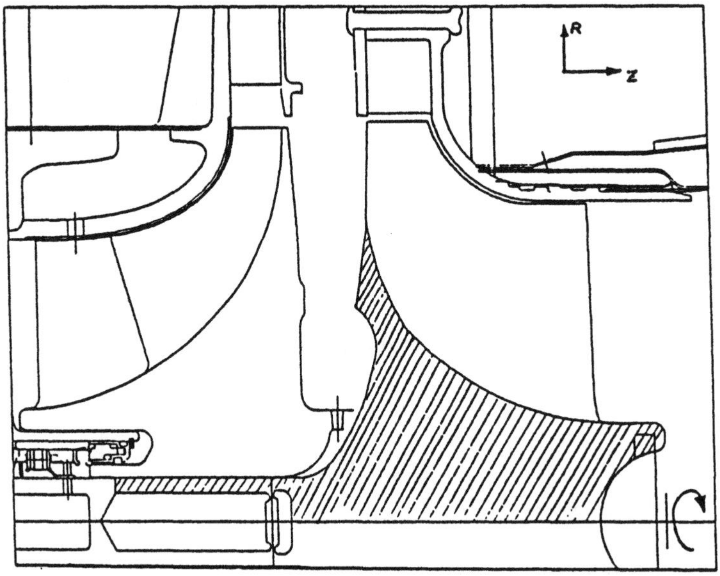

Figure 5. A design drawing from a CAD system database.
The shaded componet is selected fo BE analysis

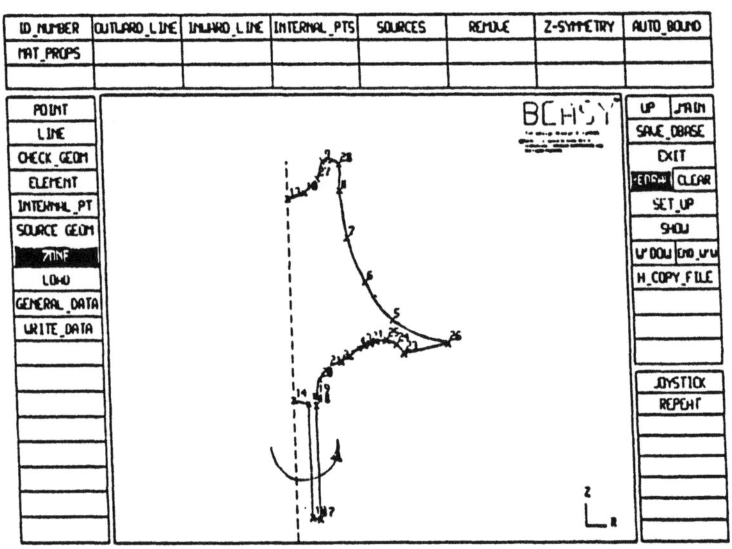

Figure 6. The IGES description of the component from fig 5.
after being transfered to BEASY interactive
modelling system

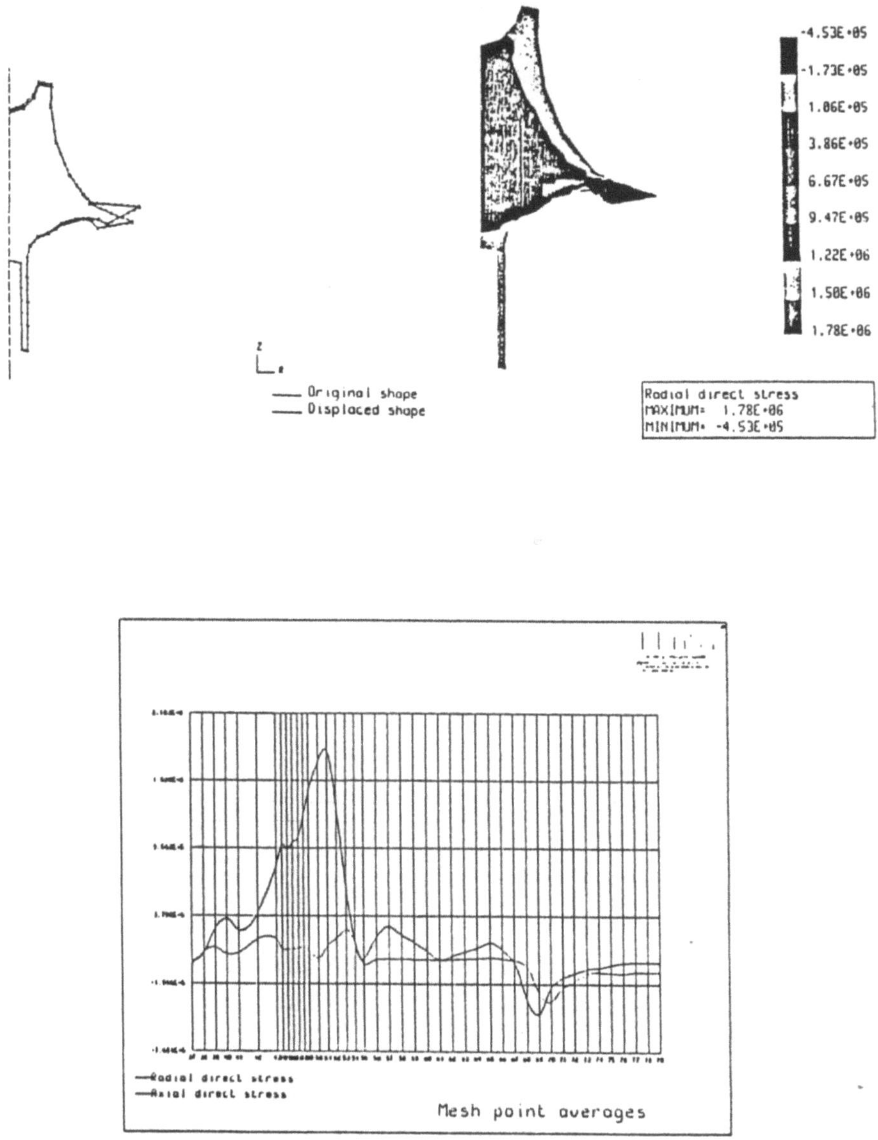

Figure 7.
Boundary results to guide the design. The peak stresses can be easily identified and geometry changes are made to reduce the peak stresses.

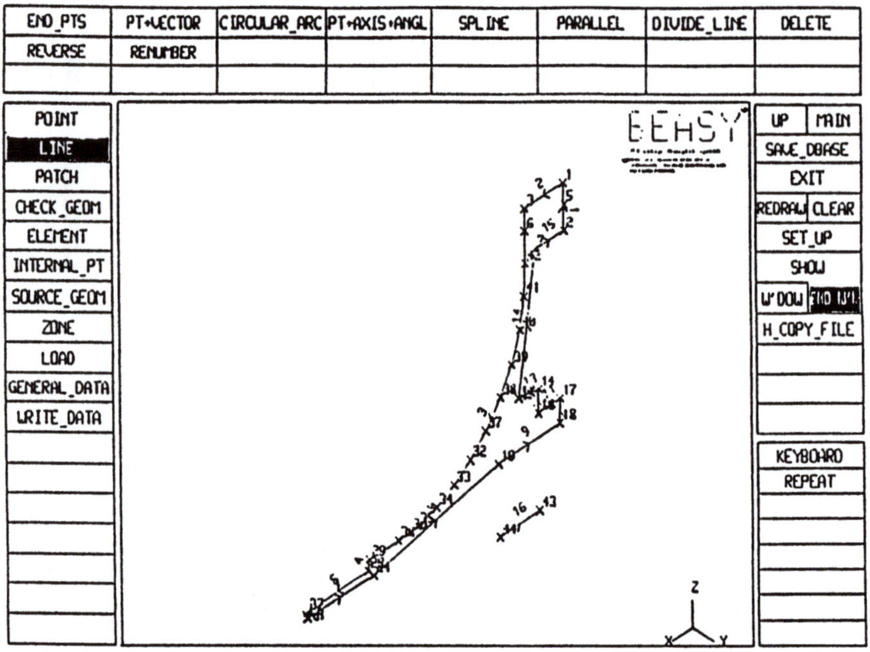

Figure 8. A cross sectional view captured from a CAD system via IGES

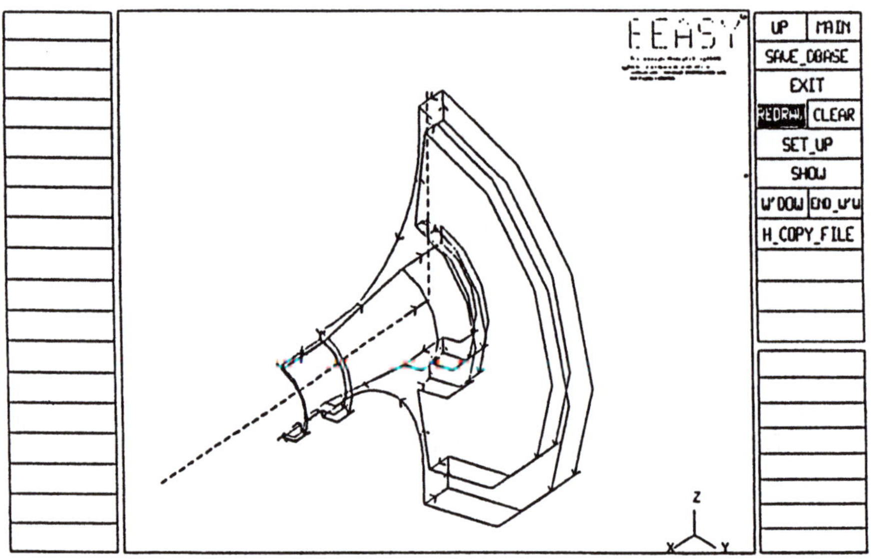

Figure 9. 3D surface model for boundary element analysis is created by the rotation of 2D geometry. Solid model is not needed for BEM.

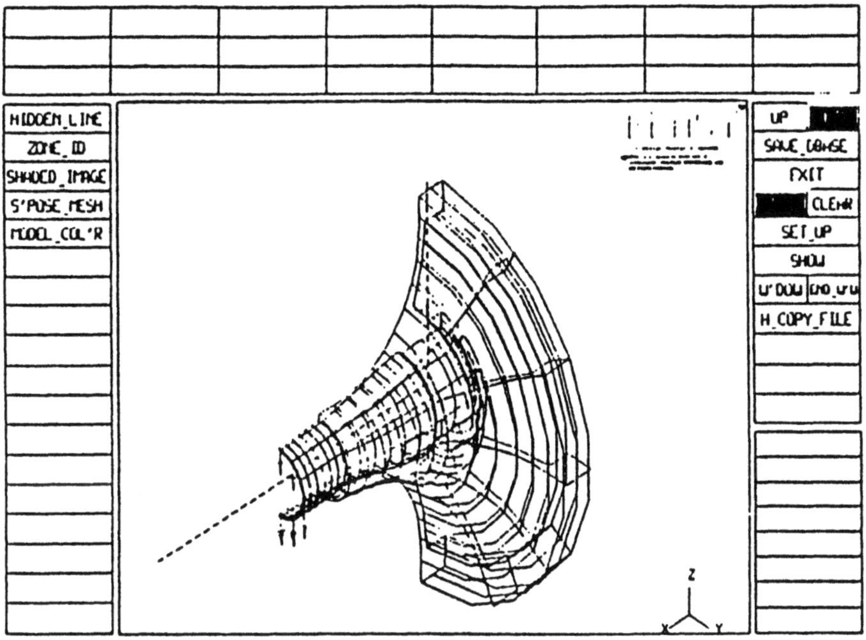

Figure 10a. Boundary elements are automatically qenetated. Only surface elements are necessary.

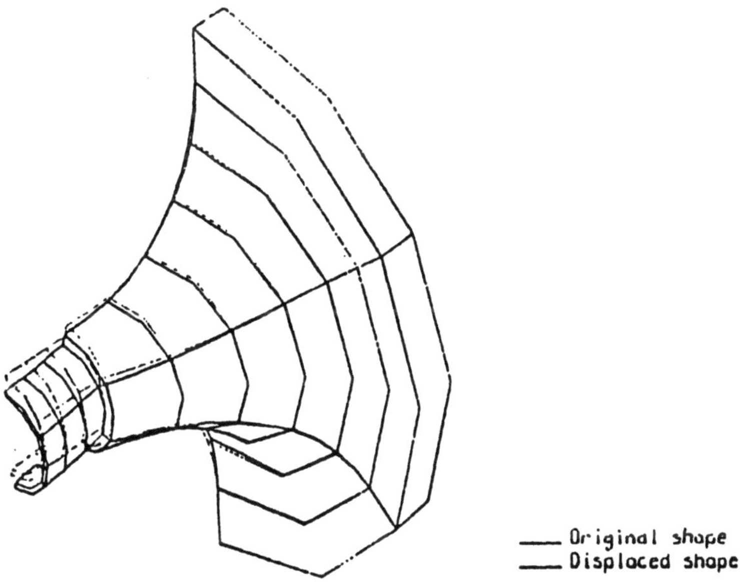

Figure 10b. Displaced shape due to end loads from BE analysis.

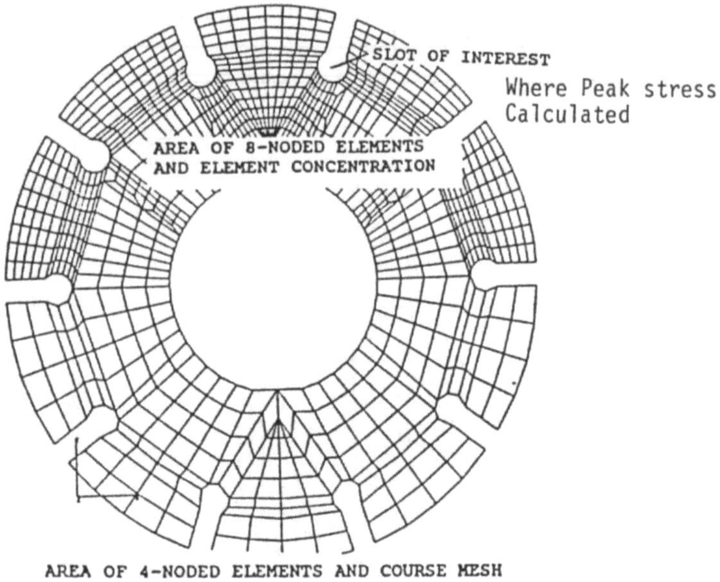

Fig 11a . FE model

ORIGINAL NASTRAN MODEL

Ref: Paul. R. Hearn Hamilton Standard Windsor Locks, CT

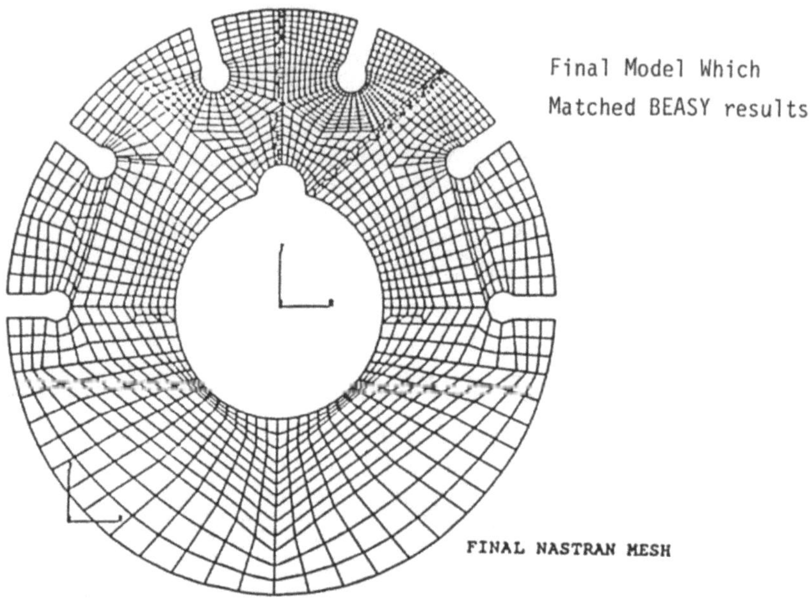

Fig 11b. Refined Model Showed 15% increase in Peak Stress

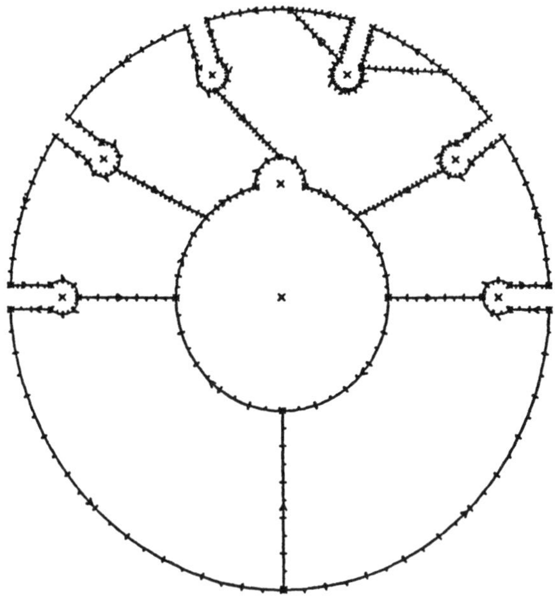

Fig 12 a Original BEASY model Which Gave the Correct Peak Stress which matched Fig 11b of FE model

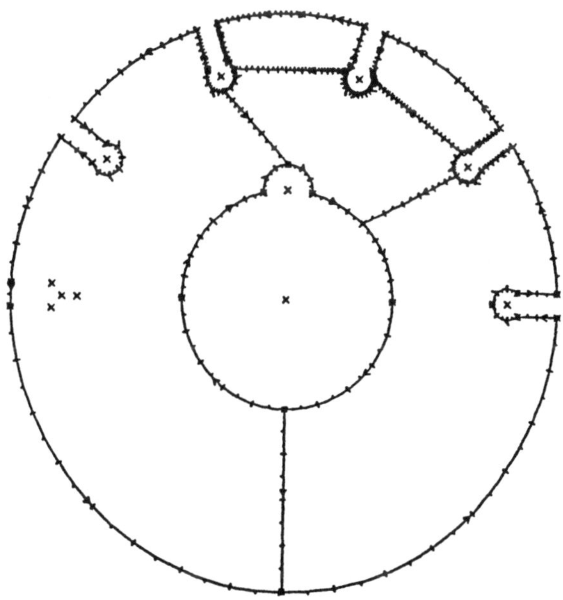

Fig 12b. Final BEASY model which gave exactly the same results as Fig 12 a.

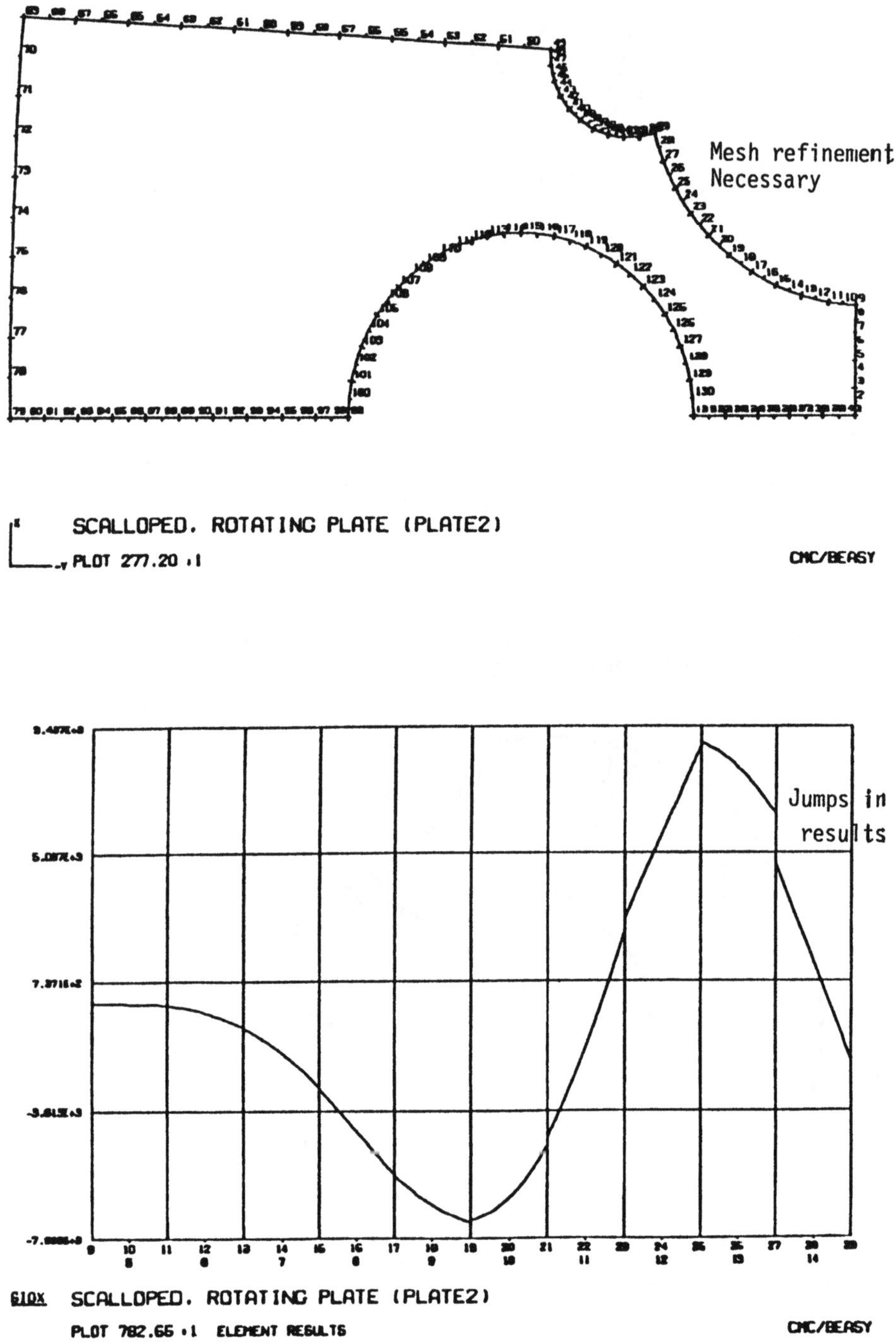

Fig 13. BEASY Results plot shows discontinuous results which indicates further mesh refinement.

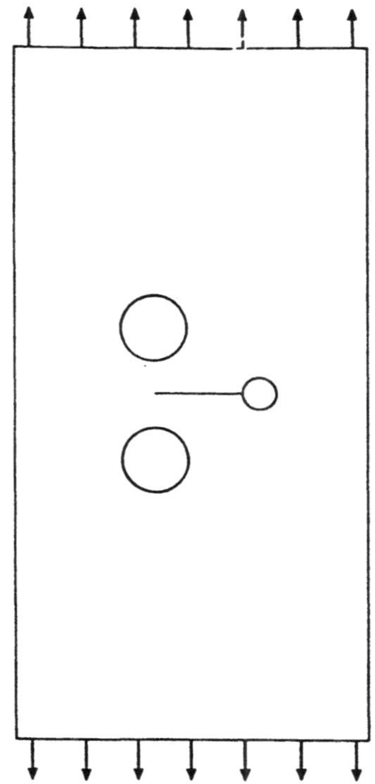

(a)

Three hole crack problem

The 2D study is the three-hole-crack problem (fig a). This was the subject of an investigation by Newman[2] at NASA Langley The problem was solved using a variety of techniques (not including boundary elements) involving different techniques for assessing the fracture properties. A model which was selected as producing accurate results for this study was a finite element mesh containing 3842 elements and using a crack-tip opening failure criterion The mesh (shown in fig b) is not a simple matter to construct, and highlights well the advantages of using BEASY for this type of problem.

A BEASY mesh to find the stress intensity factor at the crack tip is shown in fig C The elements become small around the crack tip and are graded out from there This 69-element model gave a stress intensity factor which differed by less than 5% from the 3842 element f.e. model.

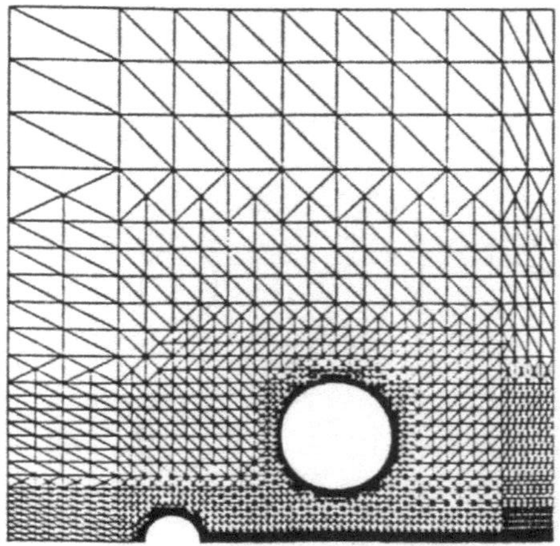

(b)
Finite element mesh for crack problem

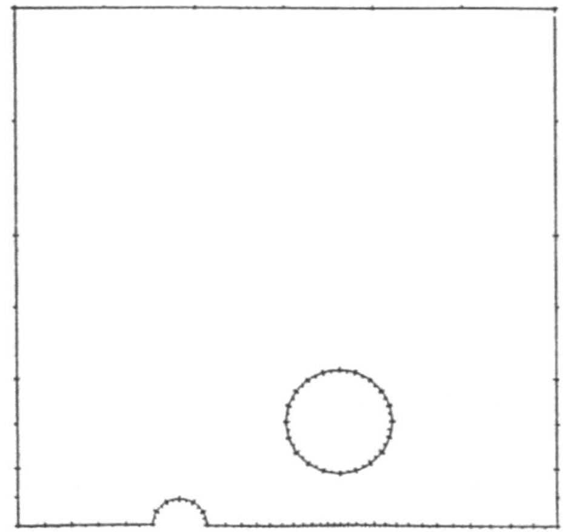

(c)
BEASY boundary element model of crack problem

Fig 13. Three Hole Crack Problem

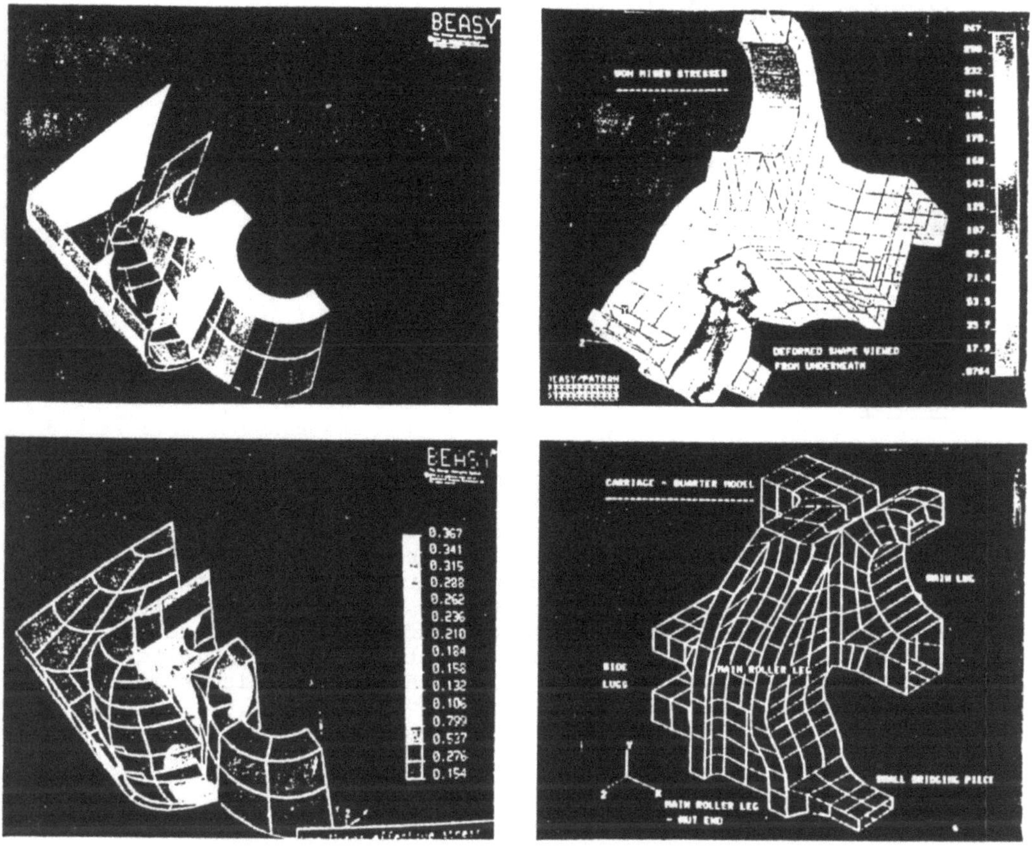

Piston Model

Flap Track Model

Fig 16. Use of BEASY for Stress concentration prediction in 3D

Coupling of Finite Element and Boundary Element Super-Element Methods

J.L. WEARING and M.A. SHEIKH

Dept of Mech. & Process Engineering
University of Sheffield,
Sheffield, U.K.

Summary

A Boundary Element Superelement Method (BESEM), based on the Indirect formulation of the Boundary Element Method (IBEM), has been developed. It produces a system of equations, for each boundary region, which can be directly merged with the Finite Element System, thus allowing the use of a combined Finite Element - Boundary Element strategy to model engineering field problems. A few examples of such models are presented. These employ the Boundary Element Superelement Method to discretize the bulk of the problems with regions of Finite Elements to represent those areas where the analysis is more complex or critical.

Introduction

The Finite Element and Boundary Element Methods are two most commonly used solution techniques in the numerical discretization of structural and other field problems. Both methods have certain advantages and disadvantages. For many linear problems the Boundary Element Method (BEM) is much more efficient due to the ease with which boundary models can be generated. It also gives more accurate results in the regions of rapid stress or flux variation. Moreover, it is ideally suited to deal with problems whose domains extend to infinity. Finite Element Method (FEM), on the other hand, is attractive if the material properties vary considerably in the domain or if the problem is nonlinear. It is sometimes advantageous to combine Finite Element and Boundary Element solutions. This requires a domain to be represented by finite elements or boundary elements, depending upon the particular geometries or boundary conditions of the problem.

One method of linking the Boundary Element Method with a Finite Element scheme is to minimise the potential energy functional so that its stationary value gives the solution of the governing differential equations in both regions while satisfying equilibrium and compatibility conditions at the interface [1]. Following this approach, a Boundary Element Superelement Method (BESEM) has been developed [2]. In this method, a boundary element region is made to appear as an equivalent finite element region or a superelement [3]. The stiffness matrix and nodal force contributions for a superelement are calculated using an Indirect Discrete Boundary Method [4], and the resulting equations have the same form as the finite element equations. These equations can therefore be directly linked with any standard Finite Element Program in the same way as the finite element substructuring strategy. The Coupled Finite Element - Indirect Discrete Method thus allows a free mixture of finite elements and boundary elements so that each method can be used to model those areas of the problem where it is most applicable.

A Fortran program based on the Boundary Element Superelement Method (BESEM) has been developed to solve problems in potential theory and linear elasticity. This program works both independently or in unison with a Finite Element program. In implementing this program a strategy similar to the one used in the Indirect Discrete Boundary Method (IDBM) is adopted to model the boundary regions [5].

In this paper the formulation of the Boundary Element Superelement Method (BESEM) and its coupling with the Finite Element Method (FEM) is presented. The numerical implementation of the coupled technique is briefly described followed by a few examples of construction and analysis of coupled Finite Element - Boundary Element models.

FE Equations

For a discretized system, the Finite Element Method produces a global system of equations:

$$\{F_n\} = [K]\{u_n\} \qquad (1)$$

where $\{F_n\}$ is the nodal force vector, $\{u_n\}$ are the nodal displacements, and [K] is the stiffness matrix given in terms of strain-displacement matrix [B] and stress-strain matrix [D] as:

$$[K] = \int_\Omega [B]^T [D][B]\, d\Omega \tag{2}$$

BE Equations

The Indirect Boundary Element Method (IBEM) is based on the use of Fundamental Solution (or Kelvin Solution) of the governing differential equation of elasticity [6]. For a homogeneous isotropic domain this solution relates the displacement at a field point $u_i(x_f)$ to a point force $s_j(x_s)$ at the source point [7] as:

$$u_i(x_f) = G_{ij}(x_f, x_s)\, s_j(x_s) \tag{3}$$

Similarly, the surface tractions $t_i(x_f)$ are given by:

$$t_i(x_f) = H_{ij}(x_f, x_s)\, s_j(x_s) \tag{4}$$

By spreading a discrete number of sources around the boundary of the domain of a given problem, the numerical approximation of Equations (3) and (4) result in the Indirect Discrete Boundary Method formulation and produces two matrix equations:

$$\{u\} = [G]\{S\} \tag{5}$$

$$\{t\} = [H]\{S\} \tag{6}$$

For a discretized system with either displacement or traction prescribed at each node on the boundary, equations (5) and (6) can be employed to assemble a system of equations relating the known boundary conditions and the source intensities as:

$$\{\bar{u}_n\} = [B]\{S\} \tag{7}$$

Equation (7) can be solved for the source intensities {S} and substituted back in equations (5) and (6) to produce the boundary solution as:

$$\{u\} = [G][B]^{-1}\{\bar{u}_n\} \tag{8}$$

$$\{t\} = [H][B]^{-1}\{\bar{u}_n\} \tag{9}$$

It is obvious that this system of equations is not directly compatible with the Finite Element System given by equation (1).

BE Superelement formulation

For a body of domain Ω with boundary Γ, the potential energy, in the absence of any body forces, is given as:

$$\Pi = \int_{\Omega} w(u)\, d\Omega - \int_{\Gamma} pu\, d\Gamma \qquad (10)$$

where the first term is a volume integral of the strain energy density w(u), and the second term is the external work in which p are the external boundary loads.

Betti's second formula is stated as:

$$\int_{\Omega} u\, A(u)\, d\Omega = 2\int_{\Omega} w(u)\, d\Omega - \int_{\Gamma} u\, t(u)\, d\Gamma \qquad (11)$$

Using this in the total potential energy expression (Eqn.10),

$$\Pi = \tfrac{1}{2} \int_{\Omega} u\, A(u)\, d\Omega + \tfrac{1}{2} \int_{\Gamma} u\, t(u)\, d\Gamma - \int_{\Gamma} pu\, d\Gamma \qquad (12)$$

If a solution is sought where the operator A(u) = 0 within Ω, then equation (12) is reduced to:

$$\Pi = \tfrac{1}{2} \int_{\Gamma} u\, t(u)\, d\Gamma - \int_{\Gamma} pu\, d\Gamma \qquad (13)$$

From the Indirect BEM equations (5) and (6), t can be expressed in terms of u as:

$$\{t\} = [H]\, [G]^{-1} \{u\} \qquad (14)$$

Substituting this in equation (14) gives:

$$\Pi = \int_{\Gamma} \{u\}^{T}[H][G]^{-1}\{u\}\, d\Gamma - \int_{\Gamma}\{u\}^{T}\{p\}\, d\Gamma \qquad (15)$$

Minimising equation (15) gives:

$$\int_{\Gamma} [H]\, [G]^{-1} \{u\}\, d\Gamma - \int_{\Gamma} \{p\}\, d\Gamma \qquad (16)$$

which can be written in matrix form as:

$$[K]\, \{u_n\} = \{F_n\} \qquad (17)$$

where $\{F_n\}$ is the vector of nodal forces and [K] is the stiffness matrix given as:

$$[K] = \int_{\Gamma} [H]\, [G]^{-1}\, d\Gamma \qquad (18)$$

Equation (17) is now in the same form as the FEM equation (1), and the stiffness matrix given by equation (18) is equivalent to a FEM superelement or substructural stiffness matrix.

Coupled FE-BE System

For a finite element domain Ω_1 (with boundary Γ_1) coupled with a boundary element domain Ω_2 (with boundary Γ_2), the total potential energy of the system is:

$$\left.\begin{aligned}\Pi = \int_{\Omega_1} w(u)\, d\Omega + \int_{\Omega_2} w(u)\, d\Omega \\ - \int_{\Gamma_1} pu\, d\Gamma \quad - \int_{\Gamma_2} pu\, d\Gamma\end{aligned}\right\} \quad \begin{aligned}\Omega = \Omega_1 + \Omega_2 \\ \Gamma = \Gamma_1 + \Gamma_2\end{aligned} \tag{19}$$

On the interface Γ_{12} (or Γ_{21}) the equilibrium condition $p_1 + p_2 = 0$ and compatibility condition $u_1 - u_2 = 0$ have to be satisfied so that:

$$\int_{\Gamma_{12}} u_1 p_1 d\Gamma + \int_{\Gamma_{21}} u_2 p_2 d\Gamma = 0 \tag{20}$$

Using this condition with equation (19) allows the potential energy of the whole domain to be split into the sum of the energies of the two subdomains, i.e.

$$\Pi_{\Omega} = \Pi_{\Omega_1} + \Pi_{\Omega_2} \tag{21}$$

Equation (21) implies that equations (1) and (17) can be merged to obtain a system based on the minimisation of total potential energy. In general, the domain can be divided into as many subdomains as required. These subdomains can be evaluated using either method before merging the final system of eqns.

Numerical implementation of Coupled FE-BE System

The analysis of a coupled FE-BE model is based on setting up and solving an overall system of equations which comes from the combination of FE and BE contributions. The FE contribution is the assembly of element stiffnesses and nodal loads given by equation (1). The BE contribution is composed of the BE Superelement stiffness matrices and nodal load vectors which are obtained by evaluating the integrals in equation (16). This is done by dividing the boundary of each superelement into a series of boundary elements. These elements are compatible with their finite element counterparts. The integration is carried out over each boundary element in turn using a gaussian quadrature scheme. The order of integration is set to be the same as used by FEM in order to maintain numerical compatibility across the FE-BE interface. H , G, u, and p are calculated at the element nodes. Their values at the gaussian integration points are obtained using the shape function interpolation, again to mimic the FEM. Equation (16) can be written in matrix form as:

$$[W]\,[H]\,[G]^{-1}\{u_n\} = [W]\{p_n\} \tag{22}$$

where [W] is the diagonal matrix of weighting factors composed of the product of shape functions, Gaussian weights and Jacobean factors. These matrices are evaluated and multiplied to to give the stiffness matrix and load vector for every superelement. These are then assembled to form the total stiffness matrix and load vector for the entire boundary element region. The final system of equations is obtained by merging this with the finite element stiffness matrix and load vector. Solution of this overall system of equations yield the displacement solution for the whole domain. These nodal displacements are then employed to calculate the stresses in the FE and BE regions by their respective programs.

Coupled FE-BE models - Examples

Model I: The quarter domain model of a plate with a circular hole (a/w=0.3) subjected to uniform pressure loading is shown in Figure. 1. It consists of 2 BE Superelements with 40 quadratic boundary elements and 20 finite elements around the hole. The Stress Concentration Factor for this model is found to be within 5% given in [8].

Model II: Figure. 2 shows a coupled FE-BE model of a crane hook which produces a region of high stress in its inner radius under static loading. It consists of 18 finite elements around the hook bend. The top part of the hook is modelled with 1 BE Superelement (13 boundary elements) and the lower arm is discretized with 5 superelements with a total of 41 quadratic boundary elements. The stress distribution around the inner radius of the bend is found to be within 2% agreement with the photoelastic results produced by Kenny [9].

Model III: Figure. 3 shows a coupled FE-BE model of the quarter domain of a tie bar which forms part of a turbine blade. It is constructed with finite element meshes around the holes with

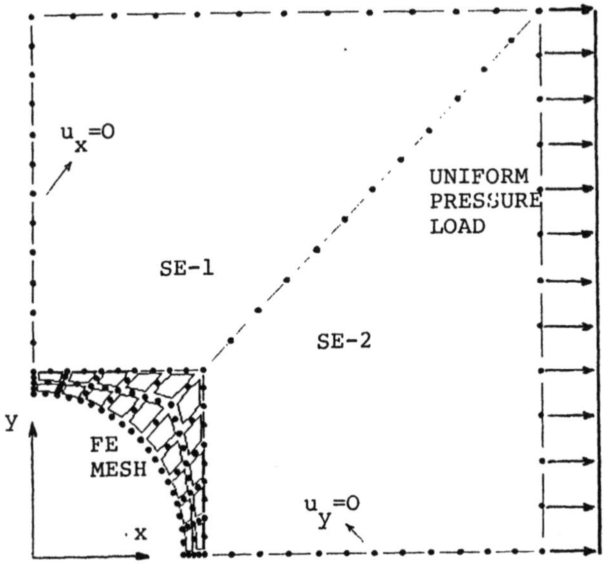

Figure.1. Coupled FE-BE model of the quarter domain of a square plate with a circular hole

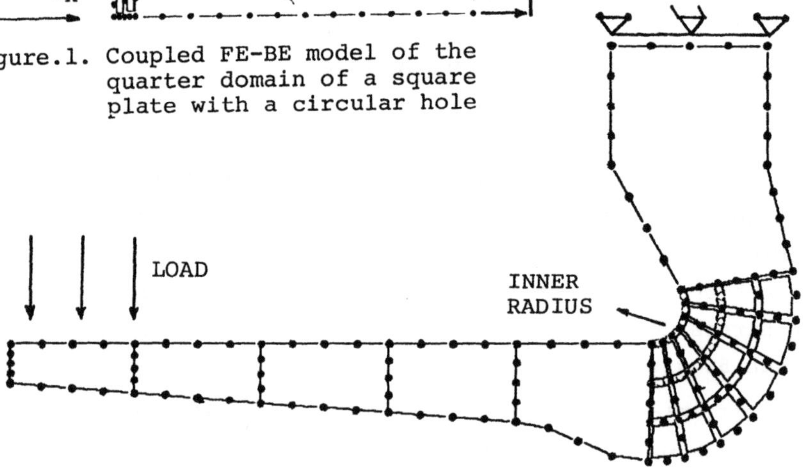

Figure.2. Coupled FE-BE model of a Crane Hook

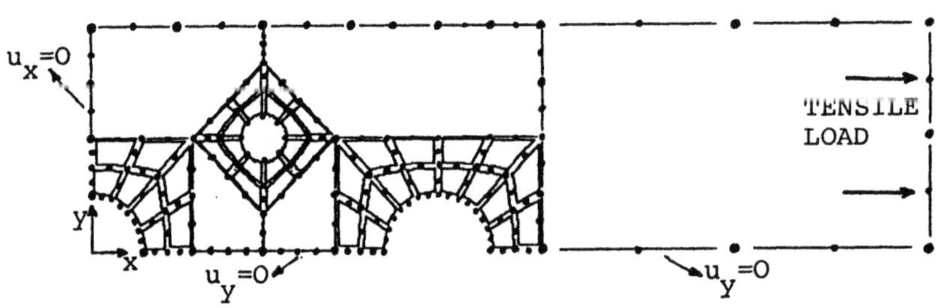

Figure.3. Coupled FE-BE model of a tie bar

5 BE Superelements linking them up. The stress concentration factors around the main holes are found to be within 5% of those calculated by FEM.

Concluding Remarks

A numerical technique has been developed which allows the stiffness matrix and nodal force contributions for a subdomain to be calculated using BEM and then link them as a substructure to a Finite Element system. The resulting unified technique, employing coupled FE-BE models, allows optimal use of both methods and thus offer greater flexibility in modelling complex problems.

Acknowledgement

The work presented in this paper has been supported by the Central Electricity Generating Board and Science & Engineering Research Council in U.K.

References

1. Kelly, D.W.; Mustoe, G.W.; Zienkiewicz, O.C.: Coupling Boundary Element Methods with other Numerical Techniques, Developments in Boundary Element Methods, Vol.I, Banerjee, P.K.; Butterfield, R. (eds.) App.Sc. Publishers, U.K, 1979.
2. Hickson, A.J.: The Combination of Finite Elements and Boundary Elements for Stress Analysis, Ph.D Thesis, University of Sheffield, U.K., 1987.
3. Wearing, J.L.; Sheikh, M.A.: A Boundary Element Superelement Method for Coupling with the Finite Element Method of Substructures, Proc. 12th Canadian Congress App. Mech., Erki, M.A; Kirkhope, J. (eds.), Graphic Services Dept, Carleton University, Ottawa, Canada, 1989.
4. Scholfield, R.P.: Development of the Indirect Discrete Boundary Method and its application to Three dimensional Design Analysis, Ph.D Thesis, University of Sheffield, U.K., 1986.
5. Noroozi, S.: Further development of the Regular Indirect Discrete Boundary Element Method, Ph.D Thesis, University of Sheffield, U.K., 1988.
6. Brebbia, C.A.: Boundary Element Method for Engineers, Pentech Press, London, 1978.
7. Banerjee, P.K.; Butterfield, R.: Boundary Element Methods in Engineering Science, McGraw Hill, Chichester, U.K., 1981.
8. Shilkrut, D. et al: Elastic Stress Concentrations in a square plate with a large hole under uniaxial loading, J. Strain Analysis, Vol.22, no.1, 1987.
9. Kenny, B.: Internal report, Dept. of Mech. Eng., University of Sheffield, U.K., 1974.

A Variational Formulation of the Boundary Element Method in Elastodynamics

G. MAIER[(1)], M. DILIGENTI[(2)], A. CARINI[(3)]

(1) Department of Structural Engineering, Technical University of Milan
Piazza Leonardo da Vinci 32, Italy
(2) Department of Mechanical Engineering, University of Brescia, Italy
(3) Department of Civil Engineering, University of Brescia, Italy

Summary

The linear dynamic analysis of elastic bodies is formulated by means of two boundary integral equations using as source distributions both traction and displacement discontinuities variable in space and time. The integral operator thus arising turns out to be symmetric in both space and time with respect to a time-convolutive bilinear form. The solution of these equations up to any instant T (i.e. the history of tractions and displacements over the boundary and the time interval $0 \le t \le T$) is shown to be characterized by a variational theorem. Discretization in space and time leads to linear equations with a symmetric coefficient matrix.

Introduction

Boundary integral equation (BIE) formulations and their approximate solution in the time domain by boundary element methods (BEM) are the subject of an abundant literature surveyed e.g. in [1-4].

The traditional approach leads, through BE discretization, to systems of linear equations with nonsymmetric coefficient matrices, to be solved for each time step.

Formulations which are symmetric or variational with respect to the solution in space only were developed for elastostatic, quasi-static elastic-plastic analysis and contact problems, see [5-11], and proposed in terms of rates for dynamic analysis [12].

In this paper the linear analysis of a homogeneous elastic body subjected to dynamic excitations is formulated by a boundary integral equation approach which leads to symmetry both in space and time over a finite time interval. We assume as unknown sources acting on the elastic space Ω_∞: distributed time-dependent discontinuities of tractions on the constrained boundary Γ_u, and discontinuities of displacements on the unconstrained boundary Γ_p. The given displacements on Γ_u and tractions on Γ_p are expressed as effects at time t due to the past history of such sources and of the given external actions. In the pair of governing integral equations thus generated, the boundary integral operator turns out to be selfadjoint with respect to a time-convolutive bilinear form (proofs, here omitted for brevity, are presented in a forthcoming paper [13]). As a consequence, it is shown that the time history of the boundary unknowns over any finite time interval T is characterized by a variational property. Discretization in both space and time through double integrations over the elements, leads to a symmetric system of linear algebraic equations in the variables governing the time histories of boundary displacements and tractions.

The conceptual foundations of the present approach basically rest on convolutive variational principles related to differential equations [14-16], which appear to have not been considered so far in the BIE-BE context. In a companion paper [17] such concepts are employed for boundary integral formulations of transient potential problems.

Governing Equations and Traditional BIE Formulation for Elastodynamics

Consider the small-amplitude motion of a linear-elastic, isotropic and homogeneous body occupying an open volume Ω with smooth boundary Γ. The relevant classical (Stokes') fundamental solution (or Green's function for the elastic space Ω_∞) will be represented herein in the form [1-2]:

$$\underline{G}_{uu}(\underline{x},\underline{\xi};t-\tau)\equiv\left[u^{*}_{i\alpha}(x_j,\xi_\beta;t-\tau);\,i,\alpha,j,\beta=1,2,3\right] \tag{1}$$

Here and henceforth: $\underline{x}=\{x_i\}$ ≡ receiver (or field) point where the displacement ($\underline{u}$) response of Ω_∞ is considered at time t; $\underline{\xi}=\{\xi_\alpha\}$ ≡ source (or load) point where the unit concentrated impulse (represented by a Dirac distribution in space and time) is applied at instant $\tau\le t$; $i,\alpha=1,\ 2,\ 3$ in a Cartesian reference; underlined symbols denote matrices (and column-vectors), a tilde marks transpose. Traction $\underline{p}$ at t in $\underline{x}$, according to an outward unit normal $\underline{n}$, due a unit impulse in $\underline{\xi}$ at τ, will be defined by matrix $\underline{G}_{pu}(\underline{x},\underline{\xi};t-\tau)$. This matrix contains Green functions obtained by those in Eq. (1) using derivatives with respect to $\underline{x}$, the elastic tensor and Cauchy equilibrium equations for the outward unit normal $\underline{n}$.

Betti's reciprocal theorem, combining the above fundamental solutions with the unknown dynamic response of the actual body, leads to the dynamic counterpart of Somigliana identity of linear elastostatics [1-2]. This equation *can be taken to the boundary* (i.e. written for $\underline{x}\longrightarrow\Gamma$) paying attention to the singularities arising in the boundary integrals for $\underline{x}=\underline{\xi}$ and $t=\tau$, [18]. Thus the boundary integral equation (BIE) is achieved which represents the starting point of the traditional direct BIE methods in elastodynamics.

Departing from the traditional approach, besides the above kind of sources, we involve here also the Green's functions $\underline{\underline{G}}_{up}$ and $\underline{\underline{G}}_{pp}$ which describe displacements $\underline{u}(\underline{x},t)$ and tractions $\underline{p}(\underline{x},t)$ due to a unit relative displacement concentrated in time (instant τ) and in space (point $\underline{\xi}$) according to a normal $\underline{\nu}$. i. e. due to a source represented by an impulsive kinematic discontinuity concentrates in $\underline{\xi}$ across a surface having through $\underline{\xi}$ normal $\underline{\nu}$ in $\underline{\xi}$. This kernels are explicity computed in [13].

Consider the homogeneous elastic space Ω_∞ embedding Ω and the surface Γ separating Ω from its complementary volume $\Omega_\infty-\Omega$. Let Γ^- and Γ^+ denote the set of points of Ω and $\Omega_\infty-\Omega$, respectively, infinitely close to Γ (Ω and $\Omega_\infty-\Omega$ are conceived as open subdomains of Ω_∞). The functions which define traction and displacement discontinuities across Γ are, respectively:

$$\Delta\underline{p}(\underline{\xi}) = \underline{p}^+ + \underline{p}^- \quad \text{on } \Gamma \quad ; \quad \Delta\underline{u}(\underline{\xi}) = \underline{u}^+ - \underline{u}^- \quad \text{on } \Gamma \tag{2a,b}$$

where superscripts $^+$ $^-$ mark values attained on Γ^+ and Γ^- near $\underline{\xi}$.

In the spirit of the *direct* BIE method, let the exterior domain $(\Omega_\infty-\Omega)$ be required to be unstressed and undeformed. This implies that:

$$\underline{p}^+=\underline{0}, \text{ hence } \Delta\bar{\underline{p}} = \underline{p}^- \equiv \underline{p} \quad ; \quad \underline{u}^+ = \underline{0}, \text{ hence: } \Delta\bar{\underline{u}} = -\bar{\underline{u}} \equiv -\underline{u} \tag{3a,b}$$

In internal points infinitely close to the *constrained* boundary $(\underline{x}\in\Gamma_u^-)$ the displacements at time t tend to coincide with the assigned ones $\bar{\underline{u}}(\underline{x},t)$ and can be expressed as superposition of effects due to the various causes acting on Ω_∞ during the time interval $[0,t]$, i.e. of the boundary discontinuities account taken of Eqs. (2) and (3) and of domain loads and initial conditions:

$$\int_0^t\int_{\Gamma_u}\underline{\underline{G}}_{uu}(\underline{x},\underline{\xi};t-\tau)\underline{p}(\underline{\xi},\tau)\, d\underline{\xi}\, d\tau - \int_0^t\int_{\Gamma_p}\underline{\underline{G}}_{up}(\underline{x},\underline{\xi};t-\tau)\underline{u}(\underline{\xi},\tau)\, d\underline{\xi}\, d\tau = \bar{\underline{f}}_u(\underline{x},t) \qquad \text{for } \underline{x}\in\Gamma_u^- \tag{4}$$

having set:

$$\bar{\underline{f}}_u(\underline{x},t) \equiv \bar{\underline{u}} - \int_0^t\int_{\Gamma_p}\underline{\underline{G}}_{uu}\,\bar{\underline{p}}\, d\underline{\xi}\, d\tau + \int_0^t\int_{\Gamma_u}\underline{\underline{G}}_{up}\,\bar{\underline{u}}\, d\underline{\xi}\, d\tau + $$

$$- \int_0^t\int_{\Omega}\underline{\underline{G}}_{uu}\,\bar{\underline{b}}\, d\underline{\xi}\, d\tau + \rho\int_\Omega\left(\left[\underline{\underline{G}}_{uu}\right]_0 \bar{\dot{\underline{u}}}_0 - \left[\frac{\partial}{\partial\tau}\underline{\underline{G}}_{uu}\right]_0 \bar{\underline{u}}_0\right) d\underline{\xi} \tag{5}$$

Here and henceforth, for brevity, $d\Gamma$ and $d\Omega$ are indicated by $d\underline{x}$ (or $d\underline{\xi}$) if the integration is with respect to $\underline{x}$ (or $\underline{\xi}$); sometimes as in Eq. (5), the arguments of functions are omitted and $[\cdot]_0$ indicates that $\cdot$ is evaluated for $t=\tau=0$.

Similarly, in interior points infinitely near to the *unconstrained* boundary ($\underline{x}\in\Gamma_p^-$), the tractions at instant t equal the given surface loads, $\bar{\underline{p}}(\underline{x},t)$ and can be expressed by superposition of effects on Ω_∞ along the interval $[0,t]$ due to the same causes as before:

$$\int_0^t\int_{\Gamma_u}\underline{G}_{pu}(\underline{x},\underline{\xi};t-\tau)\underline{p}(\underline{\xi},\tau)d\underline{\xi}d\tau-\int_0^t\int_{\Gamma_p}\underline{G}_{pp}(\underline{x},\underline{\xi};t-\tau)\underline{u}(\underline{\xi},\tau)\,d\underline{\xi}\,d\tau=\bar{\underline{f}}_p(\underline{x},t)\qquad\text{for }\underline{x}\in\Gamma_p^-\tag{6}$$

having set:
$$\bar{\underline{f}}_p(\underline{x},t)\equiv-\bar{\underline{p}}+\int_0^t\int_{\Gamma_p}\underline{G}_{pu}\,\bar{\underline{p}}\,d\underline{\xi}\,d\tau-\int_0^t\int_{\Gamma_u}\underline{G}_{pp}\,\bar{\underline{u}}\,d\underline{\xi}\,d\tau+$$
$$+\int_0^t\int_{\Omega}\underline{G}_{pu}\,\bar{\underline{b}}\,d\underline{\xi}\,d\tau+\rho\int_\Omega\left(\left[\underline{G}_{pu}\right]_0\dot{\bar{\underline{u}}}_0-\left[\frac{\partial}{\partial\tau}\underline{G}_{pu}\right]_0\bar{\underline{u}}_0\right)d\underline{\xi}\tag{7}$$

The integral equations (4) and (6) govern the dynamic response of the body occupying volume Ω over the time interval $0\le t\le T$, as an alternative formulation to the traditional BIE formulation [1-2]. Eqs. (4) and (6) can be rewritten together in the compact operatorial form:

$$\underline{L}\,\underline{y}=\bar{\underline{f}}\tag{8}$$

by setting:

$$\underline{y}\equiv\begin{Bmatrix}\underline{u}(\underline{\xi},\tau) & \underline{\xi}\in\Gamma_p\\ \underline{p}(\underline{\xi},\tau) & \underline{\xi}\in\Gamma_u\end{Bmatrix}\quad;\quad\bar{\underline{f}}\equiv\begin{Bmatrix}\bar{\underline{f}}_p(\underline{x},t) & \underline{x}\in\Gamma_p\\ \bar{\underline{f}}_u(\underline{x},t) & \underline{x}\in\Gamma_u\end{Bmatrix}\tag{9}$$

$$\underline{L}\equiv\begin{bmatrix}\int_0^t\int_{\Gamma_p}\underline{G}_{pp}(\underline{x},\underline{\xi};t-\tau)[\cdot]d\underline{\xi}\,d\tau & -\int_0^t\int_{\Gamma_u}\underline{G}_{pu}(\underline{x},\underline{\xi};t-\tau)[\cdot]d\underline{\xi}\,d\tau\\ -\int_0^t\int_{\Gamma_p}\underline{G}_{up}(\underline{x},\underline{\xi};t-\tau)[\cdot]d\underline{\xi}\,d\tau & \int_0^t\int_{\Gamma_u}\underline{G}_{uu}(\underline{x},\underline{\xi};t-\tau)[\cdot]d\underline{\xi}\,d\tau\end{bmatrix}\begin{matrix}\underline{x}\in\Gamma_p^-\\ \\ \underline{x}\in\Gamma_u^-\end{matrix}\tag{10}$$

<u>A Convolutive-Variational Principle Governing the Dynamic Response on the Boundary over a Time Interval</u>

Eq. (10) specifies the linear integral operator $\underline{L}$ on which the present formulation is centered. Consider the following scalar product with convolution with respect to time:

$$<\underline{g},\underline{y}'>\equiv\int_{\Gamma}\int_{0}^{T}\tilde{\underline{g}}(\underline{x},T-t)\underline{y}'(\underline{x},t)dtd\underline{x} \qquad (11)$$

where $\underline{g}$ and $\underline{y}'$ are any sufficiently regular functions defined over Γ and T.

Assuming $\underline{g} = \underline{L}\ \underline{y}$, Eq.(11) defines the bilinear form (convolutive with respect to t):

$$<\underline{L}\ \underline{y},\underline{y}'> = \int_{\Gamma}\int_{0}^{T} \tilde{\underline{y}}'(\underline{x},T-t)(\underline{L}\ \underline{y})_{\underline{x},t}dt\ d\underline{x} \qquad (12)$$

As a consequences of Betti's reciprocity theorem (see e.g. [7]), the Green's function matrices intervening in operator $\underline{L}$, Eq. (10), can be shown to exhibit the following simmetries in space (for any $t-\tau$):

$$\underline{G}_{hk}(\underline{x},\underline{\xi};t-\tau) = \tilde{\underline{G}}_{kh}(\underline{\xi},\underline{x};t-\tau)\ , \qquad \text{for } h=u,p\ ;\ k=u,p \qquad (13)$$

with the restriction $\underline{x}\neq\underline{\xi}$ for $\underline{G}_{up}$ and $\underline{G}_{pu}$.

Using Eq. (13), by a path of reasoning expounded in the fortcoming paper [13], it can formally proved that:

$$<\underline{L}\ \underline{y}\ ,\ \underline{y}'> = <\underline{L}\ \underline{y}',\ \underline{y}>, \qquad \text{for any } \underline{y} \text{ and } \underline{y}' \qquad (14)$$

The result expressed by Eq. (14) is an important property of operator $\underline{L}$ and can be translated into the statement: *The boundary integral operator* $\underline{L}$, Eq. (10), *is symmetric (or self-adjoint) with respect to the bilinear form* (12), *which is convolutive in time and defined over any time interval* $0\leq t\leq T$.

Such a symmetry, in space and time, of operator $\underline{L}$ leads to the following variational theorem (counterpart to the one in space given [8]):

Time histories of displacements $\underline{u}(\underline{x},t)$ *on the unconstrained boundary* Γ_p *and of tractions* $\underline{p}(\underline{x},t)$ *on the constrained boundary* Γ_u *represent the dynamic response on* Γ $(=\Gamma_p\cup\Gamma_u)$ *to the external actions acting during the interval* $0\leq t\leq T$, *if and only if they make stationary the quadratic functional* (where the given functions $\bar{\underline{f}}_u$ and $\bar{\underline{f}}_p$ containing all data are defined by Eqs.(5) and (7), respectively):

$$F(\underline{u} \text{ on } \Gamma_p,\ \underline{p} \text{ on } \Gamma_u) \equiv \frac{1}{2} <\underline{L}\ \underline{y},\ \underline{y}> - <\bar{\underline{f}},\ \underline{y}> \qquad (15a)$$

$$F = \frac{1}{2} \int_0^T\int_0^t \Big[\int_{\Gamma_p}\int_{\Gamma_p} \underline{\tilde{u}}(\underline{x},T-t)\underline{G}_{pp}(\underline{x},\underline{\xi};t-\tau)\underline{u}(\underline{\xi},\tau)\ d\underline{\xi}\ d\underline{x}\ + \tag{15b}$$

$$-\int_{\Gamma_p}\int_{\Gamma_u} \underline{\tilde{u}}(\underline{x},T-t)\underline{G}_{pu}(\underline{x},\underline{\xi};t-\tau)\underline{p}(\underline{\xi},\tau)\ d\underline{\xi}\ d\underline{x}\ +$$

$$-\int_{\Gamma_u}\int_{\Gamma_p} \underline{\tilde{p}}(\underline{x},T-t)\underline{G}_{up}(\underline{x},\underline{\xi};t-\tau)\underline{u}(\underline{\xi},\tau)\ d\underline{\xi}\ d\underline{x}\ +$$

$$+\int_{\Gamma_u}\int_{\Gamma_u} \underline{\tilde{p}}(\underline{x},T-t)\underline{G}_{uu}(\underline{x},\underline{\xi};t-\tau)\underline{p}(\underline{\xi},\tau)\ d\underline{\xi}\ d\underline{x} \Big]\ d\tau\ dt\ +$$

$$-\int_0^T\int_{\Gamma_p} \underline{\tilde{u}}(\underline{x},T-t)\underline{\bar{f}}_p(\underline{x},t)\ d\underline{x}\ dt - \int_0^T\int_{\Gamma_u} \underline{\tilde{p}}(\underline{x},T-t)\underline{\bar{f}}_u(\underline{x},t)\ d\underline{x}\ dt\ .$$

The proof of this statement rests on the traditional arguments which follow. Evaluate the first order variation of the functional in form (15a) due to variation $\delta\underline{y}$ of the variables:

$$\delta^{(1)}F = \frac{1}{2} < \underline{L}\ \underline{y},\ \delta\underline{y} > + \frac{1}{2} < \underline{L}\ \delta\underline{y},\ \underline{y} > - < \underline{\bar{f}},\ \delta\underline{y} > \tag{16}$$

By virtue of Eqs. (8) and (14) we may write for any $\delta\underline{y}$:

$$\delta^{(1)}F = < \underline{L}\ \underline{y}\ ,\ \delta\underline{y} > - < \underline{\bar{f}},\ \delta\underline{y} > = < \underline{L}\ \underline{y} - \underline{\bar{f}},\ \delta\underline{y} > = 0 \tag{17a,b,c}$$

Conversely, if Eq. (17c) holds for any $\delta\underline{y}$, the system of boundary integral equation Eq. (8) is fulfilled and, hence, $\underline{y}$ is the boundary solution.

Closing Remarks

A variational characterization was achieved in the previous section for the time histories of the unknowns on the boundary Γ over a time T. Approximate solutions may be sought by modelling the variables in space and time (over ΓxT) and by imposing the stationarity of the functional (15b) with respect to the vectors of independent (nodal) unknowns. Clearly, the algebraic equationsystem thus generated shall exhibit a symmetric coefficient matrix, computed by double integrations of kernels multiplied by interpolating functions over elements in ΓxT. The burden implied by such integrations, at least as long as they are performed analytically, is expected to be more than compensated by accuracy increase and storage savings, on the basis of what was observed for symmetric method Galerkin in elasticity and plasticity [10]. The forthcoming paper [13] contains, besides the proofs here omitted for brevity, a detailed presentation of the needed Green's functions not available in the literature and the extensions of the present variational method to multidomain problems and to elastically constrained boundary. The development of the present approach in viscoelasticity [20] and in elastic-plastic dynamics is currently being dealt with. However, an implementation and comparative assessment of computational performances of symmetric BEM outside quasi-static elastoplasticity appear to be still unexplored issues worth of future work.

Aknowledgements. This study is part of a research project sponsored by MPI (40%).

References

1. Manolis, G.D. ; Beskos, D.E.: Boundary Element Methods in Elastodynamics, Unwin Hyman Ltd, London 1988.
2. Antes, H.: Anwendungen der Methode der Randelemente in der Elastodynamik und der Fluiddynamik, B.G. Teubner Stuttgart 1988.
3. Kobayashi, S.: Elastodynamics, in D.E. Beskos (Ed.), Boundary Element Methods in Mechanics, Vol. 3, North-Holland (1987) 192-255.
4. Banerjee, P.K.; Ahmad, S.; Manolis, G.D.: Advanced elastodynamic analysis, in D.E. Beskos (Ed.)Boundary Element Methods in Mechanics, Vol. 3, North-Holland, (1987) 257-284.
5. Sirtori, S.: General stress analysis by means of integral equation and boundary elements. Meccanica 14 (1979) 210-218.
6. Costabel, M.; Wendland, W.L.: Strong ellipticity of boundary integral operators. J. Reine Angewandte Math. 372 (1986) 34-63.
7. Maier, G.; Polizzotto, C.: A Galerkin approach to boundary element elastoplastic analysis. Computer Methods in Applied Mechanics and Engineering 60 (1987) 175-194.
8. Polizzotto, C.: An energy approach to the boundary element method. Part I: Elastic solids. Part II: Elastic-plastic solids. Computer methods in applied mechanics and engineering 69 (1988) 167-184, 263-276.
9. Maier, G.; Novati, G.; Sirtori, S.: Symmetric formulation of indirect boundary method for elastic-plastic analysis and relevant extremum properties. First Joint Japan-US Symp. on Boundary Element Methods, Tokyo, 3-6 October 1988.
10. Maier, G.; Novati, G.; Sirtori, S.: On symmetric formulations of the boundary element methods in elastoplasticity. in Discretization Methods in Structural Mechanics, IUTAM/IACM Symposium, Vienna, 5-9 June 1989.
11. Panagiotopoulos, P.D.; Lazaridis, P.P.: Boundary minimum principles for the unilateral contact problems. Int. J. Solids Structures 23 (1987) 1465-1484.

12. Polizzotto, C.; Zito, M.: A variational approach to boundary element methods. First Joint Japan/US Symp. on Boundary Element Methods, Tokyo, 3-6 October, 1988.
13. Maier, G.; Diligenti, M.; Carini, A.: A variational formulation of the boundary integral equation method in elastodynamics for multidomain problems. To appear.
14. Gurtin, M.E.: Variational principles for linear initial-value problems. Quart. Appl. Math., 22 1964.
15. Tonti, E.: On the variational formulation for linear initial value problems. Annali di Matematica Pura ed Applicata 45 (1973) 331-359.
16. Reddy, J.N.; Rasmussen, M.L.: Advanced Engineering Analysis, John Wiley & Sons, New York 1982.
17. Carini, A.; Diligenti, M.; Maier, G.: Variational formulation of the boundary element method in transient heat conduction. Computers and Mathematics with Applications, Special issue on BEMs, G. Symm (Ed.). To appear.
18. Krishnasamy, G.; Schmerr, L.W.; Rudolphi, T.J.; Rizzo, F.J: Hypersingular integral formulas and wave scattering with the Boundary Element Method. Proc. 1st Japan/US Sympos. on Boundary Element Methods, Tokyo, 3-6 October, 1988.
19. Sirtori, S.; Maier, G.; Miccoli, S.; Novati, G.: Symmetric formulation and implementation of the boundary element method in elastoplasticity. To appear.
20. Carini, A.; Diligenti, M.; Maier,G.: A variational formulation of viscoelastic analysis by boundary integral equations. Nat. Conference of Mech. of Materials and Structures, Rome 25-26 October, 1989.

Progress in Applications of BEM to Inelastic Analysis of Solids

P.K. Banerjee
D.P. Henry, Jr.
G.F. Dargush

Department of Civil Engineering
State University of New York at Buffalo
Buffalo, New York U.S.A.

ABSTRACT

A number of recently developed boundary element solutions are used for solving a range of elastoplastic and thermoplastic problems. The first algorithm is the conventional volume integral formulation in which the unknown boundary solution and the initial stress (or strain) rates are found together either in an incremental iterative fashion or by a direct variable stiffness type algorithm. The second procedure which is entirely new differs from the previous one in that volume integration is not required to incorporate the nonlinear effects in the analysis. Instead, initial stress rates are introduced in the boundary element system via particular integrals.

INTRODUCTION

During the last decade the development of BEM in the inelastic analysis has accelerated. Starting from a series of first generation implementation which were not efficient but showed some potential, a second generation implementation has been done for axisymmetric, two-, and three-dimensional problems. All inelastic formulations are implemented in a general purpose, multi-region system that utilizes quadratic isoparametric shape functions to model the geometry and field variables of the body, can admit up to 15 substructured regions of different material properties and includes sophisticated inelastic algorithms.

THE INTEGRAL FORMULATION FOR ELASTOPLASTICITY

Volume Integral Formulations

The conventional boundary integral equation for displacement satisfying the governing equation for plasticity can be expressed as [1]:

$$C_{ij}(\xi)\dot{u}_i(\xi) = \int_S [G_{ij}(x,\xi)\dot{t}_i(x) - F_{ij}(x,\xi)\dot{u}_i(x)]dS(x) + \int_V B_{ikj}(x,\xi)\dot{\sigma}^o_{ij}(x)dV(x) \tag{1}$$

where G_{ij}, F_{ij} and B_{ijk} are kernel functions. The two- and three-dimensional kernel functions are given by Banerjee and Raveendra [5] and the axisymmetric kernels can be found in Henry and Banerjee [13,14].

The initial stress rate σ^o_{ij} can be defined in terms of the plastic strain rate as:

$$\dot{\sigma}^o_{ij} = D^e_{ijkl}\dot{\varepsilon}^p_{kl} \tag{2}$$

Body forces, if present, can be incorporated via a particular integral [9].

The integral equation for the strain at an interior point is found analytically by substituting equation (1) with $(C_{ij} = \delta_{ij})$ into the strain-displacement relations, where differentiation is with respect to the field point ξ:

$$\begin{aligned}\dot{\varepsilon}_{ij}(\xi) = &\int_S [G^\varepsilon_{kij}(x,\xi)\dot{t}_k(x) - F^\varepsilon_{kij}(x,\xi)\dot{u}_k(x)]dS(x) \\ &+ \int_V B^\varepsilon_{klij}(x,\xi)\dot{\sigma}^o_{kl}(x)dV(x) + J^\varepsilon_{klij}\dot{\sigma}^o_{kl}(\xi)\end{aligned} \tag{3}$$

By introducing this result into the inelastic stress-strain equation $(\dot{\sigma}_{ij} = D^e_{ijkl}\dot{\varepsilon}_{kl} - \dot{\sigma}^o_{ij})$, the stress integral equation is derived:

$$\begin{aligned}\dot{\sigma}_{ij}(\xi) = &\int_S [G^\sigma_{kij}(x,\xi)\dot{t}_k(x) - F^\sigma_{kij}(x,\xi)\dot{u}_k]dS(x) \\ &+ \int_V B^\sigma_{klij}(x,\xi)\dot{\sigma}^o_{kl}(x)dV(x) + J^\sigma_{klij}\dot{\sigma}^o_{kl}(\xi)\end{aligned} \tag{4}$$

The last integral in equations (3) and (4) is strongly singular and must be treated in the Cauchy-principle sense with additional jump terms J^σ and J^ε (Banerjee and Raveendra [5]). All kernel functions G, F and B as well as the jump terms are defined in Banerjee and Raveendra [5] and Henry and Banerjee [13,14].

The stress rate on the boundary can be obtained from the boundary traction and displacement rates without any integration by substitution of the inelastic stress-strain relation in place of Hooke's law [5].

The above equations can be then discretized and solved either iteratively [1-5,11] or by a direct stiffness type approach [6-10,12-16].

ELASTOPLASTIC BEM FORMULATION USING PARTICULAR INTEGRALS

The elastoplastic, inhomogeneous differential equation given in Banerjee and Butterfield [1] can be written in operator notation as:

$$L(\dot{u}_i) = \dot{\sigma}^o_{ij,j} \tag{5}$$

where $L(\dot{u}_i)$ is a self-adjoint, homogeneous, differential operator and $\dot{\sigma}^o_{ij,j}$ is the (assumed known) inhomogeneous quantity.

The solution of the above equation can be represented as the sum of a complementary function $\dot{u}^c_i$ satisfying the homogeneous equation

$$L(\dot{u}^c_i) = 0 \tag{6}$$

and a particular integral

$$L(\dot{u}_i^p) = \dot{\sigma}^o_{ij,j} \tag{7}$$

The total displacement rate $\dot{u}_i$ is expressed as $\dot{u}_i = \dot{u}_i^c + \dot{u}_i^p$.

The (boundary only) integral equation satisfying the homogeneous part of the differential equation represents the complementary function which can be expressed for the displacement rate at point ξ, as

$$C_{ij}(\xi)\dot{u}_i^c(\xi) = \int_S [G_{ij}(x,\xi)\dot{t}_i^c(x) - F_{ij}(x,\xi)\dot{u}_i^c(x)]dS(x) \tag{8}$$

where the $\dot{u}_i^c$ and $\dot{t}_i^c$ are the complementary functions for displacement and traction rates, respectively.

Similarly a complementary function for the interior stress rate $\dot{\sigma}^c_{ij}$ related to the homogeneous equation can be written for an interior point ξ as

$$\dot{\sigma}^c_{ij}(\xi) = \int_S [G^\sigma_{kij}(x,\xi)\dot{t}^c_k(x) - F^\sigma_{kij}(x,\xi)\dot{u}^c_k(x)]dS(x) \tag{9}$$

The total solution for displacement rate $\dot{u}_i$, traction rate $\dot{t}_i$, and stress rate $\dot{\sigma}_{ij}$ are then obtained by summing up the complementary and particular solutions [9,12,15,16].

In the present formulation, the two- and three-dimensional particular integrals for displacement rates are related to the Galerkin vector $\dot{F}_i$ via

$$\dot{u}_i^p = \frac{(1-\nu)}{\mu}\dot{F}_{i,kk} - \frac{1}{2\mu}\dot{F}_{k,ki} \tag{10}$$

where ν is Poisson's ratio.

Substituting this into equation (5) renders a relationship between the Galerkin vector and the initial stress rate function:

$$\dot{F}_{i,kkjj} = \frac{1}{(1-\nu)}\dot{\sigma}^o_{ij,j} \tag{11}$$

For convenience, a tensor function $\dot{h}_{ij}$ is introduced such that

$$\dot{h}_{ij,mmnn} = \dot{\sigma}^o_{ij} \qquad \text{and} \qquad \dot{F}_i = \frac{1}{(1-\nu)}\dot{h}_{ij,j} \tag{12}$$

Substituting this into equation (10) yields the desired particular integral for displacement rate

$$\dot{u}_i^p = \frac{1}{\mu}\dot{h}_{il,lkk} - \frac{1}{2\mu(1-\nu)}\dot{h}_{lm,ilm} \tag{13}$$

The particular integral for strain, stress and traction rates are found using the strain displacements and stress strain relations.

The tensor $\dot{h}_{ij}(x)$ can now be expressed in terms of a fictitious tensor density $\dot{\phi}_{ij}(\xi)$ as an infinite series using a global shape function $C(x,\xi)$:

$$\dot{h}_{ml}(x) = \sum_{n=1}^{\infty} C(x,\xi_n)\dot{\phi}_{ml}(\xi_n), \tag{14}$$

where $m, l = 1, 2$ for two-dimensions, $m, l = 1, 2, 3$ for three-dimensions and

$$C(x, \xi_n) = A_o^4[\rho^4 - b_n\rho^5]$$

A_o is a characteristic length, ρ is the euclidean distance between the field point x and the source point ξ_n, b_n is a constant, chosen in a manner described by Henry [12].

The unknown fictitious densities are related to the initial stress rates through equation (12). Substituting equation (14) into equation (12) leads to [12,16]:

$$\dot{\sigma}^o_{lm} = \sum_{n=1}^{\infty} K(x, \xi_n)\dot{\emptyset}_{lm}(\xi_n) \tag{15}$$

where $K(x, \xi_n) = C_{,mmnn}(x, \xi_n) = a - b\rho$, $a = 8d(d+2), b = b_n 15(d+3)(d+1)$, $d = 3$; for three-dimensional analysis, and $= 2$; for two-dimensional (plane strain) analysis.

The particular integral for the displacement rate can be found as a function of $\emptyset_{lm}(\xi_n)$ by substituting equation (15) into equation (13).

$$\dot{u}_i^p(x) = \sum_{n=1}^{\infty} D_{iml}(x, \xi_n)\dot{\emptyset}_{lm}(\xi_n) \tag{16}$$

where

$$D_{iml}(x, \xi_n) = A_o\left[(c_1 + d_1\rho)(y_i\delta_{lm} + y_m\delta_{il}) + (c_2 + d_2\rho)y_l\delta_{im} + \frac{d_1}{\rho}y_i y_l y_m\right]$$

$$y_i = [x_i - (\xi_n)_i], \quad c_1 = \frac{-8}{2\mu(1-\nu)}, \quad d_1 = \frac{15b_n}{2\mu(1-\nu)}, \quad c_2 = c_1 + \frac{8(d+2)}{\mu}, \quad d_2 = d_1 - \frac{15(d+3)b_n}{\mu}$$

The particular integral for strain, stress, and traction rates are found by substituting (16) into the strain displacements and the stress strain relations [12,16].

NUMERICAL IMPLEMENTATION

In this section, highlights of a general-purpose, state-of-the-art numerical implementation are presented. This entire implementation has been accomplished within the GP-BEST (General Purpose Boundary Element Solution Technique) analysis system, which contains a wide spectrum of capabilities in linear, non-linear static and dynamic problems of solid and fluid mechanics. As a result, many of the features and techniques used in this implementation are common to all developments.

Perhaps the most significant of these items is the capability to analyze substructured problems. This not only extends the analysis to bodies composed of several different materials, but also often provides computational efficiencies. During the integration process, the individual regions remain separate entities. The geometric modeling regions (or GMR's) are brought together for the first time at the assembly stage, where compatibility relations are enforced on the common boundary between adjacent regions. This multi-GMR assembly process produces block-banded system matrices that can be solved in an efficient manner.

As another convenience, a high degree of flexibility is provided for the specification of boundary conditions. In general, time-dependent values can be defined in either global or local coordinates, with or without sliding. Not only can generalized displacements and tractions be specified, but also spring, convection, and contact resistance relations, respectively, are available. Furthermore, a comprehensive symmetry capability is included with provisions for both planar and cyclic symmetry.

EXAMPLES

Axisymmetric Steel Pressure Vessel

The elastoplastic analysis of a vessel subjected to internal pressure is shown in Figure 1. Six regions, ninety-nine quadratic boundary elements, and twelve quadratic volume cells are used to model the body.

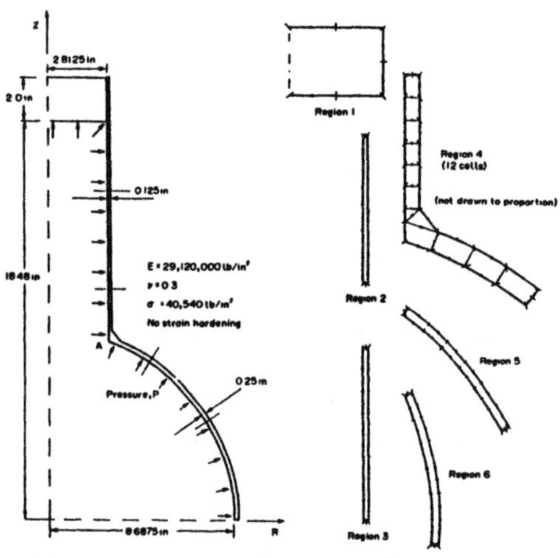

Fig. 1. Axisymmetric steel pressure vessel and discretization by region

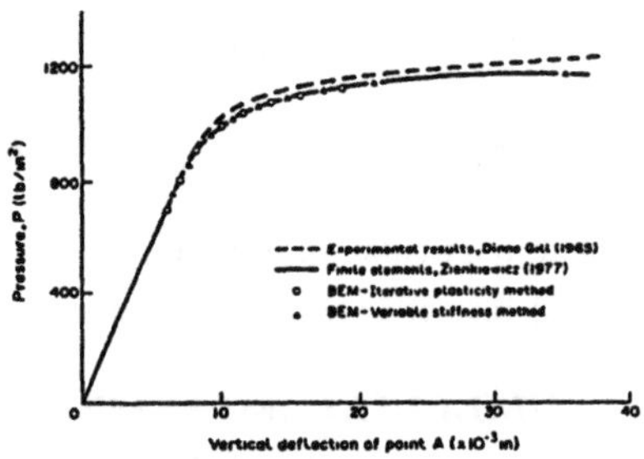

Fig. 2. Vertical deflection at point A of the pressure vessel with increasing pressure

A plot of the vertical deflection of point A with increasing pressure obtained from both the variable stiffness and iterative plasticity algorithm by volume integration is shown in Figure 2. Results obtained by others using the finite element method and experiments are also presented. The BEM results are in excellent agreement with the results obtained by FEM.

It should be noted that the variable stiffness method proceeds farther along the curve than the iterative procedure which does not converge at high load levels. The use of a multi-region system in both of the BEM analyses reduces the computational time dramatically, since each nodal equation need only be integrated over the surface in which it is contained. Furthermore, the volume integration need only be performed over the region containing cells.

Three-dimensional Analysis of a Perforated Plate

The plastic deformation of a perforated plate in tension is analyzed under plane stress conditions. The volume distribution of initial stress is represented using either twenty-noded isoparametric volume cells or the point based particular integral. Each representation is coupled with either the iterative or the variable stiffness solution process, leading to a total of four distinct algorithms. The material properties for the plate are:

$$E = 7000 kg/mm^2 \qquad \sigma_o = 24.3 kg/mm^2 \text{ (Von Mises yield criterion)}$$

$$\nu = 0.2 \qquad h = 224.0 kg/mm^2$$

The diameter of the circular hole, at the center of the plate, is one-half the width, and the thickness is one-fifth the width. A quarter of the plate is discretized in two subregions, as shown in Figure 3. The first region, containing the root of the plate, has thirty quadratic boundary elements. For the volume integral based analysis the domain of this region is discretized using nine (twenty-node) isoparametric cells. In the particular integral analysis, particular integrals are defined at points corresponding to the cell nodes used in the volume integral analysis. The second region has twenty-three boundary elements. No volume discretization or definition of particular integrals is required in this region since it remains elastic throughout the analysis. Boundary conditions on both the front and back faces are assumed traction free.

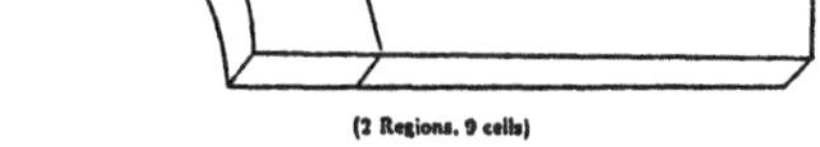

Fig. 3. Three-dimensional mesh of a perforated plate

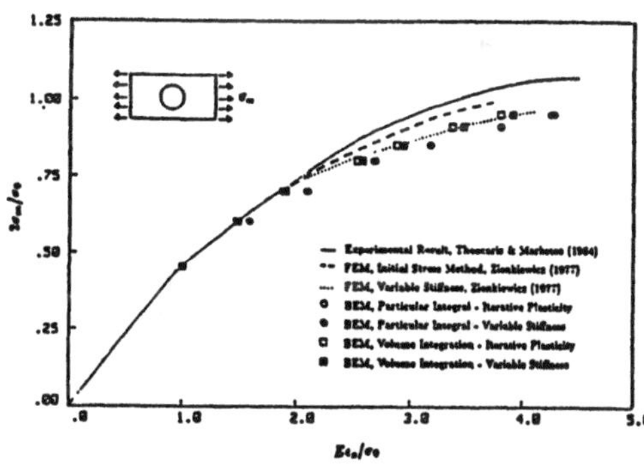

Fig. 4. Stress-strain response at the root of a 3-D perforated plate

The results obtained by the boundary element analysis is compared to FEM results in Figure 4. The results obtained using the various BEM algorithms show good agreement with one another and with the variable stiffness FEM analysis.

Two-dimensional Analysis of a Perforated Plate

The perforated plate of the previous problem is analyzed using the particular integral based, inelastic, two-dimensional, plane stress analysis. Three discretizations, shown in Figure 5, are used to model the plate. Each mesh is divided into two subregions. The initial stress distribution in the inelastic region is defined using the particular integral representation. Twenty, forty-one and sixty-five nodes are used, respectively, in these discretizations to define the particular integrals.

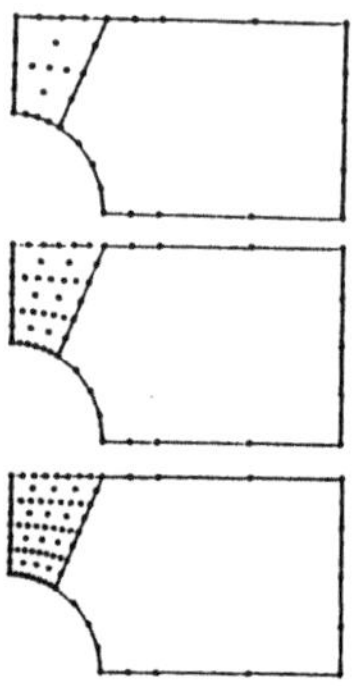

Fig. 5. Discretization of a two-dimensional perforated plate

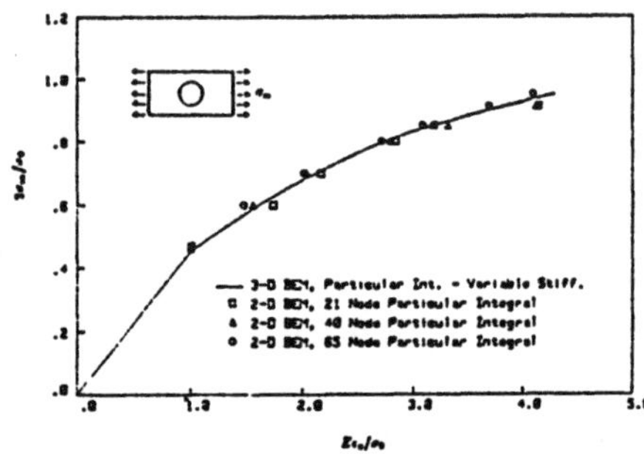

Fig. 6. Stress-strain response at the root of a perforated plate (plane stress)

A solution is obtained using the variable stiffness method, and the results are compared with the particular integral based, variable stiffness results of the previous three-dimensional analysis. The stress-strain response at the root of the plate is shown in Figure 6. The results of the more refined (65 particular integral nodes) mesh, exhibit a smoother response and follows the three-dimensional solution more closely than the other two-dimensional results.

Residual Stress in a Steel Cylinder

This example involves determination of the residual stresses resulting from the rapid cooling of a long, 2.5 in diameter steel cylinder. This cylinder is initially in a stress free state at 1250° F, and then suddenly exposed to an 80°F fluid. A film coefficient of $h = 12.25$ in-lb/sec in^2°F is specified on the cylindrical surface for the ensuing convective process. The volume integral based thermoplasticity algorithm was used [11].

The following standard thermoelastic material properties are employed for 1060 steel:

$$E = 30 \times 10^6 \text{psi} \qquad \rho c_e = 283 \text{in lb/in}^3 {}^\circ\text{F}$$

$$\nu = 0.30 \qquad k = 5.8 \text{in lb/sec in}^\circ\text{F}$$

$$\alpha = 6 \times 10^{-6}/^\circ\text{F}$$

Additionally, a temperature-dependent Von Mises thermoplastic constitutive model [11] is adopted. The yield stress at room temperature is 48 ksi reducing to zero at 1300°F, and zero strain hardening is assumed.

The boundary element surface model, displayed in Fig. 7, consists of 38 source points and 12 elements, while the complete description of the interior requires one extra node and three volume cells. After some numerical experimentation, a time step of 4 sec was selected.

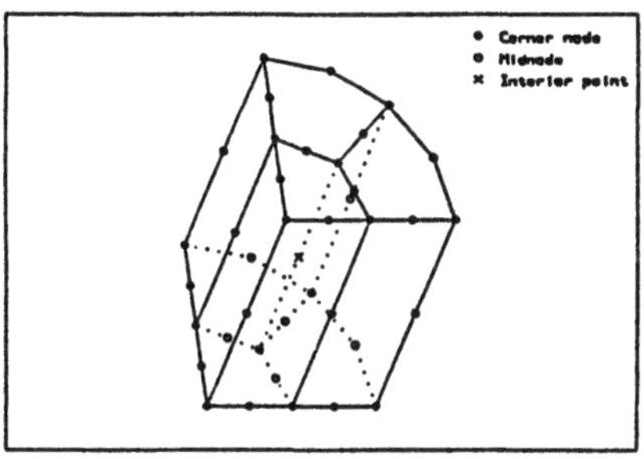

Fig. 7. BEM discretization for a segment of steel cylinder

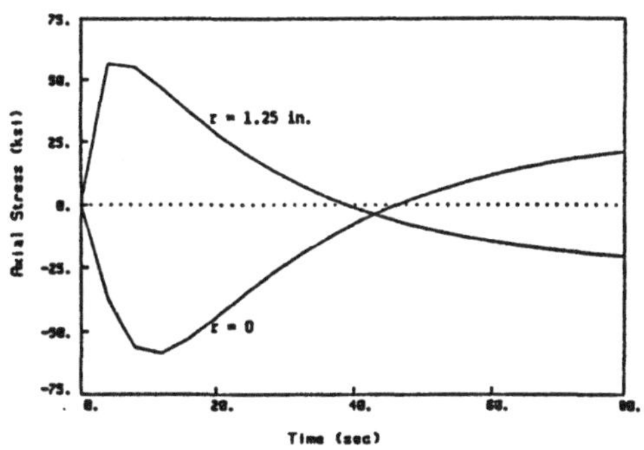

Fig. 8. Variation of axial stress with time in a steel cylinder

Figure 8 shows the axial stresses as a function of time for the surface ($r = 1.25in$) and center ($r = 0$) of the cylinder at mid-height. Notice that rapid cooling of the surface occurs during the first few seconds. This produces tensile stress in the outer portions and compression toward the center. However, throughout this initial period, the material remains quite hot, and consequently, has a relatively low yield stress, leading to significant plastic deformation.

After the beginning ten seconds, the yield stress is elevated considerably and the all important thermal gradients diminish, so that no further plastic flow takes place. Instead, elastic unloading commences. At about the forty second mark, the axial stresses change signs. Thus, the outer fibers go into compression, while tension occurs in the central portion. This trend continues to steady-state. After the cylinder cooled for 80 seconds, the axial stress on the surface and at the center are −20.5 ksi and +21.1 ksi, respectively.

CONCLUSIONS

Multi domain boundary element method has been developed for application to two- and three-dimensional and axisymmetric problems of elastoplasticity and thermoplasticity. The general purpose, state-of-the-art boundary element implementation (GP-BEST) insures efficient integration, assembly and solution, while providing capability for the modeling of complicated multi-region geometry and the specification of arbitrary time-varying boundary conditions. Additionally, facilities are provided for symmetry and the calculation of interior and boundary values. Realistic numerical examples validate this new formulation and highlight the versatility of the implementation.

ACKNOWLEDGEMENT

The work described in this paper was made possible by the NASA contract NAS3-23697 with supplemental funding from the Structures Technology Division, Pratt and Whitney, Hartford, Connecticut and is part of BEST3D and GPBEST systems. The authors wish to express their sincere gratitude to Dr. Chris Chamis, the NASA Project Manager and Dr. Edward Todd, the Pratt and Whitney Program Manager for their support and encouragement.

REFERENCES

1. Banerjee, P.K. and Butterfield, R., Boundary Element Methods in Engineering Science, McGraw-Hill, London (1981).

2. Banerjee, P.K., Cathie, D.N. and Davies, T.G., 'Two and Three-dimensional Problems of Elastoplasticity,' Chapter IV in *Developments in Boundary Elements*, **V1**, Applied Science Publishers, London (1979).

3. Cathie, D.N. and Banerjee, P.K., 'Boundary Element Methods for Axisymmetric Plasticity,' *Innovative Numerical Methods for the Applied Engineering Science*, Eds. R. Shaw, W. Pilkey, B. Pilkey, R. Wilson, A. Lakis, A. Chaudouet, and C. Marino, University of Virginia Press (1980).

4. Banerjee, P.K. and Davies, T.G., 'Advanced Implementation of the Boundary Element Method for Three-dimensional Problems of Elastoplasticity and Viscoplasticity,' Chapter I in *Developments in Boundary Elements*, **V3**, Eds. P.K. Banerjee and S. Mukherjee, Applied Science Publishers, London (1984).

5. Banerjee, P.K. and Raveendra, S.T., 'Advanced Boundary Element Analysis of Two- and Three-dimensional Problems of Elastoplasticity,' *Int. Jour. for Num. Methods in Engrg.*, **V23**, pp. 985-1002 (1986).

6. Banerjee, P.K. and Raveendra, S.T., 'A New Boundary Element Formulation for Two-dimensional Elastoplastic Analysis,' *Jour. of Engrg. Mech.*, **V113**(2), pp. 252-265 (1986).

7. Banerjee, P.K. and Henry, D.P., Jr., 'Recent Advances in the Inelastic Analysis of Solids by BEM,' *Proc. of ASME Conf. on Advances in Inelastic Analysis*, **AMD, V88**, pp. 177-192 (1987).

8. Banerjee, P.K., Wilson, R.B. and Raveendra, S.T., 'Advanced Applications of BEM to Three-dimensional Problems of Monotonic and Cyclic Plasticity,' *Int. Jour. Mech. Sciences*, **V29**(9), pp. 637-653 (1987).

9. Banerjee, P.K., Wilson, R.B. and Miller, N., 'Advanced Elastic and Inelastic Three-dimensional Analysis of Gas Turbine Engine Structures by BEM,' *Int. Jour. Num. Methods in Engrg.*, **V26**, pp 393-411 (1988).

10. Banerjee, P.K., Henry, D.P. and Raveendra, 'Advanced Inelastic Analysis of Solids by BEM,' *Int. Jour. Mech. Sciences*, **V31**, pp. 309-322 (1989).

11. Dargush, G.F., Boundary Element Methods for the Analogous Problems of Thermomechanics and Consolidation, Ph.D. Thesis, State University of New York at Buffalo (1987).

12. Henry, D.P., Jr., Advanced Development of the Boundary Element Method for Elastic and Inelastic Thermal Stress Analysis, Ph.D. Dissertation, State University of New York at Buffalo (1987).

13. Henry, D.P., Jr. and Banerjee, P.K., 'A Thermoplastic BEM Analysis for Substructured Axisymmetric Bodies,' *Jour. of Engrg. Mech.*, ASCE, **V113**(12), pp. 1880-1900 (1987).

14. Henry. D.P. and Banerjee, P.K., 'A Variable Stiffness Type Boundary Element Formulation for Axisymmetric Elastoplastic Media,' *Int. Jour. for Num. Methods in Engrg.*, **V25**, pp. 1005-1027 (1988).

15. Henry, D.P. and Banerjee, P.K., 'A New BEM Formulation for Thermoelasticity by Particular Integrals,' *Int. Jour. Num. Methods in Engrg.*, **V26**(9), pp. 2061-2078, (1988).

16. Henry, D.P. and Banerjee, P.K., 'A New BEM Formulation for Elastoplasticity by Particular Integrals,' *Int. Jour. Num. Methods in Engrg.*, **V26**(9), pp. 2079-2096 (1988).

Bifurcation Analysis of Elastic Shallow Arch by the Boundary-Domain Element Method

N. TOSAKA and M.NONAKA

College of Industrial Technology, Nihon University
Chiba,275,Japan

S. MIYAKE

Faculty of Economics, Kanto Gakuen University
Gunma,373,Japan

Summary

The nonlinear bifurcation analysis of an elastic shallow circular arch subjected to three kinds of loading is presented by using the boundary-domain element approach in the nonlinear integral equation of the problems. The resulting nonlinear system of algebraic equations of equilibrium is solved effectively by the path-following iterative solution procedure. Numerical results show a complex nonlinear behaviour such as the snap-through, the looping and the bifurcation. As the result of numerical computation, it is found newly that there exists another bifurcation point in addition to the existing bifurcation point. The obtained results are compared with the Galerkin solutions, which is shown to be in good agreement.

Introduction

A shallow circular arch is presented as the simplest structural model which exhibits such general nonlinear shell behaviour as the snap-through and bifurcation phenomena. Therefore, there exist a large number of researches on nonlinear analysis of the arch by using various solution procedures.

In this paper, the geometrically nonlinear analyses of a shallow sinusoidal arch are presented by using the boundary-domain element approach in the integral equation method. The nonlinear system of integral equations of the problem is derived with the aid of fundamental solution for the fourth order differential operator which is only a linear part of the nonlinear governing equation. The derived nonlinear system is discretized by using not only boundary elements but also domain elements, i.e., the boundary-domain element approach[7,8]. The discretized nonlinear system of algebraic equations are solved effectively by the arc length iterative solution procedure[2-5]. Concequently, the very complicated equilibrium paths, in which the limit points and the bifurcation points are included, can be obtained. As the result of our computations, the existance of two points of bifurcation can be found out newly. And various branching paths which bifurcate from the symmetric fundamental path with symmetric or asymmetric deflection modes can be also tracked. The obtained results are compared with the Galerkin solutions.

Formulation

Let us consider an elastic shallow arch of length l and cross-sectional area A subjected to a distributed normal load p or a concentrated load P. We assume this arch as a beam with a small initial curvature described with the initial centroidal locus of the arch, z=z(x). The well-known differential-integro equation governing nonlinear behavior of the arch is given in terms of the non-dimensional transverse deflection W as the following non-dimensional form:

$$\frac{d^4 w}{d x^4} - \frac{1}{2\pi}\int_0^\pi \left\{2\frac{d\beta^*}{dx}\frac{dw}{dx} - \left(\frac{dw}{dx}\right)^2\right\}dx \left(\frac{d^2\beta^*}{dx^2} - \frac{d^2 w}{dx^2}\right) = q \tag{1}$$

in which we introduce the following non-dimensional quantities defined by

$$\left.\begin{aligned} & w = \frac{w}{r}, \quad r = \sqrt{\frac{I}{A}}, \quad x = \frac{\pi}{\ell} x \\ & , \quad q = \frac{P}{EIr}\left(\frac{\ell}{\pi}\right)^4, \quad \beta^* = \frac{z}{r} \end{aligned}\right\} \tag{2}$$

where E denotes the Young's modulus and I the second moment of area. Following the our previous paper [6], we can derive the following nonlinear system of integral equations:

$$\begin{aligned} w(y) = & -\left[\frac{d^3 w}{dx^3} g^*\right]_0^\pi + \left[\frac{d^2 w}{dx^2}\frac{d g^*}{dx}\right]_0^\pi - \left[\frac{dw}{dx}\frac{d^2 g^*}{dx^2}\right]_0^\pi + \left[w\frac{d^3 g^*}{dx^3}\right]_0^\pi - \int_0^\pi q g^* dx \\ & + \frac{1}{2\pi}\int_0^\pi \left\{2\frac{d\beta^*}{dx} - \left(\frac{dw}{dx}\right)^2\right\} dx \left[\begin{aligned} & \left[\frac{d\beta^*}{dx} g^*\right]_0^\pi - \int_0^\pi \frac{d\beta^*}{dx}\frac{d^2 g^*}{dx^2} dx \\ & + \left[\frac{dw}{dx} g^*\right]_0^\pi + \int_0^\pi \frac{dw}{dx}\frac{d^2 g^*}{dx^2} dx \end{aligned}\right] \end{aligned} \tag{3}$$

$$\frac{dw}{dy}(y) = -\left[\frac{d^3w}{dx^3}\frac{dg^*}{dy}\right]_0^\pi + \left[\frac{d^2w}{dx^2}\frac{d^2g^*}{dxdy}\right]_0^\pi - \left[\frac{dw}{dx}\frac{d^3g^*}{dx^2dy}\right]_0^\pi + \left[w\frac{d^4g^*}{dx^3dy}\right]_0^\pi$$
$$-\int_0^\pi q\frac{dg^*}{dy}dx$$
$$+\frac{1}{2\pi}\int_0^\pi\left\{2\frac{d\beta^*}{dx} - \left(\frac{dw}{dx}\right)^2\right\}dx\begin{bmatrix}\left[\frac{d\beta^*}{dx}\frac{dg^*}{dy}\right]_0^\pi - \int_0^\pi\frac{d\beta^*}{dx}\frac{d^3g^*}{dx^2dy}dx \\ +\left[\frac{dw}{dx}\frac{dg^*}{dy}\right]_0^\pi + \int_0^\pi\frac{dw}{dx}\frac{d^3g^*}{dx^2dy}dx\end{bmatrix} \tag{4}$$

where we use the fundamental solution g^* which satisfies

$$\frac{d^4g^*}{dx^4} = \delta(x-y) \tag{5}$$

This fundamental solution and its derivatives are given explicitly by

$$g^* = \frac{1}{12}r^3 \tag{6}$$

$$\left.\begin{aligned} \frac{dg^*}{dx} &= \frac{1}{4}r^2\mathrm{sgn}(x-y) & \frac{dg^*}{dy} &= -\frac{1}{4}r^2\mathrm{sgn}(x-y) \\ \frac{d^2g^*}{dx^2} &= \frac{1}{2}r & \frac{d^2g^*}{dxdy} &= -\frac{1}{2}r \\ \frac{d^3g^*}{dx^3} &= \frac{1}{2}\mathrm{sgn}(x-y) & \frac{d^3g^*}{dx^2dy} &= -\frac{1}{2}\mathrm{sgn}(x-y) \end{aligned}\right\} \tag{7}$$

in which $r=|x-y|$ and sgn denotes the signum function.

Numerical Solution Procedure

In order to solve approximately the nonlinear system of integral equations (3) and (4) we adopt the boundary-domain elements discretization scheme which its validity was already verified in integral equation analyses of geometrically nonlinear problems for thin elastic body [6-8]. Applying this scheme to the nonlinear integral equations (3) and (4), we can get finally the nonlinear system of algebraic equation in terms of the field unknown vector $\boldsymbol{v}$ and loading parameter α as follows:

$$F(v, \alpha) = 0 \tag{8}$$

It is well-known that the typical nonlinear behaviour of arches exhibits snap-through, bifurcation and looping of their load-deflection curves [4]. In order to track such a higher nonlinear equilibrium path, numerous methods have been proposed in the context of an incremental solution procedure [2-5]. Especially, the path-following incremental solution procedure which contains the so-called arc length procedure may be most effective. Therefore, in this paper we apply the arc-length solution procedure developed in [2,3,5] to determine the equilibrium solution $\boldsymbol{v}(\alpha)$ from the system (8).

Numerical Examples

In this place we show the numerical results obtained for a very shallow arch subjected to three types of loading: (1) the sinusoidally distributed load (2) the uniform distributed load (3) the concentrated load. In our numerical performances, we assume the initial shapes as the simple sinusoidal one,

$$\beta = H^{*} \sin x \tag{9}$$

where H^* denotes the nondimensional rise of the arch and we adopt $H^*=8.0$ in the following calculation. And we wish to prescribe the boundary condition as the following simply supported conditions:

$$\left.\begin{aligned} &w(0) = w(\pi) = 0 \\ &\frac{d^2 w}{d x^2}(0) = \frac{d^2 w}{d x^2}(\pi) = 0 \end{aligned}\right\} \tag{10}$$

Figs.1-3 illustrate the typical non-linear load-deflection curves for each loading condition through the comparison with the Galerkin solutions. It is evident that the behaviour of the arch is dependent on the type of loading. In Fig.4 we show the fundamental and bifurcation pathes and the corresponding deflection modes of each point indicated on the curve in the case of sinusoidal distributed loading shown in Fig.1. Especially, it is noteworthy that there exists another new bifurcation point in addition to the existing bifurcation point from which the unsymmetric deflection mode bifurcates.

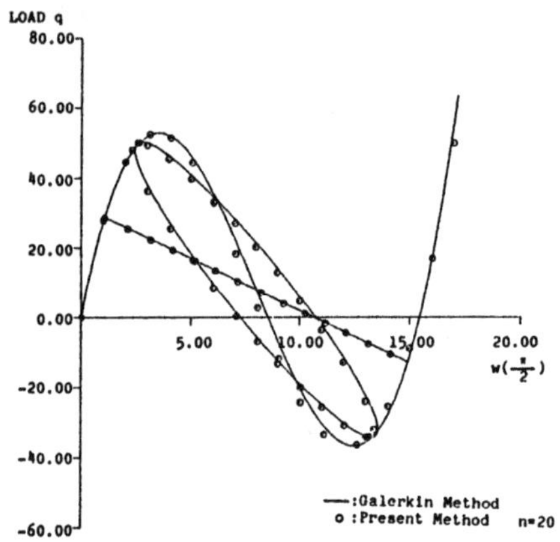

Fig.1. Load-deflection curve for the sinusoidal distributed loading

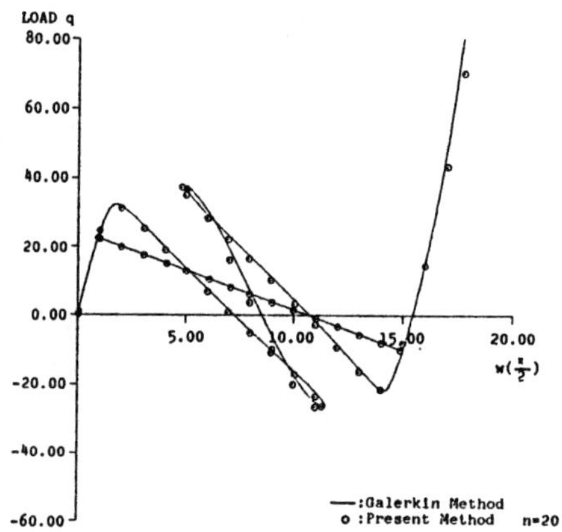

Fig.2. Load-deflection curve for the uniform distributed loading

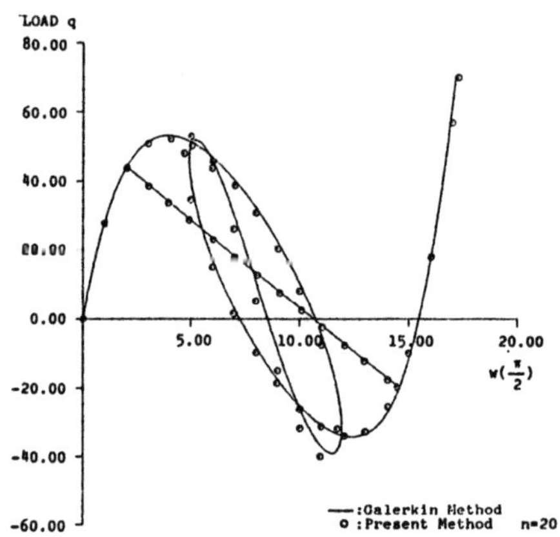

Fig.3. Load-deflection curve for the concentrated loading

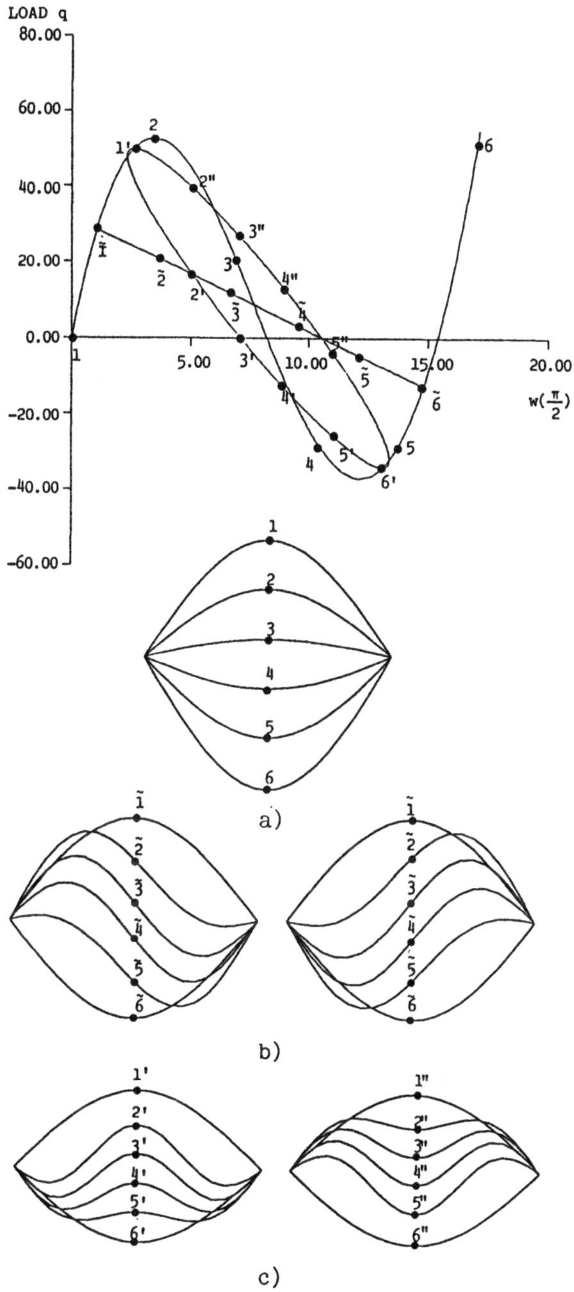

Fig.4. Deflection modes: a) fundamental symmetric modes
b)First bifurcation (a symmetric) modes c) Second bifurcation symmetric modes

Conclusions

This paper has presented an integral equation analyses of nonlinear complex behaviour of thin elastic circular arch subjected to three types of loading. The snap-through with bifurcation, and the looping in the load-deflection curves for each loading system can be tracked numerically by using the boundary—domain element approach with the arc-length solution procedure. Finally, excellent areements with the seven-modes Galerkin solutions are obtained.

References

1. Walker,A.C.: A non-linear finite element analysis of shallow circular arches, Int.J.Solids Structures, 5 (1986) 97-107

2. Wempner,G.A.: Discrete approximations related to nonlinear theories of solids, Int.J.Solids Structures, 7 (1971) 1581-1599

3. Hosono,T.: Analysis of elastic buckling problems by arc length method(in Japanese), Trans. of A.I.J, No.24 (1976) 41-49

4. Sabir,A.B. and Lock,A.C.: Large deflection, geometrically non-linear finite element analysis of circular arches, Int.J.Mech.sci.,15 (1973) 37-47

5. Riks,E.: An incremental approach to the solution of snapping and buckling problems, Int.J.Solids Structures, 15 (1979) 529-551

6. Miyake,S. and Tosaka,N.: Bifurcation analysis for thin elastic bodies by using an integral equation method, Boundary Element Methods in Applied Mechanics (Eds.M.Tanaka and T.A.Cruse), pp.483-490, Pergamon Press, 1988

7. Tosaka,N. and Miyake,S.: Geometrically nonlinear analysis of shallow spherical shell using an integral equation method, Boundary Elements VIII (Eds.M.Tanaka and C.A.Brebbia), vol.II,pp.573-576, Springer-Verlag, 1986

8. Tosaka,N. and Miyake,S. : Integral equation analysis for geometrically nonlinear problems of elastic bodies, Theory and Applications of Boundary Elements Methods (Eds. M.Tanaka and Qiughua Du), pp.251-260, Pergamon Press, 1987

Inelastic Analysis via Integral Equations: The Symmetric Approach Illustrated with Reference to the Beam on Elastic Foundation

G. NOVATI* and T. BURCZYNSKI**

* Department of Structural Engineering
Technical University (Politecnico) of Milan, Italy
** Institute of Mechanics and Fundamentals of Machine Design
Silesian Technical University, Gliwice, Poland

Summary

The problem of a beam on elastic foundation in the presence of inelastic strains is tackled by a symmetric version of the direct boundary integral equation approach. The generation of the symmetric equation system for the boundary unknowns and of the symmetric and sign-definite matrix operators which transform inelastic strains into self-stresses is described. A numerical test validates the use of "constant" and "linear" internal elements to obtain the stress response under given inelastic strains. The interest of this 1-D formulation lies in its suitability to point at the basic ideas of the symmetric approach in a simple and effective fashion.

Introduction and problem formulation

The beam on elastic foundation has already been treated via a boundary integral equation (BIE) appoach by Butterfield [1], who used it to comparatively illustrate indirect and direct methods and to suggest an algorithm for continuously inhomogeneous bodies, and by Providakis-Beskos [2] in conjunction with free and forced vibration problems.

In this paper the elastostatic analysis of a beam on elastic foundation subjected to a transversal loading and to (given) inelastic strains, is carried out using a symmetric BIE method which is the counterpart of and parallels the one first put forward for elastoplastic continua by Maier-Polizzotto [3].

Fig.1 illustrates a finite-length beam $\Omega = (a \le x \le b)$ of bending stiffness EI (= const.), supported on a bed of springs of stiffness (per unit length) K (= const.). It also defines the sign conventions for the load f(x), displacement u(x), slope $\theta(x) = \frac{du(x)}{dx}$, bending moment m(x), shear t(x) and spring reaction r(x); the sign of the curvature $\chi(x)$ and of the spring elongation e(x) are defined by $\chi(x) = \frac{d\theta(x)}{dx}$ and e(x) = u(x).

The presence of given inelastic strains $\chi^p(x)$ (in the beam) and $e^p(x)$ (in the foundation) is allowed for. The equilibrium differential equations are $\frac{dm}{dx} = -t$ and $\frac{dt}{dx} = -f + r$; the elastic constitutive laws are $m = EI(\chi - \chi^p)$ and $r = K\,(u - e^p)$. Hence, the governing differential equation is

$$EI\,\frac{d^4u}{dx^4} + K\,u = f + EI\,\frac{d^2\chi^p}{dx^2} + K\,e^p \tag{1}$$

The same problem can also be formulated by means of four Somigliana-type integral equations for the quantities u,θ,m,t. These equations, together with a fifth one in terms of the reaction r, are compactly represented by the following single equation in which symbol k runs over the set (u,θ,m,t,r)

$$\begin{aligned} k(\xi) = &- \Big[\theta(\eta)\,G_{mk}(\eta,\xi) + u(\eta)\,G_{tk}(\eta,\xi)\Big]_a^b + \\ &+ \Big[m(\eta)\,G_{\theta k}(\eta,\xi) + t(\eta)\,G_{uk}(\eta,\xi)\Big]_a^b + \int_a^b f(\eta)\,G_{uk}(\eta,\xi)\,d\eta + \\ &+ \int_a^b e^p(\eta)\,G_{pk}(\eta,\xi)\,d\eta + \delta_{pk} K e^p(\xi) + \int_a^b \chi^p(\eta)\,G_{mk}(\eta,\xi)\,d\eta \end{aligned} \tag{2}$$

In eq.(2) ξ and η are abscissas along the x axis, the quantity $[\;]_a^b = ([\;]|_{\eta=b} - [\;]|_{\eta=a})$ represents the 1-D counterpart of a boundary integral, $\delta_{pk} = 1$ for k=p and $= 0$ for $k \neq p$, and symbols G_{hk} denote influence functions for the infinitely long beam Ω_∞ ("fund. solutions") in a notation introduced by Polizzotto [4]. Precisely, $G_{hk}(\eta,\xi)$ is the response measured at η, in terms of quantity h, to a unit concentrated source placed at ξ, whose nature is singled out by subscript k as follows: to mean a certain source type, index k is set equal to the corresponding work-conjugate quantity.
Thus the standard integral equations for $u(\xi)$ and $\theta(\xi)$ (cp.[1]) involve the conventional fund. solutions $G_{\cdot u}(\eta,\xi)$ and $G_{\cdot\theta}(\eta,\xi)$ due to a unit force ($t(\xi^+) - t(\xi^-) = -1$) and a unit couple ($m(\xi^+) - m(\xi^-) = -1$) ,respectively, while the integral equations for $m(\xi)$ and $t(\xi)$ (not dealt with in [1]) imply the use of the fund. solutions $G_{\cdot m}(\eta,\xi)$ and $G_{\cdot t}(\eta,\xi)$ which are the influence functions of a unit slope discontinuity ($\theta(\xi^+) - \theta(\xi^-) = +1$) and a unit displacement discontinuity

($u(\xi^+) - u(\xi^-) = +1$), respectively. Finally the integral equation for $r(\xi)$ is also included in eq.(1) for formal completeness: the influence functions involved, of the type $G_{\cdot r}(\eta,\xi)$, turn out to represent the effects of a concentrated inelastic spring-shortening (i.e. of the inelastic strain $e^p(x) = \Delta(x-\xi)$, where Δ denotes the Dirac delta function).

Properties of the fundamental solutions and symmetric formulation of the equations for the boundary quantities

On the way towards symmetry, it is essential to recognize the following properties exhibited by the fund. solutions

$$G_{hk}(\eta,\xi) = G_{kh}(\xi,\eta) \qquad h,k = u,\theta,m,t,r \tag{3}$$

For $\eta \neq \xi$ eq.(3) is valid for any combination of subscripts (included the case h=k); eq.(3) makes sense and holds true for $\eta = \xi$ only when involves influence functions which do not exhibit a discontinuity when the field and source point coincide; for the discontinuous kernels one can write $G_{hk}(\xi^+,\xi) = G_{kh}(\xi^-,\xi)$. These properties can be proved by the (generalized) Betti theorem applied over Ω_∞ to pairs of elastic states generated by the various concentrated sources (cp. [3]).

Let us now address the determination of the four quantities at the beam ends which are unknown for any set of generic, mixed boundary conditions. There are eight available equations: they stem in pairs from the first four integral equations (2) on taking ξ to either end of the beam. The equation-selection criterion leading to symmetry , which hinges on the properties (3) of the kernels, is the following. Let quantity k be known at $x = a$ ($= b$); then pick the BIE written by "collocating" the k-conjugate source at $\xi = a^+$ ($= b^-$).

At this stage let us model the inelastic strain fields

$$\chi^p(x) = \sum_1^n{}_i \phi_i(x)\chi_i^p = \underline{\phi}^T(x)\underline{\chi}^p \qquad e^p(x) = \sum_1^m{}_i \psi_i(x)e_i^p = \underline{\psi}^T(x)\underline{e}^p \tag{4a,b}$$

where the interpolation functions ϕ_i and ψ_i (chosen, for instance, piecewise constant or linear) are conceived as defined over the whole Ω and the nodal values are given data.

The BIE system acquires the matrix form

$$\underline{A}\ \underline{X} = \underline{DF} + \underline{DM}\ \underline{\chi}^p + \underline{DR}\ \underline{e}^p \qquad (5)$$

For the case of a beam simply supported at both ends, setting $\underline{X}^T = \{\ -t(a)\ -\theta(a)\ t(b)\ \theta(b)\}$ (where superscript T means transpose) and using the above equation-selection criterion, the matrices appearing in eq.(5) turn out to have the following expressions (vector $\underline{DF}$ is not detailed for paucity of space)

$$\underset{4\times4}{\underline{A}} = \begin{bmatrix} G_{uu}(a,a) & -G_{mu}(a,a) & G_{uu}(b,a) & -G_{mu}(b,a) \\ -G_{um}(a,a) & G_{mm}(a,a) & -G_{um}(b,a) & G_{mm}(b,a) \\ G_{uu}(a,b) & -G_{mu}(a,b) & G_{uu}(b,b) & -G_{mu}(b,b) \\ -G_{um}(a,b) & G_{mm}(a,b) & -G_{um}(b,b) & G_{mm}(b,b) \end{bmatrix} \qquad (6)$$

$$\underset{4\times n}{\underline{DM}} = \underset{j=1..n}{[\underline{DM}_j]},\ \underline{DM}_j^T = \int_\Omega \{-G_{mu}(\eta,a) \mid G_{mm}(\eta,a) \mid -G_{mu}(\eta,b) \mid G_{mm}(\eta,b)\}\phi_j(\eta)\,d\eta$$

$$\underset{4\times m}{\underline{DR}} = \underset{j=1..m}{[\underline{DR}_j]},\ \underline{DR}_j^T = \int_\Omega \{-G_{ru}(\eta,a) \mid G_{rm}(\eta,a) \mid -G_{ru}(\eta,b) \mid G_{rm}(\eta,b)\}\psi_j(\eta)\,d\eta \qquad (7a,b)$$

where in eqs.(7) and henceforth the symbol $\int_\Omega$ stands for $\int_a^b$ and $\{\cdots\}$ denotes a row vector. The symmetry of the coefficient matrix $\underline{A}$, evident from properties (3), characterizes the boundary-solution phase of the present approach.

Discretized equations for the generalized stresses

The generalized stresses are defined by weighting the original stress distribution with the interpolation functions adopted in eq.(4a,b) for the corresponding work-conjugate inelastic-strain field

$$\underset{n\times1}{\underline{M}} = \int_\Omega \underline{\phi}(\xi)\ m(\xi)\ d\xi \qquad \underset{m\times1}{\underline{R}} = \int_\Omega \underline{\psi}(\xi)\ r(\xi)\ d\xi \qquad (8a,b)$$

Inserting the discretizations (4a,b) into the integral representations for $m(\xi)$ and $r(\xi)$, eq.(2), and substituting the latter into eqs.(8a,b), one obtains

$$\underline{M} = -\ \underline{MB}\ \underline{X} + \underline{MF} + \overline{\underline{MM}}\ \underline{\chi}^p + \overline{\underline{MR}}\ \underline{e}^p \qquad (9a)$$

$$\underline{R} = - \underline{RB}\ \underline{X} + \underline{RF} + \overline{\underline{RM}}\ \underline{\chi}^p + \overline{\underline{RR}}\ \underline{e}^p \tag{9b}$$

In eqs.(9a,b) matrices $\underline{MB}$ and $\underline{RB}$ depend on the beam boundary conditions; with reference to the case of simple supports at both ends they are given by

$$\underset{n\times 4}{\underline{MB}} = [\underline{MB}_i],\ \underset{i=1..n}{\underline{MB}_i} = \int_\Omega \{-G_{um}(a,\xi) \mid G_{mm}(a,\xi) \mid -G_{um}(b,\xi) \mid G_{mm}(b,\xi)\}\phi_i(\xi)\,d\xi$$

$$\underset{m\times 4}{\underline{RB}} = [\underline{RB}_i],\ \underset{i=1..m}{\underline{RB}_i} = \int_\Omega \{-G_{ur}(a,\xi) \mid G_{mr}(a,\xi) \mid -G_{ur}(b,\xi) \mid G_{mr}(b,\xi)\}\psi_i(\xi)\,d\xi \tag{10a,b}$$

The remaining four matrices, involving double integrations over the domain Ω, read

$$\underset{n\times n}{\overline{\underline{MM}}} = [\overline{MM}_{ij}],\quad \overline{MM}_{ij} = \int_\Omega\int_\Omega \phi_i(\xi)\ \phi_j(\eta)\ G_{mm}(\eta,\xi)\ d\eta\ d\xi \tag{11a}$$

$$\underset{n\times m}{\overline{\underline{MR}}} = [\overline{MR}_{ij}],\quad \overline{MR}_{ij} = \int_\Omega\int_\Omega \phi_i(\xi)\ \psi_j(\eta)\ G_{rm}(\eta,\xi)\ d\eta\ d\xi \tag{11b}$$

$$\underset{m\times n}{\overline{\underline{RM}}} = [\overline{RM}_{ij}],\quad \overline{RM}_{ij} = \int_\Omega\int_\Omega \psi_i(\xi)\ \phi_j(\eta)\ G_{mr}(\eta,\xi)\ d\eta\ d\xi \tag{11c}$$

$$\underset{m\times m}{\overline{\underline{RR}}} = [\overline{RR}_{ij}],\quad \overline{RR}_{ij} = \int_\Omega\int_\Omega \psi_i(\xi)\ \psi_j(\eta)\ G_{rr}(\eta,\xi)\ d\eta\ d\xi \tag{11d}$$

while vectors $\underline{MF}$ and $\underline{RF}$ are not detailed for paucity of space. Finally, substitution of the expression for $\underline{X}$, obtained from eq.(5), into eqs.(9a,b), yields:

$$\underline{M} = \underline{MF} + \underline{MM}\ \underline{\chi}^p + \underline{MR}\ \underline{e}^p \qquad \underline{R} = \underline{RF} + \underline{RM}\ \underline{\chi}^p + \underline{RR}\ \underline{e}^p \tag{12a,b}$$

having set

$$\underline{MF} = - \underline{MB}\ \underline{A}^{-1}\ \underline{DF} + \underline{MF} \qquad \underline{RF} = - \underline{RB}\ \underline{A}^{-1}\ \underline{DF} + \underline{RF} \tag{13a,b}$$

$$\underline{MM} = - \underline{MB}\ \underline{A}^{-1}\ \underline{DM} + \overline{\underline{MM}} \qquad \underline{RM} = - \underline{RB}\ \underline{A}^{-1}\ \underline{DM} + \overline{\underline{RM}} \tag{13c,d}$$

$$\underline{MR} = - \underline{MB}\ \underline{A}^{-1}\ \underline{DR} + \overline{\underline{MR}} \qquad \underline{RR} = - \underline{RB}\ \underline{A}^{-1}\ \underline{DR} + \overline{\underline{RR}} \tag{13e,f}$$

In view of properties (3), it turns out that $\underline{DM} = \underline{MB}^T$, $\underline{DR} = \underline{RB}^T$, $\overline{\underline{MR}} = \overline{\underline{RM}}^T$ and that matrices $\overline{\underline{MM}}$ and $\overline{\underline{RR}}$ are symmetric. Therefore, due to the symmetry of $\underline{A}^{-1}$, the matrices defined by eqs.(13c,d,e,f) which transform inelastic strains into generalized self-stresses, enjoy the following properties

$$\underline{MM} = \underline{MM}^T, \qquad \underline{RR} = \underline{RR}^T, \qquad \underline{MR} = \underline{RM}^T \tag{14}$$

Equations (12a,b) can be compactly rewritten as

$$\begin{bmatrix} \underline{M} \\ \underline{R} \end{bmatrix} = \begin{bmatrix} \underline{MF} \\ \underline{RF} \end{bmatrix} + \underline{Z} \begin{bmatrix} \underline{\chi}^p \\ \underline{e}^p \end{bmatrix}, \quad \text{with} \quad \underline{Z} = \begin{bmatrix} \underline{MM} & \underline{MR} \\ \underline{RM} & \underline{RR} \end{bmatrix} \tag{15a,b}$$

Matrix $\underline{Z}$ turns out to be symmetric and negative semidefinite. The last property stems from the fact that the quadratic form $-\frac{1}{2}\{(\underline{\chi}^p)^T, (\underline{e}^p)^T\}\,\underline{Z}\,\{(\underline{\chi}^p)^T, (\underline{e}^p)^T\}^T$ represents exactly the elastic strain energy of the system (beam + springs) due to any set of inelastic strains ($\underline{\chi}^p$, $\underline{e}^p$). These characterizations of matrix $\underline{Z}$, not shared by the analogous matrix of the standard non-symmetric approach, have important implications, discussed in [3] and [4], when elastolplastic problems are addressed.
Once the generalized stresses $\underline{M}$ and $\underline{R}$ are known, the corresponding physical quantities can be obtained assuming that they be modeled like the work-conjugate inelastic strains: $\tilde{m}(x) = \underline{\phi}^T(x)\underline{\tilde{m}}$ and $\tilde{r}(x) = \underline{\psi}^T(x)\underline{\tilde{r}}$. This yields $\underline{\tilde{m}} = (\underline{T}^{\phi})^{-1}\,\underline{M}$, with $T^{\phi}_{ij} = \int_\Omega\int_\Omega \phi_i(\xi)\phi_j(\eta)d\eta d\xi$, $\underline{T}^{\phi}$ being a diagonal matrix for piecewise constant $\phi_i(x)$ and a tri-diagonal matrix for piecewise linear, hat-shaped $\phi_i(x)$; and analoguosly $\underline{\tilde{r}} = (\underline{T}^{\psi})^{-1}\,\underline{R}$.

A numerical test

The application of the method is illustrated with reference to a steel beam, simply supported at both ends, of length l=10m and flexural rigidity EI = 7.0×10^7 Nm^2, subjected to a distribution of inelastic curvature (no inelastic spring-elongation, and no loads) which vanishes everywhere except in the three intervals $2.40<x<3.10$, $4.65<x<5.35$ and $6.90<x<7.60$ where it has a bilinear variation with $\chi^p(2.75) = \chi^p(7.25) = -0.6 \times 10^{-1}$ m^{-1} (stretching lower fibres), and $\chi^p(5.00) = +0.4 \times 10^{-1}$ m^{-1} (stretching upper fibres). For two different values of spring-bed stiffness, $K_1 = 1.5 \times 10^7$ N/m^2 and $K_2 = 0.5 \times 10^7$ N/m^2, Fig.2 shows the solution in terms of the (physical) bending moment $\tilde{m}(x)$ computed using 40 constant (discontinuos) elements and 20 linear continuous elements (4 constant el. and 2 linear el. are used in each of the above three intervals; the symmetry of the system is not exploited).

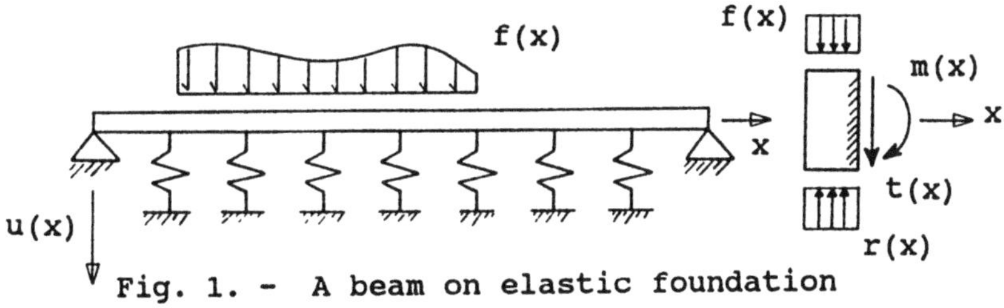

Fig. 1. - A beam on elastic foundation

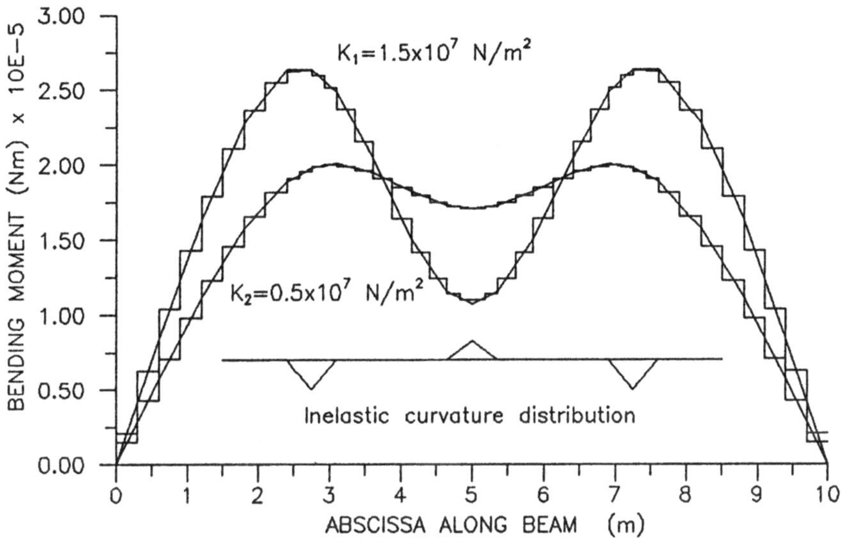

Fig. 2. - Bending moment $\tilde{m}(x)$ due to the given inelastic curvature distribution, for two different values of spring-bed stiffness (using 40 constant el. and 20 linear el.)

References

1. Butterfield, R.: New concepts illustrated by old problems. Developments in Boundary Element Methods - 1 (Banerjee, P.K. and Butterfield, R. Eds.). London: Applied Science Publishers, 1979, 1-30.
2. Providakis, C.P. and Beskos D.E.: Dynamic analysis of beams by the boundary element method. Computers & Structures 22 (1986) 957-964.
3. Maier, G. and Polizzotto C.: A Galerkin approach to boundary element elastoplastic analysis. Comp. Meth. Appl. Mech. Engng. 60 (1987) 175-194.
4. Polizzotto C.: An energy approach to the boundary element method, Part I and II. Comp. Meth. Appl. Mech. Engng. 69 (1988) 167-184, 263-276.

On Boundary Element Inelastic Analysis in the Presence of Softening

Z. Cen*, G. Maier & G. Novati

Dept. of Structural Engineering, Technical University (Politecnico) of Milan, P. Leonardo 32, Milan 20133, Italy

Summary

Two kinds of quasi-static nonlinear BE analyses are considered: elastic-plastic solids with softening behaviour described by the constitutive law; elastic bodies with cohesive-softening cracks on the boundary or on interfaces. Some consequences of local softening behaviour are pointed out and illustrated by numerical tests, namely: overall instability and mesh-dependance of the computed response in the former case; snap-back phenomena and bifurcations of equilibrium paths in the latter case.

Introduction

In plasticity, softening (or negative hardening) means reduction of strength (stress- or load-carrying capacity) due to inelastic deformation processes. Typical experimental situations where softening shows up in the behaviour of material specimens or structural elements include: compression, tension and bi- and triaxial tests on concrete and various geomaterials; bending tests on reinforced concrete beams; etc..

The essential feature of softening is that an external agency apt to impose an infinitesimal configuration change while preserving equilibrium, performs negative second-order work for some deformation path. In other terms, softening in the local behaviour is a manifestation of instability; it is a 'physical' instabilizing effect which may entail instability manifestations in the overall response of a solid or structure, even in the absence of 'geometric' instabilizing effects leading to conventional buckling. Overall instability phenomena include bifurcation of the equilibrium path, strain localization and 'internal' (uncontrollable, catastrophic or snap-back) behaviour. Softening and its important mechanical and engineering implications are the subject of an abundant,

---

* On leave of absence from Tsinghua University, Beijing, China

mostly recent literature not to be surveyed here. Representative contributions related to the results presented in this paper concern the following aspects: basic concepts of material instability as recently reviewed by Drucker[1]; overall instability with bifurcation and snap-back [2]-[5]; two-laws descriptions and 'cohesive crack' models [2] [6]; experimental investigations [7]-[9]; finite element analysis [10]-[13].

In this paper we address the following question: what is the use of boundary element methods (BEM) to predict numerically the overall behaviour in the presence of (local) softening behaviour in contexts where it is reasonable to assume small deformations (no geometric effects) and time independence (no viscous behaviour)?. Partial answers to this question and some comments are provided for two categories of situations: elastic-plastic solids with constitutive laws accommodating softening; fracture tests of concrete interpreted by means of an elastic-cohesive crack model.

Elastic-plastic BE analysis in the presence of softening

The most popular (direct, collocation) BEM for elastic-plastic analysis can be concisely outlined as follows (see e.g. [14]).The discretized boundary integral equation is generated by means of fundamental solutions (e.g. Kelvin's) for displacements, tractions and domain stresses; through suitable modelling and integrations, this gives rise to the algebraic linear equation:

$$\underline{A}\ \Delta\underline{X} = \Delta\underline{F} + \underline{D}\ \Delta\underline{\varepsilon}^{p} \tag{1}$$

where: vector $\Delta\underline{X}$ contains increments of boundary nodal unknowns (displacements $\Delta\underline{u}$ on the unconstrained boundary Γ_p; tractions on the constrained boundary Γ_u, $\Gamma_p + \Gamma_u = \Gamma$); vector $\Delta\underline{\varepsilon}^p$ contains plastic strain increments at the nodes ('strain points') in the domain Ω; $\Delta\underline{F}$ and $\Delta\hat{\underline{F}}$ in eq. (2) contain boundary and domain data concerning the load step over the (ordinative) time interval Δt; $\underline{A}$, $\underline{D}$ and $\hat{\underline{A}}$,$\hat{\underline{D}}$ in (2) are coefficient matrices.

The discretized integral equation for strains in the strain points reads:

$$\Delta\underline{\varepsilon} = \hat{\underline{A}}\ \Delta\underline{X} + \Delta\hat{\underline{F}} + \hat{\underline{D}}\ \Delta\underline{\varepsilon}^{p} \tag{2}$$

The constitutive law can be expressed for all strain points simultaneously:

$$\Delta\underline{\varepsilon}^{p} = \Delta\underline{\varepsilon}^{p}(\Delta\underline{\varepsilon}) \tag{3}$$

For solving eqs. (1)-(3), which govern the response to load increments starting from a known state at time t, one can use the following traditional algorithm ('successive substitution' or modified Newton-Raphson).

1. Calculate once for all matrices $\underline{A}$, $\underline{D}$, $\hat{\underline{A}}$, $\hat{\underline{D}}$.
2. Compute the data vectors $\Delta\underline{F}$ and $\Delta\hat{\underline{F}}$ containing the current load increments; set $\Delta\underline{\varepsilon}^{p} = \underline{0}$.
3. Solve equation (1) with respect to $\Delta\underline{X}$, substitute into equation (2) and use this to compute $\Delta\underline{\varepsilon}$.
4. Through the constitutive relations (3), compute a new vector $\Delta\underline{\varepsilon}^{p}$.
5. If the difference between new and preceding $\Delta\underline{\varepsilon}^{p}$ is negligible, enter the subsequent load increments in 2; otherwise go to 3 with the new $\Delta\underline{\varepsilon}^{p}$.

A fairly general description of elastic-plastic material behavior is as follows, e.g. [15]:

$$\underline{\sigma} = \underline{E}\,(\underline{\varepsilon} - \underline{\varepsilon}^{p}) \quad , \qquad \phi\,(\underline{\sigma}, \underline{\chi}) \le 0 \tag{4a,b}$$

$$\dot{\underline{\varepsilon}}^{p} = \frac{\partial\psi}{\partial\underline{\sigma}}\,(\underline{\sigma}, \underline{\chi})\,\dot{\lambda} \; , \qquad \dot{\underline{\eta}} = -\frac{\partial\psi}{\partial\underline{\chi}}\,(\underline{\sigma}, \underline{\chi})\,\dot{\lambda} \tag{5a,b}$$

$$\dot{\lambda} \ge 0 \quad , \qquad \phi\,\dot{\lambda} = 0 \quad , \qquad \underline{\chi} = \underline{\chi}\,(\underline{\eta}) \tag{6a,b,c}$$

where: $\underline{E}$ = matrix of elastic moduli (symmetric and positive definite); ϕ= yield function; ψ = plastic potential; $\underline{\chi}$, $\underline{\eta}$ = vectors of static and kinematic internal variables, respectively, related to each other by the evolution law (6c); dots mark time derivatives (or rates), tildes transpositions. For a yielding process ($\phi = 0$, $\dot{\phi} = 0$), by differentiating ϕ and substituting (4a) and (5) into $\dot{\phi} = 0$, solving with respect to $\dot{\lambda}$ and substituting the expression of $\dot{\lambda}$ into (5a) and (4a), one obtains the elastic-plastic stiffness matrix (such that $\dot{\underline{\sigma}} = \underline{P}\,\dot{\underline{\varepsilon}}$) in the form:

$$\underline{P} = \underline{E} - \underline{E}\,\frac{\partial\psi}{\partial\underline{\sigma}}\left(\frac{\partial\phi}{\partial\tilde{\underline{\sigma}}}\,\underline{E}\,\frac{\partial\psi}{\partial\underline{\sigma}} + H\right)^{-1}\frac{\partial\phi}{\partial\tilde{\underline{\sigma}}}\,\underline{E} \tag{7}$$

having defined the hardening modulus $H(\underline{\sigma}, \underline{\chi})$ as:

$$H \equiv \frac{\partial\phi}{\partial\tilde{\underline{\chi}}}\,\frac{\partial\underline{\chi}}{\partial\tilde{\underline{\eta}}}\,\frac{\partial\psi}{\partial\underline{\chi}} \tag{8}$$

It appears from (7) and (8) that: for H<0 (softening), matrix $\underline{P}$ may not be positive definite; there is a critical value H_c for which the scalar in parenthesis in (7) vanishes and hence, the material is no longer controllable in strains (internal instability). Multicriterion constitutive laws (more than one yield function ϕ) would lead to a generalization of equations (4)-(8) involving a hardening matrix $\underline{H}$ at 'corners' of the current yield surface.

Numerical experiments were carried out [16] for various negative H on the basis of the following specialization of equations (4)-(6): associated flow rule ($\phi = \psi$); linear isotropic hardening (H=constant); Drucker-Prager yield criterion, namely:

$$\phi = \sqrt{J_2} + \alpha\, I_1 - \left(\frac{\sqrt{3}}{2} + \alpha \right) \sigma_0^t - H\,\lambda \tag{9}$$

where: J_2 = second invariant of deviatoric stresses; I_1 = first invariant of the stress tensor; α, σ_0^t = material parameters.

The first example had been dealt with earlier for H≥0 and α=0 (plain strain, polystyrene plate with voids, Mises criterion: [17] [14]). Our results for H<0 are illustrated in Fig. 1a,b.

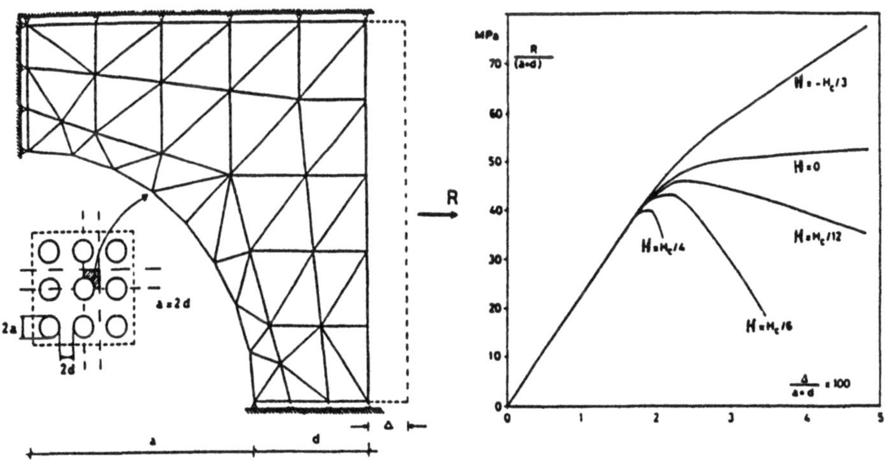

a. Perforated, plane-strain plate

b Reaction R vs. displacement Δ for different H (H_c = - 3G)

Fig. 1

The second example concerns a tensile concrete specimen [7]. The results of some BE computations are presented in Fig.2a, b and c.

a.

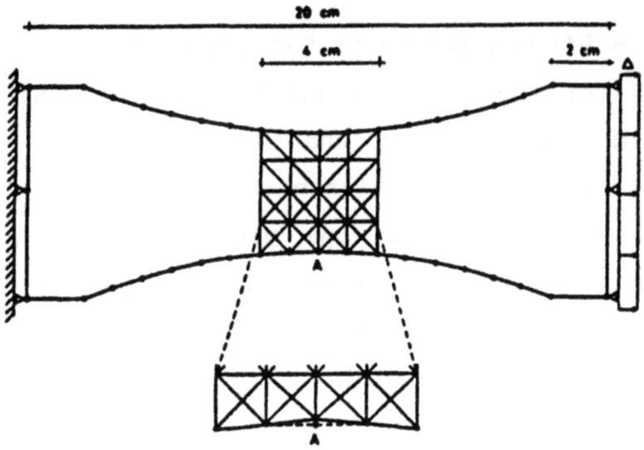

Fig. 2a. Concrete specimen with 'imperfection': BE discretization with 40 linear elements and 48 linear cells

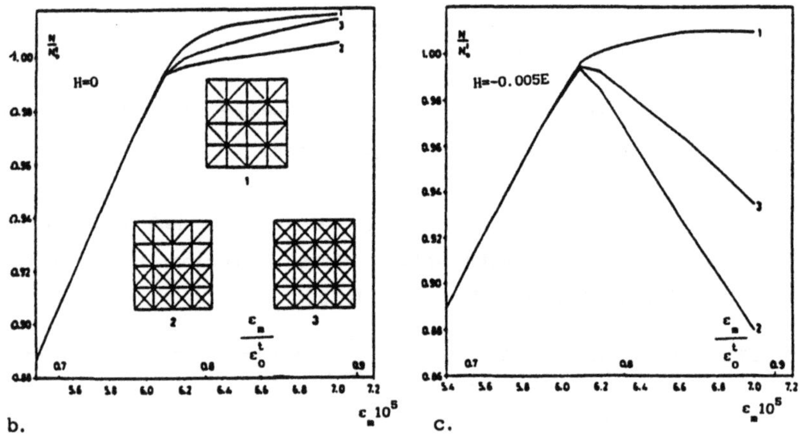

b. c.

Fig. 2b, c Axial force vs. imposed strain for H=0 and H=-0.005 E, with three different meshes

These and other numerical tests concerning BE analysis of elastic-plastic softening solids corroborate the following remarks.

(a) For increasing $|H|$ with $H<0$: the consequent reduction of load-carrying capacity is well captured; termination occurs earlier because of lack of convergence of the iterative solution to the finite increment boundary value problem. A trend toward strain localization could be observed.

(b) For $H<0$, the falling branch in the overall response (reactive resultant versus imposed displacement) exhibits a strong mesh dependence, not encountered for $H\geq 0$.

(c) The subjectivity of results noted in (b) might probably be attenuated by computational provisions basically similar to those devised in the FE context [12] [13]; however, remedies in the BE context appear to require nontrivial ad-hoc techniques, to be discussed elsewhere.

Elastic-cohesive crack BE analysis

Consider a two-dimensional elastic solid (Fig.3a) subdivided into subdomains Ω_1 and Ω_2 separated by an interface Γ_c. For each subdomain we can write a BE equation:

$$\underline{A}^1 \begin{Bmatrix} \underline{X}^1 \\ \underline{u}_c^1 \\ \underline{p}_c^1 \end{Bmatrix} = \underline{F}^1 \alpha \quad ; \quad \underline{A}^2 \begin{Bmatrix} \underline{X}^2 \\ \underline{u}_c^2 \\ \underline{p}_c^2 \end{Bmatrix} = \underline{F}^2 \alpha \tag{10}$$

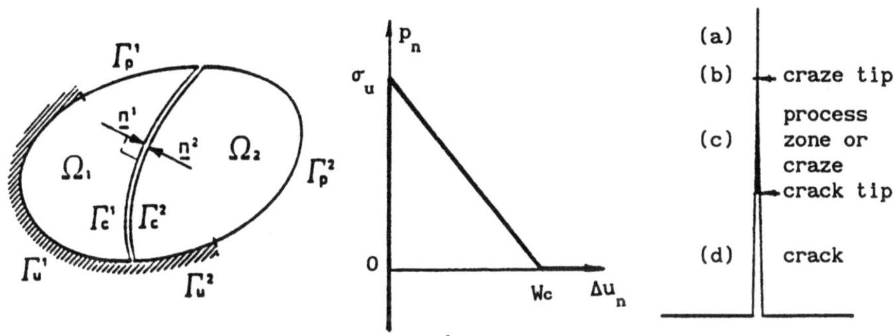

a. b. c.

Fig. 3

The symbols are the same as for eq. (1), except that the interface variables ($\underline{u}_c$, $\underline{p}_c$) are evidenced, total quantities instead of increments are used, α is a load factor and $\Delta\underline{\varepsilon}^p = 0$. Accepting for the interface Γ_c the cohesive crack model with linear softening illustrated by Fig. 3b,c [5] [6], the matching conditions to impose at each node on the normal (u_n, p_n) and tangential (u_t, p_t) components are, alternatively (σ_u= ultimate tensile stress; Wc= critical value of opening displacement discontinuity Δu_n):

(a) $\quad u_n^1 = -u_n^2 = u_n \quad ; \quad u_t^1 = -u_t^2 = u_t \quad ; \quad p_n^1 = p_n^2 = p_n \quad ; \quad p_t^1 = p_t^2 = p_t$

(b) $\quad u_n^1 = -u_n^2 = u_n \quad ; \quad u_t^1 = -u_t^2 = u_t \quad ; \quad p_n^1 = p_n^2 = \sigma_u \quad ; \quad p_t^1 = p_t^2 = p_t$

(c) $\quad u_t^1 = -u_t^2 = u_t \quad ; \quad p_t^1 = p_t^2 = p_t \quad ; \quad p_n^1 = p_n^2 = \sigma_u - \frac{\sigma_u}{Wc} \Delta u_n$

(d) $\quad p_n^1 = p_n^2 = 0 \quad ; \quad p_t^1 = p_t^2 = 0 \qquad (11)$

In eqs. (11), interface conditions (a) concern the uncracked part, (b) the 'craze' tip, (c) the craze or 'process zone', (d) the crack. The quati-static evalution of the system is conceived as driven by the location of the craze tip, which migrates monotonously from the boundary along the node sequence on the assigned Γ_c. Consequently, the load factor α becomes an unknown variable and the cohesive crack law can be regarded as reversible (holonomic) and, hence, dealt with in total (rather then incremental) terms.

The algorithm adopted to perform the analysis on the basis of the above assumptions can be outlined as follows.

1. Generate eqs. (10); eliminate non-interface variables, thus forming the interface equations.
2. For the crack growth step i, assign the node where the craze tip is located (set $p_n = \sigma_u$ there).
3. Assemble the interface equations of the subdomains and enforce conditions (11): at the first iteration r=1 for step i, by taking as normal tractions $(p_n)_r^i$ on the craze the final value of the preceding (i-1) step; at r>1, by using $(\Delta u_n)_{r-1}^i$ provided by iteration r-1.
4. Solve the assembled equations thus obtaining α_r^i. If negligible differences, go to 2 starting growth step i+1 from the next node; otherwise go to 3.

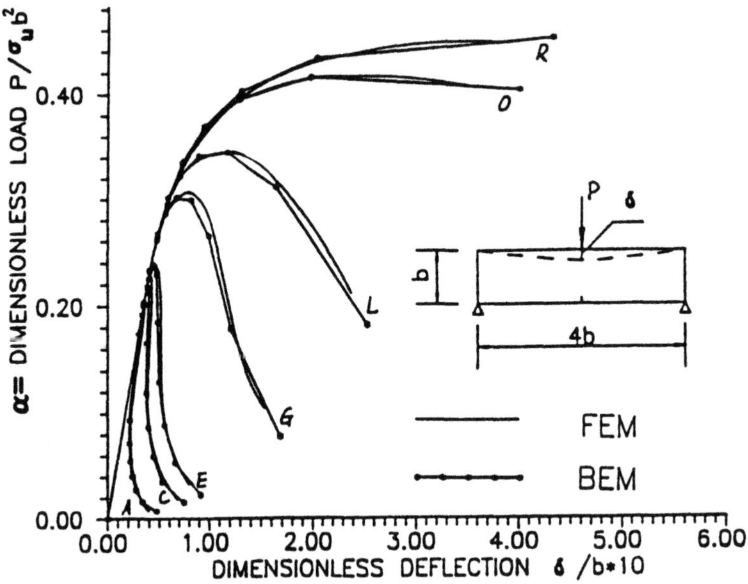

Fig. 4 Load factor vs. displacement for 3-point test

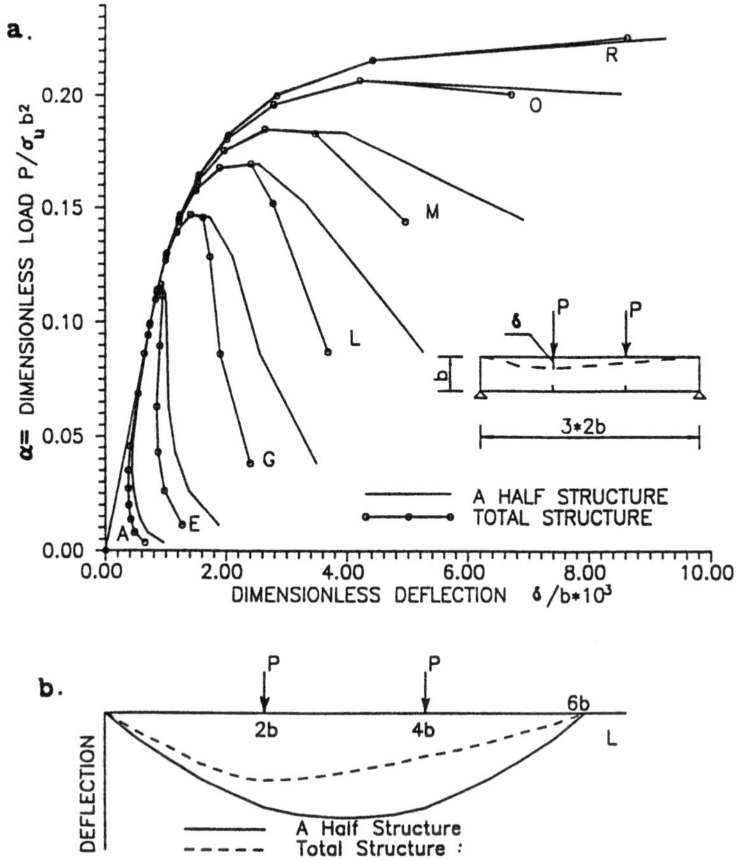

Fig. 5 Load factor vs. displacement for 4-point test (a), Symmetric and nonsymmetric deformation (b)

The results depicted in Fig.4 and 5 were obtained by assuming: $E=1.5*10^5$daN/cm^2; $\nu=0.1$; $\sigma_u=13.1$daN/cm^2; Wc=0.003cm. The 'three-point' (Fig.4) and the 'four point' (Fig.5) fracture specimens were studied as plane-stress systems discretized by quadratic conforming BE, for the same 'brittleness numbers' Wc/2b as in [5]: $2.09*10^{-5}$ graph A; $6.27*10^{-5}$ C; $1.045*10^{-4}$ E; $4.18*10^{-4}$ G; $1.04*10^{-3}$ L; $2.08*10^{-3}$ M; $6.27*10^{-3}$ O; $2.09*10^{-2}$ R. The results of Fig.4 achieved by means of 24 BE are practically coincident with those obtained by about 1000 FE in [5].

Our computational experience seem to corroborate the following conclusions.

1. When cohesive crack at interfaces and linear behaviour within subdomains represent an accepted idealization, the BEM is ideally suited and probably superior to FEM.

2. If the marching solution is controlled by the craze tip location as input variable, unstable paths and snap-back phenomena (catastrophic failure, internal overall instability [2] [5]) can easily be predicted (see Fig. 4).

3. Bifurcations of equilibrium path pointed out first in [3] can occur and can be captured, provided symmetries are dealt with properly. A symmetric system may have a nonsymmetric response, besides the symmetric one (as shown in Fig. 5).

4. If the cohesive crack propagates along an unknown itinerary, a criterion for choosing the advancement direction must be added, as in multidomain BEM for fracture mechanics (see e.g. [18]). Other cohesive crack models (see e.g. [19]) might be conveniently dealt with by BEM. These and other issues will be investigated elsewhere.

References

[1] Palmer, A.C.; Maier, G.; Drucker, D.C.: Normality relations and convexity of yield surfaces for unstable materials or structural elements, Trans. ASME, J. Appl. Mech., (1967) 464-470.

[2] Maier, G.: On softening flexural behaviour in elastic-plastic beams. Rendiconti Istituto Lombardo, Classe Sci., A, 102, (1968) 648-677. English translation in 'Studi e Ricerche', 8, Politecnico Milano, (1986) 85-117.

[3] Maier, G.; Zavelani Rossi, A.; Dotreppe, J.C.: Equilibrium branching due to flexural softening. Proc. ASCE, J. Struct. Div., 99, 4, (1973) 897-906.

[4] Bazant, Z.P.: Stable states and paths of propagation of damage zones and interactive fractures, Cracking and Damage, Mazars, J. and Bazant, Z.P. (eds.), (1989) 183-206.

[5] Carpinteri, A.: Softening and snap-back instability in cohesive solids. Int.J.Num.Meth.Engng., 28, 7, (1989) 1521-1538.

[6] Hillerborg, A. et al.: Analysis of crack formation and crack growth in concrete by means of fracture mechanics and finite elements. Cement Concrete Res., 6, (1976) 773-782.

[7] Dei Poli, S.; Iori, I.: Osservazioni e rilievi sul comportamento a trazione dei calcestruzzi: analisi di risultanze sperimentali. Studi e Ricerche, 8, (1986), Politecnico di Milano, 35-62

[8] Biolzi, L.: Flexure of beams with elastic-softening behaviour, SM Archives 13, 3, (1988) 177-191.

[9] Iori, I.: On internal instability of concrete elements, Proc. AIMETA, Conf. Mech. Materials & Struct., Rome. Oct. (1989).

[10]Pietrusczak, S.; Mroz, Z.: Finite element analysis of deformation of strain-softening materials. Int. J. Num. Meth. in Engng., 17, (1981) 327-334.

[11]Willam, K.J.; Bicanic, N.; Pramono, E.; Sture, S.: Composite fracture model for strain softening computations of concrete. Proc. Int. Conf. on Fracture Mechanics of Concrete (ed. Wittman, F.H.), Amsterdam, (1986) 149-169.

[12]Ortiz, M.; Leroy, Y.; Needleman, A.: A finite element method for localized failure analysis. Comp. Meth. in Appl. Mech. and Engng., 61 (1987) 189-214.

[13]Franchi, A.; Genna, F.: Self-adaptive model for structural softening of brittle materials. Proc. ASCE, J. of Eng. Mech. Div., 115, 7, (1989) 1543-1558.

[14]Cruse, T.A.; Polch, E.Z.: Elastoplastic BIE analysis of cracked plates and related problems, Part. 1: formulation. Int. J. Num. Meth. in Engng., 23, (1986) 429-437.

[15]Ortiz, M.;Martin, J.B.: Symmetry-preserving return mapping algorithms and incrementally extremal paths: a unification of concepts, Int. J. Num. Meth. in Engng., 28, 8, (1989) 1839-1854.

[16]Borré, G.; Maier, G.; Novati, G.: Sull'analisi elastoplastica per elementi di contorno in presenza di instabilità del materiale (softening). Cinquantenario Facoltà di Architettura Napoli: Franco Jossa e la Sua Opera, Giannini Publ., Napoli, (1988) 159-175.

[17]Brebbia, C. A.; Telles, J. C. F.; Wrobel, L. C.: Boundary Element Techniques. Springer-Verlag, Berlin, (1984).

[18]Blandford, G.E.; Ingraffea, A.R.; Liggett, J.A.: Two-dimensional stress intesity factor computations using the boundary element method. Int. J. Num. Meth. in Engng., 17, (1981) 387-404.

[19]Hui, C.; Lagoudas, D.C.; Ruina, A.: Constitutive models for crazes and their effects on crack growth in glassy polymers. Appl. Mech. Div. ASME, 85, (1988) 87-115.

An Algorithm for Vectorized Computation in Boundary Element Methods

S. Kobayashi, N. Nishimura and H. Kamiya

School of Civil Engineering
Kyoto University, Kyoto, Japan

Summary

This paper describes an efficient algorithm for high-speed computation in boundary element methods using a vector processor FACOM VP-400. The key to achieve efficient vectorization is to make the innermost DO-loop in a program as large as possible. The point of the algorithm proposed in the present paper is such that the coefficients of the algebraic equations derived from integral equations by discretization are constructed by changing the order of summations in such a manner as the summation over the node is carried out in the innermost DO-loop for a temporarily fixed Gauss point before taking the summation over the Gauss points in an element. We call this algorithm Gauss-point-wise algorithm (GPWA), while we call the ordinary algorithm for scalar processor element-wise algorithm (EWA). Computing time, for example, for the second kind exterior elastostatic problems is reduced remarkably by the GPWA used in the vector processor from that by the EWA used in a scalar processor. Even when a scalar processor is only available, the GPWA provides far more efficient computation than the ordinary EWA.

Introduction

Boundary element method (BEM) is now used as a practical tool for analysis and design in many fields of engineering. In this situation, it is highly efficient computation that is urgent to be developed. One way of improving computing time in BEM is to vectorize the computation. The key to achieve efficient vectorization is to program the innermost DO-loop as large as possible. This type of algorithm can be most efficiently applied to BEM than any other numerical techniques. In addition to this, the vectorization is suitable for 3-dimensional problems.

We will see, for instance, the efficiency of the vectorized computation applied to 3-dimensional elastostatic problems. Of course, the idea displayed in this paper can be applied to any other problems.

BEM in 3-Dimensional Elastostatics

Statement of elastostatic problems: With an assumption of the infinitesimal deformation, elastostatic problems for homogeneous linear isotropic material may be stated to find a solution of the Navier equation

$$\mu u_{i,kk}+(\lambda+\mu)u_{k,ik}+\rho b_i=0 \quad \text{in } D \tag{1}$$

with boundary conditions

$$\begin{aligned} u_i&=\bar{u}_i && \text{on } \partial D_1 \\ t_i&=\sigma_{ik}n_k=\bar{t}_i && \text{on } \partial D_2 \end{aligned} \tag{2}$$

where σ_{ik}:stress, u_i:displacement, t_i:surface traction, b_i:body force, n_k:unit outward normal vector to ∂D, λ,μ:Lamé's constants, ρ:density, and $(\bar{\ })$:prescribed boundary quantity, $(\ )_{,i}=\partial(\ )/\partial x_i$, $\partial D=\partial D_1 \cup \partial D_2$, and the summation convention rule is applied to the repeated indices.

Integral equation: Integral equation for the above-mentioned problem can be expressed, by ignoring the body force, as

$$C_{ik}u_i(\mathbf{y})=\int_{\partial D}U_{ik}(\mathbf{y},\mathbf{x})t_i(\mathbf{x})dS(\mathbf{x})-\int_{\partial D}W_{ik}(\mathbf{y},\mathbf{x})u_i(\mathbf{x})dS(\mathbf{x}) \tag{3}$$

where the first and the second integrals are called the simple layer and the double layer potentials, respectively, and the kernels U_{ik} and W_{ik} are the first and the second fundamental solutions of elastostatics, respectively, and C_{ik} is a constant tensor, such as $C_{ik}=\delta_{ik}/2$ on a smooth surface.

Discretization of the integral equation: We discretize the integral equation by use of isoparametric quadratic elements. Adopting a local coordinate system $(o;\xi_1,\xi_2,\xi_3)$, where ξ_3 is in the direction of the unit outward normal vector to the element, the global coordinate x_i and functions $u_i(\mathbf{x})$ and $t_i(\mathbf{x})$ may be transformed into the local coordinates by use of basis functions $N^\alpha(\xi)$ expressed as

$$x_i(\xi)=\sum_{\alpha=1}^{8} N^\alpha(\xi)x_i^\alpha, \quad u_i(\xi)=\sum_{\alpha=1}^{8} N^\alpha(\xi)u_i^\alpha, \quad t_i(\xi)=\sum_{\alpha=1}^{8} N^\alpha(\xi)t_i^\alpha \tag{4}$$

where the 8-node isoparametric element is assumed and $(\ )_i^\alpha$ is the quantity at node α.

The discretized version of equation (3) with N elements having 8 nodes can be expressed as follows,

$$\sum_{i=1}^{3} C_{ik}u_i(y)=\sum_{i=1}^{3}\sum_{j=1}^{N} t_i^{\alpha(j)}\int_{S(j)}U_{ik}(y,x_{(j)}(\xi))N^{\alpha}(\xi)J_{(j)}(\xi)dS(\xi)$$

$$-\sum_{i=1}^{3}\sum_{j=1}^{N} u_i^{\alpha(j)}\int_{S(j)}W_{ik}(y,x_{(j)}(\xi))N^{\alpha}(\xi)J_{(j)}(\xi)dS(\xi) \qquad (5)$$

where $S_{(j)}$, $J_{(j)}$, $\alpha(j)$ and $x_{(j)}(\xi)$ imply the surface, Jacobian, node α and the selected point of the element j, respectively.

If we apply the Gaussian quadrature formula, say 4x4-points, for numerical integration of the integrals, it can be approximated, for example, as

$$\int_{S(j)}W_{ik}(y,x_{(j)}(\xi))N^{\alpha}(\xi)J_{(j)}(\xi)dS(\xi)$$

$$=\sum_{l=1}^{4}\sum_{m=1}^{4} W_{ik}(y,x_{(j)}(\xi))N^{\alpha}(\xi)w_l w_m J_{(j)}(\xi) \qquad (6)$$

where w_l and w_m are weights of Gaussian quadrature formula corresponding to Gaussian points on ξ_1- and ξ_2-axes, respectively. Note that the singular integral must be evaluated adequately. For example, the singular part may be evaluated together with the non-integral part by use of the raw subtraction technique.

A system of algebraic equations with respect to unprescribed boundary quantities at nodal points is easily obtained by using Gaussian quadrature of the type (6) and the raw subtraction technique.

A New Algorithm for Vectorized Computation

Integration of the function-kernel product: In order to solve the system of equations, integration of the function-kernel product of the type of equation (6), i.e.,

$$K(n(y)_k,n(\alpha(j))_i)$$

$$=\sum_{l=1}^{4}\sum_{m=1}^{4} W_{ik}(y,x_{(j)}(\xi_1^{\,l},\xi_2^{\,m}))N^{\alpha}(\xi_1^{\,l},\xi_2^{\,m})w_l w_m J_{(j)}(\xi) \qquad (7)$$

must be evaluated beforehand.

In this expression, $n(\mathbf{y})$ and $n(\alpha(j))$ stand for the node number of the observation point $\mathbf{y}$, and the node number corresponding to node number α of the element j, respectively. Indices k and i imply the components in 3-dimension. When the total number of nodes of the discretization is M, the matrix size becomes 3Mx3M. Note that the singular part is excluded from the expression (7).

<u>Ordinary algorithm for scalar computation</u>:
The most widely used algorithm for computing the coefficient matrix is as follows:

i)choose the observation point $\mathbf{y}$ first,

ii)select an element j, then

iii)carry out the numerical integration independently with respect to all nodes of the element j, and components k and i.

iv)Finally, construct the matrix by adding the contributions from the ki-element $K(n(\mathbf{y})_k, n(\alpha(j))_i)$ evaluated as mentioned above.

The process may be expressed in a form

$$\mathop{\mathrm{DO}}_{n(\mathbf{y})=1}^{M}\ \mathop{\mathrm{DO}}_{j=1}^{N}\ \mathop{\mathrm{DO}}_{k=1}^{3}\ \mathop{\mathrm{DO}}_{i=1}^{3}\ \mathop{\mathrm{DO}}_{\alpha\neq\alpha(\mathbf{y})}\ \sum_{l=1}^{4}\sum_{m=1}^{4} W_{ik}(\mathbf{y},\mathbf{x}_{(j)}(\xi_1{}^l,\xi_2{}^m))N^{\alpha}(\xi_1{}^l,\xi_2{}^m)w_l w_m J_{(j)}(\xi) \quad (8)$$

where the symbol $\mathop{\mathrm{DO}}_{n(\mathbf{y})=1}^{M}\ \mathop{\mathrm{DO}}_{k=1}^{3}$ implies the operation of substituting the appropriate values into the row elements $n(y)_k$, and the symbol $\mathop{\mathrm{DO}}_{j=1}^{N}\ \mathop{\mathrm{DO}}_{i=1}^{3}\ \mathop{\mathrm{DO}}_{\alpha\neq\alpha(\mathbf{y})}$ stands for the process of introducing appropriate values into the column elements $n(\alpha(j))_i$. The symbol $\sum$ means the ordinary sum. We call this algorithm the "Element-Wise Algorithm (EWA)".

<u>A new algorithm for vectorized computation</u>: As mentioned before, the key to achieve an efficient computation by a vector processor is to make the innermost DO-loop as large as possible. A simple idea to realize the efficient computation is as follows:

i)choose a Gauss point in an element first,

ii)select an element j, and

iii)fix the observation point y.

iv)Carry out the computation independently with respect to node and components i and k and construct the matrix by adding the ki-element $K(n(\mathbf{y})_k, n(\alpha(j))_i)$.

The process may be expressed by

$$\sum_{l=1}^{4}\sum_{m=1}^{4}\underset{j=1}{\overset{N}{DO}}\ \underset{n(\mathbf{y})=1}{\overset{M}{DO}}\ \underset{k=1}{\overset{3}{DO}}\ \underset{i=1}{\overset{3}{DO}}\ \underset{\alpha\neq\alpha(\mathbf{y})}{DO}\ W_{ik}(\mathbf{y},\mathbf{x}_{(j)}(\xi_1{}^l,\xi_2{}^m))N^{\alpha}(\xi_1{}^l,\xi_2{}^m)w_l w_m J_{(j)}(\xi) \quad (9)$$

However, this is not effective since the length of the innermost DO-loop is at most the number of nodes in one element. We can easily improve this by expanding the last three DO-loops into 9 elements corresponding to W_{ik} having 8 terms of N^{α}. This is easily carried out, because they are not so large. If we adopt this idea, the length of the innermost DO-loop becomes so large as the total number of nodes of the discretized equations. We call this algorithm the "Gauss-Point-Wise Algorithm (GPWA)". Similar algorithms are also developed by Miyoshi and Hirai[1], and Ymaguchi[2].

Applications to Elastostatics

In order to evaluate the efficiency of the proposed algorithm GPWA, we solved 3-dimensional second kind exterior elastostatic problems.

Four types of computation were executed by

i)scalar processor applying the EWA,

ii)vector processor applying the EWA,

iii)scalar processor applying the GPWA, and

iv)vector processor applying the GPWA.

The results are compared in Fig. 1. The solid lines show the number of nodes versus the computing time required for constructing the coefficient matrix of the discretized equations, whereas the broken lines indicate the number of nodes versus total computing time required for obtaining the boundary displacements.

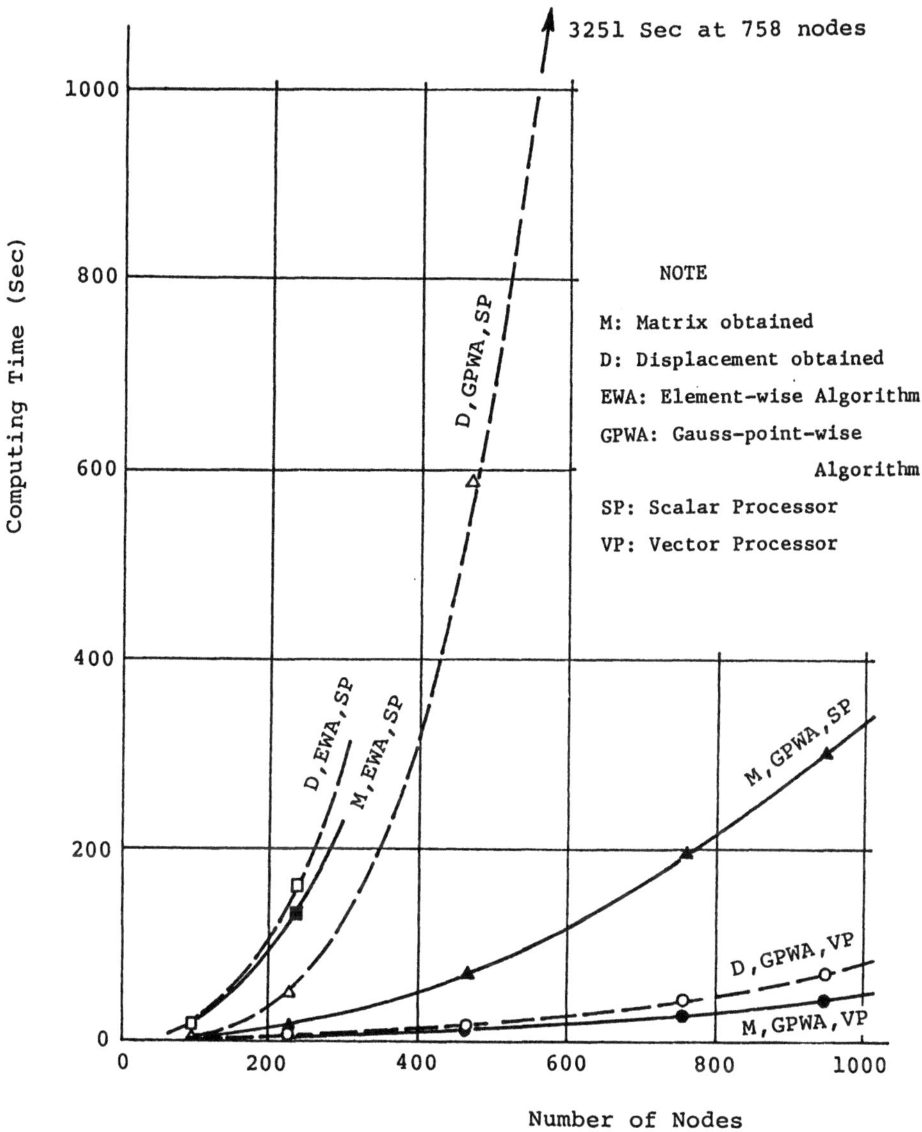

Fig.1. Number of nodes vs. computing time. Two algorithms EWA and GPWA are used in the scalar and the vector processors.

Fig.1 shows that the computing time is drastically reduced by making use of the GPWA. For example, in the case of 237 nodes, computing time is reduced to 5.4sec from 161.6sec when the vector processor is used by adopting the GPWA instead of using the scalar processor by adopting the EWA. The reduction rate of the computing time in this case is about 30. The rate becomes larger as the number of nodes increases. It is also noted that the computing time is greatly reduced by adopting the GPWA even if the scalar processor is used. For example, the computing time is reduced to 59.4sec from 161.1sec in the case of 237 nodes when the GPWA is adopted instead of using the EWA.

The use of the vector processor together with the GPWA reduces the computing time most extremely. For 231, 471 and 758 nodes, for instance, it reduces computing time down to 5.2, 17.0 and 44.2sec from 52.8, 587, 3251sec, respectively, by the scalar processor used with the GPWA. The reduction rates are 10, 34.5 and 73.6, respectively.

Roughly speaking, the computing time applying the GPWA increases as almost proportional to the number of nodes, whereas it increases with almost cubic power of the number of nodes when the EWA is adopted.

Concluding Remarks

Boundary element methods may be suited for vectorization. Therefore, a high speed computation can be performed using the GPWA when a supercomputer is available. Even when only a scalar processor is at hand, the GPWA provides far more efficient computation than the ordinary EWA. Numerical examples showed herein demonstrate the high efficiency of the GPWA. It is magnified immensely by use of the vector processor. This type of algorithm (GPWA) may be widely used for practical computations.

References

1. Miyoshi, T. and Hirai, C. (1987): Boundary Element Analysis of Surface Cracks by the Supercomputer (in Japanese), Trans. Japan Soc. Mech. Eng., vol. 53, pp.250-254.
2. Yamaguchi, K. (1988): High speed computation of BIEM by V. P., (in Japanese), MSc. Thesis, Dept. of Civil Eng., Kyoto Univ.

Overcoming Nonuniqueness in the Direct BIEM for Three-Dimensional Steady-State Exterior Elastodynamic Problems

I.R. Gonsalves[†] , D.J. Shippy[‡] and F.J. Rizzo[††]

[†] Caterpillar Inc., Tech. Center A, P.O. Box 1875, Peoria, IL 61656
[‡] Engg. Mechanics, University of Kentucky, Lexington, KY 40605
[††] Engg. Sc. and Mechanics, Iowa State University, Ames, IA 50011

Abstract

The nonuniqueness in the solution of three-dimensional steady-state elastodynamic transmission problems by the direct Boundary Integral Equation Method (BIEM) at "irregular" frequencies is investigated. The nonuniqueness is overcome by constraining the BIE with the "interior identity". Numerical examples are presented to demonstrate the effectiveness of the method.

Introduction

A difficulty encountered in the solution of steady-state (SS) exterior acoustic and elastodynamic problems by the direct BIEM is the nonuniqueness in the BIEM solution at certain discrete frequencies [1,2]. In contrast to interior problems, real eigenfrequencies are not encountered in exterior problems. Thus the failure to obtain a unique solution to the integral equation describing the exterior problem is an artifact of the integral equation rather than a difficulty with the boundary value problem itself. The irregular frequencies are therefore referred to as fictitious eigenfrequencies and the nonuniqueness in the BIEM solution is called the Fictitious Eigenfrequency Difficulty, or FED. For scattering from perfectly hard or soft inclusions, the fictitious eigenfrequencies have been identified as the natural frequencies of vibration, i.e., real eigenfrequencies, of the inclusion (when regarded as being constituted of the same material as the exterior medium) under certain boundary conditions [3]. The FED also occurs for the SS transmission problem of acoustics, i.e., for acoustically penetrable inclusions (see e.g., [4]).

The purpose of the present paper is to numerically demonstrate the FED for the SS elastodynamic transmission problem, i.e., for elastic inclusions. When the FED is encountered, it is overcome by modifying an adaptation [2] of Schenck's formulation developed for acoustics [1]. The effectiveness of the formulation is demonstrated with numerical examples.

Boundary integral equations

Consider a three-dimensional inclusion B with a bounding surface S, bonded to a full-space F. Both B and F are homogeneous, isotropic, and linearly elastic. A SS incident wave with displacement $\underline{u}^I(\underline{x},\omega)$ at $\underline{x}$ impinges on B. (The time-dependence $e^{-i\omega t}$, where $i^2=-1$, ω is circular frequency, and t is time, is dropped for clarity). The total displacements $\underline{u}(\underline{x},\omega)$ in F and $\underline{u}_B(\underline{x},\omega)$ in B are sought which satisfy the Navier-Cauchy equations of motion [5] in F and B, respectively, the radiation condition [5] on the scattered $\underline{u}^s(\underline{x},\omega) = \underline{u}(\underline{x},\omega) - \underline{u}^I(\underline{x},\omega)$ at infinity, and the transmission conditions

$$\underline{u}(\underline{x},\omega) = \underline{u}_B(\underline{x},\omega), \quad \underline{t}(\underline{x},\omega) + \underline{t}_B(\underline{x},\omega) = 0 \tag{1}$$

on S. In (1), $\underline{t}$ and $\underline{t}_B$ are total tractions on S evaluated with normals $\underline{n}$ and $\underline{n}_B$ pointing away from F and B, respectively.

The integral reformulations of the Navier-Cauchy equations, which satisfy the radiation condition are [6]:

$$\underline{C}^T(\underline{x})\underline{u}(\underline{x},\omega) = \int_S [\underline{U}^T(\underline{x},\underline{y},\omega)\underline{t}(\underline{y},\omega) - \underline{T}^T(\underline{x},\underline{y},\omega)\underline{u}(\underline{y},\omega)]dS + \underline{u}^I(\underline{x},\omega) \tag{2}$$

$$\underline{C}_B^T(\underline{x})\underline{u}_B(\underline{x},\omega) = \int_S [\underline{U}_B^T(\underline{x},\underline{y},\omega)\underline{t}_B(\underline{y},\omega) - \underline{T}_B^T(\underline{x},\underline{y},\omega)\underline{u}_B(\underline{y},\omega)]dS \tag{3}$$

where $\underline{U}$ and $\underline{T}$ are Stokes' (full-space) fundamental displacement and traction tensors for F, $\underline{C}$ is the coefficient tensor, superscript T denotes tensor transposition, and subscript B denotes quantities pertaining to the inclusion B. For $\underline{x}$ on S, (2) and (3) constitute two integral equations to be solved for $\underline{u}$, $\underline{t}$, $\underline{u}_B$ and $\underline{t}_B$, subject to the transmission conditions (1).

Numerical procedure

The BIEs (3) and (4) are first recast into forms free from Cauchy-principal-value integrals as described in [6]. The discretization of S into quadratic isoparametric quadrilateral and/or triangular elements, nodal collocation of (2) and (3) at the N nodes on S, and enforcement of (1) results in 6N linear algebraic equations

$$\begin{bmatrix} D & E \\ D_B & E_B \end{bmatrix} \begin{Bmatrix} u \\ t \end{Bmatrix} = \begin{Bmatrix} u^I \\ 0 \end{Bmatrix} \tag{4}$$

where $[D]$, $[E]$, $[D_B]$ and $[E_B]$ are 3Nx3N matrices, $\{u\}$, $\{t\}$ and $\{u^I\}$ are 3Nx1 vectors of nodal total-displacement, total-traction and incident-displacement, respectively, and $\{0\}$ is the 3Nx1 zero vector.

If ω is a regular frequency, the system (4) is solved by Gaussian elimination for $\{t\}$ and $\{u\}$. If ω is close to or is a fictitious eigenfrequency, the nonuniqueness in the solution to (2) is reflected in a near-singular matrix of coefficients, which becomes more singular with mesh refinement. Consequently, although the natural frequencies of the inclusion may not be available apriori for purposes of identifying the fictitious eigenfrequencies, numerical indications of the FED consist of (a) a marked increase in the BIEM matrix condition number (CN) for small changes in the frequency, and (b) the absence of solution convergence with mesh refinement at a given frequency.

For scattering from cavities and rigid inclusion, a number of remedies to the FED have been used in acoustics and some of these have been extended to full-space elastodynamics (e.g., [2]). In the present paper, the BIFILM (Boundary-integral Interior Equation Formulation Including Lagrange Multipliers) formulation devised in [2], which is a variation of the CHIEF (Combined Helmholtz Integral Equation Formulation, [1]) method in acoustics, is used to overcome the FED. In CHIEF and BIFILM, additional equations are generated by collocating (2) inside B. In the CHIEF method, the discrete forms of the interior equations

$$[D_c]\{u\} + [E_c]\{t\} = \{u_c^I\} \tag{5}$$

are used with (4) to form an overdetermined (rectangular) system, which is then solved with a least-squares approximation. In (5), $[D_c]$ and $[E_c]$ are $3N_c$x3N matrices and $\{u_c^I\}$ is a $3N_c$x1 vector of incident displacement at the interior points. In contrast, BIFILM uses Lagrange multipliers to form the augmented square system

$$\begin{bmatrix} D & E & D_c^T \\ D_B & E_B & E_c^T \\ D_c & E_c & 0 \end{bmatrix} \begin{Bmatrix} u \\ t \\ \lambda \end{Bmatrix} = \begin{Bmatrix} u^I \\ 0 \\ u_c^I \end{Bmatrix} \tag{6}$$

where $\{\lambda\}$ is a $3N_c$x1 vector of Lagrange multipliers. Equation (4) is then solved by Gaussian elimination.

As in the CHIEF method, the accuracy of the BIFILM solution deteriorates if N_c is large in comparison to N. Moreover, the BIFILM solution degenerates when the interior points are located on or close to the nodal surface(s) of the associated interior problem. However, as demonstrated in the next section, the CN of the BIFILM coefficient matrix can be used to assess the efficacy of the number and location of interior points.

Numerical examples

Consider the scattering of a plane dilatational incident wave from an elastic sphere (radius=a) bonded to an elastic full-space. The incident displacement is given by $\underline{u}^I = u_o e^{-i\omega x_3/\alpha} \underline{e}_3$, where u_o is the (possibly complex) amplitude of the incident displacement, α is the dilatational wave speed in F, and $\underline{e}_3$ is a unit vector in the x_3 cartesian coordinate direction. The center of the sphere is located at the center of the coordinate system. Fig.1 shows the 24- and 16-element discretizations of one-half of the sphere surface which are used to obtain the BIEM solution at $ka=\omega a/\alpha=2$. The numbers in parentheses in Fig. 1 denote the numbers of elements and nodes, respectively. Material properties chosen are those for a germanium sphere in an aluminum full-space, for which $\nu=0.35$, $\nu_B=0.1553$, $\alpha_B/\alpha=0.8061$ and $\rho_B/\rho=1.9852$, where ν and ρ are Poisson's ratio and density, respectively, of F. Fig. 2 is a polar plot of the amplitude of scattered radial displacement u_r^s on the surface of the inclusion in the x_1-x_3 plane. The axial symmetry of the 16-element discretization results in a BIEM solution which is as accurate as the 24-element solution, but with less computational effort. However, at $ka=2.1177$ (the fundamental frequency of radial vibrations of a solid aluminum sphere with a free boundary), the deterioration in the 16-element BIEM solution shown in Fig. 3 is accompanied by a jump in the matrix CN from $2.55\text{x}10^3$ at ka=2 to $1.06\text{x}10^5$ at ka=2.1177. When the finer 32-element (Fig.1) discretization was used, the BIEM solution diverged, with a matrix CN of $4.89\text{x}10^5$, thereby suggesting the FED. However, as seen in Fig.3, the use of BIFILM with four interior points located on a spherical surface of radius $r_c=a/2$ results in an

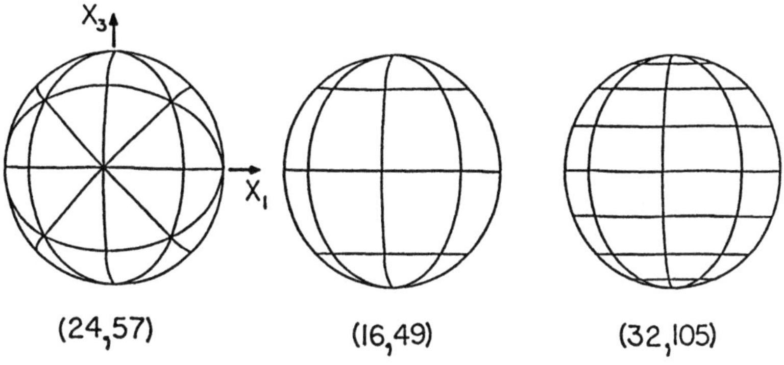

Fig.1 BIEM dicretizations of the sphere surface

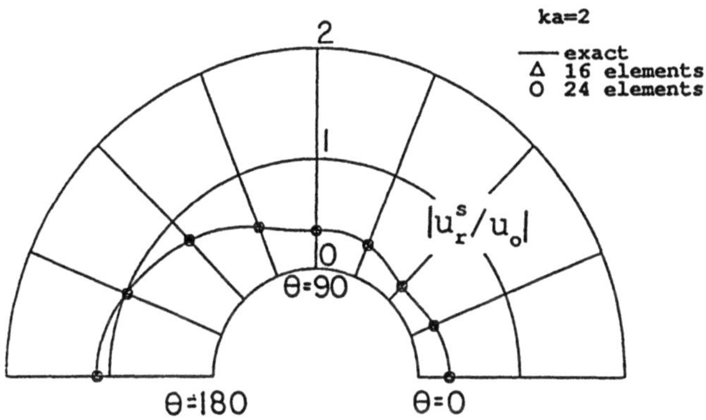

Fig.2 Amplitude of scattered radial displacement on the inclusion surface at ka-2.

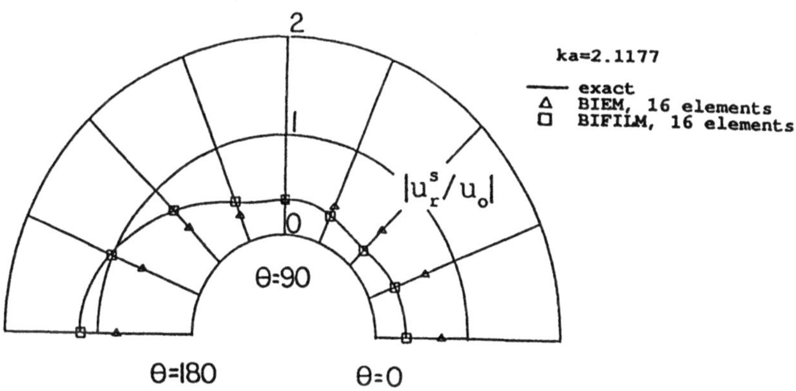

Fig.3 Amplitude of scattered radial displacement on the inclusion surface at a fictitious eigenfrequency ka-2.1177.

accurate 16-element BIFILM solution with a matrix CN=1.38×10^{3}.

Now consider a frequency ka=4.4934, which is the fundamental frequency of radial vibrations of an aluminum sphere with a fixed boundary. The accuracy of the 32-element BIEM solution at ka=5 (Fig. 4) deteriorates at the lower frequency ka=4.4934. The FED is indicated by the increase in the matrix CN from 6.24×10^{4} at ka=5 to 1.57×10^{6} at ka=4.4934. However, the BIFILM formulation (with four interior points at $r_c=a/2$) results in an accurate 32-element solution (Fig. 5) with a CN = 3.61×10^{5}.

Conclusions

The direct BIEM for the SS elastodynamic transmission problem exhibits the FED at the eigenfrequencies of the associated interior problem for Dirichlet and Neumann boundary conditions. The BIFILM procedure can be used together with the BIEM and BIFILM matrix CNs as an effective remedy to the FED.

Acknowledgements

This study was funded in part by the Office of Naval Research (contract N00014-86-K-0051) and the Center for Computational Sciences at the University of Kentucky.

References

1. Schenck, H.A.: Improved integral formulation for acoustic radiation problems. J. Acous. Soc. Amr. 44 (1968) 41-58.

2. Rezayat, M.; Shippy, D.J.; Rizzo, F.J.: On time-harmonic elastic wave analysis by the boundary element method for moderate to high frequencies. Comp. Meth. Appl. Mech. 55 (1986) 349-367.

3. Shaw, R.P.: BIE methods applied to wave problems. Developments in Boundary Element Methods - I (Banerjee, P.K.; Butterfield, R. (eds.)). London: Applied Science Publishers 1985.

4. Kleinman, R.E.; Martin, P.A.: On single integral equations for the transmission problem of acoustics. Tech. Rep. 87-1. Center for the Mathematics of Waves. University of Delaware 1987.

5. Eringen, C.A.; Suhubi, E.S.: Elastodynamics - II. New York: Academic Press 1975.

6. Rizzo, F.J.; Shippy, D.J.; Rezayat, M.: A boundary integral equation method for radiation and scattering of elastic waves in three dimensions. Int. J. Num. Meth. Engg. 21 (1985) 115-129.

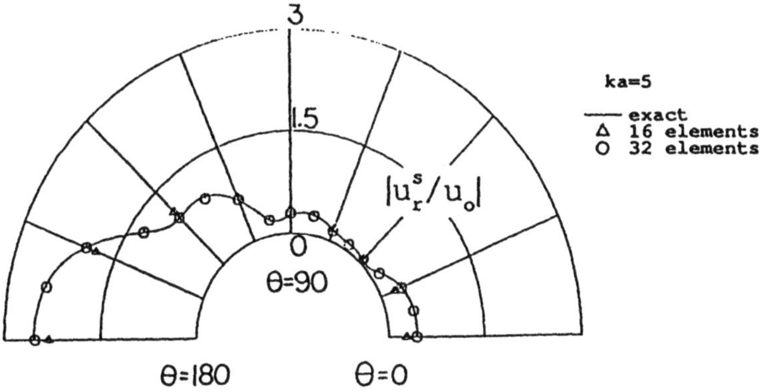

Fig.4 Amplitude of scattered radial displacement on the inclusion surface at ka=5.

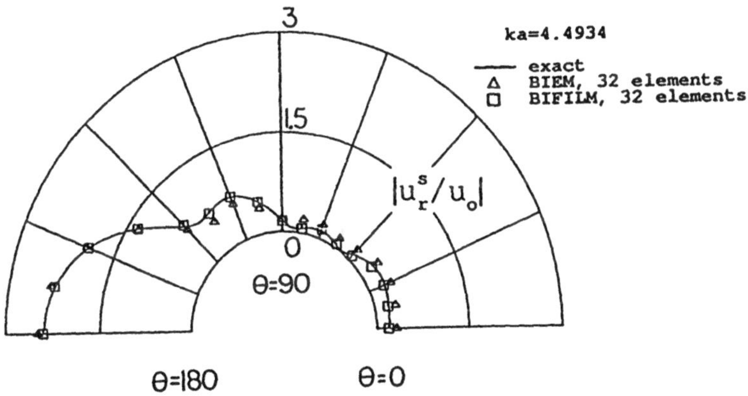

Fig.5 Amplitude of scattered radial displacement on the inclusion surface at a fictitious eigenfrequency ka=4.4934.

Efficient Time and Space Integrations of Stokes Fundamental Solutions – Applications to a Direct Time Domain BEM for Elastodynamics

DIMITRIS L. KARABALIS

Department of Civil Engineering
University of South Carolina
Columbia, SC 29208, USA

Summary

The numerical treatment of the fundamental singular solutions of the Stokes' state of quiescent past and their implementation within the framework of a 3-D time domain BEM formulation for linear elastodynamics is discussed. Of particular interest are simplified time and space integrations of the above fundamental solutions. The accuracy as well as the efficiency of such computations are tested through comparison studies against other more computationally intense BEM techniques.

Introduction

While transform domain, primarily Laplace or Fourier, BEM formulations have received a lot of attention throughout the brief BEM history, 2-D and 3-D time domain formulations have seen only limited applications. This, among other factors, is perhaps due to the formidable computational effort required in order to implement properly the time domain fundamental solutions or Green's functions into an efficient numerical scheme. In this work some simplified techniques for treating the Stokes' pair of fundamental singular solutions are discussed along with their implementation into a BEM formulation for linear elastodynamics. Although a certain degree of computational efficiency is by definition embodied into such simlified approaches their accuracy should not, of course, be assumed a priori. To this extend some comparison studies in the last section of this article aim at establishing the credibility of the proposed approaches within the range of their applicability. More extensive discussions on the subject, which in part is still under development, will be published in a forthcoming communication.

Detailed discussions on the derivation and mathematical properties of the pair of fundamental singular solutions corresponding to the Stokes' state of quiescent past can be found in Eringen and Suhubi [5]. In their general form these operators express the response of an infinite linear elastic space to a point load of arbitrary time variation. In this form they were used for the first time by Karabalis [7], and Karabalis and Beskos [9] in their treatment of 3-D SSI problems using a direct time domain BEM approach. A special form of these fundamental solutions corresponding also to a point load but with a Dirac delta function variation in time has been used by Benerjee *et al* [1,2]. The implementation of these fundamental solutions into a time domain BEM formulation requires discretizations in time and space. With regard to time discretization, in Refs. [2,9-13] a constant variation of the field variables is assumed over each time step while a linear one is considered in Ref. [1]. A uniform variation of the field variables is also assumed in Refs. [9,10,13] over boundary elements while higher order spatial discretizations, based on the work of Lachat and Watson [14], have been included in Refs. [1,2,13].

Besides, the Stokes' pair of fundamental singular solutions other sources of such functions are available for time domain elastodynamics. As an example, Johnson [6] has derived a far more complex pair of Green's functions by just considering a traction free plane surface as a boundary condition to the Stokes' state of quiescent past. Extensive reviews on these matters as well as details on a number of 2-D and 3-D BEM implementations are out of the context of this article but they can be found in, among others, Refs. [3,4,8,11].

Formulation and Numerical Solution

Assuming i) zero initial conditions, and ii) zero body forces, the solution of a 3-D linear elastodynamic problem governed by the Cauchy-Navier equations of motion can be cast into a boundary integral equation (BIE) form as

$$(0.5)\, u_k(\xi,t) = \int_B \{u_{ik}[\underline{x},t;\xi|t_{(\underline{n})i}(\underline{x},t)] - t_{(\underline{n})ik}[\underline{x},t;\xi|u_i(\underline{x},t)]\} \qquad (1)$$

where $\underline{x}$ and ξ are points on the smooth boundary B of the domain of interest, u_k and $t_{(\underline{n})i}$ are, respectively, the boundary displacements and tractions, $\underline{n}$ is the outward normal on the boundary, t is the time variable and the operators $[u_{ik},t_{(\underline{n})ik}]$ comprise the Stokes' pair of fundamental singular solutions. Analytic expressions of the above fundamental solutions and details on the derivation of Eq.(1) can be found in a number of sources, e.g. Ref. [5].

Even for relatively simple geometries and time variations a numerical solution of Eq. (1) is in order. With regard to the time variable, the time axis can be discretized into a number of equally spaced intervals within each one of which the field variables are assumed to be constant. As an example, the traction vector for the time interval n, i.e., $(n-1)\Delta t<n<n\Delta t$, can be approximated as

$$t_{(\underline{n})i}(\underline{x},t) = \quad t^n_{(\underline{n})i}(\underline{x})\, \{H[t-(n-1)\Delta t] - H(t-n\Delta t)\} \qquad (2)$$

where $H(t-\tau)$ is the Heaviside step function, and $t^{n}{}_{(\underline{n})i}$ is the magnitude of the rectangular impulse function at time $t=(n-0.5)\Delta t$, i.e.,

$$t^{n}{}_{(\underline{n})i}(\underline{x}) = t^{n}{}_{(\underline{n})i}[\underline{x},(n-0.5)\Delta t] \tag{3}$$

With regard to the space variables, the boundary of the domain of interest is also discretized into a number of quadrilateral elements over each one of which tractions and displacements assume a predescribed spatial variation. For the purposes of this work three such BEM discretizations are used: i) constant elements with a collocation point at its geometrical center, ii) linear elements with one nodal point at each corner, and iii) quadratic elements with three nodes on each side. All of these BEM discretizations converge rapidly to the "exact" solution of the problems considered, and therefore there is no need for higher order spatial elements [14] or time variations. However, the following formulation is general in the sense that it is independent of the particular BEM discretization. Under the above assumptions BIE (1), written for a number of M boundary elements with a total of Q nodal points (if constant elements are used M=Q) yields a system of 3Q linear algebraic equations for each time step N. In a matrix summation form this can be expressed as (a detailed account of this procedure is given in Refs. [9,10])

$$(0.5)\{u^{N,q}\} = \sum_{n=1}^{N} \sum_{m=1}^{M} \left([G^{n,m}]\{t^{N-n+1,m}\} - [T^{n,m}]\{u^{N-n+1}\}\right) \tag{4}$$

where $\{u^{k,q}\}$ or $\{u^{k,m}\}$ is the displacement vector of the receiver point $q=1,2,\ldots,Q$ or the source element $m=1,2,\ldots,M$, and $\{u^{k,l}\}$ and $\{t^{k,l}\}$ are, respectively, the displacement and traction vectors. The displacement tensor $[G^{n.m}]$ and the traction tensor $[T^{n,m}]$ represent the time- and space-discretized equivalents of the fundamental solutions. Physically, the 3x3 displacement tensor $G_{ik}{}^{n,m}$ represents the displacement component in the direction i at the nodal point q (receiver) and time step n due to a rectangular impulse traction component in the direction k acting at element m (source) during the first time step. A similar physical meaning is attached to the 3x3 traction tensor. For the nodal point a of the element m these tensors can be computed from the expressions

$$G_{ij}{}^{n,a} = \int_{(s)} G_{ij}{}^{n}[\underline{x}(\underline{h})]\, N^{a}(\underline{h})\, J(\underline{h})\, d(\underline{h})$$

$$T_{ij}{}^{n,a} = \int_{(s)} T_{ij}{}^{n}[\underline{x}(\underline{h})]\, N^{a}(\underline{h})\, J(\underline{h})\, d(\underline{h}) \tag{5}$$

where (s) is the area Of element m, N^{a} is the shape function corresponding to the nodal point a, and J is the Jacobian that transforms the local coordinates $\underline{x}$, which are based on the receiver nodal point, into the intrinsic coordinates $\underline{h}$. The terms $G_{ij}{}^{n}[\underline{x}(\underline{h})]$ and

$T_{ij}{}^{n}[\underline{x}(\underline{h})]$ stand for the time discretized counterparts of the fundamental solutions as given in Ref. [10].

Non-Singular Elements

For non-singular elements the spatial integrations indicated by Eqs.(5) can be carried out routinely using standard quadrature schemes. However, the accuracy in the computation of the influence matrices $[G^{n,s}]$ and $[T^{n,s}]$ of Eq. (4) depends, primarily, upon a numerical scheme that can properly take into account the causality property of the fundamental solutions. Consider, as an example, the relatively simpler case of a constant element discretization. The time and space discretized displacement and traction tensors are given in Ref. [11]

$$G_{ik}^{n,m} = \frac{1}{4\pi\rho}\left\{B_c^n \int_{(s)} \frac{1}{r^2}\left(\frac{3r_i r_k}{r^3} - \frac{\delta_{ik}}{r}\right)\left[H\left(t-\frac{r}{c_1}\right)F_1 - H\left(t-\frac{r}{c_2}\right)F_2\right] \mathrm{d}(s)\right.$$
$$\left. + \frac{B_{c_1}^n}{c_1^2}\int_{(s)} \frac{r_i r_k}{r^3}\, d(s) + \frac{B_{c_2}^n}{c_2^2}\int_{(s)}\left(\frac{\delta_{ik}}{r} - \frac{r_i r_k}{r^3}\right)\mathrm{d}(s)\right\}, \tag{6}$$

$$T_{ij}^{n,m} = T_{ijk}^{n,m} e_k\,, \tag{7}$$

$$T_{ijk}^{n,m} = \frac{1}{4\pi}\left\{B_c^n(-6c_2^2)\int_{(s)}\left(5\,\frac{r_i r_j r_k}{r^7} - \frac{\delta_{ij}r_k + \delta_{ik}r_j + \delta_{jk}r_i}{r^5}\right)\right.$$
$$\times\left[H\left(t-\frac{r}{c_1}\right)F_1 - H\left(t-\frac{r}{c_2}\right)F_2\right]\mathrm{d}(s)$$
$$- B_{c_1}^n \int_{(s)}\left[2\,\frac{c_2^2}{c_1^2}\left(6\,\frac{r_i r_j r_k}{r^5} - \frac{\delta_{ij}r_k + \delta_{ik}r_j + \delta_{jk}r_i}{r^3}\right) + \left(1 - 2\,\frac{c_2^2}{c_1^2}\right)\frac{r_k\delta_{ij}}{r^3}\right]\mathrm{d}(s)$$
$$\left. + B_{c_2}^n \int_{(s)}\left[2\left(6\,\frac{r_i r_j r_k}{r^5} - \frac{\delta_{ij}r_k + \delta_{ik}r_j + \delta_{jk}r_i}{r^3}\right) - \frac{\delta_{ik}r_j + \delta_{jk}r_i}{r^3}\right]\mathrm{d}(s)\right\} \tag{8}$$

where r is the distance from the collocation point of the receiver element q to a point on the source element m, and

$$F_\beta = \begin{cases} [t^2 - (r/c_\beta)^2]/2\,, & \text{if } (q-1)\Delta t < t - (r/c_\beta) \le q\Delta t \\ [2t(\Delta t) - (\Delta t)^2]/2\,, & \text{if } q\Delta t < t - (r/c_\beta)\,,\ \beta = 1,2\,. \end{cases} \tag{9}$$

$$B_c^n, B_{c_\alpha}^n = \begin{cases} 1\,, & \text{if } r \in B_{n,c}, B_{n,c_\alpha} \\ 0\,, & \text{if } r \notin B_{n,c}, B_{n,c_\alpha}\,, \end{cases} \qquad \alpha = 1,2\,. \tag{10}$$

with

$$B_{n,c} = \{x; x \in B \text{ and } c_2(n-1)\Delta t < r < c_1 n\Delta t\},$$
$$B_{n,c_1} = \{x; x \in B \text{ and } c_1(n-1)\Delta t < r \leq c_1 n\Delta t\}, \quad (11)$$
$$B_{n,c_2} = \{x; x \in B \text{ and } c_2(n-1)\Delta t < r \leq c_2 n\Delta t\}.$$

The coefficients $B^n{}_c$, $B^n{}_{c1}$ and $B^n{}_{c2}$ are used to specify whether or not the source element m has any influence on the receiver point q during time step n. It shown in this work, that the accuracy and the efficiency in the computation of the influence matrices $[G^{n,s}]$ and $[T^{n,s}]$ depends, primarily, not upon an elaborate spatial integration scheme (the time integration in this case has been carried out analytically) but upon a numerical scheme that can properly account for the time retarded character of the fundamental solutions and their causality property both of which are expressed by the coefficients $B^n{}_c$, $B^n{}_{c1}$ and $B^n{}_{c2}$. Using one, or maybe few more, of the collocation points within the usual BEM mesh as a control point(s) towards establishing the necessary criterion of whether the source element s in Eq. (4) is "active" or not (and if yes in which way) during a given time step, with respect to a receiver point q, can lead to a number of serious inaccuracies. As it is depicted in Fig.(1), this is due to the fact that using only one or a

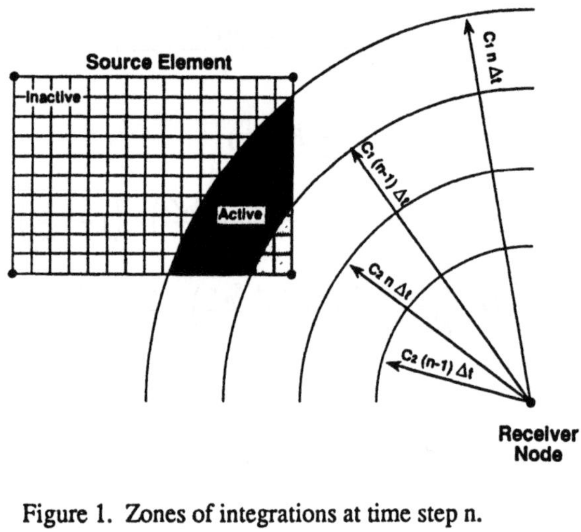

Figure 1. Zones of integrations at time step n.

limited number of control points per element no adequate information can be obtained pertaining to which portion, if any at all, of the source element is really "active" in order to perform appropriately the integration indicated by Eqs. (5). A numerically simple cure to this shortcoming has been proposed by Karabalis and Beskos [9]. According to this procedure, which has also been adopted in this work, each source element is subdivided into a number of M' source subelements and the integrations of Eqs. (5) are performed over the area (s') of the subelement m'. In this case Eqs. (5) become

$$G_{ij}^{n,a} = \sum_{m'=1}^{M'} \int_{(s')} G_{ij}^{n}[\underline{x}(\underline{h})] \, N^{a}(\underline{h}) \, J(\underline{h}) \, d(\underline{h})$$

$$T_{ij}^{n,a} = \sum_{m'=1}^{M'} \int_{(s')} T_{ij}^{n}[\underline{x}(\underline{h})] \, N^{a}(\underline{h}) \, J(\underline{h}) \, d(\underline{h}) \qquad (12)$$

In Eqs. (12) any desired number of M' control points can be placed within each source element in an effort to account for the causality property of the fundamental solutions as accurately as possible. Therefore, the possibility of only certain portions of the source element being "active" can be taken into consideration. The integrations indicated in Eqs. (12) can be carried out as usually by standard quadrature schemes. Alternatively, provided that enough subelements are used within each element, this last spatial integration indicated in Eq.(12a), and similarly for Eq.(12b), can be carried out by the most simple means using the definition of a surface integral as

$$\int_{(s')} G_{ij}^{n}[\underline{x}(\underline{h})] \, N^{a}(\underline{h}) \, J(\underline{h}) \, d(\underline{h}) = G_{ij}^{n}[\underline{h}_{m'}] \, N^{a}(\underline{h}_{m'})(s') \qquad (13)$$

where $\underline{h}_{m'}$ is the location of an assinged control point for the subelement m'. The accuracy of this simple integration scheme proves to be surprisingly good, as it is shown in the numerical results section, testifying, primarily, on behalf of the assertion made priviously that it is not the elaborate spatial integration scheme of Eq. (5) that will result to an accurate computation of the influence matrices but the appropriate incorporation into the integration scheme of the time retarded character and the causality property of the fundamental solutions. Naturally, the computational efficiency of the integration indicated by Eq. (13) is superior to any other standard integration scheme. This is, of course, a matter of serious concern as a major part of the total computational effort in a time domain BEM is spent in performing just the integrations shown in Eq. (5).

Singular Elements

The singularities of the fundamental solutions which, due to their wave propagating character, appear only at the first time step can be treated numerically as shown originally in Ref. [16] and applied in time domain problems in Ref. [1,2]. Alternatively, these singularities can be treated analytically as it is proposed by Karabalis and Beskos [9] for constant elements and is extended here to higher order elements. It is first observed that because of the traveling character of the waves emanated at the singular source nodal point q, a singularity can occur only during the first time step $(0,\Delta t)$. If the dinstances travelled by the dilatational and shear waves from the singular point are denoted by the radii R_1 and R_2, respectively, a direct application of Eq. (6) in cylindrical coordinates yields

$$
\begin{aligned}
G_{ik}^{1,m} = \frac{1}{4\pi\rho}\Bigg\{ & \frac{1}{2}\left(\frac{1}{c_2^2}-\frac{1}{c_1^2}\right)\int_0^{2\pi}\int_0^{R_2}\left(\frac{3r_{,i}r_{,k}}{r^2}-\delta_{ik}\right) dr\, d\theta \\
& + \frac{1}{2}\int_0^{2\pi}\int_0^{R_1}\frac{1}{r^2}\left[(\Delta t)^2-\left(\frac{r}{c_1}\right)^2\right]\left(\frac{3r_{,i}r_{,k}}{r^2}-\delta_{ik}\right) dr\, d\theta \\
& + \frac{1}{c_1^2}\int_0^{2\pi}\int_0^{R_1}\frac{r_{,i}r_{,k}}{r^2}\, dr\, d\theta + \frac{1}{c_2^2}\int_0^{2\pi}\int_0^{R_2}\left(\delta_{ik}-\frac{r_{,i}r_{,k}}{r^2}\right) dr\, d\theta\Bigg\},
\end{aligned}
\tag{14}
$$

Eq. (14) presents no singularities with respect to the space variables. A similar expression can be written for the traction tensor of Eq. (8). Certain simplifications for the case of relaxed boundary conditions, as well as analytic expressions for constant elements can be found in Refs. [8-11].

In most time domain BEM formulations the size of the time step is associated through the wave velocities to the dimensions of the element(s) used. This imposes certain restrictions to the accuracy and, mostly, the efficiency of the computation since a fine spatial discretization requires a very small time step, while a coarse spatial discretization a larger time step. It is rather straightforward to show that the time and space discretizations proposed in this work can be applied with far less restrictions on the required size of the time step rendering it almost completely independent from the size of the elements used in the BEM discretization. However, due to space limitations details will be given in a future article.

Numerical Examples and Discussion

A typical SSI problem is used in this section to illustrate the applicability of the proposed time domain methodology. The problem of a rigid surface foundation on a infinite elastic half-space, subject to transient external forces serves in this capacity. As only the accuracy and efficiency in computing the influence matrices are of interest in this work, only vertical forces are applied under relaxed boundary conditions in an effort to bypass certain aspects of the solution of this problem that are out of the context of this work [13,15]. The vertical response of a 5x5 square rigid, surface foundation on a half-space with: modulus of elasticity 2.58984E+09, Poisson's ratio 1/3, and mass density 10.368, in self-consistent units, is studied. The forcing function is a ractangular impulse described as

$$P_3(t) = \begin{cases} 100, & 0<t<\Delta t \\ 0, & t>\Delta t . \end{cases} \qquad (15)$$

To illustrate the issues brought forward in this work a 6x6 BEM discretization is chosen for the contact surface between the foundation and the half-space. This level of discretization has been proven to produce reliable results for the problem at hand although, finer discretizations have been suggested [9,15,17]. In Fig.(2) a study of the effect of the number of subelements upon the accuracy of the solution is shown (the number of subelements used is shown in parentheses next to the number of elements). It is apparent that the 6x6(1x1) discretization scheme produces highly oscilatory results for both the constant, Fig.(2a) and the quadratic, Fig.(2b), BEM mesh. Moreover, at this level of 6x6 element discretization, accuracy is improved at the next level of (3x3) subelement discretization, but it is the (5x5) subelement mesh that starts producing accurate results. The accuracy of the results is verified in Fig. 3. A constant suddenly applied vertical force of magnitude 100 is applied to the foundation-half-space system described previously. The results obtained via a 6x6(7x7) constant element discetization scheme reach exactly the static solution [17] of the problem at later times, as it was expected for this massless foundation problem. Surprisingly enough, it is also shown in Fig. (2) that even a 2x2(9x9) constant element discretization scheme is capable of reaching the exact solution at considerable computer time savings. The success of this last discretization is of course solely attributed to the number of subelements (81 in this case) used per element. Testistifing to this assertion is the inaccurate result obtained by a 6x6 constant element discretization scheme with only (3x3) subelement mesh per element. The simplified spatial integration scheme of Eq. (13) has been used to obtain the results shown in Fig.(3)Similar results have been obtained using linear or quadratic BEM discretization schemes. The convergence of i) a 6x6(5x5) constant element mesh using the simplified integration scheme discussed previously ii) a 6x6(5x5) constant element mesh using analytical integration for one dimension and a 4-point Gauss quadrature for the other [9], iii) a 6x6(5x5) linear element discretization with a complete 4x4-point Gauss quadrature, and iv) a 6x6(5x5) quadratic element discretization with a complete 4x4-point Gauss quadrature, is shown in Fig.(4). Finally in Fig.(5) the computer time required for producing the response of the above foundation-half-space system to the rectangular impulse force of Eq. (15) is plotted versus the number of subelements used per element. All the above computations have been performed on a IBM AT-compatible machine equiped a 16 MHz 80386 processor and a 80387 coprocessor. In view of Fig.(5), and keeping in mind the previous discussion, the substantial benefits, with regard to accuracy and computational

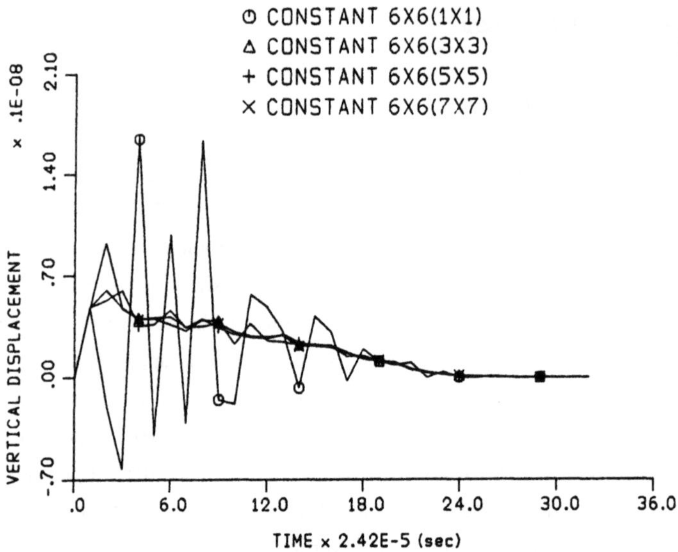

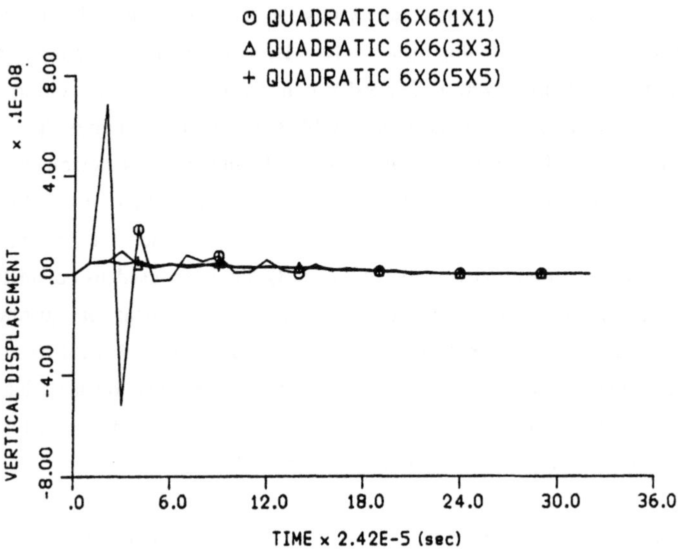

Figure 2. Effect of number of subelements upon accuracy.

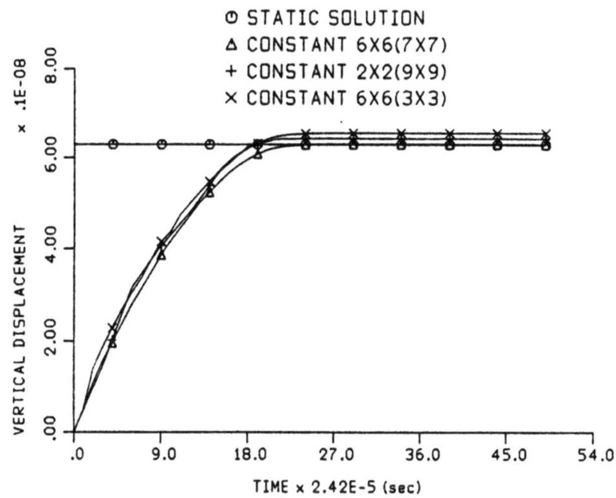

Figure 3. Accuracy versus number of elements and subelements.

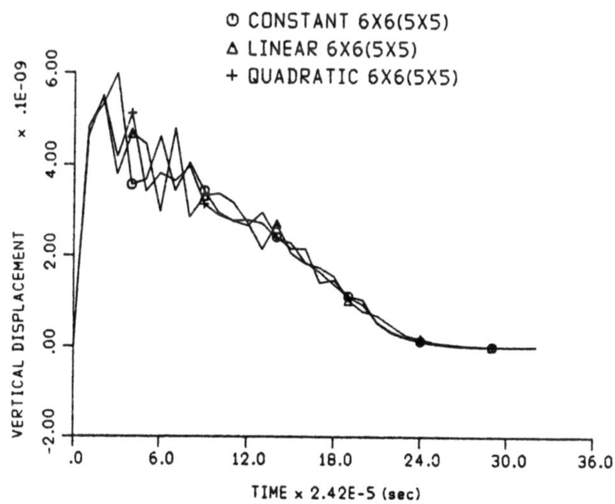

Figure 4. Convergence of different orders of boundary elements.

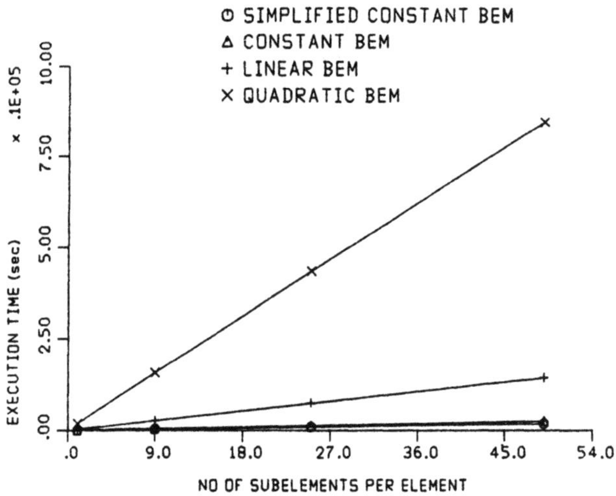

Figure 5. Computational efficiency of a number of boundary element discretizations versus number of subelements per element.

efficiency, that can result by implementing the proposed integration scheme in a time domain BEM code become apparent.

References

1. S. Ahmad and P. K. Banerjee, 'Time-Domain Transient Elastodynamic Analysis of 3-D Solids by BEM,' *Int. Journal for Numerical Methods in Engineering*, Vol. 26, 1709-1728 (1988).
2. P. K. Banerjee, S. Ahmad and G. D. Manolis, 'Advanced Elastodynamic Analysis,' Ch. 5, in: D. E. Beskos, Ed., *Boundary Element Methods in Mechanics*, North-Holland, Amsterdam, 1987.
3. D. E. Beskos, 'Boundary Element Methods in Dynamic Analysis,' *Applied Mechanics Review*, Vol. 40, No.1 (1987).
4. D. E. Beskos and G. D. Manolis, *Boundary Element Methods in Elastodynamics* (Unwin-Hyman, London, 1988).
5. A. C. Eringen and E. S. Suhubi, *Elastodynamics - Volume II, Linear Theory*, Academic Press, New York, 1975.
6. L. R. Johnson, 'Green's Function for Lamb's Problem,' *Geophysical Journal of the Royal Astronomical Society*, Vol. 37, 99-131 (1974).
7. D. L. Karabalis 'Dynamic Response of Three-Dimensional Foundations,' Ph.D Thesis, University of Minnesota, Minneapolis, Minnesota, 1984.
8. D. L. Karabalis, 'Dynamic Soil-Structure Interaction by BEM,' Proceedings of the 10th BEM International Conference, Southampton, United Kingdom, 1988.

9. D. L. Karabalis and D. E. Beskos, 'Dynamic Response of 3-D Rigid Surface Foundations by Time Domain Boundary Element Method,' *Earthquake Engineering and Structural Dynamics*, Vol. 12, 73-93 (1984).
10. D. L. Karabalis and D. E. Beskos, 'Dynamic Analysis of 3-D Rigid Embedded Foundations by Time Domain Boundary Element Method," *Computer Methods in Applied Mechanics and Engineering*, Vol. 56, 91-119 (1986).
11. D. L. Karabalis and D. E. Beskos, 'Dynamic Soil-Structure Interaction,' Ch. 11, in: D. E. Beskos, Ed., *Boundary Element Methods in Mechanics*, North-Holland, Amsterdam, 1987.
12. D. L. Karabalis and D. E. Beskos, 'Time-Domain Transient Elastodynamic Analysis of 3-D Solids by BEM,' *Int. Journal for Numerical Methods in Engineering* (in print).
13. D. L. Karabalis and C.-F. D. Huang, 'Inertial Soil-Structure Interaction by A Direct Time Domain BEM,' *Int. Journal of Computers and Mathematics* (in print).
14. J. Lachat and J. O. Watson "Effective Numerical Treatment of Boundary Integral Equation: A Formulation for Three-Dimensional Elastodynamics," *Int. Journal for Numerical Methods in Engineering*, Vol. 10, 991-1005 (1975).
15. M. Mohammadi and D. L. Karabalis, '3-D Soil-Structure Interaction Analysis by BEM: Comparison Studies and Computational Aspects,' *Soil Dynamics and Earthquake Engineering* (in print).
16. F. J. Rizzo, D. J. Shippy and M Rezayat, 'A Boundary Integral Equation Method for Radiation and Scattering of Elastic Waves in Three Dimensions, *Int. Journal for Numerical Methods in Engineering*, Vol. 21, 115-129 (1985).
17. H. L. Wong and J. E. Luco, "Dynamic Response of Rigid Foundations of Arbitrary Shape," *Earthquake Engineering and Structural Dynamics*, Vol. 4, 579-587 (1976).

Analytical Treatment of Free Term for 3-Dimensional Elastodynamic BEM

M. Kitahara
School of Marine Science and Technology
Tokai University
Shimizu, Shizuoka 424, Japan

K. Nakagawa
Total System Institute
Suginami-ku, Tokyo 166, Japan

Summary

An analytical treatment for the evaluation of the free term is shown in some detail for the three - dimensional elastodynamic boundary integral equations. For the special case of an edge or a corner with ninety degrees, there exist already the results for the analytical evaluation of the free term. For the numerical treatment of the boundary integral equations, however, it is necessary to consider the arbitrary pattern of the multiple folding of elements at a corner node. The point is to pay attention to an element which forms the corner node. In this case we can write down analytically the contribution to the free term from the element. Then we can sum up the contributions and the analytical evaluation of the free term can be obtained. The key formula used is the one for the spherical triangle.

Free Term

The boundary integral equation for the three-dimensional exterior time-harmonic elastodynamics takes the form:

$$C_{ij}(\mathbf{x})u_j(\mathbf{x}) = \int_\Gamma U_{ij}(\mathbf{x},\mathbf{y})t_j(\mathbf{y})d\Gamma_y - \int_\Gamma T_{ij}(\mathbf{x},\mathbf{y})u_j(\mathbf{y})d\Gamma_y + u_i^{(I)}(\mathbf{x}), \qquad \mathbf{x} \in \Gamma . \tag{1}$$

The coefficient of free term is defined as [1]

$$C_{ij}(\mathbf{x}) = \lim_{\varepsilon \to 0} \int_{S(\mathbf{x},\varepsilon)} T_{ij}(\mathbf{x},\mathbf{y})d\Gamma_y , \tag{2}$$

where $S(\mathbf{x},\varepsilon)$ is the part of the surface of ε-sphere (see, Fig.1), which is in the side of the exterior domain Ω. The center of the ε-sphere is the boundary point $\mathbf{x}$. In eq.(1), fundamental solution U_{ij} and T_{ij} can be separated as

$$U_{ij} = \dot{U}_{ij} + \overline{U}_{ij} \ , \qquad T_{ij} = \dot{T}_{ij} + \overline{T}_{ij} \ , \tag{3a, b}$$

where $\dot{U}_{ij}$ and $\dot{T}_{ij}$ are fundamental solutions in statics, $\overline{U}_{ij}$ and $\overline{T}_{ij}$ are regular terms [2]. Equation (2) reduces

$$C_{ij}(\mathbf{x}) = \lim_{\varepsilon \to 0} \int_{S(\mathbf{x},\varepsilon)} \dot{T}_{ij}(\mathbf{x},\mathbf{y})d\Gamma_y \tag{4}$$

from eq.(3b), where $\dot{T}_{ij}$ has the expression

$$\dot{T}_{ij} = \frac{1}{8\pi(1-\nu)r^2}\left[\{(1-2\nu)\delta_{ij} + 3r_{,i}r_{,j}\}\, r_{,k}n_k - (1-2\nu)(r_{,i}n_j - r_{,j}n_i)\right] . \tag{5}$$

In this eq.(5), the comma notation $(\cdot)$, means the differentiation with respect to the field point $\mathbf{x}$.

Free Term in Local Spherical Coordinates

Let the field point $\mathbf{x}$ be the corner point $\mathbf{x}$ as shown in Fig.2. The unit normal vector $\mathbf{n}$ points from $\mathbf{y}$ to $\mathbf{x}$. Now we introduce the local coordinates (x_1, x_2, x_3) and spherical coordinates (r, θ, ϕ) as

$$y_i - x_i = \begin{cases} r cos\phi sin\theta & i = 1 \\ r sin\phi sin\theta & i = 2 \\ r cos\theta & i = 3 \,. \end{cases} \tag{6}$$

The following relations hold

$$\begin{aligned} &(a)\quad r_{,i} = \partial r/\partial x_i = (x_i - y_i)/r = \begin{cases} -cos\phi sin\theta & i = 1 \\ -sin\phi sin\theta & i = 2 \\ -cos\theta & i = 3 \end{cases} \\ &(b)\quad n_i = (x_i - y_i)/r = r_{,i} \\ &(c)\quad r_{,k} n_k = n_k n_k = 1 \\ &(d)\quad r_{,i} n_j - r_{,j} n_i = n_i n_j - n_j n_i = 0 \,. \end{aligned} \tag{7a, b, c, d}$$

From eq.(5) and (7), eq.(4) becomes

$$C_{ij}(\mathbf{x}) = \frac{1}{8\pi(1-\nu)} \int_{S(\mathbf{x},1)} \{(1-2\nu)\delta_{ij} + 3 n_i n_j\} \, sin\theta \, d\theta \, d\phi \tag{8}$$

where $S(\mathbf{x}, 1)$ is the part of the surface of unit sphere. We set

$$\begin{aligned} \alpha &= \int_{S(\mathbf{x},1)} sin\theta \, d\theta \, d\phi = \int_{S(\mathbf{x},1)} d\Gamma_y \\ &= area\ of\ the\ part\ of\ unit\ sphere : S(\mathbf{x}, 1) \end{aligned} \tag{9}$$

$$\begin{aligned} \beta_{ij} &= \int_{S(\mathbf{x},1)} n_i n_j \, sin\theta \, d\theta \, d\phi \\ &= \int_{S(\mathbf{x},1)} n_i n_j \, d\Gamma_y . \end{aligned} \tag{10}$$

Then the C_{ij} in eq.(8) can be written

$$C_{ij}(\mathbf{x}) = \frac{1}{8\pi(1-\nu)} \{(1-2\nu)\alpha\delta_{ij} + 3\beta_{ij}\} \,. \tag{11}$$

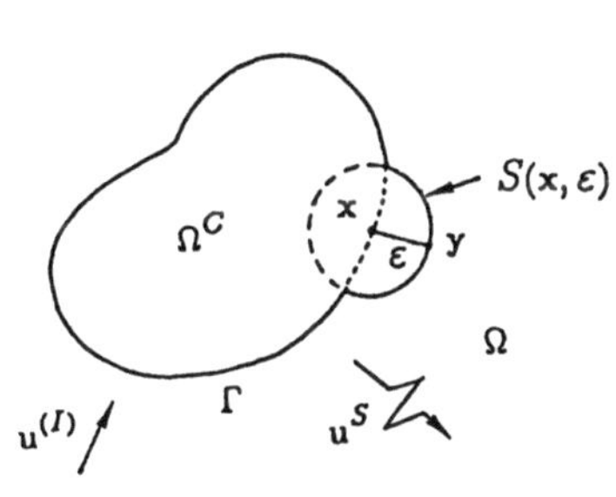

Fig.1 Part of the surface of ε-sphere $S(\mathbf{x}, \varepsilon)$

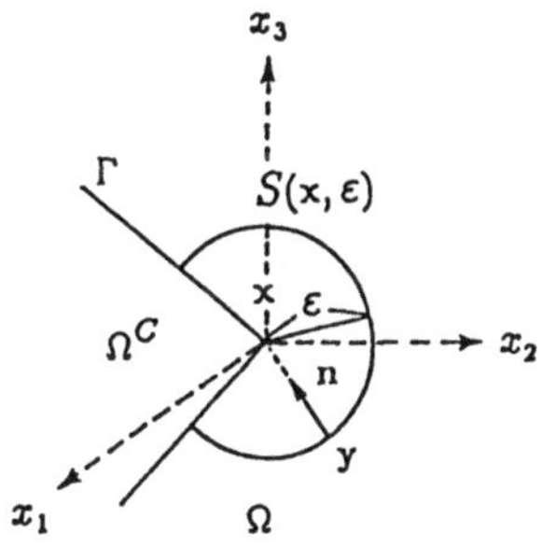

Fig.2 Corner point $\mathbf{x}$ and local coordinates

The α in eq.(9) is the coefficient of free term for the potential problem. In eq.(10), the normal vector $\mathbf{n}$ appears as the tensor product $\mathbf{nn}$. In this sense, the direction of $\mathbf{n}$ does not change the value of β_{ij}. In what follows we fix the normal $\mathbf{n}$ to be opposite direction, for convenience to evaluate β_{ij}. That is, the $\mathbf{n}$ points to the side of exterior domain Ω.

The Case of Spherical Rectangle

When the part of surface of unit sphere $S(\mathbf{x},1)$ is the special shape of spherical rectangle, the coefficient of free term $C_{ij}(\mathbf{x})$ in eq.(8) reduces a simple form. Let $S(\mathbf{x},1)$ be the spherical rectangle which was formed by two longitudes (of ϕ_1 and ϕ_2) and two latitudes (of θ_1 and θ_2), as shown in Fig.3. The $S(\mathbf{x},1)$ is the part of unit sphere in $\phi_1 \leq \phi \leq \phi_2, \theta_1 \leq \theta \leq \theta_2$. In this case, the integrations in eqs.(9) and (10) can be easily done and have the results

$$\alpha = (\phi_2 - \phi_1)(cos\theta_1 - cos\theta_2) \tag{12}$$

and

$$\begin{aligned}
\beta_{11} &= \frac{1}{8}(\phi_2 - \phi_1 + \frac{1}{2}sin2\phi_2 - \frac{1}{2}sin2\phi_1)f(\theta_1,\theta_2) \\
\beta_{12} &= \frac{1}{8}(-cos2\phi_2 + cos2\phi_1)f(\theta_1,\theta_2) \\
\beta_{13} &= \frac{1}{4}(sin\phi_2 - sin\phi_1)g(\theta_1,\theta_2) \\
\beta_{22} &= \frac{1}{8}(\phi_2 - \phi_1 - \frac{1}{2}sin2\phi_2 + \frac{1}{2}sin2\phi_1)f(\theta_1,\theta_2) \\
\beta_{23} &= \frac{1}{4}(-cos\phi_2 + cos\phi_1)g(\theta_1,\theta_2) \\
\beta_{33} &= \frac{1}{4}(\phi_2 - \phi_1)h(\theta_1,\theta_2)\ ,
\end{aligned} \tag{13a $\sim$ f}$$

where $f(\theta_1,\theta_2), g(\theta_1,\theta_2)$, and $h(\theta_1,\theta_2)$ are defined as

$$\begin{aligned}
f(\theta_1,\theta_2) &= -3cos\theta_2 + 3cos\theta_1 + \frac{1}{3}cos3\theta_2 - \frac{1}{3}cos3\theta_1 \\
g(\theta_1,\theta_2) &= sin\theta_2 - sin\theta_1 - \frac{1}{3}sin3\theta_2 + \frac{1}{3}sin3\theta_1 \\
h(\theta_1,\theta_2) &= -cos\theta_2 + cos\theta_1 - \frac{1}{3}cos3\theta_2 + \frac{1}{3}cos3\theta_1\ .
\end{aligned} \tag{14a, b, c}$$

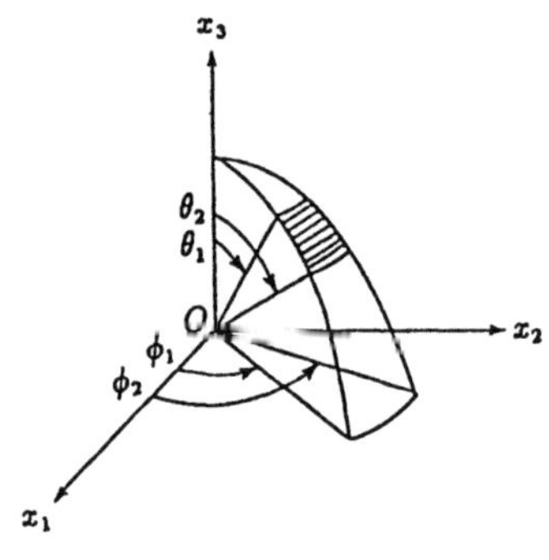

Fig.3 Spherical rectangle formed by longitudes and latitudes

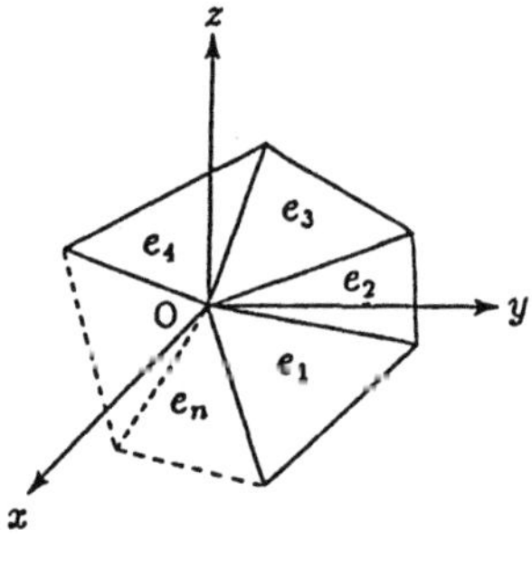

Fig.4 A corner point O with n elements

From the results in eqs.(12)~(14), the explicit expression of C_{ij} is obtained by eq.(11). A special case of this result is the case of the half sphere of $S(\mathbf{x}, 1)$. By setting $\theta_1 = 0$, $\theta_2 = \pi/2$, and $\phi_1 = 0$, $\phi_2 = 2\pi$ in eqs.(12)~(14), we have

$$\alpha = 2\pi, \quad \beta_{ij} = \frac{1}{3}(2\pi)\delta_{ij} \tag{15}$$

and well known result

$$C_{ij} = \frac{1}{2}\delta_{ij} \tag{16}$$

from eq.(11). The results for the 90° edge and for the 90° corner can also be reduced from the expressions of eqs.(11)~(14).These results of the 90° edge and the 90° corner are found in References [3] and [4].

The General Case: Spherical Triangle

Let a corner be made of n pieces of triangular plane panels as shown in Fig.4. The corner is denoted by capital O and the local coordinates (x, y, z) with the origin O is introduced. Now we pay attention to one element of n panels and calculate the contribution to α and β_{ij} in eqs.(9)~(11) from the element.

Figure 5 explains the situation. The part of one element which is contained in the unit sphere is denoted by OBC. In eqs.(9) and (10), the integrations are carried out on the spherical triangle ΔABC on the unit sphere, and they are expressed in the form

$$I = \int_{\Delta ABC} J(\theta, \phi) d\theta\, d\phi \ . \tag{17}$$

The unit vectors which point to B and C from the origin are denoted by $\hat{\mathbf{r}}_b$ and $\hat{\mathbf{r}}_c$, respectively. These vectors $\hat{\mathbf{r}}_b = (\hat{x}_b, \hat{y}_b, \hat{z}_b)$ and $\hat{\mathbf{r}}_c = (\hat{x}_c, \hat{y}_c, \hat{z}_c)$ are expressed as

$$\begin{cases} \hat{x}_b = cos\phi_1 sin\theta_1 \\ \hat{y}_b = sin\phi_1 sin\theta_1 \\ \hat{z}_b = cos\theta_1 \end{cases}, \qquad \begin{cases} \hat{x}_c = cos\phi_2 sin\theta_2 \\ \hat{y}_c = sin\phi_2 sin\theta_2 \\ \hat{z}_c = cos\theta_2 \end{cases} . \tag{18a, b}$$

From eq.(18), it follows that

$$\begin{cases} \phi_1 = tan^{-1}(\hat{y}_b/\hat{x}_b) \\ \theta_1 = sin^{-1}(\hat{x}_b^2 + \hat{y}_b^2)^{1/2} \end{cases}, \qquad \begin{cases} \phi_2 = tan^{-1}(\hat{y}_c/\hat{x}_c) \\ \theta_2 = sin^{-1}(\hat{x}_c^2 + \hat{y}_c^2)^{1/2} \end{cases}, \tag{19a, b}$$

where ϕ_i and θ_i are the azimuthal and the polar angles, and the points B and C can be expressed as

$$\hat{\mathbf{r}}_b = (\phi_1, \theta_1) \ , \quad \hat{\mathbf{r}}_c = (\phi_2, \theta_2) \ . \tag{20a, b}$$

Evaluation of α in eq.(9)

A portion of a spherical surface bounded by three arcs of great circles is called a spherical triangle[5]. The ΔABC in Fig.6 (a) is the spherical triangle. We define here the spherical moon as a portion of a spherical surface bounded by two arcs of great circles. The spherical moon is denoted by $[ABA_1, ACA_1]$ as shown in Fig.6 (b). The value of α in eq.(9) is the area of the spherical triangle ΔABC, it can be evaluated[5] as

$$\alpha = \int_{\Delta ABC} d\Gamma_y = \phi_A + \phi_B + \phi_C - \pi \ , \tag{21}$$

where ϕ_A, ϕ_B, and ϕ_C are spherical angles [5] (see, Fig.5).

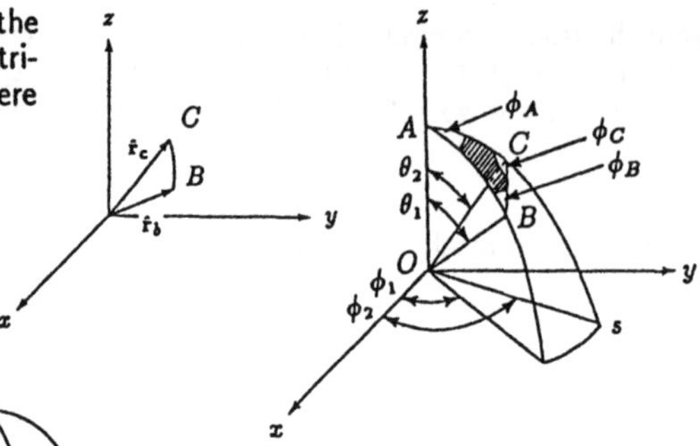

Fig.5 Part of an element OBC in the unit sphere and the spherical triangle ΔABC on the unit sphere

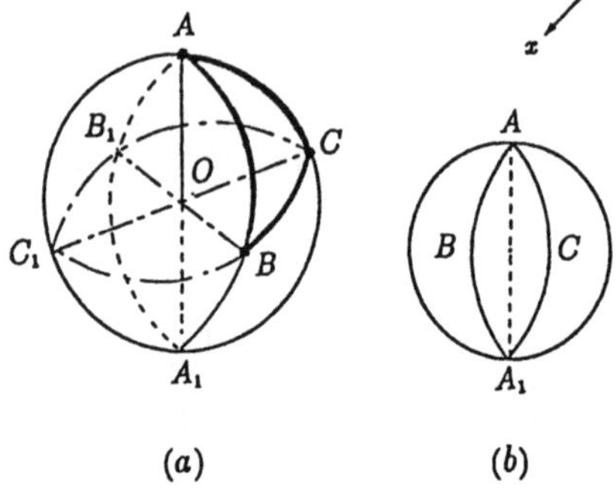

Fig.6 (a) Spherical triangle ΔABC, and (b) Spherical moon $[ABA_1, ACA_1]$

Evaluation of β_{ij} in eq.(10)

For the sake of convenience, the integral of β_{ij} in eq.(10) on the spherical triangle ΔABC is written as

$$(ABC) = \int_{\Delta ABC} n_i n_j d\Gamma_y \tag{22}$$

and the integral on the spherical moon $[ABA_1, ACA_1]$ as

$$(ABA_1, ACA_1) = \int_{[ABA_1, ACA_1]} n_i n_j d\Gamma_y \tag{23}$$

in what follows. The tensor indices i and j are abbreviated in (ABC) and (ABA_1, ACA_1). First we consider the following quantity:

$$(ABA_1, ACA_1) + (BCB_1, BAB_1) + (CAC_1, CBC_1)\ . \tag{24}$$

Then the following relations hold true (see, Fig.6)

$$(ABA_1, ACA_1) = (ABC) + (BCA_1) = (ABC) + (B_1C_1A) \tag{25a}$$
$$(BCB_1, BAB_1) = (BCA) + (ACB_1) \tag{25b}$$
$$(CAC_1, CBC_1) = (CAB) + (AC_1B)\ , \tag{25c}$$

where the last equality in eq.(25a) comes from the relation

$$(BCA_1) = (B_1C_1A) \tag{26}$$

which follows from the tensor product property of the integrand in eq.(10). Substitution of eq.(25) into eq.(24) gives

$$\begin{aligned}&(ABA_1, ACA_1) + (BCB_1, BAB_1) + (CAC_1, CBC_1)\\ &= 2(ABC) + \{(ABC) + (ACB_1) + (AB_1C_1) + (AC_1B)\}\ .\end{aligned} \tag{27}$$

From eq.(27), the important expression on β_{ij} is obtained

$$(ABC) = \frac{1}{2}[(ABA_1, ACA_1) + (BCB_1, BAB_1) + (CAC_1, CBC_1) \\ - \{(ABC) + (ACB_1) + (AB_1C_1) + (AC_1B)\}] \;. \tag{28}$$

The last term of eq.(28) is the integration on the hemisphere and has the simple result

$$(ABC) + (ACB_1) + (AB_1C_1) + (AC_1B) = \frac{1}{3}(2\pi)\delta_{ij} \tag{29}$$

from eq.(15). Three terms (ABA_1, ACA_1),(BCB_1, BAB_1), and (CAC_1, CBC_1) on the right hand side of eq.(28) can be evaluated by using the result of eq.(13).

(a) The term (ABA_1, ACA_1)

We choose the local coordinates that the z axis to be $\overline{OA}$ direction. Then, setting $\theta_1 = 0$ and $\theta_2 = \pi$ in eq.(13), we have

$$\begin{aligned} \beta_{11} &= \frac{2}{3}(\phi_2 - \phi_1 + \frac{1}{2}sin2\phi_2 - \frac{1}{2}sin2\phi_1) \\ \beta_{12} &= \frac{2}{3}(-cos2\phi_2 + cos2\phi_1), \quad \beta_{13} = 0 \\ \beta_{22} &= \frac{2}{3}(\phi_2 - \phi_1 - \frac{1}{2}sin2\phi_2 + \frac{1}{2}sin2\phi_1) \\ \beta_{23} &= 0, \quad \beta_{33} = \frac{2}{3}(\phi_2 - \phi_1) \;, \end{aligned} \tag{30}$$

where ϕ_1 and ϕ_2 have been given in eq.(19).

(b) The term (BCB_1, BAB_1)

The local coordinates $(\overline{x}, \overline{y}, \overline{z})$ are now chosen to be (see, Fig.7):

$$\overline{\mathbf{e}}_z = \hat{\mathbf{r}}_b \;, \qquad \overline{\mathbf{e}}_y = (\hat{\mathbf{r}}_b \times \hat{\mathbf{r}}_c)/|\hat{\mathbf{r}}_b \times \hat{\mathbf{r}}_c| \;, \qquad \overline{\mathbf{e}}_x = \overline{\mathbf{e}}_y \times \overline{\mathbf{e}}_z \;. \tag{31}$$

The transformation matrix $\overline{\mathbf{A}}$ is defined as

$$\mathbf{e}_i = \overline{A}_{ij}\overline{\mathbf{e}}_j \,, \tag{32}$$

where $\mathbf{e}_i$ are base vectors in the (x, y, z) coordinates. The spherical angle ϕ_B at the pole B is to be determined, and it has the form

$$\phi_B = cos^{-1}(\overline{\mathbf{t}} \cdot \overline{\mathbf{e}}_x) \;, \tag{33a}$$

where

$$\overline{\mathbf{t}} = -cos\theta_1 cos\phi_1 \mathbf{e}_x - cos\theta_1 sin\phi_1 \mathbf{e}_y + sin\theta_1 \mathbf{e}_z \tag{33b}$$

is the tangent at B along the great circle BA. Now we can use the results of eq.(13) and have

$$\begin{aligned} \overline{\beta}_{11} &= \frac{2}{3}(\phi_B + \frac{1}{2}sin2\phi_B) \\ \overline{\beta}_{12} &= \frac{2}{3}(-cos2\phi_B + 1), \quad \overline{\beta}_{13} = 0 \\ \overline{\beta}_{22} &= \frac{2}{3}(\phi_B - \frac{1}{2}sin2\phi_B) \\ \overline{\beta}_{23} &= 0, \quad \overline{\beta}_{33} = \frac{2}{3}(\phi_B) \;. \end{aligned} \tag{34}$$

The expression in the local coordinate (x, y, z) is obtained from

$$\beta_{ij} = \overline{A}_{ik}\overline{A}_{jl}\overline{\beta}_{kl} \;. \tag{35}$$

(c) The term (CAC_1, CBC_1)

The local coordinates $(\bar{\bar{x}}, \bar{\bar{y}}, \bar{\bar{z}})$ are chosen as (see, Fig.8)

$$\begin{aligned} \bar{\bar{\mathbf{e}}}_z &= \hat{\mathbf{r}}_c \\ \bar{\bar{\mathbf{e}}}_y &= (\hat{\mathbf{r}}_b \times \hat{\mathbf{r}}_c)/|\hat{\mathbf{r}}_b \times \hat{\mathbf{r}}_c| \\ \bar{\bar{\mathbf{e}}}_x &= \bar{\bar{\mathbf{e}}}_z \times \bar{\bar{\mathbf{e}}}_y \ . \end{aligned} \tag{36}$$

From this expression, the transformation matrix $\bar{\bar{\mathbf{A}}}$ is determined. Applying the results of eq.(13), we have the exactly same form of results as eq.(34), except for the new polar angle ϕ_C. The polar angle ϕ_C is determined as

$$\phi_C = cos^{-1}(\bar{\bar{\mathbf{t}}} \cdot \bar{\bar{\mathbf{e}}}_x) \ , \tag{37a}$$

where

$$\bar{\bar{\mathbf{t}}} = -cos\theta_2 cos\phi_2 \mathbf{e}_x - cos\theta_2 sin\phi_2 \mathbf{e}_y + sin\theta_2 \mathbf{e}_z \ . \tag{37b}$$

We complete the calculation by transforming the result to the coordinates (x, y, z).

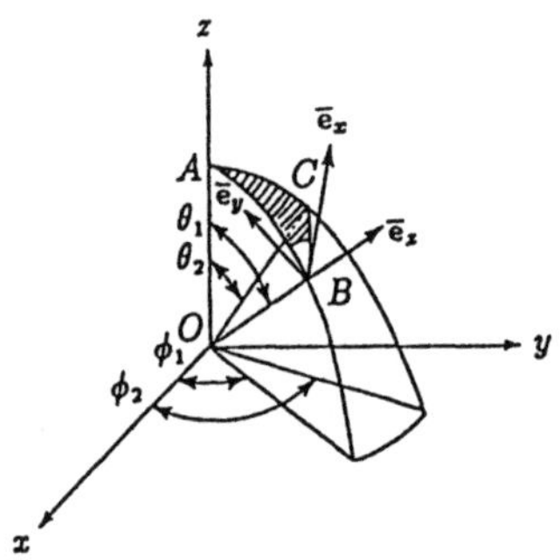

Fig.7 Local coordinates $(\bar{x}, \bar{y}, \bar{z})$

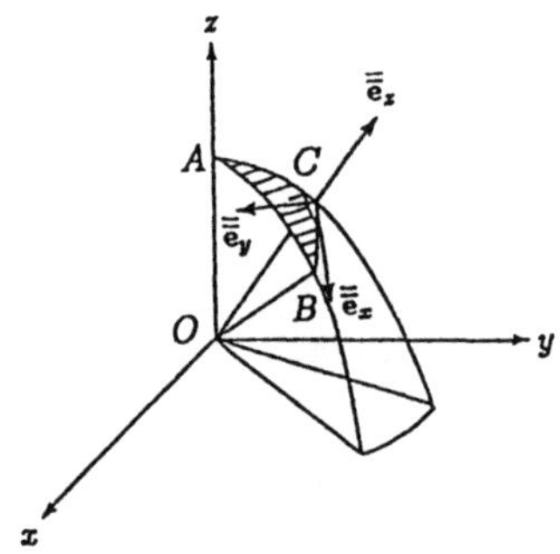

Fig.8 Local coordinates $(\bar{\bar{x}}, \bar{\bar{y}}, \bar{\bar{z}})$

Final comment

It is necessary to sum up the results for every elements in order to get the expression of C_{ij} in the local coordinate (x, y, z) whose origin is a corner point. Final form of C_{ij} is obtained from the coordinate transformation to the global coordinate.

Reference

1. Cruse, T.A.; An improved boundary-integral equation method for three dimensional elastic stress analysis, Computers & Structures, Vol.4, pp.741-754, 1974.
2. Kitahara, M., Nakagawa, K. and Achenbach, J.D.; Boundary-integral equation method for elastodynamic scattering by a compact inhomogeneity, Computational Mechanics (In press).
3. Hartmann, F.; Computing the C-matrix in non-smooth boundary points, New Developments in Boundary Element Methods, Ed. C.A. Brebbia, pp.367-379, Butterworths, 1980.
4. Hartmann, F.; Elastostatics, Progress in Boundary Element Methods, Vol.1, Ch.4, Ed. C.A. Brebbia, pp.84-167, Pentech Press 1981.
5. Gay, H.J.; Computation and Trigonometry, Ch.13-15, pp.188-228, The Macmillan Company, 1938.

Analysis of 3-D Near-interface Fractures in Bounded, Heterogeneous Domains Using the Surface Integral and Finite Element Hybrid Method

William D. Keat and Michael P. Cleary
Massachusetts Institute of Technology
Department of Mechanical Engineering
Cambridge, Massachusetts, U.S.A.

SUMMARY

Efficient numerical strategies for studying 3-D fractures in piecewise homogeneous domains have been developed based on the surface integral and finite element hybrid method. The basic procedure combines a surface integral model of the fracture in an infinite region with a finite element model of the uncracked bounded domain, leading to a coupled matrix formulation. Fractures approaching and/or intersecting planar bimaterial interfaces pose a special challenge because of the dual singularities in stress at the crack tip and at the material interfaces. The hybrid formulation was specialized for this class of problems by introducing the force dipole solutions for an infinite bimaterial region. Demonstration results which test the accuracy and generality of these procedures are presented.

INTRODUCTION

The precise characterization of near interface crack tip fields is of fundamental importance when assessing the failure strength of composite materials or when predicting the growth behavior of fractures in geotechnical environments. It represents a formidable numerical challenge owing to the stress singularities along the crack front and at the material interface, and thus has eluded practical analysis via nonspecialized techniques such as finite elements.

The objective of this research was to develop a numerical strategy that effectively models planar, arbitrarily-shaped fractures in piecewise homogeneous, bounded domains. Of special interest are geometries in which the fracture approaches and/or intersects multiple planar bimaterial interfaces. Only mode 1 type loadings are considered.

The Surface Integral and Finite Element Hybrid Method uses the principle of superposition to combine a surface integral model of a fracture in an infinite medium with a finite element model of the bounded domain without the crack. The effect of this strategy is to bring to the forefront the unique advantages of the component methods, while diminishing the impact of their known limitations. The hybrid method has been applied with success to both two- and three-dimensional problems in fracture mechanics [1,2,3,4], including those with material nonlinearities [1]. This paper draws on the influence functions for a general fracture event near a planar bimaterial interface [5], the concept of substructuring [6], and the hybrid formulation to develop a general strategy for modelling fractures in bounded domains with material inhomogeneities.

SURFACE INTEGRAL FORMULATION

A general fracture event of infinitesimal extent can be represented mathematically as a force multipole. In the case of mode 1 type crack opening, the force multipole consists of an orthogonal triad of tensile force dipoles (see e.g. [7]); a primary force dipole oriented perpendicular to the fracture surface simulates the crack opening while laterally acting dipoles serve to negate the superfluous Poisson contractions induced by the primary dipole.

The 3-D surface integral method is essentially an indirect boundary element formulation which represents the deformed crack geometry as a piecewise continuous distribution of force multipoles. The accumulated effect of one such multipole distribution on a scalar function (t) at a field point ($\vec{x}$) can be expressed by the following integral equation, written here for the special case of a fracture intersecting the planar interface of an infinite bimaterial region:

$$t(\vec{x}) = \iint_{S_1} \bar{E}_1\, \gamma_1\, \delta(\vec{\varsigma})\, dA + \iint_{S_2} \bar{E}_2\, \gamma_2\, \delta(\vec{\varsigma})\, dA \tag{1}$$

where $\delta(\vec{\varsigma})$ is crack opening; $\bar{E}_i$ are material constants relating the multipole strengths to the actual crack opening; γ_i represent the fundamental solutions (also referred to as influence functions) for a multipole in an infinite bimaterial region; S_1 and S_2 are the fracture surfaces lying to either side of the planar interface. The influence functions used in this study were derived in separate work [5,3].

When $\vec{x}$ lies on the fracture surface, as will occur during a boundary collocation, the resulting integral equation is singular at $\vec{x}$. The singular integration can be made tractable by first resolving the the influence functions into singular and nonsingular components, as expressed by $\gamma_i = \gamma_i^s + \gamma_i^{ns}$, and then subtracting a singular function from the remaining singular integrand to form a new integrand that is bounded at $\vec{x}$ and therefore integrable; the subtracted function in this case amounts to a rigid body translation. When $\vec{x}$ falls within S_1, this leads to:

$$t(\vec{x}) = \iint_{S_T} \bar{E}_1\, \gamma_1^s\, [\delta(\vec{\varsigma}) - \delta(\vec{x})]\, dA + \iint_{S_1} \bar{E}_1\, \gamma_1^{ns}\, \delta(\vec{\varsigma})\, dA + \iint_{S_2} \bar{E}_2\, \gamma_2\, \delta(\vec{\varsigma})\, dA \tag{2}$$

where the singular integration now spans the entire crack plane S_T, of which S_i is a subset. The first of these integrals can be evaluated analytically; for the remaining two nonsingular integrals, it suffices to use standard Gauss quadrature. The analogous expression for a collocation point in S_2 is found by transposing the material subscripts and replacing the influence functions.

HYBRID FORMULATION

A solution to the proposed problem can be achieved by superimposing the four models shown in Figure 1. Model I is an uncracked finite element model of the bounded domain. Models II and III are surface integral models which separately model the two bimaterial interfaces and the two respective halves of the fracture which lie closest to them. As both surface integral domains extend to infinity we must truncate the stress and displacement fields, in particular at the other's interface, so that in the resulting domains, their material properties will correspond to those of the actual geometry. These finite domains are imagined to be held in equilibrium by the tractions R^{c2} and R^{c3}, which are computed assuming that the fracture lies in an infinite region. The use of the bimaterial influence functions with these models avoids the near-singular R^c distributions that would be induced if instead we were to employ the influence functions for an infinite homogeneous region. Model IV is an uncracked finite element model of the homogeneous subregion. As we shall see, this model is necessary to ensure displacement continuity across the interfaces. A fifth model, not shown, will also be introduced in order to satisfy the displacement boundary conditions. Since the component models of Figure 1 individually satisfy equilibrium and strain compatibility within the limits of their respective formulations, all that remains to define the solution is to enforce both the prescribed boundary conditions and displacement/traction continuity at the bimaterial interfaces.

The traction boundary conditions on the external boundary, as well as traction continuity across the interfaces, are enforced by applying corrective tractions in the form of statically equivalent nodal forces to the boundary and interface nodes of model I. These corrective tractions are taken to be the negative of the external tractions holding models II through V in equilibrium. Thus the finite element equations for model I will have the standard form but with four additional load vectors on the right-hand side:

$$[\bar{K}]\{U^{FE}\} = \{R\} - \{R^{c2}\} - \{R^{c3}\} - \{R^{int}\} - \{R^{con}\} \tag{3}$$

where $[\bar{K}]$ and $[K]$, used below, represent the reduced and unreduced forms, respectively, of the finite element stiffness matrix; $\{U^{FE}\}$ is the vector of nodal displacements associated with model I; $\{R\}$ is the vector of prescribed nodal forces. With regard to the correction load vectors, $\{R^{c2}\}$ and $\{R^{c3}\}$ are computed using the surface integral equations and thus may be reexpressed in terms of crack opening displacements; $\{R^{int}\}$ and $\{R^{con}\}$ refer to nodal support reactions induced by the imposition of nodal displacements at the interface nodes of model IV and at the displacement-constrained nodes of model V, respectively. We therefore have the following expressions for the correction load vectors:

$$\{R^{c2}\}+\{R^{c3}\}=[G]\{\delta\};\quad \{R^{int}\}=[K^{sub}]\{U^{int}\};\quad \{R^{con}\}=[K]\{U^{con}\} \tag{4}$$

where $[K^{sub}]$ is the finite element stiffness matrix for the subregion. One may observe that our fifth model is in fact, like model I, an uncracked finite element model of the entire bounded domain but differs in the respect that it has not been reduced to fit the prescribed displacement boundary conditions.

The displacement constraints, and displacement continuity across the interfaces, being already assured in model I, are enforced by requiring the interface nodes of model IV and the displacement-constrained nodes of model V to displace as the negative of the surface integral results at those locations, the latter being obtained by summing the results of models II and III:

$$\{U^{int}\} = -\left[\,L^{int}\,\right]\{\delta\}; \quad \{U^{con}\} = -\left[\,L^{con}\,\right]\{\delta\} \tag{5}$$

Imposing these displacements on the finite element models effectively cancels the discontinuity in the displacement field introduced in models II and III when we truncated the surface integral domains.

The enforcement of traction boundary conditions at the crack surface is based on the surface integral equations of models II and III. Since the fracture has not been explicitely accounted for in any of the finite element models (i.e. I, IV or V), these models will produce nonzero tractions ($\{T^{c1}\}$, $\{T^{c4}\}$, and $\{T^{c5}\}$) at locations coincident with the surface of the fracure. To cancel these extraneous tractions, they are reversed in sign and applied as additional boundary conditions in the surface integral formulation:

$$\left[\,C\,\right]\{\delta\} = \{T\} - \{T^{c1}\} - \{T^{c4}\} - \{T^{c5}\} \tag{6}$$

The surface integral influence coefficients are obtained by discretizing, then integrating, equation (2) at every designated collocation point on the crack surface. The results for both surface integral models are then assembled in the coefficient matrix $[\,C\,]$. The evaluation of $\{T^{c1}\}$ is based on the equations of those finite elements (in model I) which would contain the fracture if it was to be explicitely modelled. Since the mesh topologies of finite element models I, IV and V are identical in the volume spanned by the subregion, it follows that:

$$\{T^{c1}\}=\left[\,S\,\right]\{U^{FE}\}; \quad \{T^{c4}\}=\left[\,S\,\right]\{U^{int}\}; \quad \{T^{c5}\}=\left[\,S\,\right]\{U^{con}\} \tag{7}$$

where the same $[\,S\,]$ applies to $\{T^{c1}\}$, $\{T^{c4}\}$, and $\{T^{c5}\}$. One may note that $\{T^{c2}\}$ and $\{T^{c3}\}$ of Figure 1 do not appear in equation (6) as these cross-coupling terms have been absorbed in $[\,C\,]$.

The final step in this derivation is to rewrite the above expressions as a system of equations in terms of the designated primary variables U^{FE} and δ. Two coupled systems of equations result when we substitute equations (4) and (5) into (3), and equations (5) and (7) into (6). When written in partitioned matrix form, they are expressed by:

$$\begin{bmatrix} K & G - K^{sub}L^{int} - KL^{con} \\ S & C - S(L^{int} + L^{con}) \end{bmatrix} \begin{Bmatrix} U^{FE} \\ \delta \end{Bmatrix} = \begin{Bmatrix} R \\ T \end{Bmatrix} \tag{8}$$

RESULTS

The surface of each fracture was discretized into an assemblage of linear elements (in the interior of the fracture) and tip elements (along the crack front). The latter were constructed by assuming a $\sqrt{r}$ variation of crack opening, with r being the perpendicular distance from the crack front. This form is consistent with the plane strain asymptotic solution for a homogeneous domain. In contrast, standard 8-noded brick elements were used to build the finite element models.

There being a shortage of 3-D solutions in the published literature, we have had to resort to plane strain solutions to calibrate the accuracy of this hybrid formulation. Plane strain conditions were simulated in the 3-D models by employing rectangular crack geometries possessing aspect ratios of no less than ten-to-one.

(i.) *Fractures approaching a single bimaterial interface.* Here it is assumed that the fracture lies entirely to one side of a planar bimaterial interface (see Figure 2). In the absence of external boundaries, the surface integral formulation can be used alone to obtain a solution. The long and short sides of the rectangular crack geometry were divided into ten and eight equally-spaced elements, respectively. Though coarse by most standards, this mesh yielded results (see Table 1) which compared very favorably with Cook and Erdogan's plane strain solution [9].

(ii.) *Fractures intersecting a single bimaterial interface.* The geometry shown in Figure 3 is numerically more challenging than the previous problem because, in addition to the stress singularity along the crack front, we must now also contend with the stress singularity at the bimaterial interface. For the most part, the latter singularity is taken care of in the natural course of evaluating equation (2), by choosing interpolation functions which properly account for the discontinuity in the first derivative of crack opening. The coarse fracture discretization used with Table 1 was retained here to produce the results appearing in Table 2. A comparison of the 3-D results with the corresponding plane strain solution due to Erdogan [10] yielded consistently good correlation. One may note that the error distribution in Table 2 is slightly asymmetric with respect to the value of c/a; this feature arises because of the asymmetry of the respective crack opening profiles.

The stress intensity factors for the crack fronts terminating at a material interface have not been recorded. In these cases, the stress singularity at the crack tip deviates from the expected $r^{-1/2}$ variation, which we associate with the homogeneous solution. Because of this mismatch between the actual singularity and our definition of the stress intensity factor, the stress intensity factor will tend towards ∞ or 0 depending on whether the crack tip at the interface lies embedded within the stiffer or softer material, respectively (see e.g. [11]).

(iii.) *Fractures approaching two bimaterial interfaces.* The geometry depicted in Figure 4 was analyzed using the hybrid strategy represented by Figure 1. The hybrid model consisted of the 1/8 symmetric finite element mesh shown in Figure 5 and a rectangular crack mesh with coarseness comparable to those described above. The use of symmetry might suggest that we were able to do without one of the two surface models appearing in Figure 1. But this is not the case, as both models are required in order to produce the correct boundary conditions on the planes of symmetry.

The high degree of correlation (see Table 3) between the hybrid results and Hilton and Sih's [12] plane strain solution is surprising in view of the coarseness of the finite element mesh. This is an indication that the R^c being applied to the finite mesh at the interface is smoothly varying, as was intended when this strategy was first formulated.

The rectangular mesh used in the previous example was replaced by an elliptical-shaped fracture with a two-to-one aspect ratio as an example of a realistic 3-D geometry (see Figure 6). The stress intensity factors were recorded for the elements lying nearest the material interfaces. The results shown in Figure 7 exhibited the anticipated behavior, i.e. the value of K_I approached the homogenous solution for small values of a/h, while peaking at the interface.

CONCLUSIONS

The applicability of the solution procedures presented in this paper extends considerably beyond the calibration models examined thus far. For example, the bimaterial interfaces are not required to be parallel, nor is their number limited to two. In fact, it should be possible to model fractures intersecting any finite number of parallel interfaces by introducing an additional finite element substructure for each intersected layer.

ACKNOWLEDGEMENTS

Although the majority of the 3-D hybrid capabilities were developed at the Massachusetts Institute of Technology with support from the MIT UFRAC Project, the National Science Foundation (NSF), and the Air Force Office of Scientific Research (AFOSR), this most recent work was conducted at Resources Engineering Systems Inc. (RES) under the sponsorship of the Gas Research Institute (GRI). The authors are grateful for support from all of these agencies, and also wish to express their thanks to Michael C. Larson and D. Edward Johnson , both of MIT, for their assistance with the preparation of the figures and tables.

REFERENCES

[1] Annigeri, B.S., "Surface Integral Finite Element Hybrid Method for Localized Problems in Continuum Mechanics", Sc.D. Thesis in the Department of Mechanical Engineering, M.I.T., April 1984.

[2] Annigeri, B.S. and M.P. Cleary, "Surface integral finite element hybrid (SIFEH) method for fracture mechanics", Int.'l. J. for Num. Meth. in Eng., Vol 20, 869-885, 1984.

[3] Keat, W.D., "Surface Integral and Finite Element Hybrid Method for the Analysis of Three-Dimensional Fractures", Ph.D. Thesis in the Department of Mechanical Engineering, M.I.T., April 1989.

[4] Keat, W.D., B.S. Annigeri, and M.P. Cleary, "Surface integral and finite element hybrid method for two- and three-dimensional fracture mechanics analysis", International Journal of Fracture, Vol 36, 35-53, 1988.

[5] Erguven, M.E., W.D. Keat, and M.P. Cleary, "General 3-D solution for arbitrarily-oriented fracture dipoles near bimaterial interfaces", to be published.

[6] Annigeri, B.S., W.D. Keat and M.P. Cleary, "Fracture mechanics research using the surface integral and finite element hybrid method", *Proceedings of First Joint Japan/U.S. Symposium on Boundary Element Methods*, University of Tokyo, October 1988, Pergamon Press, 191-202.

[7] Cleary, M.P., "Fundamental solutions for a fluid-saturated porous solid", Int.'l J. Solid Structures, Vol 13, 785-806, 1977.

[8] Rongved,"Force interior to one of two joined semi-infinite solids", Proceedings of Second Midwestern Conf. on Solid Mech., 1-13, 1955.

[9] Cook, T.S., and F. Erdogan, "Stresses in bonded materials with a crack perpendicular to the interface", Int. J. Engng. Sci., Vol 10, 677-697, 1972.

[10] Erdogan, F., and V. Biricikoglu, "Two bonded half planes with a crack going through the interface", Int. J. Engng. Sci., Vol 11, 745-766, 1973.

[11] Cleary, M.P., "Primary factors governing hydraulic fractures in heterogeneous stratified porous formations", presented at Energy and Petroleum Conference, Houston, November 1978, ASME Paper No. 78-Pet-47, 1978.

[12] Hilton, P.D., and G.C. Sih, "A laminate composite with a crack normal to the interfaces", Int. J. Solids Struct., Vol 7, 913-930, 1971.

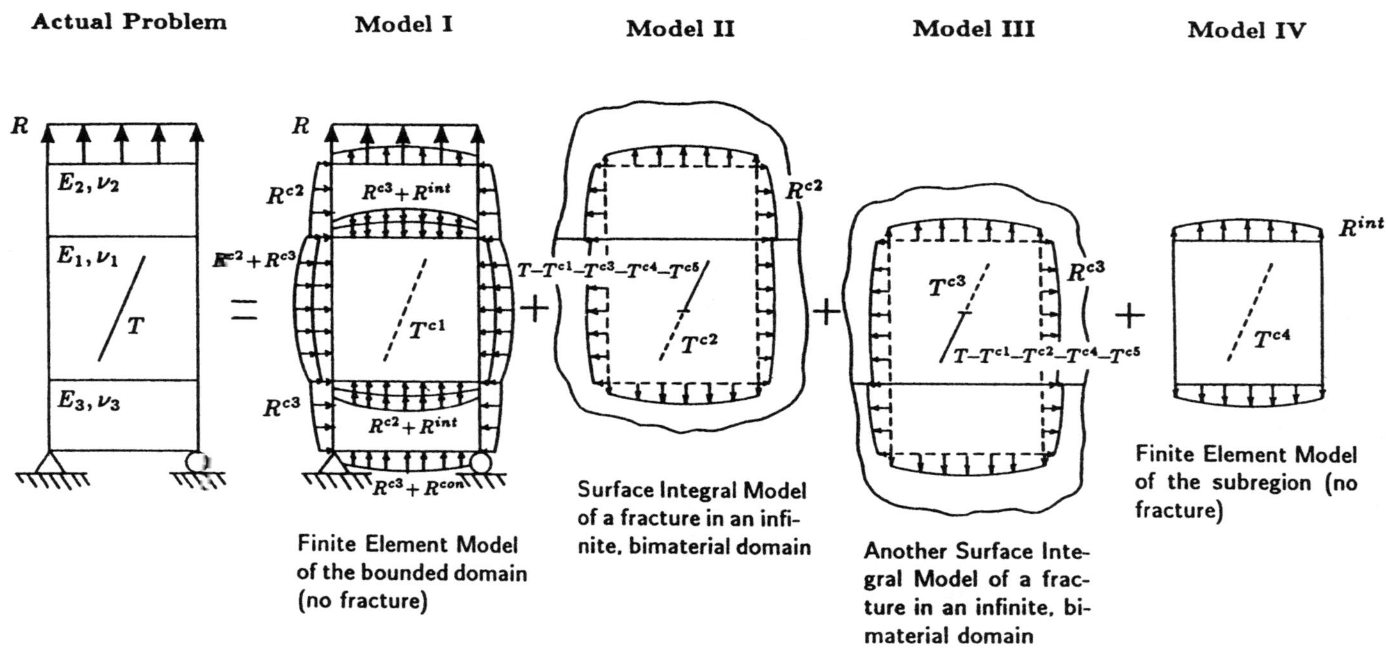

Fig.1. Component geometries used with the hybrid method to model a fracture near multiple planar interfaces in a piecewise homogeneous, bounded domain

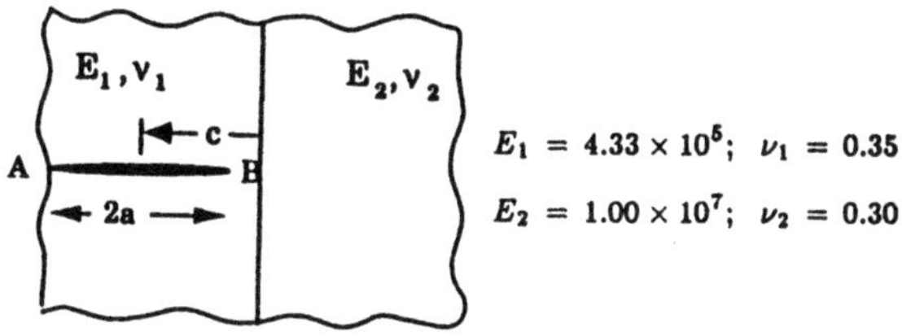

Fig.2. Fracture (under plane strain conditions) approaching a single planar bimaterial interface. See Table 1 for results.

c/a	$\left(\frac{K_I}{\sigma\sqrt{\pi a}}\right)_A$	(% Deviation)$_A$	$\left(\frac{K_I}{\sigma\sqrt{\pi a}}\right)_B$	(% Deviation)$_B$
2.0	0.9682	0.686	0.9416	0.717
1.25	0.9202	0.404	0.7887	0.625
1.15	0.9078	0.298	0.7250	1.003
1.10	0.9010	0.278	0.6787	1.693
1.00	0.8841	0.159	0.5221*	—

Table 1. Comparison of Cook and Erdogan's plane strain solution [9] with the 3-D surface integral results

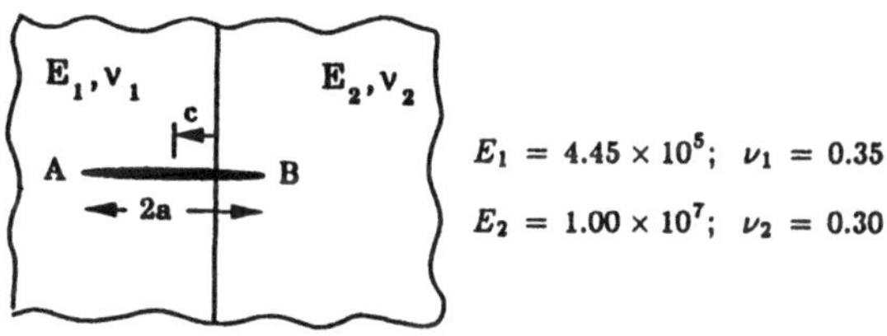

Fig.3. Fracture (under plane strain conditions) intersecting a planar bimaterial interface. See Table 2 for results.

c/a	$\left(\frac{K_I}{P_1\sqrt{\pi a}}\right)_A$	(% Deviation)$_A$	$\left(\frac{K_I}{P_2\sqrt{\pi a}}\right)_B$	(% Deviation)$_B$
1.0	0.8840	0.147	—	—
.75	0.8734	0.000	0.5597	-5.245
.50	0.9124	-0.642	0.7916	-1.395
.25	0.9984	-1.723	0.9524	-1.008
0.0	1.144	-2.97	1.083	-0.923
-.25	1.378	-4.149	1.194	-0.870
-.50	1.774	-5.203	1.287	-0.946
-.75	2.621	-5.304	1.364	-1.132
-1.0	—	—	1.360	0.354

Table 2. Comparison of Erdogan and Biricikoglu's plane strain solution [10] with the 3-D surface integral results

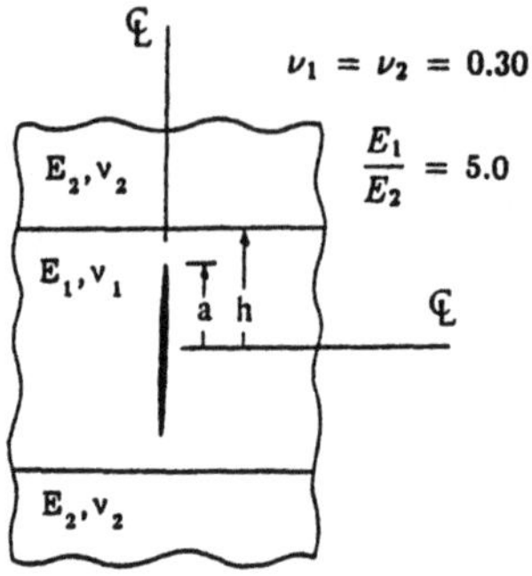

Fig.4. Fracture (under plane strain conditions) approaching two planar bimaterial interface. See Table 3 for results.

a/h	$\frac{K_1}{\sigma\sqrt{\pi a}}$	% Deviation
0.3	1.040	0.970
0.6	1.156	0.522
0.9	1.611	2.611

Table 3. Comparison of Hilton and Sih's Plane Strain Solution [11] with the 3-D hybrid results

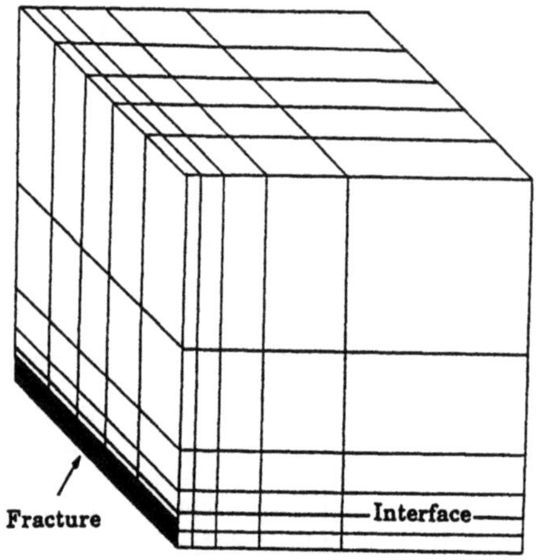

Fig.5. 1/8 symmetric hybrid model used with the double interface problem

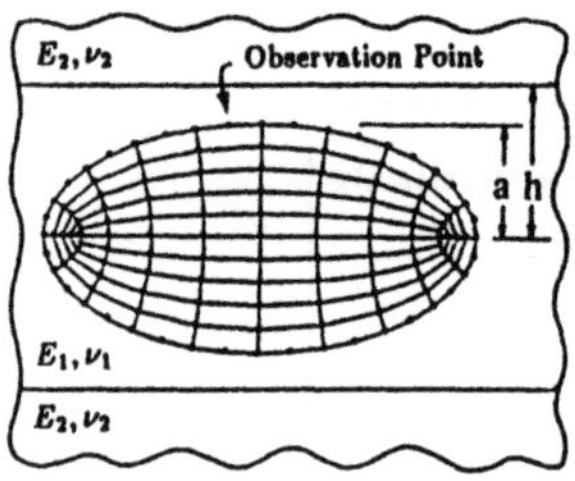

Fig.6. Elliptical fracture approaching two planar bimaterial interfaces. The fracture has been represented by its surface integral discretization.

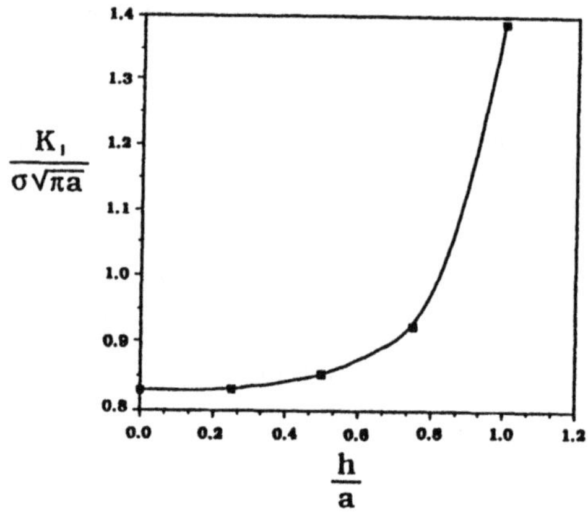

Fig.7. Dependence of K_1 on the value of a/h for a 3-D elliptical fracture. Values of K_1 were recorded at the observation point indicated in Figure 6.

Development of Yield Strip Models Using the Surface Integral and Finite Element Hybrid Method

Balkrishna S. Annigeri

United Technologies Research Center
East Hartford, Connecticut 06108
U.S.A.

SUMMARY

The further development of the Surface Integral and Finite Element Hybrid method for modelling yielding of the material at a crack tip is presented in this paper. The crack in an infinite domain is modelled using a dislocation based integral equation formulation. The uncracked finite body is modelled using finite elements and these two models are coupled using linear superposition for solving the problem of a crack in a finite domain. The yield strips are lumped models that can model small-scale plastic yielding and are useful for both plane stress and plane strain situations. Results obtained using the hybrid formulation are in good agreement with those obtained using the classical Dugdale-Barenblatt and Bilby-Cottrell-Swinden models.

INTRODUCTION

Materials that are used for the manufacture of structural components inherently contain flaws or defects such as inclusions, voids, porosities, microcracks etc. The presence of a flaw thus has to be accounted for in the design of structural components. These flaws can grow, especially under fatigue loading, and it is important to be able to predict the evolution of these flaws as a function of the loading. Fracture mechanics has been developed and applied successfully for the analysis of cracks and associated crack propagation. The Griffith-Irwin-Orowan linear elastic fracture mechanics (LEFM) theory has been adequate for modelling cracks in structural components where the elastic K (Stress Intensity Factor) fields dominate the solution [1,2]. The K dominance is appropriate when the plastic zone size is small compared to the length of the crack and other dimensions of the body. The development of LEFM has been followed by the development of elastic-plastic fracture mechanics (EPFM) with the pioneering works of Hult and McClintock [3], Dugdale [4], Barenblatt [5], Bilby, Cottrell and Swinden [6], Rice [7] and Hutchinson [8]. EPFM is applicable and needed especially for high toughness and low strength materials where the elastic K dominance may not be satisfied and has to be replaced by evaluating the energy release rate, for example, by using the J integral. The J integral is based on the existence of a strain energy functional and is strictly valid when the crack propagates under monotonic loading.

The motivation for developing the yield strip models has been provided by the need to model plasticity at the crack tip for determining parameters such as the crack tip opening displacement and the energy release rate (J integral) for predicting fatigue crack growth rates. The need for modelling plastic unloading along the crack faces during crack propagation is also important as it determines the extent of crack closure. This closure affects the rate of crack growth and is one reason for why physically "small" cracks grow faster than "long" cracks as shown in fig.1. The definition of what constitutes small and long cracks has been reported by Suresh and Ritchie [14] and relates the crack size with the plastic zone size or other relevant dimensions of the body including the grain size of the material. In this paper, the development of the yield strip models will be presented.

THE HYBRID FORMULATION WITH YIELD STRIPS

As a first order correction to the LEFM stress intensity factor solutions, Dugdale[4] and Barenblatt[5] proposed a yield strip model which is valid for plane stress situations and typically useful for thin sheets. An elastic perfectly plastic material model is used in the analysis. The Dugdale model which is for Mode I loading uses yield strips which are additional ligaments at the end of the crack tips. The Bilby, Cottrell and Swinden(BCS) model [6] is similar to the Dugdale model but it is for in-plane Mode II shear loading. These yield strips can support a maximum stress equal to the yield stress for the material. The basis of implementing the yield strip models in the Surface Integral and Finite Element Hybrid Method can be developed by considering the problem of a crack in a plate as shown in fig. 2. Details of the derivation (without the yield strips) are provided in references [9-13]. The physical length of the crack is 2a; appended to the crack tips are two yield strips whose length is c. These yield strips can sustain a maximum stress corresponding to the yield stress in tension for the opening (Mode I) case or the yield stress in shear for the shear (Mode II)case. This problem can be expressed as the sum of a finite domain without a crack and an infinite domain with a crack, using incremental linear superposition. The finite domain without the crack is modelled effectively using the finite element method; the infinite domain with the crack(s) is modelled effectively using the surface integral method which is based on dislocation theory. Since, the crack is being modelled in an infinite plate, an appropriate boundary load correction vector R^c is computed using the surface integral model and applied to the finite element model to ensure that the correct load is applied to the boundary. Similarly, the traction vector T^c in the uncracked body which is acting along the original crack location is computed using the finite element model and applied to the surface integral model. The traction vector T_y is assembled using the yield stress values in tension and shear for the material. Note that the traction applied along the yield strips is $T_y - T^c$.

The governing equation for the hybrid method is obtained as follows. The finite element equation for the uncracked plate is given by:

$$[K]\ \{U^{FE}\} = \{R\} - \{R^c\} \tag{1}$$

$[K]$ = finite element stiffness matrix of the uncracked body; $\{U^{FE}\}$ = vector of unknown finite element nodal displacements for the uncracked body; $\{R\}$ = vector of applied loads; $\{R^c\}$ = load correction vector for the boundary due to the presence of the crack.

The crack in an infinite domain is represented by using a continuous distribution of dislocations leading to a singular integral equation formulation [10]. The integral equation for an infinite domain with a crack is given by:

$$\vec{T}(x_0) = \vec{n} \cdot \int_{S_c} \Gamma(\vec{x}, \vec{x}_0) \cdot \mu(\vec{x})\, dS_c \tag{2}$$

$T(x_0)$ =traction at x_0 due to a distribution of dislocations along the crack surface; $\vec{n}$= the normal to the crack surface at x_0; $\Gamma(\vec{x}, \vec{x}_0)$= fundamental stress solution for a dislocation(kernel function); $\mu(\vec{x})$= dislocation density at $\vec{x}$; S_c= surface area of the crack. The singular integral equation is evaluated numerically in the Cauchy principal value sense and the discretized form of equation (2) is :

$$[C]\,\{F\} = \{T\} - \{T^c\} + \{T_y\} \tag{3}$$

$[C]$ = coefficient matrix for the singular integral equation; $\{F\}$ = vector of amplitudes of the dislocation density (for two dimensional analysis); $\{T\}$ = vector of applied tractions along the crack; $\{T^c\}$ = traction vector for the crack surfaces due to the applied load $R - R^c$ on the finite element model. $\{T_y\}$ = traction vector assembled using the yield stress values for the material and applied to the yield strips.

Equations (1) and (3) are fully coupled through the boundary force matrix G ($R^c = [G]\{F\}$) and the stress feedback matrix S ($T^c = [S]\{U^{FE}\}$) which leads to the following representation for the hybrid formulation:

$$\begin{bmatrix} K & G \\ S & C \end{bmatrix} \begin{Bmatrix} U^{FE} \\ F \end{Bmatrix} = \begin{Bmatrix} R \\ T + T_y \end{Bmatrix} \tag{4}$$

In equation (4) the variable U^{FE} represents the continuous displacement field due to just the finite element model. This has to be changed to the total displacement field by addition of the discontinuous field due to the crack so that the displacement boundary conditions can be enforced, as follows:

$$\{U\} = \{U^{FE}\} + \{U^{SI}\} \tag{5}$$

$\{U^{SI}\}$ = vector of unknown finite element nodal displacements due to the surface integral model. $\{U^{SI}\}$ is obtained by setting up the integral equation for displacements and evaluating the integral (Annigeri and Cleary [10]).

$$\{U^{SI}\} = [L]\ \{F\} \tag{6}$$

$[L]$ = displacement matrix. Thus equations(4), (5) and (6) can be combined to form the governing equation for the hybrid formulation:

$$\begin{bmatrix} K & G-KL \\ S & C-SL \end{bmatrix} \begin{Bmatrix} U \\ F \end{Bmatrix} = \begin{Bmatrix} R \\ T+T_y \end{Bmatrix} \tag{7}$$

Equation (7) allows imposition of arbitrary force, traction and displacement boundary conditions. The sparsity and symmetry of the stiffness matrix is taken full advantage of using a special solution scheme which has been reported in Annigeri and Cleary [11]. As is evident, the basic problem of a crack with yield strips is a nonlinear one as the length of the yield strips is dependent on the applied loading. A modified Newton algorithm with interval halving [15] has been implemented in the SAFE (Surface-Integral and Finite Element) Hybrid compter code for determining the length of the yield strips. The length of the yield strip is changed until K_I and/or K_{II} is reduced to a small number within a certain tolerance. This algorithm has worked very effectively in being able to compute the length of the yield strip. Results are presented in the next section.

RESULTS

Plate with a Center Crack - Mode I and Mode II Loading. A plate with a center crack loaded with a uniform tensile load is shown in fig. 3. The dimensions of the plate and other material parameters are also given in the figure. Plane stress conditions are used to determine the constitutive behaviour. The yield strip length and the crack tip opening displacement (CTOD) for the center crack problem in an infinite medium can be obtained using the Dugdale solution [4]. The solution obtained using the SAFE hybrid code is given in figs. 4 and 5. The SAFE results are in excellent agreement with the Dugdale solutions. The departure of SAFE predictions for a half crack length of greater than 0.3 is due to the Dugdale solution being valid only for the infinite domain. Similarly the Mode II loading situation is shown in fig. 6. The SAFE solutions are given in figs. 7 and 8. Again the SAFE results for yield strip length and the crack tip slip development (CTSD)compare very well with the analytical solutions of Bilby, Cottrell and Swinden [6]. For plane strain conditions inclined strip models at the crack tip can be used to determine the extent of the plastic zone and the CTOD or CTSD [18].

CONCLUSIONS

The development of the Surface Integral and Finite Element Hybrid method for material nonlinearity has been presented in this paper. The material nonlinearity is modelled using yield strips with a distribution of climb (opening) and/or glide(slip) dislocations for modelling mixed mode fracture. The advantage of using the yield strips is that whether the crack is physically small or long, the finite element discretization used for modelling the domain remains relatively coarse. The hybrid method has already been shown in references [9,11,12] to be a very effective procedure for modelling propagation of cracks under LEFM conditions. At present, propagation of cracks with plastic unloading behind the crack tip is being worked on using the yield strip models; the work of Budiansky and Hutchinson [17] is very relevant to this effort though it was done in the context of an infinite domain.

ACKNOWLEDGEMENTS

The research performed in this paper was supported by the United States Air Force Office of Scientific Research (AFSC) under contract F49620-86-C0095 with Lt. Col. George K. Haritos as the program manager; this support is gratefully acknowledged. Thanks are also due to the personnel of the illustrating group at the United Technologies Research Center for their assistance with the preparation of the figures.

REFERENCES

1. Rolfe, S.T.; Barsom J.M.: *Fatigue and Fracture Control of Structures*, Prentice Hall, 1977.
2. Kanninen, M.F.; Popelar, C.H.: *Advanced Fracture Mechanics*, Oxford University Press, 1985.
3. Hult, J.A.H.; McClintock, F.A.: Elastic-plastic stress and strain distributions around sharp notches under repeated shear, *Proceedings of the 9th International Congress for Applied Mechanics*, Vol.8, University of Brussels, 51-58, 1957.
4. Dugdale, D.S.: Yielding of steel sheets containing slits, Journal of the Mechanics and Physics of Solids, Vol. 8, 100-108, 1960.
5. Barenblatt, G.I.: The mathematical theory of equilibrium of a crack in brittle fracture, Advances in Applied Mechanics, Vol.7, 55-129, 1962.
6. Bilby, B.A.; Cottrell, A.H.; Swinden, K.H.: The spread of plastic yield from a notch, Proceedings of the Royal Society, Vol. A285, 22-33, 1965.
7. Rice, J.R.: Mathematical Analysis in the Mechanics of Fracture, Chapter 3 of *Fracture: An Advanced Treatise*, Leibowitz H.(ed.), Vol. 2., 191-311, Academic Press, New York, 1968.
8. Hutchinson, J.W.: Singular behavior at the end of a tensile crack in a hardening material, Journal of the Mechanics and Physics of Solids, Vol. 16, 13-31, 1968.
9. Annigeri, B.S.: Surface Integral Finite Element Hybrid Method For Localized Problems in Continuum Mechanics, Sc.D. Thesis, Department of Mechanical Engineering, M.I.T., 1984.

10. Annigeri, B.S.; Cleary M.P.: Surface integral finite element hybrid (SIFEH) method for fracture mechanics. International Journal for Numerical Methods in Engineering, Vol.20, 869-885, 1984.
11. Annigeri, B.S.; Cleary M.P.: Quasi-static fracture propagation using the surface integral finite element hybrid method. *ASME PVP Vol.85*, 1984.
12. Annigeri, B.S.: Effective modelling of stationary and propagating cracks using the surface integral and finite element hybrid method. *ASME AMD Vol.72*, 1985.
13. Annigeri, B.S.: Thermoelastic fracture analysis using the surface integral and finite element hybrid method. Presented at the ICES-88 Conference, Atlanta, Georgia, 1988.
14. Suresh S.; Ritchie R.O.: Propagation of short fatigue cracks. International Metal Reviews, Vol. 29, No. 6, 445-476, 1983.
15. Carnahan, B.; Luther, H.A.; Wilkes J.O.: *Applied Numerical Methods*, John Wiley, New York, 1969.
16. Keat, W.D.; Annigeri B.S.; Cleary, M.P.: Surface integral and finite element hybrid method for two and three dimensional fracture mechanics analysis. International Journal of Fracture,Vol.36, 35-53, 1988.
17. Budiansky B.; Hutchinson J.W.: Analysis of closure in fatigue crack growth. Journal of Applied Mechanics, Vol.45, 267-276, 1978.
18. Vitek, V.: Yielding on inclined planes at the tip of a crack loaded in uniform tension, Journal of the Mechanics and Physics of Solids, Vol. 24, 263-275, 1976.

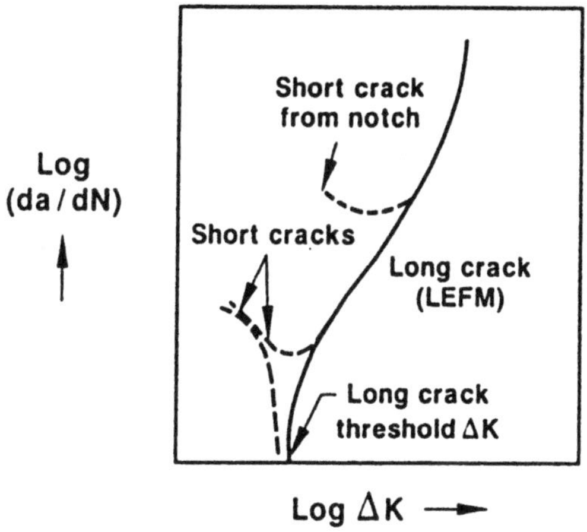

Fig.1. Typical fatigue crack propagation rates (da/dN) for short and long cracks as a function of stress intensity factor range (ΔK)

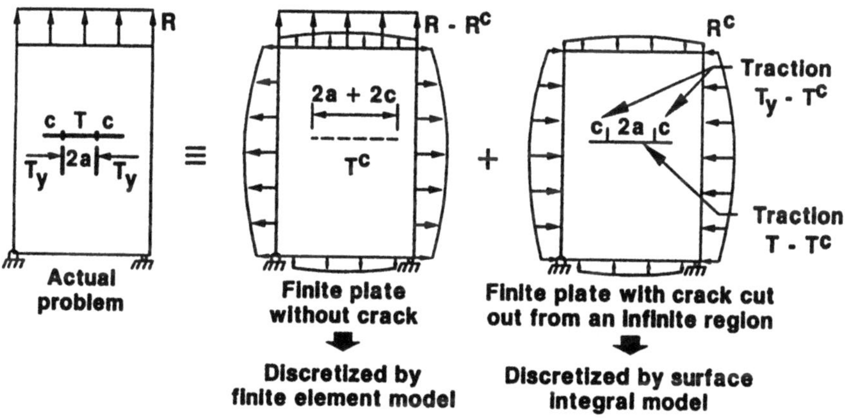

Fig.2. Superposition of the surface integral model (with yield strips) and the finite element model

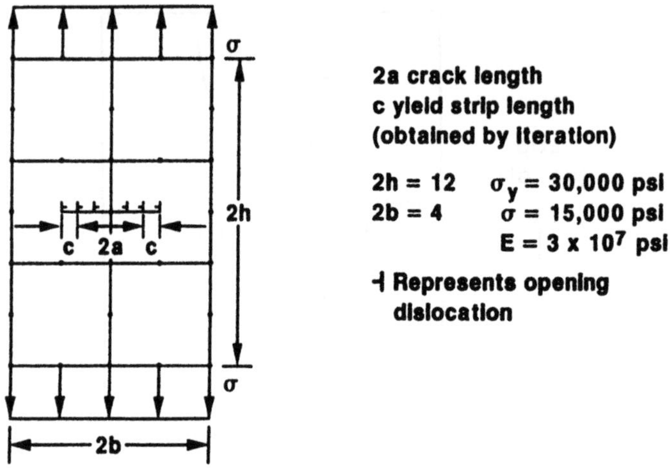

Fig.3. Hybrid model for center cracked panel under Mode I (tensile) loading

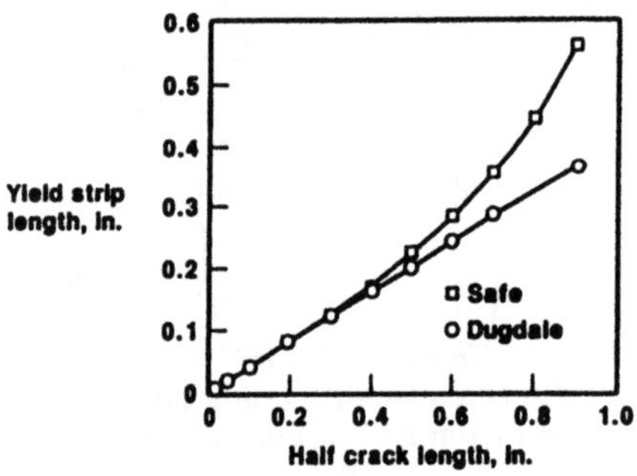

Fig.4. Yield strip lengths for Mode I loading

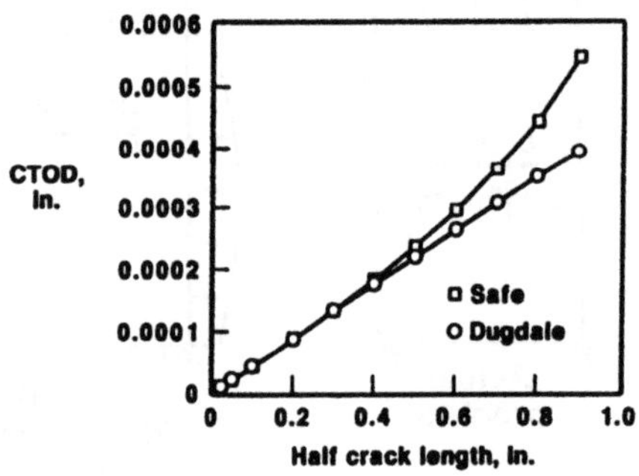

Fig.5. Crack tip opening displacement (CTOD) for Mode I loading

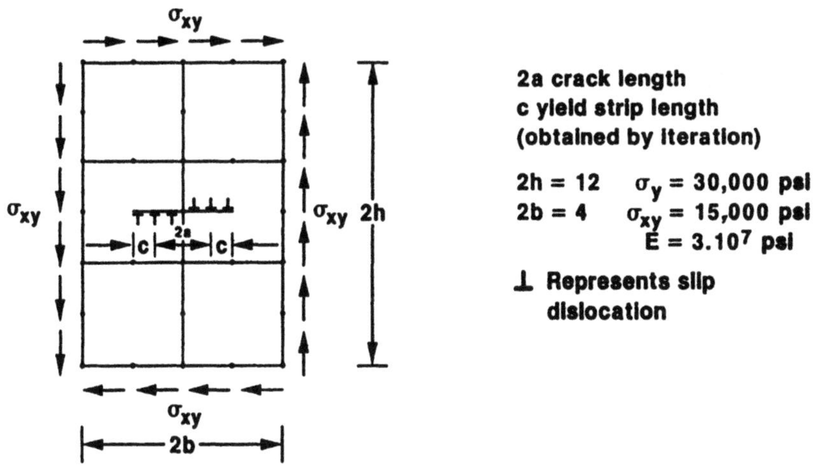

Fig.6. Hybrid model for center cracked panel under Mode II (shear) loading

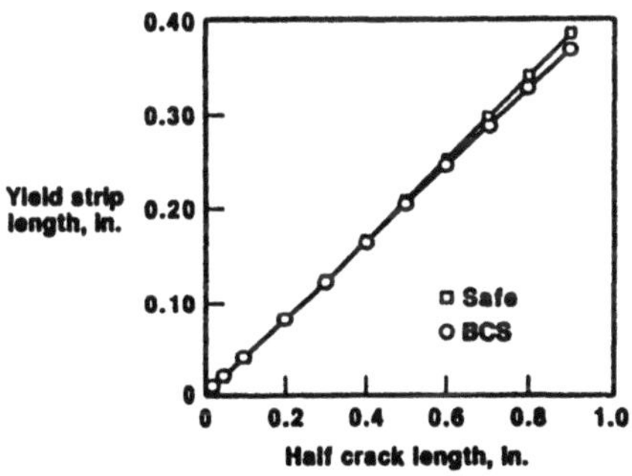

Fig.7. Yield strip lengths for Mode II loading

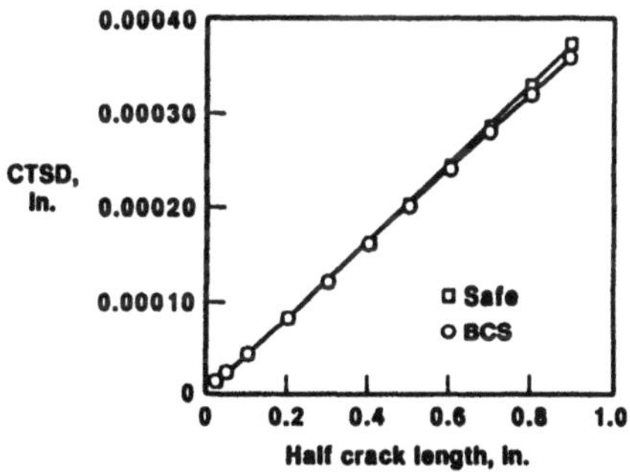

Fig.8. Crack tip slip displacement (CTSD) for Mode II loading

Time Domain Boundary Element Analysis of Two-Dimensional Crack Problems

J. DOMINGUEZ AND R. GALLEGO

Escuela Superior de Ingenieros Industriales

Universidad de Sevilla, Spain

Summary

The time domain dynamic formulation of the BEM is used in combination with quadratic spatial discretization and with singular quarter-point elements that represent the behaviour of stresses and displacements near the crack tip. Transient dynamic stres intensity factors, computed directly from the traction nodal values at the tip, are ploted versus time for two cracked plates. The obtained results are compared with those due to other authors computed by different procedures. The results are shown to be very accurate and present little dependency on the space and time discretization as opposite to Finite Element results.

Introduction

Dynamic Stress Intensity Factors (SIF) computation is an important objetive in fracture mechanics which has been study by a number of authors in the last few years. The Boundary Element Method is very well suited for crack problems and has been used, in combination with singular quarter-point (SQP) elements that include the $r^{-1/2}$ singularity of the tractions, to compute static SIF from the crack opening displacement (COD) by Blandford et al. [1]. Martinez and Dominguez [2] used the same singular boundary element and showed that the traction nodal value at the crack tip gave values of the SIF very accurate and less sensitive to the discretization. The same procedure has been used by Dominguez and Chirino [3,4] for dinamic SIF computations in the frequency domain.

In the present paper the time domain formulation of the BEM is used in combination with a quadratic space variation of tractions and displacements along the boundary elements. SQP elements are used to model the tractions and displacements in the vicinity of the crack tips.

The SIF computed in this paper for cracks in finite domains are compared with those obtained by other authors using different procedures. The computed results are very accurate, do not present oscilations and show very little dependency on the space and time discretization as opposite to FE results.

Time Domain Boundary Elements

Piecewise linear and piecewise constant time discretization are assumed for displacements and tractions, respectively. The boundary variables are written as:

$$u_j = \sum_q \sum_m \varphi^q (r)\ \eta^m (\tau)\ u_j^{mq}$$

$$p_j = \sum_q \sum_m \psi^q (r)\ \mu^m (\tau)\ p_j^{mq} \tag{1}$$

where "q" indicates the boundary node and "m" the time step, φ^q and ψ^q are the space interpolation functions, and η^m and μ^m are the time interpolation functions.

Once the integral representation has been discretized it may be written as:

$$c_{ij}u_j = \sum_{m=1}^{M} \sum_{q=1}^{Q} \left\{ \left[\int_{\Gamma_q} U_{ij}^m \ d\Gamma \right] \psi^q\ p_j^{mq} - \left[\int_{\Gamma_q} T_{ij}^m \ d\Gamma \right] \varphi^q\ u_j^{mq} \right\} \tag{2}$$

where U^m_{ij} and T^m_{ij} are obtained by integration of the fundamental solution along a time step [5].

A cracked domain is divided into two subdomains by an internal boundary that leaves one side of the crack on the boundary of each subdomain. The two elements adjacent to a crack tip are SQP elements (Figure 1). The SIF are computed by

$$K_I = p_2^k \sqrt{2\pi l}$$
$$K_{II} = p_1^k \sqrt{2\pi l} \qquad (3)$$

where p^k_1 and p^k_2 are the traction nodal values at the tip node as part of the element that goes from the tip towards the interior of the domain. The SIF can also be computed from the COD of node k-1 or of nodes k-1 and k-2 (Figure 1) using the same expressions of elastostatics [1,2].

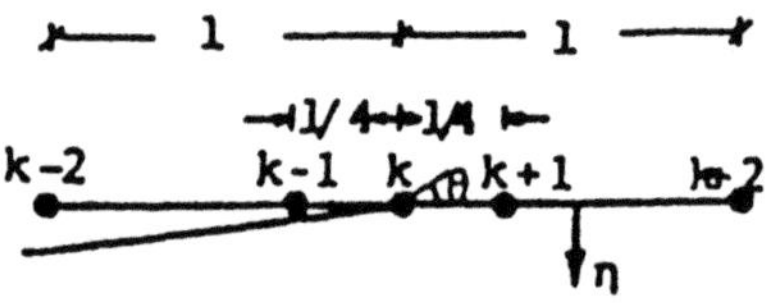

Fig.1. Quarter-point elements at the crack tip

Central Cracked Plate

Figure 2(a) shows the geometry and boundary conditions of a central cracked plate. The load is apply as a step function in time. This problem was solved by Chen [6] using finite diferences, and is frequently used as a reference to validate other methods. The material properties are: shear modulus μ= 76923 GPa, Poisson's ratio $\nu = 0.3$, and density $\rho = 5000$ Kg/m^3 .

Because of the symmetry only one quarter of the plate is discretized as shown in Figure 2(b). Two equal length elements are used to discretize one half of the crack (l/a= 0.5), the element closer to the tip being SQP. The time step Δt = 0.32 μs is such that the P-waves travel 2.4 mm per time step. This size of the time step has been chosen in relation to the size of the elements of the spatial discretization. An analysis of the effects of the changes in the size of the time steps may be found in reference [5]. Figure 2(c) shows the K_I SIF normalized by $\sigma\sqrt{(\pi a)}$, computed directly from the traction nodal values at the tip, versus time.

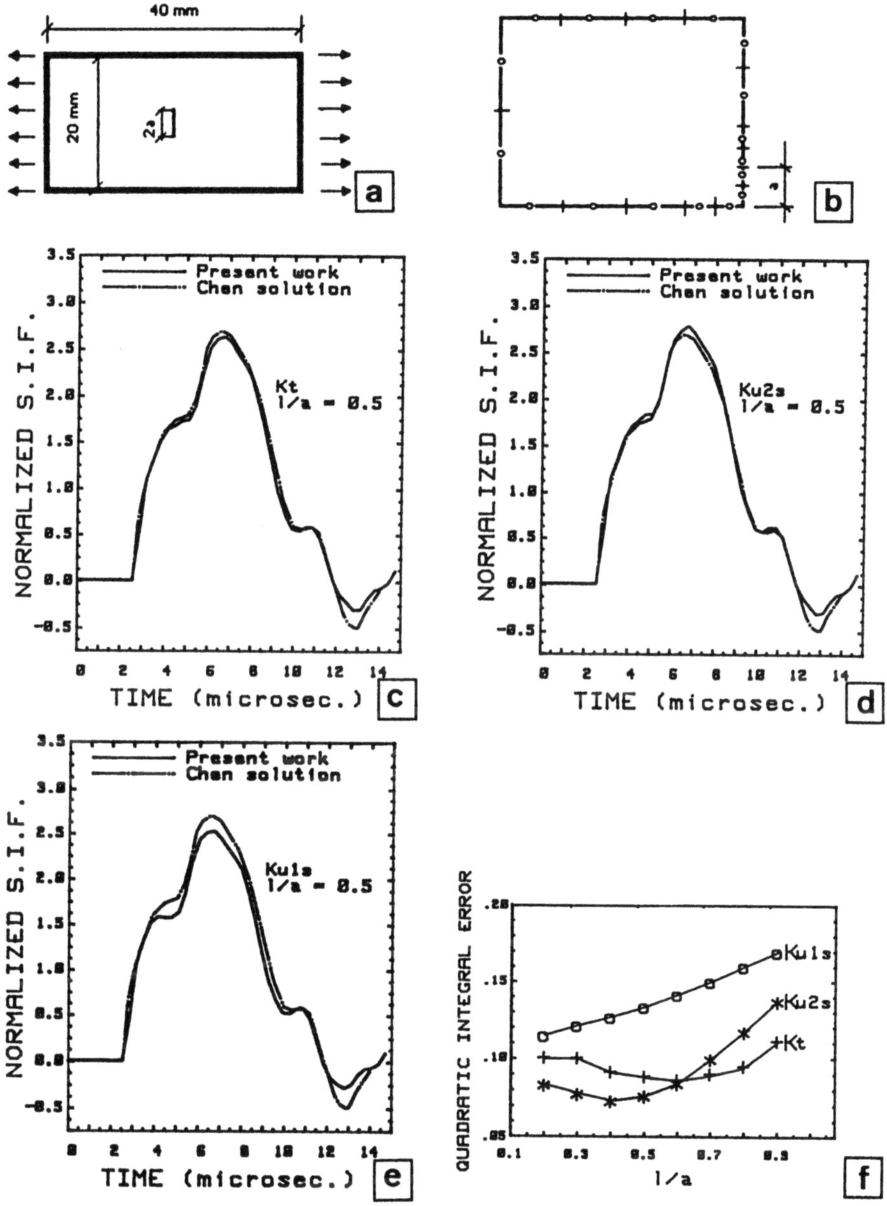

Fig.2.(a) Central cracked plate;(b) Boundary discretization;(c) SIF computed from the traction nodal values at the tip;(d) SIF computed from the COD at two point;(e) SIF computed from the COD at one point;(f) Quadratic integral of the difference between BEM and Chen's results

The agreement of the BEM results computed using that procedure (K_t) and those due to Chen is very good. Figures 2(d) and 2(e) show the SIF computed using the same discretization and two points and one point COD (K_{u2s} and K_{u1s}), respectively. The agreement with Chen's results is also very good.

In order to evaluate the spatial discretization sensitivity of the different approaches, the length of the elements adjacent to the tip was varied from 0.2a to 0.9a while the next elements were varied from 0.8a to 0.1a. Figure 2(f) represents the integrals of the square of the difference between the curves computed by the three BEM approaches and Chen's curve. The integrals are represented versus the length of the tip elements. It can be seen from Figure 2(f) that the approach that uses the traction nodal values at the tip to compute the SIF (K_t) is the less sensitive to the spatial discretization and hence the most reliable for problems which solution is not known. As a reference for comparison, Figures 3(a) and 3(b) show results obtained by Murty and Valliapan [7] for the same problem using FEM and a similar discretization over the boundaries and crack surfaces. Consistent and lamped mass matrices were used to compute the results in Figures 3(a) and 3(b), respectively. The results of Figure 3 are probably the best shown by those authors in reference [7]. Other FEM discretizations produced a more oscilatory behaviour of the solution.

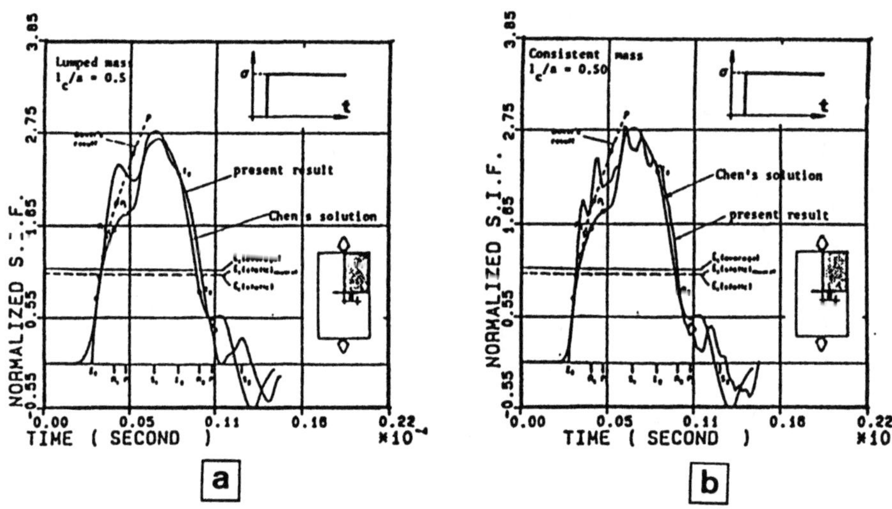

Fig.3. Transient SIF computed in reference[7] using the FEM.

Inclined Cracked Plate

The plate shown in Figure 4(a) has a 45° inclined crack. The domain is divided into two parts as shown in Figure 4(b). The load is again a uniform traction applied as a step function. The time step is $\Delta t = 3.5$ μsc. The mode-I and mode-II SIF are computed using the crack tip nodal values of the traction. Figures 4(c) and 4(d) show the two SIF normalized by $\sigma\sqrt{(\pi a_0)}$. The computed results are compared with those obtained by Aoki et al.[8] using the FEM. Results from Thau and Lu [9] which are only valid for a short period of time are also shown in the Figures. The agreement of the BEM results with those obtained using other methods can be consider as very good, in particular when one takes into account the courseness of the mesh and the complexity of the problem.

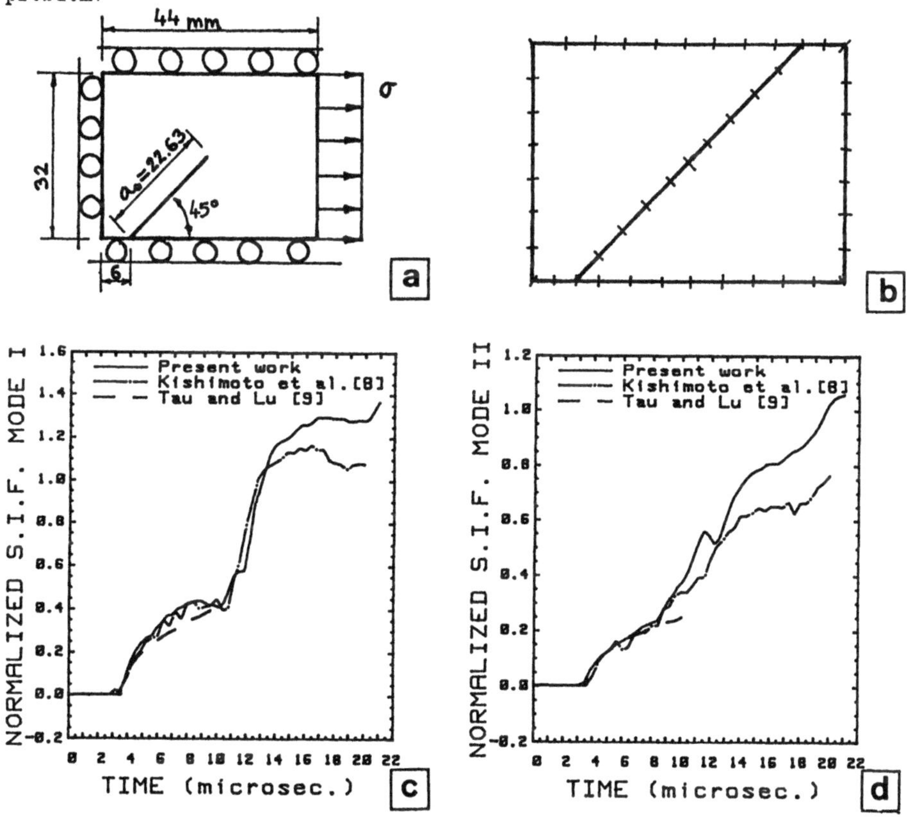

Fig.4.(a) Inclined cracked plate;(b) BE discretization;(c) Mode-I SIF;(d) Mode-II SIF

References

1. Blandford, G.E., Ingraffea, A.R. and Liggett, J.A., (1981), "Two-dimensional Stress Intensity Factor Computations Using the Boundary Element Method." Int. J. Num. Meth. Eng., Vol. 17, pp. 387-404.

2. Martínez, J. and Domínguez, J., (1984), "On the Use of Quarter-point Boundary Elements for Stress Intensity Factor Computations." Int. J. Num. Meth. Eng.. Vol. 20, pp. 1941-1950.

3. Domínguez, J. and Chirino, F. (1986), "B.E.M. for Dynamic Crack Problems in Two Dimensions" BETECH 86, J.J. Connor and C.A. Brebbia Ed., Computational Mechanics Publications, Southampton - Boston.

4. Chirino F. and Dominguez J. (1989), "Dynamic Analysis of Cracks using Boundary Element Method", Eng. Fracture Mechanics, In press.

5. Dominguez J. and Gallego R. (1989), "On the Time Domain Boundary Element Method for Elastodynamic Problems", Computer & Mathematics wit Appl., In press.

6. Chen, Y.M., (1975), "Numerical computation of Dynamic Stress Intensity Factor by Lagrangian Finite-difference Method.", Eng. Fract. Mech., Vol. 7, pp. 653-660.

7. Murti, V., adn Valliappan, S., (1986), "The Use of Quarter Point Element in Dynamic Crack Analysis." Eng. Fract. Mech., Vol. 23, pp. 505-614

8. Aoki, S., Kishimoto, K., Kondo, H., and Sakata, M., (1978), "Elastodynamic Analysis of Crack by Finite Element Method Using Singular Element", Int. J. Fract., Vol. 14. pp. 59-68.

9. Thau, S.A., and Lu, T.H.,(1971), "Transient Stress Intensity Factors for a Finite Crack in an Elastic Solid Caused by a Dilatational Wave." Int. J. Solids, and Struct., Vol.7, pp.731-750.

Fracture Analysis of Asymmetrical Cracks by the Surface Integral and Finite Element Hybrid Method

A. Lisa Burgarella
Pratt and Whitney, East Hartford, CT, USA

Balkrishna S. Annigeri
United Technologies Research Center, East Hartford, CT, USA

Eric H. Jordan
University of Connecticut, Storrs, CT, USA

ABSTRACT

A method of solution for crack problems which utilizes integral equations in the infinite domain coupled to the finite element method in the finite domain has been used to analyze asymmetrical cracks. An important feature of this method is that a single finite element discretization without crack-tip mesh refinement may be used to study almost any type of crack geometry, including curved cracks, branched cracks, and multiple cracks. Once the finite element mesh is constructed , the parametric studies needed for most fatigue life analyses can be run very quickly. Two example applications of the method will be presented. In the first application, the stress intensity factors for a single off-center crack in a rectangular plate were calculated and the hybrid method results compared with the results from finite element analyses,weight function studies, and the stress intensity handbook. In the second application, the hybrid method is used to determine the crack growth path in a thermally loaded plate. The path is found to be in good agreement with experimental results.

INTRODUCTION

The study of asymmetrical cracks by the surface integral and finite element hybrid method was initiated at Pratt and Whitney in order to evaluate the hybrid method as a tool for solving practical problems. In some jet engine components, the combustor being the prime example, cracks are present during a considerable portion of the service life. To analytically predict the durability life of such a component as accurately as possible, it is necessary to predict crack growth.

Several tools are available for analyzing crack behavior. Both finite element and weight function approaches have been used to calculate stress intensity factors, a first step in predicting crack propagation. The finite element method, however, requires crack tip mesh refinement and the generation of a number of meshes to study the crack at various lengths, making it impractical for timely fracture analysis. The weight function method, in contrast, requires only the results from a stress analysis of the uncracked body to calculate stress intensity. However, weight functions may not be readily available for the particular crack geometry under study, and, as will be shown, approximating with a weight function for a "similar" geometry may lead to large inaccuracies. To perform fracture analyses with the hybrid method, it is necessary only to create a single finite element breakup of the uncracked body. The crack, discretized by the surface integral model, is relatively independent of the finite element mesh. The hybrid method, therefore, allows the study of almost any type of crack geometry, making it very attractive as a practical fracture analysis tool.

The hybrid method was evaluated in two fracture applications. The first application is the calculation of stress intensity factors for a single off-center crack in a plate under uniform tension. The hybrid method results are compared to the results from finite element analyses, weight function calculations, and a stress intensity handbook (1). In the second application, the hybrid method is used to determine the crack growth path in a thermally loaded plate. The path is compared to experimental results.

SUMMARY OF THE SURFACE INTEGRAL AND FINITE ELEMENT HYBRID METHOD

The following section briefly outlines the theoretical formulation of the surface integral and finite element hybrid method. A more detailed presentation along with definitions of the finite element and surface integral matrices discussed below is given in the reference (2).

The surface integral and finite element hybrid method is based on the linear superpositon of the finite element and surface integral models (Figure 1) .

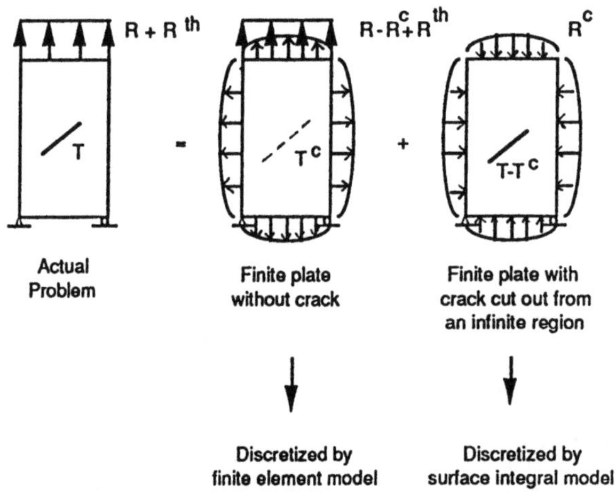

Figure 1. Linear superposition of the finite element and surface integral models

The actual problem has a load R applied on the boundary, a thermal load R^{th}, and tractions T applied to the crack face. The finite element method models the uncracked finite body with the applied boundary and thermal loads, while the surface integral method models the crack in an infinite medium, using a continuous distribution of dislocations. The two models are coupled through the tractions T^c and the load,R^c. The tractions T^c occur in the uncracked finite element model as a result of the boundary and thermal loads, and must be applied along the crack in the surface integral model as a correction so that superpostion yields the original problem. The load R^c,on the other hand, appears in the surface integral model as a result of the dislocations modeling the crack. It must in turn be applied to the boundary of the finite element model as a correction.

The finite element equations, relating the finite element displacement vector U^{FE} to the load vector R, now include the correction load R^c

$$KU^{FE} = R - R^c + R^{th} \tag{1}$$

where R^c is related to the dislocation density amplitude vector F, in the surface integral model by the G matrix.

$$R^c = GF \tag{2}$$

Combining (1) and (2) yields the finite element equations:

$$KU^{FE} + GF = R + R^{th} \tag{3}$$

Likewise, the surface integral equations, relating the dislocation density amplitude F to the tractions , now include the traction correction T^c.

$$CF = T - T^c \tag{4}$$

where T^c is related to the finite element displacements and tractions due to thermals by:

$$T^c = SU^{FE} - T^{th} \tag{5}$$

Combining (4) and (5) yields the surface integral equations:

$$SU^{FE} + CF = T + T^{th} \tag{6}$$

The combined surface integral and finite element equations, (6) and (3), are coupled:

$$\begin{bmatrix} K & G \\ S & C \end{bmatrix} \begin{bmatrix} U^{FE} \\ F \end{bmatrix} = \begin{bmatrix} R + R^{th} \\ T + T^{th} \end{bmatrix} \tag{7}$$

In equation (7) U^{FE} represents the continuous displacement field at the finite element nodes rather than the total displacement field in the actual problem. The total displacements are the sum of the finite element and surface integral displacements.

$$U_{total} = U^{FE} + U^{SI} \tag{8}$$

where

$$U^{SI} = LF \tag{9}$$

Combining (8) and (9) with the coupled matrix equation in (7) yields the matrix equation governing the hybrid method:

$$\begin{bmatrix} K & G\text{-}KL \\ S & C\text{-}SL \end{bmatrix} \begin{bmatrix} U \\ F \end{bmatrix} = \begin{bmatrix} R + R^{th} \\ T + T^{th} \end{bmatrix} \tag{10}$$

APPLICATION 1: THE STRESS INTENSITY FACTOR FOR AN ASYMMETRICAL CRACK IN A PLATE IN TENSION

In this application, the stress intensity factors for an asymmetrical crack in a plate in tension were calculated by the hybrid method and compared to the results from finite element analyses, weight function calculations, and a stress intensity factor handbook (1).

The asymmetrical or off-center crack geometry is shown in Figure 2. The parameters δ and λ, which may range from 0.0 to 1.0, are measures of the eccentricity of the crack and the proximity of the crack tip to the plate edge respectively. A large value of δ indicates a highly off-centered crack, while a large value of λ implies that a crack tip is very close to the edge of the plate. In this comparison, five crack geometries were analyzed, with δ values ranging from 0.0 (center crack) to 0.62 and λ values from 0.268 to 0.91. Three of the cases studied are fairly "extreme" cracks, with large δ and/or λ values. The remaining two cases are a center crack (δ=0.0) and slightly off-center crack (δ=.256, λ=.268).

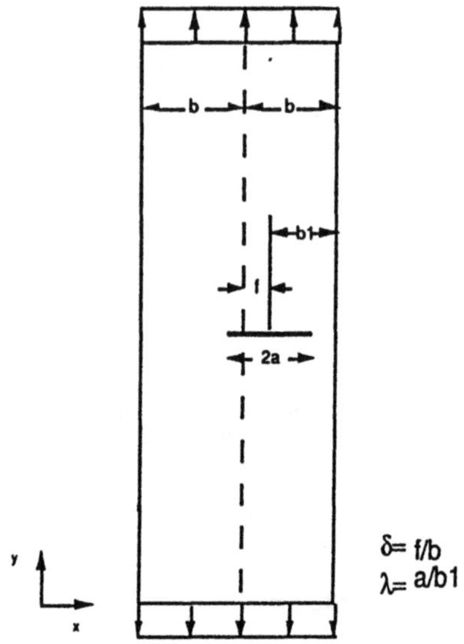

Figure 2. Asymmetrical crack geometry

To perform the finite element analyses, a mesh with refinement at the crack tip locations was generated for each of the crack geometries. Due to its symmetry in the y direction, half the plate was modeled so that each crack appears on the boundary of the finite element mesh with the crack face free of constraint. One of the five meshes is shown in Figure 3, the case of δ=0.4 and λ=0.83. The meshes were composed of 8 noded, isoparametric quadrilateral elements.

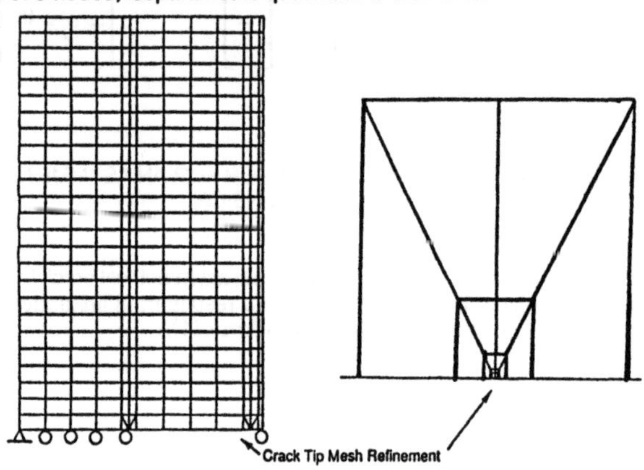

Figure 3 Finite element mesh used to calculate stress intensity, δ=0.4, λ=0.83.

In performing the weight function analyses, the weight function for the off-center crack in a finite width plate as shown in Figure 2 was not immediately available. To calculate stress intensity factors, the weight function for a crack geometry similar to that in Figure 2 was substituted. The geometry selected is shown in Figure 4, a crack approaching a bimaterial interface in an infinite body. For the plate with an off-center crack, the discontinuity is actually the plate edge, such that E_1 = the plate modulus and $E_2 = 0.0$.

The mesh used to perform the surface integral and finite element hybrid analysis, a model of the entire plate, is shown in Figure 5. The five crack geometries were analyzed with this mesh.

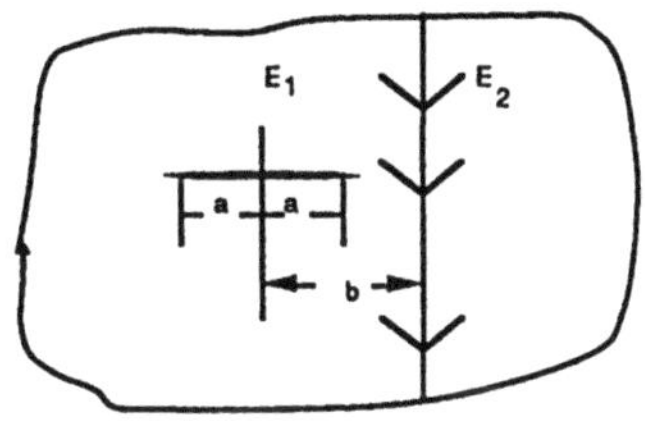

Figure 4. Crack approaching a material discontinuity

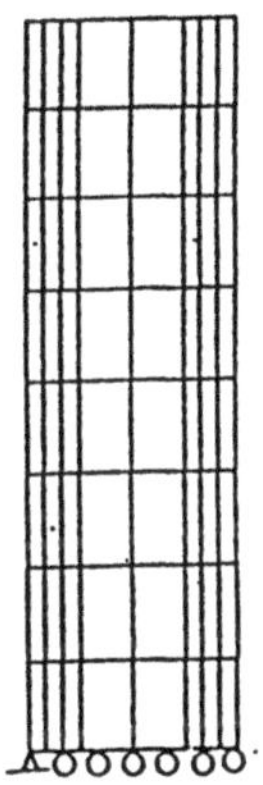

Figure 5. Mesh used to calculate stress intensity with hybrid method.

The stress intensity factors calculated by the three methods were compared to the handbook (1) results for an infinite strip in tension with an eccentric crack (accurate to within 1%). Figure 6 shows the percent error in stress intensity for each method relative to the handbook at the left crack tip, and Figure 7 shows the error at the right tip. The error in the stress intensity factors calculated by the hybrid method is less than the error with weight functions for all cases, and in the majority of the cases, less than the error with the finite element method. For the more extreme geometry of the first three cases, the error in the weight function stress intensities is greater than 10% at the left tip and greater than 15%, reaching as high as 33%, at the right tip. The hybrid method error in these cases is considerably lower, less than 5% for the left tip and as high as 10% only for the right tip in case 1. For the less extreme cracks in cases 4 and 5, the hybrid method error is less than 1%, better than the finite element method.

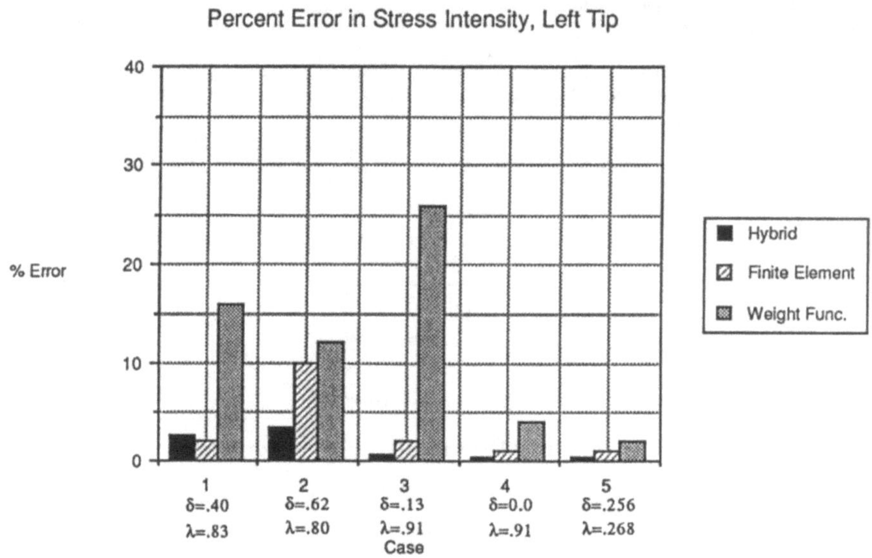

Figure 6. Percent error in stress intensity relative to handbook for left crack tip

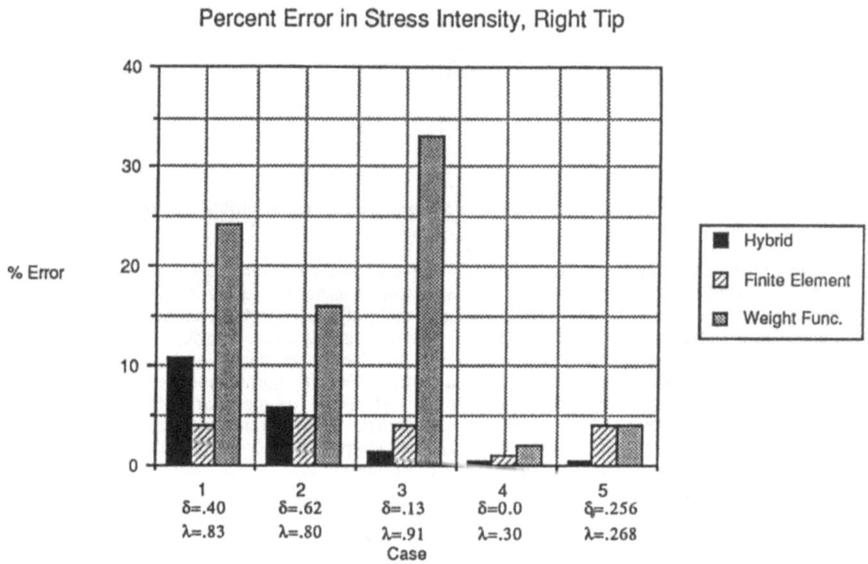

Figure 7. Percent error in stress intensity relative to handbook for right crack tip

It was concluded from this application that the hybrid method calculates stress intensity factors with accuracy comparable to the finite element method, yet is a considerably easier and faster technique to use. It is true that the accuracy of the finite element results could be improved by greater crack tip refinement -- however, this would require a further increase in engineering time. The hybrid method results, as well, could be improved by using a finer mesh. In contrast to the finite elements, the weight function approach did prove to be convenient. However, it was sufficiently accurate only for the shorter, more symmetrical cracks (cases 4 and 5). A new weight function which more closely matches the geometry under consideration would be needed to improve the weight function results for the first three cases. In this application, the hybrid method appears to be the best combination of efficiency and accuracy.

APPLICATION 2: CRACK PATH FOR AN ASYMMETRICAL CRACK IN A PLATE UNDER THERMAL LOADING

In this second application, the hybrid method was used to predict the crack path in a thermally loaded plate, and the predicted path was then compared to test results.

The test panels were cycled between a cooler, isothermal, and a hotter temperature state, with the hotter portion of the cycle including a thermal gradient. The location and direction of the crack path occurring in the thermally cycled panels is shown in Figure 8. The temperature distribution on the panel during the hot portion of the cycle was measured using an optical device, and then mapped onto the hybrid model for the analysis (Figure 9).

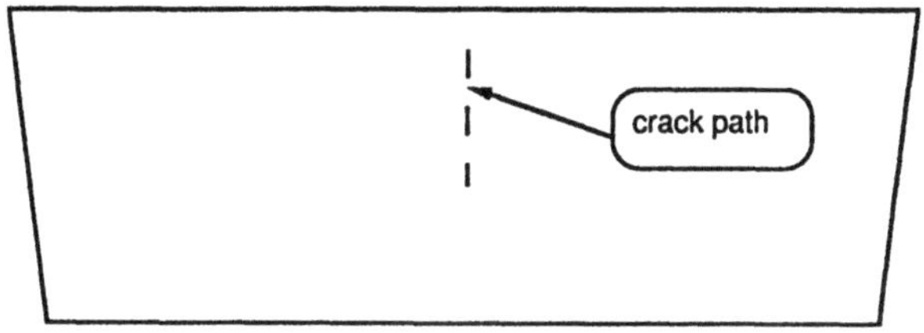

Figure 8. Crack path occurring in thermally loaded test panels

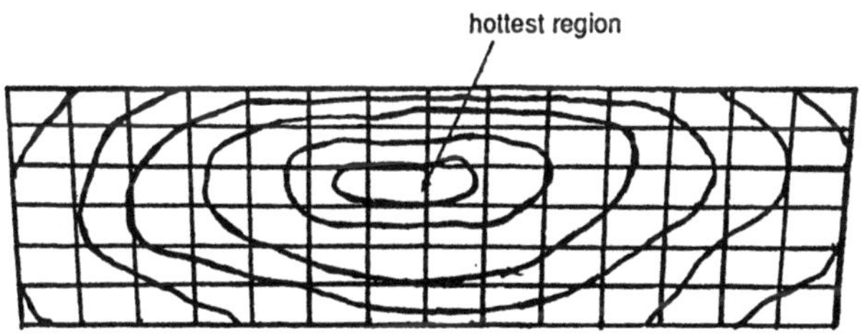

Figure 9. Hybrid model used to predict crack path with isotherms

In order to predict the crack path with the hybrid method, a small crack must be initiated by the user in the thermally loaded model, then allowed to "grow" in the direction calculated. In this study, the crack was first initiated in line with the crack observed during testing. The growth predicted by the hybrid method continued along this line, in agreement with the experimental crack. To make the comparison more rigorous, a second analysis was performed with the crack initiated 45° relative to the test crack. In this case, the predicted crack path turned to realign itself with the path in the test panel, remaining in agreement with the experimental results. In a final test, the crack in the hybrid model was initiated perpendicular to the actual crack path. In this instance, the hybrid method predicted no crack growth (Figure 10). This was the expected result, due to the thermal stress field. From this study it was concluded that the surface integral and finite element hybrid method could be used as a tool for predicting the crack growth path in thermally loaded components.

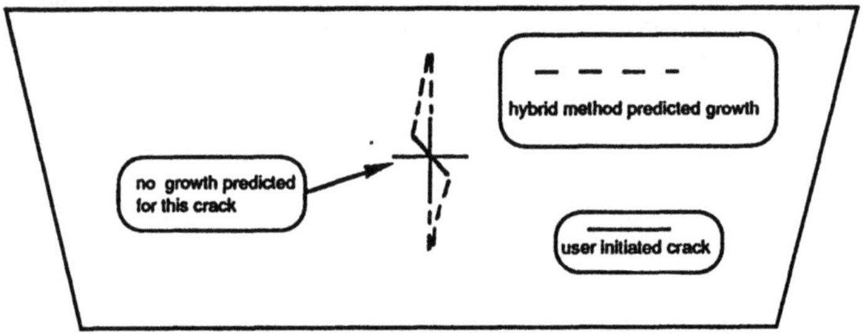

Figure 10. Crack paths predicted with the hybrid method for cracks initiated in three directions

CONCLUSIONS

The surface integral and finite element hybrid method requires a single finite element discretization of the uncracked body to study almost any type of crack geometry. This feature is important because it allows the parametric fracture studies needed to predict crack growth to be performed very quickly. The hybrid method was evaluated in two applications; calculation of stress intensity factors for asymmetrical cracks and prediction of the crack path in a thermally loaded plate. In both cases it was found to be accurate and efficient. In the first application the hybrid method calculated stress intensity factors with error less than or equal to that of the finite element method while requiring considerably less engineering time. In the second, the crack path predicted by the hybrid method matched the experimental path. The results of this evaluation indicate that the hybrid method possesses great potential as a fracture analysis tool. Work is currently in progress to use the hybrid method to predict crack growth life in thermally loaded components, and correlate this life with test results.

REFERENCES

[1] H. Tada, P. C. Paris, and G. R. Irwin, *The Stress Analysis of Cracks Handbook,* Del Research Corporation, Hellertown, Pennsylvania, 1973.

[2] B. S. Annigeri and M. P. Cleary, "Surface Integral and Finite Element Hybrid (SIFEH) Method for Fracture Mechanics", *International Journal for Numerical Methods in Engineering,* **20**, 869-885 (1984).

Regularised BIEs for Miscellaneous Elasticity Problems

N. Nishimura & S. Kobayashi

Dept. Civil Eng., Kyoto Univ., Kyoto, Japan

Various regularised integral equations in the theory of elasticity are presented. The proposed regularisation method applies not only to the hypersingular kernels of crack problems but to ordinaly double layer kernels in elastostatics and in elastodynamics. Some numerical examples are given to show the efficiency of the proposed methods.

1. Introduction

The boundary integral equation method has been shown to be useful for analysing elastic fields produced by obstacles and cracks. In particular the BIE formulation based on the double layer potential is considered to be effective in crack problems. As has been discussed, however, the double layer formulation for cracks is made practicable only if one finds effective methods of computing integrals including hypersingular kernels. In view of this the present authors have proposed a regularisation technique which uses a stress function for the fundamental solution[1]. They showed, in isotropic elastodynamics, that a repeated use of integration by parts, together with the aforementioned stress function, reduces the hypersingular kernel function of crack problems to an integrable one. As we shall see, however, this method applies not only to crack problems for isotropic materials, but to a wider class of problems. Hence in this paper we shall discuss such extended applicability of this method to various elasticity problems.

This paper begins by recapitulating the stress function regularisation method in crack problems. As has been pointed out in [1], this method is expected to work regardless of the material symmetry. We shall confirm this by presenting an expression of the relevant stress function for the transversely isotropic elastostatics in 3D. Also, we shall show that a similar approach transforms the Cauchy principal value integrals in the conventional BIE into ordinary integrals, both in elastostatics and in elastodynamics of any spatial dimension. We then provide some numerical results in order to test the performance of the proposed methods. Finally a few comments conclude this paper.

2. Crack problems

Let S be a smooth piece of curved surface in R^3 bounded by a smooth edge ∂S. Also, let $\boldsymbol{n}$ be the unit normal vector to S. The time-harmonic elastodynamic crack problem yields the following boundary value problem: *Find a solution* $\boldsymbol{u}$ *which satisfies*

$$\Delta^* \boldsymbol{u} + \rho\omega^2 \boldsymbol{u} = 0 \quad \text{in } R^3 \setminus S, \qquad (\Delta^* \boldsymbol{u} := \operatorname{div} \boldsymbol{C}[\nabla \boldsymbol{u}]) \tag{1}$$

$$\lim_{y \to x} T\boldsymbol{u}(y) = 0, \quad y \in D, \quad \boldsymbol{x} \in S, \qquad (T\boldsymbol{u} := \boldsymbol{C}[\nabla \boldsymbol{u}]\boldsymbol{n})$$

$$\lim_{\boldsymbol{x}\to\boldsymbol{x}_0} \varphi(\boldsymbol{x}) = 0, \quad \boldsymbol{x} \in S, \quad \boldsymbol{x}_0 \in \partial S,$$

and the radiation condition for $\boldsymbol{u} - \boldsymbol{u}_I$. In this statement Δ^*, T, C, ρ, ω, φ and $\boldsymbol{u}_I$ stand for the Navier operator, elastic traction operator, elasticity tensor, density, frequency, crack opening displacement and the incident field which satisfies (1) in R^3, respectiviely. The solution $\boldsymbol{u}$ to this problem has a representation given by

$$\boldsymbol{u}(\boldsymbol{x}) = \int_S \hat{\boldsymbol{\Gamma}}_I(\boldsymbol{x},\boldsymbol{y})\varphi(\boldsymbol{y})dS + \boldsymbol{u}_I(\boldsymbol{x}), \qquad \boldsymbol{x} \notin S$$

where $\hat{\boldsymbol{\Gamma}}_I$ is the double layer kernel defined in terms of the fundamental solution $\hat{\boldsymbol{\Gamma}}$ as

$$\hat{\Gamma}_{Iij} = \partial_{y_q} \hat{\Gamma}_{ip}(\boldsymbol{x}-\boldsymbol{y}) C_{jkpq} n_k(\boldsymbol{y}).$$

The crack opening displacement φ is obtained as the solution of an 'integral' equation

$$g(\boldsymbol{x}) = \text{p.f.} \int_S T\hat{\boldsymbol{\Gamma}}_I(\boldsymbol{x},\boldsymbol{y})\varphi(\boldsymbol{y})dS, \quad \boldsymbol{x} \in S, \quad g(\boldsymbol{x}) = -T\boldsymbol{u}_I(\boldsymbol{x}), \tag{2}$$

where p.f. stands for the finite part of a divergent integral. The integral equation in (2) is not suitable for numerical analyses because it has a hypersingular kernel. It can be shown, however, that the integral in (2) is made integrable with the help of the following 'stress function representation' of the stress of the double layer kernel:

$$\begin{aligned} &C_{abik}\partial_{x_k}\partial_{y_l}\Gamma_{ij}(\boldsymbol{x}-\boldsymbol{y})C_{cdjl} \\ &\quad = e_{aip}e_{bjq}e_{ckr}e_{dls}\partial_{x_i}\partial_{x_j}\partial_{y_k}\partial_{y_l}\Phi_{pqrs}(\boldsymbol{x}-\boldsymbol{y}) + \rho\omega^2\Psi_{abcd}(\boldsymbol{x}-\boldsymbol{y}) \end{aligned} \tag{3}$$

where $\boldsymbol{\Phi}$ is the stress function whose second derivatives are locally integrable, $\boldsymbol{\Psi}$ is a kernel function which has an integrable singularity at $\boldsymbol{x} = \boldsymbol{y}$, and e_{jkl} is the permutation symbol. Indeed, the use of (3) in (2) gives

$$\begin{aligned} \text{p.f.} \int_S T\hat{\boldsymbol{\Gamma}}_I(\boldsymbol{x},\boldsymbol{y})\varphi(\boldsymbol{y})dS &= \rho\omega^2 \int_S n_b(\boldsymbol{x})\Psi_{abcd}n_c(\boldsymbol{y})\varphi_d dS \\ &- \int_S (n_b(\boldsymbol{x}) - n_b(\boldsymbol{y}))e_{aip}e_{bjq}e_{dls}\partial_{x_i}\partial_{x_j}\partial_{y_l}\Phi_{pqrs}(n_c e_{ckr}\partial_{y_k}\varphi_d)dS \\ &- \text{p.f.} \int_S e_{aip}e_{dls}\partial_{x_i}\partial_{y_l}\Phi_{pqrs}n_b e_{bjq}\partial_{y_j}(n_c e_{ckr}\partial_{y_k}\varphi_d)dS \end{aligned} \tag{4}$$

where the p.f. symbol on the right hand side of (4) is due to the near tip singularily of φ. One shows that the kernels on the RHS of (4) are all integrable when S is smooth. Also, the p.f. integral on the RHS of (4) is easily converted into an ordinary integral as one finds in [1]. Hence one now uses (2), together with (4) and expressions for $\boldsymbol{\Phi}$ and $\boldsymbol{\Psi}$, to compute φ numerically. Formulae for $\boldsymbol{\Phi}$ and $\boldsymbol{\Psi}$ in isotropic elastostatics and elastodynamics are found in [1].

Notice that the existence of a stress function does not depend on the material symmetry. As a matter of fact, expressions for the Fourier transforms of $\boldsymbol{\Phi}$ and, hence, $\boldsymbol{\Psi}$ for the general 3D anisotropic case are available[2]. This general method applied to the 3D transverse isotropic elastostatics yields $\boldsymbol{\Psi} = 0$ and

$$\Phi_{ijkl}(\boldsymbol{x}) = \sum_{I=1}^{3} \frac{|x_3|\log\left[(r_I + |x_3|)/\sqrt{\kappa_I}\right] - r_I}{4\pi(D_{12}D_{33} + D_{33}D_{66}/2 - D_{13}^2)D_{44}\prod_{J\neq I}(\kappa_I - \kappa_J)}$$
$$\Big[a_{11}(\kappa_I - 1)(\delta_{ij}\delta_{kl} - J_iJ_jJ_kJ_l) + (\kappa_I a_{21} - a_{22})(\delta_{ik}\delta_{jl} + \delta_{il}\delta_{jk} - 2J_iJ_jJ_kJ_l)$$
$$+ (\kappa_I a_{31} - a_{32})(\delta_{ij}J_kJ_l + \delta_{kl}J_iJ_j - 2J_iJ_jJ_kJ_l)$$
$$+ (\kappa_I a_{41} - a_{42})(\delta_{ik}J_jJ_l + \delta_{il}J_jJ_k + \delta_{jl}J_iJ_k + \delta_{jk}J_iJ_l - 4J_iJ_jJ_kJ_l)$$
$$+ (\kappa_I a_{51} - a_{52})J_iJ_jJ_kJ_l\Big], \tag{5}$$

where

$$r_I = \sqrt{\kappa_I x_i^\alpha x_i^\alpha + x_3^2}, \quad x_i^\alpha := x_i - x_kJ_kJ_i, \quad \delta_{ij}^\alpha = \delta_{ij} - J_iJ_j, \quad J_m = \delta_{m3},$$

$\kappa_1 = D_{66}/D_{44}$ and $\kappa_{2,3}$ are the zeros of the following expression:

$$2(2D_{12}D_{33} + D_{33}D_{66} - 2D_{13}^2)\kappa^2 - 2(D_{12}D_{44} + 2D_{13}D_{66} + D_{66}D_{44})\kappa + (4D_{12} + D_{66})D_{66}.$$

Also, the constants a_{ij} are given in terms of the compliance constants D_{ij} as

$$\begin{aligned}
a_{11} &= -D_{66}D_{12},\\
a_{21} &= [(D_{66} + 2D_{12})D_{44} + 2D_{66}D_{12}]/4, \quad a_{22} = [D_{66}(D_{66} + 4D_{12})]/4,\\
a_{31} &= D_{66}D_{12} - D_{44}D_{13}, \quad a_{32} = (D_{66} - D_{44})D_{12} + (D_{12} - D_{13})D_{66},\\
a_{41} &= [D_{66}(2D_{33} - 2D_{12} - D_{44}) + 4D_{12}D_{33} - 2D_{12}D_{44} - 4D_{13}^2]/4,\\
a_{42} &= [(D_{44} - D_{66})(D_{66} + 2D_{12}) + 4D_{66}(D_{13} - D_{12})]/4,\\
a_{51} &= D_{33}D_{44}, \quad a_{52} = (D_{44} + 2D_{13})^2 - D_{33}(D_{66} + 4D_{12}),
\end{aligned}$$

where we have used Voigt's notation. With this expression and the numerical technique used in [1], or possibly with other techniques (e.g. [2],[3]), one establishes a BIEM for elastostatic crack problems for a transversely isotropic solid.

3. Ordinary problems

The regularisation method developed so far depends only on the existence of stress functions for some singular kernels. This suggests that the proposed method is not a special technique for crack problems. Indeed, this method is expected to work with any kernel function as long as it allows an interpretation as a traction of another kernel. An important example of such a kernel is the double layer potential. Consider, for example, the double layer potential for the 3D isotropic elastostatics. Since the double layer kernel $\overset{\circ}{\Gamma}_I$ is defined as the traction associated with the fundamental solution $\overset{\circ}{\Gamma}$, it admits a representation in terms of a stress function $\overset{\circ}{E}$ given by

$$\overset{\circ}{\Gamma}_{Iij}(\boldsymbol{x}, \boldsymbol{y}) := \partial_{y_l}\overset{\circ}{\Gamma}_{ik}(\boldsymbol{x} - \boldsymbol{y})C_{kljm}n_m(\boldsymbol{y}) = \partial_{y_p}\partial_{y_q}\overset{\circ}{E}_{ibc}e_{spb}e_{jqc}n_s(\boldsymbol{y}). \tag{6}$$

As a matter of fact, this expression holds only locally, because the nonzero body force of $\overset{\circ}{\boldsymbol{\Gamma}}$ gives rise to some discontinuity of $\overset{\circ}{\boldsymbol{E}}$. However, one may introduce a function $\overset{\circ}{\boldsymbol{F}}$ and modify (6) by

$$\partial_{x_l} \overset{\circ}{\Gamma}_{ik}(\boldsymbol{x}) C_{klj m} = \partial_{x_p} \overset{\circ}{F}_{ijb}(\boldsymbol{x}) e_{mpb} + c_{ijm} \delta(\boldsymbol{x} \cdot \boldsymbol{N}_1^{\perp}) \delta(\boldsymbol{x} \cdot \boldsymbol{N}_2^{\perp}) H(-\boldsymbol{x} \cdot \boldsymbol{N}), \quad (7)$$

where $\boldsymbol{N}$, and $\boldsymbol{N}_{1,2}^{\perp}$ are mutually orthogonal unit constant vectors, H is Heaviside's step function and c_{ijm} is a constant. An expression (not unique) for $\overset{\circ}{\boldsymbol{F}}$, obtained with the help of Fourier transform, is

$$\overset{\circ}{F}_{ijm}(\boldsymbol{x}) = \frac{\mu e_{ijm}}{4\pi r(\lambda + 2\mu)} - \frac{\delta_{ij} N_l e_{lpm} x_p}{4\pi r(r + \boldsymbol{N} \cdot \boldsymbol{x})} - \frac{(\lambda + \mu) x_i x_p e_{jpm}}{4\pi(\lambda + 2\mu) r^3}.$$

An application of (7) is found in the computation of the 'influence coefficients' which appear in the conventional BIEM. One usually computes the convolution of $\overset{\circ}{\boldsymbol{\Gamma}}_I$ and a shape function Ω on ∂D according to the definition, where ∂D is the boundary of a domain D. This method, however, often gives very poor numerical results near the boundary because of the singularity of the kernel. An alternative to the conventional approach is to use the following identity:

$$\int_{\partial D} \overset{\circ}{\Gamma}_{Iij}(\boldsymbol{x}, \boldsymbol{y}) \Omega(\boldsymbol{y}) dS = -\int_{\partial D} \overset{\circ}{F}_{ijm}(\boldsymbol{x} - \boldsymbol{y}) e_{kqm} n_k \partial_{y_q} \Omega(\boldsymbol{y}) dS. \quad (8)$$

The vector $\boldsymbol{N}$ is chosen so that the ray in the positive direction of $\boldsymbol{N}$ emanating from $\boldsymbol{x}$ does not intersect the support of Ω. Notice that the RHS of (8) is integrable.

The stress function method of regularisation works in dynamics as well, especially in time domain where one can write the stress functions explicitly. For example, the regularisation of the double layer potential in 2D elastodynamics uses the following decomposition of the stress of the fundamental solution $\boldsymbol{\Gamma}$:

$$\begin{aligned}
\partial_{x_l} \Gamma_{ik}(\boldsymbol{x}, t) C_{kljm} &= e_{mq} \partial_{x_q} F_{ij}(\boldsymbol{x}) + (\partial_t)^2 G_{ijm}(\boldsymbol{x}, t), \\
F_{ik}(\boldsymbol{x}, t) &= \frac{1}{2\pi} \left\{ e_{ik} \frac{c_T}{R_T} - N_p e_{pl} \frac{\boldsymbol{x} \cdot \boldsymbol{N} x_l}{(R_T^2 + (\boldsymbol{x} \cdot \boldsymbol{N})^2) R_T} \right. \\
&\quad \left. - 2 c_T^2 e_{kp} \left[\left(\delta_{ip} - 2 \frac{x_i x_p}{r^2} \right) \left(\frac{R_T}{c_T r^2} - \frac{R_L}{c_L r^2} \right) - \frac{x_i x_p}{r^2} \left(\frac{1}{c_T R_T} - \frac{1}{c_L R_L} \right) \right] \right\}, \\
G_{ikj}(\boldsymbol{x}, t) &= \frac{1}{2\pi} \left\{ \left(\frac{R_T}{c_T r^2} - \frac{R_L}{c_L r^2} \right) x_i \delta_{jk} - \frac{\delta_{ik} N_j}{c_T} \tan^{-1} \frac{R_T}{\boldsymbol{x} \cdot \boldsymbol{N}} \right\},
\end{aligned}$$

where c_L and c_T are the velocities of the longitudinal and transverse waves and

$$R_{L,T} = \left((c_{L,T} t)^2 - r^2 \right)_+^{1/2}, \quad (x)_+ := \begin{cases} x & x > 0 \\ 0 & \text{otherwise.} \end{cases}$$

The double layer potential is now regularised as $(\Gamma_{Iij} = \partial_{y_k}\Gamma_{il}C_{kljm}n_m(y))$

$$\int_0^t\int_S \Gamma_I(x,y,t-s)\Omega(y,s)dSds = -\int_0^t\int_S\Big[F_{ij}(x-y,t-s)e_{mp}n_m(y)\partial_{y_p}\Omega(y,s) + G_{ijm}(x-y,t-s)n_m(y)\left(\partial_s\right)^2\Omega(y,s)\Big]dSds, \tag{9}$$

where the shape function Ω is assumed to vanish initially. One may choose the vector N in the same manner as in the static case. Notice that the otherwise non-integrable singularities of Γ_I on the past cones are weakened to integrable singularities.

4. Numerical analysis

4.1 Crack problems The numerical method of analysis for crack problems using the present scheme of regularisation has been discussed in [1]. This method, together with the stress function given in (5) enables us to compute φ for a crack in a transversely isotropic solid. Fig. 1 shows a result of such analysis. The crack has a square shape of a side length of $2a$, is perpendicular to the plane of isotropy and is subjected to a remote normal tension of p_0. The material has elasticity constants for magnesium. The crack opening displacements given in Fig. 1 are obtained with 7×7 DOF B-spline boundary elements.

4.2 Ordinary problems We have tested integration schemes for the 3D elastostatic double layer potential using the shape function obtained as a combination of 4 four-node isoparametric square elements (Fig.2). We computed the (1,1) and (3,3) components of the integral in (8) at $x_{1,2} = 0$, $x_3 = 0$, $0.01a$, $0.1a$, $0.5a$ and a, by using both sides of (8), and compared the results in Table 1. We use Lachat & Watson's 'triangular element technique' [4] with 4 point and 8 point Gaussian integration formulae, and Poisson's ratio of 1/4. This table shows that the proposed scheme gives accurate numerical results even with small number of integration points. Also noteworthy is the fact that the new method seems to show no sign of the so called 'boundary layer effect' while such inaccuracy is evident with the conventional method. As the second example of the regularised double layer potential approach we computed the deformation of a circular hole having a radius of a in a 2D infinite medium subjected to a plane step P wave. The hole is traction free and Poisson's ratio is 1/4. We solved this problem by using a regularised direct time domain BIEM. The boundary elements used are piecewise constant spatially, and piecewise linear in time. As (9) tells, this choice saves the effort of computing one of the double integrals in (9) because the derivatives of Ω in (9) are proportional to Dirac's delta, either with respect to space or time. In addition the remaining integrals can be evaluated in closed forms for any location of x. Hence the regularisation technique has a 'side effect' of making analytical integration easier. The boundary is devided into 32 elements of an equal length Δx. The time increment Δt is $\Delta t = \Delta x/c_L$ initially, and is doubled every 50 time steps. This choice seems to stabilise the time integration in problems where the solutions approach certain steady states as $t \to \infty$. Figure 3 shows the deformation ($u - u_I$ to be precise) on the boundary for several values of t. It is seen that the analysis is quite stable for a long period of time.

Table 1. Comparison of integration schemes in 3D elastostatics

Method	regularised		conventional	
G. int. pt.	4pt.	8pt.	4pt.	8pt.
$(1,1)$, $x_3 = 0$	0.5000	0.5000	-	-
$0.01a$	0.4699	0.4616	0.1998	0.5605
$0.1a$	0.2979	0.3014	0.3359	0.3023
$0.5a$	0.0866	0.0866	0.0872	0.0866
$1.0a$	0.0286	0.0286	0.0286	0.0286
$(3,3)$, $x_3 = 0$	0.5000	0.5000	-	-
$0.01a$	0.4923	0.4885	0.0554	0.3049
$0.1a$	0.4031	0.4051	0.5609	0.3925
$0.5a$	0.2151	0.2149	0.2117	0.2148
$1.0a$	0.1091	0.1091	0.1092	0.1091

5 Concluding remarks

(1) The regularisation of the double layer potential in elastostatics introduced in (8) is related to the formulation by Ghosh and Mukherjee[5] in that they also start from an equation similar to (7) except for the delta-Heaviside function term. A difference between our method and theirs is that they formulate a new integral equation method while we propose a method of computing the conventional influence coefficients. Also their kernel function is singular on a entire line passing through the singular point x of the fundamental solution Γ, but ours is singular on a ray emanating from x.

(2) In 2D statics one may integrate (2) between $x_0, x_1 (\in S)$ to have

$$\int_{x_0}^{x_1} g_i(x) dS = \int_S e_{ia} e_{lb} \partial_{x_a} \partial_{y_d} (\Phi(x_0 - y) - \Phi(x_1 - y))\, e_{ck} n_k(y) \partial_{y_c} \varphi_l(y) dS.$$

Acknowledgement

The authors wish to express their gratitude to Messrs. N. Nakamura & T. Yamamoto for their help in numerical work.

References

1. Nishimura, N.; Kobayashi, S.: A regularized boundary integral equation method for elastodynamic crack problems. Comp. Mech. 4 (1989) 319-328.

2. A joint work of N.N. with E. Bécache and J.C. Nédélec, to be published.

3. Polch, E.Z.; Cruse, T.A.; Huang, C.-J.: Traction BIE solutions for flat cracks. Comp. Mech. 2 (1987) 253-267.

4. Lachat, J.C.; Watson, J.O.: Effective numerical treatment of boundary integral equations. Int. J. Num. Meth. Eng. 10 (1976) 991-1005.

5. Ghosh, N.; Mukherjee, S.: A new boundary element method formulation for three dimensional problems in linear elasticity. Acta Mech. 67 (1987) 107-119.

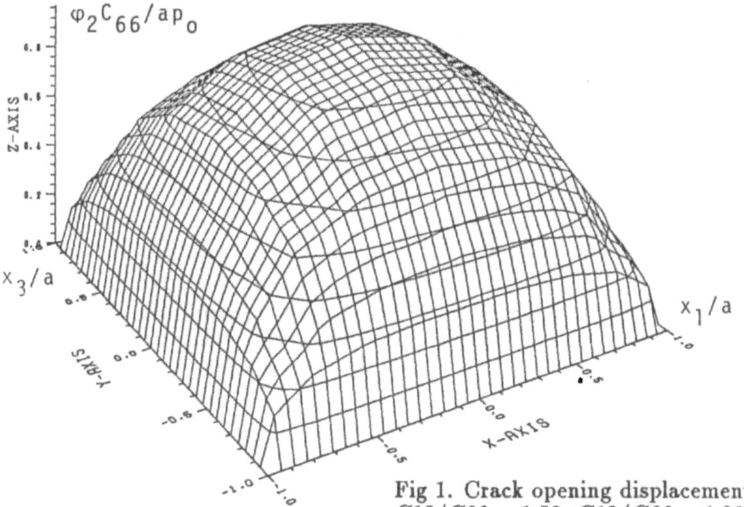

Fig 1. Crack opening displacements
$C12/C66 = 1.53$, $C13/C66 = 1.28$,
$C33/C66 = 3.67$, $C44/C66 = 0.98$

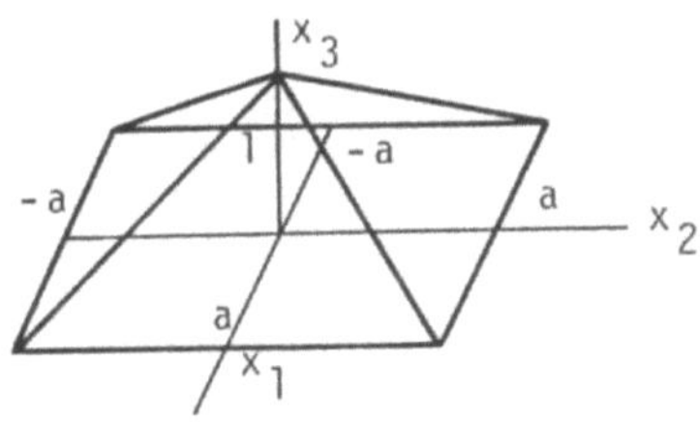

Fig. 2 Shape function

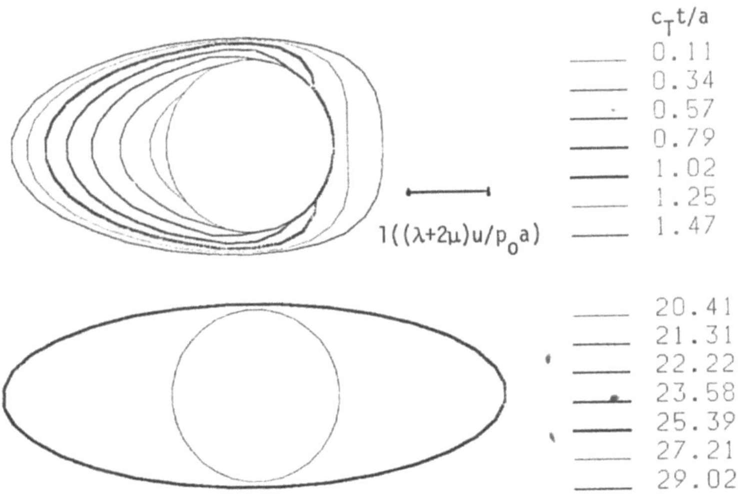

Fig. 3 Deformation of a circular hole subject to plane P wave

A B.E. Solution to Some Contact Problems with Adhesion

C. ALESSANDRI, A. TRALLI

Dipartimento di Costruzioni, Università di Firenze
Piazza Brunelleschi 6, 50121 Firenze, Italy

Summary

A B.E. numerical solution of contact problems between elastic solids in presence of adhesion (i.e. contact with finite bonding strength) is discussed and two paradigmatic examples are presented.

Introduction

A large number of papers have recently been devoted to particular unilateral contact problems characterized by a finite tensile strength. At this regard, for instance, the contact between solids made up of viscoelastic and low modulus materials can be quoted. In such cases a contact with adhesion occurs as a consequence of the action of the intermolecular forces of attraction and the traction distribution in the contact area, as well as the shape of the deformed surfaces outside this area, exhibit considerable discrepancies with those obtainable in the contact without adhesion /1-3/. Moreover, such a mechanical model can be usefully employed to study (and this is a more interesting feature) delamination problems between layers of composite elastic plates /4-6/; the results so obtained compare successfully with those of the fracture mechanics approach. As a matter of fact, the interlaminar stresses may cause debonding normally to the contact area as well as interlayer slip and, as known /7/, the first mode is crucial for the delamination of the laminate. In a number of papers /4-6,8/ the problem has been studied from a mathematical point of view in the context of the theory of the hemivariational inequalities, the existence of solutions has been established and the solution has been formulated as the minimum of the non convex Total Potential Energy. Numerical analyses have been carried out so far by using only F.E. Methods /9,10/. The numerical solution obtainable for such a class of problems by using direct B.E. Methods will be briefly discussed. As displacements and tractions on boundary surfaces are provided directly by the B.E. discretization, these Methods appear to be particularly suited to the resolution of contact problems in which the non linearity is confined to the contact area. For this type of analysis they should usually be preferred to the F.E. Methods which generally require a high discretization of the domain in order to get sufficiently accurate results. That gives rise to a high number of variables which cannot be used directly to define the behaviour of the contact area, but have to be expressed in terms of the contact ones. Suitable boundary variational formulations, previously derived /11,12/ for classical unilateral contact problems are revisited with reference to contact problems with adhesion. In Sect.2 the mechanical model of the contact between two solids in presence of finite bonding strength is shortly described, while in Sect.3 the B.E. discretization of the problem is introduced and two simple but yet significant and paradigmatic examples are presented.

Basic relationships for contact with adhesion

Let us consider, for the sake of simplicity, a plane contact problem between a linear elastic body and a rigid support. The body, subject to a state of plane stress (strain), occupies an open, bounded, simply-connected domain Ω of the plane z=0 in the Euclidean space R^3; its boundary $\Gamma\equiv\partial\Omega$ is sufficiently regular. Let us assume $\Gamma=\Gamma_o\cup\Gamma_q\cup\Gamma_c$. On Γ_o and Γ_q classical boundary conditions are defined, for instance

$$u_i = 0 \text{ on } \Gamma_o \qquad \sigma_{ij}n_j = \bar{q}_i \text{ on } \Gamma_q \qquad (i,j = 1,2) \qquad (1,2)$$

where $\underline{u}$ denotes the displacement vector, $\bar{\underline{q}}$ the prescribed external forces, $\underline{\sigma}$ the stress tensor and $\underline{n}$ the outer normal unit vector respectively; on Γ_c the body is in unilateral contact with the support. The frictionless and adhesionless case (e.g. the well known Signorini-Fichera problem) is described by the following relationships

$$w = -u_i n_i \geq 0, \qquad p = \sigma_{ij}n_j n_i \geq 0 \qquad (i,j = 1,2) \qquad (3_{1,2})$$

$$pw = 0 \qquad (3_3)$$

where w and p denote the normal components of the displacement and of the support reaction respectively. Let us remark that the solution of this simpler case can be discussed in the framework of the theory of the variational inequalities and it is usually achieved by minimizing the Total Potential Energy (in this case a convex quadratic functional) over the convex set (3_1); moreover the hypothesis $\Gamma_o\neq 0$ (rigid body motions are prevented) guarantees the existence and the uniqueness of the solution either in terms of $\underline{u}$ or in terms of p. In order to describe the finite bonding strength between the elastic solid and the support, a number of different relations have been proposed in the literature. The simplest and the most straightforward way is probably to define a relation between the support reaction p and the normal displacement w (spring model) /4,5/, such as

$$p(w) = \begin{vmatrix} kw & \text{if} & w \leq w_o & (k,\ w_o > 0) \\ 0 & \text{if} & w > w_o & \end{vmatrix} \qquad \begin{matrix}(4_1)\\(4_2)\end{matrix}$$

Even under this assumption, the strain energy of the spring defined in eqs. $(4_1,4_2)$ is not convex; consequently the solution turns out to be not unique and load path-dependent. The main critical remark which can be made about this model concerns the difficulty of defining the parameters k and w_o. An alternative approach consists of modelling the debonding process by introducing a superficial strain energy release rate, as in fracture mechanics. In particular, on the basis of thermodynamic considerations /2/, it is possible to demonstrate its coincidence with the Dupré energy of adhesion depending on the superficial and interface energies of the two solids in contact. Therefore the Total Potential Energy $\Pi(\underline{u})$ can be written (in absence of body forces) as follows

$$\Pi(\underline{u}) = \frac{1}{2}\int_{\Omega}\varepsilon_{ij}C_{ijhk}\varepsilon_{hk}d\Omega - \int_{\Gamma_q}\overline{q}_i u_i d\Gamma - \int_{\overline{\Gamma}_c \subset \Gamma_c}\overline{w}\, d\Gamma \qquad (5)$$

where $\varepsilon_{ij} = 1/2(u_{i,j} + u_{j,i})$ $(i,j = 1,2)$ represents the strain tensor components, C_{ijhk} $(i,j,h,k = 1,2)$ is Hooke's tensor of elasticity, $\overline{w}$ is the Dupré superficial energy and $\overline{\Gamma}_c \subset \Gamma_c$ denotes the surface of effective contact. The solution can be therefore formulated, for both models, as the problem of finding the minimum of the non convex Total Potential Energy on the convex set defined by inequality (3_1).
As a remarkable extension of the latter model, Frémond /8/ introduced a new kinematical variable function β, defined as intensity of adhesion, which represents the proportion of microscopic constraints still active between the body and the support. Therefore it yields

$$0 \leq \beta \leq 1 \qquad ; \qquad \beta\overline{w} = 0 \qquad (6_{1,2})$$

($\beta = 1$ denotes total adhesion, whereas $\beta = 0$ implies no adhesion). As a consequence, the energy term concerning the contact surface has to be modified into the form $\int_{\Gamma_c}\beta\overline{w}d\Gamma$ and relations $(6_{1,2})$ have to be added to (3_1) in order to define the admissible solutions. F.E. models, based on this approach, have been applied to the analysis of contact problems of viscoelastic solids in both the assumptions of small /9/ and large displacements /10/. As a final remark, the models presented so far for the contact with adhesion require the minimization of non convex and not everywhere differentiable functionals over convex and non convex sets. Moreover, the use of the trial and error procedure, often adopted in contact analysis with B.E.Ms. (see discussion in Ref.11) seems possible only by adopting relations $(4_1, 4_2)$ where the constraints on w and p are given in explicit form.

Boundary Element discretization and numerical results

Although the energy approaches to the B.E. Methods appear to be elegant and effective /13/, most of the technical literature and all the currently available codes are based on collocation procedures. Nevertheless, an estimate of the Total Potential Energy $\Pi(u)$ by means of standard B.E. techniques turns out to be quite simple and efficient. As a matter of fact, for any given effective contact area $\overline{\Gamma}_c \subset \Gamma_c$ in linear elastic solids under the hypotheses previously introduced, a trivial application of the classical Clapeyron Theorem provides the following identity

$$\frac{1}{2}\int_{\Omega}\varepsilon_{ij}C_{ijhk}\varepsilon_{hk}d\Omega - \int_{\Gamma_q}\overline{q}_i u_i d\Gamma = -\frac{1}{2}\int_{\Gamma_q}\overline{q}_i u_i d\Gamma \qquad (7)$$

Therefore, an estimate of $\Pi(\underline{u})$ can be obtained by simply evaluating the work done by the external forces.
The principal drawback of this approach is, in principle, the necessity of performing these computations for every possible contact configuration. It seems in fact difficult to apply an automatic procedure. An alternative approach, potentially useful for employing non convex non linear programming tools /14/, can be based on variational boundary formulations either in terms of di-

splacements or in terms of reactions p on Γ_c (surface of possible contact) /11,15/. In particular in References /11,12/, to which the reader is referred for details, an approach only in terms of p is developed. The following identity can be demonstrated

$$\frac{1}{2}\int_{\Omega}\varepsilon_{ij}C_{ijhk}\varepsilon_{hk}d\Omega - \int_{\Gamma_q}\overline{q}_i u_i d\Gamma = C + \tag{8}$$

$$+ \frac{1}{2}\int_{\Gamma_c}\int_{\Gamma_c} p(x) n_j G_{ij}(x,\xi) n_i p(\xi) d\Gamma d\Gamma -$$

$$- \int_{\Gamma_c}\int_{\Gamma_q} p(x) n_j G_{ij}(x,\xi)\overline{q}_i(\xi) d\Gamma d\Gamma$$

where $G_{ij}(x,\xi)$ is Green's function of the problem constrained only on Γ_o, and C a constant independent of p. The discretization is achieved by approximating Green's function with the numerical results obtained by means of standard collocation-based B.E. programs. The symmetry of the continuum problem is generally lost, whereas the coerciveness is preserved /12/. Moreover, it is possible to modify straightforwardly the reported relations in order to take into account rigid body motions and the contact between two or more deformable bodies. The numerical examples presented have been solved by an enumeration algorithm which does not require the evaluation of the derivative of the objective function. The definition of the most appropriate algorithm and the possibility of an extension to three-dimensional problems need further investigations.

The numerical applications concern the failure process of the adhesive junction between an elastic (low modulus) cylinder and a rigid foundation and the delamination problem of a rectangular plate bonded to a rigid (elastic) support and subjected to a line load along the short edge (peeling test).

Example 1. An infinitely long circular cylinder (R=100cm) resting on a rigid flat foundation is subjected, along its top, to a uniform load of variable intensity, Fig.1.a. The material is homogeneous and isotropic, with Young's modulus $E = 100\text{Ncm}^{-2}$ and Poisson's ratio $\nu = 0.33$, and the adhesive bond between the two solids is characterized by a 'work of adhesion' $\overline{w}=0.01\text{Ncm}^{-1}$. The boundary is discretized with 98 constant B.Es., closer-meshed along the supposed contact area, Fig.1.b. Under a compressive load the surfaces are pulled into contact by the adhesive forces over an area which exceeds that given by the Hertz theory. Reducing the load to zero leaves the surfaces adhering together and a tensile (negative) load is needed to make the contact width shrink further. Such a relationship between 'P' and 'a' is plotted in Fig.2 (B.E. solution for contact with adhesion) and compared with the corresponding Hertz analytical solution and B.E. unilateral solution. The tensile load can be increased up to a limit value, Fig.2, beyond which the adhesive junction is supposed to break /1/. B.E. vertical displacements of the load point are computed in both types of contact for each value of the compressive load 'P' and their variations are presented in Tab.I. In Fig.3 the surface tractions obtained numerically for a given P in the case of adhesive contact are compared with the corresponding analytical and numerical traction distributions.

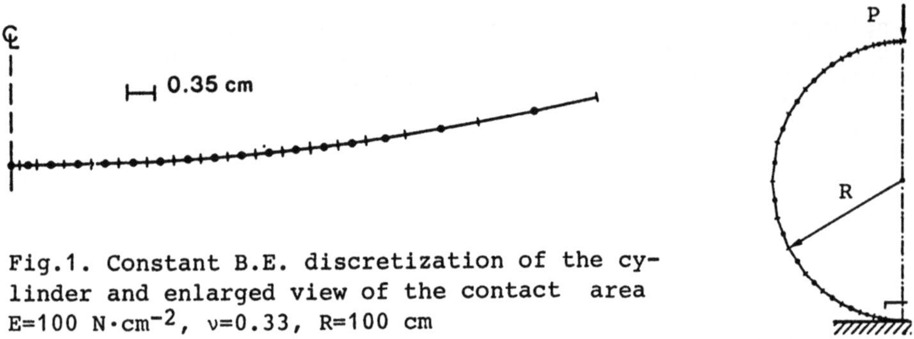

Fig.1. Constant B.E. discretization of the cylinder and enlarged view of the contact area $E=100\ N\cdot cm^{-2}$, $\nu=0.33$, $R=100$ cm

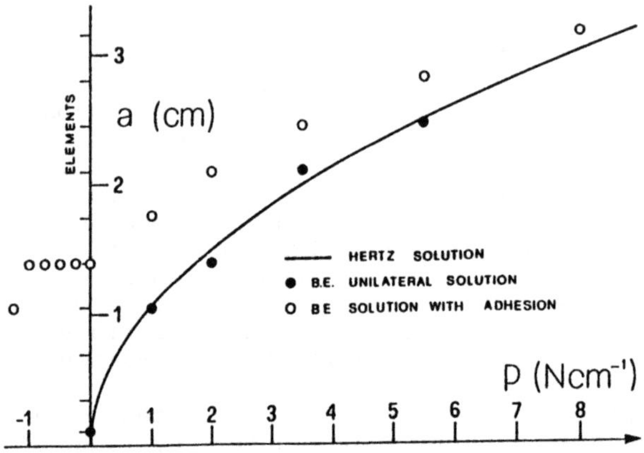

Tab.I

P(N)	$\Delta\delta\cdot 10^2$ cm
8.0	.0082
5.5	.0317
3.5	.0532
2.0	.1213
1.0	.1718
0.0	.4551

Fig.2. Variation of semi-contact width 'a' with load 'P'

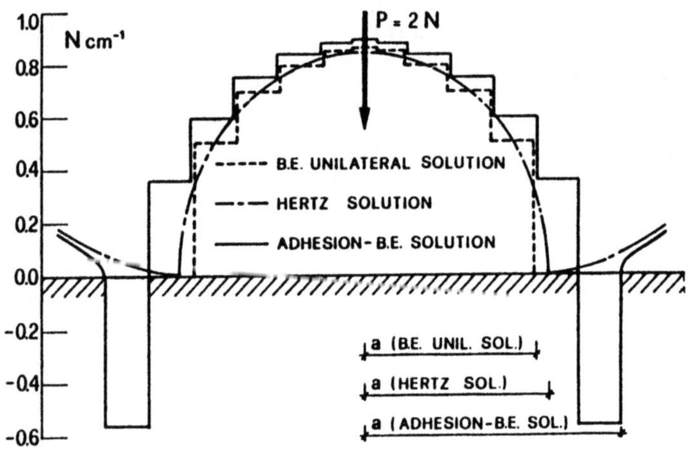

Fig.3. Surface tractions of an elastic cylinder in contact with a rigid plane

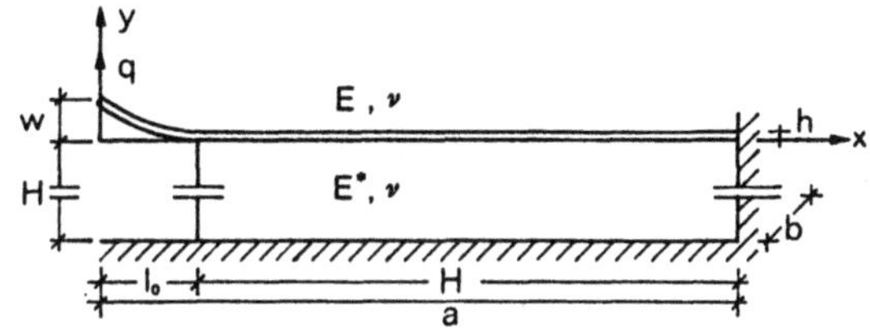

Fig.4. Delamination problem

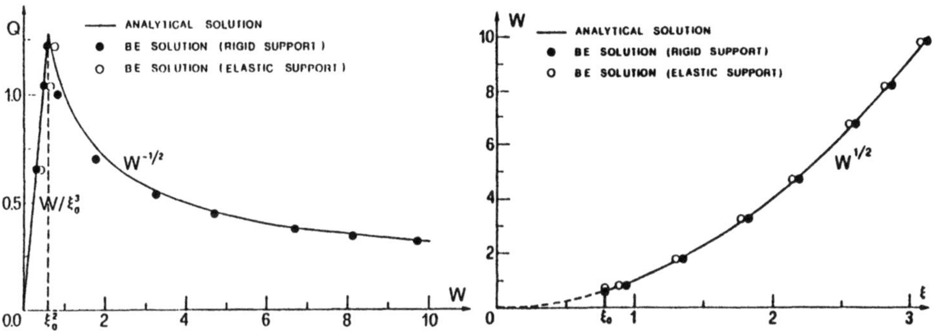

Fig.5.6. Nondimensionalized load-deflection and delamination-deflection diagrams

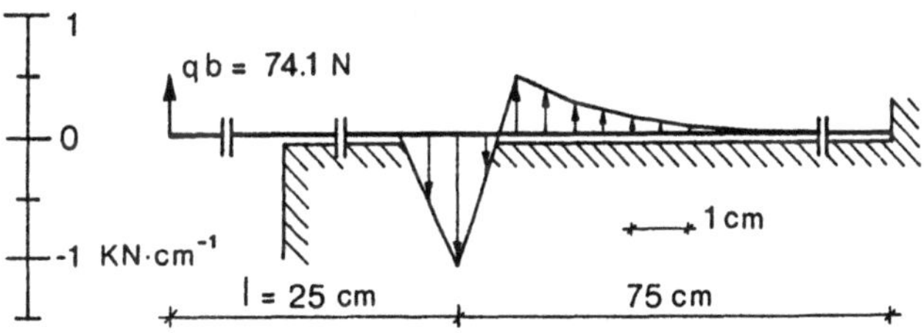

Fig.7. Elastic support: traction distribution on the contact area for a prescribed load.

	Anal. sol. Fract. Anal.		B.E. sol. rigid supp.		B.E. sol. elastic supp.		Tab.II
Q	W	ξ	W	ξ	W	ξ	
1.221	0.586	0.783	0.584	0.783	0.753	0.783	$a/h=10^3$, $a/b=10$
1.000	1.000	1.000	0.827	0.939	0.828	0.887	$a/L=5.22$
0.707	2.000	1.414	1.765	1.357	1.752	1.305	H=85 cm
0.534	3.500	1.870	3.257	1.827	3.237	1.775	$\xi_0=0.783$
0.447	5.000	2.236	4.710	2.193	4.685	2.140	$E=21*10^6 Ncm^{-2}$
0.377	7.000	2.645	6.717	2.610	6.687	2.558	$E^*=\infty$
0.343	8.500	2.915	8.114	2.871	8.081	2.819	$E^*=E*10^{-4}$
0.316	10.00	3.162	9.712	3.132	9.676	3.080	$\nu=0.3$

Example 2. A narrow plate of length 'a' and width 'b' is welded to a support, as shown in Fig.4. The load-deflection relationship is analysed when the plate is subjected to a uniform line load along the edge x=0. As the problem has usually been tackled by a fracture mechanics approach /4,5/ (mode I fracture), the numerical results are compared with the corresponding ones obtained analytically, via fracture analysis. The plate is modelled by finite one-dimensional beam elements (100 elements of constant length), whereas the base is discretized by linear B. Es. (85 for the upper edge and 17 for each of the other sides). By analogy with the fracture mechanics approach, the delamination is assumed to occur when the energy release $T = -d\Pi/d\ell$ equals the adhesive surface energy $\overline{w}$.b. By introducing the non dimensionalized quantities /4,5/

$$L = \sqrt{2D/\overline{w}} \qquad Q = q_o/\overline{w} \qquad W = 1.5\ w/L \qquad \xi = \ell/L \tag{9}$$

where D is the flexural rigidity of the plate and by assuming

$$a/b = 10 \qquad a/h = 10^3 \qquad a/L = 5.22 \qquad \xi_o = 0.783 \tag{10}$$

the nondimensionalized load-deflection and tip deflection-delamination relationships plotted in Fig. 5,6 can be obtained. The values of ξ and W, computed analytically for a prescribed Q, are compared with the corresponding ones obtained via B.Es. in Tab. II. In Fig.7 the traction distribution for a given P in the case of adhesive contact with an elastic support is presented. The computations have been performed with an Olivetti OC3250.

Acknowledgement

The financial support of the Italian Ministry of Education (M.P.I. 40% funds) is gratefully acknowledged.

References

1. Johnson, K.L.: Contact Mechanics. Cambridge University Press, (1985).
2. Maugis, D.; Barquins M., C.R. Acad. Sc. Paris., t. 287, (1978).
3. Stolarski, T.A., Int. J. Engng. Sc., 2, (1989), 169-179.
4. Reddy, J.N.; Grimaldi, A., Rep. VPJ-E-82-83, Virginia Polytechnic Inst. and State Univ., Blacksburg, VA 24061, (1982).
5. Ascione, L.; Bruno, P.; Grimaldi, A., Rep. 62, Univ. of Calabria,(1983).
6. Naniewicz, Z.; Wozniak, Cz., Ing. Archiv, 58, (1988), 403-412.
7. Jones, R.: Mechanics of Composite Materials. Mc-Graw Hill, New York , (1985).
8. Frémond, M., Journal de Mécanique Théorique et appliquée, 6, N.3,(1987), 103-407.
9. Le Tallec, P.; Lotfi, A., C.M.A.M.E., 68, (1988), 67-82.
10. Point, N., Math. Methods in the Appl. Sciences, 10, (1988), 367-381.
11. Tralli, A.; Alessandri, C.; Alliney, S., Boundary Elements X, Proc.10th Int. Conf., C.M.P., Southampton, (1988), 339-352.
12. Alliney, S.; Tralli, A.; Alessandri C., paper accepted for publication on Computational Mechanics.
13. Polizzotto, C., C.M.A.M.E., 69, (1988), 263-276.
14. Forgò, F.: Non Convex Programming, Akadémiai Kiadò, Budapest, (1988).
15. Lazaridis, P.P.; Panagiotopoulos P.D., Comp. & Struct., 25, 1, (1987), 35-49.

Calculation of Aerodynamic Sensitivities by Boundary-Integral Methods and Application to Lifting-Surface Theory

E. CARSON YATES, JR.
ROBERT N. DESMARAIS

NASA Langley Research Center
Hampton, Virginia 23665, U.S.A.

Abstract

Methodology for analytically generating the aerodynamic sensitivities (rates of change of surface pressures and aerodynamic coefficients with respect to changes in aircraft geometry), needed in aerodynamic shape optimization and multidisciplinary design synthesis, has been derived and illustrated with lifting-surface theory. The arbitrary continuous or discontinuous global or local perturbation is defined in terms of a single variable. The procedure is adaptable to surface-panel and even nonlinear integral methods without the prohibitively expensive perturbation of each panel node as proposed by others.

Introduction

Accurate and efficient computation of aerodynamic sensitivities (rates of change of surface pressures and weighted integrals of pressures with respect to changes in aircraft geometry) is of growing importance for aerodynamic shape optimization and multidisciplinary design synthesis. Such sensitivities can, of course, be obtained approximately by differencing the results of solutions for a base configuration and a geometrically perturbed configuration. Resulting sensitivity values, however, are dependent upon the size of the perturbation and are subject to cancellation errors. Calculating sensitivities by analytical methods avoids such problems and offers possibilities for significant improvements in computational efficiency.

Some capability to generate sensitivities has been developed from a steady-state surface-panel method [1], but applications have been limited to perturbations of lifting surfaces in the direction normal to the chord plane.

This paper presents much simpler methodology for generating aerodynamic sensitivities for general perturbations, including planform variations. The procedure is illustrated with subsonic unsteady nonplanar lifting-surface theory [2]. Corresponding expressions have been obtained with sonic and supersonic lifting-surface theory but are not included because of space limitation. In addition to steady-state applications, the expressions derived here can be used to track flutter and other dynamic-response characteristics in multidisciplinary design processes. The geometric perturbation is defined in terms of a single variable. Moreover, considerable simplification is accomplished without loss of generality by varying only the normalwash distribution and/or the streamwise planform coordinates. Note that aspect ratio and even spanwise location of a leading-edge or trailing-edge break point can be varied by perturbing only streamwise coordinates. The user needs to provide only simple analytical expressions or similar means for defining the continuous or discontinuous global or local perturbations of interest. In contrast, the method of [1] requires that the perturbation be made up of individual perturbations of each of the panel nodes - a computationally expensive operation. Lifting-surface-theory calculations are themselves quite fast so that computation of sensitivities by differencing is not prohibitively expensive. Even so, the analytical expressions presented here provide convenient means for computation as well as accurate benchmark criteria for assessing the accuracy of differencing so that, when needed, differencing can be used with better understanding and confidence. The primary value of the present process, however, lies in the fact that it appears to be readily adaptable to more general surface-panel methods and even to nonlinear integral methods.

Theoretical Development

The integral equation relating lifting pressure $\Delta\bar{C}_p(\xi,\eta)$ and normalwash $\bar{W}_N(x,y,z)$ distributions on a nonplanar lifting surface (Fig. 1) oscillating in subsonic flow can be solved in the form

$$\sum_n \sum_m a_{nm} F_{nm} - 8\pi \frac{b_o}{\ell} \bar{W}_N = 0 \tag{1}$$

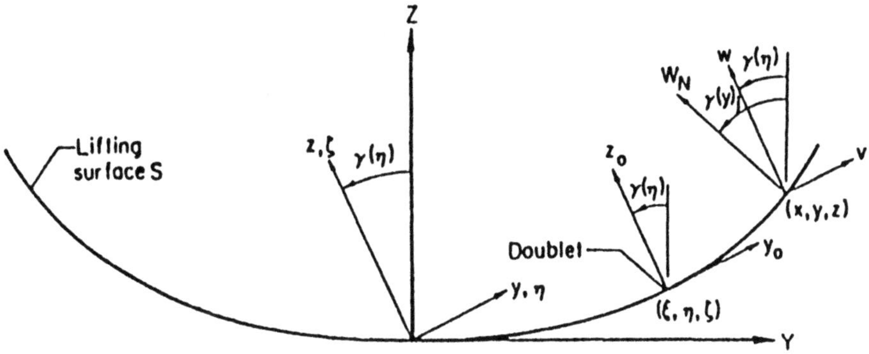

Fig. 1 Nonplanar lifting surface as seen from downstream.

where $\bar{W}_N$ is normalized with respect to freestream speed, and $\Delta\bar{C}_p$ is represented by a linear combination of chordwise $f_n(\hat{\xi})$ and spanwise $g_m(\hat{\eta})$ shape functions with coefficients a_{nm}

$$\Delta\bar{C}_p(\hat{\xi},\hat{\eta}) = \sqrt{1-\hat{\eta}^2}\sqrt{\frac{1-\hat{\xi}}{1+\hat{\xi}}}\, b^{-1}(\hat{\eta}) \sum_n \sum_m a_{nm} f_n(\hat{\xi}) g_m(\hat{\eta}) \tag{2}$$

All lengths are in units of $b_o = b(0)$, where $b(\eta)$ is semichord; ℓ is arc length of semispan in cross-stream direction. Coordinates X,x,ξ are streamwise, and the planform has been mapped into a unit square on which the coordinates are designated by (^).

$$F_{nm} = \oint_{-1}^{1}\int_{-1}^{1} \sqrt{1-\hat{\eta}^2}\sqrt{\frac{1-\hat{\xi}}{1+\hat{\xi}}}\, f_n(\hat{\xi}) g_m(\hat{\eta}) \Big\{ K_W \cos[\gamma(y)-\gamma(\eta)] - K_V \sin[\gamma(y)-\gamma(\eta)] \Big\}\, d\hat{\xi}\, d\hat{\eta} \tag{3}$$

The components of the kernel function are given by [2]

$$K_V = e^{-ikx_o} \frac{y_o z_o}{r_o^3} \left\{ k\left[P_1(k_1) + Q_1(k_1) - i2\int_0^{u_1} \frac{u e^{-ik_1 u}}{\sqrt{1+u^2}}\, du - k_1 \int_0^{u_1} \frac{e^{-ik_1 u}}{\sqrt{1+u^2}}\, du \right] + \left[\frac{x_o}{Rr_o}\left(2 + \frac{\beta^2 r_o^2}{R^2}\right) + i\,\frac{k_1}{R}\left(1 + M\,\frac{x_o}{R}\right) \right] e^{-ik_1 u_1} \right\} \tag{4}$$

$$K_W = \frac{z_o}{y_o} K_V - e^{-ikx_o} \left\{ \frac{x_o}{Rr_o^2} e^{-ik_1u_1} + \frac{k}{r_o}\left[P_1(k_1) - i\int_0^{u_1} \frac{ue^{-ik_1u}}{\sqrt{1+u^2}}\,du\right]\right\} \tag{5}$$

$x_o = x - \xi$, $y_o = y - \eta$, $z_o = z - \zeta$, $r_o = (y_o^2 + z_o^2)^{1/2}$, $R = (x_o^2 + \beta^2 r_o^2)^{1/2}$, $u_1 = (MR - x_o)/\beta^2 r_o$, $\beta^2 = (1 - M^2)^{1/2}$, M is free-stream Mach number; k is reduced frequency, and $k_1 = kr_o$. $P_1(k_1)$ and $Q_1(k_1)$ are linear combinations of modified Bessel and Struve functions $I_1(k_1)$, $K_1(k_1)$, $L_1(k_1)$. The basic aerodynamic sensitivity is

$$\frac{\partial \Delta \bar{C}_p}{\partial Q} = \sqrt{1 - \hat{\eta}^2}\sqrt{\frac{1-\hat{\xi}}{1+\hat{\xi}}}\left[\frac{\partial b^{-1}(\hat{\eta})}{\partial Q}\sum_n\sum_m a_{nm} f_n(\hat{\xi}) g_m(\hat{\eta}) + b^{-1}(\hat{\eta})\sum_n\sum_m \frac{\partial a_{nm}}{\partial Q} f_n(\hat{\xi}) g_m(\hat{\eta})\right] \tag{6}$$

Q represents the geometric perturbation of the lifting surface which, without loss of generality, will be performed by varying only the normalwash distribution $\bar{W}_N$ and/or the streamwise coordinates x and ξ in order to avoid spanwise derivatives of K_V and K_W which involve a third-order singularity at $r_o = 0$. Now $\partial b^{-1}(\hat{\eta})/\partial Q$ is purely geometrical; consequently, attention is focused on evaluating $\partial a_{nm}/\partial Q$ by differentiating (1) with respect to Q

$$\sum_n\sum_m \left(a_{nm}\frac{\partial F_{nm}}{\partial Q} + \frac{\partial a_{nm}}{\partial Q} F_{nm}\right) - 8\pi \frac{b_o}{\ell}\frac{\partial}{\partial Q}\bar{W}_N = 0 \tag{7}$$

F_{nm} and a_{nm} are obtained from solution of (1) for the base configuration, and $\partial \bar{W}_N/\partial Q$ is geometrical and known a priori. Eq. (7) shows that if perturbation Q does not involve a planform change, then $\partial F_{nm}/\partial Q = 0$, and the required values of $\partial a_{nm}/\partial Q$ may be obtained from direct solution of (1) with the normalwash perturbation $\partial \bar{W}_N/\partial Q$ input in place of $\bar{W}_N$. Differentiating (3) with respect to Q gives

$$\frac{\partial F_{nm}}{\partial Q} = \int_{-1}^{1} \int_{-1}^{1} \sqrt{1 - \hat{\eta}^2} \sqrt{\frac{1 - \hat{\xi}}{1 + \hat{\xi}}} f_n(\hat{\xi}) g_m(\hat{\eta}) \left\{ \frac{\partial K_W}{\partial x_o} \cos[\gamma(y) - \gamma(\eta)] - \frac{\partial K_V}{\partial x_o} \sin[\gamma(y) - \gamma(\eta)] \right\} \frac{\partial x_o}{\partial Q} \, d\hat{\xi} \, d\hat{\eta} \tag{8}$$

Note that $$\frac{\partial x_o}{\partial Q} = \frac{\partial (x - \xi)}{\partial Q} = \left.\frac{\partial x}{\partial Q}\right|_{x,y,z} - \left.\frac{\partial x}{\partial Q}\right|_{\xi,\eta,\zeta} \tag{9}$$

is the only quantity in (8) that depends on the particular perturbation Q. Differentiating (4), (5) gives the required expressions for $\partial K_V/\partial x_o$, $\partial K_W/\partial x_o$.

$$\frac{\partial K_V}{\partial x_o} = e^{ikx_oM^2/\beta^2} \, 3\beta^4 \frac{y_o z_o}{R^5} \left\{ \left[1 + i \frac{kMR}{\beta^2} + \frac{1}{3}\left(i \frac{kMR}{\beta^2} \right)^2 \right] e^{-i\frac{kMR}{\beta^2}} \right\} - ikK_V \tag{10}$$

$$\frac{\partial K_W}{\partial x_o} = e^{ikx_oM^2/\beta^2} \beta^2 \left\{ \frac{-1}{R^3}\left(1 + i \frac{kMR}{\beta^2}\right) + 3 \frac{\beta^2 z_o^2}{R^5} \left[1 + i \frac{kMR}{\beta^2} + \frac{1}{3}\left(i \frac{kMR}{\beta^2} \right)^2 \right] \right\} e^{-i\frac{kMR}{\beta^2}} - ikK_W \tag{11}$$

The singularities of K_V and K_W ((4) and (5)), and in particular the r_o^{-2} singularity, remain in $\partial K_V/\partial x_o$ and $\partial K_W/\partial x_o$ ((10) and (11)) only for unsteady flow. For both steady and unsteady flow, however, $\partial K_W/\partial x_o$ contains a third-order singularity (R^{-3}) at the point $x_o = y_o = z_o = 0$ where $\partial x_o/\partial Q$ is zero. Since $\partial x_o/\partial Q$ is antisymmetric, the integral in Eq. (8) is a Cauchy integral. A third-order singularity in $\partial K_W/\partial x_o$ along the line $r_o = 0$ has been avoided by not perturbing in y_o direction.

Eqs. (8) to (11) permit (7) to be solved for the required values of $\partial a_{nm}/\partial Q$ which, in turn, permit evaluation of $\partial \Delta \bar{C}p/\partial Q$ from (6). If this procedure is employed for a number of different perturbations, only the simple geometric expressions $\partial b^{-1}/\partial Q$ in (6) and $\partial x_o/\partial Q$ in (8) need be reevaluated for each. After the basic sensitivity is determined from Eq. (6), sensitivities of other aerodynamic parameters can be obtained by direct integration.

Applications

To illustrate the validity of the present procedure, it has been applied to an isolated untapered wing in steady flow at Mach number 0. The right side of Fig. 2 shows the variation of lifting pressure ΔC_p with sweep angle Λ for aspect ratio 4 and with aspect ratio A for sweep angle 40° at one point on the wing (0.25 semispan, 0.30 chord). The curves were obtained by direct calculation with the base method (Eqs. (1)-(3)). The slopes of the tangent lines are the desired sensitivities, $\partial \Delta C_p/\partial \Lambda$ and $\partial \Delta C_p/\partial A$. The left side of the figure shows these sensitivities as calculated by differencing, by spline fit to the base-method values of ΔC_p, and by the present analytical method (dash lines). For variation of sweep angle, the three results are in good agreement. For variation of aspect ratio, however, there are some differences because of the wiggly curve generated by the base method.

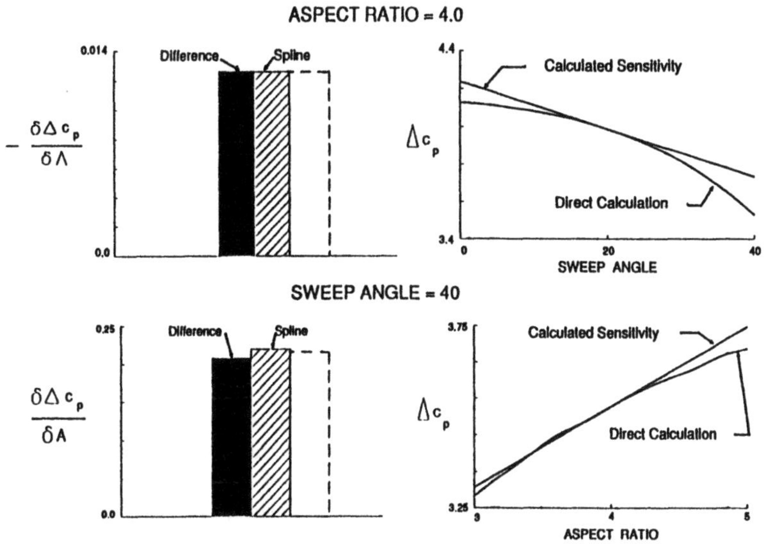

Fig. 2 Aerodynamic sensitivities for untapered wing.

Concluding Remarks

Analytical expressions for general aerodynamic sensitivities have been derived using nonplanar lifting-surface theory for steady and unsteady flow. The geometric perturbation is defined in terms of a single variable, and the user need only provide simple expressions or similar means for defining perturbations of interest. In addition to direct computational use, the method should provide benchmark criteria for assessing the accuracy of aerodynamic sensitivities obtained by approximate methods. The present process appears to be readily adaptable to more general surface-panel methods and nonlinear integral methods.

References

1. Hawk, J. D.; and Bristow, D. R., Development of MCAERO Wing Design Panel Method With Interactive Graphics Module. NASA CR-3775 (1984).

2. Yates, E. C., Jr., A Kernel-Function Formulation for Nonplanar Lifting Surfaces Oscillating in Subsonic Flow. AIAA Journal, Vol. 4, No. 8, pp. 1486-1488 (1966).

Aerodynamic Surface Element Modeling for Analysis of Aircraft Flight Dynamics

A. R. Dusto, M. A. Epton and M. F. Fitzpatrick
The Boeing Company, Seattle, Washington

1: Introduction

In this paper we describe aerodynamic boundary elements of an aeroservoelastic (ASE) analysis method used in designing aircraft flight control systems. The ASE analysis method generates flexible aircraft equations of motion that are integrated in time by a flight simulation computer program. Together, the ASE analysis method and the simulation program perform two operations required by the design process: (1) they evaluate control system performance by evaluating the controlled flight dynamics of candidate aircraft designs, (2) they generate the sensitivity of control performance to variations in control system design parameters.

The simulation program must execute in real time for pilot-in-the-loop and hardware-in-the-loop simulations and this requirement has an important influence on the operations performed by the ASE analysis method. The equations of motion must be expressed with some of their nonlinear terms evaluated in advance and stored as tabular data. This tabular data is then interpolated as the simulation program executes its numerical integration procedure. In general, the data represents the nonlinear terms as first order expansions in perturbation motion variables that describe deviations from a datum motion. As an example, assume the motions to be simulated are nearly coordinated, steady maneuvers. The simulated motion can then be described as a small deviation from a datum motion consisting of a nonrotating state of steady acceleration normal to the planform of the aircraft. The datum state is evaluated by solving a steady aeroelastic problem posed by the governing equations for a flight condition specified by: load factor, Mach number, dynamic pressure, and mass distribution (fuel and payload). The ASE analysis method formulates and solves this problem repeatedly over a set of flight conditions that form an interpolation point covering of the aircraft's flight envelope.

Nonlinear generalized aerodynamic force terms are primary quantities that must be evaluated by the ASE analysis method. At each interpolation point flight condition, the ASE analysis method must evaluate, not only the generalized aerodynamic forces in the datum state but, the coefficients appearing in first order expansions of these quantities about their values in the datum state. These evaluations incorporate the influence of the deformed shape and aerodynamics of the aircraft in the interpolation point flight conditions; in addition, they contain the effects of elastic deformation caused by the simulated motion deviating from the datum motion. Consequently, the aerodynamic coefficients must be evaluated including the effect of a small, but arbitrary, elastic deformation. The following describes a means of formulating aerodynamic boundary elements for evaluating that effect.

2: Structures Formulation

The simulation equations are derived from the principle of least action following the approach of Morino and Baillieul[1]. As a result, the elements of the rotation matrix are the generalized coordinates describing inertial orientation of an aircraft. The generalized coordinates of elastic deformation, however, are the node point generalized displacement components of a finite element model of the aircraft structure, in contrast with Morino and Baillieul. Further, elastic stiffness is expressed in terms of a stiffness matrix generated using the principle of minimum potential energy from classical theory of elasticity; hence, rotations of the structure due to elastic deformation are restricted to an infinitesimal order of magnitude. As a consequence, our equations appear to be more restrictive than those of Morino and Baillieul which admit arbitrary elastic deformations. The engineering practices employing the equations of motion, however, generate the elastic deformation in small increments, viewed in a numerical integration in time as occurring in a temporal sequence. The procedure is that commonly used in which large deformations, relative to an initial undeformed configuration, are the result of a sequence of small deformations relative to a sequence of undeformed configurations.

The choice of structural finite elements affects the accuracy of the structural/aerodynamic interface. The nature of the approximation follows from the choice of the basis functions $\mathbf{N}_j(p)$ that describe the displacement of material point p induced by a unit value of node point degree of freedom u_j. Using these basis functions, the displacement of every material point relative to the undeformed configuration is described as follows:

$$\mathbf{d}(p,t) = \sum_j \mathbf{N}_j(p)u_j(t). \tag{2.1}$$

The fidelity of the basis functions $\mathbf{N}_j$ affects the accuracy of the nodal point aerodynamic forces, given by the usual formula

$$\{p_j^A\} = -\iint_S \mathbf{p}(p,t)\cdot \mathbf{N}_j(p)dS. \tag{2.2}$$

where $\mathbf{p}(p,t)$ is the surface pressure derived from the aerodynamic solution. In section 4, we show another method for calculating p_j^A using the displacement formula (2.1) together with an alternative application of the principle of virtual work.

3: Aerodynamic Formulation, Summary of Standard Panel Technology

The current aerodynamic formulation used in our ASE analysis is based upon the panel method techniques developed in PAN AIR and A502, for both steady and unsteady flow in the subsonic and supersonic flight regimes. Those techniques are thoroughly discussed in references[2–4]. The key idea underlying this formulation is that the perturbation to the free-stream velocity field is expressed as the gradient of a potential function ϕ satisfying a linear partial differential equation with constant coefficients. A free space kernel function is found for the partial differential operator, and from this an integral representation formula is developed for ϕ. An integral equation is then obtained by imposing appropriate boundary conditions.

The governing partial differential equation for the velocity potential function ϕ is derived from the small disturbance wave equation written in the undisturbed-air coordinate system, $\{y_i\}_{i=1}^3$, by transforming that equation to a coordinate system $\{x_i\}_{i=1}^3$ moving in the $-y_1$ direction with velocity V. The resulting differential equation reads,

$$\left[\nabla_x \cdot \nabla_x - \frac{1}{a_\infty^2}\left(\frac{\partial}{\partial t} + V\frac{\partial}{\partial x_1}\right)^2\right]\phi = 0 \tag{3.1}$$

where a_∞ denotes the sound speed. The velocity of a fluid particle relative to the $\mathbf{x}$ coordinate system, neglecting atmospheric disturbance, is given by the expression: $\mathbf{V} = (V, 0, 0) + \nabla_x \phi = \mathbf{V}_\infty + \nabla_x \phi$.

Partitioning ϕ into steady and unsteady components, $\phi = \phi^S + \phi^U$, ϕ^S is found to satisfy the Prandtl-Glauert equation at Mach number $M = V/a_\infty$,

$$\left[\nabla_x \cdot \nabla_x - M^2 \frac{\partial^2}{\partial x_1^2}\right] \phi^S = 0. \tag{3.2}$$

For unsteady flows, we assume that ϕ^U has the form $\hat{\phi}_\omega e^{\iota\omega t}$; thus $\hat{\phi}_\omega$ satisfies

$$\left[\nabla_x \cdot \nabla_x - \frac{1}{a_\infty^2}\left(\iota\omega + V\frac{\partial}{\partial x_1}\right)^2\right] \hat{\phi}_\omega = 0. \tag{3.3}$$

Rather than working with the Fourier component function $\hat{\phi}_\omega$, we find it convenient to define the allied potential ϕ^* by

$$\phi^* = e^{-\iota\Omega M x}\hat{\phi}_\omega, \qquad \Omega = (\omega M)/[V(1-M^2)] \tag{3.4}$$

where Ω denotes the Helmholtz frequency. It is readily verified that ϕ^* satisfies

$$\left[(1-M^2)\frac{\partial^2}{\partial x_1^2} + \frac{\partial^2}{\partial x_2^2} + \frac{\partial^2}{\partial x_3^2} + \Omega^2(1-M^2)\right] \phi^* = 0. \tag{3.5}$$

The boundary conditions satisfied by the potential function ϕ are the usual ones of an impermeable surface condition on the aircraft's wing and body and a zero pressure jump condition on the wake surface trailing behind.

Given the partial differential equations (3.2) for steady flow (or (3.5) for unsteady flow), the standard panel techniques employ an *integral representation formula* of ϕ^S (or ϕ^*) having the following form:

$$\phi(\mathbf{r}) = -\iint_S \sigma(\mathbf{q})\psi(\mathbf{q},\mathbf{r})dS_q + \iint_S \mu(\mathbf{q})\frac{\partial\psi(\mathbf{q},\mathbf{r})}{\partial\nu_q}dS_q \tag{3.6}$$

to obtain an integral equation, or a set of integral equations, for the unknown functions σ, μ. In this equation, $\phi(\mathbf{r})$ denotes the potential function, ϕ^S for steady flow, ϕ^* for unsteady flow. S is the singularity surface across which ϕ experiences a jump of $\mu = [\phi]$ and the conormal derivative of ϕ experiences a jump of $\sigma = [\partial\phi/\partial\nu]$, where the conormal vector $\boldsymbol{\nu}$ is defined in terms of the normal vector $\mathbf{n}$ by $\boldsymbol{\nu} = [(1-M^2)n_x, n_y, n_z]$. The kernel function ψ denotes the fundamental free-space Green's function for the governing partial differential equation, (3.2) for steady flow, (3.5) for unsteady flow. There are a variety of ways in which the representation formula (3.6) can generate an integral equation. If, for the moment, we restrict our attention to potential flow around a sphere at $M = 0$, (for which $4\pi\psi = 1/||\mathbf{r}-\mathbf{q}||$), then at least four possible formulations are possible. These correspond to four separate choices about the nature of the fictitious flow *internal* to the sphere. Note that in each formulation, the integral expressions are always to be evaluated by computing the average value of the given integral with $\mathbf{r}$ taken to be slightly above and slightly below S.

Method 1 (Hess; $\mu = 0$)

$$\frac{1}{2}\sigma(\mathbf{r}) - \frac{\partial}{\partial n_r}\left(\iint_S \sigma\psi dS_q\right) = -\mathbf{V}_\infty \cdot \mathbf{n} \tag{3.7a}$$

Method 2 ($\phi = 0$ internal to the sphere, freestream inside: $\phi_L = 0$; $\sigma = -\mathbf{V}_\infty \cdot \mathbf{n}$)

$$\frac{1}{2}\mu(\mathbf{r}) - \left(\iint_S \mu\frac{\partial\psi}{\partial n_q}dS_q\right) = -\iint_S \sigma(\mathbf{q})\,\psi\, dS_q \tag{3.7b}$$

Method 3 ($\Phi = 0$ internal to the sphere: interior stagnation; $\sigma \equiv 0$)

$$\frac{1}{2}\mu(\mathbf{r}) - \left(\iint_S \mu \frac{\partial\psi}{\partial n_q} dS_q\right) = \mathbf{V}_\infty \cdot \mathbf{r} \tag{3.7c}$$

Method 4 ($\phi = 0$ interior to the sphere; $\partial\phi/\partial n = -\mathbf{V}_\infty \cdot \mathbf{n}$ exterior to the sphere.)

$$0 + \frac{1}{2}\mu(\mathbf{r}) + \left(\iint_S \sigma\psi dS_q\right) - \left(\iint_S \mu \frac{\partial\psi}{\partial n_q} dS_q\right) = 0 \tag{3.7d}$$

$$\frac{1}{2}\sigma(\mathbf{r}) + 0 - \left(\iint_S \sigma \frac{\partial\psi}{\partial n_r} dS_q\right) + \left(\iint_S \mu \frac{\partial^2\psi}{\partial n_r \, \partial n_q} dS_q\right) = -\mathbf{V}_\infty \cdot \mathbf{n}(\mathbf{r}) \tag{3.7e}$$

Notice that methods 2 and 3 involve the same integral operator applied to μ, and that method 4 generates a Hermitian pair of integral equations. There is a fair amount of folklore that method 4 produces the most accurate solutions, and it is our conjecture that this is due to a general niceness associated with Hermitian integral operators.

Once an integral equation is formulated, the task of solving it is accomplished by a collocation procedure, for which finite dimensional representations of the unknown functions are developed from the user provided panelling of S. These take the form

$$\sigma(\mathbf{q}) \approx \sum_\beta \bar{s}_\beta(\mathbf{q})\,\sigma_\beta; \qquad \mu(\mathbf{q}) \approx \sum_\beta \bar{m}_\beta(\mathbf{q})\,\mu_\beta \tag{3.8}$$

where the basis functions $\bar{s}_\beta$, $\bar{m}_\beta$ are completely determined by the networks of panels that specify the singularity surface S. When the unknown function is $\mu(\mathbf{q})$ as in formulation 2, the coefficients μ_β are determined by enforcing the integral equation (3.7b) at a number of points equal to the number of unknown values of μ_β. In the more general case (as in formulation 4), each singularity parameter σ_β, μ_β is associated with the enforcement of a specific integral equation at a specific collocation point.

In general the implementation of the collocation scheme requires the evaluation of source and doublet integrals over S having the form

$$\phi_s(\mathbf{r}) = \iint_S \bar{s}(\mathbf{q})\,\psi\, dS_q \qquad \phi_d(\mathbf{r}) = \iint_S \bar{m}(\mathbf{q})\,\partial\psi/\partial\nu_q\, dS_q \tag{3.9a}$$

$$\mathbf{v}_s(\mathbf{r}) = \nabla_r \phi_s(\mathbf{r}) \qquad \mathbf{v}_d(\mathbf{r}) = \nabla_r \phi_d(\mathbf{r}) \tag{3.9b}$$

where $\bar{s}$ and $\bar{m}$ are polynomial functions of local coordinates on flat subsurfaces $Q \subset S$. The evaluation of $\mathbf{v}_d$ presents some special challenges in the case that $M > 1$. In that case it is necessary to assume that the functions $\bar{m}$ are continuous in $\mathbf{q}$ and go to zero along the boundary of S in order to remove the nonintegrable line vortex singularity.

An interesting computational difficulty arises in the integral evaluation for supersonic unsteady flow. The kernel function in that case has the form $2\pi\psi = \cos(\Omega R)/R$, where R satisfies: $R^2 = (r_1 - q_1)^2 - (M^2 - 1)[(r_2 - q_2)^2 + (r_3 - q_3)^2]$. The difficulty is that it is quite possible for the total phase variation $\Omega(R_{max} - R_{min})$ of the cosine argument to be fairly large, say 10 radians, even if the expression ΩD, where D is the panel diameter, is bounded by 1. The reason that this can happen is that the function $R(\mathbf{r}, \mathbf{q})$ can vary quite rapidly for points $\mathbf{q}$ lying near to the boundary of a control point $\mathbf{r}$'s domain of dependence. Although this problem can occur for only a small fraction of the total number of integrals required, it must nevertheless be attended to with appropriately robust numerical techniques.

4: Approaches to interfacing structures to aerodynamics

An ASE analysis method must solve the coupled governing equation of structures and aerodynamics. In this section we explore the formulation of those equations in a way that permits substitution of alternative technologies for either the structural or the aerodynamic analysis. Of particular concern at Boeing is the requirement that the formulation interface smoothly with the transonic finite element code, TRANAIR[5]. We now investigate the formulation of the aeroelastic problem.

We suppose first that the shape of the airplane is defined by a vector of geometric parameters, denoted $\vec{u} = \{u_j\}$ (cf. equation (2.1)), and that from these parameters the complete current shape can be determined. In particular it is required that these parameters determine the current aerodynamic shape via a set of equations of the form

$$\mathbf{z}_\alpha = \mathbf{z}_\alpha^{(0)} + \sum_j \mathbf{N}_j(p_\alpha)u_j; \qquad \alpha = 1, \cdots, N_z \tag{4.1}$$

where the material points p_α correspond to the *unique, independent* mesh points of the aerodynamic surface networks. The points $\mathbf{z}_\alpha^{(0)}$ describe the location of those mesh points in the reference configuration when the structural deformation is zero ($\vec{u} = 0$). If we aggregate all of the mesh points $\mathbf{z}_\alpha$ into a single vector $\vec{z}$ (of dimension $3N_z$), then (4.1) can be written

$$\vec{z} = \vec{z}^{(0)} + T\vec{u} \tag{4.2}$$

The surface kinematics of PAN AIR uses $\vec{z} = \{\mathbf{z}_\alpha\}$ to parameterize points $\mathbf{q}$ on the singularity surface S by the formula

$$\mathbf{q}(\sigma,\tau) = \sum_\alpha \mathbf{z}_\alpha \vartheta_\alpha(\sigma,\tau) \qquad (\sigma,\tau) \subset \mathcal{P} \equiv \bigcup_{Q\subset S} P_Q \tag{4.3}$$

where the basis functions ϑ_α are linear functions of (σ,τ), a point in the parameter space $\mathcal{P}$ that, for every nondegenerate triangular subpanel Q of S, contains a copy P_Q of the standard triangle, $P_Q = \{(\sigma,\tau)\,|\,\sigma \geq 0,\ \tau \geq 0,\ 1-\sigma-\tau \geq 0\}$.

Now the aerodynamic problem is completely specified to both PAN AIR and TRANAIR by the value of $\vec{z}$, along with ancillary boundary condition information *that does not change* as $\vec{u}$ varies. Thus, once the boundary condition information has been specified along with a baseline geometry $\vec{z}^{(0)}$, the aerodynamic solution is determined by a set of aerodynamic constraints having the form

$$\Lambda(\vec{\lambda}; \vec{z}) = 0 \tag{4.4}$$

where the function Λ is a finite dimensional characterization of the aerodynamic partial differential equations and $\vec{\lambda}$ parameterizes the field variables of the aerodynamic solution. For example, for the panel methods discussed in section 3, $\vec{\lambda}$ would comprise the source parameters σ_β and the doublet parameters μ_β. Using formulation 2 given by equation (3.7b), we would obtain a function $\Lambda(\vec{\lambda}; \vec{z})$ given by

$$\Lambda(\vec{\lambda}; \vec{z}) = \begin{bmatrix} A_\sigma(\vec{z})\vec{\sigma} + A_\mu(\vec{z})\vec{\mu} \\ \cdots\cdots\cdots\cdots\cdots\cdots \\ \sigma_\alpha - (\partial\phi/\partial n)_+\big|_{\mathbf{r}_\alpha} \end{bmatrix} \tag{4.5}$$

In this equation the rows of the matrices $A_\sigma(\vec{z})$ and $A_\mu(\vec{z})$ are generated from the boundary conditions $\phi_L(\mathbf{r}_\alpha) = 0$. The integral representation formula (3.6) combined with the finite dimensional representations (3.8) allows us to write: $\phi_L(\mathbf{r}_\alpha) =$

$$\textstyle\sum_\beta \left[-\iint_S \bar{s}_\beta(\mathbf{q})\psi(\mathbf{q},\mathbf{r}_\alpha)dS_q\right]\sigma_\beta + \sum_\beta \left[\left(\iint_S \bar{m}_\beta(\mathbf{q})[\partial\psi(\mathbf{q},\mathbf{r}_\alpha)/\partial\nu_q]dS_q\right) - \frac{1}{2}\bar{m}_\beta(\mathbf{r}_\alpha)\right]\mu_\beta$$

where there is an obvious correspondence with the terms of equation (3.7b). Clearly the coefficient of σ_β (μ_β) in this equation is the entry in row α, column β of A_σ (A_μ).

The second group of equations in (4.5) specify the source strength $\sigma = \partial\phi/\partial n_+$, ϕ being identically zero on the lower surface of S. The value for $\partial\phi/\partial n_+$, which denotes the value of $\partial\phi/\partial n$ evaluated on S's upper surface, is computed from the impermeable surface boundary condition

$$[(\mathbf{V}_\infty + \mathbf{w} + \nabla\phi_+) - (\mathbf{v}_0 + \boldsymbol{\omega}\times\mathbf{q} + \partial\mathbf{q}/\partial t)]\cdot[\partial\mathbf{q}/\partial\sigma \times \partial\mathbf{q}/\partial\tau] = 0 \qquad (4.6)$$

Here $\mathbf{V}_\infty + \mathbf{w} + \nabla\phi$ is the total fluid velocity with respect to the $\mathbf{x}$ coordinate system, including the coordinate system velocity $\mathbf{V}_\infty$, the wind velocity $\mathbf{w}$ and the perturbation potential velocity $\nabla\phi$. The term $\mathbf{v}_0 + \boldsymbol{\omega}\times\mathbf{q} + \partial\mathbf{q}/\partial t$ is the velocity of a point on the surface with respect to the $\mathbf{x}$ coordinate system, consisting of a rigid body motion $\mathbf{v}_0$, a rotation rate $\boldsymbol{\omega}$ about the center of gravity at $\mathbf{x}=0$, and $\partial\mathbf{q}/\partial t$, the independent motion of the surface S. The partial derivatives of the surface function $\mathbf{q}(\sigma,\tau;t)$ are computed using the surface kinematics (4.3) in the form

$$\mathbf{q}(\sigma,\tau;t) = \sum_\alpha \mathbf{z}_\alpha(t)\,\vartheta_\alpha(\sigma,\tau); \qquad \mathbf{z}_\alpha(t) = \mathbf{z}_\alpha^{(0)} + \mathbf{d}_\alpha(t)$$

where $\mathbf{d}_\alpha(t)$ denotes meshpoint displacement functions. Notice that $[\partial\mathbf{q}/\partial\sigma \times \partial\mathbf{q}/\partial\tau]$ in (4.6) is normal to the *deformed* shape of the surface. Now, recalling the simulation program of section 1, we recognize that equations (4.5) and (4.6) express impermeability on a moving aerodynamic surface that is a deviation from the datum surface. The deviations are small, but they must be treated as arbitrary. To do this, equations (4.5) must be formulated for points on the perturbed surface using a first order expansion about points on the datum surface.

In practice, some entries of the function Λ corresponding to $\phi_L = 0$ boundary conditions are replaced with conditions that enforce the continuity of the basis functions $\bar{m}_\beta$ at network abutments. Also, some other entries of Λ are replaced by Kutta condition constraints. Nevertheless, for all PAN AIR formulations, the final result has the form (4.5). More generally, all TRANAIR formulations have the form $\Lambda(\vec{\lambda};\vec{z}) = 0$ as well. In that instance the entries of Λ correspond to the residuals of the finite dimensional representation of the Bateman variational principle that lies at the heart of TRANAIR. We re-emphasize the basic point. Once a certain set of boundary condition information has been specified, a finite dimensional characterization of the aerodynamics may be generated having the form (4.4), for which the specification of the dynamic surface (via $\vec{z}$) is sufficient to generate the whole aerodynamic solution.

Turning now to the structures part of the problem we note that the structural degrees of freedom are determined by a structural constraint equation of the form

$$(d/dt)[\partial T/\partial\dot{u}_i] - [\partial T/\partial u_i] + [\partial E/\partial u_i] = p_i^A(\vec{\lambda};\vec{z}) \qquad (4.7)$$

where T is the kinetic energy, E is the strain energy and p_i^A are the aerodynamic generalized forces. Note that this equation, along with (4.4) and (4.2), constitute the aeroelastic system of equations to be solved for the unknowns $\vec{\lambda}$, $\vec{u}$ and $\vec{z}$. Notice as well that the formulation (4.4) together with the kinematics (4.3) carefully account for the dependence of the function Λ upon the aerodynamic geometry $\vec{z}$, thus permitting analysis of variations in $\vec{z}$.

The computation of the aerodynamic generalized forces p_i^A, which constitutes the major interface problem between the structures and aerodynamics technologies, will be explored throughout the remainder of this section. We begin with the statement that the aerodynamic generalized forces p_i^A of equation (4.7) are conjugate to the generalized virtual displacements δu_i in the sense that the virtual work performed by the aerodynamic forces is given by

$$\delta W = \sum_i p_i^A\,\delta u_i \qquad (4.8)$$

Now this virtual work must be equivalent to the virtual work performed by the aerodynamic surface pressure, $(-p\mathbf{n})$ acting on the aircraft surface undergoing a virtual displacement $\delta\vec{z}$. Using the kinematics of S expressed by (4.3), that virtual work takes the form

$$\delta W = \iint_S \delta\mathbf{q}\cdot(-p\,\mathbf{n}\,dS)\ (4.9a); \quad \delta\mathbf{q}(\sigma,\tau) = \sum_\alpha \delta\mathbf{z}_\alpha\,\vartheta_\alpha(\sigma,\tau), \quad (\sigma,\tau)\in\mathcal{P}\ (4.9b)$$

The integral over S can be expressed as an integral over the parameter space by using the relation from differential geometry, $\mathbf{n}\,dS = (\partial\mathbf{q}/\partial\sigma \times \partial\mathbf{q}/\partial\tau)\,d\sigma\,d\tau = \mathbf{m}(\sigma,\tau)\,d\sigma\,d\tau$. Using this together with the formula (4.9b) for $\delta\mathbf{q}$, equation (4.9a) for δW becomes,

$$\delta W = -\sum_\alpha \delta\mathbf{z}_\alpha\cdot\iint_{\mathcal{P}} \vartheta_\alpha(\sigma,\tau)\,\mathbf{m}(\sigma,\tau)\,p\,d\sigma d\tau = \delta\vec{z}^T\vec{F}, \tag{4.10}$$

where the vector $\vec{F}$ is the aggregation into a single vector of the mesh point forces $\mathbf{F}_\alpha$,

$$\mathbf{F}_\alpha = -\iint_{\mathcal{P}} \vartheta_\alpha(\sigma,\tau)\,\mathbf{m}(\sigma,\tau)\,p(\sigma,\tau)\,d\sigma d\tau. \tag{4.11}$$

The surface pressure p is expressed as a function on $\mathcal{P}$ via the mapping $\mathbf{q}(\sigma,\tau)$: $p(\sigma,\tau) = p(\mathbf{q}(\sigma,\tau))$. Substituting the equation $\delta\vec{z} = T\delta\vec{u}$, which follows from (4.2), into (4.10), we obtain $\delta W = (\delta\vec{u}^T)T^T\vec{F}$. Writing (4.8) in the form $\delta W = \delta\vec{u}^T\,\vec{p}^A$ and equating the two forms of δW for arbitrary $\delta\vec{u}$, we conclude that

$$\vec{p}^A = T^T\vec{F}. \tag{4.12}$$

This establishes the desired form of the structural node force vector $\vec{p}^A$ by expressing it in terms of the aerodynamic mesh point forces $\vec{F}$, to be computed from equation (4.11).

In addition to providing an effective means for computing $\vec{p}^A$, equation (4.12), by comparison with (2.2), provides a means for evaluating the accuracy of the aerodynamics to structures interface. Since the pressure functions appearing in (2.2) and (4.11) are derived from the same aerodynamic solution, differences between the two methods of computing $\vec{p}^A$ are due to differences in the representation of the aerodynamic surface as expressed by (2.1) and (4.3), with the two representations connected by (4.1).

References

[1] L. Morino and J. Baillieul: A geometrically-exact non-linear Lagrangian formulation for the dynamic analysis of a flexible maneuvering airplane, Boston Univ. report CCAD-TR-87-02.

[2] L. Morino: A general theory of unsteady compressible potential aerodynamics, NASA CR-2464, 1974.

[3] A. E. Magnus and M. A. Epton: PAN AIR–A computer program for predicting subsonic and supersonic linear potential flows about arbitrary configurations using a higher order panel method, Vol. I, Theory Document (ver 1.0), NASA CR-3251, 1980.

[4] A. R. Dusto and M. A. Epton: An advanced panel method of analysis of arbitrary configurations in unsteady subsonic flow, NASA CR-152323, 1980.

[5] F. T. Johnson, S. S. Samant, M. B. Bieterman, R. G. Melvin, D. P. Young, J. E. Bussoletti and M. Madsen: Application of TRANAIR rectangular grid approach to aerodynamic analysis of complex configurations, AGARD Fluid Dynamics Panel, 64th meeting, May 1989, Loen, Norway

A Boundary Element Shape Design Sensitivity Analysis Formulation for Thermal Radiation Problems

James H. Kane and Hua Wang

Clarkson University, Mechanical & Industrial Engineering Department, Potsdam, New York 13676

SUMMARY

Effective shape optimization procedures for nonlinear thermal problems can be developed that utilize the concept of design sensitivity analysis (DSA). DSA provides an economical procedure to obtain the rates of change of all thermal response quantities involved in the problem with respect to the design variables that control the shape. In this paper, implicit differentiation of the discretized boundary element analysis (BEA) equations is performed to produce a DSA formulation that includes radiation boundary conditions. An example problem is presented to demonstrate the accuracy and efficiency of this approach.

INTRODUCTION

The BEA method is capable of predicting the thermal response of solids subjected to temperature dependent boundary conditions and with temperature dependent material properties. Textbook treatments of such techniques include [1-3]. Approaches that obviate the need for volume discretization and simulate temperature dependent material properties via the Kirchhoff transformation include Skerget and Brebbia [4], Anza et. al. [5], Bialecki and Nowak [6], and Akkuratov and Mikhailov [7]. Onishi and Kuroki [8], Butterfield [9], Banerjee [10], and Khader and Hanna [11] also discuss the solution of such thermal problems. In References [6] and [11], the convergence properties of several iterative solution strategies are studied. Recent works include Onishi and Kuroki [12] and Wrobel and Brebbia [13] dealing with transient nonlinear thermal problems, Wrobel and Azevedo [14], and Azevedo and Wrobel [15] dealing with nonlinear heat transfer in composite bodies using a multi-zone boundary element approach. These approaches involve an iterative process wherein the unknowns associated with the nonlinear effects must be evolved to their final converged values in a series of linear steps. Kane [16], and Kane, Saigal, et.. al. [17-22] have shown that a multi-zone BEA and DSA approach significantly impacts the ability to exploit the additional matrix sparsity present in the DSA step occurring during shape optimization of objects with partial geometric sensitivity. Kane and Saigal [23-24] have also used the concept of boundary element zone condensation to obtain significant computational economies in structural shape optimization. Kane, Kumar, and Gallagher [25] have shown that multi-zone capability can facilitate the effective utilization of reanalysis techniques in the continuum structural shape optimization context. Recently, Kane and Wang [26] have used the boundary element zone condensation techniques to economize the solution of nonlinear thermal problems.

In this paper, the treatment of DSA within the context of a nonlinear thermal radiation problem is discussed. This problem, however, can be regarded as typical of a larger class of nonlinear heat transfer problems. Implicit differentiation of the BEA equations is shown to yield a set of nonlinear algebraic equations that have the same left hand side matrix evolved to its converged state during a previous nonlinear thermal analysis. Thus, the sensitivities associated with this problem can be obtained with respect to design variables for little additional computational burden. An example thermal radiation problem is considered for which an exact closed form solution is obtained. The model geometry of this problem is parametrized in terms of a design variable and a material derivative is taken to yield an exact expression for the sensitivities of temperature and heat flow. These exact sensitivities are compared to those computed in the BEA approach discussed in the paper.

NONLINEAR BOUNDARY ELEMENT ANALYSIS FORMULATION

In thermal BEA, the fundamental solution, symbolized by T^*, refers to the response of an infinite conducting medium, with conductivity k, to a unit point source of thermal energy. Mathematically, this point source at location **d** is represented by a Dirac delta function $\delta(\mathbf{x}-\mathbf{d})$ at sample point **x**. A governing boundary integral equation can be derived by substituting T^* for 'g' in Green's second identity, and choosing 'f' to be the temperature response of the actual problem under consideration. In this discussion, Ω is the domain of the actual problem, Γ is the boundary of this domain and **n** is the unit normal vector on the surface. The symbol q^* denotes the normal component of the heat flux vector corresponding to the fundamental temperature solution. Use of the Fourier law of heat conduction has also been made in this derivation, along with the selection property of the delta function.

$$\int_\Omega \nabla^2 f\, g\, d\Omega - \int_\Omega \nabla^2 g\, f\, d\Omega = \int_\Gamma g\, \nabla f \cdot \mathbf{n}\, d\Gamma - \int_\Gamma f\, \nabla g \cdot \mathbf{n}\, d\Gamma \tag{1}$$

$$\nabla^2 T^* = -\frac{1}{2\pi k}\, \delta(\mathbf{x}-\mathbf{d}) \tag{2}$$

$$c\, T(\mathbf{d}) + \int_\Gamma q^*\, T\, d\Gamma = \int_\Gamma T^*\, q\, d\Gamma \tag{3}$$

Equation (3) can be discretized by breaking the surface of the object under consideration into boundary elements and approximating the behavior of the geometry, boundary conditions, and unknown response using simple interpolating functions and the values of these quantities at a finite number of nodes.

$$c\, T(\mathbf{d}) + \sum_{E=1}^{NEL} \left\{ \int_{-1}^{+1} q^* \{H\}\, J\, da \right\}^{(E)} \{T\}^{(E)} = \sum_{E=1}^{NEL} \left\{ \int_{-1}^{+1} T^* \{H\}\, J\, da \right\}^{(E)} \{q\}^{(E)} \tag{4}$$

In this expression, $\{H\}$ is a row vector of element interpolation functions associated with each node on element E, 'a' is an element intrinsic coordinate that runs from -1 to +1 in each element, and J is the Jacobian of the transformation from the differential length $d\Gamma$ in the actual coordinate system to the differential length da in the element's intrinsic coordinate system. $\{T\}^{(E)}$ and $\{q\}^{(E)}$ are column vectors of node point temperatures and normal heat flux components respectively for element E and have been taken outside the integrals to produce an algebraic expression. A singular boundary element formulation is obtained by locating the source point at each of the nodes present in the model, producing a square system of algebraic equations.

$$[F]\, \{t\} = [G]\, \{q\} \tag{5}$$

$\{t\}$ and $\{q\}$ are column vectors of nodal temperatures and normal heat flux components respectively. $\{t\}$ has an entry for each node in the problem, while $\{q\}$ may have additional entries if jumps in the heat flux occurs at any node. [F] is a square matrix, and [G] is either square of rectangular.

In a well posed boundary value problem, half of the temperature and normal heat flux components will be specified and the other half will be unknown. By transferring all knowns to the right hand side vector {q}, placing all unknown temperature and normal heat flux components in the left hand side vector {t}, exchanging corresponding columns of the respective rectangular matrices, and performing the indicated matrix-vector multiplication on the right hand side, a solvable system of equations can be produced. This matrix equation is usually solved by triangular factorization of the matrix [A] using Gauss elimination with partial pivoting, followed by forward reduction of {b} and backward substitution to determine the unknown response vector {x}.

$$[A]\{x\} = \{b\} \tag{6}$$

Note that the notation has been generalized to symbolize that {t} is the vector of unknown boundary response quantities, while {q} denotes the vector of specified boundary conditions in the problem. At each node, if q has been specified, then {t} contains T for this node. If, on the other hand, T had been specified for this node, then {t} contains q for this entry.

Radiation boundary conditions can be accounted for as an equivalent temperature dependent convection boundary condition. Convection boundary conditions relate the normal component of the heat flux vector q on the surface of an object to the surface temperature T, the free stream (or bulk) temperature T_{∞} and a convection coefficient h. Considering the ith equation in the boundary element system equations, substitution of the convection law for the convection boundary condition at node k yields the following:

$$q = h\,(T - T_{\infty}) \tag{7}$$

$$f_{i1}\,T_1 + f_{i2}\,T_2 + \cdots + f_{ik}\,T_k + \cdots + f_{in}\,T_n = g_{i1}\,q_1 + g_{i2}\,q_2 + \cdots + g_{ik}\,q_k + \cdots + g_{in}\,q_n \tag{8}$$

$$f_{i1}\,T_1 + f_{i2}\,T_2 + \cdots + f_{ik}\,T_k + \cdots + f_{in}\,T_n = g_{i1}\,q_1 + g_{i2}\,q_2 + \cdots + g_{ik}\,h_k\,(T_k - T_{\infty}) + \cdots + g_{in}\,q_n$$

The temperature at node k appears on both the left and right hand side of this equation. Different approaches can be taken to solve these equations. One approach consists of leaving the temperature on both sides of the equation and iterating to find the correct value for T_k as expressed in Equation (9).

$$\begin{aligned} &f_{i1}\,T_1 + f_{i2}\,T_2 + \cdots + f_{ik}\,T_k^{(j+1)} + \cdots + f_{in}\,T_n \\ &\quad = g_{i1}\,q_1 + g_{i2}\,q_2 + \cdots + g_{ik}\,h_k\,(T_k^{(j)} - T_{\infty}) + \cdots + g_{in}\,q_n \end{aligned} \tag{9}$$

In a second approach, the terms that multiply T_k on the right side of Equation (8) are brought to the left hand side.

$$\begin{aligned} &f_{i1}\,T_1 + f_{i2}\,T_2 + \cdots + (f_{ik} - g_{ik}\,h_k)\,T_k + \cdots + f_{in}\,T_n \\ &\quad = g_{i1}\,q_1 + g_{i2}\,q_2 + \cdots + g_{ik}\,(-h_k T_{\infty}) + \cdots + g_{in}\,q_n \end{aligned} \tag{10}$$

For cases where the convection coefficient is not a function of T, the second method characterized by Equation (10) is preferred because it results in a linear problem. When h is a function of T, both approaches have be attempted, however, there are many cases in which the approach characterized by Equation (9) does not converge. Other solution schemes have been presented [4-15] that are related to the Newton-Raphson iterative technique for simultaneous nonlinear equations that behave like the approach characterized by Equation (10). The approach characterized by Equation (9) leaves the left hand side of the overall boundary element system equations unchanged from one iteration to the next, thus allowing for the triangular factorization of [A], formed in the first iteration, to be reused in all subsequent iterations. The second method involves changing [A] in each iteration. Although this latter approach at first seems to be extremely noncompetitive, it has been shown [26] that the algorithms that employ this left hand side modification strategy converge in just a few iterations, while the algorithms that only modify the right hand side of the boundary element system equations converge at a much slower rate and may actually diverge. It has also been shown that employment of the boundary element zone condensation techniques [26] reduces the computational effort associated with the re-factorization of partially modified left hand side matrices to the point where the overall algorithm becomes superior to other approaches for this class of problems.

Radiation boundary conditions have been incorporated in thermal BEA. The law for the normal heat flux on the surface of an object due to radiation effects can be manipulated into a form that makes it appear exactly like a temperature dependent convection coefficient.

$$q = \sigma E\,(T^4 - T_r^4)\,;\ \text{where } E = V\left\{\frac{1}{\varepsilon} + \frac{1}{\varepsilon_r} - 1\right\}^{-1} \tag{11}$$

$$q = \sigma E\,(T^2 + T_r^2)\,(T + T_r)\,(T - T_r) = h_r\,(T - T_r) \tag{12}$$

In the above expressions, σ is the Stefan-Boltzmann constant, T is the surface temperature, T_r is the temperature of the known external radiation source, V is the radiation view factor, ε is the surface emissivity of the object being analyzed, and ε_r is the emissivity of the radiation source. It is therefore possible to solve problems involving radiation boundary conditions using a nonlinear BEA code that allows for temperature dependent convection coefficients that behaves like the cubic polynomial.

Solution of the nonlinear set of algebraic equations is accomplished by an iterative technique consisting of an initial guess at the surface temperature at nodes with radiation boundary conditions, computation of the equivalent h_r, and assembly of the matrix equations corresponding to Equation (10). The left hand side matrix is then factored and forward reduction and back substitution is done with the right hand side vector to obtain an updated response vector. This response is then used to repeat the entire process until it converges. The overall process can be characterized by an expression similar to Equation (6).

$$[A]^{(M-1)}\{x\}^{(M)} = \{b\}^{(M-1)} \tag{13}$$

The superscript notation is used to convey that the temperatures obtained in the M-1th iteration are used to construct the left hand side matrix [A] and the right hand side vector {b} to predict the response in iteration M.

DESIGN SENSITIVITY ANALYSIS FORMULATION

A very effective DSA formulation for this nonlinear problem can be developed by performing implicit differentiation of the converged Equation (13) with respect to the Lth design variable X_L. That is to say, during the analysis phase of the nonlinear thermal problem, the matrix $[A]^{(M)}$ is evolved to its 'correct' state, with all temperatures used to make the h_r coefficients at their converged values. The resulting sensitivity equations are or

$$\frac{\partial}{\partial X_L}\left([A]\{x\} = \{b\} \right) \Rightarrow [A]_{,L}\{x\} + [A]\{x\}_{,L} = \{b\}_{,L}$$

$$[A]\{x\}_{,L} = \left(\{b\}_{,L} - [A]_{,L}\{x\} \right) \tag{14}$$

To visualize the details symbolized in this matrix equation, a representative 3 degree of freedom example system is shown below. In this example, node 2 is the node with the radiation boundary condition.

$$\begin{bmatrix} f_{11} & (f_{12} - g_{12}\,h_r(T)) & f_{13} \\ f_{21} & (f_{22} - g_{22}\,h_r(T)) & f_{23} \\ f_{31} & (f_{32} - g_{32}\,h_r(T)) & f_{33} \end{bmatrix} \begin{Bmatrix} T_{1,L} \\ T_{2,L} \\ T_{3,L} \end{Bmatrix} = \begin{bmatrix} g_{11,L} & g_{12,L} & g_{13,L} \\ g_{21,L} & g_{22,L} & g_{23,L} \\ g_{31,L} & g_{32,L} & g_{33,L} \end{bmatrix} \begin{Bmatrix} q_1 \\ -h_r(T)\,T_\infty \\ q_3 \end{Bmatrix} \tag{15}$$

$$+ \begin{bmatrix} g_{11} & g_{12} & g_{13} \\ g_{21} & g_{22} & g_{23} \\ g_{31} & g_{32} & g_{33} \end{bmatrix} \begin{Bmatrix} 0 \\ -h_{r,L}(T)\,T_\infty \\ 0 \end{Bmatrix} - \begin{bmatrix} f_{11,L} & (f_{12,L} - g_{12,L}\,h_r(T) - g_{12}\,h_{r,L}(T)) & f_{13,L} \\ f_{21,L} & (f_{22,L} - g_{22,L}\,h_r(T) - g_{22}\,h_{r,L}(T)) & f_{23,L} \\ f_{31,L} & (f_{32,L} - g_{32,L}\,h_r(T) - g_{32}\,h_{r,L}(T)) & f_{33,L} \end{bmatrix} \begin{Bmatrix} T_1 \\ T_2 \\ T_3 \end{Bmatrix}$$

Finally, one can differentiate the relation for h_r shown in Equation (12).

$$h_{r,L}(T) = \sigma E\, T_{,L} \left(3\,T^2 + 2\,T\,T_r + T_r^2 \right) \tag{16}$$

In this expression it is assumed that the radiation view factor is insensitive to changes in the design variable. A formula involving $E_{,L}$ could also be employed in a straightforward manner.

The fundamental observation to be made regarding the solution of Equation (15) is that $T_{,L}$ appears on both sides of the equation. Thus, it is a nonlinear equation that must be solved iteratively. However, Equation (15) is far different from the nonlinear equation set solved in the previous thermal analysis. The left hand side matrix present in Equation (15) is 'correct', having been constructed using the converged values of the thermal response. This left hand side matrix has already been factored and can be saved from the last iteration of the previous thermal analysis and reused in the iterative process to evolve $T_{,L}$ to its converged state. These iterations therefore involve only forward reductions and back substitutions. Symbolically the solution of Equation (15) can be characterized as

$$[A]\{x\}_{,L}^{(M)} = \{c\}_{,L}^{(M-1)} \tag{17}$$

EXAMPLE PROBLEM

The problem of a hollow circular cylinder shown in Figure 1 is examined. The inner radius of this cylinder is maintained at a specified temperature of 200 degrees, while the outer radius is subjected to a radiation boundary condition. A quarter symmetry model is used with adiabatic conditions used along the straight sides of the model. The single zone boundary element model contains 40 nodes and 20 three-node quadratic boundary elements. The design variable in this example was chosen to be the outer radius 'b' of the cylinder. The node point were controlled by this design variable so that the geometric sensitivity was a linear function of their radial position as shown in the figure.

To appraise the accuracy of the computational procedure developed, the exact solution of this nonlinear problem was obtained and differentiated to produce an exact sensitivity solution. This axisymmetric problem is characterized by the differential equation and boundary conditions shown below.

$$\frac{1}{R}\frac{d}{dR}\left(R\frac{dT}{dR}\right) = 0;\ T\Big|_{R=25} = 200;\ T\Big|_{R=60} = \sigma E\,(T^4 - T_r^4) \tag{18}$$

The radiation boundary condition makes this problem nonlinear, and the evaluation of the constants of integration have to be determined in an iterative fashion. Likewise, the determination of the sensitivities of these constants to the design variable must be determined iteratively. The result of this process is

$$T = C_1 \ln(R) + C_2;\ T_{,L} = C_{1,L} \ln(R) + C_1 R^{-1} R_{,L} + C_{2,L}$$

$$\frac{dT}{dR} = \frac{C_1}{R};\ \left(\frac{dT}{dR}\right)_{,L} = C_{1,L} R^{-1} - C_1 R^{-2} R_{,L} \tag{19}$$

where

$$C_1 = 2056.1133;\ C_2 = 200 - C_1 \ln(25);\ C_{1,L} = -39.1405;\ C_{2,L} = -C_{1,L} \ln(25)$$

Table 1 contains a comparison of the exact and computed sensitivities at a number of sample points on the surface of the boundary element model. The agreement between the exact and computed response sensitivities is excellent. Thus, it has been demonstrated that the implicit differentiation approach for the computation of shape sensitivities is extendible to nonlinear thermal problems.

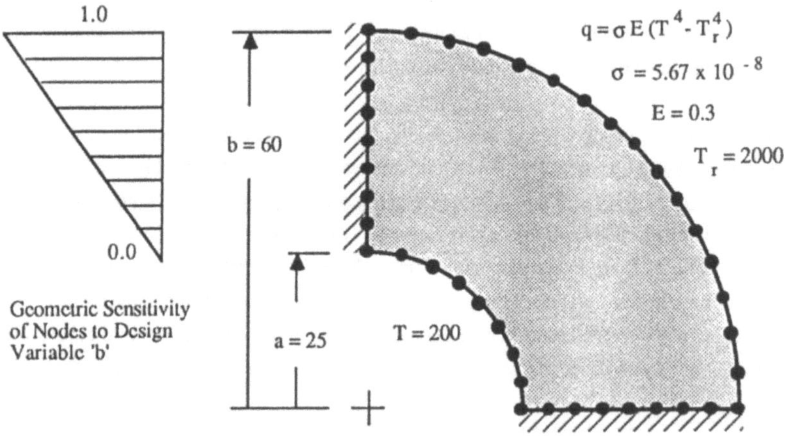

Figure 1. Hollow Cylinder Sensitivity Example Problem with Radiation Boundary Condition

Table 1. Response Sensitivities for Hollow Cylinder with Radiation Boundary Conditions.

R	$R_{,L}$	$T_{,L}$ exact	$T_{,L}$ computed	$q_{,L}$ exact	$q_{,L}$ computed
25.0	0.0	0.0	0.0	-1.5656	-1.5611
29.375	0.125	2.4372	2.4398	-1.6302	na
33.75	0.25	3.4842	3.4864	-1.6106	na
38.125	0.375	3.7069	3.7087	-1.5571	na
42.5	0.5	3.4204	3.4218	-1.4901	na
46.875	0.625	2.8107	2.8112	-1.4198	na
51.25	0.75	1.9928	1.9929	-1.3508	na
55.625	0.875	1.0404	1.0403	-1.2851	na
60.0	1.0	0.0024	0.0023	-1.2234	-1.2240

na - not available because only normal components of the heat flow vector are computed

ACKNOWLEDGEMENT

Portions of the research discussed herein have been supported by grants from the U. S. National Science Foundation (DMC-8708254) to Worcester Polytechnic Institute and (DDM-8996171) to Clarkson University, and also by grants or contracts from General Dynamics Electric Boat Division and United Technologies Corporation. Any opinions, findings, and conclusions or recommendations expressed in this publication are those of the authors and do not reflect the views of these other organizations.

REFERENCES

1. P. K. Banerjee and R. Butterfield, Boundary Element Methods In Engr. Sci., McGraw Hill Book Co. UK. London, 1981
2. C. A. Brebbia and S. Walker, Boundary Element Techniques in Engineering, Newnes Butterworths, London and Boston, 1980
3. C. A. Brebbia, J. C. F. Telles, and L. C. Wrobel, Boundary Element Techniques, Springer Verlag, Berlin and New York, 1984
4. P. Skerget and C. Brebbia, 'Nonlinear Potential Problems,' Progress in Boundary Element Methods, Vol. 2, Pentech Press, London, Springer-Verlag, NY, 1983
5. J. J. Anza, E. Ahedo, I. Da Riva, and E. Alarcon, 'A New Boundary Condition Solved With BEM,' Boundary Element Methods in Engineering, C. Brebbia Editor, Springer-Verlag, Berlin, pp. 607-618, 1982
6. R. Bialecki and A. J. Nowak, 'Boundary Value Problems For Nonlinear Material and Nonlinear Boundary Conditions,' Applied Mathematical Modelling, Vol. 5, pp. 417-421, 1981
7. Y. N. Akkuratov and V. N. Mikhailov, 'The Method of Boundary Integral Equations for the Solution of Nonliner Heat Transmission Problems,' USSR Comput. Maths. Math. Phys. Vol. 20, pp. 117-125, 1980
8. K. Onishi and T. Kuroki, 'Boundary Element Method in Singular and Nonlinear Heat Transfer,' Boundary Element Methods in Engineering, C. Brebbia Editor, Springer-Verlag, Berlin, pp. 141-155, 1982
9. R. Butterfield, 'An Application of the Boundary Element Method to Potential Flow Problems in Generally Inhomogeneous Bodies,' Recent Advances in Boundary Element Methods, C. Brebbia Editor, Pentech Press, London, pp. 123-135, 1978
10. P. K. Banerjee, 'Nonlinear Problems of Potential Flow,' Developments in Boundary Element Methods - 1, P. K. Banerjee and R. Butterfield Editors, Applied Science Publishers, London, pp. 21-30, 1979
11. M. S. Khader and D. B. Ingham, 'An Iterative Boundary Integral Numerical Solution for General Steady Heat Conduction Problems,' Journal of Heat Transfer, Trans, ASME, Vol. 103, No. 2, pp. 26-31
12. K. Onishi and T. Kuroki, 'On Nonlinear Heat Transfer Problems,' Developments in Boundary Element Methods - 4, P. K. Banerjee and J. O. Watson Editors, Elsevier Applied Science Publishers, London, pp. 191-226, 1986
13. L. C. Wrobel and C. A. Brebbia, 'The Dual Reciprocity Boundary Element Formulation for Nonlinear Diffusion Problems,' to appear in Comp. Methods Appl. Mech. Eng.
14. L. C. Wrobel and J. P. S. Azevedo, 'A Boundary Element Analysis of Nonlinear Heat Conduction,' Proc. Fourth International Conference on Numerical Methods in Thermal Problems, R. Lewis and K. Morgan Editors, Pineridge Press, Swansea, UK, 1985
15. J. P. S. Azevendo and L. C. Wrobel, 'Nonlinear Heat Conduction in Composite Bodies: A Boundary Element Formulation,' International Journal for Numerical Methods in Engineering, Vol 26, pp 19-38, 1982
16. J. H. Kane, " Shape Optimization Utilizing a Boundary Element Formulation," BETECH 86, Proceedings, 1986 Boundary Element Technology Conference, MIT, Computational Mechanics Publications, Springer-Verlag, Southampton and Boston, 1986
17. J. H. Kane and S. Saigal, "Design Sensitivity Analysis of Solids Using BEM," Journal of Engineering Mechanics, ASCE, Vol. 114, No. 10, pp. 1703-1722, October, 1988
18. S. Saigal and J. H. Kane, "A Boundary Element Shape Optimization System for Aircraft Components," AIAA Journal, Accepted for publication.
19. S. Saigal, R. Aithal and J.H. Kane, "Conforming Boundary Elements in Plane Elasticity for Shape Design Sensitivity," International Journal for Numerical Methods in Engineering, Accepted for publication, to appear.
20. S. Saigal, J. H. Kane and R. Aithal, "Semi-Analytical Structural Sensitivity Formulation using a Boundary Elements," AIAA Journal, Accepted for publication, to appear.
21. S. Saigal, J.T. Borggaard and J.H. Kane, "Boundary Element Implicit Differentiation Equations for Design Sensitivities of Axisymmetric Structures," International Journal of Solids and Structures, Vol. 25, No. 5, pp. 527-538, 1989
22. J. H. Kane and S. Saigal, "Design Sensitivity Analysis of Boundary Element Substructures," Second NASA / Airforce Symposium on Recent Experiences in Multi Disciplinary Analysis and Optimization, September, 1988, Proceedings published.
23. J. H. Kane and S. Saigal, " An Arbitrary Condensing, Noncondensing Solution Strategy for Large Scale, Multi-zone Boundary Element Analysis," Computer Methods in Applied Mechanics and Engineering, in press.
24. J. H. Kane and S. Saigal, 'An Arbitrary Multi-zone Condensation Technique for Boundary Element Design Sensitivity Analysis,' AIAA Journal, in press.
25. J. H. Kane, B. L. K. Kumar, and R. H. Gallagher, 'Boundary Element Iterative Reanalysis Techniques for Continuum Structures,' Journal of Engineering Mechanics, ASCE, submitted for review.
26. J. H. Kane, H. Wang, and B. L. K. Kumar, 'Nonlinear Thermal Analysis with a Boundary Element Zone Condensation Technique,' Computational Mechanics, submitted for review.

Design Sensitivity Analysis for Potential Problems by the Derivative Boundary Element Method

GULPERI AKSEL and SUBRATA MUKHERJEE

Department of Theoretical and Applied Mechanics
Kimball Hall, Cornell University
Ithaca, NY 14853

Summary
The subject of this paper is the determination of solutions of two-dimensional potential problems, and their sensitivities with respect to change of shape of a body, by the derivative boundary element method (DBEM). Modelling of multiple zones and corners in the body, by the DBEM, is carried out here. Numerical results are presented for a few sample problems.

Introduction

Potential theory is a classical subject on which numerous papers and many books (eg. [1]) have been written. Numerical solutions of Laplace's equations were first obtained by the BEM at least twenty five years ago. These solutions are discussed in many textbooks devoted to the BEM (eg. [2,3]).

Very recently, Choi and Kwak [4] have proposed a BEM formulation for Laplace's equation in 2-D, where the normal and tangential derivatives of the potential function (rather than the usual potential function and its normal derivative) are used as primary boundary variables. There are several potential advantages of this so called "derivative formulation" (DBEM). Perhaps the most important ones are that the kernel with the 1/r singularity (r is the distance between a source and a field point) is avoided here and that the boundary values of potential gradients can be obtained in a very accurate and consistent fashion. Also, the latter fact has important consequences in design sensitivity analysis, as will be discussed later.

After a brief review of the work of Choi and Kwak [4], this paper presents an elegant and consistent treatment of multiple zones and corners, with the DBEM. Next, design sensitivity analysis by the DBEM is discussed.

Determination of design sensitivity coefficients (DSC's) by the BEM has received considerable attention in recent years (eg.

[5-7]). DSC's are rates of change of quantities of interest, here the potential function and its derivatives, with respect to parameters that define the shape of a body. It is shown how the DBEM is very well suited for DSC analysis, particularly for the calculation of sensitivities of potential gradients on the boundary of a body.

Numerical results are presented here for a solution of Laplace's equation in a body with multiple zones and corners, and the sensitivity problem at a corner of a body.

The DBEM in Potential Theory

Choi and Kwak [4] have recently proposed a derivative BEM formulation for potential theory in which the normal and tangential derivatives $\frac{\partial\phi}{\partial n}$ and $\frac{\partial\phi}{\partial s}$, of the potential ϕ, are the primary variables on the boundary ∂B of a body B. Considering, for simplicity, a simply connected domain, their equations have the form

$$\int_{\partial B} [G(P,Q) \frac{\partial\phi(Q)}{\partial n_Q} + H(P,Q) \frac{\partial\phi(Q)}{\partial s_Q}] ds_Q = 0 \tag{1}$$

$$\int_{\partial B} [H(P,Q) \frac{\partial\phi(Q)}{\partial n_Q} - G(P,Q) \frac{\partial\phi(Q)}{\partial s_Q}] ds_Q = 0 \tag{2}$$

where the fundamental kernels are

$$G(P,Q) = - \frac{1}{2\pi} \ell n r(P,Q) \quad , \quad H(P,Q) = - \frac{1}{2\pi} \psi(P,Q) \tag{3}$$

in terms of the source and field points P and Q (with capital letters denoting points on ∂B and lower case letters denoting points inside B), r the Eucledian distance between P and Q and ψ the angle between the vector r(P,Q) and a reference direction. One can use either of the equations (1) or (2) in a BEM formulation, provided certain continuity and single valuedness conditions are satisfied for ϕ. The authors Choi and Kwak suggest use of equation (1) at certain source points and (2) for others, in order to improve conditioning of the BEM stiffness matrix. Further, a consistency condition

$$\int_{\partial B_1} \frac{\partial \phi}{\partial s} ds = \phi_2 - \phi_1 \qquad (4)$$

(where ∂B_1 is a suitable part of ∂B with ϕ_1 and ϕ_2 the values of ϕ at the start and at the end of ∂B_1) must be used for certain problems.

Once the normal and tangential deviatives of ϕ are obtained on ∂B, it is easy to compute the gradients of ϕ, at a regular point Q on it, from

$$\phi_{,i} = \frac{\partial \phi}{\partial n} n_i + \frac{\partial \phi}{\partial s} t_i \quad , \quad i = 1,2 \qquad (5)$$

where $\underset{\sim}{n}$ and $\underset{\sim}{t}$ are the unit outward normal and anticlockwise tangent vectors, respectively, to ∂B at Q.

Zones and Corners

Multiple zones have been treated in a BEM program by several authors (eg. Banarjee et.al.in the BEST program). Here, a method for a consistent treatment of zones and corners is presented in the context of the DBEM for Laplace's equation.

Figure 1 shows a region in 2-D which is subdivided into 3 zones. A material property such as the conductivity K is assumed to be piecewise constant in each zone.

The method for modelling the regions in figure 1 is outlined below. It is easy to extend these ideas to model a body with an arbitrary number of zones.

The basic idea is to treat the body as three separate bodies. BEM equations are used in each zone (relating, as usual, only the quantities on the boundary of that zone) and corner and interface equations are used for compatibility at each corner of each zone and compatibility across the interfaces, respectively.

a) Corner equations. It can be proved that if $\nabla\phi$ is bounded at a corner of ∂B, then it is continuous there. Thus, at a corner such as A in zone 1 [8] (see equation (5))

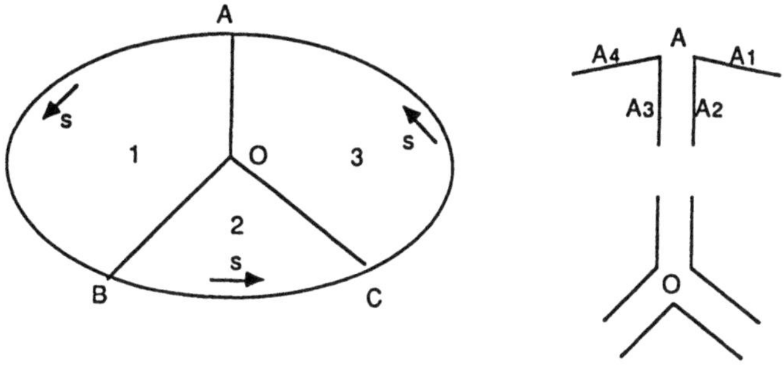

Fig. 1. Zones and corners

$$\left[\frac{\partial\phi}{\partial n}\right]^{-} n_i^{-} + \left[\frac{\partial\phi}{\partial s}\right]^{-} t_i^{-} = \left[\frac{\partial\phi}{\partial n}\right]^{+} n_i^{+} + \left[\frac{\partial\phi}{\partial s}\right]^{+} t_i^{+}, \quad i = 1,2 \qquad (6)$$

where the - superscript refers to a point just before a corner (eg. A_3) and the + superscript refers to a point just after it (eg. A_4). Of course, the quantities with superscripts can, in general, be discontinuous across a corner. Each corner is traversed in an anticlockwise sense within each zone. For the entire body , 4 corner equations can be written at each of the corners A, B and C, and 6 at the triple point 0.

b) Interface equations. The usual interface compatibility equations, say across the interface boundary between A_3 and A_2, can be written as (no sum over j or m)

$$K^{(m)} \left[\frac{\partial\phi}{\partial n}\right]^{(m)} = -K^{(j)} \left[\frac{\partial\phi}{\partial n}\right]^{(j)} \qquad (7)$$

$$\left[\frac{\partial\phi}{\partial s}\right]^{(m)} = -\left[\frac{\partial\phi}{\partial s}\right]^{(j)}$$

where the superscripts in m and j refer to the numbers of the zones on either side of the interface boundary.

This time, for the entire body in figure 1, there are 2 interface conditions near each corner A, B and C (eg. between A_3 and A_2 etc.) and 6 interface conditions near the triple point 0. Of course, other interface equations must be used across interface boundary locations away from corners. These are discussed in the next paragraph.

Let the body in figure 1 be discretized with N_E regular exterior points and N_I regular interface points. A regular point is defined as a point which lies on the boundary of a zone where it is locally smooth. Thus, for the body in figure 1, there are a total of $N_E + 2N_I + 6 + 3$ BEM equations, $N_E + 6$ external boundary conditions, $2N_I + 6 + 6$ interface equations and $12 + 6$ corner equations, for a total of $2N_E + 4N_I + 45$ equations. Here, a point such as A has 2 source points (one in each zone) but 4 field points, two external and two on the interface (eg. A_1, A_2, A_3 and A_4 for A). Finally, the total number of unknowns in this case is $2N_E + 4N_I + 3 \times 8 + 12 = 2N_E + 4N_I + 36$. The system is overdetermined, with 9 extra equations. The equations, however, are consistent.

Sensitivities

A sensitivity formulation for this problem can be derived by differentiating equations (1) and (2) with respect to, say, a shape design parameter. Following ideas discussed, for example, in references [5 - 7], one obtains, by direct differentiation (DDA) of equation (1)

$$
\begin{aligned}
0 = {} & \int_{\partial B} \left[G(b,P,Q)\,\frac{\partial \overset{*}{\phi}}{\partial n_Q}(b,Q) + H(b,P,Q)\,\frac{\partial \overset{*}{\phi}}{\partial s_Q}(b,Q)\right] ds(b,Q) \\
& + \int_{\partial B} \left[\overset{*}{G}(b,P,Q)\,\frac{\partial \phi}{\partial n_Q}(b,Q) + \overset{*}{H}(b,P,Q)\,\frac{\partial \phi}{\partial s_Q}(b,Q)\right] ds(b,Q) \qquad (8) \\
& + \int_{\partial B} \left[G(b,P,Q)\,\frac{\partial \phi}{\partial n_Q}(b,Q) + H(b,P,Q)\,\frac{\partial \phi}{\partial s_Q}(b,Q)\right] d\overset{*}{s}(b,Q)
\end{aligned}
$$

where b is a design variable and a superscribed * denotes a derivative of a quantity with respect to b,

$$
\overset{*}{G}(b,P,Q) = G_{,k}(b,P,Q)\,[\overset{*}{x}_k(Q) - \overset{*}{x}_k(P)]
$$

and similarly for $\overset{*}{H}(b,P,Q)$. Finally,

$$\overset{*}{ds}(b,Q) = [\overset{*}{x}_{k,k}(Q) - n_i n_j \overset{*}{x}_{i,j}(Q)]ds(b,Q)$$

Equation (2) can be differentiated in similar fashion. This equation is called (8$'$) and can be used in conjunction with (8) to improve conditioning of the BEM stiffnes matrix.
In order to solve the complete problem, the sensitivity equations for zones and corners must be obtained by differentiating the appropriate equations with respect to b, and these equations must be used in addition to equation (8).

Numerical Results

The above equations are discretized with straight boundary elements and linear shape functions for the unknowns over each element. It can be shown that the stiffness matrices of the BEM equations, and the sensitivity equations, are exactly the same. Of course, the sensitivity equations have a different right hand side, from the second and third integrals of equations (8) and (8$'$).
The first problem considered is the solution of Laplace's equation in the region shown in figure 2. This problem , with uniform properties in the 4 zones, has the analytical solution $\phi = x$. The BEM, corner and interface equations are solved with a total of 41 nodes on the boundaries of the four zones. The agreement between the analytical and numerical solutions is nearly perfect, with the maximum difference around 1.5%. The approach [9] is used to solve the overdetermined system of equations. This example is a test problem for the present algorithm for solving Laplace's equation in the presence of zones and corners.
A sensitivity problem for a region with a corner ($\alpha < \pi$, $\neq \pi/2$) is illustrated in figure 3. The boundary conditions are those that give the solution

$$\phi = \frac{r\sin\theta}{\sin\alpha}$$

In this case, $\nabla\phi = \underset{\sim}{j}/\sin\alpha$ and the sensitivity of $\phi_{,2}$, at the corner 0, with respect to the corner angle α is

$$\overset{*}{\phi}_{,2} = \frac{\partial}{\partial\alpha}\left[\phi_{,2}\right] = -\frac{\cos\alpha}{\sin^2\alpha}$$

The numerical result for the sensitivity at the corner, for $\alpha = 45^0$, with 13 boundary nodes, recovers the first five significant digits of the analytical solution! The numerical solution is uniformly very accurate all around the wedge, with the maximum discrepancy between the analytical and numerical solutions being about 1.6%.

Further investigations of the DBEM, in the presence of zones and corners, are in progress.

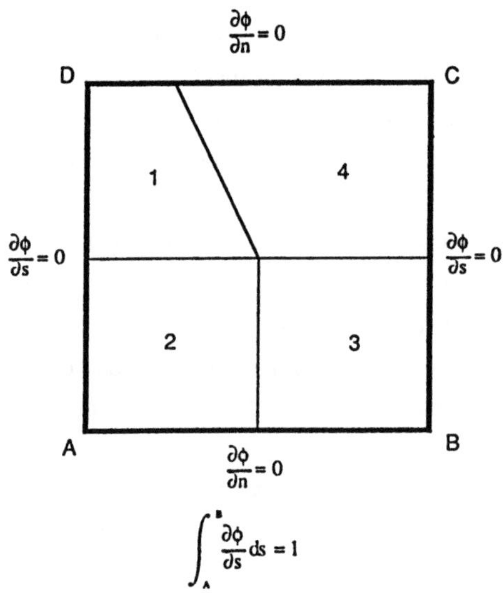

Fig. 2. An example with zones and corners

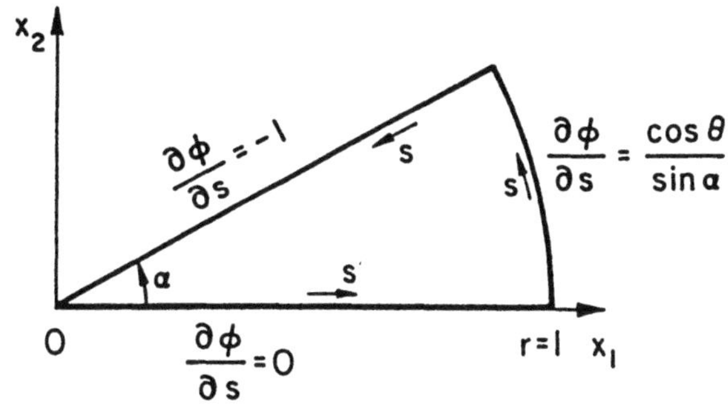

Fig. 3. A sensitivity problem in a wedge

Acknowledgement

This research has been supported by grant number MSM - 8609391 of the National Science Foundation, to Cornell University.

References

1. Kellogg, O.D.: Foundations of potential theory. Berlin: Springer Verlag 1929.
2. Banerjee, P.K.; Butterfield, R.: Boundary element methods in engineering science. U.K.: McGraw Hill 1981.
3. Brebbia, C.A.; Telles., J.C.F.; Wrobel, L.C.: Boundary element techniques: Theory and application in engineering. Berlin: Springer Verlag 1984.
4. Choi, J.H.; Kwak, B.M.: A boundary integral equation formulation in derivative unknowns for two-dimensional potential problems. ASME J. Appl. Mech. (in press).
5. Barone, M.R.; Yang, R.-J.: Boundary integral equations for recovery of design sensitivities in shape optimization. AIAA J. 26 (1988) 589-594.
6. Kane, J.H.; Saigal, S.: Design sensitivity analysis of solids using BEM. ASCE J.Eng. Mech. 14 (1988) 1703-1722.
7. Zhang, Q.; Mukherjee, S.: Design sensitivity coefficients for linear elasticity problems by the derivative boundary element method. Proceedings of IUTAM/IACM Symposium. Vienna: 1989.
8. Sladek, J.; Sladek, V.: Computation of stresses by BEM in 2-D elastostatics. Acta Technica, CSAV31 (1986) 523-531.
9. Press, W.H.; Flannery, B.P.; Teukolsky, S.A.; Vetterling, W.T.: Numerical Recipes. New York: Cambridge University Press 1986.

Consistent Sharp Corner Formulations for Boundary Element Analysis and Sensitivity Analysis

James H. Kane and B. L. Keshava Kumar

Clarkson University, Mechanical & Industrial Engineering Dept, Potsdam, New York, 13676

SUMMARY

A consistent approach to the sharp corner problem is presented for both boundary element structural analysis and sensitivity analysis. The relationships used for the consistent account of the multiple traction components at corner nodes are shown to be confined to the individual zone that contains the corner elements. This fact means that the formulation can be incorporated in existing multi-zone procedures without altering the number of degrees of freedom or the sparse blocked matrix populations present in these programs. Implicit differentiation of the sharp corner formulation is then shown to produce a consistent design sensitivity analysis formulation that allows for the use of multi-zone boundary element models with sharp corners in shape optimization problems. Example problems are presented to demonstrate the accuracy of the formulation discussed in the paper.

INTRODUCTION

In References [1-3] multi-zone boundary element analysis (BEA) is shown to be a major enhancement of the original single-zone method for a large class of problems in terms of accuracy, CPU time, and computer storage requirements. References [4-5] demonstrate that, in shape optimization, multi-zone techniques and zone condensation methods provide significant computational efficiency in problems with localized geometric sensitivity to design variables. As described in Reference [6], in order to fully exploit the benefits of these multi-zone procedures, it is necessary to be able to arbitrarily configure multi-zone boundary element models without restrictions as to how the individual zones are connected to each other. This desirable characteristic inevitably leads to the requirement for the consistent account of the internal boundary element zone sharp corners. In the structural analysis context, at such sharp corner locations, the displacements will be continuous, but one must deal with the fact that the components of the surface traction can jump in accordance with the constitutive law, the Cauchy stress transformation, the boundary element interpolation functions, and certain other geometric relationships. In this paper a relationship is developed that accounts for this effect using all of the above mentioned considerations.

As shown in References [7-9], shape optimization can be performed economically when design sensitivity analysis (DSA) is used to compute the rates of change of response quantities with respect to changes in the parameters (design variables) that control the shape. Differentiation of the sharp corner formulation is performed to produce a consistent DSA formulation that facilitates employment of multi-zone BEA models with sharp corners in shape optimization problems.

SHARP CORNERS IN MULTI-ZONE ANALYSIS

Part of a BEA model with zones that have an interior sharp corner is shown in Figure 1, where node 88 belongs to both element E1 and E2, and yet the normal to the surface of element E1 and E2 at this node are distinctly different. The displacement components at this point will be continuous, but the components of the traction vector on the surface of element E1 will be E1 and E2 at this node are distinctly different. The displacement components at this point will be continuous, but the components

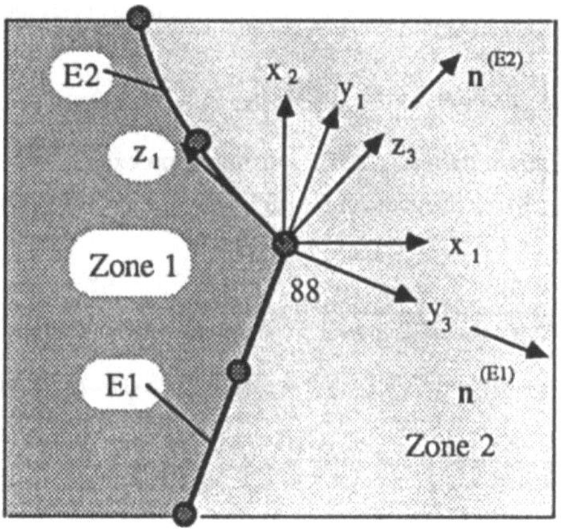

Figure 1. Interior Sharp Corner in a Mulit-zone BEA Model

of the traction vector on the surface of element E1 will be different than those onelement E2 at this node. If the sharp corner was an exterior sharp corner with specified traction boundary conditions, then the inclusion of extra traction components in the column vector {t} of nodal point tractions and the introduction of corresponding extra columns of the corresponding [G] matrix, making it a rectangular matrix, appropriately account for this effect. When the sharp corner is in the interior of the model, however, the traction components on both elements will be unknown, and some other approach to represent these components must be taken. Most multi-zone programs simply do not allow boundary element models to have sharp corners in the interior. In this paper, a relationship among the sharp corner node traction components on the surface of element E2 and the traction and displacement components in element E1 is developed. This relationship can be used to assemble the element [G] matrix contributions of element E2 in the zone matrices [F] and [G] in a systematic fashion. Although the formulation presented is three dimensional, the degeneration of these relationships for two dimensional analysis is straightforward.

Three Cartesian coordinate systems are considered. The first coordinate system is a global Cartesian system, the second system is a tangential and normal system for element E1 at the corner node, and the third system is a tangential and normal system for element E2 at the corner node. In the tangential and normal systems, the third direction is normal to the element surface at the sharp corner node. Table 1 indicates the symbols used for the components of the tensor quantities, referred to the three distinct coordinate systems. Unless stated otherwise in the expressions that follow, the range of the indices is always 1, 2, 3, and the Einstein summation convention is implied for repeated indices.

Table 1. Symbols used for the tensor quantities present in the sharp corner formulation.

System	Coordinates	Displacements	Tractions	Strain	Stress	Normals	Transformation
1	x_i	u_i	t_i	e_{ij}	s_{ij}	n_i	$x_i = \delta_{ij}\, x_j$
2	y_i	v_i	T_i	E_{ij}	S_{ij}	N_i	$y_i = L_{ij}\, x_j$
3	z_i	w_i	Γ_i	ε_{ij}	σ_{ij}	υ_i	$z_i = \beta_{ij}\, x_j$

Concentrating on quantities in the second coordinate system for an isotropic, elastic material, the stress-strain law is

$$S_{ij} = 2\,\mu\,E_{ij} + \lambda\,\delta_{ij}\,E_{kk} + S^{o}_{ij} \tag{1}$$

Solving for the normal strain term E_{33} and using it to form expressions for the components of the stress tensor that are not surface tractions on the surface with a normal in the x_3 direction yields

$$E_{33} = \frac{T_3}{(2\,\mu+\lambda)} - \frac{\lambda\,E_{11}}{(2\,\mu+\lambda)} - \frac{\lambda\,E_{22}}{(2\,\mu+\lambda)} - \frac{S^{o}_{33}}{(2\,\mu+\lambda)} \tag{2}$$

$$S_{a3} = T_a\,;\; S_{11} = aE_{11} + bE_{22} + cT_3 + f\,;\; S_{12} = dE_{12} + g\,;\; S_{22} = aE_{22} + bE_{11} + cT_3 + h \tag{3}$$

$$\text{where}\quad a = \frac{4\,\mu\,(\mu+\lambda)}{2\,\mu+\lambda}\,;\; b = \frac{2\,\mu\,\lambda}{2\,\mu+\lambda}\,;\; c = \frac{\lambda}{2\,\mu+\lambda}\,;\; d = 2\,\mu \tag{4}$$

$$f = S^{o}_{11} - \frac{\lambda\,S^{o}_{33}}{(2\,\mu+\lambda)}\,;\; g = S^{o}_{12}\,;\; h = S^{o}_{22} - \frac{\lambda\,S^{o}_{33}}{(2\,\mu+\lambda)} \tag{5}$$

For initial stresses due to thermal expansion the expressions specialize as follows:

$$S^{o}_{ij} = -\frac{E\,\delta_{ij}\,\alpha\,\theta}{(1-2\upsilon)}\,;\; f = h = -\frac{E\,\alpha\,\theta}{(1-2\upsilon)}\left(1 - \frac{\lambda}{(2\,\mu+\lambda)}\right);\, g = 0 \tag{6}$$

The tangential strains in Equation (3) can be related to the derivatives of the tangential components of the deflections. These, in turn, can be related, via a Jacobian, to derivatives of the tangential components of deflection with respect to the intrinsic surface coordinates of the element.

$$E_{11} = v_{1,1}\,;\; E_{12} = \frac{1}{2}\,(\,v_{1,2} + v_{2,1}\,)\,;\; E_{22} = v_{2,2} \tag{7}$$

where

$$E_{11} = J^{-1}_{11}\,v_{1,n1} + J^{-1}_{12}\,v_{1,n2}\,;\; E_{22} = J^{-1}_{21}\,v_{2,n1} + J^{-1}_{22}\,v_{2,n2}\,;$$

$$E_{12} = \frac{1}{2}\,\{\,J^{-1}_{21}\,v_{1,n1} + J^{-1}_{22}\,v_{1,n2} + J^{-1}_{11}\,v_{2,n1} + J^{-1}_{12}\,v_{2,n2}\,\} \tag{8}$$

and

$$v_{i,nj} = \sum_{N=1}^{NNL} h^{(N)}_{,nj}\,v^{(N)}_{i}\,;\quad J_{ji} = y_{i,nj} \tag{9}$$

J_{ij} are the entries of a 2 x 2 surface Jacobian matrix for 3D element E1, $h^{(N)}$ are the shape functions for an element with NNL nodes. The comma and subscript notation ' $h_{,nj}$ ' is used to denote the partial derivative of shape function with respect to the intrinsic coordinates, and J_{ij}^{-1} symbolizes an entry in the inverse Jacobian matrix. The range of the subscripts in these expressions is i and j = 1 , 2 , nj = n1 , n2, and a and b = 1 , 2 , 3. All these quantities are evaluated at the sharp corner node in question for element E1 and are computed in the second coordinate system. These equations completely relate the stress components ' S_{ab}' to the initial stresses, three components of traction ' T_a ' at the sharp corner node and the two tangential displacement components ' $v_i^{(N)}$ ' at each of the NNL nodes of element E1. This formulation can and is used in BEA to recover element surface stresses, given the initial stresses, and the element node point displacement and traction components.

The Cauchy stress transformation can be employed to relate the traction components at the sharp corner node on the surface of element E2 and the components of the stress tensor shown in Equation (3). To accomplish this, the components of the stress tensor must be transformed to the first coordinate system. The components (in coordinate system 1) of the normal vector to the surface of element E2, at the corner node can then be used to perform the Cauchy stress transformation.

$$t_k = s_{kp}\, n_p = (L_{ak}\, L_{bp}\, S_{ab})\, n_p = n_p\, L_{ak}\, L_{bp}\, S_{ab} \tag{10}$$

What is finally desired is a relation between the components of the traction vector on the surface of element E2 and the tractions and displacements on element E1, all referred to coordinate system 1. All that remains is to transform the tangential components of the quantities described above for element E1 into this global coordinate system and collect terms.

$$t_k^{(E2)} = n_p\, L_{ak}\, L_{bp} \left\{ I_{ab} + R_{abj}\, t_j^{(E1)} + \sum_{N=1}^{NNL} Q_{abj}^{(N)}\, u_j^{(N)} \right\} \tag{11}$$

where

$$Q_{11j}^{(N)} = a\, A_j^{(N)} + b\, B_j^{(N)}\, ;\; Q_{22j}^{(N)} = b\, A_j^{(N)} + a\, B_j^{(N)}\, ;\; Q_{12j}^{(N)} = d\, C_j^{(N)}$$

$$R_{11j} = R_{22j} = c\, L_{3j}\, ;\; R_{13j} = R_{33j} = L_{3j}\, ;\; R_{23j} = L_{2j}$$

$$I_{11} = L_{1r}L_{1m}s_{rm}^{o} - c\, L_{3r}L_{3m}s_{rm}^{o}\, ;\; I_{12} = L_{1r}L_{2m}s_{rm}^{o}\, ;\; I_{22} = L_{2r}L_{2m}s_{rm}^{o} - c\, L_{3r}L_{3m}s_{rm}^{o}$$

$$Q_{abj}^{(N)} = Q_{baj}^{(N)}\, ;\; R_{abj} = R_{baj}\, ;\; I_{ab} = I_{ba}\, ;\; \text{and all other } I_{ab},\, Q_{abj}^{(N)},\, R_{abj} = 0 \tag{12}$$

and

$$A_j^{(N)} = J_{11}^{-1}\, h_{,n1}^{(N)}\, L_{1j} + J_{12}^{-1}\, h_{,n2}^{(N)}\, L_{1j}\, ;\; B_j^{(N)} = J_{21}^{-1}\, h_{,n1}^{(N)}\, L_{2j} + J_{22}^{-1}\, h_{,n2}^{(N)}\, L_{2j}\, ;$$

$$C_j^{(N)} = \frac{1}{2} \{ J_{21}^{-1}\, h_{,n1}^{(N)}\, L_{1j} + J_{22}^{-1}\, h_{,n2}^{(N)}\, L_{1j} + J_{11}^{-1}\, h_{,n1}^{(N)}\, L_{2j} + J_{12}^{-1}\, h_{,n2}^{(N)}\, L_{2j} \} \tag{13}$$

For initial stress due to thermal expansion $I_{11} = I_{22} = f = g\, ;\; I_{12} = 0$ (14)

Equation (11) can be used to assemble the sharp corner node's element E2 contributions from the [G] matrix into the columns of the [F] and [G] matrices associated with the deflection and traction components of element E1. For example, suppose a six node triangular element is being used in the boundary element model. Then NNL will be six, and Equation (11) relates t_k for element E2 to three components of t_k, and eighteen (3 x 6) components of displacement in element E1. The 21 coefficients (3+18) can then be used to assemble the t_k column contribution from the [G] matrix of element E2 into the 21 columns of the [F] and [G] associated with element E1. Notice that the I_{ab} terms in Equation (11) are not multiplied by either traction or displacement components. These terms are used to alter the right hand side vector. This same equation can also be used to resolve problems associated with combinations of specified displacement boundary conditions that produce jumps in unknown traction components at exterior corners in boundary element models.

SHARP CORNER DESIGN SENSITIVITY ANALYSIS FORMULATION

DSA employing the implicit differentiation technique requires the development of a formulation for sensitivities of the vectors and matrices present in the BEA method. In order to develop a formulation that can account for the sharp corners that may be present on inter-zone boundaries, the sharp corner formulation developed in the previous section can be differentiated with respect to the design variable X_L. This formulation starts with Equation (11).

$$\left[t_k^{(E2)} = n_p\, L_{ak}\, L_{bp} \left\{ I_{ab} + R_{abj}\, t_j^{(E1)} + \sum_{N=1}^{NNL} Q_{abj}^{(N)}\, u_j^{(N)} \right\} \right]_{,L} \Rightarrow \left[t_k^{(E2)} = n_p\, L_{ak}\, L_{bp}\, Z_{ab} \right]_{,L}$$

$$t_{k,L}^{(E2)} = n_{p,L}\, L_{ak}\, L_{bp}\, Z_{ab} + n_p\, L_{ak,L}\, L_{bp}\, Z_{ab} + n_p\, L_{ak}\, L_{bp,L}\, Z_{ab} + n_p\, L_{ak}\, L_{bp}\, Z_{ab,L} \qquad (15)$$

The first 3 terms in Equation (15) contain response information computed in the previous analysis step (displacement and traction components) along with geometric sensitivity information (direction cosine and element normal sensitivity). The last term in this equation will involve the design's response sensitivity as shown below.

$$Z_{ab,L} = \left(I_{ab,L} + R_{abj,L}\, t_j^{(E1)} + \sum_{N=1}^{NNL} Q_{abj,L}^{(N)}\, u_j^{(N)} \right) + R_{abj}\, t_{j,L}^{(E1)} + \sum_{N=1}^{NNL} Q_{abj}^{(N)}\, u_{j,L}^{(N)} \qquad (16)$$

The term in parentheses plays a role analogous to the role that I_{ab} played in the analysis step. The last two terms, when substituted into Equation (15), yield a consistent relationship among the traction sensitivities in element E2 at node 88, the traction sensitivities in element E1 at node 88, and the displacement sensitivities on all the nodes in element E1. This relation can therefore be used to assemble the element E2 $[G]_{,L}$ contributions for design sensitivity analysis in a procedure that is entirely analogous to the assembly of the [G] contributions explained in the previous section for BEA with sharp corners. The equations that follow complete the sharp corner sensitivity formulation.

$$Q_{11j,L}^{(N)} = a\, A_{j,L}^{(N)} + b\, B_{j,L}^{(N)}\,;\; Q_{22j,L}^{(N)} = b\, A_{j,L}^{(N)} + a\, B_{j,L}^{(N)}\,;\; Q_{12j,L}^{(N)} = d\, C_{j,L}^{(N)}$$

$$R_{11j,L} = R_{22j,L} = c\, L_{3j,L}\,;\; R_{13j,L} = R_{33j,L} = L_{3j,L}\,;\; R_{23j,L} = L_{2j,L}$$

$$I_{11,L} = L_{1r,L} L_{1m} s_{rm}^{o} - c\, L_{3r,L} L_{3m} s_{rm}^{o} + L_{1r} L_{1m,L} s_{rm}^{o} - c\, L_{3r} L_{3m,L} s_{rm}^{o} + L_{1r} L_{1m} s_{rm,L}^{o} - c\, L_{3r} L_{3m} s_{rm,L}^{o}$$

$$I_{12,L} = L_{1r,L} L_{2m} s_{rm}^{o} + L_{1r} L_{2m,L} s_{rm}^{o} + L_{1r} L_{2m} s_{rm,L}^{o}$$

$$I_{22,L} = L_{2r,L} L_{2m} s_{rm}^{o} - c\, L_{3r,L} L_{3m} s_{rm}^{o} + L_{2r} L_{2m,L} s_{rm}^{o} - c\, L_{3r} L_{3m,L} s_{rm}^{o} + L_{2r} L_{2m} s_{rm,L}^{o} - c\, L_{3r} L_{3m} s_{rm,L}^{o}$$

$$Q_{abj,L}^{(N)} = Q_{baj,L}^{(N)}\,;\; R_{abj,L} = R_{baj,L}\,;\; I_{ab,L} = I_{ba,L}\,;\; \text{and other } I_{ab,L},\, Q_{abj,L}^{(N)},\, \text{and } R_{abj,L} = 0 \qquad (17)$$

and

$$A_{j,L}^{(N)} = J_{11,L}^{-1}\, h_{,n1}^{(N)}\, L_{1j} + J_{12,L}^{-1}\, h_{,n2}^{(N)}\, L_{1j} + J_{11}^{-1}\, h_{,n1}^{(N)}\, L_{1j,L} + J_{12}^{-1}\, h_{,n2}^{(N)}\, L_{1j,L}$$

$$B_j^{(N)} = J_{21,L}^{-1}\, h_{,n1}^{(N)}\, L_{2j} + J_{22,L}^{-1}\, h_{,n2}^{(N)}\, L_{2j} + J_{21}^{-1}\, h_{,n1}^{(N)}\, L_{2j,L} + J_{22}^{-1}\, h_{,n2}^{(N)}\, L_{2j,L}$$

$$C_j^{(N)} = \frac{1}{2} \{ J_{21,L}^{-1}\, h_{,n1}^{(N)}\, L_{1j} + J_{22,L}^{1}\, h_{,n2}^{(N)}\, L_{1j} + J_{11}^{-1}\, h_{,n1,L}^{(N)}\, L_{2j} + J_{12L}^{-1}\, h_{,n2}^{(N)}\, L_{2j} \}$$

$$+ \frac{1}{2} \{ J_{21}^{-1}\, h_{,n1}^{(N)}\, L_{1j,L} + J_{22}^{-1}\, h_{,n2}^{(N)}\, L_{1j,L} + J_{11}^{-1}\, h_{,n1}^{(N)}\, L_{2j,L} + J_{12}^{-1}\, h_{,n2}^{(N)}\, L_{2j,L} \} \qquad (18)$$

For initial stress due to thermal expansion the formulae for $I_{ab,L}$ simplify to

$$S_{ij,L}^{o} = -\frac{E\, \delta_{ij}\, \alpha\, \theta_{,L}}{(1-2\upsilon)}\,;\; I_{11,L} = I_{22,L} = -\frac{E\, \alpha\, \theta_{,L}}{(1-2\upsilon)} \left(1 - \frac{\lambda}{(2\,\mu + \lambda)} \right);\; I_{12,L} = 0 \qquad (19)$$

SHARP CORNER EXAMPLE PROBLEM

The two zone pressurized cylinder model shown in Figure 2 is used to demonstrate the effectiveness of the sharp corner DSA formulation. This model contains 17 elements in zone 1, 43 elements in zone 2, and a total of 99 nodes. Node 21 is the node that is located at the sharp corner on the interface between Zone 1 and Zone 2. In this example problem, the model with modulus of elasticity 30 x $10^{6\,psi}$ and Poisson's ratio 0.3 was subjected to the boundary conditions corresponding to an internal pressure and quarter symmetry. The design variable in this example was chosen to be the inner radius 'a' of the cylinder and the nodes in the model had their geometric sensitivity controlled as depicted in Figure 2. The analysis and DSA results for the response of this model both with and without the sharp corner formulation are presented in Table 2. Notice that the errors in the stress components predicted by the BEA without the sharp corner formulation are considerable near the sharp corner. The displacements were reasonably accurate for both cases. The stress components and sensitivities predicted by the BEA method with the sharp corner formulation, on the other hand, agree closely with the analytical results.

Table 2. Analysis and DSA Results for the Pressurized Cylinder Example Problem

		Analysis			Design Sensitivity Analysis		
Node	Catagory	σ_{rr}	$\sigma_{\theta\theta}$	$\sigma_{r\theta}$	σ_{rr}	$\sigma_{\theta\theta}$	$\sigma_{r\theta}$
21	exact	-0.21875	0.30208	0.0	-0.065104	0.10850	0.0
21	corner	-0.21816	0.30226	0.00054	-0.059256	0.12412	-0.000008
21	no-corner	-0.10077	0.13381	0.13380	-0.035290	0.04122	0.041209

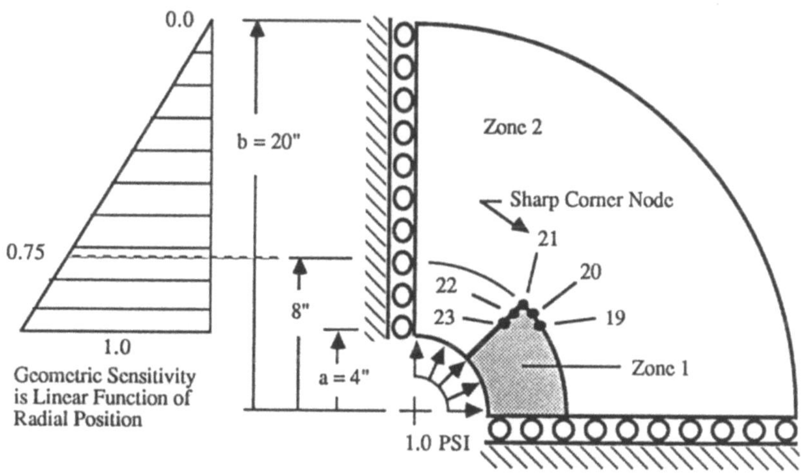

Figure 2. Pressurized Cylinder Example Problem Two-zone BEA Model

ACKNOWLEDGEMENT

Portions of the research discussed herein have been supported by grants from the U. S. National Science Foundation (DMC-8708254) to Worcester Polytechnic Institute and (DDM-8996171) to Clarkson University, and also by grants or contracts from General Dynamics Electric Boat Division and United Technologies Corporation. Any opinions, findings, and conclusions or recommendations expressed in this publication are those of the authors and do not reflect the views of these other organizations.

REFERENCES

1. J. H. Kane and S. Saigal, "Numerical Integration and Sparse Blocked Equation Solution Techniques for Large Scale Boundary Element Analysis," Proceedings: Two-Day Symposium on the Solution of Super Large Problems in Computational Mechanics, Mystic Connecticut, October 1988, (J. H. Kane and A. D. Carlson, Editors, Plenum Press Publishers)

2. J. C. Lachat, and J. O. Watson, "A Second Generation Boundary Integral Program for Three Dimensional Elastic Analysis," Chapter in, Boundary Integral Equation Method: Computational Applications in Applied Mechanics (Eds. T.A. Cruse and F. J. Rizzo) Appl. Mech. Div., ASME Vol 11, New York, 1975.

3. J. M. Crotty, "A Block Equation Solver for Large Unsymmetric Matrices Arizing in the Boundary Element Method," International Journal for Numerical Methods in Engineering, Vol 18, pp 997-1017, 1982

4. J. H. Kane and S. Saigal, 'An Arbitrary Multi-zone Condensation Technique for Boundary Element Design Sensitivity Analysis,' AIAA Journal, in press.

5. J. H. Kane, " Shape Optimization Utilizing a Boundary Element Formulation," BETECH 86, Proceedings, 1986 Boundary Element Technology Conference, MIT.

6. J. H. Kane and S. Saigal, " An Arbitrary Condensing, Noncondensing Solution Strategy for Large Scale, Multi-zone Boundary Element Analysis," Computer Methods in Applied Mechanics and Engineering, in press.

7. J. H. Kane and S. Saigal, "Design Sensitivity Analysis of Solids Using BEM," Journal of Engineering Mechanics, ASCE, Vol. 114, No. 10, pp. 1703-1722, October, 1988

8. S. Saigal, J.T. Borggaard and J.H. Kane, "Boundary Element Implicit Differentiation Equations for Design Sensitivities of Axisymmetric Structures," International Journal of Solids and Structures, Vol. 25, No. 5, pp. 527-538, 1989

9. J. H. Kane, B. L. K. Kumar, and M. Stabinsky, "Transient Thermoelasticity and Other Body Force Effects in Boundary Element Shape Sensitivity Analysis," International Journal for Numerical Methods in Engineering, submitted for review.

A Direct and Indirect Approach to Shape Optimization Using the Boundary Element Method

R. L. West
Mechanical Engineering Department
Virginia Polytechnic Institute and State University

E. Sandgren
School of Mechanical Engineering
Purdue University

Abstract
The positive and negative aspects of using the boundary integral method for shape optimization are reviewed. Aspects such as shape representation, geometric compatibility and ease of coupling the analysis and optimization modules are discussed. Two fundamentally different approaches to shape optimization with the boundary integral method are presented. The first approach involves a conventional coupling between a nonlinear programming code and a BEM code. The geometric representation is based on B-spline functions. The second approach involves a formulation of the shape optimization problem as a variational statement. Examples of both approaches are presented considering two-dimensional applications. Extensions to other problem classes are discussed.

Introduction

Shape optimization is a design process that seeks to make trade-offs in geometry to maximize design performance while maintaining design constraints. Typically the problem is characterized by complex implicitly defined relationships between the design goals and constraints. In general, the design goals and constraints are nonlinear functions.

Traditionally, the shape optimization problem has been approached from two perspectives, nonlinear programming (NLP) methods and variational methods. The NLP methods have the advantage that the formulation is general, easy to implement and there is a lot of experience using mathematical programming methods. The disadvantage of the NLP approach is that it is an inherently discrete formulation. The variational approach has the disadvantage of a more difficult and mathematically involved formulation, with the difficulties associated with obtaining solutions. The advantages are that it is a continuous formulation that provides for interaction between the objective function and constraints, it provides insight into the problem and due to the analytic form of the optimality criteria does not lose information from discretization. There is a great volume of literature on these two formulations of the shape optimization problem, the most representative review of the work can be found in the following references [1-7].

This paper discusses the investigation of the shape optimization problem from both formulations. The NLP method is the more conventional approach which couples the boundary element method to determine constraints on static response. The variational approach derives the optimality criteria and uses the boundary element method to solve for the shape directly.

NLP Formulation

The shape optimization problem has often been formulated as a NLP problem. NLP methods seek to directly minimize the objective function subject to constraints imposed on the design and bounds on the design variables. The general NLP is written as

$$\text{Minimize } f(x); \quad x=[x_1,x_2,\ldots,x_N]^T \tag{1}$$

$$\text{Subject to: } h_k(x)=0 \qquad k=1,2,\ldots,K \tag{2}$$

$$g_j(x)\geq 0 \qquad j=1,2,\ldots,J \tag{3}$$

$$x_i^l \leq x_i \leq x_i^u \qquad i=1,2,\ldots,N \tag{4}$$

In the structural shape optimization literature, the design variables have often been degrees-of-freedom associated with the nodal coordinates of a finite element model. Due to the discrete nature of NLP techniques this has often resulted in problems such as the generation of large numbers of design variables, poor representation of part shape and response error due to inadequate maintenance of the finite element mesh during optimization.

This work incorporates the concept of the design patch that was introduced by Iman [8], and Braibant and Fleury [9]. The design patch reduces the number of design variables and makes use of geometric modeling methods to better represent the shape. This also reduces the discrete effects of the design variables through the coupling of the geometric relationships. In this study, the geometry of the design patch is represented as B-spline curves. The mathematical form of the B-spline curve is

$$P(u)=\sum_{j=0}^{n} p_j B_{j,k}(u) \tag{5}$$

where p_j is the j th control point and $B_{j,k}(u)$ is the j th B-spline blending function of order k [10]. The B-spline curve has several nice properties; the control points affect the curve shape naturally, the B-spline curve is variation diminishing and independent of coordinate system. Most importantly, the B-spline basis functions exhibit the property of local control, and can be extended, with no loss of continuity by adding control points. The use of B-spline curves to represent the design patch naturally leads to using the control points as the design variables.

The difficult issue of mesh refinement and associated error is made easier by using the direct boundary element method to compute the static structural response. In the absence of body forces, the boundary element method requires only the discretization of the part boundary. This provides a natural and easy coupling between the optimization and the required structural analysis. In this case, the displacement field can be evaluated by equation (6).

$$C_{ij}(x^p)u_j(x^p)=\sum_{k=1}^{M}\sum_{\alpha=1}^{n}\left\{ \int_{\Delta S_k}\left\{G_{ij}\left[x^p,x^q(\eta_m)\right]N^\alpha(\eta_m)J(\eta_m)ds(\eta_m)\right\}t_i^\alpha - \int_{\Delta S_k}\left\{F_{ij}\left[x^p,x^q(\eta_m)\right]N^\alpha(\eta_m)J(\eta_m)ds(\eta_m)\right\}u_i^\alpha\right\} \tag{6}$$

The matrix form of the boundary element equations are

$$[H]\{U\}=[G]\{T\} \tag{7}$$

where

$$[H]=[C]+[F]. \tag{8}$$

Substructuring is also used in conjunction with the design patch to reduce the computational effort. The total structure is partitioned into component substructures. For each substructure, the nodal values are partitioned into interfacing and noninterfacing coordinates. The resulting system of equations for a substructure is given in partitioned form by equation (9).

$$\begin{bmatrix} H_{ii} & H_{ib} \\ H_{bi} & H_{bb} \end{bmatrix} \begin{Bmatrix} u_i \\ u_b \end{Bmatrix} = \begin{bmatrix} G_{ii} & G_{ib} \\ G_{bi} & G_{bb} \end{bmatrix} \begin{Bmatrix} t_i \\ t_b \end{Bmatrix} \tag{9}$$

Subscripts i and b denote noninterfaced and interfaced coordinates respectively. The resulting matrix equation written in terms of the unknown interfaced displacements u_b and interfaced tractions t_b is given in equation (10)

$$[A]\{u_b\} = [B]\{t_i\} + \{t_b\} \tag{10}$$

where

$$[A] = \left[[G_{bb}] - [H_{bi}][H_{ii}]^{-1}[G_{ib}]\right]^{-1}\left[[H_{bb}] - [H_{bi}][H_{ii}]^{-1}[H_{ib}]\right] \tag{11}$$

$$[B] = \left[[G_{bb}] - [H_{bi}][H_{ii}]^{-1}[G_{ib}]\right]^{-1}\left[[G_{bi}] - [H_{bi}][H_{ii}]^{-1}[G_{ii}]\right]. \tag{12}$$

Once equation (10) has been composed for each subregion, the subregions of the structure can be assembled using the conditions of compatibility and continuity. The resulting assembled structure matrices can be rewritten after applying the known boundary conditions to form a set of linear algebraic equations. Solution of the matrix equations yields the unknown nodal boundary displacements and tractions along with the interfaced displacements. In turn the interfaced tractions can be computed for each substructure along with desired interior stresses and displacements.

The region of the structure to be modified during optimization is defined by one or more design patches which contain a portion of the boundary elements. The change in geometry within a boundary element of the design patch for the two-dimensional problem is given in equation (13).

$$\begin{Bmatrix} x' \\ y' \end{Bmatrix} = \begin{Bmatrix} x \\ y \end{Bmatrix} + \begin{Bmatrix} \Delta x \\ \Delta y \end{Bmatrix} = \sum_{i=1}^{I} N_i(\eta) \begin{Bmatrix} x_i \\ y_i \end{Bmatrix} + \sum_{j=0}^{J} B_{j,k}(u) \begin{Bmatrix} v_{xj} \\ v_{yj} \end{Bmatrix} \tag{13}$$

Where x,y are the initial coordinates of the boundary, x',y' are the updated coordinates of the boundary, x_i,y_i are the i th nodal coordinates of each boundary element. The corresponding interpolation functions are N_i, η is the local coordinate of each element and v_{xj},v_{yj} are the variations in the design variables in x and y respectively. Once the new shape is calculated the boundary element mesh may be updated along with the current values of the stress and displacement. Gradients of the stress and displacement are evaluated by finite difference methods using the results of the boundary element analysis.

Application

The design of a hook for lifting hot-metal ladles is considered. The minimum design load for the hook is taken to be 150 tons. In this example the design goal is to minimize the frontal area of the hook in order to get the optimal bight section. Stress constraints are imposed which require the stress along the inside curve of the hook to be less than 12,500 psi.

The hook is modeled as two substructures. Substructure 1 is discretized into 47 quadratic boundary elements. Substructure 2 is approximated by 46 quadratic boundary elements. The curve between points A and B defines the design patch which contains 17 boundary elements, 35 nodes. The design patch is modeled geometrically by a seventh order B-spline curve which has nine control points, giving 18 design variables for the problem. An NLP code implementing the Generalized Reduced Gradient (GRG) method, OPT [11], was used to perform the optimization.

Figure 1 The Boundary Element Mesh for the Ladle Hook

The problem formulation follows as:

Minimize Area

$$\text{Subject to} \quad \left[\sigma_0 - \left(\sigma_j + \sum_{i=1}^{l} \frac{\partial \sigma_j}{\partial v_i} \delta v_i\right)\right] \Big/ \sigma_0 \geq 0 \quad j = 1,2,\ldots,n \tag{14}$$

and $c_i \geq v_i \geq d_i$

Where Area is the frontal area of the hook, σ_0 is the specified stress limit (12,500 psi), and σ_j are the stresses at the j th node of the boundary elements. The move limits on the design variables are c and d and the total number of nodes on the design patch are $n = 17$.

The optimum shape of the bight section on the hook is shown in Figure 2. The area of the bight section was reduced by 244.82 in.^{2} with a maximum stress of 12,380 psi.. The effect of substructuring saved two-thirds of the computer time required for each boundary element analysis. Details of this study are presented in Wu [12] and Sandgren and Wu[13].

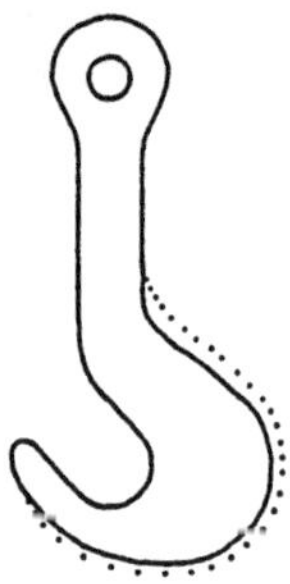

Figure 2 The Optimum Shape for the Ladle Hook

A Constrained Variational Approach

The underlying conceptual and theoretical basis of this shape optimization method is rooted in the variational principles of mechanics[14] and the calculus of variations[15]. The resulting formulation leads to a direct variational statement of the shape optimization problem. The class of functional under consideration are of the form

$$\Pi = \int_{t_1}^{t_2} L(x, y, x_t, y_t, \lambda_1, \ldots, \lambda_n)\, dt \tag{15}$$

The augmented Lagrangian, L, is a function of the components of the curve, $x(t), y(t)$ their associated first derivatives, $x_t(t) = \frac{dx}{dt}$, $y_t(t) = \frac{dy}{dt}$ and the Lagrange multipliers, λ_i. The first variation of the functional yields the optimality criteria composed of the Euler-Lagrange equations, (16-17), the equality constraint set, (18)

$$\frac{\partial L}{\partial x} - \frac{d}{dt}\left(\frac{\partial L}{\partial x_t}\right) = 0 \tag{16}$$

$$\frac{\partial L}{\partial y} - \frac{d}{dt}\left(\frac{\partial L}{\partial y_t}\right) = 0 \tag{17}$$

$$\sum_{i=1}^{N} \left[\int_{t_1}^{t_2} \frac{\partial L}{\partial \lambda_i} dt - Const_i = 0 \right] \tag{18}$$

and the associated boundary conditions.

Due to the coupled nonlinear character of the Euler-Lagrange equations and the constraint set, a linear approximation about the current shape is applied. For this class of functionals, the linearized Euler-Lagrange equations result in a set of second order linear differential equations. These equations can be algebraically rewritten in Poisson's equation form. The effect of the transformation to Poisson's equation form is that the coupling between the differential equations show up as source terms based on the current shape. Additionally, the Poisson's differential equation can now be written in integral equation form with the use of the known fundamental solution. The resulting integral equations in x, y are

$$x(\xi) = \left[\frac{dG(\xi,t)}{dt} x(t) - G(\xi,t) \frac{dx(t)}{dt} \right]_{t_1}^{t_2} + \int_{t_1}^{t_2} G(\xi,t) \Psi_x(t)\, dt \tag{19}$$

$$y(\xi) = \left[\frac{dG(\xi,t)}{dt} y(t) - G(\xi,t) \frac{dy(t)}{dt} \right]_{t_1}^{t_2} + \int_{t_1}^{t_2} G(\xi,t) \Psi_y(t)\, dt \tag{20}$$

where $G(\xi,t)$ is the one-dimensional fundamental solution for Laplaces's equation [16]. The integral equations give an expression for x and y at any point on the curve given the boundary conditions at the endpoints. In this approach the curve is discretized into a series of nodes and curve elements. The domain is represented by the curve itself, the endpoints of the curve define the boundary.

The approximate optimality criteria in matrix form can be written by assembling the system matrices from the boundary element formulation and the constraint set.

$$[P]\{x\} - [G]\{x_t\} + [G\Psi M_x]\{\lambda\} + \{G\Psi C_x\} = \{0\} \tag{21}$$

$$[P]\{y\} - [G]\{y_t\} + [G\Psi M_y]\{\lambda\} + \{G\Psi C_y\} = \{0\} \tag{22}$$

$$[C_x]\{x\} + [C_y]\{y\} + [\varnothing]\{\lambda\} = \{Const\} \tag{23}$$

A solution for the next shape estimate can be written as linear functions of the Lagrange multipliers. Substitution of these relations into the constraint set results in a linear system of equations that can be solved for estimates of the Lagrange multipliers that satisfies the constraint set. Back-substitution of the Lagrange multiplier estimates yields the calculated shape that extremizes the approximate optimality criteria and simultaneously satisfies the linear constraint set. The updated shape estimate, $x^{(\tau+1)}$, is obtained from the current shape, $x^{(\tau)}$, the calculated shape estimate, $x_c^{(\tau)}$, and the fractional step size, α, by the update scheme given in equation (24).

$$x^{(\tau+1)} = x^{(\tau)} + \alpha(x_c^{(\tau)} - x^{(\tau)}) \tag{24}$$

Once the updated shape estimate has been computed, the region properties for the curve and the boundary element system matrices are recomputed. The solution is iterated upon until convergence is achieved.

Illustration of the Method

The formulation of a two-dimensional shape optimization problem can be illustrated by determining the minimum perimeter curve with a prescribed enclosed area. This classical problem was chosen for the purpose of evaluating the method since the analytical solution and the interpretation of the Lagrange multiplier is known. The solution of the problem is that of a circle and the interpretation of the Lagrange multiplier is that of the curvature[15]. The functional for this problem is

$$\Pi = \int_{t_1}^{t_2} [x_t^2 + y_t^2]^{\frac{1}{2}} dt + \lambda\left[\frac{1}{2}\int_{t_1}^{t_2} (xy_t - yx_t)dt - Area\right]. \tag{25}$$

Requiring stationarity of the functional results in the following optimality criteria without the boundary conditions.

$$\frac{d^2x}{dt^2} = x_{tt} = \lambda\left[x_t^2 + y_t^2\right]^{\frac{1}{2}} y_t = \lambda \Psi_x \tag{26}$$

$$\frac{d^2y}{dt^2} = y_{tt} = -\lambda\left[x_t^2 + y_t^2\right]^{\frac{1}{2}} \dot{x}_t = \lambda \Psi_y \tag{27}$$

$$\frac{1}{2}\int_{t_1}^{t_2} (xy_t - yx_t)dt - Area = 0 \tag{28}$$

The discrete model is composed of 9 nodes, 2 boundary (point) elements, 8 linear domain (line) elements using Lagrangian interpolation functions. There are two prescribed boundary conditions ; $x_1 = x_9 = 2.0$, $y_1 = y_9 = 0.0$. The fractional step size is specified as, $\alpha = 0.5$ and the convergence epsilon is given to be, $\varepsilon = 0.001$. The initial configuration of the discretized curve is shown along with the first iteration shape estimate in Figure 3. Points generated from the DBEM solution are plotted against the analytic solution in Figure 4. It is encouraging to see that all the points lie on the circle. Table 1 presents the results of each iteration. Details can be found in West [17] and West and Sandgren [18].

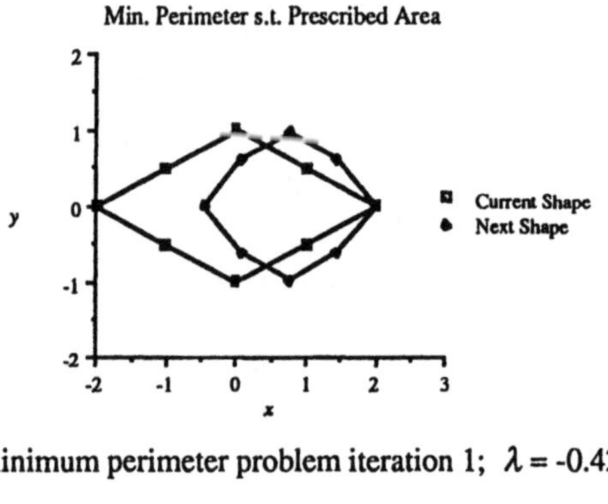

Figure 3 Minimum perimeter problem iteration 1; λ = -0.4215.

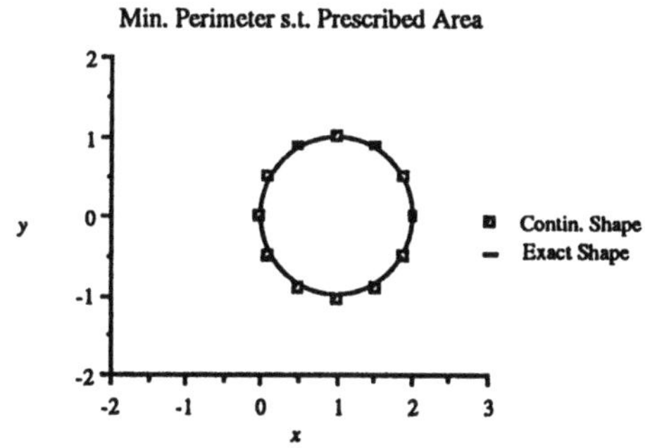

Figure 4 DBEM solution vs the exact solution.

Table 1 Minimum Perimeter Problem Iteration History

Iteration	Perimeter	Area	Constraint	Multiplier
0	5.657	2.000	-1.1416	---
1	6.353	2.994	−0.1480	−1.6661
2	6.281	2.966	−0.1758	−1.0197
3	6.276	2.970	−0.1712	−1.0562
4	6.276	2.972	−0.1698	−1.0561
5	6.277	2.972	−0.1691	−1.0560
Analytic	2π	π	0.0	−1.0

Conclusions

The discrete nature of the NLP approach to the shape optimization problem can be attenuated by the design patch concept and the use of the boundary element method. The design patch concept is effective in reducing the number of design variables and the discrete representation of the shape. The boundary element method is ideal for use in shape optimization problems because of the ease in coupling the optimization and the analysis of static response. Additional benefits are obtained from reducing the dimension in the analysis and better accuracy in the stresses on the boundary. Further improvements were made in the computational effort and the accuracy of the solution via substructuring.

A constrained variational method for two-dimensional shape optimization with equality constraints is realizable. Insight into the interaction between the design goals and constraints can be obtained from the optimality criteria. The ill-effects of linearization are reduced by linearizing the governing integral equations over the boundary and domain elements. A standard solution procedure is obtained for a class of problems that results in the use of a known fundamental solution. An expression for the optimal shape is given by the direct boundary element method and was shown to be in good agreement with the known analytic solution. The method seems stable provided enough degrees-of-freedom are used to "adequately" represent the shape. An important characteristic of this approach is that the framework established looks to be extendable to classes of three-dimensional problems.

It is clear that the accuracy of results for the continuous shape optimization problem are greater than those required for parameter shape optimization. As a result, the continuous shape optimization problem is especially demanding in terms of problem formulation. Mathematical theory will play an important role in qualifying formulations and providing more appropriate frameworks. It is also important that the numerical techniques be consistent with the problem formulation. Lastly, Further development will likely require better representation of the constraints.

References

1. Schmit, L.A., "Structural Optimization - Some Key Ideas and Insights", New Directions in Optimum Structural Design, E. Atrek, R.H. Gallagher, K.M. Ragsdell and O.C. Zienkiewicz, (Eds.), Wiley, 1984, pp 1-45.
2. Vanderplaats, G.N., "Numerical Methods for Shape Optimization: An Assessment of the State of the Art", New Directions in Optimum Structural Design, E. Atrek, R.H. Gallagher, K.M. Ragsdell and O.C. Zienkiewicz, (Eds.), Wiley, 1984, pp. 89-103.
3. Haftka, R.T., Grandhi, R.V., "Structural Shape Optimization-A Survey", Computer Methods in Applied Mechanics and Engineering, Vol. 57, 1986, pp.91-106.
4. Atrek, E., Gallagher, R. H., Ragsdell, K. M., Zienkiewicz, O. C., New Directions in Optimum Structural Design, Wiley, 1984.
5. Bennett, J. A., Botkin, M., E., The Optimum Shape: Automated Structural Design, Plenum Press, 1986.
6. Haug, E. J., "A Review of Distributed Parameter Structural Optimization Literature", Optimization of Distributed Parameter Structures, Sijthoff & Noordhoff, 1981, Vol. 1, pp. 3-74.
7. Haug, E.J. and Cea, J., Optimization of Distributed Parameter Structures, Sijthoff & Noordhoff, 1981.
8. Iman, M. H., "Three-dimensional Shape Optimization", International Journal for Numerical Methods in Engineering, 18, 1982, pp. 661-673.
9. Braibant, V., Fleury, C., "Shape Optimal Design an Approach Matching C.A.D. and Optimization Concepts", Optimization in Computer-Aided Design, J.S. Gero, (Ed.), Elsevier(North Holland), 1985, pp. 231-269.
10. Newman, W. M. and Sproull, R. F., "Principles of Interactive Computer Graphics", Addison-Wesley, New York, 1984.
11. OPT, (Rolla, MO: K, M, Ragsdell, Design Productivity Center, University of Missouri-Rolla).
12 Wu, S., J., "Applications of the Boundary Element Method for Structural Shape Optimization", Ph.D. Dissertation, Department of Mechanical & Aerospace Engineering, University of Missouri-Columbia, 1986.
13. Sandgren, E., Wu, S. "Shape Optimization Using the Boundary Element Method with Substructuring", International Journal for Numerical Methods in Engineering, Vol. 26, No. 19, 1988, pp. 1913-1924.
14. Lanczos, C., The Variational Principles of Mechanics, Dover, 1986.
15. Weinstock, R., Calculus of Variations with Applications to Physics and Engineering, McGraw-Hill, 1952.
16. Banerjee, P.K. and Butterfield, R., Boundary Element Methods in Engineering Science, McGraw-Hill, 1981.
17 West, R., L., "A Constrained Variational Shape Optimization Method", Ph.D. Dissertation, Department of Mechanical & Aerospace Engineering, University of Missouri-Columbia, 1988.
18. West, R. L., and Sandgren, E., "A Constrained Variational Method for Two-Dimensional Shape Optimization", Advances in Design Automation 1989, Proceedings of the 1989 ASME Design Automation Conference, 1989, Vol. 2, pp. 201-207.

A Variational Approach for the Sensitivity of Stress Constraints Using Boundary Elements

Sunil Saigal, Carnegie Mellon University, Pittsburgh, PA 15213
R. Aithal, University of Puerto Rico, Magauyez, PR

Abstract

An adjoint approach for the shape design sensitivity analysis (SDSA) of plane elasticity problems with a direct boundary integral equation formulation is presented. The objective function is expressed in an integral form and augmented by incorporating the elasticity equations via adjoint functions. The variation of the augmented function is taken by performing its material derivative to obtain the sensitivity of the objective function. The adjoint functions are obtained through the solution of the adjoint problem resulting from setting the local variations of the displacement, velocity, and traction fields equal to zero. An approximate procedure for the application of concentrated adjoint load in the boundary element framework is suggested. This allows the computation of displacement, traction, and stress sensitivities at discrete nodal points. The present formulation includes the treatment of body forces in sensitivity calculations. Numerical examples demonstrate the validity of the present approach.

Introduction

Shape Design Sensitivity Analysis (SDSA) of an object involves the prediction of the effect of a shape change on the performance of the object. In general, the methods used for the determination of design sensitivities are: the Finite Difference Approach (FDA), the Direct Differentiation Approach (DDA), and the Adjoint Structure Approach (ASA). Considerable effort has been devoted to developing efficient techniques for the evaluation of shape design sensitivities in the context of the finite element method, for example, Refs. [1,2]. The Boundary Element Method (BEM) has recently emerged as another powerful numerical discretization technique. Its advantages for SDSA over other analysis techniques have been given by, among others, Refs. [3,4]. As noted by Barone and Yang [3] although ASA is a straightforward technique, computational difficulties arise when evaluating displacement and stress sensitivities at discrete points. This is because the adjoint solutions for these two cases correspond to a concentrated force and moment, respectively and these solutions using BEM give rise to unbounded integrals. This problem was circumvented by Mota Soares and Choi [5] and Choi and Kwak [6] for stress sensitivity calculation by representing the von Mises stress over an element (or over an area) in an average sense using an averaging characteristic function m_p. The procedure does not provide the displacement and stress sensitivities at discrete nodal points and the results depend upon the choice of m_p. An improved formulation for SDSA with BEM including the effect of body forces using ASA is presented in this paper. This approach allows the evaluation of displacement and traction sensitivities at discrete locations on the object. The sensitivity of stress constraints at discrete locations is determined by the direct differentiation of the boundary element surface stress recovery expression.

Shape Sensitivity Analysis: Elastostatics

The shape optimization process involves the minimization of a general integral performance criterion I. A typical functional I usually involves both the domain integral and the boundary integral, such as,

$$I = \int_{\Omega} \phi_1 d\Omega + \int_{\Gamma} \phi_2 d\Gamma \tag{1}$$

where ϕ_1 and ϕ_2 are both smooth functions which are continuously differentiable with respect to the corresponding independent variables. A perturbation in the boundary Γ of an object also causes a movement of the domain Ω. The variation of the functional I due to this movement represents the sensitivity of the performance functional and is given by its Material Derivative (MD) represented here as $\dot{I}$. For 2D or 3D elastostatics analysis, the functions ϕ_1 and ϕ_2 appearing in equation (1) are generally expressed as functions of the stress σ_{ij}, the displacement u_i, the traction t_i, and the location coordinates x_i. These stresses, displacements, and tractions satisfy the equilibrium equations and the corresponding boundary conditions which define the *primary problem* and are given as

$$\sigma_{ij,j} + f_i = 0 \qquad \text{on } \Omega \tag{2}$$

$$u_i = \bar{u}_i \quad \text{on } \Gamma_1, \qquad t_i = \bar{t}_i \quad \text{on } \Gamma_2 \tag{3}$$

where $t_i = \sigma_{ij} n_j$; f_i are the body forces; Ω is the domain of the object, assumed to have no discontinuity in it; $\Gamma = \Gamma_1 + \Gamma_2$ is the boundary of the domain; the overbar represents specified values and n_j is the outward normal to the surface.

Design Sensitivities

The variation of the integral I requires the variation of various quantities, such as stresses, displacements, tractions, etc. The formulation for obtaining these variations is given below.

(a) Sensitivity of the Boundary Stress Components: Consider the local coordinate system (x_1', x_2') of a boundary element for 2D elastostatics analysis. The surface stresses are recovered from the boundary displacements and tractions obtained in a BEM analysis. Starting from the Hooke's law in the local coordinate system,

$$\sigma_{22}' = \frac{\nu}{1-\nu} \sigma_{11}' + \frac{E}{1-\nu^2} u_{2,s}' \tag{4}$$

where σ_{ij}' and $u_{2,s}'$ are the stress and strain in the local coordinate system, respectively. Substituting the interpolations for quadratic boundary elements in equation (4) we obtain

$$\sigma_{22}' = \frac{\nu}{1-\nu} \sum_{j=1}^{3} h^j t_i^j n_i + \frac{E}{1-\nu^2} J^{-1} \sum_{j=1}^{3} \frac{dh^j}{d\xi} u_i^j s_i \tag{5}$$

where h_i are the interpolation functions for node i ; n_j and s_j are the unit normal and tangential vectors at the boundary; E is the Young's modulus; ν is the Poisson's ratio; u_j^i and t_j^i are the nodal displacements and tractions, respectively, at node i; J is the Jacobian of transformation from global to local coordinate system; and ξ is the non-dimensional coordinate of the intrinsic element. Equation (5) is differentiated with respect to the design variable to obtain the stress sensitivity in the local coordinate system as

$$\dot{\sigma}'_{22} = \frac{\nu}{1-\nu}\sum_{i=1}^{3}\left(h^i \dot{t}_j^i n_j + h^i t_j^i \dot{n}_j\right) + \frac{E}{1-\nu^2} J^{-1} \sum_{j=1}^{3} \frac{dh^j}{d\xi}\left(\dot{u}_j^i s_j + u_j^i \dot{s}_j - u_j^i s_j V_{K,s} s_K\right) \tag{6}$$

where the relationship $\dot{J} = -J / V_{K,s} s_K$ has been used in arriving at equation (6). The dot denotes the derivative (sensitivity) of the quantity with respect to the design variable. The quantities $\dot{t}_j$ and $\dot{u}_j$ in equation (6) need to be determined. It should be noted that equation (6) allows the evaluation of stress sensitivities at any location within the element unlike the treatment by Choi and Kwak [6] where a value of stress sensitivity averaged over the entire element is obtained.

(b) Nodal Traction and Nodal Displacement Sensitivities: Consider a general functional I expressed as a surface integral as

$$I = \int_\Gamma f(u_i, t_i) d\Gamma \tag{7}$$

where f is any function of the displacements u_i and tractions t_i. The nodal displacement and traction sensitivity can be obtained by suitably defining the function f and then obtaining the MD of the functional I. The function f is selected as $f = \delta(x_i - \hat{x}_i)u_i$ for displacement sensitivity, and $f = \delta(x_i - \hat{x}_i)t_i$ for traction sensitivity. Then,

$$I^u = \int_\Gamma \delta(x_i - \hat{x}_i) u_i d\Gamma \quad ; \quad I^t = \int_\Gamma \delta(x_i - \hat{x}_i) t_i d\Gamma \tag{8}$$

where I^u and I^t are the functionals used for obtaining the nodal displacement and traction sensitivity, respectively; $\hat{x}_i$ are the coordinates of the node point under consideration; and δ is the Dirac-delta function. The sensitivity of I is now determined by following the steps outlined by Meric [7] as: (i) augment the performance criterion I by including the equations of motion using adjoint functions, (ii) Integrate by parts in Ω and Γ, and (iii) take the MD of the augmented functional to give

$$\dot{\hat{I}} = \dot{I} + \int_\Omega \left(-u^*_{i,j}\sigma'_{ij} - \sigma_{ij}u'^*_{i,j} + f_i u'^*_i\right) d\Omega + \int_\Gamma \left(-u^*_{i,j} - \sigma_{ij} + f_i u^*_i\right) V^n d\Gamma$$

$$+ \int_\Gamma \left(t_i u'^*_i + u^*_i t'_i + \left(t_i u^*_i\right)_{,j} V_j + t_i u^*_i V_{j,s} s_j\right) d\Gamma \tag{9}$$

(iv) substitute the partial derivative forms of the constitutive equations and kinematic conditions, (v) integrate by parts again to give

$$\dot{\hat{I}} = \int_\Omega \left\{ \left(\sigma_{ij,j} + f_i\right) u_i'^{*} + \left(\sigma^{*}_{ij,j} u_i'\right) \right\} d\Omega + \int_\Gamma \left\{ \left(-\sigma_{ij} u^{*}_{i,j} + f_i u_i\right) V^n + \left(t_i u_i^{*}\right)_{,j} V_j + \left(t_i u_i^{*}\right) V_{j,s} s_j \right\} d\Gamma$$

$$+ \int_\Gamma \left\{ \left(\delta(x_i - \hat{x}_i) \frac{\partial f}{\partial u_i} - t_i^{*} \right) u_i' + \left(\delta(x_i - \hat{x}_i) \frac{\partial f}{\partial t_i} - u_i^{*} \right) t_i' \right\} d\Gamma$$

$$+ \left\{ \frac{\partial f}{\partial u_i} u_{i,j} V_j + \frac{\partial f}{\partial u_i} t_{i,j} V_j \right\}_{x_i = \hat{x}_i} \tag{10}$$

where σ^{*}_{ij} is the adjoint stress tensor, (vi) substitute the MD forms of the boundary and initial conditions, (vii) set the coefficients of local variations of u_i and t_i to be equal to zero to obtain the *adjoint system* as

$$\sigma^{*}_{ij,j} = 0 \quad \text{in } \Omega \tag{11}$$

with the adjoint boundary conditions

$$u_i^{*} = -\frac{\partial f}{\partial t_i} \quad \text{on } \Gamma_1, \qquad t_i^{*} = -\frac{\partial f}{\partial u_i} \quad \text{on } \Gamma_2 \tag{12}$$

(viii) Finally, the sensitivity of the performance criterion can be obtained as

$$\dot{I} = \int_\Gamma \left\{ \left(-\sigma_{ij} u^{*}_{i,j} + f_i u_i\right) V^n d\Gamma + \int_\Gamma \left(\left(t_i u_i^{*}\right)_{,j} V_j + \left(t_i u_i^{*}\right) V_{j,s} s_j \right) d\Gamma \right\}$$

$$+ \left[\frac{\partial f}{\partial u_i} u_{i,j} V_j + \frac{\partial f}{\partial t_i} t_{i,j} V_j \right]_{x_i = \hat{x}_i} + \int_{\Gamma_1} \left(\frac{\partial f}{\partial u_i} - t_i^{*} \right) \left(\bar{u}_i^1 + \left(\bar{u}_{i,k} - u_{i,k}\right) V_k \right) d\Gamma$$

$$+ \int_{\Gamma_2} \left(\frac{\partial f}{\partial t_i} - u_i^{*} \right) \left(\bar{t}_i^1 + \left(\bar{t}_{i,k} - t_{i,k}\right) V_k \right) d\Gamma \tag{13}$$

Solutions of the primary problem as given by equations (2) and (3) and the adjoint problem as given by equations (11) and (12) are employed in evaluating the sensitivity given by equation (13). To obtain these solutions equations (2) and (11) are first cast in the boundary integral form and then solved in the manner explained in Ref. [8]. To obtain the solution of the adjoint problem, a unit force acting in the direction of the displacement component for which the sensitivity is required must be applied in the BEM framework. Since it is not possible to compute the integrals due to the application opf a concentrated load [3], a statically equivalent traction must be applied. Figure 1 shows the equivalent tractions applied for continuous, quadratic boundary element for the cases when the actual concentrated load is at an end node and a middle node, respectively. In Figure 1, J_N denotes the Jacobian of element N. It is noted that the above expressions are general and include the effect of body forces on the objects for which the sensitivities are required.For the solution of the adjoint problems using BEM only the right hand side force vector changes for each such problem while the left hand side coefficient matrix needs to be factorized only once for the solution of the primary problem and its factorization can be saved for reuse in solving the adjoint problems.

Numerical Examples

The above developments were used for the SDSA of a number of example problems. The results were compared to the exact sensitivity results when available and to the results obtained using the DDA. A numerical example of a circular ring with a concentrated load of 400 lb. as shown in Figure 2 with $E = 3 \times 10^7$ psi and $\upsilon = 0.3$ is presented here to demonstrate the accuracy of the present approach. Due to the symmetry in the problem, only a quarter of the ring is modeled using 18 quadratic boundary elements and the node reflection property described by Banerjee, Ahmad and Chen [9]. Both the inside face and the outside face are discretized using nine boundary elements, respectively. The boundary elements on each of these two faces are of equal lengths. The design variable, which is the inner radius of the ring, has an initial value of 1.5 inch and it is perturbed by 0.001 inch to obtain the deformation velocity field. This perturbation results in a unit normal velocity on the inner boundary and a velocity of zero on the rest of the boundary of the ring. The results for displacement sensitivity and von Mises stress sensitivity at a number of locations along the inner face of the ring using the DDA, the ASA by Choi and Kwak [6], and the present approach are shown in Table I. For the present approach, the stress sensitivities at the midnode of the element are reported. The results obtained using the approach due to Choi and Kwak [6] are based on the characteristic function m_p prescribed as a unit step function. It is seen from the Table 1 that the present approach yields more accurate results compared to the approach due to Choi and Kwak [6].

Conclusions

An improved formulation based on the BEM for the sensitivity analysis of plane elasticity problems using the material derivative concept and the adjoint variable approach is presented. By developing a consistent approach for the application of concentrated adjoint loads, this formulation allows the computation of nodal displacement and traction sensitivities. The stress recovery expression for the boundary element method is directly differentiated to provide the nodal stress sensitivities in terms of nodal displacement and traction sensitivities. Previous similar sensitivity studies provide the sensitivity of stress constraint averaged over a region and their results depend upon the selected averaging characteristic function. The effect of body forces is also included in this treatment. The numerical implementation of the proposed method is discussed and numerical examples demonstrate the accuracy of sensitivity results obtained.

Acknowledgements

Sunil Saigal was supported by the National Science Foundation grant DMC-8708254 with Dr. Tony Woo as the project monitor. R. Aithal was supported by funds provided by the United Technologies Research Center.

References

1. Dems, K., and Mroz, Z. (1984). "Variational Approach by Means of Adjoint Systems to Structural Optimization and Sensitivity Analysis II, Structure Shape Variation", *Int. J. Solids Struct.*, 20 (6), 527-552.

2. Bennett, J.A., and Botkin, M.E. (1986). *The Optimum Shape: Automated Structural Design*, Plenum Press, New York.

3. Barone, M. and Yang, R.J. (1988). "Boundary Integral Equations for Recovery of Design Sensitivities in Shape Optimization", *AIAA Jl.*, 26, 589-594.

4. Kane, J.H., and Saigal, S. (1988). "Design Sensitivity Analysis of Solids Using BEM", *J. Engng. Mech.*, ASCE, 114 (10), 1703-1722.

5. Mota Soares, C.A., and Choi, K.K. (1986). "Boundary Elements in Shape Optimal Design of Structures", *The Optimum Shape: Automated Structural Design* (Edited by J.A. Bennett and M.E. Botkin) Plenum Press, New York, NY, 199-231.

6. Choi, J.H. and Kwak, B.M. (1988). "Boundary Integral Equation Method for Shape Optimization of Elastic Structures", *Int. J. Num. Meth. Engng.*, 26, 1579-1595.

7. Meric, R.A. (1988). "Shape Design Sensitivity Analysis of Dynamic Structures", *AIAA J.*, 26 (2), 206-212.

8. Pape, D.A., and Banerjee, P.K. (1987). "Treatment of Body Forces in 2D Elastostatics BEM Using Particular Integrals", *J. App. Mech.*, ASME, 54, 866-871.

9. Banerjee, P.K., Ahmad, S., and Chen, K. (1988). "Application of BEM to Wave Barriers in Multi-Layered Three-Dimensional Soil Media", *Earthquake Engineering and Structural Dynamics*, 16, 1041-1060.

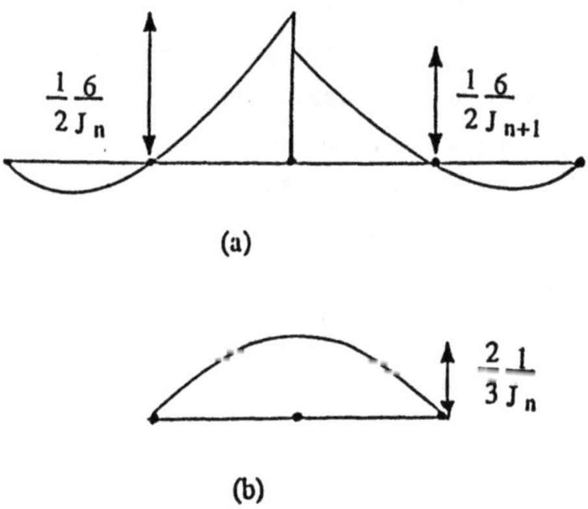

Figure 1. Statically Equivalent Tractions.
(a) Load at Element End-node
(b) Load at Element Mid-node

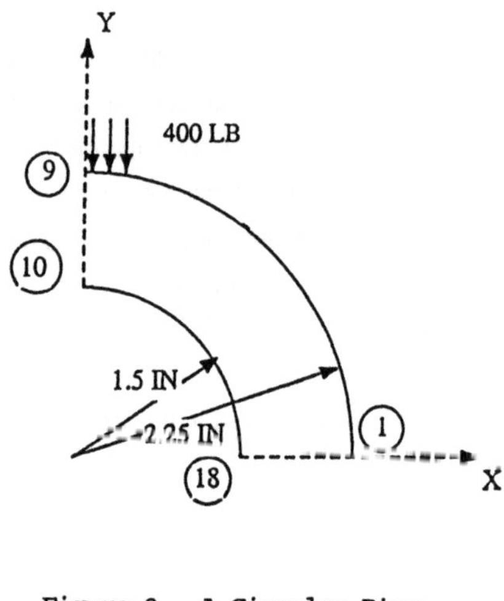

Figure 2. A Circular Ring Under Concentrated Load.

Table I: Sensitivity Analysis for a Circular Ring Under Concentrated Load.

Coordinate Location (inch)		DISPLACEMENT SENSITIVITY			
		$\dot{u}_x$		$\dot{u}_y$	
X	Y	DDA	ASA (This Study)	DDA	ASA (This Study)
0.0	1.5	0.0	0.0	4.6165×10^{-4}	4.6036×10^{-4}
0.1307	1.4943	-1.8956×10^{-5}	-1.9329×10^{-5}	4.5871×10^{-4}	4.5986×10^{-4}
0.2605	1.4772	-3.6583×10^{-5}	-3.5517×10^{-5}	4.4077×10^{-4}	4.3923×10^{-4}
0.3882	1.4489	-5.7970×10^{-5}	-5.8700×10^{-5}	4.1980×10^{-4}	4.2106×10^{-4}
0.5130	1.4095	-7.6723×10^{-5}	-7.5173×10^{-5}	3.8656×10^{-4}	3.8474×10^{-4}
0.4226	1.3595	-1.0210×10^{-4}	-1.0118×10^{-4}	3.5513×10^{-4}	3.5386×10^{-4}
0.75	1.2990	-1.2484×10^{-4}	-1.2656×10^{-4}	3.1477×10^{-4}	3.1664×10^{-4}

Element No.	DDA	ASA	
		Choi and Kwak (m_p = unit step function)	Present Method
10	-3288.00	-3075.04	-3170.40
11	-2312.50	-2185.10	-2259.30
12	-1238.40	-1185.50	-1227.80
13	-312.60	-309.80	-315.85
14	525.9	490.86	507.40
15	1240.9	1175.60	1214.70
16	1798.20	1709.50	1758.80
17	2179.00	2074.10	2130.40
18	2372.20	2259.04	2317.20

Some Observations on the Boundary Element Method

FRANK RIZZO

Department of Engineering Science and Mechanics
Iowa State University
Ames, Iowa 50011, USA

Summary

This is a qualitative look at the boundary element method (BEM) first with comments on problem types subject to BEM analysis. Next, there are some comments on BEM issues in general which, in the author's view, deserve more attention and clarification. There is a call for additional interaction with the mathematics community and with other disciplines such as physics and electromagnetics. Finally, there are a few remarks about what has been most appealing to the author and distinctive about the BEM over the last twenty-five years.

Terminology

The present discussion has to do with the boundary integral equation (BIE) method of formulating problems in engineering, mathematics, and the applied sciences, with the boundary element method of solution of the integral equations. Often the term BEM is used generically to include BIE as well as BEM ideas, and though sometimes imprecise, that practice will be used here for brevity.

Problem Types/Differential Equations

It appears that most problems to which the BEM has been applied over the years are either elliptic (time independent) to start with, or if parabolic or hyperbolic, they are converted to elliptic problems for computation using an integral transform on the time variable. For elliptic problems which are governed by linear differential equations with constant coefficients, the BEM machinery is most readily applied. Further, if any body force is present and is simple enough to be represented by equivalent boundary integrals, the BEM requires only spatial discretization.

Considerable success, however, has been achieved with transient problems via both time-stepping procedures and transform inversion techniques. Moreover nonlinear problems have been solved (iteratively) via BEM with great success as well, wherein the defining differential operator has a (hopefully dominant) linear part, which can be made the basis for the required reciprocal relation.

Inhomogeneity of the zoned type is less formidable than inhomogeneity of the continuous type which involves variable coefficients in the differential equations. The latter class of problems is, to the author's knowledge, among the most ill-suited for BEM. This is due to the difficulty in constructing an explicit fundamental solution for such problems (see ref. [1] Ch. III).

Examples and in-depth discussion of many of the topics described in this paper may be found in the texbooks [4] through [12].

Problem Types/Geometrical Features

The BEM seems to have a nice advantage over field-discretization methods, such as finite element (FEM) and finite difference (FDM), for unbounded domains, e.g., domains exterior to a simple closed surface, since only that surface needs be discretized with BEM. Even for semi-infinite domains, such as a half-space (assuming the half-space Green's function is not used) where surface discretization needs be truncated, the effects of such truncation seem to be less troublesome than truncation with FEM or FDM.

For finite regions, the BEM seems at best advantage for bulky rather than thin-shell or platelike bodies unless, of course, shell or plate theories with fundamental integral formulas exclusive to those theories are used. Similarly, layered and composite media are most readily treated via BEM if the appropriate Green's functions can be described for the geometry. Otherwise, discretization requirements for such media can be formidable.

Moving boundary problems, contact problems, and problems in general with nonlinear boundary conditions seem to be treated with greater success each year via BEM.

Finally, regions with stress-raisers, such as notches and cracks, have always been handled well via BEM. Indeed, the ability to efficiently solve crack(like) problems is one of the most important features of BEM. Issues surrounding surface-breaking and interlaminar cracks, rough cracks, crack growth, scattering from cracks, and inelastic fracture mechanics in general all are far from settled, however, as is the best form of integral equation to use in crack computations.

BIE/BEM Issues

Because of the appeal of dealing with physical variables, the comparatively nice behavior of these variables near notches and corners, and the ease of formulation for problems with mixed boundary data (the type most often encountered in practical problems it seems), the DIRECT as opposed to the INDIRECT BIE formulation has received the most attention in the last two decades. Nevertheless, there is reason to believe the indirect approach offers both theoretical and computational advantages for certain problems. As a practical matter, only one integral rather than two appears in indirect formulations, and prescribed data go into the problem directly rather than as the density in a (known) integral. Indeed, the indirect approach is the more classical one usually leading to integral equations with better-understood mathematical properties. The mathematical literature on integral equations plus the papers in this volume and associated references could be profitably studied on this matter.

It is possible to avoid VARIATIONAL ideas in BIE formulations (for linear problems at any rate) to get a relevant boundary integral equation and then use a straightforward collocation-BEM scheme to solve it. This approach stems directly from Green's or Gauss' theorems and, in this author's view, is the simplest way to present the BEM to new audiences. However, this may not be enough if one is concerned with the character of solutions to the integral equations, bounds on solutions, convergence, error analysis, etc. Variational ideas are rich with information and are perhaps essential to such important issues. This author regretfully speaks as an outsider on such matters and has been no help to them.

There seems to be a resurgence of interest in nonsingular integral equation formulations obtained by collocating off of the domain boundary, i.e. on an auxiliary boundary perhaps, to avoid dealing with Cauchy principal values (and the like) present in conventional BIE's. Such activity has advantages as noted very early by Kupradze [2]. It is also useful in avoiding certain nonuniqueness in formulations for scattering problems (cf. [3]). However, where to collocate off the boundary to ensure good conditioning (maximum conditioning seems to occur by collocating on the boundary) is the troubling question.

Regarding singular integrals, Cauchy-singular ones and now finite-part integrals, the question arises: Do we regularize such integrals or simply use or devise schemes for numerically evaluating them directly? Most effort seems to be devoted to regularization. However, reliable schemes do exist and more are being developed to compute such integrals directly, so that meaningful alternatives to regularization are available. In any case, this author believes that the singularities in the fundamental solutions at the heart of the BEM are important and good, and they provide what is most distinctive about the BEM.

In general terms, how to solve the BIE's one chooses to use is an active area. How practical an alternative to BEM-collocation are say least-squares and Galerkin methods? How sophisticated should elements be? Should they always be isoparametric? Is quadratic representation of geometry sufficient with lower or higher order variation for the functions defined on that geometry a proper strategy? Is Gaussian quadrature to be avoided in favor of simpler (but perhaps more) elements? Error analysis and self-adaptive mesh generation with BEM have a long way to go compared with FEM. This is perhaps to be expected if one is to believe those who refer to the BEM as a "quasi-analytical" technique--a term which is new to this author.

This volume contains several papers on BEM-FEM combinations, and never having done any work in this area, this author is impressed with how difficult the combination seems to be. However, will the day come when difficulties are resolved and the identity of the BEM is lost in combination? Will all that is distinctive about BEM be irrelevant someday and be regarded by users of large codes as merely a type of super finite element? This author thinks not. BEM ideas are too elegant and beautiful and rich with understanding about a given problem to be relegated completely to such status. Moreover, future generations of students with ready access to vector and parallel computing may form BEM (and BEM-FEM) codes, from scratch, which are vastly more efficient than those obtainable by vectorizing existing codes.

Interactions

The fields of physics, chemistry, chemical engineering, electromagnetics, biomechanics, and the newer field of nondestructive evaluation (NDE) are perhaps just a few with which the BEM community could have more fruitful interaction. Boundary-initial value problems involving difficult geometries abound in all of these areas, and numerical computation is in order. Interaction, such as it has already taken place, in the author's own experience and that of a few others known to him, has been especially rewarding. For example, in NDE there is considerable need for computational capabilities for unbounded-domain problems with composite structures containing cracks. The role to be played by BEM is considerable in the inverse problems of target identification and defect detection and characterization. The methods of artificial intelligence for these and other such problems have need for accurate data obtainable best via the BEM.

The most important interaction, at least for the BEM community in understanding its own "machinery", perhaps, is with the mathematicians. This author has always enjoyed, and indeed found indispensable, the collaboration of mathematicians in his research. Certain barriers of language always need to be overcome in such interaction. However, the barriers seem to be more formidable than ever today for some reason--a time when the body of mathematical knowledge about the BEM is especially rich. This author does not know how to solve this barrier-of-language problem, but he is frustrated and concerned that the body of knowledge in the school of mathematical activity, so ably represented by Professor Wendland and his colleagues, seems to exist largely in parallel with the more engineering-oriented BEM activity. There is reason to believe the same is true with other mathematical schools. This mathematical knowledge needs somehow to be "integrated" and become a more vital part of the engineering-oriented BEM activity. Any and all suggestions on this matter would be welcome.

Closing Remarks

This author wishes to emphasize that the issues specifically discussed in this paper are not necessarily more important than some BEM issues that have been omitted. The intent here was to limit the discussion to issues which the author has at least some acquaintance, and which seem to "surface" with some regularity over time.

He would like to close by expressing appreciation for the UNITY inherent in the representation integrals such as Green's integral for Laplace's equation and Somigliana's integral for the Navier-Cauchy equations of elasticity. To say just a bit more about these integrals, in them we have expressions for the fields (potential, elastic displacement) in terms of boundary integrals involving twice as much boundary data as can be prescribed in a well-posed problem. Then, taking the field point to the boundary, with due attention to "jumps" and singularities, a boundary integral equation is obtained. This BIE is a constraint among these data and is the vehicle whereby the unprescribed boundary data can be obtained from those which are prescribed. Then going back into the domain with the field point, the original representation integral, with all boundary data now known, yields the desired solution everywhere by simple quadrature.

While theoretical investigation and numerical solution of integral equations have existed for a long time, expression of the VERY IDEA above, i.e., the direct BIE, was simultaneous with its first serious numerical exploitation in the 1960's. While it is no doubt true that numerical solutions to singular integral equations would have achieved popularity by now, no matter what, this author likes to think that the direct-BIE concept, the "twice-as-much-data" concept mentioned above, has played a key role in bringing this whole BEM-business to the computational world.

Finally, this author likes the idea that the mathematics involved in the BIE is old and was developed for a different purpose than the BEM. It is nice to believe that Green, Gauss, Kelvin, Stokes, Somigliana, Navier and others would be pleased to see numbers forthcoming for PRACTICAL problems from their formulas. This bridge with the past, not to mention the bridge now formed with other computational communities, especially perhaps the FEM, is important to this writer.

Acknowledgement

Thanks are due the Office of Naval Research, under Contract No. N00014-89-K-0109, for support of this work, and to the National Science Foundation and Air Force Office of Scientific Research for earlier support. This author is grateful to his students and colleagues, over more than two decades, whose collaboration provides whatever substance is present in the above remarks.

References

1. John, F.: Plane waves and spherical means. New York. Springer-Verlag 1955.
2. Kupradze, V. D.: Potential methods in the theory of elasticity. Jerusalem. Israel Program for Scientific Translations 1965.
3. Burton, A. J.; Miller, G. F.: The application of integral equation methods to the numerical solution of some exterior boundary-value problems. Proc. Roy. Soc. Lond. A. 323 (1971) 201-210.
4. Jaswon, M. A.; Symm, G. T.: Integral equation methods in potential theory and elastostatics. London. Academic Press 1977.
5. Banerjee, P. K.; Butterfield, R.: Boundary element methods in engineering science. McGraw-Hill 1981.
6. Crouch, S. L.; Starfield, A. M.: Boundary element methods in solid mechanics. London. George Allen & Unwin 1983.
7. Liggett, J. A.; Liu, P. L-F.: The boundary integral equation method for porous media flow. London. George Allen & Unwin 1983.

8. Brebbia, C. A.; Telles, J. C. F.; Wrobel, L. C.: Boundary element techniques. Berlin. Springer-Verlag 1984.

9. Mukherjee, S.: Boundary element methods in creep and fracture. London. Applied Science Publishers 1985.

10. Cruse, T. A.: Boundary element analysis in computational fracture mechanics. Dordrecht. Kluwer Academic Publishers 1988.

11. Manolis, G. D.; Beskos, D. E.: Boundary element methods in elastodynamics. London. Unwin Hyman 1988..

12. Hartmann, F.: Introduction to boundary elements. Berlin. Springer-Verlag 1989.

Higher Order Elements and Element Enhancement by Combined Regular and Hypersingular Boundary Integral Equations

T. J. Rudolphi

Department of Engineering Science and Mechanics
Iowa State University

Summary

The direct, collocation boundary element method is used with both the regular and tangent derivative equation to formulate boundary elements with additional degrees of freedom. Use of the tangent derivative equation allows collocation at the same nodes as the regular constraint equation when the derivatives are continuous at the collocation points. Nonconforming elements with degrees of freedom corresponding to tangent derivatives of the primary function are introduced into both equations through Hermitian shape functions. The formulation is given for potential problems in two dimensions with a verification example.

Introduction

Boundary constraint equations are obtained through the use of a fundamental solution or source and the function being sought in an appropriate reciprocal identity to first produce an integral representation. Then a limit process on the representation, in which the source point is taken to the boundary, yields the constraint equation. This equation is rendered singular when, in the limit process, the "principal part" of the integral is extracted by a local analysis around the limit point, leaving the integral equation singular and subject to an interpretation set by how the principal part was extracted - generally in the sense of a Cauchy principal value. Partitioning the equation into a principal and a singular integral part may be formally elucidating, but obscures the fact that the equation itself is independent of the interpretation interposed by the analysis to extract the principal part. In fact, the constraint equation can be derived and expressed in a way that Cauchy integrals need not be introduced; all integrals being completely regular.

Likewise, the so-called hypersingular equation results when the directional derivative of the representation is taken, followed by the same limit process to extract the principal part, but yields unbounded terms. These divergent terms are a consequence of the limited local analysis and ultimately are canceled by identical terms from the remaining integral [2] to leave a bounded total. If the divergent terms are subtracted from the local analysis to leave a bounded principal part, then the like terms must also be subtracted from the remaining integral, leaving both parts, and hence the sum, finite. Like the Cauchy principal value for weaker singularities, the Hadamard [1] finite part is associated with the bounded remaining integral when the divergent terms are extracted. The finite part can thus be introduced into the equation by associating an appropriately defined integral with a particular local analysis. Also, as for regular constraint equations, it is possible to express the derivative equation without the finite part concept [2,3].

The normal derivative is the commonly used hypersingular equation since it involves the same functions as the regular constraint equation. They have been employed, most notably in crack problems where the regular equation is degenerate. For other problems, the normal derivative and the regular equation are not independent, but combinations of the two have been used [4], by collocating each at different points, to improve the system conditioning.

The tangent derivative equation has singularities similar to the normal derivative equation, but has apparently not been exploited. It introduces, however, tangent derivatives of the boundary variables but, as is shown herein, can be coupled with the regular constraint equation by collocation at common nodes. This doubles the number of available equations, and hence nodal degrees of freedom (DOF), which allows alternate element formulations.

Early computational algorithms for strongly singular equations were ad hoc and precarious, probably due to the separation of the principal part from the finite part integral, which then required a direct computation of the finite part. Through careful interpretation of the limit process in the derivation of the normal derivative equation for crack problems, Rudolphi et.al. [1] and Krishnasamy, et.al. [2] have introduced a regularization procedure that avoids the introduction of hypersingular integrals. Alternately, this author [5] has generalized the "rigid motion" solution to avoid the direct computation of finite part integrals in the normal derivative equation. This process is followed here for the tangent derivative equation.

The direct, collocation boundary element method evaluates the constraint equation at points where the essential variables are at least continuous. Secondary or natural quantities, such as flux or tractions, may have discontinuities as may occur at corners or where data types change. Polynomial based, conforming boundary elements provide this degree of continuity, even at element intersections, but cannot be used in the derivative equations with collocation at element intersections, since the essential variable needs to be Holder continuous at the collocation point. Nonconforming elements [6] provide the required continuity, although the collocation points are inside the element. A variant of these - a nonconforming Hermitian element - is used herein to couple the regular and tangent derivative equations.

Formulation

For the harmonic function $u(\vec{\xi})$ on a domain bounded externally by S with unit outward normal $\vec{n}$, the regular boundary constraint equation can be written [6]

$$\int_{S-S_0} T(\vec{x}_0;\vec{x})u(\vec{x})dS(\vec{x}) + \int_{S_0} T(\vec{x}_0;\vec{x})[u(\vec{x})-u_0]dS(\vec{x})$$

$$-\int_{S-S_0} T(\vec{x}_0;\vec{x})dS(\vec{x})u_0 = \int_{S} U(\vec{x}_0;\vec{x})q(\vec{x})dS(\vec{x}) \qquad (1)$$

where $U(\vec{\xi};\vec{x}) = -(1/2\pi\kappa)\ln r(\vec{\xi};\vec{x})$ and $T(\vec{\xi};\vec{x}) = \kappa\nabla_{\vec{x}}U(\vec{\xi};\vec{x})\cdot\vec{n}(\vec{x})$ are the fundamental solution and its flux, $q = \kappa\partial u/\partial n$, $u_0 = u(\vec{x}_0)$, κ is a material property and r is the distance between $\vec{\xi}$ and $\vec{x}$. The boundary S has been divided into the two parts, $S-S_0$ and S_0, with the collocation point $\vec{x}_0 \in S_0$. The above is equivalent to the more familiar form in which the principal part $c(\vec{x}_0)u(\vec{x}_0)$ has been explicitly extracted. Here the coefficient $c(\vec{x}_0)$ has been implicitly determined by imposing the solution $u = constant$ onto that form of the equation. Equation (1) is readily suited for computation since the second integral on the left is regular when $u(\vec{x})$ is continuous at $\vec{x}_0$, the third or "diagonal" term is determined by the sum of off-diagonal terms and that on the right is weakly singular.

Although not ordinarily done, the weakly singular part of eqn. (1) can also be removed. To do so, the integral on S is again divided into $S-S_0$ and S_0 and the simple linear solution $u=\vec{r}\cdot\vec{n}_0$, $\vec{n}_0=\vec{n}(\vec{x}_0)$, is used to eliminate the singular part on S_0. The result of this provides the completely regular result

$$\int_{S-S_0} TudS + \int_{S_0} T(u-u_0)dS - \int_{S-S_0} TdSu_0 = \int_{S-S_0} UqdS$$
$$+ \int_{S_0} U[q-(\vec{n}\cdot\vec{n}_0)q_0]dS - \int_{S-S_0} U\vec{n}dS\cdot\vec{n}_0 q_0 + \int_{S} \frac{1}{\kappa} T\vec{r}dS\cdot\vec{n}_0 q_0 \qquad (2)$$

where the explicit arguments have now been dropped. This manner of eliminating the weak singularity is comparable to that used later for the derivative equation with stronger singularities.

Similarly, through the imposition of both a constant and the linear solution $u=\vec{r}\cdot\vec{s}_0$ onto the limit form of the directional derivative of the representation, the tangent derivative equation can be expressed without Cauchy or finite part integrals. Following the notation and development of Ref. [5], one has

$$\int_{S-S_0} XudS + \int_{S_0} X(u-u_0-u_0'(\vec{r}\cdot\vec{s}_0))]dS - \int_{S-S_0} XdSu_0$$
$$- \int_{S-S_0} X\vec{r}dS\cdot\vec{s}_0 u_0' + \int_{S} \kappa Y\vec{n}dS\cdot\vec{s}_0 u_0' = \int_{S-S_0} YqdS$$
$$+ \int_{S_0} Y[q-(\vec{n}\cdot\vec{n}_0)q_0]dS - \int_{S-S_0} Y\vec{n}dS\cdot\vec{n}_0 q_0 + \int_{S} \frac{1}{\kappa} X\vec{r}dS\cdot\vec{n}_0 q_0 \qquad (3)$$

where $u_0'=(\partial u/\partial s)(\vec{x}_0)$, $\vec{s}_0$ is a unit tangent at $\vec{\xi}_0$ and the kernels Y and X are the directional derivatives of U and T, respectively; $Y(\vec{\xi};\vec{x})=\nabla_{\xi}T(\vec{\xi};\vec{x})\cdot\vec{s}(\vec{\xi})$ and $X(\vec{\xi};\vec{x})=\nabla_{\xi}T(\vec{\xi};\vec{x})\cdot\vec{s}(\vec{\xi})$.

In the above form, all terms are completely regular. The first integrals on each side are the "off-diagonal" terms, the second are regularized terms on the collocation element while the remaining are the "diagonal" terms and are calculated from integration of off-diagonal elements or non-singular integrals. Some are determined by the sum of off-diagonal terms as in the regular constraint equation while others are moments of the off-diagonal terms with respect to the collocation point.

It is noted that eqn. (3) can be derived directly by taking the tangential derivative of the regular constraint eqn. (1), while again imposing the two simple solutions to eliminate the derivatives of integrals on S_0.

Numerical Implementation

A boundary discretization is made in the usual way, with elements terminating at points of discontinuous boundary data or geometric discontinuities. On each element, the functional behavior for the boundary data is approximated by shape functions and nodal values. Since the derivative eqn. (3) involves the tangential derivatives $u' = \partial u/\partial s$, it is appropriate to use them as nodal values. Also, since Holder continuous first derivatives are required at the collocation points, it is natural to use nonconforming elements. Accordingly, Hermitian types of nonconforming elements are appropriate for approximating u and q on the elements of S.

A simple element that could be used is a linear, two DOF element with nodal values of the function and its slope. A more sophisticated element, and that used here, is the four DOF, cubic Hermitian element. It is presumed that, on each element with intrinsic coordinate $-1 \leq \xi \leq +1$, the two collocation nodes are to be taken at the locations $\xi = \pm 1/2$ and that

$$[u \;\; q] = [H_1 \;\; H_2 \;\; H_3 \;\; H_4] \begin{bmatrix} u_1 & q_1 \\ u_1' & q_1' \\ u_2 & q_2 \\ u_2' & q_2' \end{bmatrix} = [H][\{u\} \;\; \{q\}] \tag{4}$$

where $u_1' = \partial u/\partial s$ at node 1 on the element, etc., and the shape functions $H_i(\xi)$ are

$$H_1 = (1-2\xi)(1-2\xi)(1+\xi)/2 \qquad H_2 = (1-2\xi)(1-2\xi)(1+2\xi)/8$$
$$H_3 = (1+2\xi)(1+2\xi)(1-\xi)/2 \qquad H_4 = -(1+2\xi)(1+2\xi)(1-2\xi)/8 \qquad (5)$$

The tangent derivatives of u and q are then determined by $u' = [H']\{u\}$ where $[H'] = (d[H]/d\xi)(d\xi/ds)$ and similarly for $q' = dq/ds$. For curved elements, the usual quadratic (three nodes) interpolation scheme is used for the geometry so the present formulation is not isoparametric.

Substitution of the interpolation functions and the appropriate derivatives into the constraint equations (2) and (3), one has

$$\int_{S-S_0}\begin{Bmatrix}T\\X\end{Bmatrix}[H]dS\{u\} + \left\{\int_{S_0}\begin{Bmatrix}T([H]-[H_0])\\X([H]-[H_0]-\zeta[H_0'])\end{Bmatrix}dS - \int_{S-S_0}\begin{Bmatrix}T\\X\end{Bmatrix}dS[H_0]\right.$$
$$\left. - \left[\int_{S-S_0}\begin{Bmatrix}0\\X\zeta\end{Bmatrix}dS - \int_{S}\begin{Bmatrix}0\\\kappa Y\epsilon\end{Bmatrix}dS\right][H_0']\right\}\{u_0\} = \int_{S-S_0}\begin{Bmatrix}U\\Y\end{Bmatrix}[H]dS\{q\}$$
$$+\left\{\int_{S_0}\begin{Bmatrix}U\\Y\end{Bmatrix}([H]-\delta[H_0])dS - \left[\int_{S-S_0}\begin{Bmatrix}U\\Y\end{Bmatrix}\delta dS - \int_{S}\frac{1}{\kappa}\begin{Bmatrix}T\\X\end{Bmatrix}\nu dS\right][H_0]\right\}\{q_0\} \qquad (6)$$

where $\nu = \vec{r}\cdot\vec{n}_0$, $\zeta = \vec{r}\cdot\vec{s}_0$, $\delta = \vec{n}\cdot\vec{n}_0$, $\epsilon = \vec{n}\cdot\vec{s}_0$ and $[H_0'] = d[H]/ds(\vec{x}_0)$. Equations (6) can then be collocated at each of the two nodes on each element with eight associated values of the vectors $\{u\}$ and $\{q\}$, one of whose components are specified on each segment.

Demonstration Problem

As a demonstration problem for the combined approach, a simple problem of a square region $0 \le x \le 4$, $0 \le y \le 4$ with flux boundary conditions consistent with the harmonic field $u = x^2 - y^2$ was modeled with eight, equal length, straight elements. To eliminate an arbitrary potential, the element on $0 \le x \le 2$ was assigned essential conditions. The solution for the remaining boundary is given in Figure 1, showing the exact solution, the solution by the regular boundary element method and the combined approach.

The combined method is nearer the exact solution than the regular, as might be expected, since cubic elements are used in the combined method while the regular method used linear, conforming elements. Condition numbers for both solutions were small and approximately equal. This simple example is certainly not one from which to draw general conclusions, but serves to merely verify the algorithm.

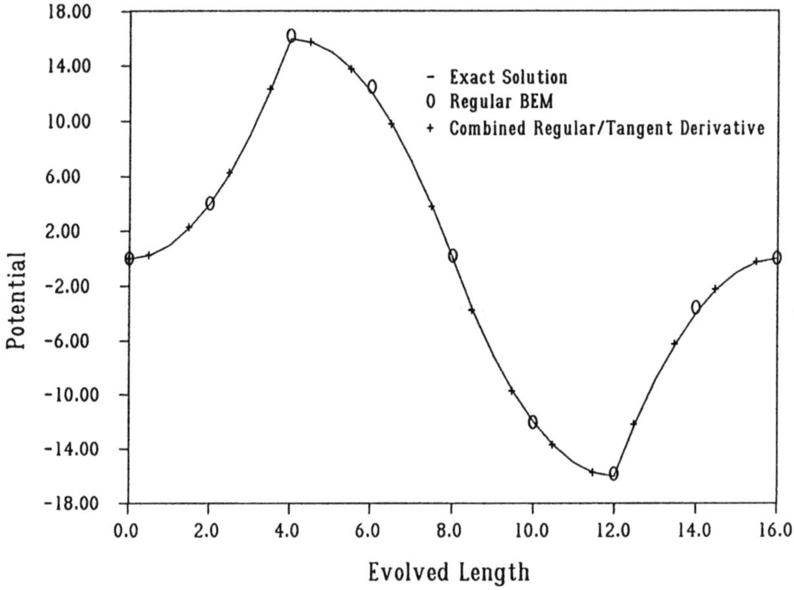

Figure 1. Potential on a Square Region

Discussion and Conclusions

The given formulation and computational approach for the simple potential problem has shown that a general algorithm for other problems is feasible and suggests several possible advantages. There is the obvious ability to collocate the two equations at the same nodes so that collocation points can be kept farther apart while retaining higher DOF elements. Since tangent derivative quantities are directly involved, they are retrieved immediately as part of the boundary solution as opposed to post-calculation with the regular method. In the potential problem, the tangential flux quantities are directly determined. For a vector problem, such as elasticity, the tangent derivative quantities would be displacement gradients, or in effect the surface strains. This means that all components of the surface stresses would be directly available.

Collocation of both equations at the same point as demonstrated here is also not necessary. Mixed approaches could be implemented. For example, if it were desired to retain conforming elements, the regular constraint equations could be collocated at element intersections while the tangent derivative equation or both might be collocated at interior points. In this fashion, elements with enriched degrees of freedom would result.

The normal derivative equation has previously been used [4,5] as an alternative to the regular boundary constraint equation. Here it has been shown that the tangent derivative equation can be simultaneously used with the regular constraint equation in boundary element solutions. The method of regularization of the strong singularities through the use of simple solutions is quite general and can be extended to other problems.

References

1. Hadamard, J.: Lectures on Cauchy's problem in linear partial differential equations. Dover Publications 1952.

2. Rudolphi, T.J.; Krishnasamy, G.; Schmerr, L.W.; Rizzo, F.J.: On the use of strongly singular integral equations for crack problems. Boundary Elements X, Brebbia, C.A. (ed.), Computational Mechanics Publications, Springer-Verlag, 3 (1988) 249-263.

3. Krishnasamy, G.; Schmerr, L.W.; Rudolphi, T.J.; Rizzo, F.J.: Hypersingular boundary integral equations: Some applications in acoustics and elastic wave scattering, Accepted for publication in Journal of Applied Mechanics.

4. Ingber, M.S.; Rudolphi, T.J.: Solution of potential problems using combinations of the regular and derivative boundary integral equations. Submitted to Appl. Math. Modelling.

5. Rudolphi, T.J.: The use of simple solutions in the regularization of hypersingular boundary integral equations. Accepted in Computers and Math. with Appl.

6. Patterson, C.; Sheikh, M.A.: Interelement continuity in boundary element analysis. Topics in Boundary Element Research, Brebbia, C.A. (ed.), Springer-Verlag, Berlin (1984) 123-141.

Uniform Convergence of Boundary Element Solutions Using the Collocation Methods

Yuusuke ISO

Research Institute for Mathematical Sciences
Kyoto University
Kitashirakawa-Oiwake Sakyo Kyoto 606, JAPAN

ABSTRACT
The author proved rigorously the uniform convergence theorem of boundary solutions for Laplace's equation with the Neumann condition[1]. The aim of the present paper is to show his results. Furthermore he gives the methods to construct the more acculate scheme than usual.

§1 Introduction

The boundary element method(=BEM) is widely used in the various engineering fields. Especially to solve boundary value problems of elliptic partial differential equations(i.e. Laplace's equation, Poisson's equation etc.), this method is more powerful than other ones. But from a mathematical point of view, BEM is far behind beside the finite difference method and the finite element method. Analysis for convergence of BEM numerical solutions has been rarely investigated. Furthermore some of such mathematical results are based on Galerkin methods, and almost all of them avoid, in their analysis, approximation of the boundaries by polygons(or polyhedra). But the most of engineers who use BEM adopt collocation methods to construct numerical schemes and approximate boundaries by polygons(or polyhedra) to integrate numerically. The author proved the uniform convergence theorem for the BEM scheme with the collocation method in which approximation of the boundary was taken into account[1]. The equation considered is Laplace's equation with the Neumann condition.

The same approach to the heat equation is also shown in [3],[4].

§2 Boundary Integral Equation

Let Ω be a bounded domain in R^2. We denote spatial points by $x=(x_1,x_2)\in R^2$. Let Γ be a smooth boundary of Ω. Here we assume that a curvature of Γ is positive. This assumption is necessary in estimation of a coefficient matrix.

We consider the following Neumann problem for Laplace's equation;

$$\Delta u(x) = 0 \quad \text{in } \Omega , \tag{2.1}$$

$$\frac{\partial}{\partial n} u(x) = q(x) \quad \text{on } \Gamma , \tag{2.2}$$

where $q(x)$ is a given smooth function defined on Γ , and satisfies

$$\int_\Gamma q(y)\, d\sigma_y = 0.$$

And $\frac{\partial}{\partial n}$ denotes an outward normal derivatives on Γ . Let $E(x)$ be the fundamental solution of Laplace's equation;

$$E(x) = -\frac{1}{2\pi} \log|x|, \tag{2.3}$$

where $|x| := ((x_1)^2 + (x_2)^2)^{\frac{1}{2}}$. The boundary integral equation derived from (2.1) and (2.2) is as follows;

$$\frac{1}{2} u(x) + \text{p.v.} \int_\Gamma \frac{\partial}{\partial n_y} E(x-y) u(y)\, d\sigma_y = \int_\Gamma E(x-y) q(y)\, d\sigma_y, \tag{2.4}$$

where $x \in \Gamma$,and p.v. means the Cauchy principal value of integrals. Our aim is to find a continuous function on Γ, which corresponds to the Dirichlet data for (2.1)and(2.2) through solving (2.4).

In order to guarantee unique solvability , we impose an additional integral equation

$$\int_\Gamma u(y)\, d\sigma_y = 0. \tag{2.5}$$

Hence our problem is to find a continuous function $u(x)$ such that $u(x)$ satisfies a system of integral equations (2.4) and (2.5). Mathematical arguments on unique solvability are shown in [1].

§3 Discretization under the collocation method

To give an acculate scheme, we must pay much attention to approximation of the outward normal derivatives on Γ .

Let $\{z_{2k}\}_{k=1}^{N}$ be a set of N nodal points on Γ. And let Γ_k be a closed line segment between the two points z_{2k-1} and z_{2k+1}. Set $\tilde{\Gamma} := \bigcup \Gamma_k$, then it is a set of polygonal lines which approximate Γ. We define z_{2k} by the middle point of Γ_k. Let $\vec{n}_z$ be a unit outward normal vector to Γ at z, and let $\vec{\nu}_k$ be a unit outward normal vector to $\tilde{\Gamma}$ on Γ_k. Set $h := \max|\Gamma_k|$, where $|\Gamma_k|$ denotes the length of Γ_k. We call h as a mesh size. We assume the next property on redivision.

Assumption A

There exists a positive constant M which is independent of h such that $\max|\Gamma_k|/\min|\Gamma_k| < M$.

Let us define a set of vectors $\{\vec{\eta_k}\}_{k=1}^{N}$ by

$$\vec{\eta_k} := \frac{|\Gamma_k|\overrightarrow{\nu_{k-1}} + |\Gamma_{k-1}|\vec{\nu_k}}{\| \, |\Gamma_k|\overrightarrow{\nu_{k-1}} + |\Gamma_{k-1}|\vec{\nu_K} \, \|_2} \qquad (1 \le k \le N), \qquad (3.1)$$

where $\| \; \|_2$ is defined by $\|\vec{a}\|_2 := ((a_1)^2 + (a_2)^2)^{\frac{1}{2}}$ for $\vec{a} = (a_1, a_2)$. we have the following estimates for them.

Lemma 1

There exists a positive constant C which is independent of h such that

$$\| \vec{\eta_k} - \overrightarrow{n_{z_{2k-1}}} \|_\infty < Ch^2, \qquad (3.2)$$

where $\| \; \|_\infty$ is defined by $\|\vec{a}\|_\infty := \max(|a_1|, |a_2|)$ for $\vec{a} = (a_1, a_2)$.

The set $\{\vec{\eta_k}\}_{k=1}^{N}$ enables us to construct a pseud-normal vector on $\tilde{\Gamma}$ which approximates $\vec{n_z}$ in $O(h^2)$. Define a vector $\overrightarrow{m_{z'}}$ by

$$\overrightarrow{m_{z'}} := \left(1 - \frac{2\lambda}{|\Gamma_k|}\right)\vec{\nu_k} + \frac{2\lambda}{|\Gamma_k|}\vec{\eta_k}, \qquad (3.3)$$

where λ denotes the distance between z' and z_{2k} (see the next figure).

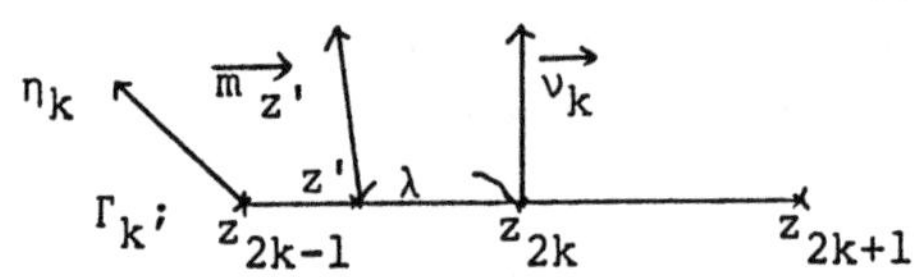

In some cases, $\{\vec{\nu_k}\}$ are adopted as approximation for $\vec{n_z}$, though piecewise-linear or higher base functions are used to approximate functions. In these cases, the leading term of error estimates is arisen from approximation for $\vec{n_z}$. Thus numerical solutions may not be improved, even if numerical integrations and approximation for given functions are improved.

Let us define a set of piecewise-functions $\{\varphi_k\}_{k=1}^{N}$ so that they satisfy the following property.

Assumption B

(i) each φ_k is a linear function on Γ_j $(1\le j,k\le 2N)$,

(ii) each φ_k is continuous on $\tilde{\Gamma}$ $(1\le k\le 2N)$,

(iii) $\varphi_k(z_j) = \delta_{k,j}$ $(1\le j,k\le 2N)$.

Let us define another set of piecewise-linear functions $\{\Phi_k\}_{k=1}^{2N}$ such that

$$\Phi_k := \varphi_k + \frac{1}{2}(\varphi_{k-1} + \varphi_{k+1}) \qquad 1\le k\le 2N,$$

where φ_0 and φ_{2N+1} coincide with φ_{2N} and φ_1 respectively.
Let us define finite dimensional function spaces V_h^1 and V_h^2 on $\tilde{\Gamma}$ by

$$V_h^1 := \text{linear span of } \langle \Phi_1, \Phi_3, \cdots, \Phi_{2N-1}\rangle,$$

$$V_h^2 := \text{linear span of } \langle \Phi_2, \Phi_4, \cdots, \Phi_{2N}\rangle .$$

In other words, $v_h^m \in V_h^m$ $(m=1,2)$ is expressed as follows;

$$v_h^1 = \sum_{1\le j\le N} a_j^1 \Phi_{2j-1}, \qquad v_h^2 = \sum_{1\le j\le N} a_j^2 \Phi_{2j} .$$

Let P_h be an operator defined by

$$P_h : \begin{array}{ccc} C^0(\Gamma) & \longrightarrow & V_h^1 \\ \Psi & & \Psi \\ f(x) & \longrightarrow & \sum_{1\le j\le N} f(z_{2j-1})\Phi_{2j-1}. \end{array} \tag{3.3}$$

We call this operator P_h as a collocation operator. Above preparations enable us to discretize the boundary integral equation. Let us express our aimed numerical solution in

$$u_h := \sum_{1\le k\le N} u_h^k \Phi_{2k}.$$

We can regard u_h both as an element of V_h^2 and as an element of R^N. Hence we are sufficient to give a system of algebraic linear equations corresponding to (2.4) and (2.5). Let us define a matrix $A_h = (a_{i,j})$ by

$$\alpha_{i,j} := \frac{1}{2}\delta_{i,j} + \text{p.v.}\int_{\tilde{\Gamma}} (\nabla_x E(z_{2i}-y))\cdot\vec{m}_y \,\Phi_{2j}(y)\, d\sigma_y \quad (1\le i,j\le N) \tag{3.4}$$

$$a_{i,j} := \alpha_{i,j} - \frac{1}{N}\sum_{1\le k\le N} \alpha_{i,j} \quad (1\le i,j\le N), \tag{3.5}$$

where $\nabla_x = (\frac{\partial}{\partial x_1}, \frac{\partial}{\partial x_2})$.
The following estimates , which play an imprtant role in the mathematical discussion, hold for this matrix A.

Lemma 2

There exist positive constants C_1, C_2, C_3 and C_4 such that

$$\text{(i)} \qquad -C_1 h \le a_{i,j} \le -C_2 h \quad (1\le i,j\le N,\ i\ne j) \tag{3.6}$$

(ii) $$C_3 \leq a_{i,i} \leq C_4 \qquad (1\leq i\leq N), \tag{3.7}$$

(iii) $$\sum_{1\leq j\leq N} a_{i,j} = 0, \tag{3.8}$$

where C_1, C_2, C_3, and C_4 are independent of h.

Remark 1

In Lemma 1, we can choose the constants C_3 and C_4 approximately 1/2.

The next result can be derived immadiately from Lemma 2.

Lemma 3

$$\operatorname{rank}(A_h) = N-1.$$

This lemma suggests to us how to give discretization of the left hand side of (2.4). Define a function $r(x)$ on $\tilde{\Gamma}$ by

$$r(x) := \int_{\tilde{\Gamma}} E(x-y)(P_h q)(y)\, d\sigma_y. \tag{3.9}$$

Set $r_h^k := r(z_{2k})$ for $1\leq k\leq N-1$. And let $\mu = (\mu_1, \mu_2, \cdots\cdots, \mu_N) \in R^N$ be a vector such that

$$A_h \mu^T = 0, \qquad \mu_N \neq 0.$$

Set

$$r_h^N := -\frac{1}{\mu_N}\sum_{1\leq k\leq N-1} \mu_k r_h^k .$$

Then, obviously , $r_h := (r_h^1, r_h^2, \cdots\cdots, r_h^N) \in R(A_h)$, where $R(A_h)$ denotes the range space of a matrix A_h. Hence the discretized equation of (2.4) is given by

$$A_h u_h = r_h. \tag{3.10}$$

Let $\{d_k\}_{k=1}^N$ be defined by

$$d_k := \int_{\tilde{\Gamma}} \phi_{2k}(y)\, d\sigma_y \qquad (1\leq k\leq N).$$

Then the discretized equation of (2.5) is given by

$$\sum_{1\leq k\leq N} d_k u_h^k = 0. \tag{3.11}$$

Therefore the system of linear equations which approximate (2.4) and (2.5) are (3.10) and (3.11).

Remark 2

From a practical point of view, the system of linear equation to be solved is as follows;

$$a_{i,1}u_h^1 + a_{i,2}u_h^2 + \cdots\cdots + a_{i,N}u_h^N = r_h^i \quad (1 \leq i \leq N-1), \qquad (3.12)$$

$$d_1u_h^1 + d_2u_h^2 + \cdots\cdots + d_Nu_h^N = 0. \qquad (3.13)$$

But in some cases, the linear equations (3.12) are replaced by

$$\alpha_{i,1}u_h^1 + \alpha_{i,2}u_h^2 + \cdots\cdots + \alpha_{i,N}u_h^N = r_h^i \quad (1 \leq i \leq N-1). \qquad (3.14)$$

This replacement may yield numerical instability in solving them by computers.

§4 Convergence Tneorem

Our aimed numerical scheme is to solve the system of linear equations(3.10) and(3.11). And we have the following mathematical results for it.

Theorem 1

The system of linear equation (3.10)and (3.11) has a unique solution in V_h (also in R^N).

Theorem 2

Let u be the exact solution of (2.4) and (2.5), and let u_h be the exact solution of (3.10) and (3.11). Then, there exists a positive constant C which is independent of h such that

$$\max_{z \in \tilde{\Gamma}} | u_h(z) - (P_h u)(z) | \leq Ch.$$

Remark 3

Theorem 2 suggests that the BEM solution obtained by(3.10) and (3.11) converges to the exact solution by O(h). The complete proof is written in [1]. Furthermore this error estimation can be improved [2].

REFERENCES

[1] Iso,Y. Uniform Convergence of Boundary Solutions for Laplace's Equation. Publ.Res.Inst.Math.Sci. vol.25, 21-43 (1989)

[2] Hayakawa,K.&Iso,Y. High-Order Uniform Convergence Estimates for Boundary Solutions for Laplace's Equation (in preparation)

[3] Iso,Y.&Onishi,K.et.al Numerical Convergence of Boundary Solution in Heat Conduction Problems. Topics in Boundary Element Research. vol.3,ch.1, Springer-Verlag (1987)

[4] Iso,Y. Convergence of the Boundary Element Solutions for the Heat Equation (to appere in Publ.Res.Inst.Math.Sci.)

An Algorithm for Computation of Delay Times in Subsonic and Supersonic Rotary Propeller Flow

Marvin I. Freedman,

Department of Mathematics, Boston University, 111 Cummington St., Boston, MA 02215

Abstract

This paper deals with the issue of accurately computing the delay times that appear in the blade fixed Green's Function for rotary propeller flow. This delay corresponds to the roots of a certain transcendental equation.

It is found that for subsonic flow only one root exists while in the supersonic regime one will encounter an odd number ≥ 3 of roots.

The algorithm developed is capable of computing the values of these roots independently of how many there are. As such, even though the main interest here is in the supersonic regime, the algorithm may be usefully applied in the case of subsonic rotary flow to accurately compute the one delay time arising in that case as well.

1. Preliminaries

In applying the integral equation formulation of the Green's Function method to a rotating moving blade one must transform the Free Space Green's Function for the 3-dimensional wave equation

$$G = \frac{\delta(t - t_* + \rho/a_\infty)}{4\pi\rho}$$

to blade fixed coordinates. (See [1],[2] and [4].)

If air fixed coordinates are denoted by $\underline{\xi}$ and blade fixed coordinates are denoted by $\underline{X}$ then the equation $\underline{\xi} = U(\omega t)\underline{x} - \underline{V}_\infty t$ where

$$U(\omega t) = \begin{bmatrix} \cos\omega t & -\sin\omega t & 0 \\ \sin\omega t & \cos\omega t & 0 \\ 0 & 0 & 1 \end{bmatrix}$$

describes the relation between the two coordinate systems. Here $\underline{V}_\infty$ represents the contribution due to a uniform motion of the blade while $U(\omega t)\underline{x}$ takes into account blade rotation about the z-axis.

Thus if $\underline{\xi}, \underline{\xi}_{\infty}$, $\underline{x}, \underline{x}_*$ represent the receiving and sending points respectively in the two coordinate systems we have

$$\underline{\rho} = \underline{\xi} - \underline{\xi}_* = U(\omega t)\underline{x} - U(\omega t_*)\underline{x}_* - \underline{V}_\infty(t - t_*)$$
$$\rho = \|\underline{\xi} - \underline{\xi}_*\| = \|U(\omega t)\underline{x} - U(\omega t_*)\underline{x}_* - \underline{V}_\infty(t - t_*)\|$$

It is well known that if we define $g(t) = t - t_* + \rho/a_\infty$ then the Green's Function in the blade fixed coordinate system will take the form

$$G = \frac{1}{4\pi} \sum_{\substack{Roots\ of \\ g(t_i)=0}} \frac{\delta(t-t_i)}{\hat{\rho}_i} \quad \text{where} \quad \hat{\rho}_i = \left(\rho \cdot |g'(t)|\right)\Big|_{\substack{t=t_i\ roots\ of \\ g(t_i)=0}}$$

(Assuming $g'(t) = 0$ has only simple roots).

Our object is to study the number and location of roots t_i to the equation $g(t_i) = 0$ and additionally to find usable expressions for the corresponding $\hat{\rho}_i$.

An interesting observation is that if we define

$$\gamma(t) = a_\infty^2 (t - t_*)^2 - \|U(\omega t)\underline{x} - U(\omega t_*)\underline{x}_* - \underline{V}_\infty (t - t_*)\|^2$$

Then the roots t_i of $\gamma(t_i) = 0$ with $t_i < t_*$ correspond exactly to the roots of $g(t) = 0$ with $t_i < t_*$.

To proceed further let us assume that the uniform velocity $\underline{V}_\infty$ lies along the axis of rotation i.e. the k axis, and that $\underline{x} - \underline{x}_* \perp \underline{V}_\infty$.

We now rescale our equations by introducing $M_\infty = \frac{|V_\infty|}{a_\infty}$, $\underline{X} = \frac{\underline{x}}{a_\infty}$, $\underline{X}_* = \frac{\underline{x}_*}{a_\infty}$
Assuming $M_\infty < 1$ i.e. the uniform flow is subsonic, we can introduce

$$\hat{\theta} = \theta\sqrt{1 - M_\infty^2}, \ \Omega = \frac{\omega}{\sqrt{1 - M_\infty^2}}$$

After some calculation we find that we are faced with finding roots $\hat{\theta}_i \geq 0$ of the equation

$$\gamma(\hat{\theta}) = \hat{\theta}^2 - \|U(-\Omega\hat{\theta})\underline{X} - \underline{X}_*\|^2 = 0 \tag{1}$$

and computing

$$\hat{\rho} = \frac{a_\infty \sqrt{1 - M_\infty^2}}{2} \frac{d\gamma}{d\hat{\theta}} \tag{2}$$

at the corresponding $\hat{\theta} = \hat{\theta}_i$.

2. Description of Figures

In the figures provided we have studied graphically the solutions $\hat{\theta}_i$ of equation (1)

$$\gamma(\hat{\theta}) = 0 \text{ where } \gamma(\hat{\theta}) = \hat{\theta}^2 - \|U(-\Omega\hat{\theta})\underline{X} - \underline{X}_*\|^2$$

In the above we can relate our normalized parameters to the original system parameters as described via the transformations given in Section 1.

Specifically we have, $X = \frac{x}{a_\infty}$, $X_* = \frac{x_*}{a_\infty}$, and $\Omega = \frac{\omega}{\sqrt{1-M_\infty^2}}$ where M_∞ is the Free Stream Mach number. In addition the solutions $\hat{\theta}_i$ of $\gamma(\hat{\theta}) = 0$ are related via

$$\theta_i = \frac{\hat{\theta}_i}{\sqrt{1 - M_\infty^2}}$$

to the actual unnormalized delay times θ_i of the underlying problem.

We assume throughout this paper that $|M_\infty| < 1$ i.e. the free stream velocity is subsonic but the effect of sufficiently high rotational velocity ω can lead to supersonic behavior as reflected in certain influence functions.

The parameter Ω takes into account both the free stream and rotational velocities and is the most significant parameter in determining the underlying behavior.

Indeed as we shall see, for small enough values of Ω equation (1) will have only one root corresponding to subsonic behavior while for larger values of Ω, multiple but generically an odd number of roots of (1) can exist, corresponding to supersonic behavior.

We wish to study the behavior of equation (1) as Ω grows, more carefully. In each of Figures 1-6 the parabola

$$z = \hat{\theta}^2 \text{ and the sinusoidal function } z = \|U(-\Omega\hat{\theta})\underline{X} - \underline{X}_*\|^2$$

have been plotted on the same axes (for $\hat{\theta} \geq 0$).

The intersections of these two curves (for $\hat{\theta} \geq 0$) correspond to roots $\hat{\theta}_i$ of $\gamma(\hat{\theta}) = 0$. In Figures 1-6 we have chosen $\underline{X} = \begin{bmatrix} 1 \\ 0 \\ 0 \end{bmatrix}$ and $\underline{X}_* = \begin{bmatrix} .5 \\ 0 \\ 0 \end{bmatrix}$ and we have increased Ω from $\Omega = 1$ in stages up to $\Omega = 25$. In Figures 1 and 2 with $\Omega = 1, 2$ respectively, we see that the situation is clearly subsonic with one intersection $\hat{\theta}$. In Figure 3 with $\Omega = 6$ we do however see the sinusoid approaching the parabola near $\hat{\theta} = 1.5$-getting ready to cross as Ω continues to increase. In Figure 4 with $\Omega = 7$ this crossing has occurred and we see three distinct intersections corresponding to three roots $\hat{\theta}_1, \hat{\theta}_2$ and $\hat{\theta}_3$ of $\gamma(\hat{\theta}) = 0$. Clearly at $\Omega = 7$ we are in the supersonic regime. In Fig. 5 with $\Omega = 10$ we again see three intersections, with the indication that as Ω continues to increase another crossing near $\hat{\theta} = 1.5$ may occur.

The last figure, Figure 6 with $\Omega = 25$ shows the presence of 9 roots and indicates how as Ω tends to ∞ we can obtain more and more crossings i.e. the number of roots will also tend to ∞ but *not* monotonically. Thus the number of roots will always be odd except for an exceptional discrete set of Ω's, corresponding to transitions in the number of roots.

3. The Quadratic Approximation for $U(\lambda)$

In Section 4 we will derive a general algorithm for accurately finding all roots $\hat{\theta}_i$ of $\gamma(\hat{\theta}) = 0$ as given by equation (1). Our algorithm will be based on the use of a newly developed quadratic approximation for $U(\lambda)$.

In this section we will derive a useful and somewhat surprising approximation for the unitary operators $U(\lambda)$ over the interval $|\lambda| \leq \pi/4$.

We remark that with

$$A = \begin{bmatrix} 0 & -1 & 0 \\ 1 & 0 & 0 \\ 0 & 0 & 0 \end{bmatrix}, \qquad \text{so that } A^2 = \begin{bmatrix} 1 & 0 & 0 \\ 0 & 1 & 0 \\ 0 & 0 & 0 \end{bmatrix}$$

the usual quadratic approximation is

$$U(\lambda) = e^{A\lambda} \sim I + \lambda A + \frac{\lambda^2}{21}A^2 \qquad i.e.\ U(\lambda) \sim \begin{bmatrix} 1-\lambda^2/2 & -\lambda & 0 \\ \lambda & 1-\lambda^2/2 & 0 \\ 0 & 0 & 1 \end{bmatrix}$$

which corresponds to the approximation $\cos\lambda \sim 1 - \lambda^2/2$ and $\sin\lambda \sim \lambda$.

Now for $|\lambda| \leq \pi/4 = .7854\ldots$ the approximation $\cos\lambda \sim 1 - \lambda^2/2$ satisfies

$$\left|\cos\lambda - (1 - \lambda^2/2)\right| \leq .0155\ldots \tag{3}$$

However the approximation $\sin\lambda \sim \lambda$ is much poorer.

Indeed
$$\max_{|\lambda|\leq\pi/4} |\sin\lambda - \lambda| = .0783\ldots$$

The question arises as to whether the approximation to $\sin\lambda$ can be easily improved over $-\pi/4 \leq \lambda \leq \pi/4$.

Since $f(\lambda) = \sin\lambda$ is an odd function, our approximating function should also be odd. Since, in addition, we restrict our attention to functions that are at most quadratic we are reduced to considering the following approximation problem

$$\sin\lambda \sim \alpha\lambda \qquad |\lambda| \leq \pi/4.$$

We wish to find the constant α that does the best job. Clearly $0 \leq \alpha \leq 1$

Theorem: *Consider the function*

$$E(\alpha) = \max_{|\lambda|\leq\pi/4} |\sin\lambda - \alpha\lambda| = \max_{0\leq\lambda\leq\pi/4} |\sin\lambda - \alpha\lambda|$$

This function is minimized for $\alpha = .925\ldots$ and the value of $E(\alpha)$ for this α is $.0194\ldots$ i.e.

$$|\sin\lambda - .925\lambda| \leq .0194\ldots$$

for $|\lambda| \leq \pi/4$ (a lot better than .0783).

After some computation we obtain

$$\max_{|\lambda|\leq\pi/4} \|U(\lambda) - U_{approx}(\lambda)\| \leq \sqrt{(.0155)^2 + (.0194)^2} \leq .0248 \tag{4}$$

A consequence of the foregoing is that for $\underset{\sim}{X}, \underset{\sim}{X}_*$ with $\|\underset{\sim}{X}\|, \|\underset{\sim}{X}_*\| \leq 1$

$$\left| < U(\lambda)\underset{\sim}{X}, \underset{\sim}{X}_* > - < U_{approx}(\lambda)\underset{\sim}{X}, \underset{\sim}{X}_* > \right| \leq .0248\ldots \text{ For } |\lambda| \leq \pi/4 \tag{5}$$

4. Algorithm:

Step 1: Compute

$$K_1 = \frac{2\Omega\|\underline{X} - \underline{X}_*\|}{\pi} - 1/2 \quad \text{and} \quad K_2 = \frac{2\Omega\|\underline{X} + \underline{X}_*\|}{\pi} + \frac{1}{2}$$

Step 2: For each integer n, $K_1 \leq n \leq K_2$ succesively set

$$\underline{Z}_{*n} = U(\frac{n\pi}{2})\underline{X}_*, \quad \underline{r}_n = \underline{X} - \underline{Z}_{*n}, \quad \text{and} \quad r_n = \|\underline{r}_n\|$$

Then set $\underline{M}_x = \Omega A \underline{X}$, $\underline{M}_{z*n} = \Omega A \underline{Z}_*$, $M_{G,n} =< \underline{M}_x, \underline{M}_{z*,n} >^{1/2}$ and $M_{av,n} = \frac{\underline{M}_x + \underline{M}_{z*,n}}{2}$. Next let $b_n = \alpha\Omega < M_{av,n} \cdot \underline{r}_n > + \frac{n\pi}{2}$, $c_n = \frac{n^2\pi^2}{4} - \Omega^2 \underline{r}_n^2$ and compute

$$\lambda_n = \frac{-b_n \pm \sqrt{b_n^2 - (1 - M_{G,n}^2)c_n}}{1 - M_{G,n}^2} \tag{6}$$

Step 3: If $|\lambda_n| \leq \pi/4$ (check both roots)

$$\text{define} \quad \hat{\theta} = \frac{\frac{n\pi}{2} + \lambda_n}{\Omega} \tag{7}$$

and $\hat{\theta}$ will be a desired approximate root of $\gamma(\hat{\theta}) = 0$.

If $|\lambda_n| > \pi/4$ increase index n by 1 and continue.

Remark 1. In the case of subsonic flow once the first acceptable λ_n is found compute $\hat{\theta} = \frac{\frac{n\pi}{2} + \lambda}{\Omega}$ and *stop* as only one root is possible. But note that this root may not necessarily occur when $n = 0$.

Remark 2. In the preceding description the value $\alpha = .925$ used in the approximation $\sin\lambda \sim \alpha\lambda$ appears only in the coefficient b_n. In fact since the approximation $\sin\lambda \sim \lambda$ (i.e. $\alpha = 1$) has error $\leq .01$ for $|\lambda| \leq \frac{\pi}{8}$ one may wish to modify Step 3 so that if $|\lambda_n|$ turns out to be $\leq \pi/8$ then step 2 is redone by recomputing b_n with $\alpha = 1$ to give a more accurate value of λ_n and thus of $\hat{\theta}$. (Care must be taken to avoid a loop if $|\lambda_n| \sim \pi/4$ though.)

Given the computation of the roots $\hat{\theta}_i$ of $\gamma(\hat{\theta}_i) = 0$ as described in the foregoing, we find that with the parameters b_n, $M_{G,n}$ and c_n as described, for each $\hat{\theta}_i$ we can compute

$$\hat{\rho} = \frac{a_\infty\sqrt{1 - M_\infty^2}}{\Omega}\sqrt{b_n^2 - (1 - M_{G,n}^2)c_n} \tag{8}$$

5. References

[1] Morino, L., Freedman, M.I., and Tseng, K., (1987): Unsteady Compressible Potential Aerodynamics of Rotors. Proceedings AHS Specialists Meeting on Aerodynamics and Aeroacoustics, Arlington, Texas Feb. 25-27, 1987.

[2] Tseng, K., and Freedman, M.I., (1987): A First Order Green's Function Method for Supersonic Oscillatory Flows. Boston University, Department of Mathematics, TR-87-01.

[3] Tseng, K. (1986): A Time-Domain Integral Formulation for Unsteady Compressible Potential Flows About Aircraft or Propeller. United Technologies Research Center. Technical Report R86-152663-1. June 1986.

[4] Morino, L., Bharadvaj, B.K., and Freedman, M.I. (1987): Boundary Integral Equation for Wave Equation with Moving Boundary and Applications to Compressible Potential Aerodynamics of Airplane and Helicopters. Submitted for publication, *Computational Mech. J.*

6. Legend for Figures 1-6

Figure #	$\underline{X}$	$\underline{X}_*$	Ω
1	(1,0,0)	(.5,0,0)	1
2	(1,0,0)	(.5,0,0)	2
3	(1,0,0)	(.5,0,0)	6
4	(1,0,0)	(.5,0,0)	7
5	(1,0,0)	(.5,0,0)	10
6	(1,0,0)	(.5,0,0)	25

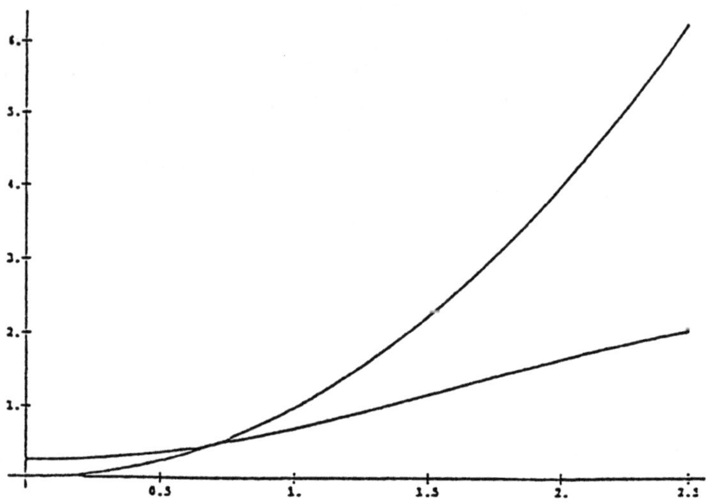

Fig. 1. $\Omega = 1$

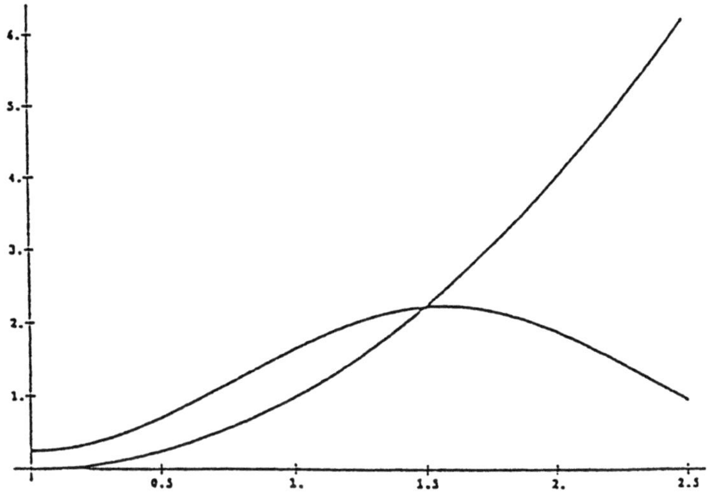

Fig. 2. $\Omega = 2$

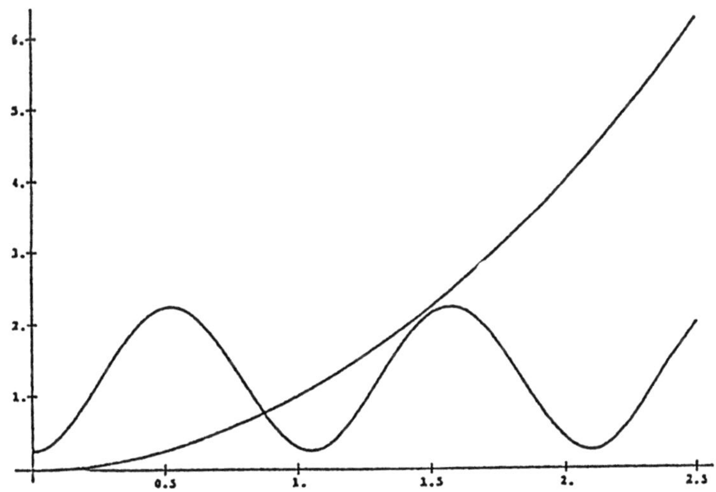

Fig. 3. $\Omega = 6$

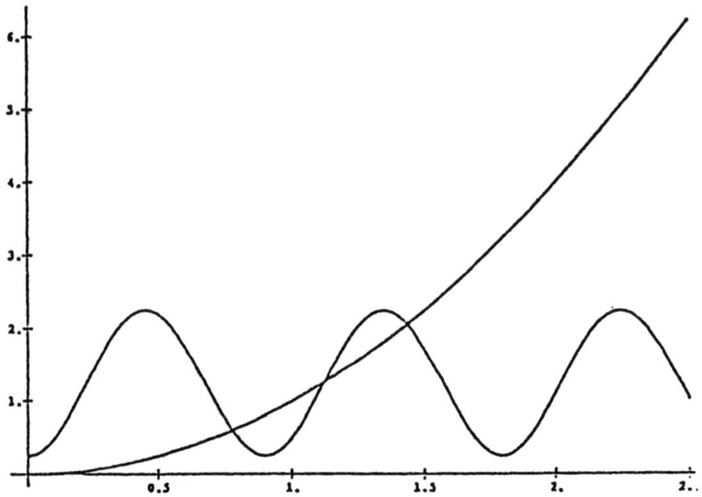

Fig. 4. $\Omega = 7$

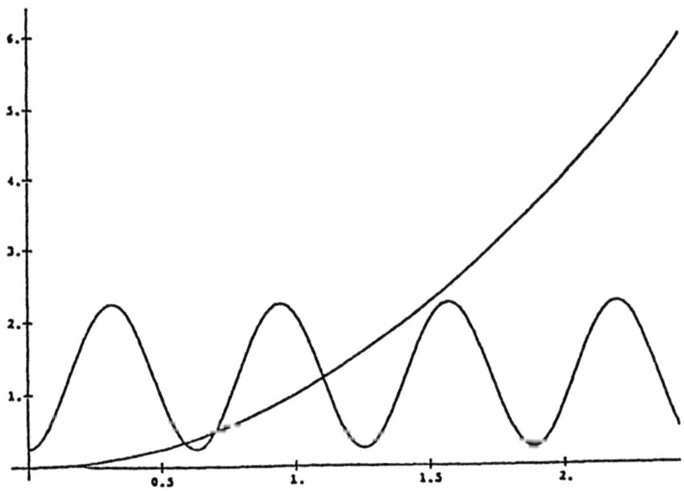

Fig. 5. $\Omega = 10$

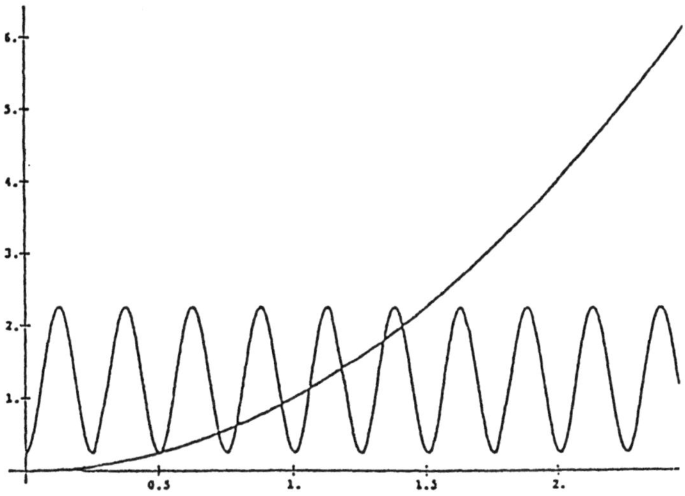

Fig. 6. $\Omega = 25$

An Efficient Nonlinear Transformation for the Numerical Computation of the Singular Integrals Appearing in the 2-D Boundary Element Method

E. ALARCON(#), M. DOBLARE(+), J. SANZ-SERNA(*)

(#) Departamento de Ingeniería de Construcción. U. P. Madrid (Spain)
(+) Departamento de Ingeniería Mecánica. Universidad de Zaragoza (Spain)
(*) Departamento de Matemática Aplicada y Computación. Universidad de Valladolid (Spain)

Summary

We discuss several methods, based on coordinate transformations, for the evaluation of singular and quasisingular integrals in the direct Boundary Element Method. An intrinsic error of some of these methods is detected. Two new transformations are suggested which improve on those currently available.

Introduction

An essential aspect of the implementation of the direct Boundary Element Method (B.E.M.) is the accurate computation of the integrals appearing in the coefficients of the resulting algebraic system. Special techniques are required, due to the inherent singularities (weak or Cauchy type) of the integrands. If constant or linear straight boundary elements are used, it is feasible to perform the integrations analytically [1]. This is not possible in general with curved elements. Semianalytical techniques can be employed with an acceptable accuracy. These techniques require an initial analytical effort to find the coefficient of the singularity and depend on the type of the singularity, so that different segments of code must be employed for different problems, with a lost of systematization [2][3].

An alternative consists in using quadrature rules tailored to the specific type of the singularity [3][4]. Again different quadratures would be required for different problems. The commonest alternative in standard B.E. codes computes singular terms by imposing special conditions (constant potentials or rigid body movements). In some cases (axisymmetric problems), it is difficult or even impossible to program these special "strain" cases.

This aspect of the B.E.M. has become even more critical by the recent inclusion in B.E. problems of adaptive techniques [5][6] with approximation functions of high degree. This imply more approximation nodes (i.e. more singularities) and nonstandard placement of the nodes within the elements.

Furthermore, quasisingular integrals appear when the collocation point is very close to the element of integration, which happens when computing variables at an internal point very close to the boundary or when two adjacent elements have very dissimilar lengths. This problem is aggravated in the p-adaptive version, where the collocation point can be very near the edges of the element.

In order to solve these problems, different authors [7][8][9] have recently suggested the use of nonlinear coordinate transformations of the integrand in order to smooth it. The aim of the present wiork is to study these transformations. We prove that some of them, although asymptotically correct as the number of Gauss points increases, incorporate a hidden error source, that can be easilly removed. Also two consistent nonlinear transformations are sugested which show a very efficient behaviour.

Nonlinear transformations for the computation of 2-D singular integrals

We begin by revising the transformations suggested in [7][8][9]. The three of them transform the integration variable (η) into a new variable (ξ). The transformations in [8] and [9], subdivide the standard integration interval (-1,1) into two subintervals $(-1,\eta_s)$ $(\eta_s,1)$ and each of them is then mapped onto the interval [ξ: (-1,1)]). The transformation [11] keeps the initial interval so that [η: (-1,1)]) transforms into [ξ: (-1,1)]).The following transformations are then defined:

$$\eta(\xi) = a\xi^3 + b\xi^2 + c\xi + d \quad \text{for transformations [7] and [8]}$$

$$\eta(\xi) = a\xi + b.tg[\phi(\xi)] + c \quad \text{for transformation [9]} \tag{1}$$

with $$\phi(\xi) = n.\frac{\pi}{4}.(\xi S + 1)$$

and S = +1, If the integration is performed along the right hand side of η_s.
S = -1, If the integration is performed along the left hand side of η_s.

The boundary conditions that are imposed to the two above transformations are defined as:

CUBIC [7]	BICUBIC [8]	BITANGENT [9]
$\eta\,(\xi = S) = S$	$\eta\,(\xi = S) = S$	$\eta\,(\xi = S) = S$
$\eta\,(\xi = -S) = \eta_s$	$\eta\,(\xi = -S) = \eta_s$	$\eta\,(\xi = -S) = \eta_s$
$\left(\frac{\partial\eta(\xi)}{\partial\xi}\right)_{\xi=-S} = 0$	$\left(\frac{\partial\eta(\xi)}{\partial\xi}\right)_{\xi=-S} = 0$	$\eta(\xi = -S\xi_a) = -\eta_s + S\varepsilon$
$\left(\frac{\partial^2\eta(\xi)}{\partial\xi^2}\right)_{\xi=-S} = 0$	$\eta(\xi = -S\xi_a) = -\eta_s + S\varepsilon$	

(2)

with ξ_a the abcissa of the last Gauss integration point, and ε the distance from η_s, to $\eta(-S\xi_a)$. In the case of [8], accurate results are obtained for ε values of

$$\varepsilon = \frac{1 - |\xi_a|}{R} \qquad R = 100 \text{ to } R = 1000 \tag{3}$$

while the bitangent transformation is much more stable for smaller Rs.

In the case of a logarithmic singularity, the integrand may be always expressed as:

$$I = a.\ln r + I_r \tag{4}$$

with I_r a regular funcion. Let us consider the effect of a transformation with a null Jacobian at the singularity. Taylor expansion near the singularity $\eta(\xi) = \eta_s + A_1(\xi - \xi_s)^2 + A_2(\xi - \xi_s)^3 + ...$ leads after transformation to the new integrand

$$I(\xi) = a.\ln r(\xi).2A_1(\xi - \xi_s) + ... + I_r(\xi).2A_1(\xi - \xi_s) + ... \tag{5}$$

which is a regular funcion. This is the reason of the good performance of all the transformations in (2): they impose the condition of null jacobian at the singularity (Telles and Cerrolaza) or this condition is approximately satisfied when using small ε (Alcantud).

In a similar form, the integrand of a Cauchy Principal Value Integral (CPVI) in two dimensions may be wrtitten as

$$I = \frac{a}{r} + I_r \tag{6}$$

with I_r regular. It is then obvious that it is sufficient to investigate the performance of the transformations of the integral

$$I = \int_{-1}^{1} \frac{d\eta}{\eta - \eta_s} \qquad \text{con} \quad -1 < \eta_s < 1 \tag{7}$$

Let us now consider a double nonlinear transformation on each side of the singularity (as in [8],[9]), with Taylor developments around η_s of the form:

$$\begin{aligned} \eta_1 &= \eta_1(\xi) = \eta_s + a_1(\xi - 1) + b_1(\xi - 1)^2 + c_1(\xi - 1)^3 + ... \\ \eta_2 &= \eta_2(\eta) = \eta_s + a_2(\xi + 1) + b_2(\xi + 1)^2 + c_2(\xi + 1)^3 + ... \end{aligned} \tag{8}$$

Taking these expressions to (7) and using the definition of CPVI it is possible to get

$$I = \int_{-1}^{\eta_1^{-1}(\eta_s - \varepsilon)} \frac{d\xi}{\xi - 1} + \int_{\eta_2^{-1}(\eta_s + \varepsilon)}^{1} \frac{d\xi}{\xi + 1} + I_r = I_s + I_r \tag{9}$$

with I_r again a regular integral. If we now perform the numerical integration (with the same quadrature rule for both sides) of I_s along the interval ξ:(-1,1) the result is zero by symmetry. However, the true value of I_s is

$$I_s = \ln \frac{|\eta_1^{-1}(\eta_s - \varepsilon) - 1|}{|\eta_2^{-1}(\eta_s + \varepsilon) + 1|} \quad \text{and in the limit } \varepsilon \to 0 \qquad \lim_{\varepsilon \to 0} I_s = \ln \frac{|a_2|}{|a_1|} \tag{10}$$

If both $a_i = 0$, (10) becomes the logarithmic of the modulus of the ratio between b_i etc. With this, it is clear that the analytical value of I_s will be the same as the numerical one when the dominant coefficients of both transformations are the same. If this is not the case, there will be an intrinsic error in the numerical computation of (7) due to the exclusion of the term (10). The transformations [7][8] and [9] include this error in their formulation, but the authors in [8] and [9] present results of the computation of CPVI with a very acceptable accuracy. We must then conclude that in these cases the term (10) must be small.

For transformation [8], the ratio of the dominant terms is given by

$$\frac{b_2}{b_1} = \frac{8\varepsilon + (1+\eta_s)(\xi_a-1)^3}{-8\varepsilon + (1+\eta_s)(\xi_a-1)^3} \tag{11}$$

Therefore, the modulus of this ratio tends to 1 when ξ_a tends to 1, and this happens very rapidly when the order of the Gauss-Legendre quadrature is increased. This explains the good performance of this transformation for high-order quadratures, despite the intrinsic error included. Note that larger values of ε imply a faster reduction on the intrinsical error. However, for large ε (R<100) this effect is offset by the poor performance in the integration of the regular part due to the important distorsion introduced by the transformation away from the singularity.

With respect to the bitangent formula, the ratio of the dominant coefficients is

$$\frac{a_2}{a_1} = \frac{\varepsilon \mathrm{tg}\, n\frac{\pi}{2} + (\xi_a-1)(1-\eta_s)\mathrm{tg} n\frac{\pi}{2} + (1-\eta_s)\mathrm{tg}[n\frac{\pi}{4}(\xi_a-1)]}{-\varepsilon\, \mathrm{tg}\, n\frac{\pi}{2} + (\xi_a-1)(1+\eta_s)\mathrm{tg} n\frac{\pi}{2} + (1+\eta_s)\mathrm{tg}[n\frac{\pi}{4}(\xi_a-1)]} \tag{12}$$

and a similar discussion applies. In this case it is possible to increase substantially the value of ε without important distorsion of the regular part. Finally (12) also explains why values of n closer to 1 give rise to better results, becouse the value of $\varepsilon \mathrm{tg} n\pi/2$ is also larger.

<u>Suggested transformations</u>

It is now clear that a transformation should satisfy the following requirements: 1) zero Jacobian at the singularity; 2) If two transformations are used they should have the same dominant terms (alternatively a single transformation may be used if it transforms η_s in $\xi_s = 0$); 3) The regular terms are not greatly distorted.

There is only one single cubic transformation satisfying 1)-2), namely $\eta(\xi) = \eta_s(1-\xi^2) + \xi^3$. However, 3) is not satisfied.In fact this transformation does not perform well in practice. An alternative is to use a single transformation of degree 4. Conditions 1)-2) leave one free parameter that can be chosen to get an almost monotonic transformation, thus implying 3). We have tested the transformation

$$\eta(\xi) = \eta_s(1-\xi^4) + \xi^3 \tag{13}$$

which has shown a very good behaviour, even in the case $\eta_s > 0.75$ where a maximum appears within the integration interval (Fig.1a).

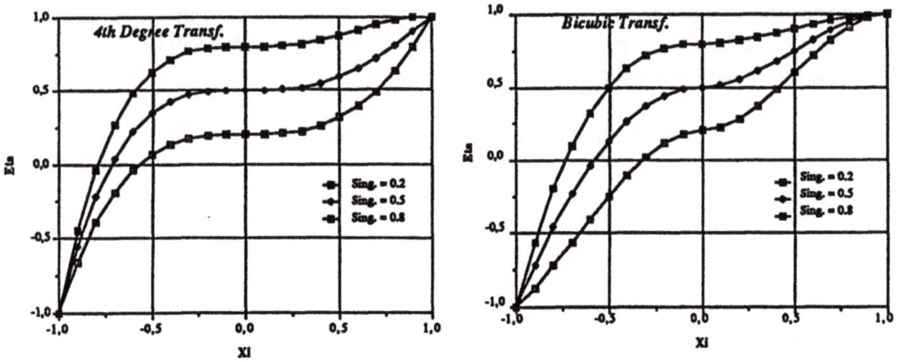

Fig. 1. Proposed nonlinear transformations

Another possibility is to use a bicubic transformation as [8][9] but with equal dominant terms. Subject to condition 1), these are the form

$$\eta_1(\xi) = \eta_s - \frac{1+\eta_s}{4}(\xi - 1)^2 + \frac{B}{2}(\xi^2 - 1)$$
$$\eta_2(\xi) = \eta_s + \frac{1-\eta_s}{4}(\xi + 1)^2 + \frac{B}{2}(1 - \xi^2) \tag{14}$$

where B is a free parameter. The transformation is monotonic if $B \geq 0$ and $B \leq \frac{3}{4}(1 - |\eta_s|)$. Good results have been obtained with $B = \frac{3}{4}[\,1 - |\eta_s|\,]$.

Results

Fig. 2a shows the results obtained when integrate an integral with a logarithmic singularity, using the different transformations included here, with 8 evaluations (4 Gauss points for each interval for the bicubic and bitangent transformations and 8 for the rest).

From this, it can be observed that good results are obtained in this case with any of these transformations, as was expected due to the condition of null jacobian at the singularity that all of them "include". However, a greater stability may be seen when using bicubic transformations, due perhaps to the lower degree of the jacobian of the transformation.

In the same way, Fig. 2b shows the results corresponding to the use of these transformations (except [7] which gives rise to error 2 or 3 order greater, as was explained before) for the numerical computation of an integral with a singularity of type 1/r. As it was expected, there appears an important error for transformations [8][9], which is avoided when using the transformations proposed here. These last ones are very similar, although the 4th degree one gives slightly better results. It is remarkable the high accuracy got with only 8 evaluations, with maximum errors of about 0,1 %, which are reduced to 0,0002 % when using 16 evaluations.

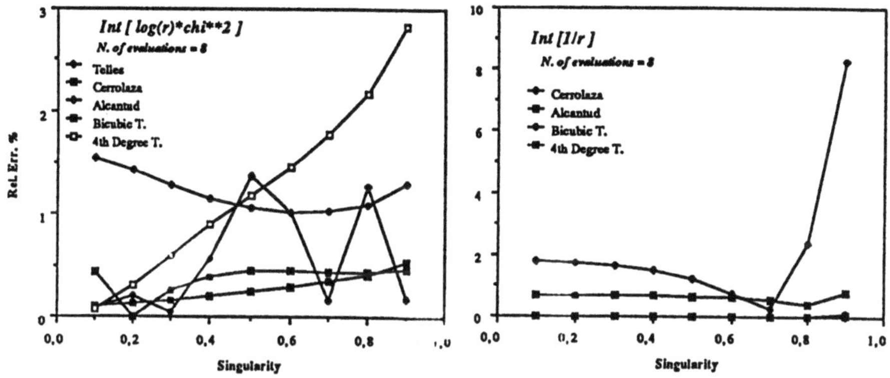

Fig. 2. Computation of singular integrals

Finally, Fig. 3a and 3b show the relative errors obtained when the proposed 4th degree transformation is used to integrate quasisingular integrals with the same type of integrands. The results are again very acurate, and of course less dependent of the position of the closest point of the interval to the collocation point, and now very dependent of the distance between these two points. Anyway, with only 8 evaluations maximum errors of 2% are obtained for distances around 1% of the length of the interval. If the quadrature is improved (16 evaluations) this error is reduce to only 0.4%.

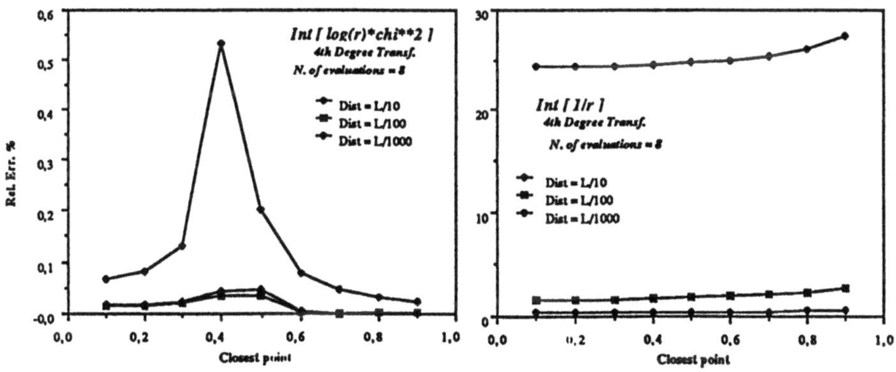

Fig. 3. Computation of quasisingular integrals

Conclusions

The above results along with extensive numerical experimentation suggest the following conclusions:

- All the nonlinear transformations considered here are well adapted to the computation of singular integrals of logarithmic type.
- For singularities of type 1/r the transformation suggested in [7] does not work.
- The bicubic [8] and bitangent [9] transformations produce similar results, although this latter is more accurate and robust. They both need high order quadratures rules (8+8 nodes) to get an acceptable accuracy. The best results with [9] are obtained when R is between 50 and 100, while [8] works best with R around 1000. In the bitangent transformation high values of n (near 0.9) are best, except when the singularity is near the edges. The large n produces an unwelcome distorsion.
- A quartic and a bicubic transformations have been suggested that, with fewer evaluations (8), produce good results when applied to the singular and quasisingular B.E.M. integrals.
- For collocation points placed at the edge of the elements it is sufficient to consider the two adjacent elements to this node as a unique interval and work over both elements at the same time with the same transformation between the coordinate s and the integration coordinate ξ.
- The methods discussed here render it possible the accurate computation of B.E. integrals in a type-dependent way, a feature that reduces coding time, coding errors and, perhaps, execution time.

References

1. Cruse, T.A.: Mathematical Foundations of the B.I.E.M. in Solid Mechanics. Report AFOSR-TR-77-1002. Pratt & Whitney Aircraft Group. 1977

2. Brebbia, C.A.; Telles, J.C,; Wrobel, L.C. : Boundary Element Techniques: Theory and Applications. Springer Verlag. 1984

3. Doblaré, M.: Computational Aspects of the B.E.M. in Topics in B.E. Research III, (Ed. C.A. Brebbia). Springer Verlag. 1987

4. Kutt, H.R.: On the Numerical Integration of Principal Value Integrals by Finite Part Integration Numerische Mathematik, 24 (1975) 205-210

5. Rank, E.: Adaptive Boundary Element Methods. Proc. IX Int. Conf. on B.E.M. (Ed. W.L. Wendland and C.A. Brebbia). Springer Verlag 1987

6. Cerrolaza, M.; Alarcón, E.: P-Adaptive Boundary Elements for Three-Dimensional Potential Problems". Communications in Appl. Num. Meth., 3 (1987) 335-345.

7. Telles, J.C.: A Selfadaptive Coordinate Transformation for the Efficient Evaluation of General Boundary Element Integrals Int. J. of Num. Meth. in Eng. 24 (1987) 937-959

8. Cerrolaza, M. Alarcón, E.: A Bicubic Transformation for the Numerical Evaluation of the Cauchy Principal Value Integrals in Boundary Elements. Int. J. Num. Meth. in Eng. 28, (1989) 987-999

9. Alcantud, M.: A bitangent transformation for the evaluation of the B.E. singular integrals (in spanish). Private communication 1989

Block LU Decomposition on the Connection Machine System

John L. Richardson
Thinking Machines Corporation
245 First Street
Cambridge MA 02142-1214
(617)-876-1111
john@think.com

October 2, 1989

Abstract

We describe the solution to *very large* systems of dense linear equations using the CM-2/DATAVAULT system. The block LU decomposition algorithm is presented and the implementation of this algorithm on the CM-2 is described. It is shown that the computation time is dominated by matrix multiplication. Performance results for the factorization of a dense complex 20k by 20k double precision matrix are given.

1 Introduction

Very large simulation problems require the numerical solution to systems of linear equations that are too large to fit into the memory of the Connection Machine[1]. The only way to solve such systems is to store results in secondary storage devices such as DATAVAULTS. We will discuss the implementation of block LU decomposition in this document[2].

The fundamental idea in partitioning large matrix problems is to divide a very large matrix into a set of smaller submatrices, each of which fit into the CM-2's memory. At the outer level, our algorithms are serial in that we sequentially operate on a few submatrices at a time. Submatrices are transferred from the DATAVAULT mass storage system to CM-2 memory, operated on, and then written back to the DATAVAULT.

At the inner level our algorithms are highly parallel and utilize the many thousands of processors in the CM-2 as well as its very high speed interprocessor communication system. The parallelism is isolated to a few optimized matrix kernels which operate on a few matrices at a time. We will find that most of the time is dominated by matrix multiplication.

In the following sections we will describe an algorithm we have implemented on a CM-2 with DATAVAULT. A detailed timing analysis is also provided.

2 Block LU Decomposition

Consider the system (real or complex) of linear equations

$$AX = B \tag{1}$$

where A is a non-singular N by N matrix and both B (the right hand sides) and X (the desired solutions) are N by K matrices. Given the right hand side B we would like to know the solution vector X. When A is so large that it cannot fit into the memory of the Connection Machine we must partition A into n^2 m by m blocks where $n = \lfloor N/m \rfloor$. Each m by m submatrix $a_{i,j}$ of A may be used in the block LU decomposition of A.

The algorithm consists of three loops and is described as follows:

LU decomposition algorithm

for $k = 0$ to $n-1$
    read $a_{k,k}$
    invert $a_{k,k}$
    write $(a_{k,k})^{-1}$
    for $i = k+1$ to $n-1$
        read $a_{i,k}$
        $m := a_{i,k}(a_{k,k})^{-1}$
        $a_{i,k} := m$
        write $a_{i,k}$
        for $j = k+1$ to $n-1$
            read $a_{i,j}$
            read $a_{k,j}$
            $a_{i,j} := a_{i,j} - m a_{k,j}$
            write $a_{i,j}$
        end j
    end i
end k

The outer loop is executed p times, the intermediate loop is executed n times and the innermost loop is executed $\frac{n(n-1)}{2}$ times. By examining the above algorithm we can again see how the execution time is divided between computation and DATAVAULT communication. This is displayed in the following table:

Forward Elimination		
operation	count	time
matrix multiply	$\frac{pn(n-1)}{2}$	t_M
disk read	$pn(n-1)$	t_R
disk write	pn	t_W

The total runtime for forward elimination of p block right hand sides is then

$$t_F = \frac{pn(n-1)}{2} t_M + pn(n-1)t_R + pnt_W \quad . \tag{3}$$

4 Back Substitution

After forward elimination the desired solution is obtained through a subsequent back substitution. The (in place) back substitution algorithm also requires three loops and is described by:

Back Substitution algorithm

```
for k = 0 to p − 1
        for i = n − 1 down to 0
                read b_{i,k}
                s := b_{i,k}
                for j = i + 1 to n − 1
                        read a_{i,j}
                        read b_{j,k}
                        s := s − a_{i,j} b_{j,k}
                end j
                read a_{i,i}
                b_{i,k} := a_{i,i} s
                write b_{i,k}
        end i
end k
```

The outer loop is executed n times, the intermediate loop is executed $\frac{n(n-1)}{2}$ times and the innermost loop is executed $\frac{n(n-1)(2n-1)}{6}$ times. By examining the above algorithm we can easily see how the execution time is divided between computation and DATAVAULT communication. This is displayed in the table:

LU decomposition		
operation	count	time
matrix multiply	$\frac{n(n^2-1)}{3}$	t_M
disk read	$\frac{(n-1)(4n^2+n+6)}{6}$	t_R
disk write	$\frac{(n-1)(n^2+n+3)}{3}$	t_W
matrix inversion	n	t_I

The total runtime for matrix factorization is given by

$$t_{LU} = \frac{n(n^2-1)}{3} t_M + \frac{(n-1)(4n^2+n+6)}{6} t_R + \frac{(n-1)(n^2+n+3)}{3} t_W + n t_I \ . \tag{2}$$

3 Forward Elimination

So far we have only discussed the LU factorization of A and have not discussed the forward elimination of B followed by the subsequent back substitution to obtain X. If there are K right hand sides then B and the solutions X are both of block size n by p where $p = \lfloor K/m \rfloor$. The (in place) forward elimination algorithm consists of three loops and is described as follows:

Forward Elimination algorithm

```
for k = 0 to p − 1
        for i = 0 to n − 1
                read b_{i,k}
                s := b_{i,k}
                for j = 0 to i − 1
                        read a_{i,j}
                        read b_{j,k}
                        s := s − a_{i,j} b_{j,k}
                end j
                b_{i,k} := s
                write b_{i,k}
        end i
end k
```

The outer loop is executed p times, the intermediate loop is executed n times and the innermost loop is executed $\frac{n(n-1)}{2}$ times. By examining the above algorithm we can again see how the execution time is divided between computation and DATAVAULT communication. This is displayed in the following table:

Back Substitution		
operation	count	time
matrix multiply	$\frac{pn(n+1)}{2}$	t_M
disk read	pn^2	t_R
disk write	pn	t_W

The total runtime for back substitution of p block right hand sides is then

$$t_S = \frac{pn(n+1)}{2} t_M + pn^2 t_R + pnt_W \quad . \tag{4}$$

5 Performance

Consider the case where A is a complex double precision matrix of size 20k by 20k so $N = 20,480$. If the block size $m = 2048$ then $n = 10$. The total runtime for the LU factorization algorithm is

$$t_{LU} = 330t_M + 624t_R + 339t_W + 10t_I \quad . \tag{5}$$

A 64k CM-2 can perform a double complex 2k by 2k matrix multiply at 4GFlop. Since there are $4 \times 2(2k)^3 = 64$GFlops in this complex matrix multiply, $t_M = 16$ seconds.

A 2048 by 2048 double complex matrix consists of 64 MBytes of data, that can be striped across two DATAVAULTS. Two striped DATAVAULTS average above 45 MBytes per second reading or writing the matrix so $t_R = t_W = 1.42$ seconds. t_I is about 100 seconds. So $t_{LU} = 7648$ seconds or 2.2 hours. This Algorithm has been implemented on a 16k CM-2 using 1k blocks and a single DATAVAULT and measured run times were under 10 hours.

6 Huge Systems

For systems that are larger than a DATAVAULT a mass tape storage system(s) that trickle feeds and drains the the DATAVAULT at 3MByte/sec per tape is employed. A super-block strategy minimizes tape transfers.

References

[1] Connection Machine Model CM-2 Technical Summary, Thinking Machine Technical Report TR-89-1

[2] Golub, Gene H.,& Van Loan, Charles F., Matrix Computations, Second Edition, The John Hopkins University Press (1989)

Parallel Implementation on the ICAP/3090 of a Boundary Element Method Formulation for Fractional Operator Modeled Viscoelastodynamic Structures

Robert D. Ciskowski
Numerically Intensive Computing Center
International Business Machines Corporation
Kingston, New York 12401 U.S.A.

Vijay Sonnad
Scientific and Engineering Computations Department
International Business Machines Corporation
Kingston, New York 12401 U.S.A.

Ke-jun Xie
Department of Mechanical and Aerospace Engineering
North Carolina State University
Raleigh, North Carolina 27695-7910 U.S.A.

Abstract

A formulation of the boundary element method (BEM) for dynamic analysis of 3-D viscoelastic structures modeled by the fractional operator constitutive equations is reviewed. Then, the parallel numerical implementation of the BEM formulation on a large grain parallel system, the ICAP/3090, is discussed.

Introduction

In the classical theory of viscoelasticity, the differential operator formulation is popularly employed to describe the mechanical behavior of materials [3]. By this approach, the stress field and strain field are related by differential operators of integer order. While this formulation enjoys wide use, it yields restricted functional representations.

The constitutive relation employing the fractional integrodifferential operator has recently been found to be attractive as a different approach for modeling the viscoelastic behavior of materials [1,2]. The fractional operator representation employs differential operators of fractional order to describe the mechanical behavior of materials. Mathematically, the use of fractional operators extends the number of functional families which may be used to represent material behavior. Thus, this approach produces more accurate analytic representations for viscoelastic problems.

The fractional operator representation of viscoelasticity has been incorporated into finite element models [1] and, most recently, BEM models as first reported by Xie, Ciskowski, and Royster [10]. This BEM formulation for the transient dynamic analysis of three-dimensional viscoelastic structures modeled by the fractional operator constitutive equations will be reviewed in this paper.

The past decade has seen the introduction of new types of hardware architecture for parallel processing systems which offer increased computer performance and are fast becoming an essential part of the engineering analysis computing environment [9]. BEM is a numerically intensive computational method and so it is only natural to implement these codes on faster architectures [5,8]. This paper describes the implementation of the BEM formulation for the analysis of fractional operator modeled viscoelastic structures on a research machine for large grain parallelism, the ICAP/3090.

Governing Equations

Consider a viscoelastic body which occupies an open region V in Euclidean 3-space E and has a boundary surface S which is piecewise smooth. With the absence of body forces the balance of linear momentum gives

$$\sigma_{ij,j} = \rho \frac{\partial^2 u_i}{\partial t^2} \tag{1}$$

subject to initial and boundary conditions. In this equation, σ_{ij} is the stress tensor, u_i the displacement vector, t_i the boundary traction vector, ρ the mass density of the body, n_i the outward pointing normal vector to the boundary and $\vec{x} = \{x_1, x_2, x_3\}^T$ the position vector. The comma denotes spatial differentiation and the summation convention is used.

Let a bar over a function denote the Laplace transformation of that function with respect to time, then the Laplace transformation applied to the governing equation (1), assuming zero initial conditions, gives

$$\bar{\sigma}_{ij,j} = s^2 \rho \bar{u}_i. \tag{2}$$

The fractional operator constitutive equation for a linear, homogeneous and isotropic viscoelastic body under an isothermal state suggested by Bagley and Torvik [2] is of the form

$$\sum_{n=0}^{N} \eta_n D^{\gamma_n} \sigma_{ij} = \delta_{ij} \sum_{n=0}^{N} \lambda_n D^{\alpha_n} \epsilon_{kk} + 2 \sum_{n=0}^{N} \mu_n D^{\beta_n} \epsilon_{ij} \tag{3}$$

where ϵ_{ij} is the strain tensor, δ_{ij} is the Kronecker delta and $\lambda_n, \alpha_n, \mu_n, \beta_n, \eta_n$, and γ_n are material constants. The fractional operator D^γ is defined by the Riemann-Liouville integral,

$$D^\gamma f(t) = \frac{1}{\Gamma(1-\gamma)} \int_0^t \frac{f(t-\tau)}{\tau^\gamma} d\tau, \tag{4}$$

in which $\Gamma()$ is the gamma function.

Applying the Laplace transformation to the fractional operator constitutive equation (3) and noting the convolutional property of equation (4) gives

$$\sum_{n=0}^{N} \eta_n s^{\gamma_n} \bar{\sigma}_{ij} = \delta_{ij} \sum_{n=0}^{N} \lambda_n s^{\alpha_n} \bar{\epsilon}_{kk} + 2 \sum_{n=0}^{N} \mu_n s^{\beta_n} \bar{\epsilon}_{ij}. \tag{5}$$

Equation (5) can be rewritten as

$$\bar{\sigma}_{ij} = \delta_{ij} s \bar{\lambda}(s) \bar{\epsilon}_{kk} + 2 s \bar{\mu}(s) \bar{\epsilon}_{ij} \tag{6}$$

where

$$\bar{\lambda}(s) = \frac{\sum_{n=0}^{N} \lambda_n s^{\alpha_n}}{s \sum_{n=0}^{N} \eta_n s^{\gamma_n}} \text{ and } \bar{\mu}(s) = \frac{\sum_{n=0}^{N} \mu_n s^{\beta_n}}{s \sum_{n=0}^{N} \eta_n s^{\gamma_n}}. \tag{7}$$

Under the assumption of infinitesimal deformation and with the introduction of equation (6) into equation (2), we obtain the governing equation in terms of transformed displacements,

$$\mu_s \bar{u}_{i,kk} + (\mu_s + \lambda_s)\bar{u}_{k,ki} = s^2 \rho \bar{u}_i, \tag{8}$$

in which

$$\mu_s = s\bar{\mu}(s) \text{ and } \lambda_s = s\bar{\lambda}(s). \tag{9}$$

BEM Formulation

The boundary integral equation can be obtained by making use of the form of Betti's theorem appropriate to the Laplace transformed viscoelastodynamic state, and then taking the limit as an internal point approaches a boundary point $\vec{y}_0$. This procedure gives

$$\begin{aligned} C_{ij}(\vec{y}_0)\bar{u}_j(\vec{y}_0, s) &= \int_S \bar{U}_{ij}(\vec{x}, \vec{y}_0, s)\bar{t}_j(\vec{y}_0, s)dS \\ &\quad - \int_S \bar{T}_{ij}(\vec{x}, \vec{y}_0, s)\bar{u}_j(\vec{y}_0, s)dS. \end{aligned} \tag{10}$$

In equation (10), $\bar{U}_{ij}$ and $\bar{T}_{ij}$ represent the transformed displacements and boundary tractions in the j direction at the field point $\vec{x}$ due to a unit point force acting in the i direction applied at load point $\vec{y}$ respectively. They are given by

$$\bar{U}_{ij} = \frac{1}{4\pi\rho c_2^2}(\bar{\psi}\delta_{ij} - \bar{\chi} r_{,i} r_{,j}) \tag{11}$$

and

$$\begin{aligned} \bar{T}_{ij} &= \frac{1}{4\pi}[(\frac{d\bar{\psi}}{dr} - \frac{\bar{\chi}}{r})(\delta_{ij}\frac{\partial r}{\partial n} + r_{,j} n_i) - \frac{2}{r}\bar{\chi}(r_{,i} n_j - 2 r_{,i} r_{,j}\frac{\partial r}{\partial n}) \\ &\quad - 2\frac{d\bar{\chi}}{dr} r_{,i} r_{,j}\frac{\partial r}{\partial n} + (\frac{c_1^2}{c_2^2} - 2)(\frac{d\bar{\psi}}{dr} - \frac{d\bar{\chi}}{dr} - 2\frac{\bar{\chi}}{r}) r_{,i} n_j] \end{aligned} \tag{12}$$

where

$$\bar{\chi} = (\frac{3c_2^2}{s^2 r^2} + \frac{3c_2}{sr} + 1)\frac{e^{-sr/c_2}}{r} - \frac{c_2^2}{c_1^2}(\frac{3c_1^2}{s^2 r^2} + \frac{3c_1}{sr} + 1)\frac{e^{-sr/c_1}}{r}, \tag{13}$$

$$\bar{\psi} = (\frac{c_2^2}{s^2 r^2} + \frac{c_2}{sr} + 1)\frac{e^{-sr/c_2}}{r} - \frac{c_2^2}{c_1^2}(\frac{c_1^2}{s^2 r^2} + \frac{c_1}{sr})\frac{e^{-sr/c_1}}{r}, \tag{14}$$

$$c_1^2 = \frac{\lambda_s + 2\mu_s}{\rho} \text{ and } c_2^2 = \frac{\mu_s}{\rho}. \tag{15}$$

For the purpose of obtaining a numerical solution, the boundary S is discretized into a finite number, N, of boundary elements, which in this investigation are constant triangular elements. Equation (10) becomes

$$0.5\delta_{ij}\bar{u}_j^n + \sum_{k=1}^{N} \bar{u}_j^k \int_{S_k} \bar{T}_{ij}(\vec{x}, \vec{y}_n, s) dS_k = \sum_{k=1}^{N} \bar{t}_j^k \int_{S_k} \bar{U}_{ij}(\vec{x}, \vec{y}_n, s) dS_k \tag{16}$$

where $k, n = 1, 2, .., N$.

After the application of the boundary conditions, a matrix equation results from equation (16). For transient problems formulated in the Laplace transform domain, this matrix equation is solved for a sequence of values of s the and resulting solution is inverted numerically to obtain the time domain solution using the algorithm of Durbin [6].

The ICAP/3090 System

The IBM ES/3090[1] family of computers with optional Vector Facility (pipelined arithmetic unit) provides for one thru six processors (depending upon the machine configuration) sharing a single global memory, thus creating a powerful parallel vector computer for engineering and scientific processing [7].

The ICAP/3090 is a research computer that further increases the level of parallelism offered by the IBM 3090 family. This is accomplished by coupling together clusters of IBM 3090 multiprocessors to form a system that, while not massively parallel, is modular and can be expanded to match the degree of parallelism required for a given set of applications.This coupling is extendable so that, as faster versions of IBM multiprocessor systems are developed, they may be coupled together in a similar fashion. The justification for coupling together more processors than are currently available in a single system is supported by the large degree of parallelism that exists in many engineering and scientific applications. Initial efforts in permitting multiple clusters to work collectively on a single problem exhibiting large grain parallelism have proven successful. For a more detailed discussion of the ICAP/3090 system hardware and software, see reference [4].

Although the ICAP/3090 is strictly a research machine, IBM recently announced the availability of Clustered FORTRAN which was developed in cooperation with the Cornell National Supercomputer Facility. This is the first of a new product category called Supercomputing Systems Extensions (SCSE). With this product two ES/3090 processor complexes may be joined into a cluster for the purpose of exploiting application parallelism with up to 12 processors.

Parallel Implementation

The set of algebraic equations resulting from the discretization of the continuum integral equation must be solved for various values of the transform parameter s. We note that the formation and solution of the matrix equation for a given value of s, are completely independent of the solution of the equations for any other value of s. This leads to a particularly simple scheme for parallel solution of the entire problem: we divide the given values of s evenly among the processors, and allow each processor to have access to all the input and auxiliary data. Each processor then forms the global equation and solves it for all values of s that have been assigned to it.

To recover the time domain solution, an inverse Laplace transformation is performed numerically using the FFT [6]. We note that the procedure is also readily implemented in parallel since the transformation at each node is independent of all the others. The solutions for the values of s assigned to each processor are gathered into a single processor and redistributed by nodes. This scheme has a very low ratio of communication time to computation time, and is therefore well suited to large grain parallel systems such as those provided by the 3090 family, SCSE Clustered FORTRAN, and the ICAP/3090 system.

[1]ES/3090 is a registered trademark of the International Business Machines Corporation.

Numerical Examples

The BEM formulation presented has been used to predict the transient responses of a viscoelastic column of dimension 1.0 *in* × 1.0 *in* × 4.0 *in* subject to external excitations. The column is fixed at one end and subject to an axial loading at the other end. The rest of the boundary of the column is free of tractions.

The column is made of E-A-R[2] C-3200 viscoelastic damping material with a constant Poisson's ratio ($\nu = 0.36$). The mass density of the material is $\rho = 0.226 \times 10^{-4}\ lbf\ s^2/in^4$. A uniaxial five parameter fractional operator viscoelastic model was utilized to represent the mechanical behavior of the material.

The boundary of the column was discretized into 144 constant triangular elements. The predicted end traction of the column subject to a step external loading at the free end is presented in Fig. 1 along with the 1-D analytical solution. For the material with $\nu = 0.0$, the response of the column is confined to the axial direction and the 3-D BEM predicted response should agree with the 1-D analytical solution. It is seen that the predicted response compares favorably. The response for the material with $\nu = 0.36$ (E-A-R C-3200) is slightly delayed in comparison to that for $\nu = 0.0$ because of additional energy loss in the lateral directions.

Parallel Performance

The parallel scheme described previously has been implemented on the ICAP/3090 using two 3090/600 processors (one a Model S and the other a model E). Minor adjustments were necessary in the implementation because the processor speeds were not identical; the model S performed about 20 percent faster on the given problem. Load balancing was achieved by distributing the work between processors in inverse proportion to the processor speeds. The table below gives the timings for various numbers of processors. The model S was used for the 1 processor and 4 processor runs.

Number of Processors	*Elapsed Time (minutes)*	*Parallel Efficiency E_n (max.=1.0)*
1	60.9	1.0
4	16.4	0.93
8	9.0	0.92
12	6.5	0.87

The parallel efficiency is the ratio of the actual speedup to the ideal speedup for a given number of processors. In calculating this number when using a combination of 3090 systems (8 and 12 processors), we define a "mean" processor with a performance averaged between the S and E systems, and use that single processor number to calculate the ideal speedup. We note that the efficiency remains high even when data is transmitted across systems using relatively slow communication hardware; this is possible due to the high level of parallelism inherent in the transform domain analysis with the boundary element method.

[2]E-A-R is a registered trademark of the E-A-R Division of the Cabot Corp., Indianapolis, IN.

Summary

A three-dimensional BEM formulation incorporating the fractional operator constitutive relation for periodic and transient viscoelastodynamic analysis was reviewed. The algorithm was implemented on the ICAP/3090 using 12 processors where it displayed a high degree of parallel efficiency. The transient response of a numerical example predicted by this formulation was presented and satisfactory results were obtained.

References

[1] Bagley R.L. (1979). "Application of Generalized Derivatives to Viscoelasticity," Ph.D. Thesis, Air Force Institute of Technology.

[2] Bagley R.L. and P.J. Torvik (1986). "On the Fractional Calculus Model of Viscoelastic Behavior," *J. Rheol.*, 30(1), 133-155.

[3] Bland, D.R. (1960). *The Theory of Linear Viscoelasticity*, Pergamon Press, Oxford.

[4] Clementi, E., Logan, D., and Saarinen, J. (1988). "ICAP/3090: Parallel Processing for Large-scale Scientific and Engineering Problems," *IBM Sys. J.*, 27(4), 475-507.

[5] Davies, A.J. (1988). "Quadratic Isoparametric Boundary Elements: An Implementation on the ICL DAP," in *Boundary Elements X: Vol. 3* edited by C.A. Brebbia, CM Pub., Southampton, 657-666.

[6] Durbin, F. (1974). "Numerical Inversion of Laplace Transforms: an Efficient Improvement to Dubner and Abate's Method," *Comp. J.*, 7, 371-376.

[7] Gibson, D.H., Rain, D.W., and Walsh, H.F. (1986). "Scientific Processing on the IBM 3090," *IBM Sys. J.*, 25(1), 36-50.

[8] Hebeker, F.K. (1989). "Experiences with the Vectorization of a Boundary Element Code for the Neumann Problem of the Laplacean," *Applications of Supercomputers in Engineering: Fluid Flow and Stress Analysis Applications*, Computational Mechanics Publications, Southampton.

[9] Melli, P. and Brebbia, C.A., editors (1989). *Supercomputing in Engineering Structures*, Computational Mechanics Publications, Southampton.

[10] Xie, K.J., R.D. Ciskowski, and L.H. Royster (1988). " An Investigation of Wave Propagation in Viscoelastic Media Modeled by Fractional Derivative Using the Boundary Element Method," in *Boundary Element Methods in Applied Mechanics*, edited by M. Tanaka and T.A. Cruse, Pergamon Press, New York, 523-531.

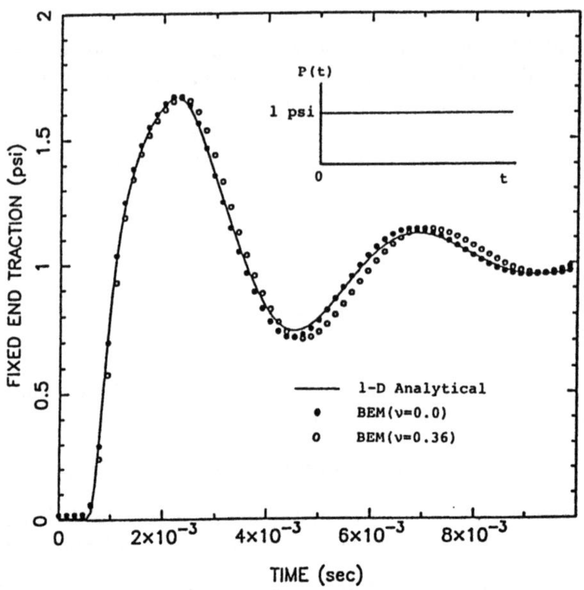

Fig. 1 Fixed End Traction in Axial Direction Due to a Step Function Input Excitation

Helicopter Free-Wake Analysis on a Massively Parallel Computer

T. Alan Egolf, Senior Principal Engineer
United Technologies Research Center
East Hartford, Connecticut 06108

Abstract

An algorithm for the efficient implementation of a helicopter rotor free-wake analysis on the massively parallel Connection Machine (CM-2) is described whose computational performance approaches 10 times that of an efficient vectorized implementation on a Cray-2 processor.

1 Introduction

Historically the prediction of distorted rotor wake geometry using vortex lattice methods has required many simplifying assumptions. In particular, single tip-vortex distortion models. Recent work has demonstrated the need to predict the distortion of the complete wake surface for some conditions (Refs. 1-3). New designs which alter the blade trailing vorticity distribution may require this type of model for all flight conditions. Advances in computer architectures and computational speeds make these types of predictions practical (Ref. 4). Figure 1 is an example of a fully distorted wake surface model predicted by UTRC's new free-wake code (Ref. 1). This code makes few assumptions concerning the wake geometry, other than inviscid incompressible flow and the discretization of the wake surface. The wake convection and the blade bound circulation solution (spanwise airloads) are computed with this code. In this implementation, the bound circulation calculation is based on lifting-line theory and includes non-linear airfoil characteristics and blade motions and controls. If desired, more sophisticated methods such as lifting surface, panel, or finite-difference methods could replace the lifting-line solution. This code has been recently implemented by UTRC using FORTRAN with the proposed 8X array extensions on the massively parallel Connection Machine (CM-2), developed by Thinking Machines Corporation (TMC). This paper describes the key features of the algorithms used to produce an efficient code. Recent success on the CM-2 with a finite-difference viscous flow solver for helicopter blade flows, requiring this distorted wake influence, is given in Ref. 5. Other scientific applications by UTRC on this new supercomputer are given in Ref. 6.

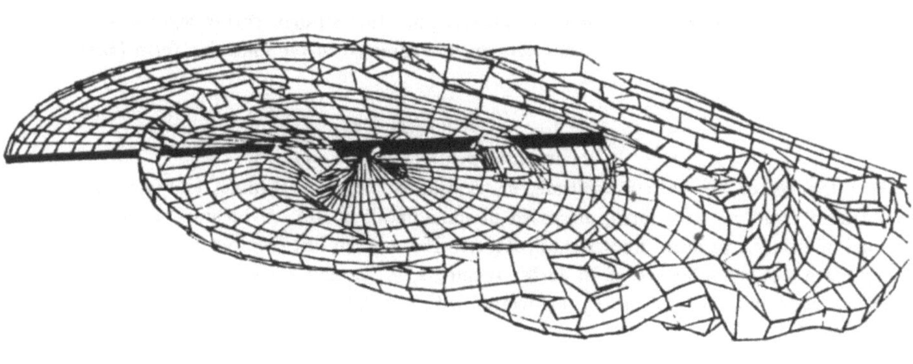

Figure 1: Predicted Distorted Wake Surface

2 The Free-Wake Model

Free-wake methods are generally based on the solution of the inviscid fluid flow equations. They rely on an assembly of discrete singularity elements distributed on surfaces representing bodies or fluid discontinuities. The associated induced influence is a function of the elements strength distribution and geometry relative to the influenced points. The strengths are generally determined by solving a boundary value problem at each time step on these body surfaces and convecting the strengths along the wake surfaces. In the free-wake method, satisfaction of zero pressure across the wake surface is generally obtained by a time-stepping solution process in which the wake moves under the combined freestream and self-induced velocity field. The free-wake method used herein is based on the use of vortex-box elements consisting of four, straight, constant-strength vortex segments (Fig. 2) combined to form a wake lattice structure of the complete wake surface (Fig. 1). The combination of these vortex boxes models represents a discrete model of radially and azimuthally varying surface of vorticity.

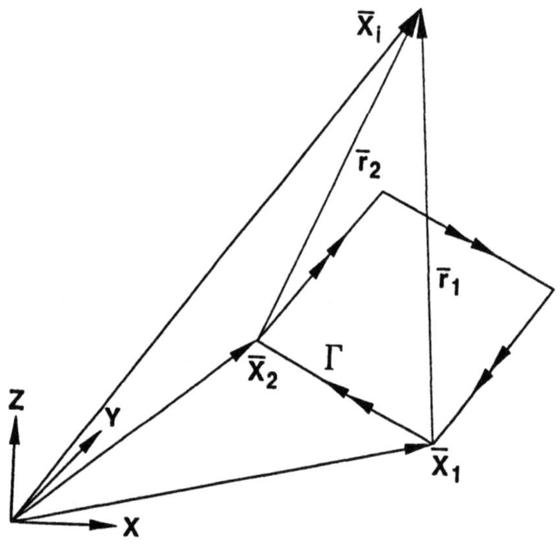

Figure 2: Vortex-Box Geometry

The induced velocity, $\mathbf{v}$, at a point in space, $\mathbf{x}$, due to a straight vortex segment of constant strength, Γ, with endpoints at $\mathbf{x}_1$ and $\mathbf{x}_2$ can be expressed in the following form (Biot-Savart Law),

$$\mathbf{v}(\mathbf{x}) = \frac{\Gamma}{4\pi} \frac{\mathbf{r}_1 \times \mathbf{r}_2 (r_1 + r_2)}{max(r_c^2, r_1 r_2 (r_1 r_2 + \mathbf{r}_1 \cdot \mathbf{r}_2))} \tag{1}$$

where the vector, $\mathbf{r}_i$, with magnitude, r_i, is defined from the segment endpoint, $\mathbf{x}_i$ to the influenced point $\mathbf{x}$. This form, discussed in some detail in Ref. 1, includes a "pseudo" core model of size, r_c, to avoid numerical problems. For a given lattice geometry, the appropriate combination of the influence of all vortex segments, $\mathbf{v}_i$, on any point, is expressed in terms of the discrete bound circulation, Γ_j, associated with the influence coefficients of the vortex-box elements, $\mathbf{A}_j$.

$$\mathbf{v}_i = \sum_{j=1}^{N} \Gamma_j \mathbf{A}_j \tag{2}$$

There are two basic portions to the code, the calculation of the blade bound circulation (airloads) and the calculation of the wake convection. For the bound circulation solution, the wake-induced influence is calculated at the relatively few blade lifting-line collocation points, on the order of 100. For the wake convection process, the wake self-induced influence is calculated at each node point of the vortex lattice wake structure. The number of node point calculations is proportional to the number of elements squared and dominants the computational cost of the problem.

3 The Connection Machine

The Connection Machine Model CM-2 is an integrated system of hardware and software for data parallel computing (Refs. 7 and 8). The hardware consists of a SIMD parallel processing unit containing up to 65536 physical processors (64k) in blocks of 8k ("k" stands for 1024). Each processor has its own local memory of 2048 32-bit words (expandable to 8196 words). The system can have one to four front-end computers. The CM-2 system software is designed to use existing programming languages and environments. UTRC uses a DEC 6320 UNIX based front-end and FORTRAN. Parallel data structures are spread across the data processors, with a single element stored in each processors's memory. When the number of parallel data elements exceeds the number of physical processors (the normal case), the hardware operates in virtual processor mode, effectively presenting the user with a larger number of processors with correspondingly smaller memory. The ratio of virtual to physical processors is known as the virtual-processor ratio (vp ratio). The CM-2 parallel floating-point hardware can perform at an aggregate peak of about 4000 single precision MFlops (4 GFlops) on a full machine (64k) using a high-level parallel language, such as CM FORTRAN.

The CM-2 system provides two forms of inter-processor communication. The general mechanism, the router, allows any processor to communicate in parallel with any other processor with an average time to send data of sixty (60) integer add operations. A local communication mechanism allows processors to pass data in parallel in a regular multi-dimensional pattern with adjacent processors in about six (6) integer add operations. A special communication process can make optimal use of the underlying communication hardware, when the distance between communicating processors is an integer power-of-two.

4 CM FORTRAN

Connection Machine FORTRAN (CM FORTRAN) consists of the incorporation of proposed FORTRAN 8X Array Extensions (Ref. 9) to the ANSI FORTRAN 77 standard. The most important enhancement is the treatment of entire arrays as objects. On the CM-2 the use of an array without explicit indexing presumes a parallel operation. Most of these constructs can be categorized as: elemental, reduction, complex, and communication operations. The following code illustrates several of these constructs by the parallel calculation of the z-component of induced velocity from Eqs. 1 and 2 for a single vortex filament comprised of N-1 connected segments. In this example, each processor is associated with a vortex-segment endpoint (N values). In addition, each processor is associated with a segment (N-1) and is used to calculate the induced velocity of that segment, thus the last processor has no associated segment.

```
PARAMETER (N=128,PI4=12.56637)
REAL R1X(N),R1Y(N),R1Z(N),R1(N),R2X(N),R2Y(N),R2Z(N),R2(N),R1DR2(N),R1CR2Z(N)
COMMON/FIELDPOINT/X,Y,Z           ! scalars
COMMON/DATA/ X(N),YW(N),ZW(N),GAM(N),RCSQ(N),VZ(N)
LOGICAL SEGMENT(N)
```

```
SEGMENT    = .TRUE.
SEGMENT(N) = .FALSE.   ! No vortex-segment at N, non-parallel code

WHERE (SEGMENT)
  R1X     = X - XW
  R1Y     = Y - YW
  R1Z     = Z - ZW
  R1      = SQRT( R1X**2 + R1Y**2 + R1Z**2 )
  R2X     = X - CSHIFT( XW, 1, +1 )
  R2Y     = Y - CSHIFT( YW, 1, +1 )
  R2Z     = Z - CSHIFT( ZW, 1, +1 )
  R2      = SQRT( R2X**2 + R2Y**2 + R2Z**2 )

  R1DR2  = R1X * R2X + R1Y * R2Y + R1Z * R2Z
  R1CR2Z = R1X * R2Y - R1Y * R2X

  VZ    = GAM * R1CR2Z * (R1+R2) / MAX( (R1*R2*(R1*R2+R1DR2)),RCSQ) / PI4
ELSEWHERE
  VZ = 0.0          ! no segment
ENDWHERE

VTZ  =  SUM ( VZ )
```

The first section of code defines a logical mask (SEGMENT) used as the argument to a WHERE block. The mask is an array of logical bits, each bit associated with a single processor, whose context is true or false. The WHERE block is a logical construct which controls the calculation of the relationships expressed inside the WHERE block for each processor depending on the value of its associated mask element (SEGMENT).

The first section of code inside the WHERE statement calculates the components of the vector, $\mathbf{r}_1$, and its magnitude. This vector is defined from the first endpoint of each segment $\mathbf{x}_1$ to the influenced point, $\mathbf{x}$. These geometric calculations are then performed for the second endpoint of each segment ($\mathbf{x}_2$). Note, that the coordinates of the first endpoint of each segment are associated with the same processor as the vortex segment, while the coordinates of the second endpoint of each segment are the coordinates of first endpoint of the neighboring processor (in the positive index sense). In this example, the new FORTRAN command CSHIFT is used to get, in parallel, the specified data from all processors one positive index away. The next section of code calculates the scalar product and the z-component of the vector product of the two vectors of each segment. The last section of code inside the first part of the WHERE block calculates the z-component of induced velocity for each vortex-segment. Since there is no segment associated with the last processor a value of zero is assigned to the z-component for this processor following the ELSEWHERE statement. The WHERE block is then terminated with the ENDWHERE statement. The final section of code uses the parallel reduction operator (SUM) to sum the z-components of induced velocity for each vortex-segment into one scalar value (VTZ). Note that all executable lines of code are parallel constructs with the exception of the explicit statement which sets the last element of the logical mask to a false value in the beginning of the code.

5 Code Implementation And Performance

5.1 Vector Implementation

Before describing the massively parallel implementation, the regular vectorized FORTRAN implementation as used in Ref. 1 is discussed. In the computationally intensive kernel portion of the FORTRAN code, the geometric relationships in Eq. 1 between each segment endpoint and a specified influenced point are precomputed and stored in memory to avoid the cost of re-computing the common geometric quantities. These common geometric quantities occur because of the common vortex segments of adjacent vortex boxes. For each influenced endpoint, the calculation is performed using a multiple DO LOOP structure over number of blades, spanwise position along the blade, and finally over the wake age of the influencing wake segment endpoints (the longest loop). Using the precomputed geometric quantities, the induced influence coefficients of the individual vortex segments are then computed (Eq. 1) using a similar loop structure. This loop structure over all of the vortex boxes is then used to calculate the vortex-box induced influence due to the appropriately combined individual segment influence coefficients and vortex-box circulation strengths. This process is repeated for each influenced endpoint. The performance of this code is about 140 MFlops (0.14 GFlops) on the Cray-2.

5.2 Parallel Implementation

In the full code of Ref. 1, the blade circulation solution requires the calculation of the wake influence at the blade collocation points (kernel 1) and cannot use the algorithm of Ref. 4 (kernel 2 programmed in LISP), while the wake convection portion can.

5.2.1 Kernel 1

The first kernel is a direct translation of the vectorized algorithm into FORTRAN 8X array extensions. Performance of this kernel applied to the wake convection process is compared with the vectorized performance on the Cray-2 and Cray-YMP processors are provided in Table I. The results are presented in terms of the speed-up relative to the Cray-2 performance for several combinations of number of wake elements and processors on the CM-2. Note that the 16k wake element case, run using 8k processors represents a vp ratio of two, while all other cases are for vp ratios of one. The actual Cray-2 times for the 8k and 16k cases were 47 and 186 seconds per time step, respectively. (The algorithm was also run on all 8 processors of the Cray-YMP with only a small increase on total time for all eight processors, 4 and 1 percent for the 8k and 16k cases respectively.) The estimated times, denoted by *, for the 64k wake elements and processors assume that the problem scales with the square of the number of elements on the Cray-2 and linearly on the CM-2.

Table I. Relative Computational Performance of Kernel 1 - Speed-up

Wake Elements	Cray-2	Cray-YMP	CM-2(8k)	CM-2(16k)	CM-2(64k)
8k	1.0	1.26	0.71	–	–
16k	1.0	1.44	1.00	1.41	–
64k	1.0	–	–	–	5.64*

Although the results are impressive considering the relative cost differences of a Cray-2 and 64k processor CM-2, even better performance is possible. The kernel 1 implementation requires broadcasting of the data associated with the point being influenced to all influencing processors which always requires long-range communication. This algorithm does not efficiently use the CM-2 processor communication architecture, but is required for the blade circulation solution. The second kernel uses the communication more efficiently and is preferable for the wake convection process. It was originally programmed (Ref. 4) in LISP and has now been implemented using the FORTRAN 8X array extensions.

5.2.2 Kernel 2

In the kernel 2 implementation, each processor is associated with a corner of the wake-lattice structure, shown conceptually in Fig. 3. Each processor represents two vortex segments, west and south. The combination of all of these corner influences is equivalent to the total of all of the vortex-box elements as performed in the vectorized FORTRAN code (the kernel 1 implementation). Special treatment is required for the western and southern edges of the lattice. The details of this kernel are described in Ref. 4. The key aspect of this implementation is that long-range communication becomes local.

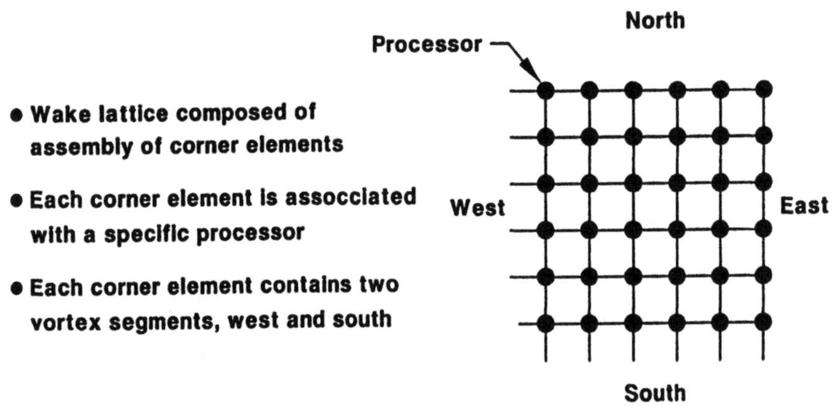

Figure 3: Computational Wake Lattice/Processor Structure

This second kernel performs the calculation in a different order than the kernel 1 implementation and requires additional memory, but has the same end result. Since all of the corner endpoints associated with each processor represent all of the points to be influenced, an algorithm which moves these coordinate values, assigned to different arrays, from processor to processor can be constructed. At each step, the induced influence due to the influencing corner elements can be calculated and summed into arrays representing the total induced influence. These total induced influence arrays are then moved along with the arrays representing the influenced endpoint coordinates from one processor to the next. At the end of one complete cycle, the movement to all influencing corner points produces the same total induced influence as kernel 1. The number of steps is the same, but the communication is local. Note that in kernel 2, there is redundant calculation of the relative geometry, unlike kernel 1. This redundancy could be removed, but would require increased local communication. Since any form of communication is significantly more costly than a few floating point calculations, this has not been done. Table II presents the relative computational performance of this kernel. The communication cost for the 16k wake element case using 16k processors was found to be 18 percent of the total time. The performance of this kernel is about 1700 MFlops (1.7 GFlops) on a 64k CM-2, including the redundant geometry calculations.

Table II. Relative Computational Performance of Kernel 2 - Speed-up

Wake Elements	Cray-2	Cray-YMP	CM-2(8k)	CM-2(16k)	CM-2(64k)
8k	1.0	1.26	1.33	–	–
16k	1.0	1.44	1.80	2.70	–
64k	1.0	–	–	–	10.8*

5.2.3 Full Code

As noted earlier, the full code uses both kernels, kernel 1 for the blade circulation solution and kernel 2 for the wake convection process. In addition, there are many small portions of the code that are not parallelizable or are parallelizable, but have not been implemented on the CM-2. These portions run on the front-end computer of the CM-2. For this problem, the impact of these portions of the code are small but noticeable, while they are insignificant on the Cray's. Performance comparisons for the full code, including the work done on the front-end are provided in Table III for the 16k wake element cases. The actual time for the Cray-2 for this case was 185 seconds per time step.

Table III. Relative Computational Performance of Full Code - Speed-up

Wake Elements	Cray-2	Cray-YMP	CM-2(8k)	CM-2(16k)	CM-2(64k)
16k	1.0	1.43	1.57	2.43	-
64k	1.0	-	-	-	9.73*

6 Concluding Remarks

A helicopter free-wake prediction code has been successfully implemented on the massively parallel Connection Machine (CM-2) using the proposed ANSI FORTRAN 8X array extensions. The performance of the full code on the CM-2 (65536 processors) compared with vector supercomputers has been shown to approach 10 times that of a Cray-2 processor and demonstrates the potential of massively parallel computer systems to provide significant computational performance for this type of problem.

7 Acknowledgments

The use of the Cray-2 and Cray-YMP of the Numerical Aeronautics Simulation Program (NAS) at NASA Ames Research Center is gratefully acknowledged.

8 References

1. Egolf, T.A.,"Helicopter Free Wake Prediction of Complex Wake Structures Under Blade-Vortex Interaction Operating Conditions," Proceedings of the 44th Annual Forum of the American Helicopter Society, June 16-18, 1988.
2. Bliss, D.B., L. Dadone, and D.A. Wachpress,"Rotor Wake Modeling For High Speed Flows," 43rd Annual Forum Proceedings of the American Helicopter Society, May 18-20, 1987.
3. Berry, J.D.,"A Multi-Element Vortex Lattice Method for Calculating the Geometry and effects of a Helicopter Rotor wake in Forward Flight," AIAA 26th Aerospace Sciences meeting, AIAA-88-0664, Reno NV, January 1988.
4. Egolf, T.A., and J.P. Massar,"Helicopter Free Wake Implementation On Advanced Computer Architectures," Presented at The 2nd International Conference on Basic Rotorcraft Research, University of Maryland, College Park, Maryland, February 16-18, 1988.
5. Wake, B.E. and T.A. Egolf,"Implementation of a Rotary-wing Three-Dimensional Navier-Stokes Solver on a Massively Parallel Computer," AIAA 89-1939, AIAA 9th Computational Fluid Dynamics Conference. June 13-15, 1989.
6. Egolf, T.A.,"Scientific Applications of the Connection Machine at the United Technologies Research Center," Proceedings of the Conference on Scientific Applications of the Connection Machine, NASA Ames Research Center, Sept. 12-14, 1988.
7. Hillis, W.D.,"The Connection Machine," *Scientific American*, pp. 108-115, June 1987.
8. Connection Machine Model CM-2 Technical Summary, Thinking Machines Technical Report HA87-4, April 1987.
9. Metcalf, M. and J. Reid, *Fortran 8x Explained*, Clarendon Press, Oxford, 1987.

The Influence of Massively Parallel Processing in Boundary Element Computations

B.L. Keshava Kumar[1], J.H. Kane[2], A.V. Srinivasan[3] and R.B. Wilson[4]

1 Graduate Student, Clarkson University
2 Associate Professor, Clarkson University
3 Manager, Applied Mechanics Research, United Technologies Research Center
4 Research Engineer, Pratt and Whitney, Commercial Engines

Abstract

Computational efficiencies in executing boundary element elasto-static problems on a massively parallel processor system (Connection Machine) are examined by making comparisons with the corresponding results obtained on a supercomputer (CRAY XMP). The problem chosen is characteristic of the size and type encountered in CPU intensive phases of a typical calculation i.e. calculation of integrals and the solution of the resulting matrix.

The surface elements used in these calculations are three dimensional, eight-noded quadrilateral, isoparametric patches. The solver is based on a Block LU decomposition procedure. The results indicate that significant reductions in CPU time can be obtained for those aspects of the procedure that are computationally intensive. Thus, parallel computation of boundary element analysis has the potential to enhance the ability of engineers to solve realistic large size problems economically.

Introduction

Computations based on boundary element analysis typically involve three major steps: (1) numerical calculation of integrals for each element/node pair (2) assembling a matrix of coefficients on the basis of prescribed boundary conditions and (3) numerical solution of the resulting equation system. A realistic problem from a typical jet engine application, for example, may involve as many as 1500-2000 elements leading to the calculation of several million integrals. Experience in conducting stress analysis of turbine blades has shown that it is simply uneconomical to run a 3D boundary element analysis of the problem size needed by the analyst even when these codes are exercised on a CRAY system. Further, with very optimistic assumptions on efficiency improvements which might be obtained by optimizing a three-dimensional elasto-static boundary element analysis for a serial computer with vector processing capabilities, the computing requirements for analysis of entire components would still be unrealistic as can be seen from fig. 1. The projections in the figure assume an improvement in integration time by a factor of ten and the ability to solve linear equations out-of-core with no more penalty than a factor of two over the vectorized LINPACK routine. This is a serious limitation, because it not only restricts the design space but also the possible extent of parametric studies needed to make decisions leading to an optimum component configuration.

An important feature of the boundary element approach is that the calculations of the kernel function/shape function integrals which form the system matrix are conducive to calculations that can be performed simultaneously for all nodes over a single element. This feature lends itself in a natural way to processing data in a parallel computing environment. The particular machine on which the computations reported in this paper were made is the Connection Machine with a capacity of 16384 physical processors. A full machine can have up to 65,536 physical processors, each containing 8 kilobytes of bit-addressable local memory. The Connection Machine is a SIMD (Single Instruction Multiple Data) machine in which each processor executes the same instruction set on a separate data packet. Data structures are spread in parallel across the processors and the machine can also be configured to operate in a virtual processor mode with a smaller memory but a larger number of processors than the full size machine. A high speed routing network establishes communications among the processors. A front-end provides the programming environment for the system allowing program storage there. More details pertaining to the computing environment are provided in the subsequent section.

The advantages of developing boundary element codes for the analysis of realistic problems in a parallel environment are established in this paper. While computational advantages appear possible in all the three steps outlined above, the focus in this paper is on the calculations, made in parallel, of integrals only. Computational efficiencies in regard to the solution of general systems of linear equations are cited from the results of Reference 1.

The Computational Environment

Figure 2 shows the computing environment which includes the front end and the massively parallel processors i.e. the CM-2 connected to its data vault. A front-end computer serves as a gateway to the Connection Machine system. It provides software development tools, software debugging tools, and a program execution environment familiar to the user. From the point of view of the user, the Connection Machine environment appears to be an extended version of the normal front-end environment. The front end also contains specialized hardware, called a Front-End Bus Interface which allows communication with the Connection Machine. The front end used in the present work is a VAX6320.

A brief description of the CM-2 is provided below and is drawn from Ref.2. The interested reader is directed to Ref.2 and the references cited there-in. The connection machine model CM-2 is an integrated system of hardware and software for data parallel computing. This type of computing associates one processor with each data element to exploit the natural computational parallelism inherent in many data-intensive problems. The system software is based upon the operating environment of the front-end computer.

The system implements data parallel programming constructs directly in hardware. The hardware consists of 65536 physical processors each with its own local memory. The machine architecture is classified as a SIMD type i.e. all the processors perform the same instruction on the data at the same time, although the individual data is different for each processor, and that there exists more than one path to access information from the other processors. Parallel data structures are spread across the data processors, with a single element stored in each processor memory. When parallel data structures have more then 65,356 data elements (the normal case), the hardware operates in a virtual processor mode, presenting the user with a larger number of processors, each with a correspondingly smaller memory. The ratio of the number of virtual processors to the number of actual physical processors is known as the Virtual Processor ratio (VP ratio).

A CM-2 system can be configured as 64K, 32K, or 16K data processors (here "K" stands for 1024) representing a full, half, or quarter of a machine. Each data processor has 64K bits (2048 32-bit words) of bit-addressable local memory and an arithmetic logic unit that can operate on variable-length operands. Each data processor can access its memory at a rate of at least 5 megabits per second. A fully configured CM-2 thus has 512 megabytes of memory that can be read or written at about 300 gigabits per second. When 64K processors are operating in parallel, each performing a 32-bit integer addition, the CM-2 parallel processing unit operates at about 2500 Mips. The CM-2 parallel processing unit also has an optional parallel floating point processor per 32 data processors that currently perform at 3500 MFlops (single precision) or 2500 MFlops (double precision) using the parallel machine language, PARIS. Even higher performance is possible by micro-coding these floating point processors (special cases have execeded 20 GFlops).

The CM-2 system provides two forms of communication within the parallel processing unit. The more general mechanism is known as the router, which allows any processor to communicate with any other processor. One may think of the router as allowing every processor to send a message to any other processor, with all messages being sent and delivered at the same time. Alternatively, one may think of the router as allowing every processor to access any memory location within the parallel processing unit, with all processors making memory accesses at the same time; in effect, the router allows the local memories of the data processors to be treated as a single large shared memory. The messages (data) may be of any length. The throughput of the router depends on message length and the pattern of accesses; typical values are 8 million to 250 million 32 bit-accesses per second. On the average, the time to send data through the router is equivalent to sixty add operations. The CM-2 also provides a more structured, somewhat faster communication mechanism called NEWS grid.

The underlying hardware used by these communication systems is based on a hyper-cube system and a local neighbor processor communication system at the chip level.

However, this is of no direct importance to the general user at the normal interface level, such as FORTRAN.

High speed transfers between peripheral devices and CM-2 memory take place through the CM-2 I/O system. All processors, in parallel, pass data to and from I/O buffers. The data is then moved between the buffers and the peripheral devices. CM-2 high-speed peripherals include the DataVault mass storage system and the CM-2 graphics display system.

Problem formulation

A three dimensional elasto-static problem modelled by a boundary element procedure can be shown to lead to the following integral equation (see References 3 and 4) :

$$C_{ij}(\xi)u_i(\xi) = \int_S \{t_i(x)G_{ij}(x,\xi) - F_{ij}(x,\xi)u_i(x)\}dS \tag{1}$$

where G_{ij} is the fundamental solution for displacement, F_{ij} is the fundamental solution for traction, S is the surface over which integration is being performed, u_i and t_i are the actual displacements and tractions on the surface S.

$$C_{ij}(\xi) = \frac{\delta_{ij}}{2} \textit{ on a smooth surface,}$$

$$G_{ij} = \frac{1}{16\pi\mu R(1-\nu)}\{R_{,i}R_{,j} + (3-4\nu)\delta_{ij}\}, \tag{2}$$

$$F_{ij} = \frac{-1}{8\pi R^2(1-\nu)}\{R_{,k}N_k[(1-2\nu)\delta_{ij} + 3R_{,i}R_{,j}] + (1-2\nu)[R_{,i}N_i - R_{,j}N_i]\}, \tag{3}$$

$$R = \left(\sqrt{(R_1^2 + R_2^2 + R_3^2)}\right), \tag{4}$$

$$R_{,j} = \frac{R_j}{R}, \tag{5}$$

$$R_j = x_j - d_j, \tag{6}$$

x_j and d_j are the coordinates of the load point and sample point respectively

In the numerical solution of the above equation the entire boundary S is modelled by quadratic isoparametric patches or elements. The integrals extend over the entire boundary, and the basic operation in their numerical evaluation is the integration of an element/node point pair. Thus in a typical sequential calculation a term such as $t_i(x)G_{ij}(x,\xi)$ is calculated for a given x and ξ pair where ξ is fixed and x varies over all of the Gauss points used in the integration over an element. The power of parallel computation is evident when we recognize that each processor calculates the integrals of eq.1 so that the entire calculation for one element and for all nodes can be performed *SIMULTANEOUSLY* and either written to the datavault or copied onto the front end for further use. Each processor on the CM is associated with a node point and the front end interacts with the CM for the element to be integrated. The front end computes the shape functions and Gauss point information for a given number of Gauss points and this information will be used by all the processors which are calculating integrals without subdividing the element (normal integrals). In case of singular or nearly singular integrals the element is subdivided into eight sub-elements and each sub-element is associated with a processor. These processors compute the Gauss point information and the shape functions necessary for computing the integrals. Thus the computations for singular or nearly singular integrals can be done in parallel with the normal integrals. These calculations are repeated for all the elements. Thus the saving in time is clearly evident. It is to be noted that there is no inter-processor communication in this scheme while the integrals are being computed.

Results and Conclusions

The parallel algorithm described above was implemented on the Connection Machine, first for the normal (non-subdivided) integrals and then for the complete integration, including the singular and near-singular cases. The time required to carry out the normal integration on the Connection Machine increases linearly with the number of surface elements (figure 3), compared to the quadratic variation found on the serial computers (including those with vector processing capability). While the addition of the singular and near-singular integration (figure 4) increased the computing time required, it did not alter the linear relationship between computing time and the model size. It should be noted that the code used in obtaining these results has not been optimized either in terms of computational efficiency or in terms of its Connection Machine implementation. Such optimization, together with anticipated enhancements to the Connection Machine compiler can be expected to lead to significant reductions in the times shown in this paper. Work reported by John Richardson (Reference 1) indicates very clearly that full, nonsymmetric matrices such as those produced in the boundary element method can be solved efficiently on the Connection Machine. Typically the system matrix for a three dimensional boundary element elastic analysis has a dimension which is very close to $9N$, where N is the number of surface elements in the model. Preliminary calculations based on Richardson's work indicate that such system matrices can be solved in less time than required for the integration process, up to model sizes of at least 3000 elements. In conclusion, it is clear from the results of this paper and of Reference 1 that the three-dimensional boundary element method can be effectively implemented in a massively parallel computing environment. This will lead to reduced analysis cost for problems which can be done using current serial computers. Much more importantly, however, feasible problem sizes will expand to accommodate realistic models of components.

REFERENCES

1. John L. Richardson; Block LU decomposition on the Connection Machine System, ISBEM 1989,East Hartford, CT 06108.
2. Wake, B.E., and Egolf, T.A.; Implementation of a Rotary-Wing Three-Dimensional Navier-Stokes Solver on a Massively Parallel Computer, AIAA Paper 89-1939-CP, AIAA 9th Computational Fluid Dynamics Conference, Buffalo, NY, 1989.
3. Bannerjee, P.K. and Butterfield, R.; Boundary Element Methods in Engineering Science, McGraw-Hill Book Company, 1981.
4. Brebbia C.A., Telles J.C.F. and Wrobel L.C.; Boundary Element Techniques, Springer-Verlag, 1984.

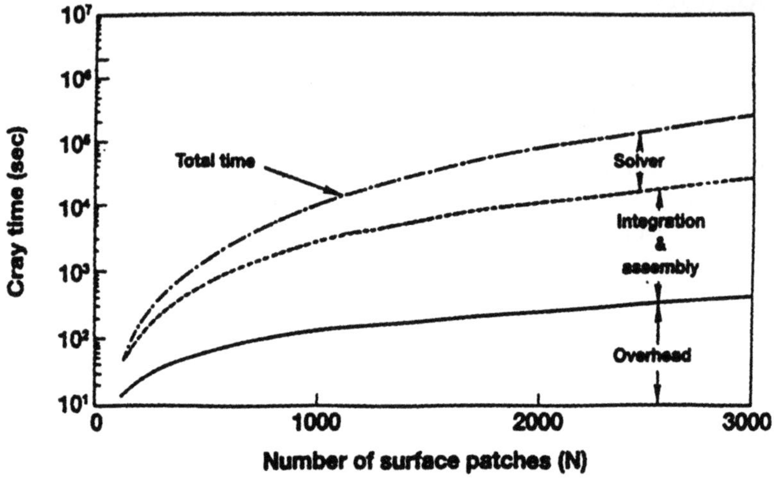

Fig. 1. Projected computing time (CRAY XM-P) requirements for optimized elastic three-dimensional boundary element program.

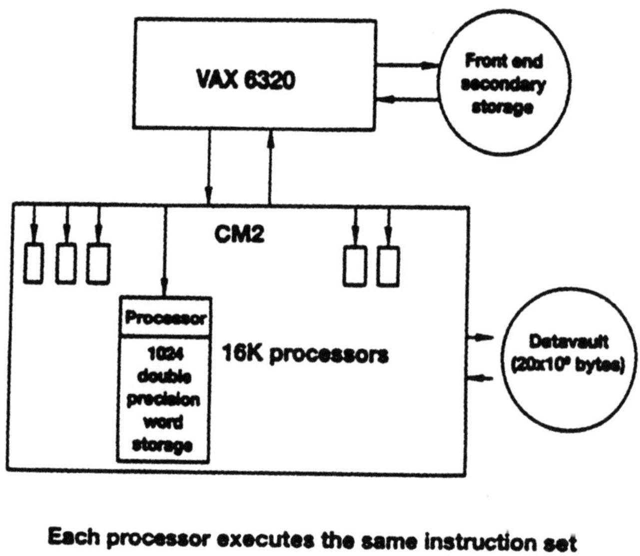

Fig. 2. Schematic of Connection Machine architecture.

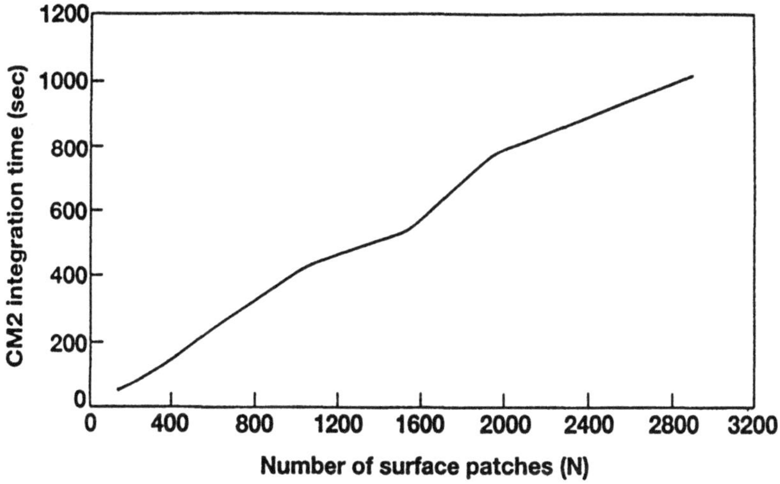

Fig. 3. Computing time required for normal integration on Connection Machine.

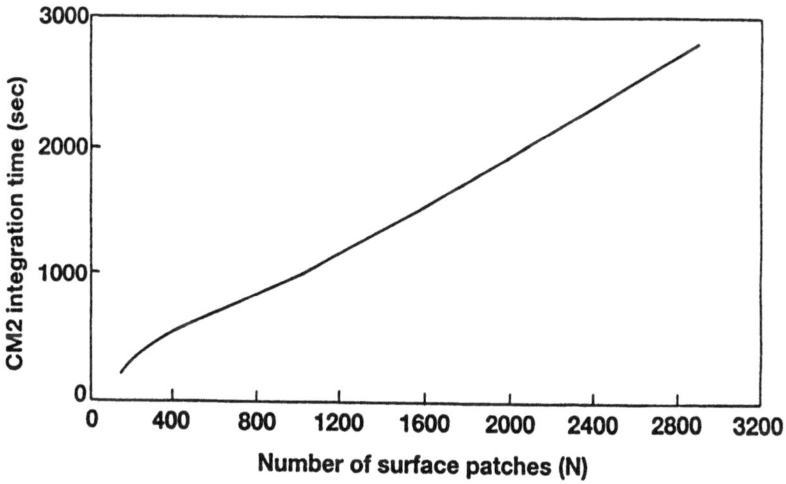

Fig. 4. Computing time required for full integration on Connection Machine.

High Speed Propeller Acoustics and Aerodynamics – A Boundary Element Approach

F. FARASSAT

NASA Langley Research Center
Hampton, Virginia

M. K. MYERS

The George Washington University (JIAFS)
Hampton, Virginia

M. H. DUNN

Planning Research Center
Hampton, Virginia

Summary

The Boundary Element Method (BEM) is applied in this paper to the problems of acoustics and aerodynamics of high speed propellers. The underlying theory is described based on the linearized Ffowcs Williams-Hawkings equation. The surface pressure on the blade is assumed unknown in the aerodynamic problem. It is obtained by solving a singular integral equation. The acoustic problem is then solved by moving the field point inside the fluid medium and evaluating some surface and line integrals. Thus the BEM provides a powerful technique in calculation of high speed propeller aerodynamics and acoustics.

Introduction

High speed propellers with advanced blade geometry are currently the subject of considerable interest because of their fuel efficiency. Both the problems of aerodynamics and acoustics of these propellers pose challenges to the theoreticians because of the unusual blade geometry and operating conditions. The classical propeller aerodynamic theory is not directly applicable. The acoustic theory can run into difficulties with mathematical and real singularities which require careful analysis. One is also faced with the problem of choosing, from many possible formulations, one which is suitable for implementation on a computer.

Almost from the beginning, the choice of working in the time or frequency domain must be made. In acoustics, the frequency domain has traditionally been preferred to the time domain with the exception of transient analysis. In linear unsteady aerodynamics, the frequency domain is also often preferred to the time domain because of the similarity of the governing equation to that of the acoustics. It is neither more difficult nor in any way less general to use time domain method for both the acoustics and the aerodynamics. The fundamental solution of the wave equation is particularly simple. In addition, one can easily overcome the problem of working with the additional dimension of time in interpreting the resulting formulas.

In this paper, an overview of the application of BEM to the acoustics and aerodynamics of high speed propellers is presented. The derivation of results is not stressed here. Reference to the papers where the derivations may be found is given. In the next section, the governing wave equation used in both acoustics and aerodynamics is given. In the following sections the acoustics and aerodynamic formulations are presented. The method of application of these results is also discussed.

The Governing Equation

Let $f(\vec{x},t)=0$ be the equation of a moving surface which in our problems is the blade surface. This function is defined so that f>0 outside the blade with unit outward normal $\vec{n}=\nabla f$. The blade is assumed thin so that small perturbation in fluid is generated by the blade motion and linear analysis can be applied. Let $p=c^2\tilde{\rho}$, where $\tilde{\rho}$ is the perturbation density ρ-ρ_o. Here ρ_o and c are the density and speed of sound of the undisturbed medium. Because of the assumption of linearity, p can be taken as pressure. The governing equation for both acoustics and aerodynamics is the Ffowcs Williams-Hawkings (FW-H) equation [1]:

$$\frac{1}{c^2}\frac{\partial^2 p}{\partial t^2}-\nabla^2 p=\Box^2 p=\frac{\partial}{\partial t}[\rho_o v_n \delta(f)]-\frac{\partial}{\partial x_i}[p n_i \delta(f)] \tag{1}$$

Here v_n is the local normal velocity on the blade surface. The Dirac Delta function is denoted $\delta(f)$.

The appearance of v_n and p on the right side of Eq. (1) suggests that the boundary conditions are included in the governing equation. Indeed this is so, and by doing so considerable algebraic manipulations are avoided. The idea behind the inclusion of the boundary conditions in a differential equation is explained in reference [2]. Equation (1) will be used as the basis for deriving an aerodynamic integral equation based on the acceleration potential. The acoustic pressure is then described by some integrals all whose integrands are known once the aerodynamic problem is solved. Our method is, thus, a boundary element method.

The Aerodynamic Problem

We can define the thickness and lifting problems of aerodynamics as follows. By taking the blade as infinitely thin, we can breakdown Eq. (1) as two equations as follows:

$$\Box^2 p_T = 2\frac{\partial}{\partial t}[\rho_o \bar{v}_n \delta(f)] \tag{2-a}$$

$$\Box^2 p_L = \frac{\partial}{\partial x_i}[\Delta p n_i \delta(f)] \tag{2-b}$$

where the subscripts T and L stand for thickness and lifting, respectively and we define

$$\bar{v}_n = \frac{1}{2}\left[(v_n)_{lower} + (v_n)_{upper}\right] \tag{3-a}$$

$$\Delta_p = (p)_{lower} - (p)_{upper} \tag{3-b}$$

with obvious reference to the lower and upper surfaces of the blade. This classification corresponds to the usual definition of the thickness and loading problems [3]. Of the two problems, the lifting case is more important and interesting since it leads to an integral equation. The thickness problem has a well-defined source depending on blade geometry and kinematics as seen from Eq. (2-a). We will not discuss this case further. See reference [4] for more details.

The lifting problem is described by Eq. (2-b). The fundamental solution of the wave equation is $\delta(g)/4\pi r$ where

$$g = \tau - t + r/c. \tag{4}$$

Here $r = |\vec{x} - \vec{y}|$ and $(\vec{x}, t)$ and $(\vec{y}, \tau)$ are the observer (field) and source space-time coordinates. Thus, the formal solution of Eq. (2-b) is

$$4\pi p(\vec{x}, t) = \frac{\partial}{\partial x_i}\int \frac{n_i}{r}\Delta p\,\delta(f)\delta(g)\,d\vec{y}\,d\tau \tag{5}$$

Using the following relation

$$\frac{\partial}{\partial x_i}\left[\frac{\delta(g)}{r}\right] = -\frac{1}{c}\frac{\partial}{\partial t}\left[\frac{\hat{r}_i\delta(g)}{r}\right] - \frac{\hat{r}_i\delta(g)}{r^2}, \tag{6}$$

where $\hat{r}_i = (x_i - y_i)/r$, Eq. (5) can be written as

$$4\pi p(\vec{x}, t) = -\frac{1}{c}\frac{\partial}{\partial t}\int \frac{\Delta p\cos\theta}{r}\delta(f)\delta(g)\,d\vec{y}\,d\tau - \int \frac{\Delta p\cos\theta}{r^2}\delta(f)\delta(g)\,d\vec{y}\,d\tau \tag{7}$$

In this equation, $\cos\theta = n_i\hat{r}_i$, i.e. θ is the angle between the radiation direction $\vec{\hat{r}}$ and the normal to the blade (pointing to the suction side of the airfoil section).

The various equivalent interpretations of the integrals in Eq. (7) are given in reference [5]. We take one based on the acoustic planform which is defined by $F(\vec{y}; \vec{x}, t) = [f(\vec{y}, \tau)]_{ret} = f(\vec{y}, t - r/c) = 0$. We call this surface Σ. Equation (7) is then written as

$$4\pi p(\vec{x}, t) = -\frac{1}{c}\frac{\partial}{\partial t}\int_{F=0} \frac{1}{r}[\Delta p\cos\theta]_{\tau*}\,d\Sigma - \int_{F=0} \frac{1}{r^2}[\Delta p\cos\theta]_{\tau*}\,d\Sigma \tag{8}$$

where τ^* is the emission time based on the solution of

$$\tau^* - t + \frac{1}{c}\left|\vec{x} - \vec{y}(\vec{\eta}, \tau^*)\right| = 0 \tag{9}$$

In Eq. (9) $\vec{\eta}$ is a Lagrangian variable describing points on the blade. When the field point is moved to the surface, it can be shown that Eq. (8) does not lead to a useful integral equation [6]. Another approach must be sought.

Let $\phi(\vec{x}, t)$ be the velocity potential. The following relation holds between p and ϕ:

$$p = -\rho_0 \frac{\partial \phi}{\partial t} \tag{10}$$

Using Eq. (10) on the left side of Eq. (9), integrating with respect to time from $-\infty$ to t and finally taking derivative of both sides of the resulting equation in the direction normal to the airfoil gives

$$4\pi\rho_0 v_n(\vec{x}, t) = \frac{1}{c}\frac{\partial}{\partial n}\int_{F=0}\frac{1}{r}\left[\Delta p \cos\theta\right]_{\tau'^*} d\Sigma + \frac{\partial}{\partial n}\int_{-\infty}^{t}\int_{F=0}\frac{1}{r^2}\left[\Delta p \cos\theta\right]_{\tau'^*} d\Sigma\, dt' \tag{11}$$

Here τ'^* is solution of Eq. (9) with t' replacing t and v_n is the local normal velocity of the blade mean surface. The right side of Eq. (11) must be interpreted as a limit when $\vec{x}$ approaches the blade surface.

Equation (11) is now a legitimate integral equation. Following the results of reference [5], we can write Eq. (11) in several different equivalent forms depending on local airfoil Mach number which may be subsonic or transonic/supersonic. We mention here that for numerical computation, the blade can be divided into panels and assuming Δp is uniform on each panel, Eq. (11) can be written as

$$4\pi\rho_0 v_n\left(\vec{x}_j, t\right) = \sum_i \Delta p_i \left\{\frac{1}{c}\frac{\partial}{\partial n}\int_{F_i=0}\frac{1}{r}\left[\cos\theta\right]_{\tau^*} d\Sigma \right.$$

$$\left. + \frac{\partial}{\partial n}\int_{-\infty}^{t}\int_{F_i=0}\frac{1}{r^2}\left[\cos\theta\right]_{\tau'^*} d\Sigma\, dt' \right\} = 4\pi\rho_0 \sum_i \Delta p_i K_{ij} \tag{12}$$

The function K_{ij} is known as the aerodynamic kernel function. It is the normal velocity induced at the jth panel by the i-th panel.

The kernel function can be found analytically for a panel in rectilinear uniform motion [7]. For general motion of a panel, it can be shown that the normal derivative of the integrals in Eq. (12) can be found analytically [6] resulting in both surface and line integrals over the edge of the acoustic planform. These integrals must then be evaluated numerically. The Galerkin method can then be used as in Reference [7]. The Kutta condition then can be applied to the basis functions used in Galerkin method for airfoil sections moving at subsonic speeds.

The Acoustic Problem

The acoustic problem is relatively easier since the solution of equation (1) can be written in various forms using the fundamental solution of the wave equation. It is necessary however that two separate formulations be used for the subsonic and transonic/supersonic regions of the blade. For subsonic regions, from the formal solution of Eq. (1) and using Eq. (6), the acoustic pressure can be written as

$$4\pi p(\vec{x},t) = \frac{1}{c}\frac{\partial}{\partial t}\int_{f=0}\left[\frac{\rho_o v_n + p\cos\theta}{r(1-M_r)}\right]_{\tau^*} dS + \int_{f=0}\left[\frac{p\cos\theta}{r^2(1-M_r)}\right]_{\tau^*} dS \tag{13}$$

Here M_r is the Mach number in the radiation direction.

For numerical computation, it is preferable that the time derivative in Eq. (13) be taken under the integral. From Eq. (9), we can show that

$$\frac{\partial}{\partial t} = \left[\frac{1}{1-M_r}\frac{\partial}{\partial \tau}\right]_{\tau^*} \tag{14}$$

Using this equation in Eq. (13), we get

$$4\pi p(\vec{x},t) = \int_{f=0}\left\{\frac{1}{c}\left[\frac{1}{1-M_r}\frac{\partial}{\partial \tau}\frac{\rho_o v_n + p\cos\theta}{r(1-M_r)}\right] + \left[\frac{p\cos\theta}{r^2(1-M_r)}\right]\right\}_{\tau^*} dS \tag{15}$$

This equation is highly suitable for numerical calculations when the blade section Mach number is subsonic. Note that Eq. (15) develops singularities when $M_r \to 1$. This only happens when the blade section moves supersonically.

The transonic/supersonic acoustic problem involves considerably more manipulation. Although it is possible to manipulate the formal solution of Eq. (2) to get the required result [8], a more direct way would be through the manipulation of the source terms of Eq. (2). The source terms of this equation can be written as follows using results from generalized function theory and differential geometry [9]. We first give some definitions. Assume that the leading and trailing edges are sharp and that these edges are described by implicit equations $f = \tilde{f} = 0$, i.e. the intersection of two surfaces. Also let the inward unit geodesic normal to the edge curve be denoted by $\vec{\nu}$. The geodesic normal is the vector tangent to the surface and normal to the edge. Note that at the LE and TE there are two geodesic normals at each point corresponding to the upper and lower surfaces, respectively. Let $v_\nu = \vec{v}\cdot\vec{\nu}$ where $\vec{v}$ is the local velocity of the edge. We define

$$\psi_1 = \rho_o\left(\dot{v}_n - \vec{v}_t\cdot\nabla v_n\right) - 2H_f p \tag{16-a}$$

$$\psi_2 = -\rho_o v_n^2 - p \tag{16-b}$$

$$\psi_3 = -\rho_o (v_n v_\upsilon)_{upper} - \rho_o (v_n v_\upsilon)_{lower} \tag{16-c}$$

Here $\dot{v}_n$ is rate of change of v_n for an observer moving with the blade and $\vec{v}_t$ is the tangential component of blade velocity. Also H_f is the mean curvature of the blade surface.

Equation (2) can now be written as

$$\Box^2 p = \psi_1 \delta(f) + \hat{\psi}_2 \delta'(f) + \psi_3 \delta(f)\delta(\tilde{f}) \tag{17}$$

where $\hat{\Psi}_2$ stands for the restriction of Ψ_2 to the surface f=0 [9]. We note that the first two terms on the right of Eq. (17) are surface sources while the last term is a line source on the LE and TE. The solution of Eq. (17) can be written as [9]

$$4\pi p(\vec{x},t) = \int_{F=0} \frac{1}{r}[Q_F]_{ret} d\Sigma + \int_{F=0} \frac{1}{r^2}[Q_N]_{ret} d\Sigma + \int_{\substack{F=0\\ \tilde{F}=0}} \frac{1}{r}[Q_o]_{ret} d\gamma \tag{18}$$

where Q_F,, Q_N and Q_o are complicated expressions involving terms in Ψ_1, Ψ_2, Ψ_3 and other geometric parameters [8]. In this equation $\tilde{F} = [\tilde{f}]_{ret}$ and $d\gamma$ is the element of length of the edge curve of the acoustic planform generated by the blade planform.

Both equations (15) and (18) have been coded for prediction of the noise of advanced propellers. The blade is divided into panels and a decision is made in the code which of the two equations should be used for each panel. The procedure is carried out entirely in the time domain and the result of computation is the acoustic pressure signature. The process is explained fully in reference [10]. It must be mentioned that the blade surface pressure on the right of Eqs. (15) and (18) is often supplied to the acoustic code from nonlinear aerodynamic codes rather than from linear analysis.

Concluding Remarks

The boundary element method gives a unified approach to both the aerodynamics and acoustics of high speed propellers. This method is generally much faster to execute on a computer than nonlinear analysis based on Euler's equation, for example, particularly when unsteady blade surface pressure is involved. The method as described here applies to thin blades because of the assumption of linearity. It can be extended to the nonlinear regime by a field panel method using unknown quadrupole sources in the volume around the blade. In this case, the full Ffowcs Williams-Hawkings equation should be used in place of Eq. (2). Such a method has not been programmed yet but shows good promise based on past experience.

One of the chief advantages of BEM as applied here, i.e., using a frame fixed to the undisturbed fluid medium, is the simplicity of the fundamental solution of the wave equation. There is no restriction on the motion of the moving body and no assumption in the acoustic results concerning the position of the observer in the near or far field. In fact, motion of the observer can also easily be included in numerical computations [10].

This paper has shown that much analytic work can be done in application of the BEM to the aerodynamics and acoustics of high speed propellers. In fact, it should be clear that a change of formulation to one of the equivalent forms can save considerable computation time as shown in the case of equations (15) and (18), for low and high speed motions of a panel on the blade, respectively.

References

1. J. E. Ffowcs Williams and D. L. Hawkings, "Sound Generation by Turbulence and Surfaces in Arbitrary Motion," Philosophical Transactions of the Royal Society of London, Vol. A264, 1969, 321-342.

2. F. Farassat, "Theory of Noise Generation from Moving Bodies with an Application to Helicopter Rotors," NASA TR R-451, 1975.

3. H. Ashley and M. Landahl, "Aerodynamics of Wings and Bodies," Addison-Wesley, 1965 (also published by Dover publications).

4. L. Long, "The Compressible Aerodynamics of Rotating Blades Based on An Acoustic Formulation," NASA TP-2197, December 1983.

5. F. Farassat, "Linear Acoustic Formulas for Calculation of Rotating Blade Noise, "AIAA Journal, Vol. 19, No. 9, 1981, 1122-1130.

6. F. Farassat and M. K. Myers, "Aerodynamics Via Acoustics: Application of Acoustic Formulas for Aerodynamic Calculations," AIAA 86-1877, 1986.

7. R. Milliken, "A New Lifting Surface Method: An Acoustic Approach," M. Thesis, The George Washington University, 1986.

8. F. Farassat, "Theoretical Analysis of Linearized Acoustics and Aerodynamics of Advanced Supersonic Propellers, AGARD-CP-366, (10), 1-15, 1985.

9. F. Farassat, "Lectures on the Aeroacoustics of Rotating Blades in Time Domain," Lectures delivered in the Dept. of Mechanics and Aeronautics, University of Rome "La Sapienza," July 1989.

10. F. Farassat, S. L. Padula and M. H. Dunn, "Advanced Turboprop Noise Prediction Based on Recent Theoretical Results," Journal of Sound and Vibration, Vol. 119(1), 1987, 53-79.

A Unified Approach for Aerodynamics and Aeroacoustics of Rotors in Compressible Potential Flows

M. Gennaretti and L. Morino
Università di Roma "La Sapienza", Rome, Italy

The objective of this work is to present a new unified approach to both aerodynamics and aeroacoustics, with applications to hovering helicopter rotors and propellers in subsonic potential flows. This approach consists of solving the velocity-potential equation by a boundary-integral-equation technique and then applying Bernoulli's theorem. The solution for the velocity potential is obtained on the basis of a formulation introduced by Morino, Freedman, Deutsch and Sipcic (1985) and used by Morino, Bharadvaj, Freedman and Tseng (1989) in order to analyze the flow around a hovering helicopter rotor. In this work the numerical implementation has been refined, partially removing the low-Mach-number approximation used by Morino, Bharadvaj, Freedman and Tseng (1989). In addition, a novel approach has been used for the aeroacoustic problem. Specifically, since the late sixties, the work in this field is primarily based on the equations introduced by Ffowcs Williams and Hawkings (1969). These allow one to evaluate the aeroacoustic pressure, once the pressure distribution on the body surface and Lighthill's stress tensor in the field have been evaluate by an independent process (theoretical or experimental). With the present methodology, a single formulation is used to evaluate the velocity potential, on the body and in the field (and then the pressure in the field using Bernoulli's theorem).

1. Formulation for Aerodynamics and Aeroacoustics

This work deals with fluids which are compressible, inviscid and non-conducting. Starting from the equations of conservation of mass, momentum and energy, adding the hypotheses of initial isentropicity and irrotationality, we obtain that the flow is isentropic and irrotational at all times, except for those points that come in contact with solid boundaries in the field (wake). The above hypotheses, in a frame of reference fixed with undisturbed air (air frame), yield the following wave equation for velocity potential[1]

$$\nabla^2_\xi \phi - \frac{1}{a^2_\infty}\frac{\partial^2 \phi}{\partial \tau^2} = \chi \qquad \boldsymbol{\xi} \text{ outside } S(\tau) \qquad (1)$$

where S indicates a surface outside which the flow is potential (*i.e.*, any surface surrounding the body and the wake) and χ represents the non-linear terms (which are relevant only in the transonic regime). The following boundary conditions complete the problem: (*i*) $\partial\phi/\partial n = \mathbf{v}_B \cdot \mathbf{n}$ on S_B; (*ii*) $\Delta(\partial\phi/\partial n) = 0$ on the wake surface S_W; (*iii*) $\Delta\phi$ =constant, following a wake point; (*iv*) $\phi = 0$ at infinity; and (*v*) $\Delta\phi_W = \Delta\phi_B$ at trailing edge (see Morino, Bharadvaj, Freedman and Tseng, 1989).

Equation 1 may be recast as a boundary integral equation in several ways; in the present formulation the boundary integral representation is first obtained in the air frame, and then is rewritten in a moving frame of reference (body frame). In this paper we are concerned with propellers and hovering helicopter rotors in subsonic flows. Therefore, we may assume that the body frame moves in uniform helicoidal motion, *i.e.*, its constant velocity $\mathbf{v}_0$ is parallel to its constant angular velocity $\boldsymbol{\omega}$, and the transformation relating the frames of reference is given by $\boldsymbol{\xi}(\mathbf{x}, t) = \mathbf{U}(t)\,\mathbf{x} + \mathbf{v}_0 t$, $\tau = t$, where $\mathbf{U}(t)$ is a uniform-rotation matrix. Note that $d\mathbf{U}/dt = \mathbf{AU}$, where $\mathbf{A}$ is such that, for any $\mathbf{z}$, $\mathbf{Az} = \boldsymbol{\omega} \times \mathbf{z}$. If the z-axis is chosen to be in the direction of $\mathbf{v}_0$

[1] *In this work, Greek letters are used for the air frame and Latin for the body frame.*

and ω, we have $\mathbf{v}_0 = v_0\mathbf{k}$, $\boldsymbol{\omega} = \omega\mathbf{k}$,

$$\mathbf{U}(t) = \begin{pmatrix} \cos\omega t & -\sin\omega t & 0 \\ \sin\omega t & \cos\omega t & 0 \\ 0 & 0 & 1 \end{pmatrix} \quad \text{and} \quad \mathbf{A} = \begin{pmatrix} 0 & -\omega & 0 \\ \omega & 0 & 0 \\ 0 & 0 & 0 \end{pmatrix} \tag{2}$$

Next, consider a surface S surrounding body and wake, and rigidly connected with the body frame of reference. The integral formulation for Eq. 1, for subsonic flows and for $\chi = 0$, is given by (Morino, Bharadvaj, Freedman and Tseng, 1989)

$$E(\mathbf{x}_*)\phi(\mathbf{x}_*, t_*) = \oint\!\!\!\oint_S \left[\frac{-1}{4\pi\hat{\varrho}}\frac{\partial\phi}{\partial\check{n}} - \frac{\partial}{\partial\hat{n}}\left(\frac{-1}{4\pi\hat{\varrho}}\right)\phi + \frac{-1}{4\pi\hat{\varrho}}\frac{\partial\phi}{\partial t}\left(\frac{\partial\theta}{\partial\hat{n}} + 2\frac{\mathbf{M}\cdot\mathbf{n}}{a_\infty}\right)\right]^\theta dS \tag{3}$$

where $\partial/\partial\check{n} = \partial/\partial n - \mathbf{M}\cdot\mathbf{n}\ \mathbf{M}\cdot\nabla$ and $\partial/\partial\hat{n} = -\mathbf{n}_*\cdot\nabla_* + \mathbf{M}\cdot\mathbf{n}\ \mathbf{M}_*\cdot\nabla_*$, with $\mathbf{M} = (\boldsymbol{\omega}\times\boldsymbol{\xi} + \mathbf{v}_0)/a_\infty$ and $\mathbf{M}_* = (\boldsymbol{\omega}\times\boldsymbol{\xi}_* + \mathbf{v}_0)/a_\infty$, whereas $[...]^\theta$ indicates evaluation at $t = t_* - \theta$. In Eq. 3 the time-delay $\theta \geq 0$ is evaluated as the root of

$$\left(a_\infty^2 - v_0^2\right)\theta^2 + 2\mathbf{x}^T\left[\mathbf{U}(\theta) - \mathbf{I}\right]\mathbf{x}_* + 2(\mathbf{x} - \mathbf{x}_*)^T\mathbf{v}_0\theta - r^2 = 0 \tag{4}$$

where $r = |\mathbf{x} - \mathbf{x}_*|$, and the $\hat{\varrho}$ is defined as

$$\hat{\varrho} = \varrho|1 + M_\varrho| = (1 + M_0^2)a_\infty\theta + \frac{1}{a_\infty}\mathbf{x}^T\mathbf{A}\mathbf{U}(\theta)\mathbf{x}_* + \mathbf{M}_0^T(\mathbf{x} - \mathbf{x}_*) \tag{5}$$

where $M_0 = v_0/a_\infty$, $\varrho = |\boldsymbol{\xi} - \boldsymbol{\xi}_*|$, $M_\varrho = \mathbf{M}\cdot\boldsymbol{\varrho}/\varrho$ and the matrix $\mathbf{A}$ is defined above.
Note that $E(\mathbf{x}_*) = 1$ if $\mathbf{x}_*$ is outside S, $E(\mathbf{x}_*) = 0$ if $\mathbf{x}_*$ is inside S, and $E(\mathbf{x}_*) = 1/2$ if $\mathbf{x}_*$ is a smooth point of S. Thus, if $\mathbf{x}_*$ is in the fluid region, Eq. 3 is an integral representation of $\phi(\mathbf{x}_*, t_*)$, in terms of ϕ and $\partial\phi/\partial n$ on S. On the other hand, if $\mathbf{x}_*$ is on the surface S, Eq. 3 is a compatibility condition between the values of ϕ and $\partial\phi/\partial n$ on S, *i.e.*, an integral equation of the second kind that may be used to solve for ϕ on S if $\partial\phi/\partial n$ on S is known. Equation 3 may be used to study helicopter rotors in hover, ascent and descent, as well as propellers in axial flow. In this case, the problem is made more complicated by the presence of the wake. This is dealt with by letting the two sides of the surface S approach the wake surface. As a results the wake is modeled as doublet layer (for compressible flow, there exists a source layer which is typically negligible, see Morino, Bharadvaj, Freedman and Tseng, 1989, for details).
In order to solve the aerodynamic and aeroacoustic problems, the first step is to solve the integral equation, Eq. 3, for the unknown ϕ on the surface S. Once the potential on the body surface has been evaluated, it is possible to determine both the velocity and the pressure anywhere, on the surface of the body (aerodynamics) or in the field (aeroacoustics): the velocity is obtained by taking the gradient of potential, the pressure by applying Bernoulli's theorem.
As mentioned above, a simplified formulation was used by Morino, Bharadvaj, Freedman and Tseng (1989). This is obtained by observing that the matrix $\mathbf{U}(\theta)$ may be expressed as $\mathbf{U}(\theta) = \sum \mathbf{A}^n\theta^n/n!$ and neglecting the terms of order θ^3. This yields $\mathbf{U}(\theta) = \mathbf{I} + \mathbf{A}\theta + \mathbf{A}^2\theta^2/2$ and consequently simplified expressions for θ and $\hat{\varrho}$ result. Here, the above expressions are used as an initial guess for the evaluation of θ and $\hat{\varrho}$ (see Gennaretti, 1989, for details). The discretization used is still a zeroth-order boundary-element formulation.

2. Numerical Results

In this section some numerical results are presented for a hovering rotor and a propeller. The results are based on the work of Gennaretti (1989), where additional results are presented. The flow is assumed to be subsonic, and a fixed-wake analysis is used. Both aerodynamic and aeroacoustic results have been obtained for cases that are steady-state in the rotor frame of reference.

First, consider the numerical results obtained for subsonic flows around a hovering rotor for which experimental results are available from Caradonna and Tung (1981). The rotor has a NACA 0012 section, radius of 45 *in*, constant chord of 7.5 *in* and no twist. The case considered here is that with a collective pitch of 8°. In obtaining the results, the wake geometry is prescribed to be close to that measured experimentally by Caradonna and Tung (1981) (see Morino, Bharadvaj, Freedman and Tseng, 1989, for details). For all the aerodynamic results presented here, a three-spiral wake has been considered: a previous convergence analysis on the number of wake spirals, showed that the three-spirals solution is close to convergence. The values of the sectional lift coefficient, C_l, and of the pressure coefficient, c_p, at different blade sections, have been computed for two angular speeds: 1,250 rpm ($M_{tip} = 0.439$) and 2,050 rpm ($M_{tip} = 0.727$)[2]. Figures 1a and 1b present the sectional lift coefficient as a function of r/R (for $M_{tip} = 0.439$ and 0.727, respectively), with comparisons to the experimental results of Caradonna and Tung (1981). Figures from 2a to 2d, show the pressure coefficients for $M_{tip} = 0.439$, at four different radial sections, again with comparison to the experimental results of Caradonna and Tung (1981). Similar results are presented in Figs. from 3a to 3d, for $M_{tip} = 0.727$. It may be noted that the difference between numerical and experimental results is similar in both cases. This indicates that the effects of the compressibility are captured correctly by the proposed algorithm, and that the difference between the numerical and experimental results is probably due to the use of a prescribed-wake analysis (see Morino, Bharadvaj, Freedman and Tseng, 1989, for an analysis of this effect). In order to emphasize this point, the difference between the numerical and the experimental results are presented in Figs. 4a to 4d: the agreement between experimental and numerical results is good, especially at the section near the tip, where the effects of the compressibility are more pronounced.

Next, consider the aeroacoustic analysis. We found that the number of spirals required for convergence in the aeroacoustic analysis is much higher than that for the aerodynamic calculations (*i.e.*, several hundred spirals for the acoustic analysis *vs.* three for the aerodynamic one). In order to reduce the computational time, the following algorithm was used: for the first 100 spirals, the surface integral in Eq. 3, has been computed by the trapezoidal rule, whereas the rest has been computed analytically by approximating the mean value of the integrand as a third-degree polynomial of $1/\alpha$, where α is the azimuthal angle along the wake. The case considered is that of a three-bladed general aviation propeller, considered by Farassat and Succi (1980). The rotor radius is 1.295 m, (for the variation of the thickness and of the angle of attack with r/R, see Farassat and Succi, 1980). The speed of rotation is 1,760 rpm and the advancing velocity is 55 m/s, which correspond to M_{tip}=0.71. Figure 5a presents the pressure *vs.* time at a point in the plane of propeller, 7.28 m from the axis of rotation. Also included in Fig. 5a are the experimental results of Magliozzi (1977), as well as the numerical results by Farassat and Succi (1980), obtained by using a code based on the Ffowcs Williams and Hawkings (1969) equations, It should be pointed out that the phase angle between the our results and those of Farassat and Succi (1980), has been chosen arbitrarily. Nonetheless the result obtained appears to be very satisfactory. Finally, for the same test case, Fig. 5b shows the contributions due to source and doublet integrals (thickness and loading noises), as evaluated by the present analysis and a comparison with thickness noise computed by Succi (1979).

3. Conclusions

A unified integral formulation for the aerodynamic and aeroacoustic analyses of unsteady, compressible flows around rotors in uniform helicoidal motion has been presented. The emphasis has been on loads on blade and, above all, on a new approach for aeroacoustic analysis. Good numerical results have been obtained.

[2] *The sectional lift coefficient is equal to the sectional lift divided by the product of the tip dynamic pressure times the chord, the pressure coefficient is obtained with reference to the local dynamic pressure.*

Current work on the aerodynamic formulation includes higher-order discretization and free-wake analysis, as well as application to forward flight. On the aeroacoustic side, the analysis requires improvements in the evaluation of the contribution of the wake. Several approaches are currently under investigation. One of them is outlined in Appendix A, where the relationship of the present formulation and that of Ffowcs Williams and Hawkings (1969) is also discussed.

References

Caradonna, F. X., Tung, C., (1981): "Experimental and Analytical Studies of a Model Helicopter Rotor in Hover," USAAVRADCOM TR-81-A-23.

Farassat, F., (1977): "Discontinuities in Aerodynamics and Aeroacoustics: the Concept and Applications of Generalized Derivatives," Journal of Sound and Vibration, Vol. 55, pp. 165-193.

Farassat, F., Succi, G. P., (1980): "A Review of Propeller Noise Prediction Technology with Emphasis on Two Current Methods for Time Domain Calculation," Journal of Sound and Vibration, Vol. 71, pp. 399-419.

Ffowcs Williams, J. E., Hawkings, D. L., (1969): "Sound Generated by Turbulence and Surfaces in Arbitrary Motion," Philosophical Transaction of the Royal Society, A 264, pp. 321-342.

Gennaretti, M., (1989): "Metodo delle Equazioni Integrali al Contorno per Analisi Aerodinamica e Aeroacustica di Rotori in Flussi Potenziali, Compressibili, Non Stazionari," Tesi di Laurea, Università di Roma "La Sapienza", Rome, Italy.

Magliozzi, D., (1977): "The Influence of Forward Flight on Propeller Noise," NASA CR145105.

Morino, L., Bharadvaj, B. K., Freedman, M. I., and Tseng, K., (1989): "Boundary Integral Equation for Wave Equation with Moving Boundary and Applications to Compressible Potential Aerodynamics of Airplanes and Helicopters," Computational Mechanics, Vol. 4, pp. 231-243.

Morino, L., Freedman, M. I., Deutsch, D. J., and Sipcic, S. R., (1985): "An Integral Equation Method for Compressible Potential Flows in an Arbitrary Frame of Reference," Ed.: L. Morino, Computational Methods in Potential Aerodynamics, Computational Mechanics Publications, Southampton, UK.

Succi, G. P., (1979): "Design of Quiet Efficient Propellers," S. A. E. Paper 790584.

A. Relationship with the FW-H Formulation

In this appendix we discuss the relationship between the present formulation and that of Ffowcs Williams and Hawkings (1969). Starting from the continuity equation $\partial\rho/\partial t + \partial(\rho v_j)/\partial x_j = 0$ and the dynamic equilibrium equation $\partial(\rho v_i)/\partial t + \partial(\rho v_i v_j)/\partial x_j = -\partial p/\partial x_i + \partial V_{ij}/\partial x_j$, where V_{ij} is the viscous stress tensor, one obtains the following equation for the acoustic pressure $p' = a_\infty^2(\rho - \rho_\infty)$

$$\Box^2 p' = \frac{1}{a_\infty{}^2}\frac{\partial^2 p'}{\partial t^2} - \nabla^2 p' = \frac{\partial^2 T_{ij}}{\partial x_i \partial x_j} \tag{6}$$

where $T_{ij} = (p - p')\delta_{ij} + \rho v_i v_j - V_{ij}$ known as the Lighthill stress tensor. Next, introduce the function $\tilde{p}' = Ep'$ (with E defined above). Combining with Eq. 6, one obtains

$$\Box^2 \tilde{p}' = E\frac{\partial^2 T_{ij}}{\partial x_i \partial x_j} + \frac{1}{a_\infty}\left[\frac{\partial E}{\partial t}\frac{\partial p'}{\partial t} + \frac{\partial}{\partial t}\left(p'\frac{\partial E}{\partial t}\right)\right] - \left[\frac{\partial E}{\partial x_i}\frac{\partial p'}{\partial x_i} + \frac{\partial}{\partial x_i}\left(p'\frac{\partial E}{\partial x_i}\right)\right] \tag{7}$$

Noting that

$$\frac{\partial^2 (ET_{ij})}{\partial x_i \partial x_j} = E\frac{\partial^2 T_{ij}}{\partial x_i \partial x_j} + \frac{\partial E}{\partial x_i}\frac{\partial T_{ij}}{\partial x_j} + \frac{\partial}{\partial x_i}\left(T_{ij}\frac{\partial E}{\partial x_j}\right) \tag{8}$$

and combining with the Eq. 7, one obtains

$$\Box^2 \hat{p}' = \frac{\partial^2 (ET_{ij})}{\partial x_i \partial x_j} + \frac{1}{a_\infty}\left[\frac{\partial E}{\partial t}\frac{\partial p'}{\partial t} + \frac{\partial}{\partial t}\left(p'\frac{\partial E}{\partial t}\right)\right] - \frac{\partial E}{\partial x_i}\frac{\partial Q_{ij}}{\partial x_j} - \frac{\partial}{\partial x_i}\left(Q_{ij}\frac{\partial E}{\partial x_j}\right) \tag{9}$$

where $Q_{ij} = p\delta_{ij} + \rho v_i v_j - V_{ij}$. From this equation, as shown by Farassat (1977), it is possible to derive the Ffowcs Williams and Hawkings equations. Note that Eq. 7, with p' replaced by ϕ (and $\partial^2 T_{ij}/\partial x_i \partial x_j$ replaced by χ), is the starting point to obtain Eq. 3. Thus the two formulations are closely related. Note that the small-perturbation formulations based on Eqs. 7 and 9 are not equivalent (although Eqs. 7 and 9 are). For, in the process of linearizing Eqs. 7 and 9 one neglects the terms $\partial^2 T_{ij}/\partial x_i \partial x_j$ and $\partial^2(ET_{ij})/\partial x_i \partial x_j$, respectively, which are related by Eq. 8. Finally, the solution of Eq. 3, with ϕ replaced by p' (*i.e.*, the integral formulation obtained from Eq. 7), is now under investigation. This approach has the advantage that, within a small-perturbation approximation, the wake contribution is negligible.

Acknowledgments: This work was partially supported by the Army Research Office (Contract No. DAAL-03-87-K-0022 to Boston University).

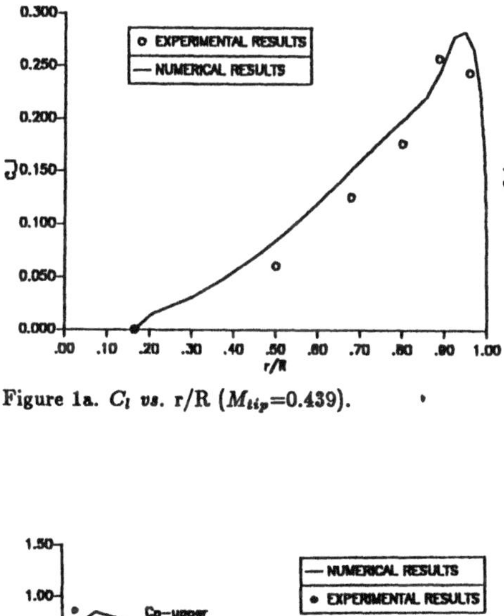

Figure 1a. C_l *vs.* r/R (M_{tip}=0.439).

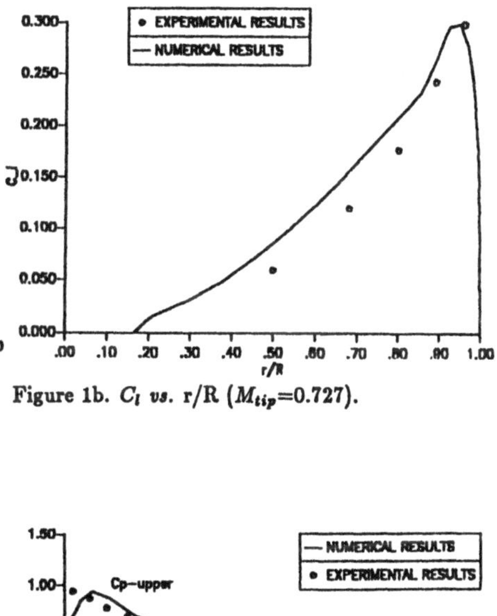

Figure 1b. C_l *vs.* r/R (M_{tip}=0.727).

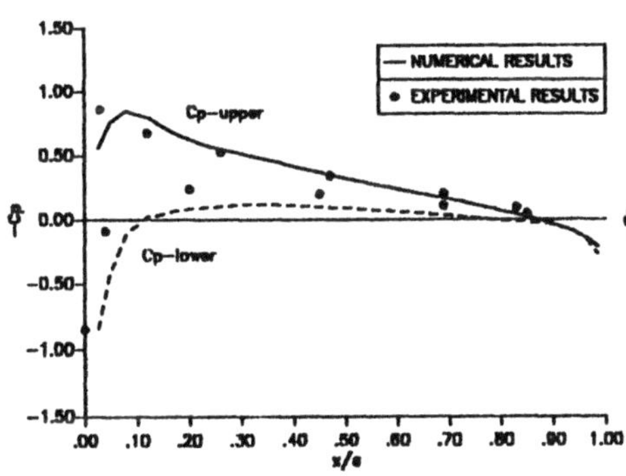

Figure 2a. c_p *vs.* x/c at r/R=0.5 (M_{tip}=0.439).

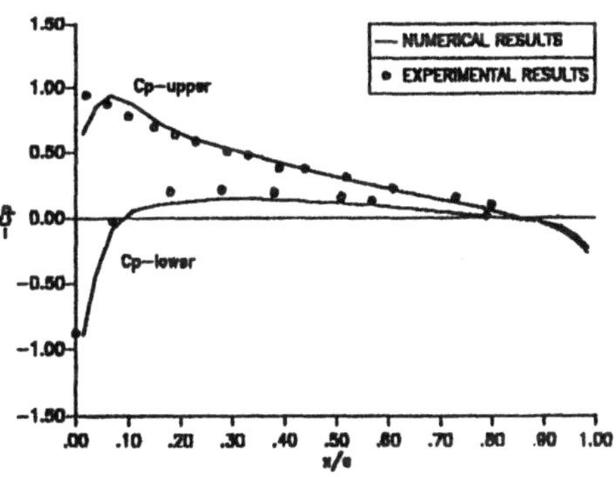

Figure 2b. c_p *vs.* x/c at r/R=0.68 (M_{tip}=0.439).

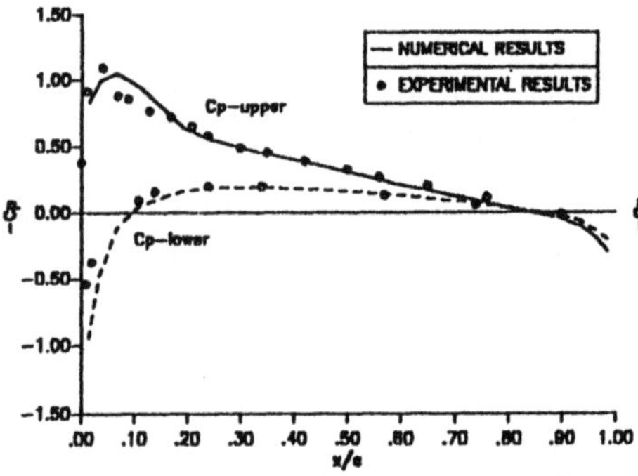

Figure 2c. c_p *vs.* x/c at r/R=0.8 (M_{tip}=0.439).

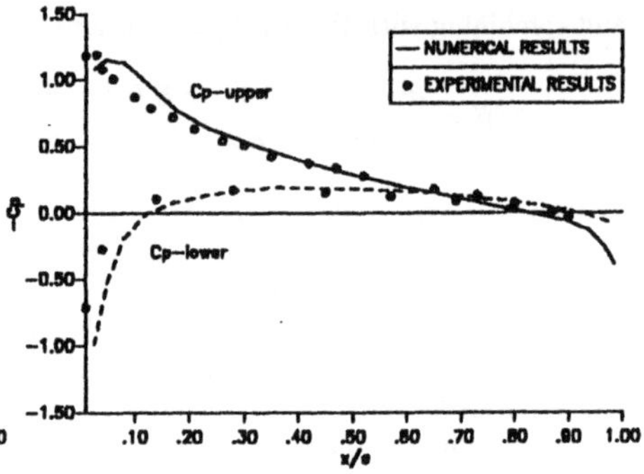

Figure 2d. c_p *vs.* x/c at r/R=0.89 (M_{tip}=0.439).

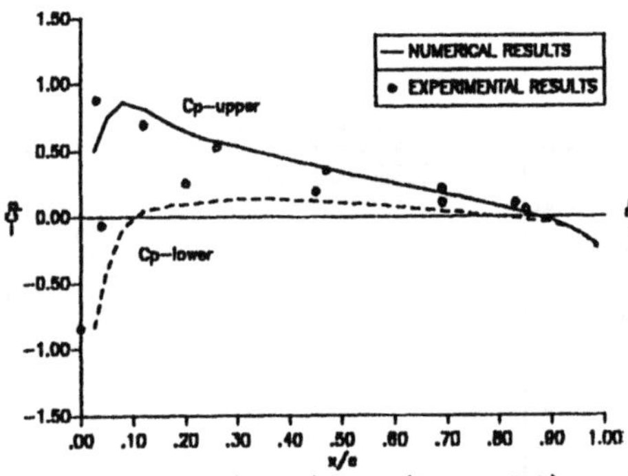

Figure 3a. c_p *vs.* x/c at r/R=0.5 (M_{tip}=0.727).

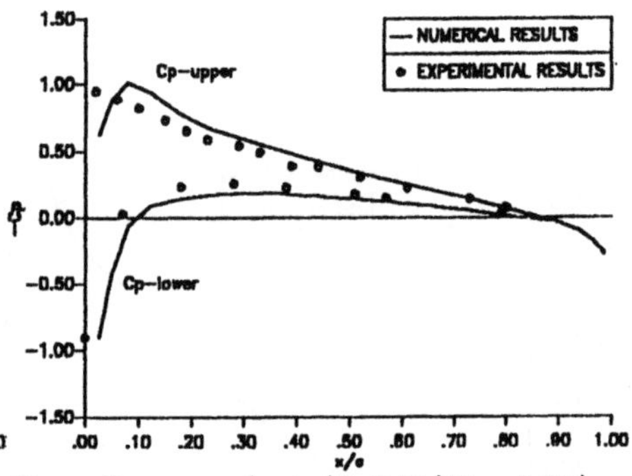

Figure 3b. c_p *vs.* x/c at r/R=0.68 (M_{tip}=0.727).

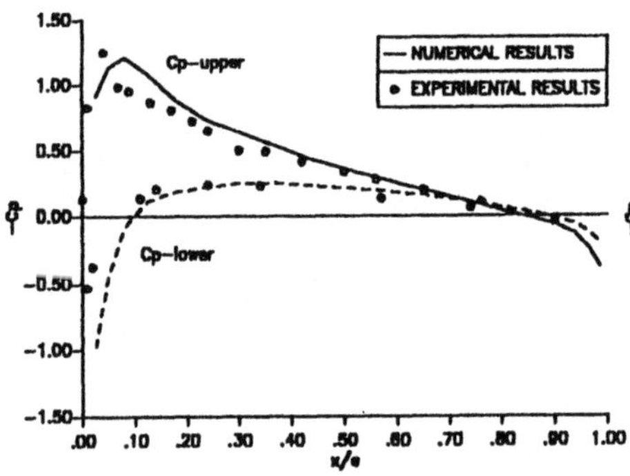

Figure 3c. c_p *vs.* x/c at r/R=0.8 (M_{tip}=0.727).

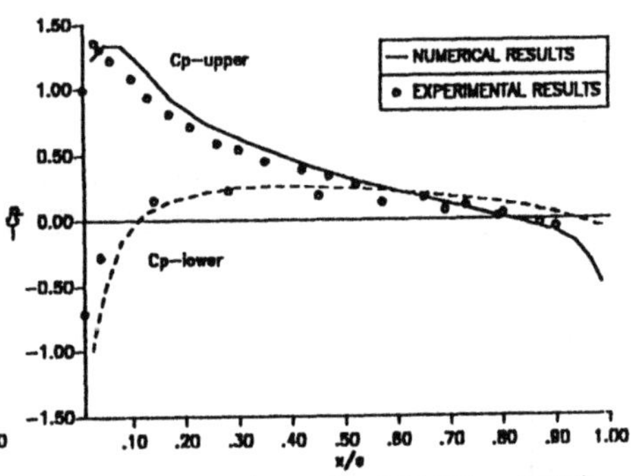

Figure 3d. c_p *vs.* x/c at r/R=0.89 (M_{tip}=0.727).

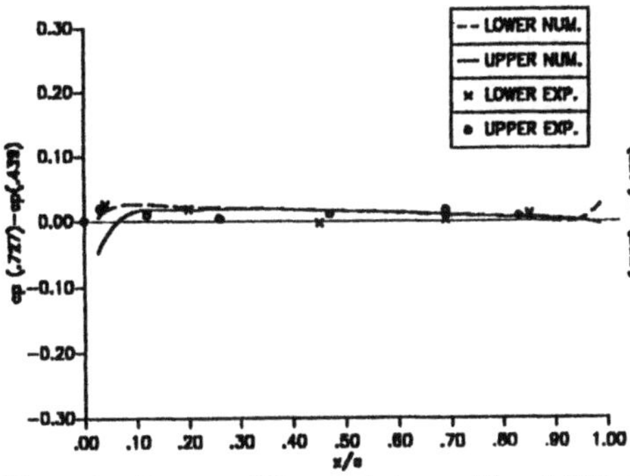

Figure 4a. Pressure difference between M_{tip}=0.727 and M_{tip}=0.439 (r/R=0.5).

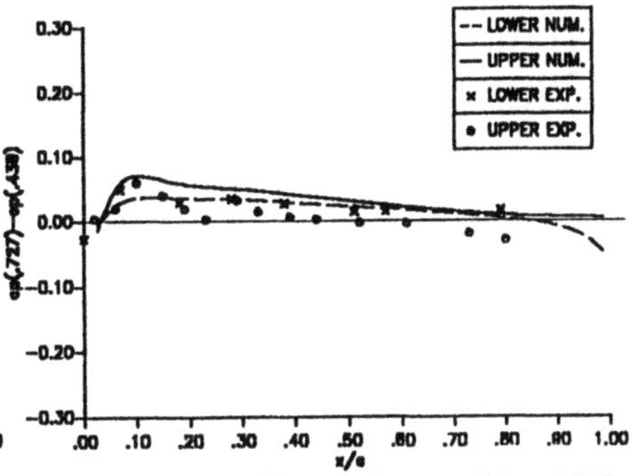

Figure 4b. Pressure difference between M_{tip}=0.727 and M_{tip}=0.439 (r/R=0.68).

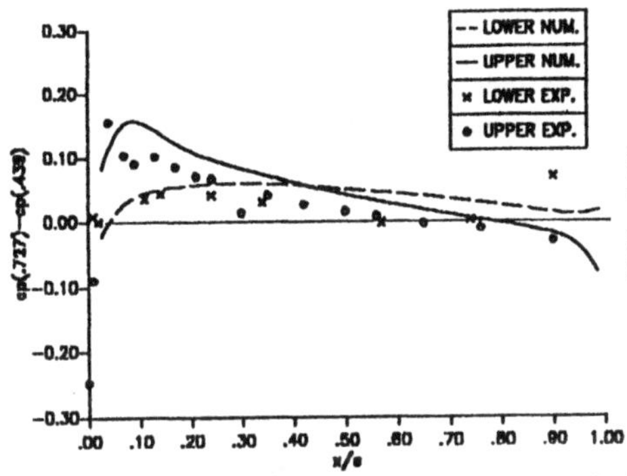

Figure 4c. Pressure difference between M_{tip}=0.727 and M_{tip}=0.439 (r/R=0.8).

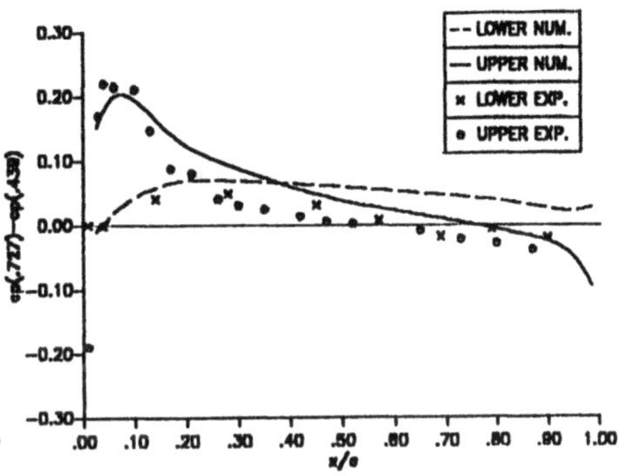

Figure 4d. Pressure difference between M_{tip}=0.727 and M_{tip}=0.439 (r/R=0.89).

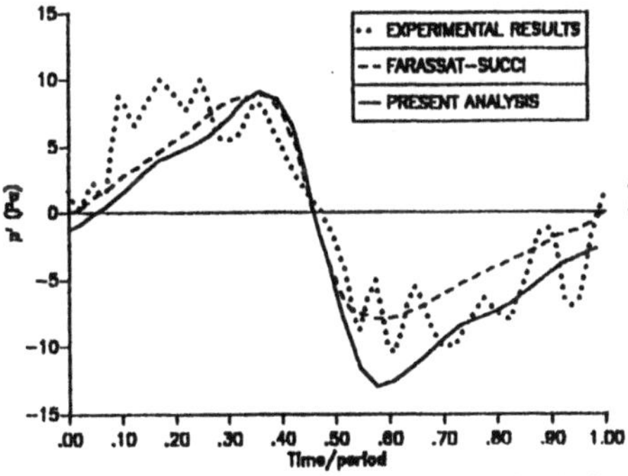

Figure 5a. Pressure signature for general aviation propeller; period=11.36 m/s.

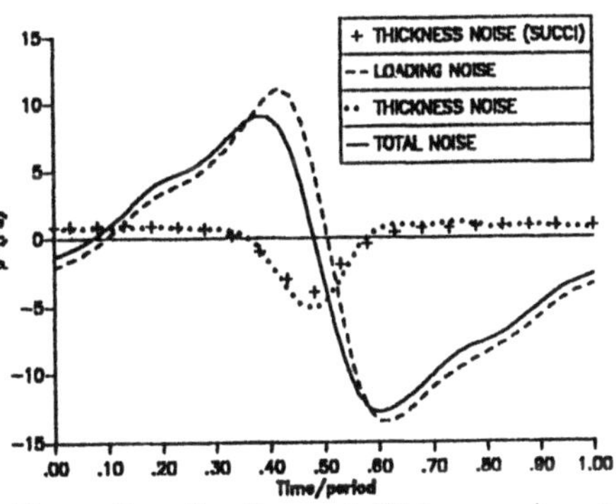

Figure 5b. Loading and thickness noise; period=11.36 m/s.

On the Possibility of Singularities in the Acoustic Field of Supersonic Sources When BEM is Applied to a Wave Equation

E. De Bernardis

CIRA
Italian Aerospace Research Center
81043 Capua (CE) — Italy

F. Farassat

NASA
Langley Research Center
Hampton, VA 23665 — U.S.A.

Summary

Using a time domain method based on the Ffowcs Williams-Hawkings equation, a reliable explanation is provided for the origin of singularities observed in the numerical prediction of supersonic propeller noise. In the last few years Tam and, more recently, Amiet have analyzed the phenomenon from different points of view. The method proposed here offers a clear interpretation of the singularities based on a new description of sources, relating to the behaviour of lines where the propeller blade surface exhibits slope discontinuity.

Introduction

Modern advanced propellers for the propulsion of future airliners have thin, highly twisted and swept blades. They operate at transonic and supersonic speed over a large portion of the blade span. The accurate prediction of the noise of these propellers is a significant problem of research both from the point of view of fuselage design and community noise impact studies. One of the most useful methods of noise prediction of these propellers is based on the acoustic analogy which is in fact a boundary element method. The specific approach we study here is the time domain method based on the Ffowcs Williams-Hawkings (FW-H) equation [1]. Several computer codes have been developed at Langley Research Center based on this approach [2,3].

In the numerical implementation of the BEM to the supersonic propeller noise problem, sharp pulses are sometime observed in the acoustic pressure signature. The peaks of these pulses increase as the time mesh size is decreased indicating possible singularities in the solution of the governing wave equation. Tam has shown that such singularities are possible by studying the convergence of the Fourier spectrum of the noise [4]. In a more recent paper, Amiet analyzes the origin of these singularities based on the structure of the acoustic planform [5]. It will be shown here from a more direct and general method that the origin of these singularities for airfoil with sharp leading and trailing edges are line sources along these edges that can be clearly identified in the sources of the governing wave equation. Furthermore a clear geometrical interpretation of the condition of occurrance of singularities can be given by this method. This paper is on the mathematical aspects of the problem of singularities in the application of BEM to supersonic surfaces.

Mathematical statement of the problem

We will consider the thickness noise problem of propellers. The loading problem is identical and will not be touched upon here. Let the blade surface be given by the equation $f(\mathbf{x},t) = 0$ such that $\nabla f = \mathbf{n}$, where $\mathbf{n}$ is the unit outward normal. Assume that the blade has sharp leading and trailing edges with no other slope discontinuities on the upper and lower surfaces of the airfoil.

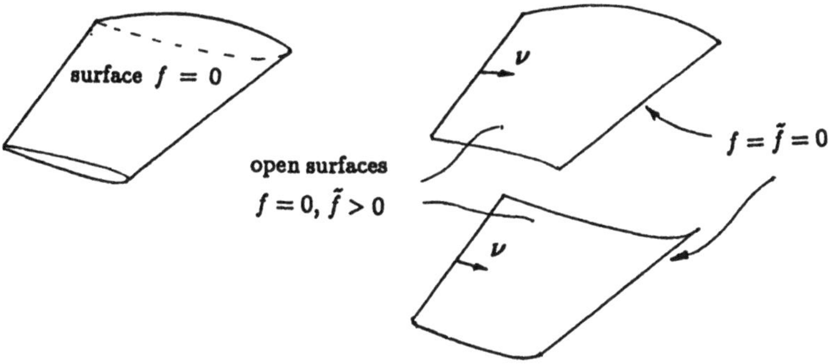

Figure 1: partitioning of the blade surface into upper and lower surfaces joining along the leading and trailing edges.

Then, from the FW-H equation, the thickness noise problem is described by [1]:

$$\left(\frac{1}{c^2}\frac{\bar{\partial}^2}{\partial t^2} - \bar{\nabla}^2\right) p' = \bar{\Box}^2 p' = \frac{\bar{\partial}}{\partial t}[\rho_0 v_n \delta(f)] \tag{1}$$

Here p' is the acoustic pressure in the far field, ρ_0 and c are the density and the speed of sound in the undisturbed medium and v_n is the local normal velocity of the blade. In the above equation $\delta(f)$ is the Dirac delta function and a bar on a differential operator signifies generalized differentiation [6]. We assume here that $v_n < c$ over the entire blade surface.

To study the effect of the slope discontinuities at the edges of the blade, we break up the surface $f = 0$ into the upper and lower surfaces which are now open (*i.e.* not closed) surfaces. These two surfaces join along the leading and trailing edges to form the blade, as shown in figure 1. We will concentrate on one of these surfaces. To be precise, we assume that the edge of each open surface is described by $f = \tilde{f} = 0$ where $\tilde{f} > 0$ on the surface. We thus study the wave equation

$$\bar{\Box}^2 p' = \frac{\bar{\partial}}{\partial t}[\rho_0 v_n H(\tilde{f})\delta(f)] \tag{2}$$

where $H(\tilde{f})$ is the Heaviside function. To obtain the edge source, we must take the time derivative inside the square brackets. First we state more precisely the variables of $\tilde{f}$. On the open surface we use variables (u^1, u^2) as surface coordinates then extend these along normals to $f = 0$. We also take $u^3 = f$ which gives $\mathbf{u} = \mathbf{u}(\mathbf{x}, t)$. Using this frame, we define $\tilde{f} = \tilde{f}(u^1, u^2)$.

The right-hand side of eq.(2) can be written as

$$\frac{\bar{\partial}}{\partial t}[\rho_0 v_n(u^1, u^2, u^3, t)H(\tilde{f})\delta(u^3)] = \frac{\bar{\partial}}{\partial t}[\rho_0 v_n(u^1, u^2, 0, t)H(\tilde{f})\delta(u^3)] \tag{3}$$

where, due to the presence of $\delta(u^3)$, all remaining functions within square brackets can be thought of as depending on surface variables (u^1, u^2). Taking the time derivative inside the square brackets gives rise to three terms whose meaning will be analyzed in the following. In fact one has:

$$\begin{aligned}\frac{\bar{\partial}}{\partial t}[\rho_0 v_n H(\tilde{f})\delta(u^3)] &= \rho_0 \frac{\partial}{\partial t}[v_n(u^1, u^2, 0, t)]H(\tilde{f})\delta(u^3) \\ &+ \rho_0 v_n \frac{\bar{\partial}}{\partial t}[H(\tilde{f})]\delta(u^3) \\ &+ \rho_0 v_n H(\tilde{f})\frac{\bar{\partial}}{\partial t}[\delta(u^3)]\end{aligned} \tag{4}$$

The generalized derivative has been removed from the first term in the right-hand side, v_n being continuous on the open surface $f = u^3 = 0$, $\tilde{f} > 0$. In this term one has:

$$\begin{aligned}\frac{\partial v_n}{\partial t} &= \frac{d}{dt}\{v_n[\mathbf{y}(\mathbf{u},t),t]\} - \nabla v_n \cdot \frac{\partial \mathbf{y}}{\partial t} \\ &= \dot{v}_n - \mathbf{v_T} \cdot \nabla_2 v_n \end{aligned} \tag{5}$$

where $\dot{v}_n$ is time rate of change of v_n as measured by an observer moving with the blade, $\mathbf{v_T}$ is tangential velocity of the blade and ∇_2 is the surface gradient.

In the second term the derivative of the Heaviside function can be calculated giving:

$$\frac{\bar{\partial}}{\partial t}[H(\tilde{f})] = H'(\tilde{f})\frac{\partial \tilde{f}}{\partial t} = -\delta(\tilde{f})v_\nu \tag{6}$$

The symbol v_ν stands for the component of the tangential velocity normal to the edge pointing into the surface, *i.e.* $v_\nu = \mathbf{v_T} \cdot \boldsymbol{\nu}$ where $\boldsymbol{\nu}$ is the inward unit geodesic normal of the edge.

In the third term at the right-hand side of eq.(4) one simply has:

$$\frac{\bar{\partial}}{\partial t}[\delta(u^3)] = \delta'(u^3)\frac{\partial u^3}{\partial t} = -\delta'(u^3)v_n \tag{7}$$

Accounting for eqs.(5), (6) and (7), eq.(4) can be written as

$$\begin{aligned}\frac{\bar{\partial}}{\partial t}[\rho_0 v_n H(\tilde{f})\delta(u^3)] &= \rho_0(\dot{v}_n - \mathbf{v_T} \cdot \nabla_2 v_n)H(\tilde{f})\delta(u^3) \\ &- \rho_0 v_n v_\nu \delta(\tilde{f})\delta(u^3) \\ &- \rho_0 v_n^2 H(\tilde{f})\delta'(u^3)\end{aligned} \tag{8}$$

Here the first term is a source extending to the open surface $f = u^3 = 0$, $\tilde{f} > 0$ and leads to nonsingular surface integrals. The second term, which involves the product of delta functions $\delta(\tilde{f})\delta(u^3)$, is a line source which does result in singularities in the acoustic field. The third term can be shown to result in forms similar to the previous ones [7].

We will thus study the generic wave equation with a line source of the following type

$$\bar{\Box}^2 p' = Q(\mathbf{x},t)\delta(\tilde{f})\delta(f) \tag{9}$$

where we have taken $Q(\mathbf{x},t) = -\rho_0 v_n v_\nu + (\rho_0/\Lambda)v_n^2 m_n \hat{\mathbf{r}} \cdot \boldsymbol{\nu}$ and replaced u^3 by f again. In the above expression of the source, $\hat{\mathbf{r}}$ is the unit vector in the source-observer direction and $\Lambda = \{1 + m_n^2 - 2m_n \cos\theta\}^{\frac{1}{2}}$, with $m_n = v_n/c$, and θ the angle between $\hat{\mathbf{r}}$ and $\mathbf{n}$. In the next section we will give the solution and discuss the possibility of singularities of the solution of eq.(9).

Solution of wave equation with line sources

Several forms of solution of the FW-H equation have been proposed by Farassat over the last fifteen years. The method based upon the collapsing sphere technique will be exploited here. The solution of eq.(9) can be written in general

$$4\pi p'(\mathbf{x},t) = \int \frac{Q(\mathbf{y},\tau)}{r}\delta(f)\delta(\tilde{f})\delta(g)\, d\mathbf{y} d\tau \tag{10}$$

Here the free-space Green's function for the wave equation has been used. The four-dimensional space-time integral can now be reduced to a simple time integral working on the Dirac delta functions. To this aim the function $\tilde{f}(u^1, u^2)$ will be extended. The function $\tilde{f}_e$ is defined in such a way that $\tilde{f}_e = 0$ represents a moving surface whose intersection with $f = 0$ is the edge of

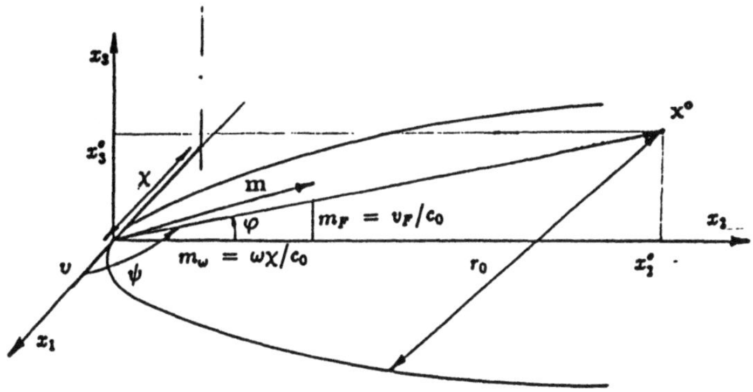

Figure 2: Behaviour near the condition $\cos\psi = 0$: configuration referring to the instant when the condition occurs.

the open surface described by $f = 0$, $\tilde{f} > 0$. The extension can be defined to meet the following conditions

$$\tilde{f}_e(u^1, u^2, 0) = \tilde{f}(u^1, u^2), \quad \nabla\tilde{f}_e \cdot \nabla f = \boldsymbol{\nu} \cdot \mathbf{n} = 0, \quad |\nabla\tilde{f}_e| = 1 \tag{11}$$

Considering the above conditions the transformation $(y_1, y_2, y_3) \rightarrow (f, \tilde{f}_e, g)$ can be applied, for which

$$\left|\frac{\partial(y_1, y_2, y_3)}{\partial(f, \tilde{f}_e, g)}\right| = \frac{1}{|\nabla f \times \nabla\tilde{f}_e \cdot \nabla g|} = \frac{c}{|\hat{\gamma} \cdot \hat{r}|} \tag{12}$$

In the above equation $\hat{\gamma}$ is the unit vector along the edge $f = \tilde{f}_e = 0$. If ψ is the angle between $\hat{\gamma}$ and the source-observer direction, substituting the far right-hand side of eq.(12) in eq.(10) one finally obtains:

$$4\pi p'(\mathbf{x}, t) = \int_{-\infty}^{t} \left[\frac{cQ(\mathbf{y}, \tau)}{r|\cos\psi|}\right]_{f=\tilde{f}_e=g=0} d\tau \tag{13}$$

The above equation describes the contribution of line integrals to the perturbation at $\mathbf{x}$ and t as a result of the overlapping of signals emitted by points of the curve $f = 0$, $\tilde{f} = 0$ at the corresponding emission times, that is when the points of that curve touch the collapsing sphere. The singularity of the function under the integral in equation (13) takes place when the condition $\cos\psi = 0$ occurs: at the points where this happens, the vector $\hat{\gamma}$ is perpendicular to $\hat{\mathbf{r}}$ that is the $f = \tilde{f} = 0$ curve is tangent to the collapsing sphere.

Analysis of singularities

In order to understand the behaviour of the integral (13) in the vicinity of the condition $\cos\psi = 0$, the function $\cos\psi$ will be studied in a short time interval including the instant where the above condition occurs [8]. A simple model will be studied: the discontinuity line is represented by a segment of a straight line in steady helical motion with forward speed $\mathbf{v_F}$ and angular velocity ω. For the sake of simplicity, the line will be assumed to pass through the axis of rotation and perpendicular to the axis.

Figure 2 shows the configuration at the instant τ_0 where, for a point of the $f = \tilde{f} = 0$ curve, the condition $\cos\psi = 0$ occurs. The position occupied at τ_0 by the above point is chosen as the origin of the frame of reference that we use to perform the analysis. It has the x_1-axis coinciding with the direction of the $f = \tilde{f} = 0$ curve at τ_0, while x_3 has the direction of the forward velocity $\mathbf{v_F}$ and the axis of rotation is represented by: $x_1 = -\chi$, $x_2 = 0$, χ being a positive number. The observer is placed at $\mathbf{x}^o \equiv \{0, x_2^o, x_3^o\}$ and r_0 is its distance at τ_0 from the $f = \tilde{f} = 0$ curve.

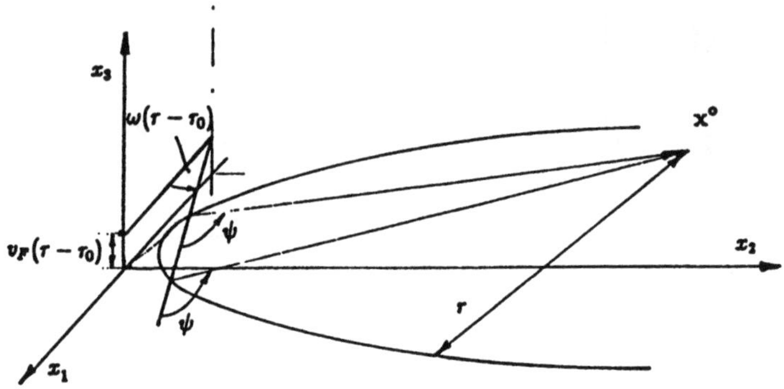

Figure 3: Behaviour near the condition $\cos\psi = 0$: configuration referring to an instant shortly after the condition occurred.

Figure 3 refers to an instant τ, shortly after τ_0: the $f = \tilde{f} = 0$ line has moved according to the helical motion and intersects the collapsing sphere in two (distinct) points; the radius of the sphere has reduced to $\tilde{r} = c(\tau - \tau_0)$. The coordinates of the above points will be found in order to calculate the current (*i.e.* at the source time τ) values of $\cos\psi$, whose behaviour near the singular point will be determined as the limit when τ tends to τ_0.

At time τ the $f = \tilde{f} = 0$ line is represented by the equations:

$$x_2 = \tan\left[\omega(\tau - \tau_0)\right]\,(x_1 + \chi) \tag{14}$$

$$x_3 = v_F(\tau - \tau_0) \tag{15}$$

while the equation of the collapsing sphere is:

$$x_1^2 + (x_2 - x_2^o)^2 + (x_3 - x_3^o)^2 = r^2 = (r_0 - \tilde{r})^2 \tag{16}$$

Letting now $m_F = v_F/c$, $m_\omega = \omega\chi/c$, and observing from the figure that $x_2^o = r_0\cos\varphi$, $x_3^o = r_0\sin\varphi$, one obtains from (14), (15) and (16):

$$\begin{aligned} x_1^2 + \tan^2(m_\omega\tilde{r}/\chi)(x_1+\chi)^2 &+ m_F\tilde{r}^2 - 2r_0\cos\varphi\tan(m_\omega\tilde{r}/\chi)(x_1+\chi) \\ &- 2r_0\sin\varphi\, m_F\tilde{r} + 2r_0\tilde{r} - \tilde{r}^2 = 0 \end{aligned} \tag{17}$$

Solving for x_1 intersections between the sphere and the $f = \tilde{f} = 0$ curve at time τ are determined.

Since $\tau \to \tau_0$ implies $\tilde{r} \to 0$ one can assume, in the limit, $\tan(m_\omega\tilde{r}/\chi) \simeq (m_\omega\tilde{r}/\chi)$. Neglecting now terms of order higher than one in $\tilde{r}$, equation (17) becomes:

$$x_1^2 - 2r_0\cos\varphi(m_\omega\tilde{r}/\chi)x_1 - 2r_0\cos\varphi\, m_\omega\tilde{r} - 2r_0\sin\varphi\, m_F\tilde{r} + 2r_0\tilde{r} = 0 \tag{18}$$

Considering the relation:

$$m_\omega\cos\varphi + m_F\sin\varphi = m_r$$

where $m_r = v_r/c = (\mathbf{v}\cdot\hat{\mathbf{r}})/c$, one finally obtains:

$$x_1^2 - 2r_0\cos\varphi(m_\omega\tilde{r}/\chi)x_1 + 2r_0\tilde{r}(1 - m_r) = 0 \tag{19}$$

whose solutions are expressed by:

$$x_1^* = r_0\cos\varphi(m_\omega\tilde{r}/\chi) \pm \left\{\left[r_0\cos\varphi(m_\omega\tilde{r}/\chi)\right]^2 - 2r_0\tilde{r}(1 - m_r)\right\}^{\frac{1}{2}} \tag{20}$$

Neglecting the term of $O(\tilde{r}^2)$ in the square brackets, when $\tilde{r}(m_r - 1) > 0$, the above solutions represent two distinct (and real) intersections. They can be written in the form:

$$x_1^* = A\tilde{r} \pm B\tilde{r}^{\frac{1}{2}}$$

Approximating r with r_0, from figure 3 it is easily seen that:

$$\cos\psi \simeq \frac{1}{r_0}\left[\cos(m_\omega\tilde{r}/\chi)(x_1^* - x_1^o) + \sin(m_\omega\tilde{r}/\chi)(x_2^* - x_2^o)\right] \tag{21}$$

Since $\tau \to \tau_0$ one has $\cos(m_\omega\tilde{r}/\chi) \to 1$ and $\sin(m_\omega\tilde{r}/\chi) \to (m_\omega\tilde{r}/\chi)$. Then the above expression leads to:

$$\cos\psi = O(\tilde{r}^{\frac{1}{2}}) = O\left[(\tau - \tau_0)^{\frac{1}{2}}\right]$$

This shows that when $\tilde{r}(m_r - 1) > 0$, then $1/\cos\psi$ is integrable (with respect to τ) in the neighbourhood of τ_0. The integral will behave like $(t - t_0)^{\frac{1}{2}}$ near $t_0 = \tau_0 + r_0/c$, *i.e.* at the instant when the observer receives the perturbation generated at τ_0.

It is easily seen from equation (20) that, in the case where the condition $m_r = 1$ occurs along with $\cos\psi = 0$, to the first order in $\tau - \tau_0$ one of the solutions turns out to be $x_1^* = 0$. This shows that τ_0 represents a double (at least) solution to $g = 0$, *i.e.* a double emission time (as it is obvious considering that $1 - m_r = \partial g/\partial\tau$). To the first order in $\tau - \tau_0$, the second solution of equation (20) is given by:

$$x_1^* = r_0 \cos\varphi(m_\omega\tilde{r}/\chi)$$

Inserting this in the expression for $\cos\psi$, one shows that, in this case:

$$\cos\psi = O(\tilde{r}) = O(\tau - \tau_0)$$

In the vicinity of the reception time t_0 one then has, for the integral (13):

$$4\pi p'(\mathbf{x}^o, t) = \int [\ldots]\frac{d\tau}{|\tau - \tau_0|} = C\ln|t - t_0| \tag{22}$$

One can go on with the analysis considering terms of order $\tilde{r}^2$ in the left-hand side of equation (17) in the case $m_r = 1$, neglecting some small coefficients of x_1 and x_1^2. This equation becomes:

$$x_1^2 - 2r_0\cos\varphi(m_\omega\tilde{r}/\chi)x_1 + m_\omega^2\tilde{r}^2 + m_F^2\tilde{r}^2 - \tilde{r}^2 = 0 \tag{23}$$

whose solutions, keeping in mind that $m_F^2 + m_\omega^2 = m^2$, are given by:

$$x_1^* = r_0\cos\varphi(m_\omega\tilde{r}/\chi) \pm \left\{[r_0\cos\varphi(m_\omega\tilde{r}/\chi)]^2 + (1 - m^2)\tilde{r}^2\right\}^{\frac{1}{2}} \tag{24}$$

The above expression shows that when the condition $m = 1$ occurs along with $\cos\psi = 0$ and $m_r = 1$, then the solution $x_1^* = 0$ exist up to the order $(\tau - \tau_0)^2$ (according to $\partial^2 g/\partial\tau^2 = 1 - m^2$, which holds true for any steady motion). In this case τ_0 is a triple root to $g = 0$, a fact that is of crucial importance in coding procedures for emission time calculation at points on blades moving at transonic speed.

Concluding remarks

The derivation of a new form of the inhomogeneous wave equation used in supersonic propeller noise prediction is presented. The expression of the source term leads to a useful interpretation of the role played by lines of slope discontinuity on the blade surface in generating sound at high propeller speed. The major result is a clear kinematic explanation of the singularity that is generally observed in numerical implementation of BEM to supersonic propeller noise problem.

References

1. FFOWCS WILLIAMS, J.E.; HAWKINGS, D.L.: Sound Generation by Turbulence and Surfaces in Arbitrary Motion, *Philosophical Transactions of the Royal Society* **A264** (1969), 321–342

2. BRENTNER, K.: Prediction of Helicopter Rotor Discrete Frequency Noise (User's Manual), *NASA* **TM 87721**, 1986

3. FARASSAT, F.; PADULA, S.L., DUNN, M.H.:Advanced Turboprop Noise Prediction Based on Recent Theoretical Results, *Journal of Sound and Vibration* **119** (1987), 53–79

4. TAM, C.W.K.: On Linear Acoustic Solutions of High Speed Helicopter Impulsive Noise Problems, *Journal of Sound and Vibration* **89** (1983), 119–134

5. AMIET, R.K.: Thickness Noise of a Propeller and its Relation to Blade Sweep, *Journal of Fluid Mechanics* **192** (1988), 535–560

6. GEL'FAND, I.M.; SHILOV, G.E.: *Generalized Functions, vol. 1: Properties and Operations.* New York: Academic Press 1964

7. FARASSAT, F.: Lectures on the Aeroacoustics of Rotating Blades in Time Domain, Dipartimento di Meccanica e Aeronautica, Università La Sapienza, Roma 1989

8. DE BERNARDIS, E.: On a New Formulation for the Aeroacoustics of Rotating Blades (in Italian), Ph.D. Thesis, Università La Sapienza, Roma 1989

Boundary Element Methods in Probabilistic Acoustic Scattering Problems

Raymond P. Daddazio and Mohammed M. Ettouney

Weidlinger Associates, Inc.
Applied Science Division
333 Seventh Avenue
New York, NY 10001

Abstract

Boundary Element Methods (BEM) are suited to a wide range of engineering problems, especially those of a semi-infinite nature. Examples of such problems can be found in the fluid-structure interactions of acoustic radiation and scattering problems and in the soil-structure interactions of earthquake and machine foundation problems. The required input parameters, dynamic loads, and system properties for such problems are not in general well defined and can be considered random variables. Probabilistic structural analysis through the use of the BEM has been introduced by Ettouney et al. [8,9]. In this work, we extend the use of this probabilistic approach to area of fluid-structure interaction by applying this technique to the problem of acoustic scattering from structures.

Boundary Element Approach

The basic idea of Boundary Element Methods is to reduce a problem in n dimensions to one in $n-1$ dimensions by using integral equation techniques to express the solution over a domain in terms of its surface values. This approach has been applied to problems in acoustic radiation and scattering by among others Schenck [17], Meyer et al. [13], Rizzo et al. [15], Everstine et al. [10], Vasudevan and McCormick [18] and Seybert et al. [16]. Additional details can be found in Banerjee and Butterfield [2], and Brebbia, et al. [6]. We briefly describe its application to the exterior fluid problem.

For steady state harmonic motion at angular frequency ω, the propagation of acoustic waves through a homogeneous fluid of density ρ and wave speed c is described by the Helmholtz equation. Applying Green's Theorem together with the appropriate boundary conditions, the pressure and velocity can be related through the Helmholtz integral equation (Baker and Copson, [1]). Following the notation of Everstine and Quezon [11], this is written as

$$\int_S p(\underline{x})\frac{\partial D(\underline{r})}{\partial \underline{n}}\,dS - \int_S q(\underline{x})D(\underline{r})\,dS = \begin{cases} p(\underline{x}')/2 - p_I(\underline{x}'), & (\underline{x}') \text{ on } S \\ p(\underline{x}') - p_I(\underline{x}'), & (\underline{x}') \text{ in } E \end{cases} \tag{1}$$

where S and E denote the surface and exterior fluid points, respectively, $p(\underline{x})$ is the pressure on S, p_I is the pressure of an acoustic wave incident upon the closed surface S and $\underline{r}$ is the vector defined by $\underline{x} - \underline{x}'$ (Figure 1) in which $\underline{x}$ is the location of a point on S and $\underline{x}'$ is either on S or in E, and $k = \omega/c$ is the acoustic wave number. The expression $D(\underline{r}) = \frac{e^{-ikr}}{4\pi \underline{r}}$ is the three dimensional free space Green's function for equation (1), and $\partial/\partial\underline{n}$ denotes differentiation in the outward normal direction ($\underline{n}$) on S at $\underline{x}$. At any point $\underline{x}$ on S, the relation between the pressure and velocity is

$$q(\underline{x}) = \frac{\partial p(\underline{x})}{\partial \underline{n}} = -i\omega\rho v_n(\underline{x}) \tag{2}$$

where $v_n(\underline{x})$ is the outward normal surface velocity.

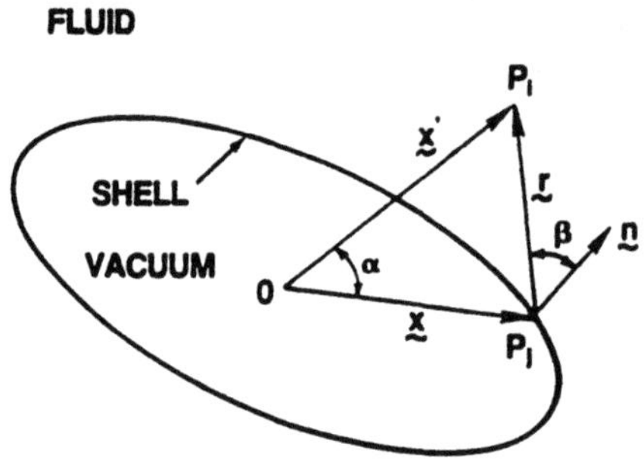

Figure 1. Notation for Helmholtz integral equation

Substitution of $D(\underline{r})$ and equation (2) into equation (1) leads to the following integral equation relating the total pressure, $p(\underline{x})$ and normal velocity $v_n(\underline{x})$ on S .

$$\frac{1}{2}p(\underline{x}') - \int_S p(\underline{x})\left(\frac{e^{-ik\underline{r}}}{4\pi\underline{r}}\right)\left(ik + \frac{1}{r}\right)\cos\beta \, dS = i\omega\rho \int_S v_n(\underline{x})\left(\frac{e^{-ik\underline{r}}}{4\pi\underline{r}}\right) dS + p_I(\underline{x}') \tag{3}$$

Once surface values are properly defined, the problem specification is complete. We assume that surface pressures are specified over certain portions of the surface and that normal velocities are prescribed over the remaining portions of S. Equation (3) is used in the probabilistic the analysis.

A numerical solution can be obtained by discretizing the surface and approximating the integrals in equation (3) by a weighted sum of functions evaluated at the surface points. If m is the total number of points on the surface, then equation (3) is approximated by a set of m linear algebraic equations:

$$\mathbf{E}\mathbf{p} = \mathbf{C}\mathbf{v} + \mathbf{p}_\mathbf{I}. \tag{4}$$

where $\mathbf{E}$ and $\mathbf{C}$ are square matrices whose elements are computed by numerical integration of the kernel functions with due regard being given to integrable singularities (Benthien and Barach, [5] and Everstine and Quezon, [11]). The elements of the vectors $\mathbf{p}$, $\mathbf{v}$ and $\mathbf{p}_\mathbf{I}$ are the surface pressures, the normal surface velocities and the prescribed incident pressures at the m surface points, respectively. Solution of these equations yields approximate values for the unknown surface pressures and normal velocities. Once the surface pressures and normal velocities are known on the entire surface, the value of the pressure anywhere in the exterior fluid can be calculated using equation (3).

Uncertainty Modeling

A case of considerable interest in acoustic scattering from structures is one where the statistics of the material parameters c, ρ and ω are available so that their mean values are known and deviations relative to the mean are not too large. The unknown pressures and normal velocities on the surface then can be expanded in a Taylor series about the mean values. This technique has been used in probabilistic finite element methods. Benaroya and Rehak [4] presented a well documented summary of its applications to structural analysis. Ettouney, et al. [8,9] used this approach in probabilistic BEM as applied to structural analysis. We will use the same method for the purpose of this study.

Perturbation Expansions

The dependence on c, ρ and ω in equation (4) for the case of scattering from a rigid body is indicated explicitly by writing

$$\mathbf{E}(c,\rho,\omega)\mathbf{p}(c,\rho,\omega) = \mathbf{C}(c,\rho,\omega)\mathbf{v} + \mathbf{p_I}(c,\rho,\omega) \tag{5}$$

where $\mathbf{v}$ is specified, and therefore has no c,ρ,ω dependence. Expansions in Taylor series of $\mathbf{p}$ about the mean values $\overline{c}$,$\overline{\rho}$ and $\overline{\omega}$, with $\varepsilon_c = (c-\overline{c})$, $\varepsilon_\rho = (\rho-\overline{\rho})$ and $\varepsilon_\omega = (\omega-\overline{\omega})$, yields

$$\mathbf{p}(c,\rho,\omega) = \overline{\mathbf{p}} + \varepsilon_c\overline{\mathbf{p}}_c + \varepsilon_\rho\overline{\mathbf{p}}_\rho + \varepsilon_\omega\overline{\mathbf{p}}_\omega + \frac{1}{2}\varepsilon_c^2\overline{\mathbf{p}}_{cc} + \frac{1}{2}\varepsilon_\rho^2\overline{\mathbf{p}}_{\rho\rho} + \frac{1}{2}\varepsilon_\omega^2\overline{\mathbf{p}}_{\omega\omega} + \varepsilon_\rho\varepsilon_\omega\overline{\mathbf{p}}_{\rho\omega} + \cdots \tag{6a}$$

where the subscripts on $\overline{\mathbf{p}}$ denote partial differentiation and the overline denotes evaluation of the functions at the respective mean values. Taylor series expansions of $\mathbf{E}(c,\rho,\omega)$, $\mathbf{C}(c,\rho,\omega)$ and $\mathbf{p_I}(c,\rho,\omega)$ in terms of the respective c,ρ and ω *sensitivity functions.* follow.

$$\mathbf{E}(c,\rho,\omega) = \overline{\mathbf{E}} + \varepsilon_c\overline{\mathbf{E}}_c + \varepsilon_\rho\overline{\mathbf{E}}_\rho + \varepsilon_\omega\overline{\mathbf{E}}_\omega + \frac{1}{2}\varepsilon_c^2\overline{\mathbf{E}}_{cc} + \frac{1}{2}\varepsilon_\rho^2\overline{\mathbf{E}}_{\rho\rho} + \frac{1}{2}\varepsilon_\omega^2\overline{\mathbf{E}}_{\omega\omega} + \cdots \tag{6b}$$

$$\mathbf{C}(c,\rho,\omega) = \overline{\mathbf{C}} + \varepsilon_c\overline{\mathbf{C}}_c + \varepsilon_\omega\overline{\mathbf{C}}_\omega + \varepsilon_\rho\overline{\mathbf{C}}_\rho + \frac{1}{2}\varepsilon_c^2\overline{\mathbf{C}}_{cc} + \frac{1}{2}\varepsilon_\rho^2\overline{\mathbf{C}}_{\rho\rho} + \frac{1}{2}\varepsilon_\omega^2\overline{\mathbf{C}}_{\omega\omega} + \cdots \tag{6c}$$

$$\mathbf{p_I}(c,\rho,\omega) = \overline{\mathbf{p}}_\mathbf{I} + \varepsilon_c\overline{\mathbf{p}}_{\mathbf{I}c} + \varepsilon_\rho\overline{\mathbf{p}}_{\mathbf{I}\rho} + \varepsilon_\omega\overline{\mathbf{p}}_{\mathbf{I}\omega} + \frac{1}{2}\varepsilon_c^2\overline{\mathbf{p}}_{\mathbf{I}cc} + \frac{1}{2}\varepsilon_\rho^2\overline{\mathbf{p}}_{\mathbf{I}\rho\rho} + \frac{1}{2}\varepsilon_\omega^2\overline{\mathbf{p}}_{\mathbf{I}\omega\omega} + \cdots \tag{6d}$$

Substitution of equation (6) into equation (5), and equating terms of the same order in ε_c, ε_ρ and ε_ω, produces the usual set of recursion relations which result from perturbation expansion techniques (Bellman, [3]). Ten terms will be needed to include up to second order perturbations. The resulting ten equations corresponding to (5) which must be solved are:

$$\overline{\mathbf{E}}\overline{\mathbf{p}} = \overline{\mathbf{C}}\overline{\mathbf{v}} + \overline{\mathbf{p}}_\mathbf{I} \tag{7a}$$

$$\overline{\mathbf{E}}\overline{\mathbf{p}}_\rho = -(\overline{\mathbf{E}}_\rho\overline{\mathbf{p}} + \overline{\mathbf{C}}_\rho\overline{\mathbf{v}}) + \overline{\mathbf{p}}_{\mathbf{I}\rho} \tag{7b}$$

$$\overline{\mathbf{E}}\overline{\mathbf{p}}_c = -(\overline{\mathbf{E}}_c\overline{\mathbf{p}} + \overline{\mathbf{C}}_c\overline{\mathbf{v}}) + \overline{\mathbf{p}}_{\mathbf{I}c} \tag{7c}$$

$$\overline{\mathbf{E}}\overline{\mathbf{p}}_\omega = -(\overline{\mathbf{E}}_\omega\overline{\mathbf{p}} + \overline{\mathbf{C}}_\omega\overline{\mathbf{v}}) + \overline{\mathbf{p}}_{\mathbf{I}\omega} \tag{7d}$$

$$\overline{\mathbf{E}}\overline{\mathbf{p}}_{\rho\rho} = -(2\overline{\mathbf{E}}_\rho\overline{\mathbf{p}}_\rho + \overline{\mathbf{E}}_{\rho\rho}\overline{\mathbf{p}} + \overline{\mathbf{C}}_{\rho\rho}\overline{\mathbf{v}}) + \overline{\mathbf{p}}_{\mathbf{I}\rho\rho} \tag{7e}$$

$$\overline{\mathbf{E}}\overline{\mathbf{p}}_{cc} = -(2\overline{\mathbf{E}}_c\overline{\mathbf{p}}_c + \overline{\mathbf{E}}_{cc}\overline{\mathbf{p}} + \overline{\mathbf{C}}_{cc}\overline{\mathbf{v}}) + \overline{\mathbf{p}}_{\mathbf{I}cc} \tag{7f}$$

$$\overline{\mathbf{E}}\overline{\mathbf{p}}_{\omega\omega} = -(2\overline{\mathbf{E}}_\omega\overline{\mathbf{p}}_\omega + \overline{\mathbf{E}}_{\omega\omega}\overline{\mathbf{p}} + \overline{\mathbf{C}}_{\omega\omega}\overline{\mathbf{v}}) + \overline{\mathbf{p}}_{\mathbf{I}\omega\omega} \tag{7g}$$

$$\overline{\mathbf{E}}\overline{\mathbf{p}}_{\rho c} = -\frac{1}{2}(\overline{\mathbf{E}}_\rho\overline{\mathbf{p}}_c + \overline{\mathbf{E}}_c\overline{\mathbf{p}}_\rho + \overline{\mathbf{E}}_{\rho c}\overline{\mathbf{p}} + \overline{\mathbf{C}}_{\rho c}\overline{\mathbf{v}}) + \overline{\mathbf{p}}_{\mathbf{I}\rho c} \tag{7h}$$

$$\overline{\mathbf{E}}\overline{\mathbf{p}}_{\rho\omega} = -\frac{1}{2}(\overline{\mathbf{E}}_\rho\overline{\mathbf{p}}_\omega + \overline{\mathbf{E}}_\omega\overline{\mathbf{p}}_\rho + \overline{\mathbf{E}}_{\rho\omega}\overline{\mathbf{p}} + \overline{\mathbf{C}}_{\rho\omega}\overline{\mathbf{v}}) + \overline{\mathbf{p}}_{\mathbf{I}\rho\omega} \tag{7i}$$

$$\overline{\mathbf{E}}\overline{\mathbf{p}}_{c\omega} = -\frac{1}{2}(\overline{\mathbf{E}}_c\overline{\mathbf{p}}_\omega + \overline{\mathbf{E}}_\omega\overline{\mathbf{p}}_c + \overline{\mathbf{E}}_{c\omega}\overline{\mathbf{p}} + \overline{\mathbf{C}}_{c\omega}\overline{\mathbf{v}}) + \overline{\mathbf{p}}_{\mathbf{I}c\omega} \tag{7j}$$

The recursive system provides an efficient procedure for calculating the terms in equation (7) since the coefficient matrix $\overline{\mathbf{E}}$ is the same for all orders. Once the elements of this matrix have been computed and the matrix has been factorized into lower and upper triangular matrices, most of the computational work has been done , and the equations can be solved quickly. It is relatively easy to assess the rate of convergence of the series and, as expected, the expansion converges rapidly for physically reasonable uncertainties in the values of c,ρ and ω.

For second order perturbation analysis, equations (7) are solved recursively for the unknown components of $\overline{\mathbf{p}}$, $\overline{\mathbf{p}}_\rho$, $\overline{\mathbf{p}}_c$, $\overline{\mathbf{p}}_\omega$, $\overline{\mathbf{p}}_{\rho\rho}$, $\overline{\mathbf{p}}_{cc}$, $\overline{\mathbf{p}}_{\omega\omega}$, $\overline{\mathbf{p}}_{\rho c}$, $\overline{\mathbf{p}}_{\rho\omega}$, $\overline{\mathbf{p}}_{c\omega}$, and substituted appropriately into equations (6a), which is used below to evaluate the first and second order moments (mean and standard deviation) of $\mathbf{p}$.

Expectations of a Model with Three Random Variables

Equation (6a) provides the basis for the probabilistic analysis of the response of acoustic systems. The mean values of the pressures at any point can be approximated by

$$\begin{aligned} E\{p(c,\rho,\omega)\} \cong\; & \overline{p} + E\{\varepsilon_\rho\}\overline{p}_\rho + E\{\varepsilon_\omega\}\overline{p}_\omega + E\{\varepsilon_c\}\overline{p}_c \\ & + \frac{1}{2}E\{\varepsilon_\rho^2\}\overline{p}_{\rho\rho} + \frac{1}{2}E\{\varepsilon_\omega^2\}\overline{p}_{\omega\omega} + \frac{1}{2}E\{\varepsilon_c^2\}\overline{p}_{cc} \\ & + E\{\varepsilon_c\varepsilon_\rho\}\overline{p}_{c\rho} + E\{\varepsilon_\rho\varepsilon_\omega\}\overline{p}_{\rho\omega} + E\{\varepsilon_c\varepsilon_\omega\}\overline{p}_{c\omega} \end{aligned} \tag{8}$$

and the mean-square values, correct to second order, by

$$\begin{aligned} E\{p(c,\rho,\omega)^2\} \cong\; & \overline{p}^2 + 2E\{\varepsilon_c\}\overline{p}\,\overline{p}_c + 2E\{\varepsilon_\rho\}\overline{p}\,\overline{p}_\rho + 2E\{\varepsilon_\omega\}\overline{p}\,\overline{p}_\omega \\ & + E\{\varepsilon_c^2\}(\overline{p}\,\overline{p}_{cc} + \overline{p}_c^2) + E\{\varepsilon_\rho^2\}(\overline{p}\,\overline{p}_{\rho\rho} + \overline{p}_\rho^2) + E\{\varepsilon_\omega^2\}(\overline{p}\,\overline{p}_{\omega\omega} + \overline{p}_\omega^2) \\ & + 2E\{\varepsilon_c\varepsilon_\rho\}(\overline{p}\,\overline{p}_{c\rho} + \overline{p}_c\,\overline{p}_\rho) + 2E\{\varepsilon_\omega\varepsilon_\rho\}(\overline{p}\,\overline{p}_{\omega\rho} + \overline{p}_\omega\,\overline{p}_\rho) \\ & + 2E\{\varepsilon_c\varepsilon_\omega\}(\overline{p}\,\overline{p}_{c\omega} + \overline{p}_c\,\overline{p}_\omega) \end{aligned} \tag{9}$$

In order to evaluate these expressions, we need to make probabilistic statements regarding the random variables c, ρ and ω. For the purpose of this study, a simple uniform probability density distribution was assumed for all the random variables. Thus the evaluation of $E\{\varepsilon_c\}$, $E\{\varepsilon_\rho\}$, $E\{\varepsilon_\omega\}$, $E\{\varepsilon_c^2\}$, $E\{\varepsilon_\rho^2\}$, $E\{\varepsilon_\omega^2\}$ $E\{\varepsilon_c\varepsilon_\rho\}$, $E\{\varepsilon_\omega\varepsilon_\rho\}$ and $E\{\varepsilon_c\varepsilon_\omega\}$ can easily be performed using the well known statistical relations for the uniform distribution case, (Papoulis [14]).

Once equations (8) and (9) are used to approximate $E\{p(c,\rho,\omega)\}$, $E\{p(c,\rho,\omega)^2\}$ it is useful to evaluate the variance (σ^2) and standard deviation (σ) of surface pressure:

$$\sigma_p = +\sqrt{\sigma_p^2} = \sqrt{E\{p^2\} - E^2\{p\}} \tag{10}$$

Given the mean and variance, it is possible to establish confidence bounds on specific realization of p.

Case Study

We will now apply the outlined methodology to the solution of the problem of the acoustic scattering from a rigid sphere which is irradiated by a plane wave. The sphere has a center at $R = 0$ and a radius at $R = a$. The exact solution for the resulting surface pressure on the sphere may be found in Junger and Feit [12].

Analysts and systems designers in the field of structural acoustics are aware of the uncertainties in the material properties of the exterior fluid. If the acoustic medium is seawater, the density and sound speed vary as a function of temperature, salinity and pressure (Del Grosso [7]). In our study, we will assume that the only random variable is the sound speed, c. The sound speed is taken to be a uniformly distributed random variable with a range of $0.95\bar{c} \leq c \leq 1.05\bar{c}$. This range of variability is a typical realistic situation.

A boundary element model for the pulsating sphere has been developed. The accuracy of the boundary element model is shown in Figure 2a. The discretization is judged to be adequate for the purposes of this study. The problem was solved for a range of frequencies from $0 < ka \leq 2.5$. At each frequency, the maximum and minimum values for the real and imaginary part of the surface pressure with a probability of non-exceedance of 84% and 97% are calculated at the point of first impact of the plane wave with the rigid sphere. These non-exceedance levels are chosen since they correspond to one and two standard deviation from the mean in a normal density function, hence they are widely used in engineering applications.

Figures 2b through 2d show the results of the probabilistic analysis. These figures indicate the deterministic solution for the real, imaginary, and amplitude of the surface pressure along with the probabilities of non-exceedance. Figure 2b and figure 2c indicate the uncertainty in the real and imaginary part of the surface pressure, respectively. An important point to be made in the interpretation of these results is that because of the normalization of the x-axis in terms of ka, the deterministic response is *independent* of the parameter c. Hence the deterministic zone has zero width (Figure 2a). The solution of the problem at an upper and lower limit of c will yield *no* information on the uncertainty in a response parameter to a random value of c. Only probabilistic approaches can yield quantitative measures of the sensitivity of the results (surface pressure in this case) to the variability of the acoustic medium. Figure 2d indicates the uncertainty in the magnitude of the pressure. The uncertainties in the real and imaginary part contribute to the uncertainty in the amplitude. Therefore, the uncertainty for a 97% probability of non-exceedence at about ±20% in the low frequency range and increases to ±60% at $ka = 2.5$.

It should be reiterated that the purpose of the example problem is to illustrate the application of the probabilistic analysis procedure presented herein. Other factors not addressed here that need be investigated are but are not limited to the effect of multiple random variables (ρ and ω in addition to c) and the correlation between them, the interaction of the random media with an elastic structure and the effects of the random media on the predictions of far field sound pressure levels.

Summary of Present Work

A probabilistic formulation for acoustic scattering problems using the boundary element method is introduced. The formulation assumes that the sound speed, mass density and the angular frequency of oscillation are random variables. Using a Taylor series technique, a perturbation approach is used to solve the problem. The statistical properties of the response measures, such as the mean and mean square are then evaluated. These statistical measures are used to evaluate the response measures at different confidence levels.

The problem of acoustic scattering from a rigid sphere was studied in order to demonstrate the use and value of the proposed method. Estimates of the uncertainty in the real and imaginary parts of the surface pressure as well as the amplitude of the surface pressure as a function of frequency in a fluid with random sound speed, are given. It is established that the probabilistic approach is needed in order to ascertain the degree of accuracy to which the response measures can be determined.

We have demonstrated the feasibility and practicality of this approach. It has been presented as an analysis and design tool, since we believe that an analytical method without close coupling to simple and clear design usage is incomplete. We have tried to demonstrate such usage with our application.

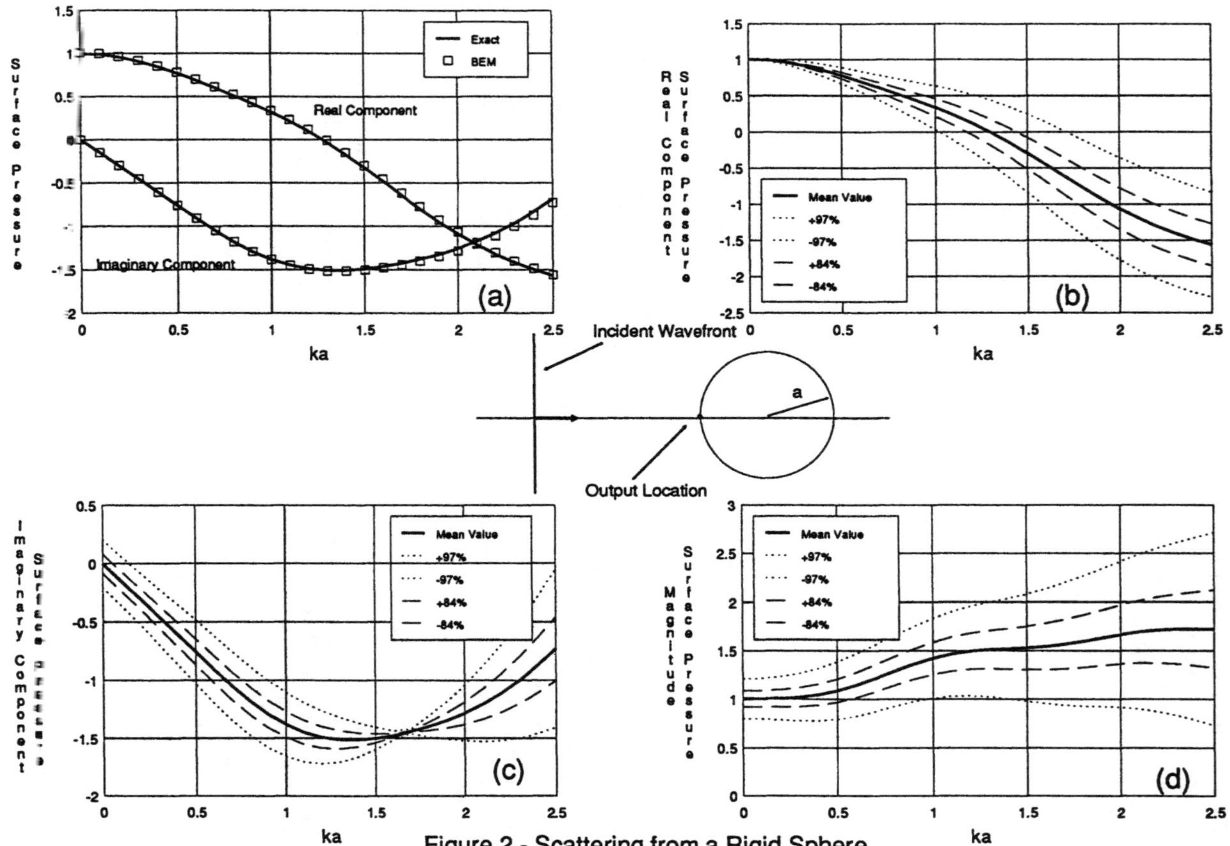

Figure 2 - Scattering from a Rigid Sphere

References

1. Baker, B., and Copson, E.,*The Mathematical Theory of Huygens' Principle*, Chelsea Publishing Company, New York, NY, Third Edition, 1987.
2. Banerjee, P., and Butterfield, R., *Boundary Element Methods in Engineering Science*, McGraw–Hill UK, 1981.
3. Bellman, R., *Perturbation Techniques in Mathematics, Physics, and Engineering*, Holt, Rinehart and Winston, 1964.
4. Benaroya, H., and Rehak, M., "Finite Element Methods and Probabilistic Structural Analysis - A Selective Review," ASME *Applied Mechanics Reviews*, Vol 41, No 5, May 1988.
5. Benthien, G.W., and Barach, D., "CHIEF Users Manual," *Technical Report*, NOSC TD 970, Naval Ocean Systems Center, San Diego, CA, pp. A2–1–A3–4, 1987.
6. Brebbia,C. A., Telles, J., and Wrobel,L., *Boundary Element Techniques*, Springer–Verlag, 1984.
7. Del Grosso, V., "New Equations for the Speed of Sound in Natural Water (with comparisons to other equations)," *J Acoust Soc Am* **56**(4),pp.1084–1091, 1974.
8. Ettouney, M., Benaroya, H., and Wright, J., "Probabilistc Boundary Element Methods (PBEM)," *Computational Mechanics in Probabilistic and Reliability Analysis*, W. K. Liu and T. Belytschko, Editors, ELMPRESS International, 1989.
9. Ettouney, M., Benaroya, H., and Wright, J., "Boundary Element Methods in Probabilistic Structural Analysis (PBEM)," *Applied Mathematical Modelling*, Vol. 13, July 1989, pp. 423–441.
10. Everstine, G., Henderson, F., Schroeder, E., and Lipman, R., "A General Low Frequency Acoustic Radiation Capability for NASTRAN," *Technical Report* NASA CP–2419, Fourteenth NASTRAN User's Colloquium, National Aeronautics and Space Administration, Washington, D.C., 1986, pp. 293–310.
11. Everstine, G., and Quezon, A., "User's Guide to the Coupled NASTRAN/Helmholtz Equation Capability (NASHUA) for Acoustic Radiation and Scattering," *Technical Report* DTRC CMLD–88/03, David Taylor Research Center, Bethesda, MD, 1988, pp. 7–9.
12. Junger M., and Feit, D., *Sound, Structures, and Their Interaction*, MIT Press, Cambridge, MA, Second Edition, 1986, pg 321.
13. Meyer, W., Bell, W., Zinn, B., and Stallybrass, M., "Boundary Integral Solutions of Three Dimensional Acoustic Radiation Problems," *Journal of Sound and Vibration*, **59**(2), 1978, pp. 245–262.
14. Papoulis,A., *Probability, Random Variables, and Stochastic Processes*, McGraw–Hill, 1965.
15. Rizzo, F., Shippy, D., and Rezayat, M., "A Boundary Integral Equation Method for Radiation and Scattering of Elastic Waves in Three Dimensions," *Int J Engng Sci*, Vol. 21, 1985, pp. 115–129.
16. Seybert, A., Wu, T., and Wu, X., "Radiation and Scattering of Acoustic Waves from Elastic Solids and Shells Using the Boundary Element Method," *J.Acoust.Soc.Am.* **84**(5), 1988, pp. 1906–1912.
17. Schenck, H., "Improved Integral Formulation for Acoustic Radiation Problems," *J Acoust Soc Am* **44**(1), 1968, pp. 41–58.
18. Vasudevan, R., and McCormick, J., "Substructuring Techniques for Acoustic Radiation from Complex Structures," Accepted for publication, ASME *Journal of Vibration, Acoustics, Stress and Reliability Design*, 1989.

Structural Acoustics Applications of the BEM and the FEM

A. F. Seybert, T. W. Wu and W. L. Li

Department of Mechanical Engineering, University of Kentucky
Lexington, KY 40506-0046 U.S.A.

SUMMARY

In this paper we discuss how the boundary element method (BEM) may be used by itself or in conjunction with the finite element method (FEM) for the solution of problems in structural acoustics. Problems in this category include: the scattering of sound from an elastic solid or shell submerged in a heavy fluid such as sea water, and the sound field generated within a vehicle passenger compartment or aircraft cabin due to vibration of the surrounding structure. In the latter case, the modes of the structure and the acoustic fluid may be coupled even if the fluid is air. Two approaches to structural acoustics modeling are considered. In one approach, the BEM is used to model both the elastic and the acoustic parts of the problem. In the second approach, the FEM is used to model the structure, while the BEM is used to model the fluid.

INTRODUCTION

The interaction of sound and vibration is an important subject in acoustics. If a light fluid is contained within a vibrating structure, such as the air within a vehicle passenger compartment, there may be significant coupling between the structural vibration and the sound field. In such cases, the coupled structural/acoustics problem must be solved. When a structure is submerged in a heavy fluid, such as sea water, the vibration of the structure will be affected by the surrounding sound pressure, and the whole system becomes a coupled structural/acoustics problem. There have been many papers published using a variety of analytical methods to examine structural/acoustic interaction [1-5]. However, when the structure has an arbitrary shape, methods amenable to numerical implementation are usually more desirable than approximate analytical methods.

In this paper we use two numerical techniques to solve problems in structural acoustics. In the first approach, the coupled elastic/acoustic boundary value problem is reformulated into a pair of integral equations, one for the interior (elastic) domain and one for the exterior (acoustic) domain. In its numerical form the integral equation method is referred to

as the boundary element method (BEM). The coupled elastic/acoustic technique used herein is referred to as the BEM/BEM approach.
The second numerical technique uses a finite element method (FEM) for the structure, but retains the BEM for the fluid. We refer to this method as the FEM/BEM approach. The FEM/BEM approach is preferred for structural acoustics problems in which the structural geometry is complex or when the structure is a thin shell. In the latter case, a BEM model of the structure would not be practical because the elements would be very small to avoid the near-singular behavior caused by the thin structural domain.

FORMULATION

For acoustic radiation or scattering from an elastic structure, we must solve a boundary value problem involving coupling between the elastic structure B_2, bounded by the surface S, and the surrounding acoustic fluid B_1, see Fig. 1. (Similarly, we may pose a problem consisting of an acoustic fluid, bounded by a surface S, contained within an elastic structure B_2 bounded on the exterior by a surface s). For scattering problems, we denote the prescribed incident wave as p_i. The elastic structure is assumed to be homogeneous, isotropic and linearly elastic, and the fluid is assumed to be inviscid. Body forces are assumed zero. Boundary conditions may be specified along the surfaces s and S. Along s, where B_1 and B_2 are not in contact, we may impose surface tractions,

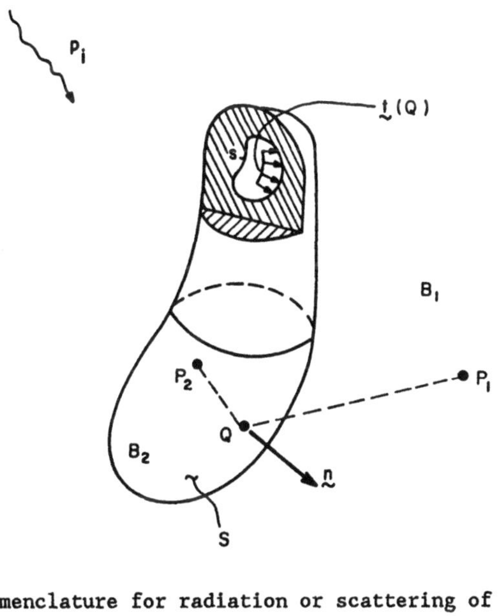

Fig. 1. Nomenclature for radiation or scattering of acoustic waves from an elastic structure.

surface displacements, or a linear combination of the two, corresponding to a well-posed boundary value problem. The appropriate boundary conditions at the interface S are continuity of normal displacement of the structure and the particle displacement of the fluid and equilibrium of the normal stress on the structure and the acoustic pressure. (A complete statement of the problem may be found in Reference 6).

BEM/BEM FORMULATION

To reformulate the problem in Fig. 1 in terms of integral equations valid at the boundary S + s, we use the boundary integral equation for the exterior acoustics problem and a companion interior integral equation for the elastic structure. The resulting boundary integral equation formulation for our coupled problem is [6]:

$$\underset{\sim}{C}^T(P_2)\underset{\sim}{u}(P_2) = \int_{S+s} [\, \underset{\sim}{U}^T(P_2,Q)\underset{\sim}{t}(Q) - \underset{\sim}{T}^T(P_2,Q)\underset{\sim}{u}(Q)] \, dS(Q) \,, \tag{1}$$

which is valid for P_2 in B_2 or on S + s and

$$C(P_1)p(P_1) = \int_S [\, p(Q) \frac{\partial\psi}{\partial n}(P_1,Q) - \frac{\partial p}{\partial n}(Q)\, \psi(P_1,Q)] \, dS(Q) + 4\pi p_i(P_1) \,, \tag{2}$$

which is valid for P_1 in B_1 or on S. In the above equations, $\underset{\sim}{U}$ and $\underset{\sim}{T}$ are the Stokes' displacement and traction tensors, respectively, $\psi(P_1,Q)$ is the free-space Green function, $\underset{\sim}{u}$ and $\underset{\sim}{t}$ are the displacement and the stress, respectively, in the elastic structure and p and $\partial p/\partial n$ are the sound pressure and the normal derivative of the sound pressure, respectively, in the acoustic fluid. Special care has to be used in integrating Eqs. (1) and (2) numerically due to the singularities in the integrands. However, it is noted that both integrals exist.

A numerical solution to Eqs. (1) and (2) can be achieved by discretizing the boundary into a number of surface elements. In this paper, we use isoparametric quadratic elements [6]. In numerical form Eq. (1) becomes

$$[H]\{u\} = [\tilde{G}]\{t\} \quad , \tag{3}$$

where H and $\tilde{G}$ are the assembled coefficient matrices resulting from the numerical integration. The momentum equation is used to replace the normal derivative of the pressure in Eq. (2) with displacement, and the resulting numerical form of Eq. (2) is

$$[T]\{u\} = [D]\{p\} + 4\pi\{p_i\} \quad , \tag{4}$$

where T and D are the assembled coefficient matrices. The coupled elastic/acoustic system of equations can be obtained by combining Eqs. (3) and (4) and replacing the tractions on the surface with pressure. The resulting linear system of equations can be solved for the nodal values of displacements and acoustic pressure on the elastic/acoustic interface as well as the displacements (or tractions) on the inner solid surface.

FEM/BEM FORMULATION

Although, theoretically, there is no limitation on the closeness between the outer surface S and the inner surface s, the BEM/BEM approach will require a large number of integration points and elements if the elastic structure is an extremely thin shell. An alternative way to model an elastic thin shell is the finite element method [7]. If we use the FEM to model the elastic thin shell and the BEM to model the acoustic field, the resulting matrix representation is

$$\begin{bmatrix} K - \omega^2 M & L \\ T & -D \end{bmatrix} \begin{bmatrix} u \\ p \end{bmatrix} = \begin{bmatrix} f \\ 4\pi p_i \end{bmatrix} \tag{5}$$

where K is the stiffness matrix, M is the mass matrix, f is the forcing vector due to the applied loading, and L takes care of the loading from the sound pressure at the interface.

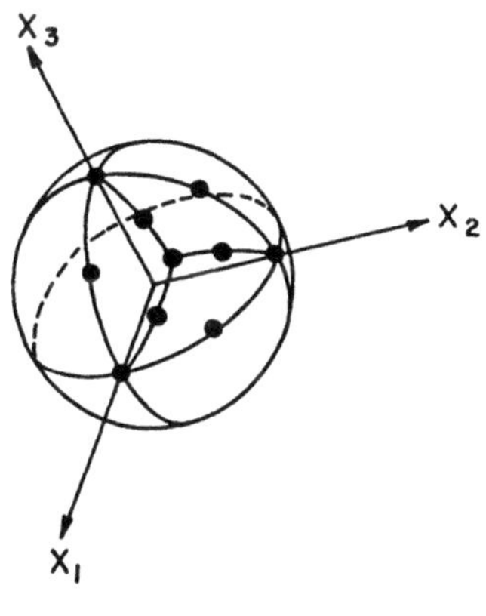

Fig. 2. One octant of the outer surface of the sphere showing nodes and element configuration.

EXAMPLE PROBLEMS

We tested the BEM/BEM and the FEM/BEM methods by solving several problems in structural acoustics having spherical geometry. A typical mesh used in these problems is shown in Fig. 2. The first problem that we consider is the radiation from a hollow steel sphere in sea water with inner radius b = 0.5 m and outer radius a = 1 m. The inner surface is loaded with a uniform time-harmonic pressure of amplitude P_0. As this is a thick, homogeneous shell, the BEM/BEM technique is used. Figure 3 shows the sound pressure at a distance of 3a from the center of the hollow sphere plotted versus ka where k is the wave number in the fluid. The data in Fig. 3 show the rapid convergence that may be expected with the BEM/BEM. The numerical solution, however, is not as good at the resonance frequency (ka=8) as it is at other frequencies. Our numerical study showed that the error at ka=8 will be reduced to 0.7 percent if a mesh of 386 nodes is used.

Next, we consider a BEM/BEM test case consisting of the scattering of an incident plane wave of amplitude P_i by the hollow sphere described above. Figure 4 shows the ratio of the scattered pressure to the incident pressure at a distance of 3a from the center of the sphere for a sphere of various composition when ka = 1. The plane wave is incident on the hollow sphere from the right in Fig. 4. The results for the rigid sphere are also shown

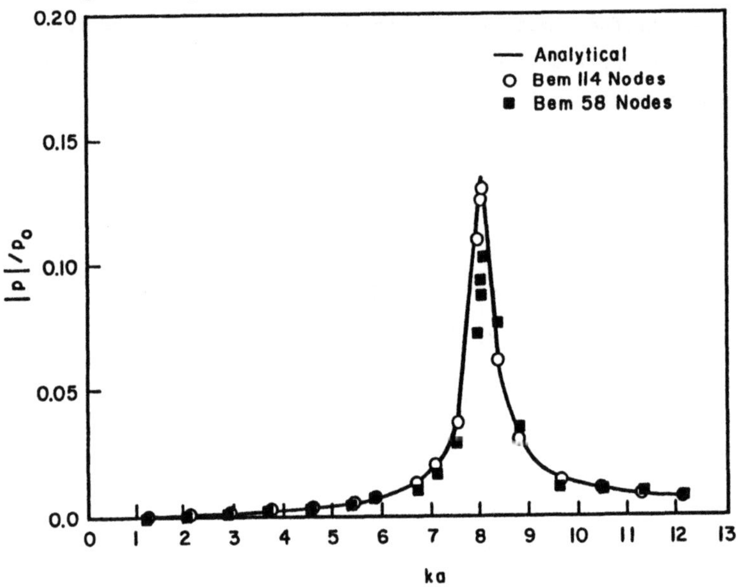

Fig. 3. Normalized sound pressure at a distance of 3a from the center of a hollow steel sphere in sea water excited by an uniform internal pressure P_0.

in Fig. 4 for comparison. All of the scattering data were obtained with the 114 node BEM/BEM model discussed above. The effect of material composition on the scattered pressure may be seen quite clearly in Fig. 4.

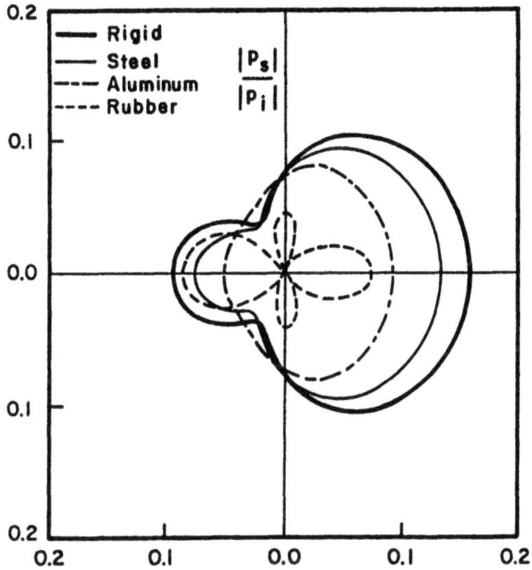

Fig. 4. Ratio of the scattered to the incident pressure at a distance of 3a from the center of a hollow sphere of various compositions in sea water when ka = 1.

AN INTERIOR STRUCTURAL/ACOUSTICS PROBLEM

As a test of the FEM/BEM technique, consider an air-filled spherical aluminum shell of radius 1 m and thickness 5 mm excited externally by a uniform pressure of 1 Pa (94 dB sound pressure level). Figure 5 shows the sound pressure level at the center of the spherical cavity as a function of frequency. The frequency regions where the sound pressure level is high correspond to the resonances of the air cavity within the shell.

DISCUSSION AND SUMMARY

Two numerical methods have been used to solve structural acoustics problems. In the BEM/BEM method, we have reformulated the acoustic radiation and scattering problems for elastic structures in terms of a pair of integral equations, one for the elastic structure, the other for the host fluid and both valid at the interface. These integral equations were reduced to a numerical form using quadratic isoparametric elements.

The BEM/BEM formulation was tested with good results for several radiation

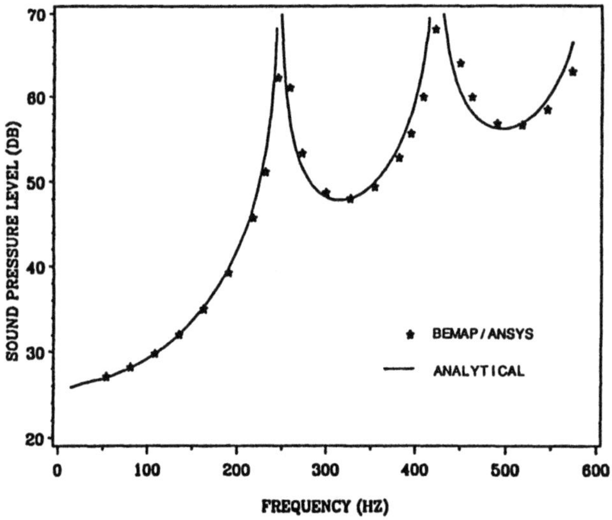

Fig. 5. Sound pressure level at the center of a air-filled aluminum shell of radius 1 m and thickness 5 mm due to a uniform external pressure of 94 dB.

and scattering problems of practical importance. For elastic thin shells, the BEM/BEM approach is still quite reliable if more integration points are used to integrate the integral equations for very close elements. However, a coupled FEM/BEM technique also provides excellent results for thin shell problems. In addition, the FEM/BEM method is more practical in cases where the structure has a complex geometry. A test problem using the FEM/BEM approach showed good results.

1. J. J. Faran, Jr., "Sound Scattering by Solid Cylinders and Spheres," J. Acoust. Soc. Am. 23, 405-418, (1951).

2. M. C. Junger, "Sound Scattering by Thin Elastic Shells," J. Acoust. Soc. Am. 24, 366-373 (1952).

3. R. Hickling, "Analysis of Echoes from a solid Elastic Sphere in Water," J. Acoust. Soc. Am. 34, 1582 (1962).

4. K. Yosioka and Y. Kawasima, "Acoustic Radiation Pressure on a Compressible Sphere," Acustica 5, 167-173 (1955).

5. T. Hasegawa and K. Yosioka, "Acoustic-Radiation Force on a Solid Elastic Sphere," J. Acoust. Soc. Am. 46, 1139-1143 (1969).

6. A. F. Seybert, T. W. Wu and X. F. Wu, "Radiation and Scattering of Acoustic Waves from Elastic Solids and Shells using the Boundary Element Method," J. Acoust. Soc. Am. 84, 1906-1912 (1988).

7. D. T. Wilton, "Acoustic Radiation and Scattering from Elastic Structures," Int. J. Num. Meth. Eng.. 13, 123-138 (1978).

A B.E. - F.E. Coupled Acoustic Element for Fluid-Structure Interaction Problems

C. Rajakumar
Ashraf Ali
Shah M. Yunus

Swanson Analysis Systems, Inc.
Johnson Road
Houston, PA 15342-0065

Summary

An acoustic element capable of coupling the boundary element (BE) and finite element (FE) discretizations has been formulated in 2-D for fluid-structure interaction problems. The Helmholtz equation governing the acoustic pressure in a fluid is discretized using the BE method and coupled to the FE discretizations of a vibrating structure that is in contact with the fluid. The coupled element is validated by verifying the computed results of a simple problem whose theoretical solution is known. Since the BE method naturally maps the infinite fluid domain into finite node points on the fluid-structure interface, the formulation is especially useful for problems where the fluid domain extends to infinity. The vibration response of a submerged cylindrical shell has been computed and compared with the results from an entirely finite element formulation.

Introduction

Engineering applications of acoustic pressure computations often require modeling of both the structure that generates the acoustic pressure wave and the fluid medium that transmits the wave. Zienkiewicz and Newton [1] first presented the finite element (FE) discretization of the acoustic wave equation to solve this fluid-structure interaction problem. Whenever a structure is in contact with fluid medium, the formulation is effectively applied in the literature [1-3] to predict the acoustic pressure and structural displacement response to harmonic excitations induced in either or both the parts of the coupled problem.

The FE procedure is quite effective for problems involving bounded fluid domain. However, when the fluid region is unbounded, such as a vibrating structure placed in an infinite medium of air or water, the FE formulation will require approximations to represent the infinite domain by a finite region of fluid. In the boundary element (BE) discretization of the Holmholtz equation, the inherent incorporation of the radiation boundary condition of the acoustic pressure allows the infinite extent of fluid to be modeled without any approximating assumptions. Therefore, the BE formulation is well suited to model the fluid region that extends to infinity. For the structure that is present in the fluid the FE discretizations are extensively available in general purpose codes such as the ANSYS® program. Therefore a coupled BE-FE formulation, where the acoustical equation is discretized using BE and the structure by FE, would find a wider application in solving fluid-structure interaction problems involving both finite and infinite fluid domain. In this paper we present the coupled formulation for 2-D problems which can be extended to 3-D case.

The idea of using BE-FE coupled analysis to solve the acoustic fluid-structure interaction problem is found in the literature [4-5] and we review them briefly here. The paper by Mathews [4] describes two ways of solving the coupled acoustic BE-structural FE problems. One of the methods, called the structural variable methodology, requires the elimination of the unknown surface pressures at the fluid-structure interface to cast the coupled equations in terms of the unknown structural displacements. The second method, namely the fluid variable methodology, involves the elimination of the structural displacements to get an equation in terms of the unknown pressures. These two procedures involve two step computations in order to get the complete solution to the fluid-structure interaction problem. In the structural variable method, the fluid pressures will have to be computed by substituting the displacement solution in the acoustic BE equation, and vice versa in the fluid variable method.

Wilton [5] presents a coupled BE-FE analysis procedure where the fluid variable method is used to study the scattering of pressure waves incident upon elastic spherical shells. Sound radiated from partially submerged elastic cylindrical shells has been studied by Iwasaki et al [6] using coupled structural FE-acoustic BE discretizations. The structural displacement and the fluid velocity potential have been solved in a single step by setting up the fluid-structure coupled matrix equation. Cunefare and Kooperman [7] show the nonuniqueness of solutions from a direct BE discretization of the exterior domain of a rigid sphere at frequencies that match with the resonant frequencies of the internal domain. A method of eliminating this nonuniqueness difficulty of the BE formulation for exterior fluid domain is described in the paper.

In the present paper, the discretized equations of the fluid domain and the structure are set up in one augmented matrix equation, and solved in a single step to get the acoustic pressure and structural displacement solutions. We have considered the acoustic Helmholtz equation in 2-D to demonstrate the BE-FE coupled analysis of the fluid-structure problem. The formulation presented has been tailored for easy adoption into existing FE codes where an extensive library of structural finite elements are available.

BE Discretization of Helmholtz Equation

The pressure amplitude of harmonically varying acoustic pressure waves in a compressible fluid medium is governed by the Helmholtz equation given by

$$\nabla^2 P + k^2 P = 0 \tag{1}$$

The wave number k is given by $k = \omega/C$ where ω is the circular frequency of oscillations and C is the speed of sound in the fluid. Equation (1) is now considered in 2-D where the Laplacian operator is given by $\nabla^2 = \partial^2/\partial x^2 + \partial^2/\partial y^2$. Au and Brebbia [11] have formulated constant, linear and quadratic boundary elements using an approximate expansion of the Hankel function, which constitutes the fundamental solution to the 2-D Helmholtz equation. In the present work, 2-D linear boundary element formulation of the Helmholtz equation is carried out with the exact series representation of Hankel function. The procedure is computationally efficient at the same time. The well known boundary integral equation in terms of pressure P and presure gradient $Q = \partial P/\partial n$ is

$$\alpha P_i(\xi) + \int_\Gamma P\, Q^*\, d\Gamma(x) = \int_\Gamma P^*\, Q\, d\Gamma(x) \tag{2}$$

α is a geometric parameter at the source point ξ, x is a field point and Γ is the boundary of the fluid domain. The fundamental solution and its normal derivative P^* and Q^*, respectively, of the Helmholtz equation are given by

$$P^* = \frac{j}{4} H_0^1 (kr) = \frac{j}{4} [J_0(kr) + j\, Y_0(kr)] \tag{3}$$

$$Q^* = - \frac{jk}{4} H_1^1 (kr) \frac{\partial r}{\partial n} = - \frac{jk}{4} [J_1(kr) + j\, Y_1(kr)] \frac{\partial r}{\partial n} \tag{4}$$

H^1 is the Hankel function of the first kind. J and Y are the Bessel functions of the first and second kind respectively. The subscripts 0 and 1 stand for the zeroeth and first order. r is the distance between ξ and x and $j = \sqrt{-1}$. Using the isoparametric concept, the problem variables P and Q , and the geometry S are approximated by linear shape functions

$$P = \sum_{m=1}^{2} \phi_m P_m, \quad Q = \sum_{m=1}^{2} \phi_m Q_m, \quad S = \sum_{m=1}^{2} \phi_m S_m \tag{5}$$

over each segment, Γ_e on the boundary (Fig. 1). Substituting equations (3), (4) and (5) into equation (2) and

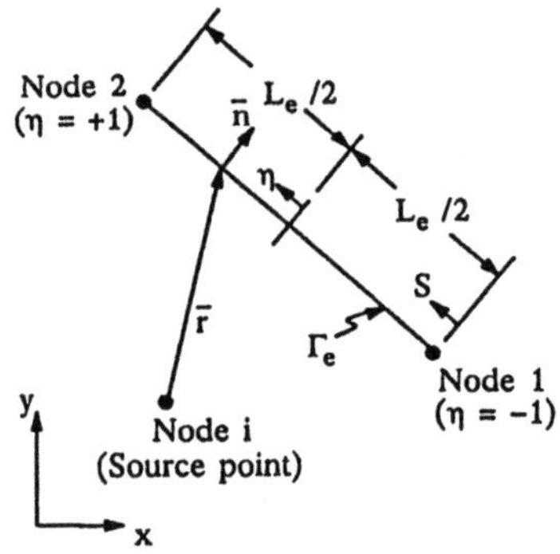

Fig. 1. A linear boundary element (Shape functions: $\phi_1 = (1-\eta)/2$, $\phi_2 = (1+\eta)/2$)

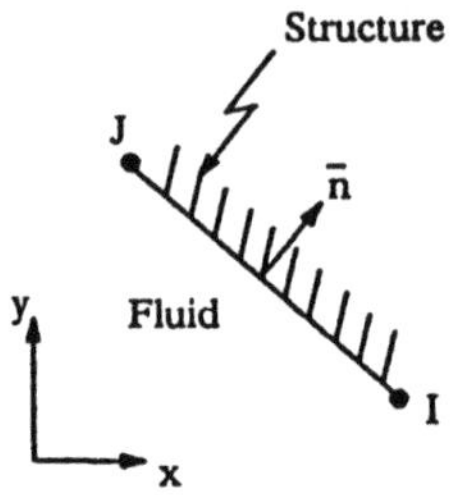

Fig. 2. 2-D BE-FE coupled acoustic element

performing the integrations over all of the segments in the boundary for i = 1, 2, ... n (n = number of boundary nodes), the BE matrix equation of the fluid domain is formed.

$$[H]\{P\} = [G]\{\partial P/\partial n\} \tag{6}$$

When the source point i lies on the segment Γ_e, the resulting singular integrals were evaluated analytically. For nonsingular elements, four point Gaussian quadrature was used to integrate.

FE Matrix Structural Dynamic Equation

Let us now consider a vibrating structure adjacent to the acoustic fluid domain as shown in Fig. (2). The structure is routinely discretized using FE procedures in order to solve for the structural displacement. In this paper, we have used the 2-D finite elements available in the ANSYS® program [8]. The matrix equation of a harmonically vibrating structure is written down as follows:

$$(-\omega^2[M] + j\omega\,[C] + [K])\,\{u\} = \{F_p\} + \{F\} \tag{7}$$

[M], [C] and [K] are the mass, damping and stiffness matrices of the FE discretized structure. {u} is the nodal displacement vector consisting of two component displacements u_x and u_y for the 2-D problem. $\{F_p\}$ is the pressure force exerted by the fluid on the structure at the fluid-structure interface and {F} represents all other nodal forces that may be acting on the structure. The pressure force at the nodes on the interface boundary of the structure is given by:

$$\{F_p\} = \sum_e \int_{\Gamma e} \Psi_i \; P \; \bar{n} \; dA \tag{8}$$

Ψ_i is the shape function employed in the FE discretization of the structure. $\bar{n}$ is the unit outward normal at the interface boundary of the structure. Σ_e represents summation over all the elements on the interface boundary. The shape functions employed in the BE discretization of the acoustic Holmholtz equation is used to represent the pressure in equation (8). Substituting for P from equation (5) into (8) and in turn into equation (7), we rewrite the FE matrix equation of the structure as follows:

$$(-\,\omega^2[M] + j\omega\,[C] + [K])\;\{u\} - [R]\;\{P\} = \{F\} \tag{9}$$

Matrix [R] of the interface is formed by the FE process and is given by

$$[R] = \sum_e \int_{\Gamma e} \Psi_i \; \phi_j \; \bar{n} \; dA \tag{10}$$

Coupled BE-FE Equation

In the BE discretized equation (6), we can express the pressure gradient at the boundary of the fluid domain. In terms of the normal structural displacement $\{u_n\}$ using the following equilibrium condition at the interface.

$$\{\frac{\partial P}{\partial n}\} = \rho\omega^2 \; \{u_n\} = \rho\omega^2 \langle n_x \;\; n_y \rangle \begin{Bmatrix} u_x \\ u_y \end{Bmatrix} \tag{11}$$

ρ is the fluid density. u_x and u_y are the x and y displacement components. n_x and n_y are the x and y components of the unit normal. Substitution of equation (11) into (6) results in an equation relating the nodal pressures and displacements at the fluid boundary.

$$[H]\{P\} = \rho\omega^2 \; [n_x G \;\vdots\; n_y G] \begin{Bmatrix} u_x \\ u_y \end{Bmatrix} \tag{12}$$

Absorbing the unit normal components into the BE matrix [G], we can redefine [G] as a rectangular matrix.

$$[G] = [n_x G \;\vdots\; n_y G] \tag{13}$$

The acoustic BE matrix equation is then written down in a compact form from equation (12):

$$[H]\{P\} - \rho\omega^2 [G]\{u\} = 0 \tag{14}$$

Now, we can combine equations (9) and (14) in order to form the BE-FE coupled matrix equation as follows:

$$\begin{bmatrix} (-\omega^2 M + j\omega C + K) & -R \\ -\rho\omega^2 G & H \end{bmatrix} \begin{Bmatrix} u \\ P \end{Bmatrix} = \begin{Bmatrix} F \\ 0 \end{Bmatrix} \tag{15}$$

Equation (15) represents a typical interface element matrix equation where the coupling of fluid and structure is brought in by the off-diagonal submatrices -[R] and $-\rho\omega^2[G]$. The top row of submatrices is FE-based and can be assembled by FE assembly procedure to form the global matrices of a fluid-structure problem. However, the bottom row of matrices are BE-based and need to be formed in a special way, as indicated in the next section, in order to take advantage of the FE assembly algorithm.

Implementation of the BE-FE Coupled Acoustic Element

Let us consider a two-noded interface element shown in Fig. 2. Using linear shape functions for both the pressure and displacements we write the matrix equation of the interface element as,

$$\begin{bmatrix} 0 & -R_e \\ -\rho\omega^2 G_e & H_e \end{bmatrix} \begin{Bmatrix} u_I \\ u_J \\ P_I \\ P_J \end{Bmatrix} = \begin{Bmatrix} F_I \\ F_J \\ 0 \\ 0 \end{Bmatrix} \tag{16}$$

The top left hand submatrix will be filled when a structural element is assembled with the interface element at nodes I and J. Coupling matrix R_e in this case is a 4x2 matrix. Matrices G_e and H_e are from BE discretization and hence will have contribution from all the nodes on the fluid boundary. Therefore, if there are n boundary nodes, they will be rectangular matrices of size nx4 and nx2, respectively. For example, the matrix H_e of an interface element will look like,

$$H_e = \begin{bmatrix} h_{1I} & h_{1J} \\ h_{2I} & h_{2J} \\ \vdots & \vdots \\ h_{nI} & h_{nJ} \end{bmatrix} \tag{17}$$

As is evident from equation (17), in a finite element sense, all of the nodes on the boundary connect to the interface element IJ.

In order to fit into the highly modular arrangement of FE codes, where the 'element' nodes connect only to those which share a common node point, we use a three-node definition for the interface element. The interface element is defined by three nodes namely I, J and K, the third node K being the 'source node' in BE terminology. So, to form the matrices G_e and H_e of the interface element IJ, we define n such three-node elements by letting K=1, 2...n as shown in Fig. 3. It is easy to see that each of the three-node elements will contribute submatrices of size 1x4 and 1x2 respectively of the interface element matrices G_e and H_e. By this procedure the FE assembly algorithm can be used to assemble and solve the coupled BE-FE equation (15) of the fluid-structure interaction problem.

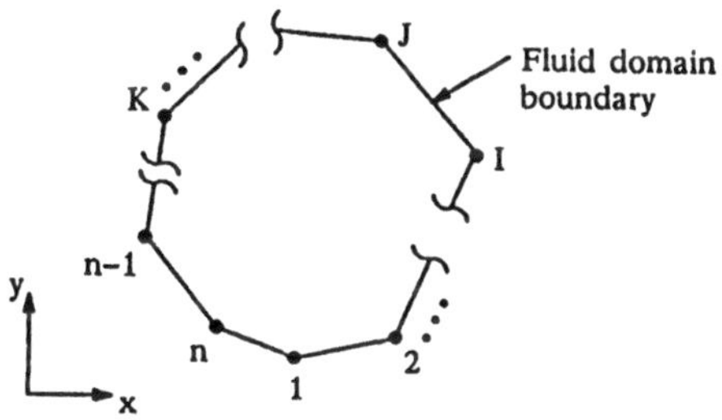

Fig. 3. Three node definition of the interface element I J (K = 1, 2...n)

Applications of BE-FE Coupled Element

The coupled element presented here has been tested by solving problems with known solutions. The example problems demonstrate the applicability of the BE-FE coupled acoustic element to both external and internal domain problems. The fluid-structure coupling is demonstrated by solving an impedance tube problem with an oscillating piston at one end of the tube. In addition, the element is applied to compute the vibration response of a cylindrical shell submerged under water.

Impedance Tube Problem:

The impedance tube shown in Fig. 4 is taken from the reference by Rajakumar [9]. For a known piston velocity of V_x = 1.27 m/s at x = 0, the magnitude of the pressure on the piston surface is plotted in Fig. 5 for frequencies ranging from f = 100 to 1000 Hz. The solid line represents the theoretical solution to the 1-D problem, which is given by

$$(P)_{x=0} = -j \rho C V_x \cot (kL) \tag{18}$$

Finite element solutions using 14 elements along the length of the tube are plotted for comparison. The same number of elements have been used in the BE discretization. The BE-FE coupled elements were used to represent the fluid-structure interface on the piston. At higher frequencies, the BE-FE solutions stay close to the theoretical values while the FE-FE solutions deviate. In order to improve the FE-FE results, the number of elements along the length of the tube will have to be increased at higher frequencies.

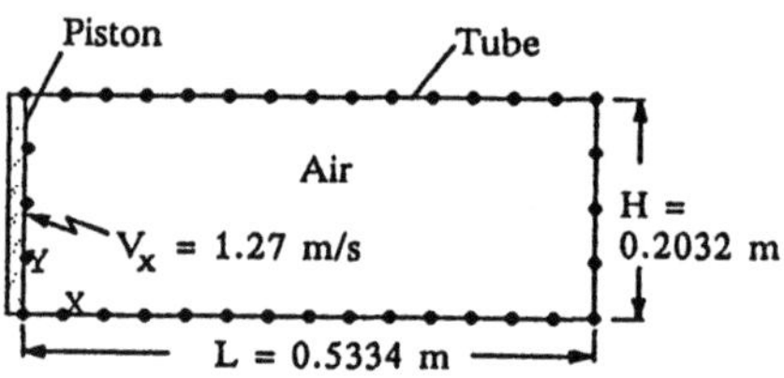

Fig. 4. Impedance tube (Air: Density ρ = 1.12 kg/m^3, Speed of sound C = 344.424 m/s)

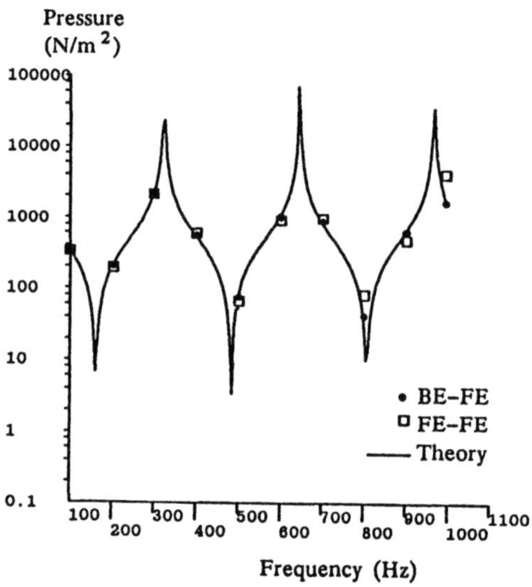

Fig. 5. Pressure magnitude on the piston surface of the impedance tube

Dynamic Response of a Submerged Cylindrical Shell:

A practical application of the BE-FE coupled element for infinite fluid domain problems is the dynamics of submerged structures. Here we study the dynamic response of a circular cylindrical shell submerged in an infinite body of water. Schroeder and Marcus [10] have presented the eigenfrequencies of the problem in plane 2-D assuming the shell to be infinitely long. The same dimensions and material properties of the shell used by them have been employed in the vibration response computations in order to validate the results obtained. The cross-section of the steel cylindrical shell is shown in Fig. 6. It is discretized using 32 finite elements around the circumference. Plane 2-D elements of the ANSYS program has been used to model the shell. The elements representing plane strain conditions allow bilinear variation of the u_x and u_y displacements. The BE discretization of the external fluid domain consists of 32 nodes representing the fluid pressure on the outer surface of the shell. The BE-FE acoustic elements used on this interface boundary, therefore, couple the shell displacement and fluid pressure.

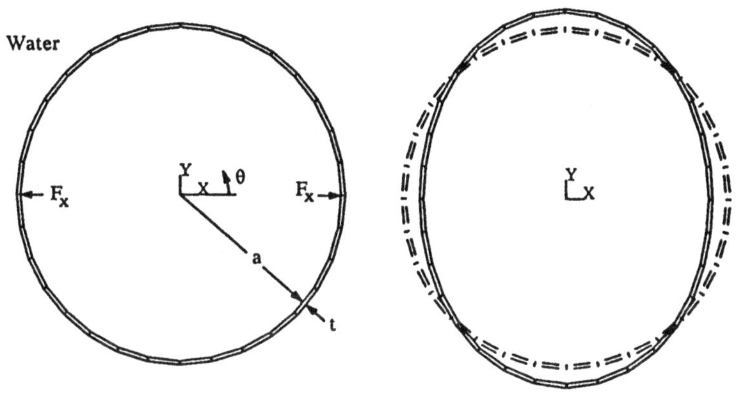

(a) Shell Cross Section

(b) Mode 2

Fig. 6. Cylindrical shell and mode investigated (Shell: a = 0.254 m, t = 0.00635 m, Young's modulus E = 2.068E11 N/m^2, Density ρ_s = 7929 kg/m^3, Poisson's ratio ν = 0. Water: Speed of sound C = 1460 m/s, Density ρ_w = 1030 kg/m^3)

In order to excite the shell into its first fundamental circumferential mode, designated as mode 2 in reference [10], two diametrically opposing forces were applied as shown in Fig. 6. For frequencies in the neighborhood of mode 2 resonance, the radial displacement at the outer radius at $\theta = 0$ location is plotted in Fig. 7. For comparison, the acoustic FE-structural FE solutions considering finite extents of the water surrounding the ring and vacuum beyond are plotted. In addition, the FE solution of the ring in vacuum is also plotted. Concentric fluid boundaries placed at radii R = 0.559 m and 0.828 m are the two cases of finite fluid extent that were modeled by the linear acoustic fluid finite element of the ANSYS program. The same level of circumferential discretization as in BE-FE solution was used. In the fluid domain, 4 and 6 radial divisions were used respectively, for R = 0.559 and 0.828 m. Zero pressure was specified on the outer fluid boundary to represent vacuum beyond radius R. From the graphs in Fig. 7, the resonant peaks of the submerged ring shift to the left as the radial extent of the fluid domain is increased in the acoustic FE solution. The response plotted from the BE-FE coupled formulation represents the solution to the problem as posed when the ring is submerged in a fluid extending to infinity. Resonant frequency obtained from this frequency response result is f = 35.75 Hz and it closely matches with the analytically found eigenfrequency of 35.76 Hz given by Shroeder and Marcus [10]. Although, with increasing R, the resonant frequency of FE solution converged to within one percent of the infinite fluid domain frequency, the response plots in Fig. 7 show a much larger error in the displacement of the ring, especially close to the resonance frequency.

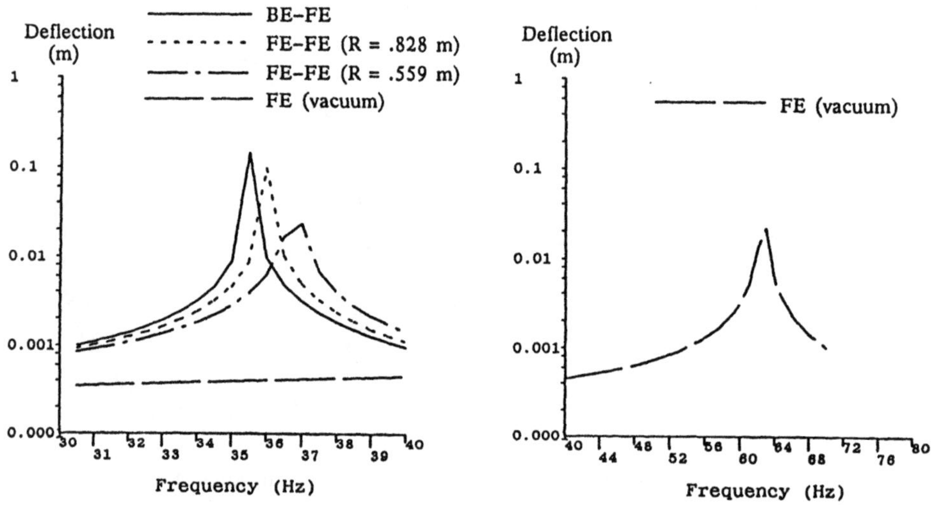

(a) Shell in water (b) Shell in vacuum

Fig 7. Response of the cylindrical shell (Radial deflection at outer radius at θ = 0).

Conclusion

The example problems presented show the validity of the BE-FE coupled formulation for both internal and external acoustic fluid domain problems. For external domain problems, the method is quite powerful when compared with the FE-FE method as pointed out in the submerged ring problem. When modeling the infinite fluid domain, the acoustic FE matrices quickly grow in size resulting in a prohibitively high computational cost, unless some form of 'infinite' finite elements are used. Furthermore, for arbitrary shaped submerged structures, there will be a degree of uncertainty about the response computed using a finite extent of the infinite fluid domain. The BE-FE coupled formulation, on the other hand, is very efficient both in terms of the computational effort and the accuracy of the computed solution.

The formulation of the 2-D BE-FE coupled acoustic element is easy to implement in existing FE codes and also can be extended to 3-D case. The issue of the BE discretization of exterior domain failing to yield a unique solution at the eigenfrequencies of the interior domain Dirichlet problem will have to be addressed to make the procedure robust. There are methods found in the literature, such as in reference [7], which can be incorporated in the BE formulation to overcome this difficulty. The acoustic BE-structural FE coupled element will then be a powerful tool for a wide range of acoustic fluid-structure problems.

Acknowledgement

The help of Michael P. Johnson of Westinghouse Electric Corp., Cleveland, OH in providing us with the relevant literature is gratefully acknowledged. Also the assistance from Jackie Williamson and Charmaine Grunick of Swanson Analysis Systems in preparing the manuscript is appreciated.

References

1. Zienkiewicz, O. C.; Newton, R. E.: Coupled vibrations of a structure submerged in a compressible fluid. Proc. Int. Symp.on Finite Element Techniques, Stuttgart (1969) 359-379.

2. Craggs, A.: A finite element model for acoustically lined small rooms. J. of Sound and Vibration 108(2) (1986) 327-337.

3. Zienkiewicz, O. C.; Bando, K.; Bettes, P.; Emson, C.; Chiam, T. C.: Mapped infinite elements for exterior wave problems. Int. J. of Numer. Methods in Eng. 21 (1985) 1229-1251.

4. Mathews, I. C.: Numerical techniques for three-dimensional steady-state fluid-structure interaction. J. of Acoustical Society of America 79(5) (1986) 1317-1325.

5. Wilton, D. T.: Acoustic radiation and scattering from elastic structures. Int. J. of Numer. Methods in Eng. 13 (1978) 123-138.

6. Iwasaki, Y.; Kawabe, H.; Bessho, M.: The underwater sound scattering problem from the floating elastic shell. Boundary Elements IX, Vol: 3 Fluid Flow and Potential Applications, Edited by Brebbia, C. A.; Wendland, C. A.; Kuhn, G.; Springer-Verlag, London (1987) 53-64.

7. Cunefare, K. A.; Kooperman, G.: A boundary element method for acoustic radiation valid for all wave numbers. J. of Acoustical Society of America 85(1) (1989) 39-48.

8. Kohnke, P.: ANSYS® Engineering Analysis System Theoretical Manual, Rev. 4.4, Swanson Analysis Systems, Inc., Houston, PA, (1989).

9. Rajakumar, C.: A comparison of acoustic finite element solutions to theory. ANSYS *News®*, Fourth Issue, Swanson Analysis Systems Inc., Houston, PA 15342, (1988)

10. Schroeder, E. A.; Marcus, M. S.: Finite element solution of fluid-structure interaction problems. 46th Shock and Vibration Symp., San Diego, CA (October 1975) 1-18

11. Au, M. C.; Brebbia, C. A.: Diffraction of water waves for vertical cylinders using boundary elements. Appl. Math. Modeling 7 (1983) 106-114.

Thermal Aspects of Metal Cutting: A Boundary Element Approach

A. Chandra and C. L. Chan

Department of Aerospace and Mechanical Engineering
University of Arizona, Tucson AZ 85721

Abstract

In this paper, the boundary element method (BEM) approach is used to analyze the thermal aspects of steady state metal cutting processes. Particular attention is paid to modelling of the boundary conditions at the tool-chip and the chip-workpiece interfaces. Since the velocities in each of the regions are different, the heat transfer within the tool, the chip, and the workpiece are first calculated separately. A complete model for heat transfer during steady state turning is then obtained by matching the boundary conditions across the primary and the secondary shear zones. An exact expression for matching is developed to avoid any iterations. The temperature fields within the workpiece, the chip and the tool for various processing conditions are obtained and presented. Design sensitivities of temperature distributions to various shape and process parameters are considered next. The BEM design sensitivity formulation is based on direct differentiation (DDA) of the relevant BEM equations. It retains the advantages of the BEM regarding accuracy and efficiency while avoiding strongly singular kernels. The numerical results obtained by the BEM are also compared to existing FEM results reported in the literature. The BEM is found to be very efficient and robust for this class of machining problems and provides a new avenue for analysis and design of the thermal aspects of various machining processes.

Introduction

In machining operations, the elevated temperature fields generated by the severe inelastic deformations in the shear plane along with the frictional conditions at the chip-tool interfaces, play crucial roles in determining the chip formation mechanics. Accordingly, the elevated temperature fields have profound effects on the surface quality of the finished product and the process efficiency. Tool life, as limited by cutting tool wear, depends to a large extent on the temperature in the vicinity of the cutting edge, which in turn imposes a practical limit on the rate of material removal. Excessive temperatures may lead to various types of surface damage. The shear zone temperatures in metal cutting influence the deformation processes, the occurrence of instabilities and the behavior of free machining inclusions. There exists various analytical and finite element analyses of heat conduction with moving or stationary heat sources together with kinematic, geometric and energetic aspects of the metal cutting process (e.g. [1-4]).

Tanaka, et. al. [5] have developed a BEM formulation for steady state conduction-convection problems. Chan and Chandra [6] have extended the BEM approach to investigate the thermal aspects of metal cutting operations. They have paid particular attention to the interface conditions at the tool-chip and chip-workpiece interfaces, and obtained a complete heat transfer model for steady state turning by matching the boundary conditions across the interfaces.

These analyses provide crucial insights into the thermal aspects of machining processes. On the one hand, numerical simulation may be used to gain insight into the physics of various aspects of the problem and on the other hand, computation methods may be used by the designer to predict the characteristics of alternative designs for heat transfer. Assuming that one has the ability to analyze the problem at hand, the question then arises of how to modify the design to improve the performance. Accordingly, accurate and efficient determination of design sensitivity is also of crucial importance.

A large amount of literature exists on evaluation of design sensitivity coefficients (DSC's) in linear problem of solid mechanics (e.g. [7]). In general, two methods emerge as the most powerful ones for the determination of design sensitivities. These are DDA and the adjoint structure approach (ASA). In the present work, DDA in conjunction with the BEM [8] is used to obtain the DSC's. It is important to note that the differentiation procedures developed by this technique do not increase the singularities of the relevant kernels.

This paper begins with a BEM formulation for analyzing the heat transfer problem during steady state turning operations. Design sensitivities are considered next. Numerical results are presented. The salient features of temperature distributions in machining processes are discussed in light of these results.

Boundary Element Formulation

A brief review of the BEM formulation for the steady state conduction-convection problem is presented in this section. Special attention is paid to modelling of the boundary conditions in turning operations along with interface conditions at the tool-workpiece, the tool-chip and the chip-workpiece interfaces. The heat transfers within the workpiece, the tool and the chip are first calculated separately due to the fact that the velocities in each of these regions are different. A complete model for heat transfer during steady state turning is then obtained by matching the boundary conditions across the interfaces. An exact expression for matching is developed to satisfy the matching boundary conditions without any iteration. The BEM equations are then differentiated to obtain the design sensitivities of temperature and flux distributions.

As shown by Chan and Chandra [6] , a boundary integral equation for steady state conduction-convection problem may be written as,

$$c(P)\,T(P) = \int_{\Gamma} \left\{ \left[\frac{\partial G(P,Q)}{\partial n_Q} T(Q) - G(P,Q) \frac{\partial T(Q)}{\partial n_Q} \right] + G(P,Q)\,T(Q)\,v_i^{(s)}\,n_i(Q) \right\} d\Gamma \qquad (1)$$

Here, P and Q are boundary source and field points respectively. The coefficient c(P) in general depends on the local geometry at P. If the boundary is locally smooth at P, c(P) = 1/2. Otherwise, it may be evaluated indirectly [9,10]

The heat generated at the shear plane due to plastic deformation is the sum of the heat flux going into the workpiece and the chip. Similarly, at the tool-chip interface, the total heat generated due to frictional condition and plastic deformation is dissipated into the chip and the tool. Further, the temperature continuity must be enforced at all the interfaces. Based on these interface conditions, an exact expression for the determination of the interface temperatures can be derived as [6].

At the chip-workpiece interface,

$$\sum_{\ell=1}^{k_2-k_1+1}\left(\frac{\partial q^{*(n)}_{c_{j-1+k_1}}}{\partial T^{*}_{c_{\ell-1+k_1}}}+\frac{\partial q^{*(n)}_{w_{\ell_2-j+1}}}{\partial T^{*}_{w_{\ell_2-\ell+1}}}\right)(T^{*}_{c_{\ell-1+k_1}}-T^{*(o)}_{c_{\ell-1+k_1}})$$

$$+\sum_{\ell=k_2-k_1+2}^{k_4-k_3+k_2-k_1+3}\frac{\partial q^{*(n)}_{c_{j-1+k_1}}}{\partial T^{*}_{c_{\ell-k_2-k_1+2+k_3}}}(T^{*}_{c_{\ell-k_2-k_1+2+k_3}}-T^{*(o)}_{c_{\ell-k_2-k_1+2+k_3}})$$

$$=1-q^{*(n)(o)}_{c_{j-1+k_1}}-q^{*(n)(o)}_{w_{\ell_2-j+1}} \qquad (2)$$

Here, nodes ℓ_1 to ℓ_2 belong to the workpiece and nodes k_1 to k_2 belong to the chip at the chip-workpiece interface, while nodes k_3 to k_4 belong to the chip at the chip-tool interface. A superscript (*) implies dimensionless qunatities. Also, superscript (o) denotes an arbitrary interface tempearture and the corresponding flux. Similarly, an equation for the chip-tool interface may also be derived.

Differentiating equation (1) with respect to a typical design parameter (designated here as b), after some algebraic manipulations we obtain the BEM design sensitivity equation as,

$$0=\int_\Gamma\left\{\left[\frac{\partial G(P,Q)}{\partial n_Q}-G(P,Q)\,v_i^{(s)}\,n_i(Q)\right][\tilde{T}(Q)-\tilde{T}(P)]-G(P,Q)\frac{\partial \tilde{T}(Q)}{\partial n_Q}\right\}d\Gamma$$

$$+\int_\Gamma\left\{\left[\widetilde{\frac{\partial G(P,Q)}{\partial n_Q}}-\tilde{G}(P,Q)\,v_i^{(s)}\,n_i(Q)-G(P,Q)\,\tilde{v}_i^{(s)}\,n_i(Q)\right.\right.$$

$$\left.\left.-G(P,Q)\,v_i^{(s)}\,\tilde{n}_i(Q)\right][T(Q)-T(P)]-\tilde{G}(P,Q)\frac{\partial T(Q)}{\partial n_Q}\right\}d\Gamma$$

$$+\int_\Gamma\left[\frac{\partial G(P,Q)}{\partial n_Q}-G(P,Q)\,v_i^{(s)}\,n_i(Q)\right][T(Q)-T(P)]-G(P,Q)\frac{\partial T(Q)}{\partial n_Q}\,d\tilde{\Gamma} \qquad (4)$$

where a superposed tilda (~) denotes a derivative with respect to b.

Numerical Results and Discussion

Steady state turning operations are considered in this section. A schematic diagram of the machining process is shown is Figure 1. In all of the following cases, the thermophysical properties are assumed to be the same in all three regions. This assumption does not change the physics of the problem but simplify the numerical implementation. These calculations provide an indication of the speed and efficiency of the method. The boundaries of the workpiece, the chip, and the tool are discretized into 62, 47 and 50 segments respectively. The segments are divided in such a way that fine scales are used to resolve the high gradient region. The calculation is done on a Sun Microsystem 3/260. The CPU time for all cases is approximately 0.4 second.

The dimensionless surface temperature ($Pe\ (T - T_\infty)\ k_w / q' \ell$) along the primary shear plane for several different Peclet numbers ranging from 1 to 40 are plotted in Figure 2. The abscissa is the dimensionless distance along the shear plane (see Figure 1) measured from A (x = 0) to B (x = 1). It can be observed that the temperature increases gradually, attains a maximum and decreases. For Pe = 1, the maxima occurs at x = 0.64. As the Peclet number increases the location of the maxima shifts to the right and for Pe = 40, the maxima occurs at x = 0.91. Its magnitude also changes from 0.637 to 0.051 as the Peclet number goes from 1 to 40. The comparisons between the present BEM results and those obtained by Dawson and Malkin [4] using FEM are presented in Figure 3. The Peclet number is 45 and the shear angle is 30 degrees. For this case, the maximum discrepancy between the two is 20 percent. It should be noted that the FEM solution presented by Dawson and Malkin [4] accounts for the workpiece only, where as, the present BEM solution accounts for all three regions.

The dimensionless temperatures along the chip-tool interface for five different Peclet number are plotted in Figure 4. The abscissa is the dimensionless distance measured from B (x = 0) to C (x = 1) along the tool-chip interface (see Figure 1). For Pe = 1, the maximum dimensionless temperature is 0.598 and occurs at x = 0.1. As the Peclet number increases the temperature peak shifts to the right and for Pe = 40, a temperature peak of 0.087 occurs at x = 0.5. The maximum temperature is believed to be responsible of the crater wear in the tool.

The variation of the distribution of the total heat flux (primary and secondary) dissipated into the chip, the tool and the workpiece with respect to the Peclet number is plotted in Figure 5. At Pe = 1, about 25 percent of the total heat generated is carried into the chip while about 50 percent goes into the workpiece. As the Peclet number increases, the portion of the heat carried by the chip increases rapidly. For Pe = 50, about 90 percent of the heat generated dissipates through the chip, while only 4 percent goes to the workpiece.

The shear plane geometry is varied as a parabola ($x = \eta$; $y = b(\eta - 1/2)(\eta + 1/2)$). The nominal configuration is a straight line (b = 0). Figure 6 shows the temperature and its sensitvity along a shear plane (nominal configuration) for Pe = 10 and 30 at a shear angle of 45°. The abscissa is the dimensionless distance along the shear plane (see Figure 1) measured from A (x = 0) to B (x = 1). The sensitivity of the temperature decreases with the Peclet number and As Pe goes from to 10 to 30, the maximum temperature sensitivity goes from -0.06 to -0.025. The position of the maxima also shifts toward point A (in Figure 1).

The wear of the tool nose with machining is considered next. The nose curve is parameterized as ($x = \eta$; $y = a(b \exp(-|\eta| /b) + |\eta|)$). The parameter b = 0 represents a perfectly sharp tool and the blunting of the nose is repesented by increasing b. The parameter a is the cotangent of the tool wedge angle. Figure 7 shows the temperature sensitivity along the chip-tool interface. The abscissa is the distance measured from point B in Figure 1. In this case, the nominal configuration is of crucial importance. The temperature sensitivity increases significantly as the tool wears out. The maximum temperature sensitivity occurs at the point B and goes from 0.021 to 0.129 as b goes from 0.01 to 0.02.

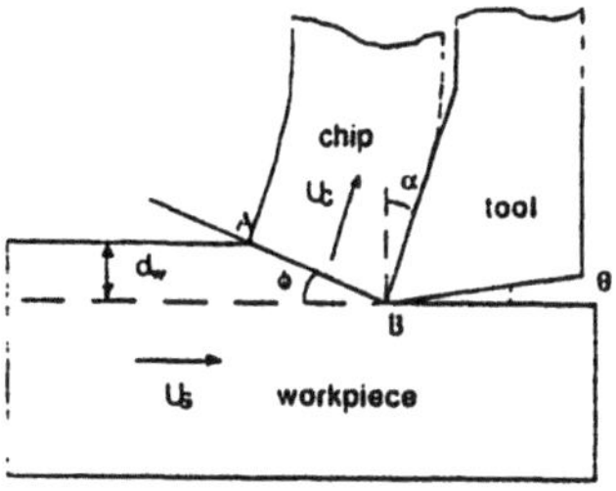

Figure 1 Schematic Diagram of Machining

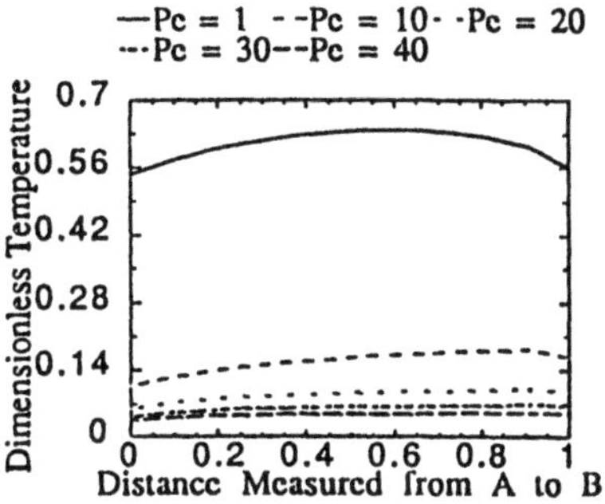

Figure 2 Temperature Along the Shear Plane

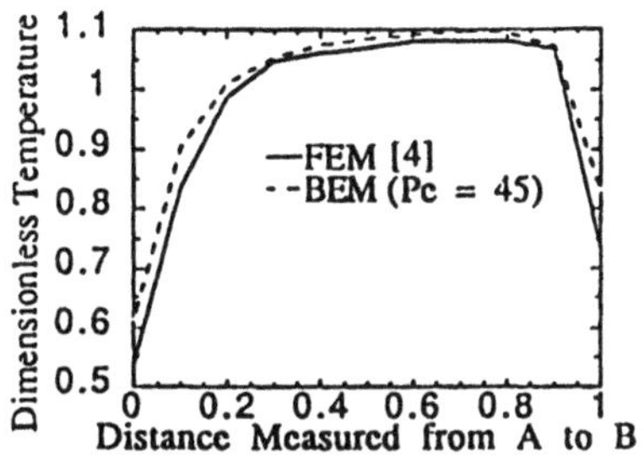

Figure 3 Comparisions of BEM and FEM

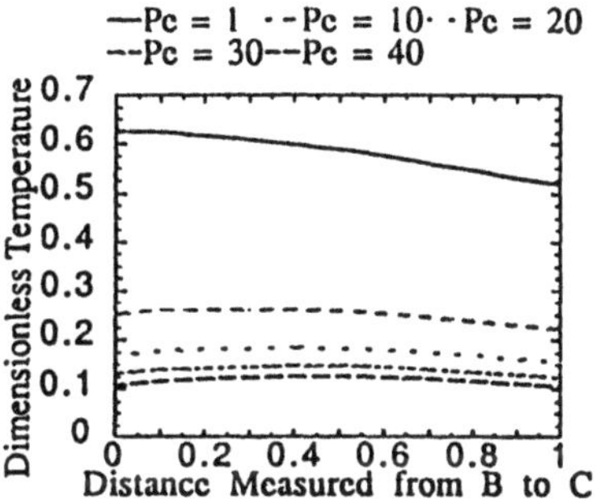

Figure 4 Temperature Along Contact Surface

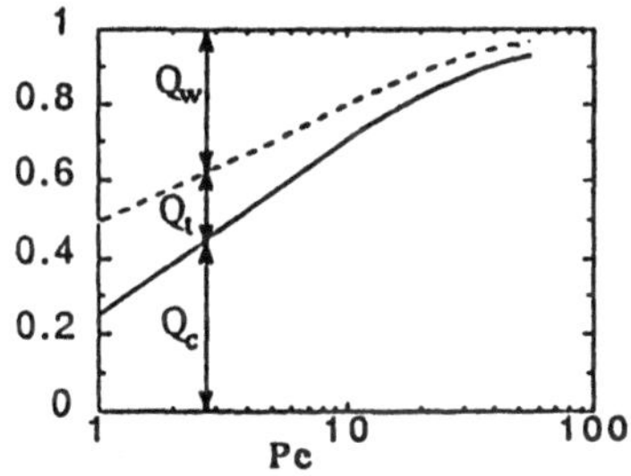

Figure 5 Distribution of Total Heat Generation

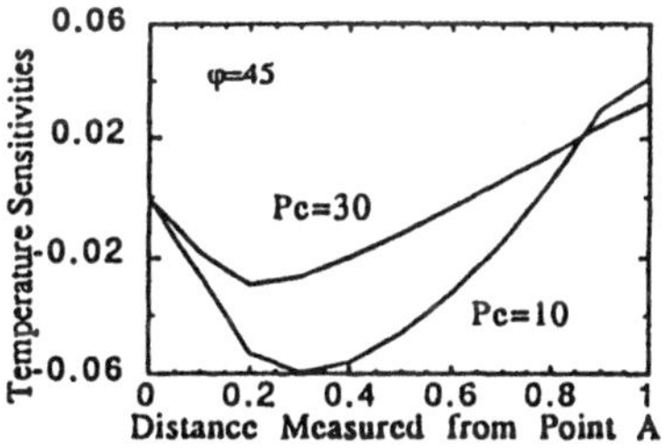

Figure 6 Temperature Sensitivities Along the Shear Plane

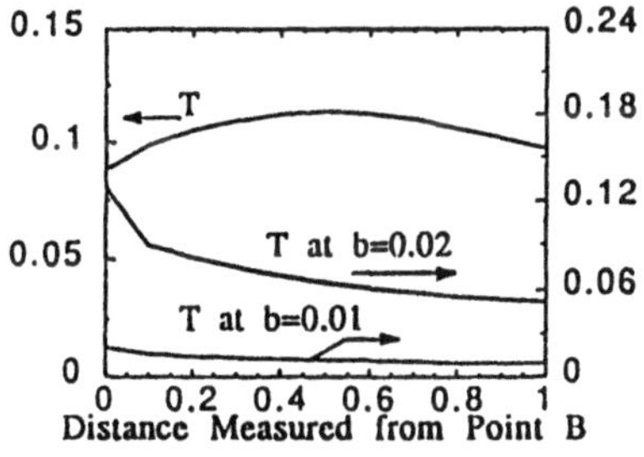

Figure 7 Temperature Sensitivities At Chip-Tool Interface

Conclusion

This paper is a first attempt at the prediction of the temperature and its sensitivity during steady state machining processes. An exact alggraic equation is derived to satisfy the matching boundary conditions which obviates the need for any iteration. The BEM solution is found to be stable regardless the Peclet number and does not show any false diffusion. The temperature sensitivity is also obtained by the DDA approach.

Acknowledgement

A. Chandra's contribution to this research has been supported by Grant No DMC-8657345 of the U.S. National Science Foundation and by Hughes Aircraft Company. C. L. Chan's contribution to this research is supported by Engineering Foundation Grant No RI-A-88-7.

References

1. Shaw, M. C., Metal Cutting Principles , Clarendon Press, Oxford, 1984.

2. Bhattacharyya, A., Metal Cutting Theory and Practice , Central Book Publisher, Calcutta, 1984.

3. Stevenson, M. G., Wright, P. K. and Chow, J. G., "Further Developments in Applying the Finite Element Method to the Calculation of Temperature Distributions in Machining and Comparisons with Experiment," J. Eng. Ind., ASME, 105, 1983, pp 149-154.

4. Dawson, P. R. and Malkin, S., "Inclined Moving Heat Source Model for Calculating Metal Cutting Temperatures," J. Eng. Ind., ASME, 106, 1984, pp 179-186.

5. Tanaka, Y., Honma, T. and Kaji, I., "On Mixed Element Solutions of Convection-Diffusion Problems in Three Dimensions," Appl. Math. Modelling, 10, 1986, pp 170-175.

6. Chan, C. L. and Chandra, A., "A Boundary Element Method Analysis of the Thermal Aspects of Metal Cutting Processes," submitted for publication in the Journal of Industry for Engineering.

7. Haug, E. J., Choi, K. K., and Komkov, V., Design Sensitivity Analysis of Structural Systems , Academic Press, New York, 1986.

8. Mukherjee, S. and Chandra, A., "A Boundary Element Formulation for Design Sensitivities in Materially Nonlinear Problems," Acta Mechanica, 78, 1989, pp243-253.

9. Banerjee, P. K. and Butterfield, R., Boundary Element Methods In Engineering Science , McGraw Hill, London, 1981.

10. Mukherjee, S., Boundary Element Methods in Creep and Fracture , Elsevier Applied Science Pub., Barking, Essex, 1982.

Summary of Modern Nodal Integral Methods in Fluid Flow and Heat Transfer

G.L. WILSON
R.A. RYDIN

Dept. of Nuclear Engineering and Engineering Physics
University of Virginia, Charlottesville

Summary
Modern nodal integral methods are a product of original applications in neutron diffusion and neutron transport calculations. These methods have recently been applied to two-dimensional dynamic heat transfer and fluid flow problems, where they have been shown to be independent of Fourier and Courant conditions. Applications in these areas are discussed and future extensions of the methodology are noted.

Static Nodal Integral Methods

Modern Coarse-Mesh Methods were initiated by Dorning [1] in the mid-1970's, with the introduction of the Partial Current Balance Method (PCBM). The PCBM was applied to neutron diffusion, employing locally defined two-dimensional Green's functions to convert the partial differential equation into a local integral equation [2]. The investigation of the PCBM quickly led to new ideas based on multiple one-dimensional Green's functions, such as the Nodal Green's Function Method (NGFM) [3] and the Hybrid NGFM (HNGFM) [4]. The NGFM introduced the concept of transverse integrations to decouple the neutron diffusion equation in its spatial variables. The HNGFM is an improved version of the NGFM using two different locally defined Green's functions (different local boundary conditions) for each volume element.

In the first use of modern integral methods for fluid dynamics, the Nodal Green's Tensor Method (NGTM) [5] was applied to two-dimensional, isothermal, incompressible, steady-state fluid flow. The transverse-integrated continuity and momentum equations are treated as a coupled set, leading to a Green's function tensor, defined locally over each node, which is used to produce a set of coupled local integral equations. The NTGM is historically important because it provides a transition from nodal Green's function methods to nodal integral methods. Nodal Integral Methods (NIM) form a subclass of modern coarse-mesh computational methods that offer high accuracy and efficiency, without the introduction of Green's functions. Azmy and Dorning [6] first applied the NIM to the same problems done with the NGTM, but the methods were quickly extended to static two-dimensional Benard natural convection.

Applications of the NIM in neutron transport have come the closest to solving practical engineering problems. The most ambitious effort to date is the recent work by Wagner on the three-dimensional multigroup neutron transport problem in hexagonal-Z geometry [7]. This static solution is currently being implemented into the quasi-static neutron kinetics code CASSANDRA by DeVooght and Mund [8]. Neutron diffusion is also the most advanced in terms of geometry. Azmy demonstrated NIM methods in cylindrical geometry for two-dimensional neutron diffusion [9]. Rajic and Ougouag recently demonstrated the ability to use within-node regional property variations [10]. Finally, Kim and Dorning [11] have explored the use of boundary-fitted coordinates in two-dimensional neutron transport problems, thus taking the first steps towards using generalized geometry.

The Dynamic Nodal Integral Method

Wilson and Rydin [12,13] developed a generalized formulation of the NIM, also extending these methods to the time variable. This approach to solving differential equations involves a series of analytical steps designed to incorporate the full homogeneous solution and an approximate particular solution. Nonlinear terms can be incorporated into source terms or may be dealt with through more subtle means. A fairly large set of approximations to various types of nonlinear terms has been developed and applied. All approximations, when made, are at the node-integral level so that the error is minimized thus preserving the integral behavior of the dependent variable.

The Time-Dependent Nodal Integral Method (TDNIM) can be illustrated using a simple model problem:

$$\alpha \frac{\partial^2 P}{\partial x^2} = \frac{\partial P}{\partial t} \tag{1}$$

where α is a constant and P is the quantity being conserved. Transverse integration of Eq.(1) leads to two ordinary differential equations (ODEs):

$$\frac{dP_t}{dt} = S_t \tag{2}$$

and

$$\alpha \frac{d^2 P_x}{dx^2} = S_x \tag{3}$$

The source terms S_x and S_t incorporate quantities resulting from the transverse-integration process plus any nonlinear convective terms present in the balance equations.

During the derivation process for the spatially dependent variables, the ODEs are treated by performing multiple integrals from each direction. This process makes the final equations invariant and stable independent of the direction of flow. Preservation of the integral time information ensures that the method is not constrained by a Courant condition. The complete homogeneous solution, even though not explicitly introduced, is implicitly present but is eliminated in favor of face-average values that are arbitrary until the boundary and initial conditions are applied. Problems solved using TDNIM techniques are characterized by generalized sparse matrices and/or sparse banded matrices. To date, the most efficient matrix inversion action has been obtained by making use of generalized matrix inversion routines to symbolically invert the Jacobian matrix.

Application of the Dynamic Nodal Integral Method

Modern dynamic formulations of the NIM have been applied to the solution of nonlinear problems in heat transfer [12] and fluid flow [13]. In general, dynamic NIM solutions have been found to be about an order of magnitude more accurate on the same mesh as quadratic finite elements. Comparisons using these methods in heat conduction have demonstrated a computational efficiency advantage over finite elements of a factor of six per dimension for spatial calculations. Because the accuracy of the comparative nonlinear finite element method (which employed a modified Crank-Nicholson method in time) was constrained by a Fourier condition, the NIM was clearly superior. Wilson and Rydin [12] demonstrated the ability to take nearly arbitrary time-step sizes, and to arbitrarily change the time-step size within a calculation (see Table I).

TABLE I. Temperature as a function of time for several TDNIM time step sizes

	Time Step Size			
Time	0.02	0.2	0.5	2
0.02	0.1999	---	---	---
0.20	1.5206	1.5205	---	---
0.30	2.0733	---	---	---
0.40	2.5596	2.5596	---	---
0.50	2.9989	---	2.9986	---
0.60	3.4026	3.4026	---	---
0.70	3.7781	---	---	---
0.80	4.1308	4.1307	---	---
0.90	4.4642	---	---	---
1.00	4.7812	4.7812	4.7812	---
1.10	5.0841	---	---	---
1.20	5.3746	5.3746	---	---
1.30	5.6540	---	---	---
1.40	5.9236	5.9236	---	---
1.50	6.1844	---	6.1843	---
2.00	7.3800	7.3800	7.3799	7.3791
3.00	9.4011	9.4011	9.4011	---
4.00	11.1144	11.1144	11.1144	11.1144
5.00	12.6284	12.6284	12.6283	---

Note: Test temperatures are expressed in degrees Kelvin.

Applications of the TDNIM to fluid flow [13] have included the investigation of the double-glazing benchmark problem, the closed-cavity Benard problem, and flow in a channel, plus natural convection near a fluid density maximum [14]. Using the TDNIM in conjunction with a static nodal integral method derived from the dynamic formulation, higher-order critical Rayleigh loci as a function of aspect ratio β have been investigated for the two-dimensional Benard problem [15]. Results of interest include the discovery of multiple equilibrium three-vortex solutions forming a figure-eight locus [16], and the convergence of the first two critical Rayleigh values with increasing aspect ratio towards the analytic solution of 1708 for an infinite cavity [17] (see Figure 1).

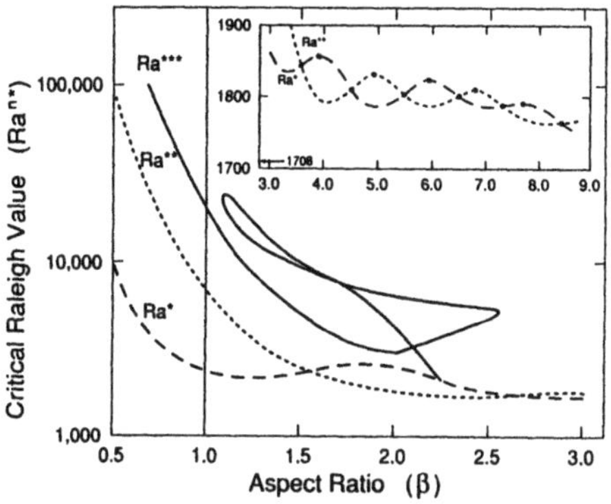

Fig. 1. The first three critical Ra curves vs. β

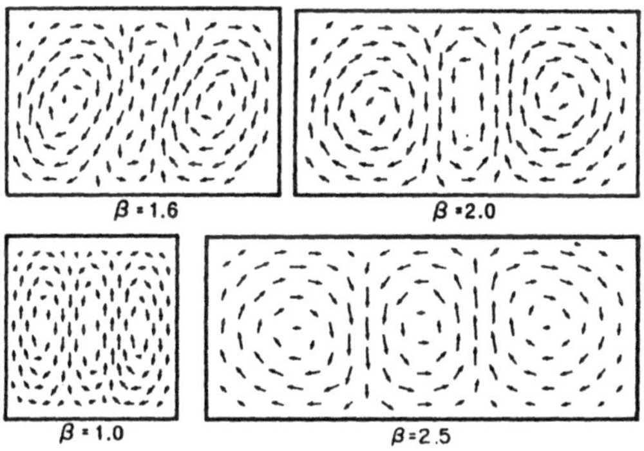

Fig. 2. Three vortex solutions for Ra = 23000 at various β

The search for the Ra^{3*} solution begins by taking a Ra value above Ra^{3*} and an initial flow distribution. Sample flow solutions for several values of β are given in Figure 2. As the aspect ratio β is increased using $\Delta\beta$ = 0.01 to 0.05, abrupt transitions occur near the turning point at $\beta \approx 1.1$, near the crossover at $\beta \approx 1.82$, and near the turning point at $\beta \approx 2.5$ as the character of the solution changes. These difficulties could only be avoided by carefully following the unfolding by a path-dependent process.

Extensions of the TDNIM Methodology for Fluid Flow Applications

The TDNIM formulation of the Navier-Stokes equations and the energy equation thus far carries all terms at high-order except for the advection terms. We estimate that the advection terms are effectively approximated at slightly less than second order, while the diffusion terms are effectively approximated at better than third order. However, in some applications the precision of these terms may need to be increased. To do this, it is necessary to define some auxiliary moment variables such that the advection terms no longer appear entirely in the source terms, but rather are explicitly carried in the homogeneous equations. For example, define the auxiliary variables

$$
\begin{aligned}
A &= u\frac{\partial u}{\partial x} + \beta v\frac{\partial u}{\partial y} \\
B &= u\frac{\partial v}{\partial x} + \beta v\frac{\partial v}{\partial y} \\
C &= u\frac{\partial T}{\partial x} + \beta v\frac{\partial T}{\partial y}
\end{aligned}
\qquad (4)
$$

The net effect is to modify the homogeneous forms of the original TDNIM equations slightly, change the source terms, and add the auxiliary coupled ODEs and boundary conditions. Hence, the problem can be attacked at a higher level of accuracy, but only at the cost of a larger set of equations that needs to be solved.

Although the TDNIM is accurate and efficient relative to competing methods such as quadratic finite elements, it is currently a significant user of computer memory resources because of the direct Newton-Raphson inversion of a large coupled set of equations. However, we note that the method is inherently in a form that encourages the use of parallel processing. The primary variables are face averages which are coupled in only one dimension. Therefore, it should be possible to sweep all of the surface equations simultaneously and then construct the nodal average values in a sequential step. Such a process has been carried out successfully by Kirk and Azmy for the neutron diffusion equations [18]. Replacement of direct inversion by iterative methods would significantly advance the TDNIM as a general purpose tool for fluid dynamics.

References

1. Dorning, J.J. : Modern coarse-mesh methods -- a development of the '70's. Proc. Topl. Mtg. Computational Methods of Nuclear Engineering, Williamsburg Virginia, 23-25 April 1979, vol. 1, p. 3-1, American Nuclear Society (1979).

2. Burns, T.J. and Dorning, J.J.: The partial current balance method: a new computational method for the solution of multidimensional neutron diffusion problems. Contributions to the NEACRP and CSNI Specialist's Meeting on New Developments in Three-Dimensional Neutron Kinetics and Review of Kinetics Benchmark Calculations, Garching, Munich (1975).

3. Lawrence, R.D. and Dorning, J.J.: A nodal Green's function method for multidimensional neutron diffusion computations. Nuclear Science and Engineering, 76 (1980) 218-231.

4. Ougouag, A.M.: A coarse-mesh nodal method for multigroup, multidimensional, neutron diffusion computations. Master's thesis in Nuclear Engineering, University of Illinois (1981).

5. Horak, W.C. and Dorning, J.J.: A nodal Green's tensor method for the efficient numerical solution of laminar flow problems. Proc International Conference on Numerical Methods in Laminar and Turbulent Flow, Venice (1981).

6. Azmy, Y.Y. and Dorning, J.J.: Numerical studies of bifurcations in the confined Benard problem. Proc. 10th Int. Conf. Numerical Methods in Fluid Dynamics, Beijing, China, 23-27 June 1986, F.G. Zhung, Ed., Springer-Verlag, Berlin, FRG (1986).

7. Wagner, M.R.: Three dimensional nodal diffusion and transport theory methods for hexagonal-Z geometry. Proc. Topl. Mtg. Reactor Physics, Jackson Hole, Wyoming, 10-22 Sept. 1988, American Nuclear Society (1988).

8. Devooght, J. and Mund, E.: Generalized quasi-static method for nuclear reactor space-time kinetics. Nuclear Science and Engineering, 76 (1980) 10-17.

9. Azmy, Y.Y.: A nodal integral method for the neutron diffusion equation in cylindrical geometry. ANS Transactions, vol. 54, 7-11 June 1987, Dallas, Texas American Nuclear Society (1987).

10. Rajic, H.L. and Ougouag, A.M.: A higher order nodal method with spatially dependent material properties. Proc. Topl. Mtg. Advances in Nuclear Engineering Computation and Radiation Shielding, vol. 1, 9-13 April 1989, American Nuclear Society (1989).

11. Kim, S.T. and Dorning, J.J.: Discrete nodal and S_n transport methods for boundary-fitted coordinate geometries. Proc. Topl. Mtg. Advances in Nuclear Engineering Computation and Radiation Shielding, vol. 1, 9-13 April 1989, American Nuclear Society (1989).

12. Wilson, G.L, Rydin, R.A. and Orivuori, S. A.: Comparison of two efficient nonlinear heat conduction methodologies using a two-dimensional time-dependent benchmark problem. Nuclear Technology, 82 (July 1988) 94-105.

13. Wilson, G.L., Rydin, R.A. and Azmy, Y.Y.: Nonlinear time-dependent nodal integral methods with applications in natural convection, fluid flow and bifurcation phenomena. Nuclear Science and Engineering, 100 (Dec. 1988) 414-425.

14. Wilson, G.L. and Rydin, R.A.: Time-dependent nodal integral methods in natural convection near a fluid density maximum. Computational Mechanics 4 (1989), 345-360.

15. Wilson, G.L. and Rydin, R.A.: Nodal integral method solutions for Benard natural convection. International Journal for Numerical Methods for Fluids, to appear in 1989.

16. Wilson, G.L. and Rydin, R.A.: Multiple equilibrium solutions to the Benard problem at the third critical Rayleigh number. Numerical Heat Transfer, 15, part B (1989) 117-126.

17. Lord Rayleigh: On convection currents in a horizontal layer of fluid, when the higher temperature is on the under side. Philos. Mag., 32, (1916) 529-547.

18. Kirk, B.L, and Azmy, Y.Y.: A parallel approach to the nodal method solution of the two-dimensional diffusion equation. Proc. Topl. Mtg. on Advances in Nuclear Engineering Computation and Radiation Shielding, vol. 1, 9-13 April 1989, American Nuclear Society (1989).

The Principle of "Limit Absorption" for Water Wave Diffraction

S. ALLINEY

Department of Mathematics and Computer Science
University of Udine, Italy

Introduction

Water wave models based on potential theory are widely used in engineering applications, mainly because of their computational efficiency. In our opinion, however, the main drawback of such models resides in the impossibility of taking into account any form of energy dissipation, with possible overestimation of certain significant quantities. In this contribution, we propose a new model, which considers water waves, that are exponentially decreasing at infinity. More precisely, we introduce an absorption condition (instead of pure reflection) at the bottom; as a consequence, the wave number becomes complex. In a forthcoming paper, we will give all the details of our developments; here, because of space limitations, we present only a synthetic account of the theory and a few numerical results.

A short account of potential theory

As well-known [1-3], the hypothesis of irrotational motion allows to define a velocity potential function $\Psi(x,y,z;t)$ such that $\underline{v}=\text{grad}\,\Psi$. If we seek a solution, that is simple harmonic in time, the potential function can be given the form $\Psi(x,y,z;t) = \Phi(x,y,z)\, e^{-i\omega t}$, ω being the angular frequency. Besides that, in the usual model further assumptions are made (the fluid is ideal, homogeneous and incompressible; the wave amplitude is infinitesimally small), so as to obtain one linear equation, together with linear boundary conditions, for the unique (complex-valued) unknown $\Phi(x,y,z)$:

$$\nabla^2\Phi(x,y,z) = 0 \qquad ; \text{ inside the domain} \qquad (1)$$

$$\Phi_z - (\omega^2/g)\,\Phi = 0 \qquad ; \text{ at the free surface} \qquad (2)$$

$$\Phi_n = 0 \qquad ; \text{ at the bottom } z=-h \qquad (3)$$

Some other boundary conditions must be added, according to the problem at hand. The most frequent case is the following: a plane wave of prescribed direction is incoming from infinity, and is scattered at certain vertical obstacles. Therefore, we need reflection conditions ($\Phi_n=0$) at the inner boundaries (the obstacle walls) and some regularity condition at infinity. If we split the velocity potential $\Phi(x,y,z)$ into two parts, the incoming wave potential and the disturbance potential, the usual choice is to impose a Sommerfeld radiation condition on the latter.

A diffraction model with attenuation

A considerable simplification of the problem is obtained, when the variation of the bottom profile can be neglected: $h(x,y) \cong$ constant. As a consequence, the normal derivative in (3) can be represented as $\Phi_n=-\Phi_z$. In that case, separation of variables is allowed and, after a simple computation, we obtain the classical diffraction model [1-3]. At this point, we will introduce the main novelty of our approach: we propose to substitute equation (3) with an "absorption condition"

$$\Phi_n = i\gamma\Phi \qquad ; \quad \text{at } z=-h \tag{4}$$

Equation (4) can represent rather well the interaction of the fluid with a sandy, irregular bottom, whose mean depth is -h.

THEOREM: For $\gamma<0$, in absence of sources, the fluid loses its energy at the rate

$$\frac{dE}{dt} = \varrho\gamma\omega \cdot \iint \mathrm{Im}(\ \Phi(x,y,z))^2 \quad dS \tag{5}$$

where the integral is extended over the whole bottom surface.

This means that a stationary motion must be sustained from outside (e.g. by the wind at the free surface); in practice, we can represent such a situation by considering a plane incoming wave, which is not attenuated, and a disturbance rapidly decreasing far from the obstacles.

Now, we will proceed with the variable separation procedure :

$$\Phi(x,y,z) = Z(z)\cdot\varphi(x,y) \tag{6}$$

With that position, equation (1) is equivalent to

$$\nabla^2\varphi(x,y) - a^2\varphi(x,y) = 0 \tag{7}$$

$$Z''(z) + a^2 Z(z) = 0 \tag{8}$$

together with the conditions at the free surface and the bottom

$$Z'(z) - (\omega^2/g)\cdot Z(z) = 0 \quad ; \quad z=0 \tag{9}$$

$$Z'(z) + i\gamma\cdot Z(z) = 0 \quad ; \quad z=-h \tag{10}$$

The dispersion relationship

The general solution of equation (8) is clearly $Z(z)=A\cdot\cos(az)+$ $+B\cdot\sin(az)$, with A and B constants to be determined. Upon substitution of such expression into conditions (9) and (10), a simple computation yields the generalized dispersion relationship for the complex wave number a :

$$(a-i\gamma\frac{\omega^2}{g}\cdot a^{-1})\cdot\tan(ah) + (\frac{\omega^2}{g}+i\gamma) = 0 \tag{11}$$

It is not difficult to prove that

THEOREM: For $\gamma < 0$ and $0 < h < +\infty$, any solution of equation (11) needs to be a complex number. Furthermore, the limit solutions are $a = \pm i\omega^2/g$ when $h \to +\infty$ (i.e., the bottom influence becomes unrelevant), and $a = \pm ik$ when $\gamma \to 0$.

Clearly, for $\gamma = 0$ equation (7) reduces to the standard Helmholtz equation and therefore from equation (11) we find again the usual dispersion relationship

$$-k\cdot\tanh(kh) + \frac{\omega^2}{g} = 0 \tag{12}$$

THEOREM: For $\gamma < 0$, equation (11) has at least two solutions having opposite signs, one of them with Re(a)>0 and Im(a)>0.

In practical applications, we have to find a numerical solution to equation (11). We point out, however, that such equation is of the type f(a)+w=0, with f(.) a complex valued, but analytical function. Therefore, f(.) admits a derivative and we can use a simple Newton algorithm; examples are shown in Table 1.

The principle of limit absorption

Now, we will focus our attention on equation (7); as we noted before, we need some additional boundary conditions. At the reflecting walls of the vertical obstacles, we impose

$$\frac{\partial\varphi}{\partial n} = -\frac{\partial\varphi_i}{\partial n} \tag{13}$$

assuming that φ represents the disturbance and φ_i the incoming wave potential. As far as the conditions at infinity are concerned, it is sufficient to recall [4-6] that:

THEOREM: If $Re(a) > 0$, equation (7) has a unique solution, which is exponentially decreasing at infinity. Furthermore, when the real part of a vanishes, such a solution tends to a function satisfying Sommerfeld's radiation condition

$$\lim_{r \to \infty} \sqrt{r} \cdot (\partial\varphi/\partial r \pm ik\varphi) = 0 \tag{14}$$

In other words, that means that for $\gamma \to 0$, and therefore for $Re(a) \to 0$, the attenuated solution tends exactly to the solution of the corresponding Helmholtz equation with a radiation condition at infinity. That is the content of the so-called "principle of limit absorption" for our particular problem.

Fundamental solution

When dealing with a problem defined over an unlimited domain, the standard boundary element technique consists of "bringing back" at a finite boundary the asymptotic condition, by means of a suitable integral equation. To do that in our case, we need the fundamental solution of equation (7), which is exponentially decreasing at infinity. Following a general method described in [6], we can prove that:

THEOREM: The unique fundamental solution of (7), which is exponentially decreasing at infinity is $(-1/2\pi)\cdot K_0(ar)$, that is the modified Bessel function of the second type, of order zero.

Keeping in mind the well-known [7] equivalence between $K_m(\xi)$ and the first type Hankel function $H_m(i\xi)$, we can state the limit behaviour of the fundamental solution and its derivative:

$$(-1/2\pi)\cdot K_0(ar) \to (i/4)\cdot H_0(kr) \quad \text{as } Re(a) \to 0 \tag{15}$$

$$(a/2\pi)\cdot K_1(ar) \to (ik/4)\cdot H_1(kr) \quad \text{as } Re(a) \to 0 \tag{16}$$

$$(-1/2\pi)\cdot K_0(ar) \doteq (1/2\pi)\cdot \log(ar) + E \quad \text{for } r \ll 1 \tag{17}$$

$$(a/2\pi)\cdot K_1(ar) \doteq (1/2\pi)\cdot r^{-1} \quad \text{for } r \ll 1 \tag{18}$$

Here, E denotes Euler's constant. We note that, when Re(a) vanishes, we find again the Sommerfeld fundamental solution, as we could expect according to the Theorem stated above.

Series solution

When we have an "external problem", defined over a domain extending to infinity and having a circular internal boundary, an alternative approach is to resort to a Fourier series solution. In that case, it is convenient to use polar coordinates (r,ϑ) with origin at the center of the circle; then, the potential φ can be expressed as

$$\varphi(r,\vartheta) = \sum_m c_m(r)\, e^{im\vartheta} \qquad (19)$$

Upon substitution into equation (7), with the Laplacian represented in polar coordinates, we obtain for the m-th coefficient

$$r^2 c_m''(r) + r\cdot c_m'(r) - (a^2r^2+m^2)\cdot c_m(r) = 0 \qquad (20)$$

whose solution bounded at infinity is

$$c_m(r) = c_m \cdot K_m(ar) \qquad (21)$$

The undefined constants c_m will be determined according to the conditions prescribed at the internal boundary. We will present later a simple application of that method.

Numerical evaluation of Bessel functions

Since the values of $K_0(.)$ and $K_1(.)$ are not tabulated when their argument is a complex number, we need a stable computational method for their evaluation. We will give a short report of the technique employed for $K_0(.)$, the other being very similar. According to the definition [7], we can write

$$K_0(z) = -\log(z/2)\cdot I_0(z) + R_0(z) \qquad (22)$$

$$I_0(z) = \sum_{k=0}^{\infty} (z/2)^{2k}/(k!) \qquad (23)$$

$$R_0(z) = \sum_{k=0}^{\infty} (z/2)^{2k}\, \Psi_0(k) \qquad (24)$$

where $\Psi_0(k)$ is related to the Euler's function, being

$$\Psi_0(k) = (-E + \sum_{h=1}^{k} \frac{1}{h})/k! \qquad (25)$$

The main difficulty remains in the evaluation of $\Psi_0(k)$, the numerator being divergent as k increases, whereas a reduced sum for $I_0(.)$ and $R_0(.)$ can be accurately computed using a Horn-type scheme. Anyway, the values of $\Psi_0(k)$ can be deduced from the following (stable) recursion:

$$\delta(k) = \delta(k-1)/k \qquad ; \ \delta(0)=1 \qquad (26)$$

$$\Psi_0(k) = (\Psi_0(k-1)+\delta(k))/k \qquad ; \ \Psi_0(0)=-E \qquad (27)$$

A similar recursion can be found also for $K_1(.)$. Once the first two functions are computed, the following can be evaluated either by extending the present scheme, or by well-known [7] recurrence relationships. In order to assess the precision of our method, we reported in Table 2 some values for z real, compared with the ones, tabulated in [8].

An example : diffraction around a circular pile

Finally, we will present some results for a circular pile of radius R, centered at the origin of the coordinate system. The incoming wave potential is assumed as a plane wave

$$\varphi_i(r,\vartheta) = e^{ikr\cdot\cos\vartheta} \qquad (28)$$

where $k<0$ is the imaginary part of a satisfying (11). The disturbance potential will be represented as

$$\varphi(r,\vartheta) = \sum_m c_m K_m(ar)\, e^{im\vartheta} \qquad (29)$$

where a is the solution of (11) with $Im(a)>0$. Using the Fourier series expansion for the incoming wave potential

$$\varphi_i(r,\vartheta) = \sum_m (i)^m J_m(kr)\, e^{im\vartheta} \qquad (30)$$

the condition $\partial\varphi/\partial r = -\partial\varphi_i/\partial r$ at the vertical wall r=R can be imposed term by term. The final result is

$$\varphi_{tot} = \varphi_i + \varphi = \sum_m (i)^m \{[J_m(kr)K_m'(aR) - -(k/a)\cdot J_m'(kR)K_m(ar)]/K_m'(aR)\}\, e^{im\vartheta} \qquad (31)$$

that for a=-ik reduces to the known result in absence of attenuation (see for example [9]). In Figures 1 and 2 we reported the computed amplitude of φ_{tot} at the wall.

References

1. Stoker, J.J.: Water waves. New York: Interscience 1957.
2. Whitham, G.B.: Linear and nonlinear waves. New York: Wiley 1974.
3. Berkhoff, C.W.: Mathematical models for simple harmonic linear water waves. Delft University of Technology Publications 163 (1976) 5-46.

4. Kato, T.: Growth properties of solutions of the reduced wave equation with a variable coefficient. Comm. Pure Appl. Math. 12 (1959) 403-425.

5. Eidus, D.M.: The principle of limit absorption. Amer. Math. Soc. Transl. (2) 47 (1965) 157-191.

6. Vainberg, B.R.: Principles of radiation, limit absorption and limit amplitude in the general theory of partial differential equations. Russian Math. Surveys 21 (1966) 115-193.

7. Gradshteyn, I.S.; Ryzhik, I.M.: Tables of integrals, series and products. New York: Academic Press 1984.

8. Abramowitz, M.; Stegun, I. (eds.): Handbook of mathematical functions. New York: Dover 1970.

9. Alliney, S.: Water wave diffraction around cylindrical obstacles. Appl. Math. Modelling 5 (1981) 237-240.

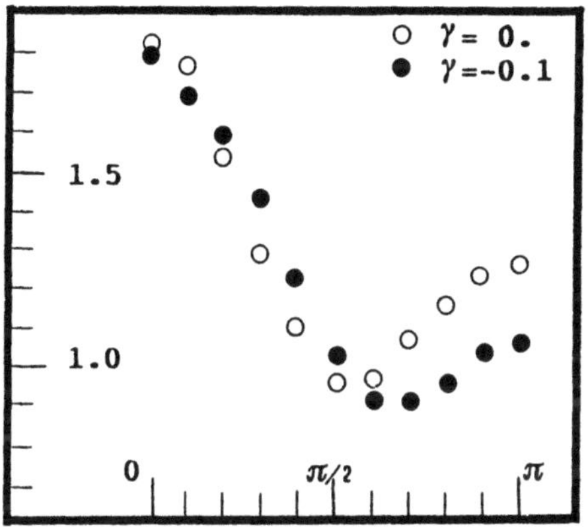

Fig.1 : Potential amplitude at the wall (R=5, h=10, ω=1).

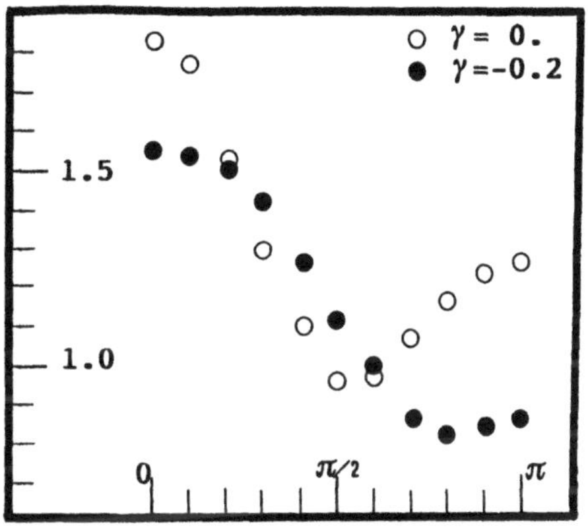

Fig.2 : Potential amplitude at the wall (R=5, h=10, ω=1).

γ	h=5. ω=1.	h=10. ω=1.
0.0	.000,.156	.000,.122
-0.1	.051,.156	.023,.115
-0.2	.096,.151	.037,.100
-0.5	.182,.116	.040,.066
-1.0	.218,.066	.031,.047
γ	h=5. ω=.5	h=10. ω=.5
0.0	.000,.073	.000,.053
-0.1	.091,.102	.065,.064
-0.2	.145,.119	.101,.060
-0.5	.243,.107	.132,.031
-1.0	.284,.062	.137,.016

TABLE 1 : Complex wave numbers.

x	Computed $e^xK_0(x)$	Tabulated
.1	2.68232632	2.68232610
.2	2.14075732	2.14075732
.5	1.52410924	1.52410938
1.	1.14446294	1.14446307
2.	0.84156817	0.84156821
5.	0.54775137	0.54780756
x	Computed $e^xK_1(x)$	Tabulated
.1	10.89018154	10.89018268
.2	5.83338547	5.83338603
.5	2.73100972	2.73100970
1.	1.63615346	1.63615348
2.	1.03347683	1.03347684
5.	0.60040343	0.60027385

TABLE 2 : Values of K_0 and K_1.

A Laplace Transform Boundary Element Method for Axisymmetric Diffusion Problems

A. H-D. Cheng, Y. Abousleiman and T. Badmus
Department of Civil Engineering, University of Delaware
Newark, Delaware, USA

Abstract

This paper presents a Boundary Element algorithm for solving axisymmetric diffusion equation in the Laplace transform domain. The free-space Green function of the governing equation is not available in closed form. The problem is resolved by using a combination of methods, including series expansion, interpolation from numeric tables, and direct numerical integration. The solution in the time domain is found by an approximate Laplace inversion.

Introduction

The purpose of this paper is to develop a Laplace transform Boundary Element Method for the diffusion equation under axisymmetry. The axisymmetric BEM in the time-domain has been implemented by Wrobel and Brebbia [1]. The Laplace transform version however has so far eluded a satisfactory treatment. The difficulty lies in the inability in obtaining the axisymmetric Green function in closed form. In general, two approaches have been taken: one uses the brutal force of numerically integrating the Green function expression [2], and the other expands it into a series [3]. Neither methods however is satisfactory because the former is highly demanding in CPU time, while the latter works only for solutions in the large time (small Laplace transform parameter) range.

The paper presents an innovative idea for overcoming the above difficulties. It is proposed that the Green function be evaluated by interpolation from pre-calculated numeric tables. This method, as simplistic as it may sound, turns out to be highly efficient and also with satisfactory accuracy. The idea is explained in more detail in the following.

Boundary Element Formulation

The diffusion equation in three-dimension is expressed as

$$\nabla^2 \Phi = \frac{1}{c}\frac{\partial \Phi}{\partial t} \tag{1}$$

where c is the diffusivity constant. Applying the Laplace transform, we obtain the modified Helmholtz equation

$$\nabla^2 \tilde{\Phi} = \frac{s}{c}\tilde{\Phi} \tag{2}$$

in which $\tilde{}$ denotes the Laplace transform, and s is the transform parameter. Making use of Green theorem, the boundary integral equation is

$$\alpha\tilde{\Phi}(x_i, y_i, z_i, s) = \int_\Gamma \left(g\frac{\partial \tilde{\Phi}}{\partial n} - \tilde{\Phi}\frac{\partial g}{\partial n} \right) dx \tag{3}$$

In the above, (x_i, y_i, z_i) denotes the 'base point', dx represents integration with respect to Cartesian coordinates (x, y, z), Γ is the solution boundary, n the unit outward normal of Γ, α

a constant respectively equal to 0, 4π and the solid angle, for base point located outside and inside of the domain, and on Γ, and g is the three-dimensional free-space Green function

$$g(x-x_i, y-y_i, z-z_i, s) = \frac{e^{-R_3\sqrt{s/c}}}{R_3} \tag{4}$$

where $R_3 = \sqrt{(x-x_i)^2 + (y-y_i)^2 + (z-z_i)^2}$ is the radial distance between the base and the field point.

In the case of axisymmetry, all quantities in (3), except for g, are independent of the θ parameter of the cylindrical coordinate system. The integration with respect to θ can be carried out to yield the following equation in the r-z plane

$$\alpha\tilde{\Phi}(r_i, z_i, s) = \int_\gamma r\left(G\frac{\partial\tilde{\Phi}}{\partial n} - \tilde{\Phi}\frac{\partial G}{\partial n}\right)\, dx \tag{5}$$

In the above, γ is the intersection of Γ with the r-z plane, dx corresponds to integration with respect to (r, z), α equals 0, 4π or twice the plane interior angle, and

$$G(r-r_i, z-z_i, s) = \int_0^{2\pi} \frac{e^{-R_3\sqrt{s/c}}}{R_3}\, d\theta \tag{6}$$

is the axisymmetric Green function as the result of the distribution of a ring source around the z-axis.

Another way of expressing G is to directly solve the axisymmetric governing equation with a delta function on the right hand side. Utilizing the Hankel transform, we arrive at the following expression [4]

$$G(r-r_i, z-z_i, s) = 2\pi \int_0^\infty \frac{\beta J_0(\beta r_i) J_0(\beta r)}{\sqrt{\beta^2 + s/c}}\, e^{-\sqrt{\beta^2+s/c}\,|z-z_i|}\, d\beta \tag{7}$$

where J_0 is the Bessel function of the first kind of order zero, and β the Hankel transform parameter. Neither of the above integrations (6) and (7) can be executed in closed form.

Pullan's Expansion

Pullan [3] expanded the exponential function in (4) into a Taylor series

$$g = \frac{1}{R_3} - \sqrt{\frac{s}{c}} + \frac{R_3}{2}\frac{s}{c} - \frac{R_3^2}{6}\left(\frac{s}{c}\right)^{3/2} + \cdots \tag{8}$$

Equation (6) can be integrated term by term to give

$$G = 4\sum_{k=0}^{\infty} \frac{(-1)^k s'^k}{k!\sqrt{a+b}}\, I_k \tag{9}$$

where

$$I_k = \int_0^{\frac{\pi}{2}} (1 - m\sin^2\phi)^{(k-1)/2}\, d\phi \tag{10}$$

and

$$\begin{aligned} a &= r^2 + r_i^2 + (z-z_i)^2 \\ b &= 2rr_i \\ m &= \frac{2b}{a+b} = \frac{2b}{R^2+2b} \\ R &= \sqrt{(r-r_i)^2 + (z-z_i)^2} \\ s' &= \sqrt{(a+b)s/c} \end{aligned} \tag{11}$$

Integrating (10) we found [5]

$$\begin{aligned} I_0 &= \mathrm{K}(m) \\ I_1 &= \pi/2 \\ I_2 &= \mathrm{E}(m) \\ I_3 &= \frac{\pi(2-m)}{4} \\ I_k &= \frac{(k-2)(2-m)}{k-1} I_{k-2} - \frac{(k-3)(1-m)}{k-1} I_{k-4}; \quad k \geq 4 \end{aligned} \tag{12}$$

In the above $\mathrm{K}(m)$ and $\mathrm{E}(m)$ are respectively the complete elliptic integral of the first and the second kind [6]. It is of interest to remark that the leading term in the series (9), $4\mathrm{K}(m)/\sqrt{a+b}$, is precisely the axisymmetric Green function of the Laplace equation [7]. The singularity of G is contained in I_0 as $m \to 1$ ($m' = 1 - m \to 0$). We therefore find that

$$\lim_{R \to 0} G = -\frac{2}{r_i} \ln R \tag{13}$$

The normal derivative of Green function, which was not given by Pullan [3], is derived as

$$\frac{\partial G}{\partial n} = \sum_{k=0}^{\infty} \frac{(-1)^k s'^k}{k!(a+b)^{3/2}} \left\{ (k-1) \left(2r_i n_r + R \frac{\partial R}{\partial n} \right) I_k + 2 \left[2(1-m) r_i n_r - mR \frac{\partial R}{\partial n} \right] I'_k \right\} \tag{14}$$

In the above $I'_k = dI_k(m)/dm$ is also evaluated in a recurrence relation

$$\begin{aligned} I'_0 &= \frac{1}{2m} \left(\frac{I_2}{1-m} - I_0 \right) \\ I'_1 &= 0 \\ I'_2 &= \frac{1}{2m} (I_2 - I_0) \\ I'_3 &= -\frac{\pi}{4} \\ I'_k &= \frac{(k-2)}{k-1} [(2-m) I'_{k-2} - I_{k-2}] - \frac{(k-3)}{k-1} [(1-m) I'_{k-4} - I_{k-4}]; \quad k \geq 4 \end{aligned} \tag{15}$$

The singularity of the normal derivative is

$$\lim_{R \to 0} \frac{\partial G}{\partial n} = \frac{n_r}{r_i^2} \ln R - \frac{2}{r_i} \frac{\partial R}{\partial n} \frac{1}{R} \tag{16}$$

The above series looks quite attractive because of the recurrence nature which allows an efficient computer evaluation. The series however breaks down as s becomes large. This is in fact not surprising if one examine the behavior of the alternate series $e^x = 1 - x + x^2/2 - x^3/3! - \cdots$. Although theoretically this expansion is uniformly convergent, for $x \gg 1$ the terms in the series quickly diverge before they start to converge. In the computer evaluation, the final answer is overwhelmed by the roundoff error.

Numerical Integration

Equations (6) and (7) can be numerically integrated by quadrature rules. To facilitate the tabulation as discussed in the next section, they are expressed as functions containing the minimum number of dimensionless parameters. For example, equation (6) can be written as

$$G = \frac{4}{\sqrt{a+b}} f_1(s', m') \tag{17}$$

where

$$f_1(s',m') = \int_0^{\frac{\pi}{2}} \frac{1}{\sqrt{1-m\sin^2\phi}}\, e^{-s'\sqrt{1-m\sin^2\phi}}\, d\phi \tag{18}$$

and $m' = 1-m$. The parameters s' and m' are dimensionless, and m' has the limits of $0 \leq m' \leq 1$. Differentiating (17) with respect to the outward normal n we obtain

$$\frac{\partial G}{\partial n} = -\frac{4}{(a+b)^{3/2}}\left[R\frac{\partial R}{\partial n} f_2(s',m') + 2r_i n_r f_3(s',m')\right] \tag{19}$$

in which $\partial R/\partial n$ is evaluated as $\partial R/\partial n = [(r-r_i)n_r + (z-z_i)n_z]/R$, n_r and n_z are the directional cosines of n, and

$$f_2(s',m') = \int_0^{\frac{\pi}{2}} \frac{1+s'\sqrt{1-m\sin^2\phi}}{(1-m\sin^2\phi)^{3/2}}\, e^{-s'\sqrt{1-m\sin^2\phi}}\, d\phi \tag{20}$$

$$f_3(s',m') = \int_0^{\frac{\pi}{2}} \frac{1+s'\sqrt{1-m\sin^2\phi}}{(1-m\sin^2\phi)^{3/2}}\, e^{-s'\sqrt{1-m\sin^2\phi}}\, \cos^2\phi\; d\phi \tag{21}$$

The functions f_1, f_2 and f_3 can be evaluated using the subroutine DCADRE of the IMSL scientific subroutine package [8]. The Green function can thus be accurately determined for any values of s.

Tabulation

The direct integration is very demanding in CPU time. It is therefore suggested that numerical tables be prepared such that the Green function values can be looked up using interpolation. Although higher order schemes such as Chebyshev polynomials can be used, we choose the simple bilinear interpolation for its apparent efficiency.

In order for the tabulation to be successful, the functions must behave smoothly over the specified parameter range. It is unfortunately not the case for the current functions which are actually singular. To overcome the difficulty, the functions are distorted both in the vertical and the horizontal scales to obtain smooth surfaces. Constant grids, which allow an efficient searching process, are overlain. Once the tables are available, for any given values of s' and m', a 'box' is located in the grid and the bilinear interpolation can be carried out.

Figures 1a and b depict the variation of f_1 under distorted scales. To optimize the representation, we actually use two tables to cover different ranges. Figure 1a plots $f_1/(1-\ln m')$ versus $\log_{10} s'$ and m', for the ranges of $1 \leq s' \leq 100$ and $0.1 \leq m' \leq 1$. Figure 1b shows $f_1/(1-\ln m')$ versus $\log_{10} s'$ and $\log_{10} m'$, for $1 \leq s' \leq 10^3$ and $1 \times 10^{-10} \leq m' \leq 0.1$. It is clear that these surfaces are quite smooth. A low resolution mesh, say 50×50, can adequately represent the function. Similarly, f_2 and f_3 can be distorted as $m' f_2$ and $f_3/(1-\ln m')$. They are respectively plotted as Figures 2a, b and 3a, b, with the same horizontal scales as the above.

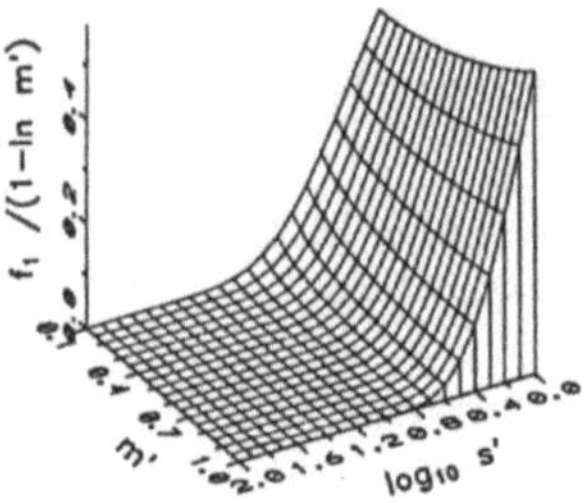

Figure 1a: $f_1(s',m')/(1-\ln m')$ versus $\log_{10} s'$ and m'.

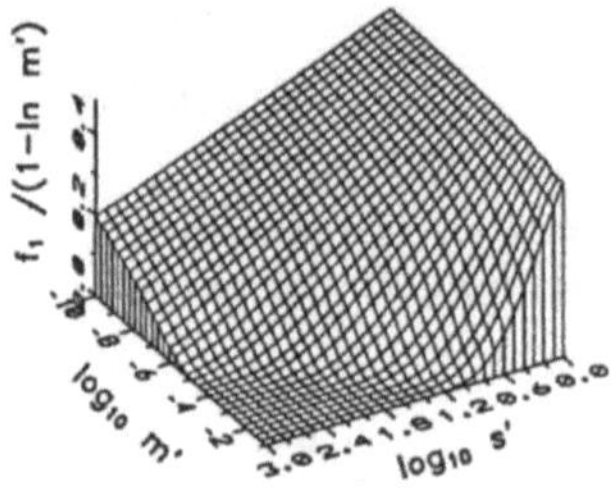

Figure 1b: $f_1(s',m')/(1-\ln m')$ versus $\log_{10} s'$ and $\log_{10} m'$.

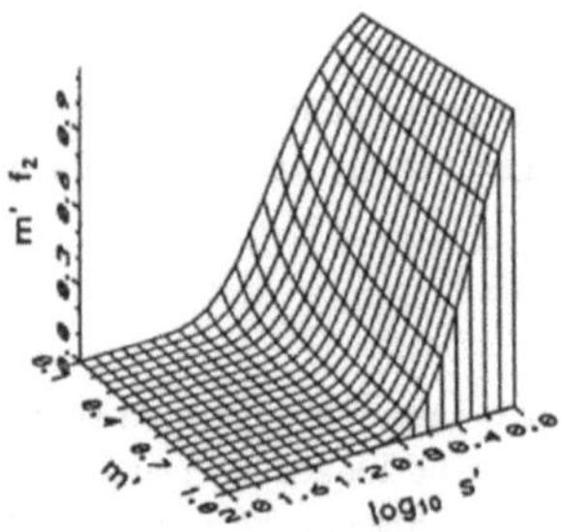

Figure 2a: $m' f_2(s',m')$ versus $\log_{10} s'$ and m'.

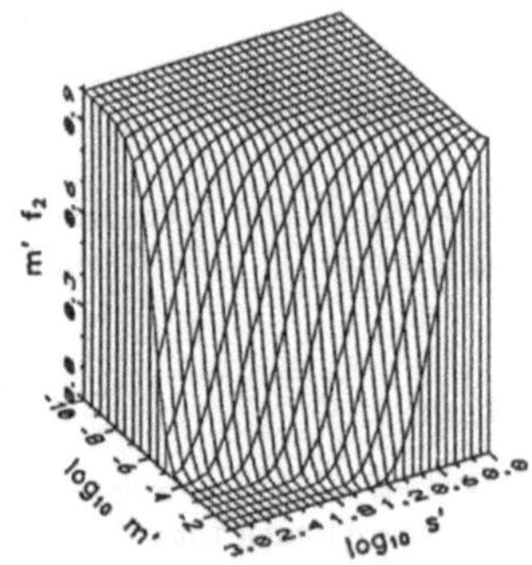

Figure 2b: $m' f_2(s',m')$ versus $\log_{10} s'$ and $\log_{10} m'$.

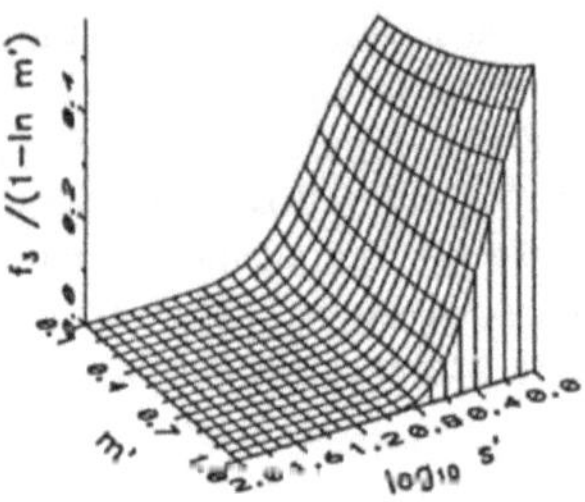

Figure 3a: $f_3(s',m')/(1-\ln m')$ versus $\log_{10} s'$ and m'.

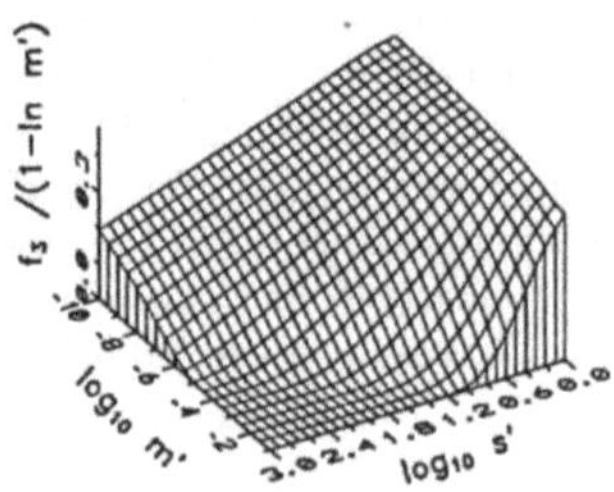

Figure 3b: $f_3(s',m')/(1-\ln m')$ versus $\log_{10} s'$ and $\log_{10} m'$.

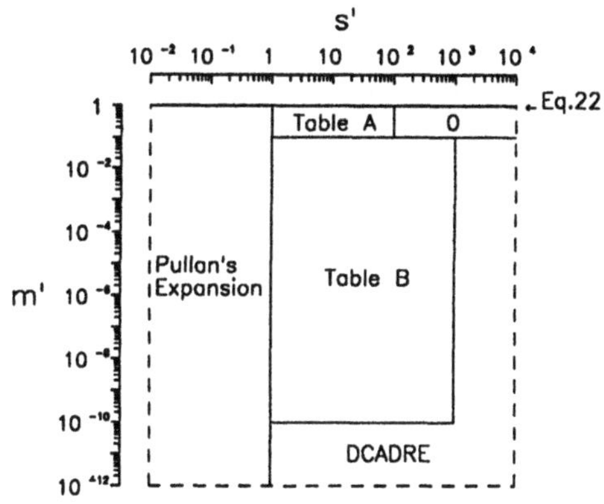

Figure 4: Map for Green function evaluation schemes.

Combined Method

To optimize the evaluation of Green function, the above methods are combined according to the layout shown in Figure 4 (solid lines indicate closed boundaries and dashed lines represent open boundaries). In the indicated range ($s' < 1$) in the map, Pullan's series always converges to a relative accuracy of 10^{-12} within in 16 terms. When m' equals to 1, the following exact formulae are used

$$\begin{aligned} G &= \frac{2\pi}{\sqrt{a}} e^{-s'} \\ \frac{\partial G}{\partial n} &= -\frac{2\pi}{a^{3/2}} \left(R\frac{\partial R}{\partial n} + r_i n_r \right) (1+s')e^{-s'} \end{aligned} \tag{22}$$

To minimize the size of the tables, Table 'A' (Figures 1a, 2a, etc.) and 'B' (1b, 2b, etc.) each covers a finite area. For $0.1 \leq m' \leq 1$ and $s' > 100$, the functions are less than 10^{-20}, hence are arbitrarily set to zero. For typical problems examined, we found it rare that s' will exceed 10^3 and m' be less than 10^{-10}. However, as a precaution, the values that are not covered by one of the above schemes are then computed by direct integration using the DCADRE subroutine. (In fact, for the sample problems investigated in the section below, the DCADRE subroutine is never invoked.) The combined method hence covers all the possible values of s' and m' for the Green functions.

Examples

Two numerical examples are presented for verification. In the first example, the diffusion of heat in a circular cylinder initially at zero temperature everywhere, and then suddenly heated from the top is examined. The geometry and the boundary conditions are illustrated in Figure 5. This is essentially a one-dimensional problem, but solved as a two-dimensional axisymmetric problem by the BEM. The three sides of the boundary are subdivided into a total of 7 quadratic elements. The centerline is not discretized because it does not contribute to the integration. A Stehfest inversion [9] with six terms in the series is used to bring the solution to the time domain. The temperature distribution in the z-direction at various dimensionless times $t^* = ct/L^2$ is plotted

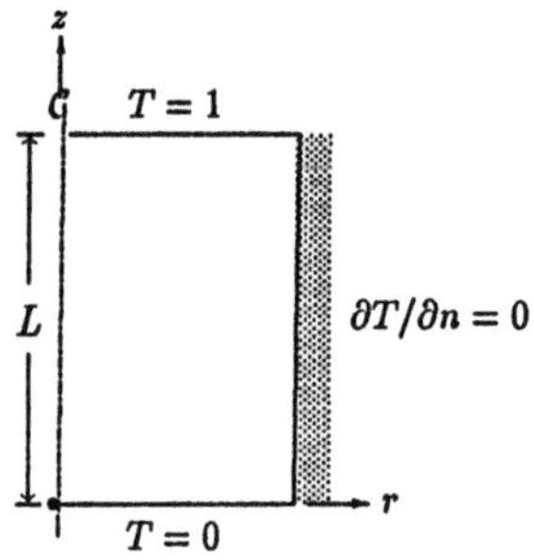

Figure 5. Circular cylinder heated from the top.

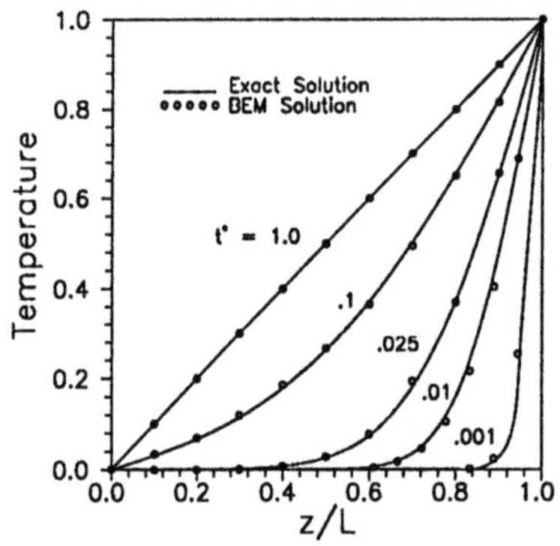

Figure 6: Temperature profile for circular cylinder.

in Figure 6 in symbols. Also shown is the exact solution as solid lines [10]. Close agreement is observed.

The second example is adopted from Wrobel and Brebbia [1]. It has the geometry of a prolate spheroid defined by the parametric equations

$$\begin{aligned} r &= L \sin \phi \\ z &= 2L \cos \phi \end{aligned} \tag{23}$$

where ϕ is the apex angle of the spherical coordinate system (measured from the z-axis). The body is initially set at zero temperature. At $t = 0^+$ the surface is suddenly heated and kept at a constant temperature $T = 1$. The temperature history at the center of the prolate is plotted in Figure 7 and compared with the analytical solution [11]. Again, excellent agreement is found.

Conclusion

The complete BEM solution for axisymmetric diffusion equation in the Laplace transform domain is presented in this paper. To overcome the difficulty of the unavailability of closed form Green function, the method of numerical tabulation is introduced. This approach has been

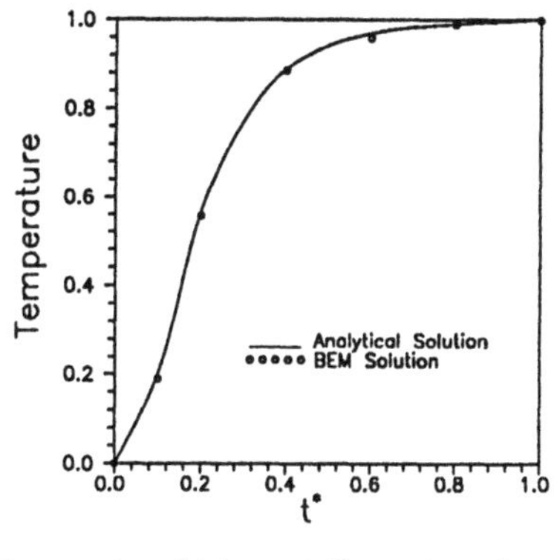

Figure 7: Temperature history at the center of prolate spheroid.

proven to be efficient and accurate. This idea of tabulating Green function is in fact fairly general. It not only applies for the current problem, but also can be extended to any other problems in which the Green function can not be obtained in closed form.

References

[1] Wrobel, L. and Brebbia, C.A., "A formulation of the boundary element method for axisymmetric transient heat conduction", Int. J. Heat Mass Transfer, **24**, 843-850, 1981.

[2] Tsuchimoto, M. and Honma, T., "New expressions of the fundamental solution of axisymmetric Helmholtz-type equations," in *Boundary Elements X. Vol. 1 Mathematical and Computational Aspects*, ed. C.A. Brebbia, Comp. Mech. Publ., 97-110, 1988.

[3] Pullan, A., "Boundary element solutions of quasilinearised time-dependent infiltration", Appl. Math. Modelling, **12**, 9-17, 1988.

[4] Stakgold, I., *Green's Function and Boundary Value Problems*, John Wiley & Sons, 1979.

[5] Gradshteyn, I.S. and Ryzhik, I. M., *Table of Integrals, Series, and Products*, Academic Press, 1980.

[6] Abramowitz, M. and Stegun, I.A., *Handbook of Mathematical Functions*, Dover, 1972.

[7] Lennon, G., Liu, P.L.-F. and Liggett, J.A., Boundary integral equation solution to axisymmetric potential flows, 1. Basic formulation", Water Resour. Res., **15**, 1102-1106, 1979.

[8] *IMSL Library, FORTRAN Subroutines for Mathematics and Statistic*, IMSL Inc.

[9] Stehfest, H., "Numerical inversion of Laplace transforms", Comm. ACM, **13**, 47-49 and 624, 1970.

[10] Churchill, R., *Operational Mathematics, 2nd ed.*, McGraw-Hill, 1958.

[11] Haji-Sheik, A. and Sparrow, E.M., "Transient heat conduction in a prolate spheroidal solid," J. Heat Transfer, Trans. ASME, **88C**, 331-333, 1966.

MIX
Papier aus verantwortungsvollen Quellen
Paper from responsible sources
FSC® C105338

If you have any concerns about our products,
you can contact us on
ProductSafety@springernature.com

In case Publisher is established outside the EU,
the EU authorized representative is:
Springer Nature Customer Service Center GmbH
Europaplatz 3, 69115 Heidelberg, Germany

Printed by Libri Plureos GmbH
in Hamburg, Germany